HERBICIDES, THEORY AND APPLICATIONS

Edited by **Sonia Soloneski**
and **Marcelo L. Larramendy**

Herbicides, Theory and Applications
Edited by Sonia Soloneski and Marcelo L. Larramendy

Published by InTech

Janeza Trdine 9, 51000 Rijeka, Croatia

Publishing Process Manager Ana Nikolic

Technical Editor Teodora Smiljanic

Cover Designer InTech Design Team

Additional hard copies can be obtained from orders@intechopen.com

Herbicides, Theory and Applications, Edited by Sonia Soloneski
and Marcelo L. Larramendy
p. cm.
ISBN 978-953-307-975-2

For EU product safety concerns:
IN TECH d.o.o., Prolaz Marije Krucifikse Kozulić 3, 51000 Rijeka, Croatia,
info@intechopen.com or visit our website at intechopen.com.

Meet the editors

Sonia Soloneski holds a PhD in Natural Sciences and is Assistant Professor of Molecular Cell Biology at the Faculty of Natural Sciences and Museum of La Plata, National University of La Plata, Argentina. Her pre-graduate studies in Finland enabled her to develop part of her doctoral thesis work at the Department of Medical Genetics, University of Finland, Finland. Member of the National Scientific and Technological Research Council (CONICET) of Argentina in Genetic Toxicology field since 2005. Presently, member of the Latin American Association of Environmental Mutagenesis, Teratogenesis and Carcinogenesis (ALAMCTA), the Argentinean Society of Toxicol- ogy (ATA) and the Argentine Society of Genetics (SAG). She has co-authored more than 35 scientific publications in reviewed scientific journals and 90 abstracts of research papers, increasingly as senior author. Most have been published in leading journals in the field of the Environmental Mutagenesis. She is a regular lecturer at the International A. Hollaender courses by the International Associa- tion of Environmental Mutagenesis Societies (IAEMS).

Marcelo L. Larramendy, PhD, Professor of Molecular Cell Biology at La Plata National University (UNLP), Argentina. Researcher at the National Scientific and Technological Research Council (CONICET) of Argentina since 1981. Presently Member of the Executive Committee of the Latin American Association of Environmental Mutagenesis, Teratogenesis and Carcinogenesis (ALAMCTA). Author of more than 265 scientific papers in the field. Recipient of several national and international awards. Regular Lecturer at The International A. Hollaender Courses by the International Association of Environmental Mutagenesis Societies (IAEMS). Former guest scientist at NHI, Bethesda, USA and University of Helsinki, Finland, among others. Expert in Molecular Cytogenetics and Genetic Toxicology.

Contents

Preface

Weeds have always represented one of the main limiting factors in crop production. For the first time in human history, we are technically able to produce as much food as needed for the ever increasing world population, thus theoretically eliminating the risk of famine. These are mainly observed in cases of wars and poor management. Herbicides have revolutionized weed control worldwide. Since weeds are responsible for a loss of over 14% of global harvests, they have been rapidly adopted worldwide. This is no small concern, since most of the largest producers of commodities are located in the so called developing world or emerging economies. The popularity of these chemicals derives from the fact that they are the most reliable and least expensive method of weed control available today. Weeds are different from other pests in crop production because they are relatively constant, while outbreaks of insects and disease pathogens are sporadic. Apart from quantitative damages caused by weeds due to competition for nutrients, light and water, they are able to cause indirect qualitative damages due to crop yield reduction, contamination of seeds, harvesting practices, and soil degradation.

In the past 60 years, agrochemical companies have successfully discovered and marketed a wide array of selective herbicides. Their success is largely responsible for the abundant and sustained food production demanded by national governments. The use of herbicides has simplified crop management attempting to keep weed population at acceptable levels. Herbicides and other agrochemicals have provided, year after year, tools to grow the most profitable crops on the same fields. Thus, reliance upon herbicides as the primary method of weed control in crop management systems is understandable. Unfortunately, they are not free from posing serious environmental risks and substantial health dangers to the population. Residues on food, groundwater contamination, as well as occupational exposure to farm workers are not to be disregarded.

In our industrialized society, the common feeling about herbicides is often indifference. In agreement with this concept, several surveys carried out by herbicides manufacturers claim that less than 10% of the interviewed consider herbicides dangerous for man and the environment. This social acceptance is most probably due to the communication gap existing between the scientific community and society. Misinformation and disinformation are also to be included in this context. Society is not usually fully aware of "the price" to be paid in order to provide an abundant and uninterrupted food production chain. Before registration of a new herbicide, rigorous testing is mandatory, surfactants and inert ingredients present in commercial formulations are included as well. These tests include animal toxicity, namely acute toxicity, carcinogenicity and teratogenicity bioassays, effects on non-target organisms, and different modes of environmental degradation.

Several excellent papers within the complex herbicide field came out in the last decade. A simple search in a databank as PubMed, displays more than 3,500 reports published in scientific journals only during 2010. As developments in this field have been quite rapid, we believe the writing of a new book scoping the subject is fully justified. To tackle among others, related geopolitical, economical and population issues in our modern, internet-economy connected societies, we aim to present a more holistic approach of the matter, in order to appreciate the full scope of the question.

The content selected in *Herbicides, Theory and Applications* is intended to provide researchers, producers and consumers of herbicides an overview of the latest scientific achievements. Although we are dealing with many diverse and different topics, we have tried to compile this "raw material" into three major parts in search of clarity and order. First, in *Weed Control and Crop Management*, readers will find twelve chapters with background information about the effects of herbicides on the undesired plants that grow and reproduce aggressively in crops as well as their management and several empirical methodologies for study. Second, in *Analytical Techniques of Herbicide Detection*, we have included seven chapters dealing with specific analytical procedures used to identify, quantify and characterize different types of herbicides. Finally, *Herbicide Toxicity and Further Applications* encloses eight chapters related to the usage of conventional and non-conventional cellular bioassays for estimating herbicide toxicity as well as the putative indications of these agrochemicals as antiparasitic compounds outside their classical, recognized herbicide use.

Many researchers have contributed to the publication of this book. Given the fast pace of new scientific publications shedding light on the matter, this book will probably be outdated very soon. We regard this as a positive and healthy fact. The editors hope that this book will continue to meet the expectations and needs of all interested in the methodology of use of herbicides, weed control as well as problems related to their use, abuse and misuse.

Sonia Soloneski and Marcelo L. Larramendy

Faculty of Natural Sciences and Museum,
National University of La Plata
Argentina

Part 1

Weed Control and Crop Management

1

Weed Control in Conservation Agriculture

Andrew Price[1] and Jessica Kelton[2]
[1]*United States Department of Agriculture*
[2]*Auburn University*
United States

1. Introduction

Prior to the introduction of the selective herbicide, 2,4-D (2,4-dichlorophenoxyacetic acid), in the 1940's, weed control in agricultural crops was primarily achieved through mechanical cultivation of the soil. Since that time, an increasing number of effective herbicide options, paired with tillage operations, have allowed agricultural producers in developed countries to significantly increase crop yields while reducing labor demands. Continuation of these practices that rely on intense soil disturbance, however, have helped fuel concerns over agricultural sustainability in light of the severe soil degradation that occurs under these conditions. In response to continued soil depletion and other environmental impacts from agricultural production, conservation agriculture has been promoted as a means of maintaining high crop productivity and increasing economic potential while preserving natural resources and limiting future environmental damage. To achieve goals proposed with conservation agriculture, innovative weed control strategies including chemical methods have and will continue to be an essential component in the development of sustainable agricultural practices.

An understanding of the fundamental components of conservation agriculture is imperative in order to appreciate the necessity for weed control strategies in these practices as well as the difficulties associated with their development. To that aim, our purpose, in part, is to identify the key components of conservation systems and the evolution of herbicide needs within these practices. Secondly, we present the strategy of high-residue cereal cover crop implementation that can be utilized in conjunction with chemical weed control methods to address the changes in weed control requirements in agricultural settings. Finally the research synopses detail recent and ongoing efforts to ensure the availability of effective herbicide applications within conservation agriculture.

2. Defining conservation agriculture

As the global population expands, food demands placed on agricultural production systems will test the capabilities of current agriculture practices. Moreover, adequate food production in the future can only be achieved through the implementation of sustainable growing practices that minimize environmental degradation and preserve resources while maintaining high yielding, profitable systems. To this end, conservation agriculture is a system designed to achieve agricultural sustainability by improving the biological functions of the agroecosystem with limited mechanical practices and judicious use of chemical inputs (FAO, 2010).

Three core elements of conservation agriculture make possible the objectives of this system including minimal tillage operations, permanent vegetative residue for soil cover, and rotation of primary crops (FAO, 2010). From these components, a more narrow focused system has been defined as conservation tillage which seeks to reduce, although not necessarily eliminate, tillage practices and increase residual soil covering, which may not be permanently maintained, to achieve similar goals as conservation agriculture (Hobbs, 2007). While sometimes mistakenly used synonymously, it is the less intensive conservation tillage system that has become more recognized, and adopted, within the agricultural community.

A host of benefits can be achieved through employing components of conservation agriculture or conservation tillage, including: reduced soil erosion and water runoff, increased productivity through improved soil quality, increased water availability, increased biotic diversity and reduced labor demands (Steiner *et al.*, 2000; Hobbs, 2007). Despite both environmental and production advantages offered through conservation systems, adoption rates have previously lagged in many countries due to several factors including: availability of required equipment, lack of information, producer mindsets, and, initially, weed control issues (Kells and Meggitt, 1985; Derpsch and Friedrich, 2009). However, recent estimates of global adoption rates of no-tillage systems have reported a substantial increase in hectares (ha) under these practices up to 105 million ha in 2008 worldwide from 45 million ha a decade ago (Derpsch and Friedrich, 2009). This increase in conservation practices can partially be attributed to increased awareness of benefits provided through conservation systems and the growing need for agricultural sustainability, but recent technological advancements and refined implementation strategies have also afforded growers an opportunity to adopt these practices with greater confidence and ease.

Presently, research in conservation practices continues to offer innovative strategies for applying conservation systems in many landscapes, climates, and crop settings, in developed or developing countries. Continued efforts to improve adoption rates as well as address current issues, such as herbicide resistance, are necessary to ensure that the global agricultural productivity can be maintained for future generations.

3. Herbicide requirements in conservation systems

The shift from conventional tillage practices, where the soil is turned prior to planting, to conservation practices, where tillage is reduced to a minimum, can be particularly difficult with respect to weed control. Successful weed control requires a producer's attention throughout the season in order to achieve an optimal harvest. In systems with intense tillage operations, growers can obtain early season weed control through turning of the soil which disrupts weed seed germination and seedling growth through burial (Steckel *et al.*, 2007). The use of selective herbicide applications over the top of the crop at a later date can, most often, sufficiently reduce weed pressure until the end of the season. In cases where there is a history of a difficult to control weed species emerging, producers have the option to use a preemergent, soil applied herbicide with residual efficacy to further reduce weed germination. Although weed control in tilled systems is no small task, conservation systems have presented an even greater challenge to achieve the same results until recently.

Many weed species within agricultural settings are able to flourish when intense tillage operations are minimized. Therefore, conservation systems have been characterized by greater weed densities than conventionally tilled agricultural productions (Cardina *et al*,

2002; Sosnoskie *et al.*, 2006). With reduced tillage practices, producers have increasingly relied on herbicide control options to obtain satisfactory crop yields; however, the initial availability of effective herbicide formulations was limited for conservation tillage. With a reduction in tillage, producers lose weed control offered from seed burial as well as the option to incorporate soil applied preemergent herbicides. Moreover, soil applied herbicides that do not require incorporation can have reduced persistence and efficacy in the presence of plant residue that may intercept and bind the chemical before it reaches the soil surface (Potter *et al.*, 2008). This loss of control options has forced producers wishing to adopt conservation practices to be primarily dependent upon postemergent chemical applications which, oftentimes, fail to provide adequate weed control. To further complicate attempts to adopt conservation practices, growers initially face shifts in weed population dynamics due to altered distribution of weed seed within the soil (Buhler, 1997); perennial weed species also thrive in reduced-tillage settings and can be difficult to control with available postemergent herbicide options (Swanton *et al.*, 1993). Although studies report that, over time, the weed seedbank, or viable weed seed within the soil, will be reduced and/or easier to manage with chemical controls due to increased selection pressures and increased uniform germination, initial weed control strategies have remained challenging for agricultural lands being switched to conservation tillage practices (Murphy *et al.*, 2006; Swanton *et al.*, 2008).

The introduction and advances with herbicide-tolerant crops made in the last 15 years have greatly altered the herbicide needs in conservation systems for those who use these technologies; however in developing regions with limited access to herbicide options or in areas where herbicide-resistant weed species have compromised the use of herbicide-tolerant crops in conservation systems, early weed management tactics and control issues in reduced tillage practices remain relevant for growers.

4. Introduction of herbicide resistant crops

In the 1990's when transgenic, herbicide-tolerant crops were first introduced, reduced-tillage, in the United States at least, became a viable option for many producers. The availability of transgenic crops with resistance to a nonselective herbicide, such as glyphosate, has provide the means for effective postemergent herbicide control of a broad spectrum of weed species while reducing labor demands and repeated herbicide applications. By combining this crop technology with conservation tillage, producers have been able to further reduce labor expenses and boost profitability. Partly through the combination of these practices, conservation tillage has been implemented on over 26 million hectares to date in the United States alone (Derpsch and Friedrich, 2009).

The process of developing and commercializing a transgenic crop cultivar is a complex and costly endeavor which has limited commercial availability of genetically modified crop varieties. For the development of a transgenic crop, particularly an herbicide-tolerant crop, to be pursued, several factors must be investigated including: spectrum of weed control provided by the herbicide, safety risks to humans and the environment, yield performance of genetically modified crop, and economic value of the crop (Devine, 2005). Currently, only a select few herbicide-tolerant crops have been fully developed, marketed, and remain commercially available although the technology exists to produce tolerant varieties for many major and minor crops throughout the world (Devine, 2005)(Table 1).

	Herbicide	
	Glufosinate	Glyphosate
Crop	Year Commercialized	
Canola	1995	1996
Corn	1997	1998
Cotton	2004	1997
Soybean	2009	1996

Table 1. Currently available transgenic crops by herbicide tolerance and year available.

Since identifying selective herbicide compounds that are active on weed species and not on a particular crop can be a difficult process, conferring herbicide tolerance of a non-selective herbicide to a crop can be tremendously valuable for effective weed control (Mazur and Falco, 1989). From the non-selective herbicides available for use, two key herbicides have been the focus for herbicide-tolerant crops: glufosinate and glyphosate (Devine, 2005).

The broad spectrum herbicide, glyphosate (*N*-(phosphonomethyl)glycine), works through the inhibition of 5-enolpyruvylshikimate-3-phosphate synthase (EPSPS), an enzyme required for the production of aromatic amino acids which are necessary for subsequent production of plant hormones and structural components (Schönbrunn *et al.*, 2001; Dill, 2005). The means for conferring glyphosate resistance to crops is achieved through the insertion of a resistant transgene, referred to as CP4-EPSPS, which allows the plant's shikimate pathway to continue to function in the presence of glyphosate applications (Funke *et al.*, 2006). Since the release of glyphosate-tolerant soybean (*Glycine max* L.) in 1996, adoption of this technology has soared in several industrialized countries, such as the United States, Argentina, and Brazil, and represents a majority of the soybean being produced in these areas (Dill, 2005). The introduction of other major crops with glyphosate tolerance soon followed with successful adoption due to the weed control efficacy, ease of use, and lower production costs from reduced herbicide applications.

Glufosinate, or L-Phosphinothricin, also a non-selective herbicide, acts as an inhibitor of glutamine synthetase which impedes the production of amino acids and inhibits photosynthesis (Dröge-Laser *et al.*, 1994; Ross and Lembi, 1999). Glufosinate-tolerant plant varieties are produced through the encoding for phosphinothricin acetyltransferase (PAT) proteins which detoxify glufosinate through *N*-acetylation (Dröge *et al.*, 1992; Hérouet *et al.*, 2005). Glufosinate-tolerant canola (*Brassica napus* L.) was introduced in Canada in 1995 with relative success (Devine, 2005; Duke, 2005). Other tolerant crop varieties have been successfully released since that time but have yet to gain as large of a market share as glyphosate-tolerant varieties potentially due to economic advantages not being realized and the lack of translocation of glufosinate which can limit its efficacy for certain weed species (Duke, 2005).

With the availability of effective broad-spectrum weed control without tillage operations and repeated use of herbicides, conservation tillage saw substantial increases after the introduction of herbicide-tolerant crop varieties. Employing the use of herbicide-tolerant crops with conservation systems offered growers even greater costs savings than utilizing either practice alone and continue to do so today. The adoption of glyphosate-tolerant crops was especially suited to conservation systems since glyphosate can effectively control many perennial species that appear when tillage practices are reduced (Ross and Lembi, 1999). Glyphosate-tolerant crops provided such effective control, the use of glyphosate, due, in part, to biotechnology, has become the predominant herbicide used globally (Baylis, 2000). Unfortunately, the cost effectiveness and weed control advantages, paired with limited herbicide choices (primarily in conservation tillage), of glyphosate-tolerant technology have compelled some growers in conventional as well as reduced-tillage systems to rely solely on this herbicide for agricultural productivity. Because of this, in some regions, the sustainability of both conservation tillage and glyphosate use has been threatened due to this overdependence and development of glyphosate resistance in multiple weed species.

5. Herbicide resistance in weed species

As early as the 1950's, shortly after widespread herbicide use began, concerns were being voiced about the possibility of herbicide-resistant weed biotypes appearing as a result of repeated exposure to one herbicide (Appleby, 2005). It was not until 1970, however, that the first case of herbicide resistance was formally documented in triazine-resistant common groundsel (*Senecio vulgaris* L.) (Ryan, 1970). Since that time, 346 herbicide-resistant weed biotypes have been reported worldwide and continue to demand considerable research attention to control existing resistance as well as to combat the further spread of resistant populations (Appleby, 2005; Heap, 2010).

Although almost all herbicide modes of action have seen resistance development, the introduction of glyphosate-tolerant crops has been a prominent factor to the development of glyphosate-resistant weed species for this herbicide (Powles and Yu, 2010). The steady adoption rate of herbicide-tolerant crops has been met with a simultaneous increase in the use glyphosate applications, particularly in conservation systems where minimal herbicide alternatives exist (Askew and Wilcut, 1999; Dill, 2005; Duke and Powles, 2008). The initial success of this weed control strategy has lead many producers to rely exclusively on this single herbicide mode of action to maintain acceptable weed control year after year (Green, 2007). Unfortunately, repeated exposure to glyphosate has greatly increased selection pressure for resistant weed biotypes among affected populations resulting in agricultural weed infestations with limited or no known control options at present (Culpepper, 2006). Rapid development of herbicide resistance is evident in the number of confirmed cases of glyphosate resistance since 1996 which has appeared in 18 different weed species and on all agriculturally productive continents (Figure 1).

From the beginning of glyphosate use in 1974 as a nonselective herbicide in nonagricultural settings, it was believed that resistance development would be highly unlikely or very slow in appearance if it did occur (Bradshaw *et al.*, 1997). At the time, naturally resistant weed species had not been identified, modifications to confer resistance resulted in low levels that could not survive glyphosate applications or reduced the plant's fitness level, and typical use patterns did not increase selection pressure for resistant biotypes (Dyer, 1994; Padgette *et al.*, 1995; Bradshaw *et al.*, 1997). Indeed, glyphosate use successfully utilized without incident for over 2 decades before a resistant biotype of rigid ryegrass (*Lolium rigidum* Gaudin) was identified in 1996 (Powles *et al.*, 1998). However, since the release of herbicide-tolerant crops, several resistant weed biotypes have been reported in glyphosate-tolerant systems in as little as 3 years (Green, 2007; Duke and Powles, 2008).

Mechanisms for herbicide resistance development vary greatly depending on many factors such as weed species and herbicide in use. When resistance emerges, it can be classified as either target-site resistance, where modifications to the active site for an herbicide limits its toxicity, or non-target-site resistance, where herbicide movement to the active site is limited in some fashion (Powles and Yu, 2010). Identified resistance mechanisms for glyphosate are comprised of both classifications of resistance, including target-site modifications, gene amplification, as well as reduced herbicide translocation (Lee and Ngim, 2000; Lorraine-Colwill *et al.*, 2003; Gaines *et al.*, 2010). It is likely that new mechanisms for glyphosate resistance will continue to be discovered within current herbicide practices which will require intensified research in order to develop innovative management practices that preserve glyphosate use in many agricultural settings (Powles and Yu, 2010).

Managing for herbicide resistance remains a key component in current developments for weed control. Proactive weed control practices that reduce initial resistance development are vital for herbicide viability in the future (which is necessary for sufficient agricultural production). In order to ensure sustainability in herbicide-tolerant crop production and conservation practices, many weed control techniques are currently being employed and evaluated.

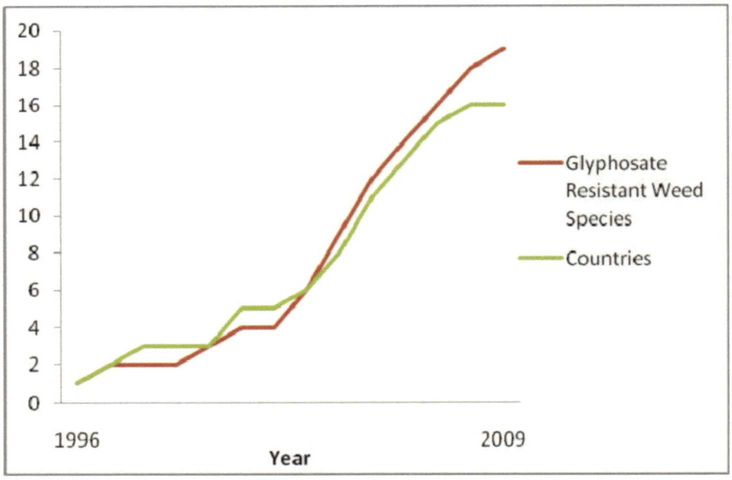

Fig. 1. Global development and spread of glyphosate resistant weed species. Adapted from Heap 2010.

6. High-residue cereal cover cropping

The continued appearance of herbicide-resistant weed species following the introduction of herbicide-resistant crops has been detrimental to conservation tillage systems where feasibility relies on this technology. In an effort to preserve conservation tillage and, subsequently, the benefits realized through these practices, current research has been directed at identifying alternative weed control strategies that can be employed in herbicide-resistant crop settings, which remain a valuable asset to conservation tillage practices, while minimizing favorable conditions for resistance development. Options for weed control must reduce selection pressure for herbicide resistance as well as provide season long weed suppression in order to be viable components of a new weed control strategy. Alternatives to repeated use of a single herbicide include: crop rotation, improved management practices (such as weed scouting and herbicide application timing), alternative herbicide chemistries, and use of high-residue cover crop systems (Price *et al.*, 2009). Ideal alternative growing practices would incorporate several of these strategies to protect against resistance development, however, the use of a particular practice, high-residue cover cropping, is proving to be an exceptional weed control technique in conservation systems.

Recent evaluations of cereal cover crop use, particularly high residue systems that achieve approximately 4,500 kilograms of residue per hectare (Reiter *et al.*, 2008), have demonstrated a promising method for battling herbicide dependence and providing a sustainable approach for the continuation of conservation tillage (Figure 2). The high-residue cover crop system, when utilized in a reduced-tillage system, can further reduce soil erosion and water runoff, improve soil fertility, and, with legume covers, reduce nitrogen input needs. These winter cereal crops or legumes can also improve weed control through both physical barriers of plant residue and chemical interference in the form of allelopathic compounds released by the cover crop. Although concerns have been raised about the potential for reduced crop yields and increased need for more herbicide applications, research continues to show that these practices can be successfully implemented into conservation tillage (Balkcom and Reeves, 2005; SAN, 2007; Price *et al.*, 2008; Reiter *et al.*, 2008).

A variety of options are available to a producer wishing to utilize a winter cover crop between the primary growing season depending on the crops grown, climate, and needs of the system. For example, legume cover crops, such as clover (*Trifolium sp.*) and vetch (*Vicia sp.*), can potentially reduce nitrogen requirements for the following crop but may not provide season-long weed control; cereal cover crops, like rye (*Secale cereale* L.) and oat (*Avena sp.*), can provide longer weed control throughout the season can also reduce available soil nitrogen (SAN, 2007). Besides legume and cereal crops, brassica and mustard species are also of interest for cover crop use due to their potential for pest management.

Cover crops can reduce weed numbers physically and chemically while actively growing or after termination. Prior to termination, cover crops can compete with weed species for necessary resources such as light, water, and nutrients; cover crops can also release allelochemicals into the soil which may be detrimental to nearby competing weed species, particularly for small-seeded weeds (Weston, 1996; Foley, 1999; Price *et al.*, 2008). After termination, weed suppression occurs by physical impedance of weed species with cover crop residue as well as continued leaching of allelochemicals into the soil (Weston, 1996). These characteristics allow cover crops to offer early weed control as well as weed suppression into the growing season (depending on the rate of decomposition).

Although the use of cover crops in production systems is a viable option for producers, there are still concerns that must still be investigated. High-residue cover crops, before termination, can deplete soil moisture needed by the primary crop (SAN, 2007); conversely, dense plant residue can retain excessive amounts of moisture during periods of high rainfall (Fernandez *et al.*, 2008). Lower soil temperatures, increased plant pest populations, as well as planting operation interferences, such as poor soil-to-seed contact, have also been attributed to high levels of cover crop residue (Fernandez *et al.*, 2008; Kornecki *et al.*, 2009). Additionally, high levels of plant residue are thought to impede herbicide movement to the soil surface through interception and sorption leading to reduced weed control under cover crop systems (Johnson *et al.*, 1989; Gaston *et al.*, 2003). Future adoption of these practices will be dependent upon continued research in many areas but especially in determining effective herbicide strategies to be employed in combination with high-residue systems.

Fig. 2. Cotton (*Gossypium hirsutum* L.) planted into a soil cover of black oat (*Avena strigosa* Schreb.).

7. Development of effective weed control strategies for use in conservation systems

The growing adoption rate of conservation practices and interest of high-residue cover crops has spawned an increase in research geared toward understanding the fit of herbicides into the sustainable agricultural landscape. Case studies described here illustrate recent projects in three major crop systems designed to determine the most effective production practices, including herbicide choice, that can be employed for successful adoption of high residue cover crop systems which will ultimately aid in reducing herbicide resistance and preserve sustainable agricultural practices for the future.

Peanut (*Arachis hypogaea L.*), cotton, and soybean comprise a substantial portion of the agricultural hectarage in the southeastern United States. A large percentage of growers in this region utilize some form of conservation tillage due to its economic benefits such as reduced labor and fuel expense. Use of cover crops managed for high-residue in these systems remains largely untried due to grower concerns about herbicide input requirements and poor yields (Schwab *et al.*, 2002). To investigate these concerns, a three year study was conducted in Headland, Alabama, in the southeastern US, to determine the effects of a winter cereal cover crop on primary crop yield as well as to identify effective herbicide practices (Price *et al.*, 2005; Reeves *et al.*, 2005; Price *et al.*, 2007).

In order to achieve high-residue stands, three winter covers, rye, wheat (*Triticum aestivum* L.), and black oat (*Avena strigosa* Schreb.), were established in November at a rate of 120 kg/ha with 56 kg ammonium nitrate and terminated three weeks prior to planting of the primary crop in the spring of the following year. A winter fallow system of annual weeds was included as a comparison and as a representation of the common conservation practice by regional producers. After termination with glyphosate as a burndown herbicide at a rate of 1.12 kg ae/ha, cover crops were rolled flat on the surface with a mechanical roller-crimper.

Crops were planted into a strip-tilled bed that limits tillage to a small area, approximately 30 cm, for seed placement. Three herbicide treatments were included for evaluation and consisted of a preemergence (PRE) herbicide application only (low input system), a PRE plus a postemergence (POST) herbicide application (high input system), or no herbicide application (no input system). Peanut and cotton received pendimethalin [*N*-(1-ethylpropyl)-3,4-dimethyl-2,6-dinitrobenzenena-amine] as a PRE at a rate of 1.12 kg ai/ha and soybean received this herbicide treatment at a rate of 0.84 kg ai/ha. Additional PRE herbicides included metribuzin [4-amino-6-(1,1-dimethylethyl)-3-(methylthio)-1,2,4-triazin-5(4*H*)] at 0.43 kg ai/ha for soybean and fluometuron [1,1-dimethyl-3-(a,a,a-triluoro-m-toly) urea] at 1.7 kg ai/ha for cotton. Postemergent herbicides included: paraquat (1,1′-Dimethyl-4,4′-bipyridynium dichloride) (0.14 kg ai/ha), bentazon [3-(1-methylethyl)-1H-2,1,3-benzothiadiadiazin-4(3H)-one 2,2-dioxide] (0.56 kg ai/ha), 2,4-DB [4-(2,4-dichlorophenoxy) butyrate] (0.22 kg ai/ha), and chlorimuron ethyl {ethyl 2-[[[[(4-chloro-6-methoxypyrimidin-2yl)]carbonyl]amino]sulfonyl]benzoate} (0.14 kg ai/ha) for peanut; DSMA (Disodium methanearsonate) (1.7 kg ai/ha), lactofen {2 ethoxy-1-methyl-2-oxoethyl 5-[2-chloro-4-(trifluoromethyl)phenoxy]-2-nitrobenzoate} (0.2 kg ai/ha) and cyanazine {2-[[4-chloro-6-(ethylamino)-1,3,5-triazin-2-yl]amino]-2-methylpropanenitrile} (0.84 kg ai/ha) for cotton; and chlorimuron ethyl(8.75 g ai/ha) for soybean. Herbicide options reflect current herbicide use in the respective crop.

At cover crop termination, biomass samples were weighed for each cover. First year results averaged 5,450 kg/ha for black oat, 5,130 kg/ha for rye, 5,100 kg/ha for wheat, and 1,410 kg/ha for fallow systems with the predominant weed species being cutleaf evening primrose (*Oenothera laciniata* Hill) and chickweed [*Stellaria media* (L.) Vill.]. Biomass residue levels for all cover crops in this year exceeded amounts considered the minimum for high-residue systems. Averages for cover crop weights were in line with previously reported biomass weights (Bauer and Reeves, 1999).

Visual weed control ratings (as a percentage of control) for year 1 are presented in Table 2. Dominant weed species in research plots for all crops included: large crabgrass [*Digitaria sanguinalis* (L.) Scop.], Texas panicum (*Panicum texanum* Buckl.), nutsedges (*Cyperus esculentus* L. and *Cyperus rotundus* L.), sicklepod [*Senna obtusifolia* (L.) Irwin and Barnaby], and Palmer amaranth [*Amaranthus palmeri* (S.) Wats.]. Analysis of data revealed significant effects on weed control from both cover crop and herbicide input treatments. Although no cover crop provided optimum season-long weed control without herbicide applications, black oat and rye did provide substantial weed control in peanut and soybean without herbicides; low herbicide input treatments, however, had acceptable weed control in peanut and soybean, especially in black oat and rye covers. In years 2 and 3 (data not shown), weed control was reduced in all covers, particularly black oat, due to below-average winter temperatures inhibiting cover crop growth and biomass production. Crop yields in cover crop systems were increased over fallow systems for all crops, however, yields were greatly reduced in systems without any herbicide application (Table 3). Increases in yield noted in cover crops systems over fallow systems are attributed to reduced weed pressure as well as other benefits from conservation systems that are amplified when high-residue cover crops are employed such as increased water infiltration and increased soil quality.

Results from this study show that high-residue cover crop systems can be effectively utilized in conservation systems with increased yield potential and possible reductions in herbicide inputs for adequate weed control. Reduced herbicide dependence, without yield decrease, can ultimately aid in reduced herbicide-resistance development and sustain conservation tillage practices well into the future. For high-residue cover crops to be more widely adopted, research continues to be necessary to fully understand the benefits, and potential drawbacks, of their use at a regional level as well as to define the most effective cover crop choices for producers in a variety of systems.

	Cotton			Peanut			Soybean		
	Herbicide input system			Herbicide input system			Herbicide input system		
Cover crop	High	Low	None	High	Low	None	High	Low	None
	----Weed control (%)---			----Weed control (%)---			----Weed control (%)---		
Fallow	94	86	13	91	88	24	92	85	29
Black oat	95	91	35	93	94	70	95	95	86
Rye	94	89	26	94	93	61	95	95	83
Wheat	94	87	14	94	93	43	95	91	61

Table 2. Weed control for year 1 in cotton, peanut, and soybean by percent control for four cover crop options and three herbicide inputs (by intensity) where 100 is total control and 0 is no control.

	Cotton			Peanut			Soybean		
	Herbicide input system			Herbicide input system			Herbicide input system		
Cover crop	High	Low	None	High	Low	None	High	Low	None
	---Seed cotton (kg/ha)----			---Peanut (kg/ha)---			---Soybean (kg/ha)---		
Fallow	3660	3010	0	4280	4100	2030	4031	4031	1344
Black oat	3840	3630	0	4760	4740	3190	6719	7391	6047
Rye	3980	3350	0	4690	4850	3460	6047	6719	6047
Wheat	3970	3120	0	4670	4420	2500	6719	6719	4703

Table 3. Crop yield for year 1 as affected by three herbicide inputs and four cover crop options. No yield could be collected for cotton without herbicide input.

8. Conclusions

Conservation systems are necessary to preserve agricultural productivity and meet future global food demands. To implement these systems, adequate weed control is crucial in their success. Herbicide use has been a valuable asset when adopting conservation practices, however, prudent use of chemical weed control is essential to fulfilling the goals of conservation agriculture, reducing detrimental environmental impact, and reducing herbicide resistance development. Further development and testing of alternative weed management practices that can be utilized along with herbicide applications must be pursued in order for conservation practices to remain successful.

9. References

Appleby, A.P. (2005) A history of weed control in the United States and Canada - a sequel. *Weed Science,* 53, 762-768, ISSN 0043-1745.

Askew, S.D. and J.W. Wilcut (1999) Cost and weed management with herbicide programs in glyphosate-resistant cotton (*Gossypium hirsutum*). *Weed Technology,* 13, 308-313, ISSN 0890-037X.

Balkcom, K.S. and D.W. Reeves (2005) Sunn-hemp utilized as a legume cover crop for corn production. *Agronomy Journal,* 97, 26-31, ISSN 0002-1962.

Bauer, P.J. and D.W. Reeves (1999) A comparison of winter cereal species and planting dates as residue cover for cotton grown with conservation tillage. *Crop Science,* 39, 1824-1830, ISSN 0011-183X.

Baylis, A.D. (2000) Why glyphosate is a global herbicide: strengths, weaknesses and prospects. *Pest Management Science,* 56, 299-308, ISSN 1526-4998.

Bradshaw, L.D., S.R. Padgette, S.L. Kimball, and B.H. Well (1997) Perspectives on glyphosate resistance. *Weed Technology,* 11, 189-198, ISSN 0890-037X.

Buhler, D.D. (1997) Effects of tillage and light environment on emergence of 13 annual weeds. *Weed Technology,* 11, 496-501, ISSN 0890-037X.

Cardina, J., C.P. Herms and D.J. Doohan (2002) Crop rotation and tillage system effects on weed seedbanks. *Weed Science,* 50, 448-460, ISSN 0043-1745.

Culpepper, A.S. (2006) Glyphosate-induced weed shifts. *Weed Technology,* 20, 277-281, ISSN 0890-037X.

Derpsch, R. and T. Friedrich (2009) Development and current status of no-till adoption in the world. Proceedings on CD, 18th Triennial Conference of the International Soil Tillage Research Organization (ISTRO), June 15-19, 2009, Izmir Turkey.

Devine, M.D. (2005) Why are there not more herbicide-tolerant crops? *Pest Management Science,* 61, 312-317, ISSN 1526-4998.

Dill, G. M. (2005) Glyphosate-resistant crops: history, status, and future. *Pest Management Science* 61, 219-224, ISSN 1526-4998.

Dröge W., I. Broer, and A. Pühler (1992) Transgenic plants containing the phosphinothricin-*N*-acetyltransferase gene metabolize the herbicide phosphinothricin (glufosinate) differently from untransformed plants. *Planta,* 187, 142-1551, ISSN 1432-2048.

Dröge-Laser, W., U. Siemeling, A. Pühler, and I. Broer (1994) The metabolites of the herbicide L-Phosphinothricin (Glufosinate). *Plant Physiology,* 105, 159-166, ISSN 1532-2458.

Duke, S.O. and S.B. Powles (2008) Glyphosate: a once-in-a-century herbicide. *Pest Management Science,* 64, 319-325, ISSN 1526-4998.

Dyer, W.E. (1994) Resistance to glyphosate, In: *Herbicide Resistance in Plants,* S.B. Powles and J.A.M. Holtum (Eds.), 229-241, Lewis Publishing, ISBN 0873717139, Boca Raton, FL, USA.

Fernandez, R., A. Quiroga, E. Noellemeyer, D. Funaro, J. Montoya, B. Hitzmann, and N. Peinemann (2008) A study of the effect of the interaction between site-specific conditions, residue cover and weed control on water storage during fallow. *Agricultural Water Management,* 95, 1028-1040, ISSN 0378-3774.

Foley, M.E. (1999) Genetic approach to the development of cover crops for weed management, In: *Expanding the Context of Weed Management,* D.D. Buhler (Ed.), 77-93, The Haworth Press, ISBN 9781560220633, New York, NY, USA.

Food and Agriculture Organization of the United Nations (FAO) (2010) Available online at http://www.fao.org/ag/ca/index.html (verified 20/7/10).

Funke, T., H. Han, M. L. Healy-Fried, M. Fischer, and E. Schönbrunn (2006) Molecular basis for the herbicide resistance of Roundup Ready crops. *Proceedings of the National Academy of Sciences,* 35,13010-13015, ISSN 1091-6490.

Gaines, T.A., W. Zhang, D. Wang, B. Bukun, S.T. Chisholm, D.L. Shaner, S.J. Nissen, W.L. Patzoldt, P.J. Tranel, A.S. Culpepper, T.L. Grey, T.M. Webster, W.K. Vencill, R.D. Sammons, J. Jiang, C. Preston, J.E. Leach, and P. Westra (2010) Gene amplification confers glyphosate resistance in *Amaranthus palmeri. Proceedings of the National Academy of Sciences,* 107, 1029-1034, ISSN 1091-6490.

Gaston, L.A., D.J. Boquet, and M.A. Bosch (2003) Pendimethalin wash-off from cover crop residues and degradation in a Loessial soil. *Communications in Soil Science and Plant Analysis,* 34, 2515-2527, ISSN 0010-3624.

Green, J.M. (2007) Review of glyphosate and ALS-inhibiting herbicide crop resistance and resistant weed management. *Weed Technology,* 21, 547-558, ISSN 0890-037X.

Heap, I. (2010) The international survey of herbicide resistant weeds. Available online at http://www.weedscience.com (verified 28/7/10).

Hérouet, C., D.J. Esdaile, B.A. Mallyon, E. Debruyne, A. Schulz, T. Currier, K. Hendrickx, R. van der Klis, and D. Rouan (2005) Safety evaluation of the phosphinothricin acetyltransferase proteins encoded by the *pat* and *bar* sequences that confer tolerance to Glufosinate-ammonium herbicide in transgenic plants. *Regulatory Toxicology and Pharmacology,* 41, 134-149, ISSN 0273-2300.

Hobbs, P.R. (2007) Conservation agriculture: what is it and why is it important for future sustainable food production? *Journal of Agricultural Science,* 145, 127-137, ISSN 0021-8596.

Johnson, M.D., D.L. Wyse, and W.E. Lueschen (1989) The influence of herbicide formulation on weed control in four tillage systems. *Weed Science,* 37, 239-249, ISSN 0043-1745.

Kells, J.J. and W.F. Meggitt (1985) Conservation tillage and weed control, In: *A Systems Approach to Conservation Tillage,* F.M. D'Itri (Ed.), 123-129, Lewis Publishing, ISBN 087371024X, Chelsea, MI, USA.

Kornecki, T.S., A.J. Price, R.L. Raper, and F.J. Arriaga (2009) New roller crimper concepts for mechanical termination of cover crops in conservation agriculture. *Renewable Agriculture and Food Systems,* 24, 165-173, ISSN 1742-1713.

Lee, L.J. and J. Ngim (2000) A first report of glyphosate-resistant goosegrass (*Eleusine indica* (L) Gaertn) in Malaysia. *Pest Management Science,* 56, 336-339, ISSN 1526-4998.

Lorraine-Colwill, D.F., S.B. Powles, T.R. Hawkes, P.H. Hollinshead, S.A.J. Warner, and C. Preston (2003) Investigation into the mechanism of glyphosate resistance in *Lolium rigidum. Pesticide Biochemistry and Physiology,* 74,62-72, ISSN 0048-3575.

Mazur, B.J. and S.C. Falco (1989) The development of herbicide resistant crops. *Annual Review of Plant Physiology and Plant Molecular Biology,* 40, 441-470, ISSN 1040-2519.

Murphy, S.D., D.R. Clements, S. Belaoussoff, P.G. Kevan, and C.J. Swanton (2006) Promotion of weed species diversity and reduction of weed seedbanks with conservation tillage and crop rotation. *Weed Science,* 54, 69-77, ISSN 0043-1745.

Padgette, S.R., K.H. Kolac, X. Delannay, D.B. Re, B.J. Lavallee, C.N. Tinius, W.K. Rhodes, Y.I. Otero, G.F. Barry, and D.A. Eicholtz (1995) Development, identification, and characterization of glyphosate-tolerant soybean line. *Crop Science,* 35, 1451-1461, ISSN 0011-183X.

Potter, T.L., C.C. Truman, T.C. Strickland, D.D. Bosch, and T.M. Webster (2008) Herbicide incorporation by irrigation and tillage impact on runoff loss. *Journal of Environmental Quality,* 37, 839-847, ISSN 0047-2425.

Powles, S.B., D.F. Lorraine-Colwill, J.J. Dellow, and C. Preston (1998) Evolved resistance to glyphosate in rigid ryegrass (*Lolium rigidum*) in Australia. *Weed Science,* 46, 604-607, ISSN 0043-1745.

Powles, S.B. and Q. Yu (2010) Evolution in action: plants resistant to herbicides. *Annual Review of Plant Biology,* 61, 317-347, ISSN 1543-5008.

Price, A.J., D.W. Reeves, and M.G. Patterson (2005) Evaluation of weed control provided by three winter cereals in conservation-tillage soybean. *Renewable Agriculture and Food Systems,* 21, 159-164, ISSN 1742-1713.

Price, A.J., D.W. Reeves, M.G. Patterson, B.E. Gamble, K.S. Balkcom, F.J. Arriaga, and C.D. Monks (2007) Weed control in peanut grown in a high-residue conservation-tillage system. *Weed Science,* 34, 59-64, ISSN 0043-1745.

Price, A.J., M.E. Stoll, J.S. Bergtold, F.J. Arriaga, K.S. Balkcom, T.S. Kornecki, and R.L. Raper (2008) Effect of cover crop extracts on cotton and radish radical elongation. *Communications in Biometry and Crop Science,* 3, 60-66, ISSN 1896-0782.

Price, A.J., D.W. Reeves, and D.A. Lamm (2009) Glyphosate resistant Palmer amaranth-a threat to conservation tillage. Proceedings of the Beltwide Cotton Conference, January 5-8, San Antonio, TX, USA.

Reiter, M.S., D.W. Reeves, C.H. Burmester, H.A. Torbert (2008) Cotton nitrogen management in a high-residue conservation system: Cover crop fertilization. *Soil Science Society of America Journal,* 72, 1321-1329, ISSN 1435-0661.

Reeves, D.W., A.J. Price, and M.G. Patterson (2005) Evaluation of three winter cereals for weed control in conservation-tillage nontransgenic cotton. *Weed Technology,* 19, 731-736, ISSN 0890-037X.

Ross, M.A. and C.A. Lembi (1999) *Applied Weed Science,* Prentice-Hall, ISBN 0137540035, Upper Saddle River, NJ, USA.

Ryan, G.F. (1970) Resistance of common groundsel to simazine and atrazine. *Weed Science,* 18, 614-616, ISSN 0043-1745.

Schönbrunn, E., S. Eschenburg, W.A. Shuttleworth, J.V. Schloss, N. Amrhein, J.N. Evans, and W. Kabsch (2001) Interaction of the herbicide glyphosate with its target enzyme 5-enolpyruvylshikimate 3-phosphate synthase in atomic detail. *Proceedings of the National Academy of Science 98,* 1376-1380, ISSN 1091-6490.

Schwab, E.B., D.W. Reeves, C.H. Burmester, and R.L. Raper (2002) Conservation tillage systems for cotton in the Tennessee Valley. *Soil Science Society of America Journal,* 66, 142-153, ISSN 0361-5995.

Sosnoskie, L.M., C.P. Herms and J. Cardina (2006) Weed seedbank community composition in a 35-yr-old tillage and rotation experiment. *Weed Science,* 54, 263- 273, ISSN 0043-1745.

Steckel, L.E., C.L. Sprague, E.W. Stoller, L.M. Wax, and F.W. Simmons (2007) Tillage, cropping system, and soil depth effects on common waterhemp (*Amaranthus rudis*) seed-bank persistence. *Weed Science,* 55, 235-239 ISSN 0043-1745.

Steiner, J.L., H.H. Schomberg, P.W. Unger, and J. Cresap (2000) Biomass and residue cover relationships of fresh and decomposing small grain residue. *Soil Science Society of America Journal,* 64, 2109-2114, ISSN 1435-0661.

Sustainable Agriculture Network (SAN) (2007) *Managing Cover Crops Profitably.* 3rd edition. A. Clark (Ed.), ISBN 978888626124, Beltsville, MD, USA.

Swanton, C.J., D.R. Clements, and D.A. Derksen (1993) Weed succession under conservation tillage: A hierarchical framework for research and management. *Weed Technology,* 7, 286-297, ISSN 0890-037X.

Swanton, C.J., K.J. Mahoney, K. Chandler, and R.H. Gulden (2008) Integrated weed management: Knowledge-based weed management systems. *Weed Science,* 56, 168-172, ISSN 0043-1745.

Weston, L.A. (1996) Utilization of allelopathy for weed management in agroecosystems. *Agronomy Journal,* 88, 860-866, ISSN 0002-1962.

2

Weed Management Systems for No-Tillage Vegetable Production

S. Alan Walters
Department Plant, Soil, and Agricultural Systems,
Southern Illinois University, Carbondale, Illinois
USA

1. Introduction

No-tillage (NT) is the extreme form of conservation tillage where the soil is left undisturbed before planting and crops are just planted into residues left on the soil surface (Morse 1999a). The direct seeding of large-seeded vegetables has generally provided successful crop establishment in NT production systems, although many vegetable crops can easily be transplanted in NT production systems including broccoli (*Brassica oleraceae* var. *italica*), cabbage (*Brassica oleraceae* var. *capitata*), cucumber (*Cucumis sativus* L.), pumpkin (*Cucurbita pepo* L.), squash (*Cucurbita pepo* L.) and tomato (*Solanum lycopersicum* L.) (Table 1). Planting is normally achieved in a narrow seedbed or slot created by a NT planter (Hebblethwaite 1997). No-tillage planters are typically equipped with coulters that cut through the mulch residue and disk openers that slice open the soil, with the seed or transplants then placed at proper depths in the soil profile. The opened soil is then closed with some type of device that presses the soil back together after the seed or transplants have been placed into the soil.

	Planting method	
Vegetable Crop	Direct seeded	Transplanted
Broccoli	Low	High
Cabbage	Low	High
Cucumber	High	High
Pumpkin	High	High
Summer squash	High	High
Tomato	Low	High

Low = poor likelihood of success using this method and High = great likelihood for success using this method.

Table 1. The likelihood of success for vegetable crop planting method in no-tillage production systems (adapted from Morse, 1999a)

1.1 Effects of No-Tillage systems

No-tillage systems are gaining increased attention by growers as a practical way to produce vegetables while improving soil quality parameters at the same time. Many vegetable growers are interested in NT production practices for many different reasons besides just decreasing the overall cost of production. No-tillage provides several advantages to conventional tillage (CT; e.g., plowing and disking of soil) including reduced soil and wind erosion, soil water conservation and the addition of organic matter to the soil (Barnes & Putnam 1983, Blevins et al. 1971). However, an adverse effect of NT vegetable production is that organic mulches or residues on the soil surface significantly decrease soil temperatures, especially during the spring, which can significantly reduce seedling emergence and vigor, and delay vegetable crop maturity. This is especially important since the earliness of production is often important in the marketing and profitability of many vegetable crops. Walters & Young (2008) found that soil covered with winter rye (*Secale cereale* L.) residues used for NT zucchini squash production will generally provide about a 5 to 6°C reduction in soil temperature compared to bare soil. Walters et al. (2007) found that the use winter rye enhanced early cucumber yields under drought conditions, but early yields were suppressed when cooler weather conditions prevailed. Thus, early yields of many vegetables can be significantly affected when using cover crop residues depending on seasonal growing conditions.

1.2 Importance of weed control in No-Tillage systems

Although the loss of vegetable crop earliness can be a problem, the adoption of NT practices by commercial vegetable growers has been limited primarily due to the lack of effective weed management in this type of production system. Although all vegetation is often herbicide-killed before planting to achieve a weed-free, stale-seedbed in NT systems, vegetable growers have been reluctant to use NT production practices due to the problem of weed control. Several researchers (Hoyt & Monks 1996, Hoyt et al. 1996, Masiunas et al. 1995, Shelby et al. 1988, Walters & Kindhart 2002) indicate that NT significantly increases weed problems in vegetable production. Weeds often become the primary problem in NT production systems, since between-row tillage is often used to reduce weed populations that preemergence (PRE) herbicides fail to control in CT systems. Weedy fields have been a major deterrent to the adoption of NT vegetable crops, especially when growers failed to achieve uniformly distributed, high-residue mulched fields (Morse 1999b, Morse et al. 2001). Although NT provides greater overall weed control compared to conventional tillage, which most likely results from tillage promoting weed germination by exposing dormant seeds to oxygen and sunlight (Teasdale et al. 1991), effective weed control soon after crop emergence is essential during the early part of the growing season to prevent weeds from suppressing vegetable crop yield and quality. Current weed control programs in NT systems recommend the use of nonselective herbicides to kill the cover crop or control emerged weeds in residues, followed by an application of PRE and then postemergence (POST) herbicides. No-tillage systems often require more herbicides than do CT programs (Wallace & Bellinder 1992). Although herbicides are typically used as the primary means for weed control in NT vegetable systems, those available for most vegetable crops do not provide season-long weed control. Thus, effective weed management systems must be developed for NT vegetable production before NT will be widely utilized by growers.

1.3 Effects of No-Tillage on weed species and population densities

The use of NT production practices has a definite influence on the weed species composition and density. Tillage practices have a significant influence on weed population and species composition by modifying the soil environment (Putnam et al. 1983). Weed species, soil seed density, seed production, and surface residue can all affect weed population dynamics under different tillage systems (Teasdale et al. 1991). Although cultivation generally stimulates weed emergence compared to NT production systems, the effects of NT on weed populations have been highly variable depending on environmental conditions as well as herbicide performance. Putnam (1986) and Putnam & DeFrank (1983) indicated that grasses and perennial weed species tend to dominate weed communities in reduced tillage systems over time, although shorter term studies have had mixed results (Masiunas et al. 1995, Teasdale et al. 1991). Research has indicated that although weed control by cover crop mulches lasts from about 30 to 75 days (Barnes & Putnam 1983, Creamer et al. 1996, Masiunas et al. 1995, Moore et al. 1994), weed species respond differently to cover crop mulches. Cover crop residues can also influence weed populations in NT cropping systems due to the proximity of the residue to the site of seed germination on the soil surface. Teasdale et al. (1991) indicated that tillage and cover crops can influence weed populations, the rate of population growth, and species composition. Seedling emergence of several weeds in NT was consistently suppressed by winter rye and wheat (*Triticum aestivum* L.) surface residues compared to CT (Blum et al. 1997). Generally, surface residues prevent germination of small seeded annual species that require light for germination, such as common lambsquarters (*Chenopodium album* L.) and various species of *Amaranthus* (e.g., redroot pigweed, *Amaranthus retroflexus* L.), whereas surface mulches do not generally prevent the germination of large-seeded annuals and perennials. Morse (1999a) indicated that fields planted to NT should not have serious perennial weed problems such as nutsedge (*Cyperus* spp.), quackgrass [*Agropyron repens* (L.) Beauv.], johnsongrass [*Sorghum halepense* (L.) Pres.], or morningglory (*Ipomoea* spp.). Furthermore, Masiunas et al. (1997) and Derken et al. (1993) indicated that prostrate pigweed (*Amaranthus blitoides* S. Wats.) was more common in CT than in cropping systems with surface residues. In NT and high residue mulch systems there tends to be more wind dispersed species [e.g., dandelion (*Taraxacum officinale* Weber)] than in CT (Bottenburg et al. 1997). Although NT cover crop systems reduce weed populations, the use of a residual herbicide is still required to prevent weed populations from increasing to severe field infestation levels. Kruidhof et al. (2009) indicated that the optimal cover crop residue management strategy for weed suppression depends on the cover crop species and the target weed species.

2. Production systems for No-Tillage vegetables

Several different production systems have been evaluated for NT vegetables. Although the vegetable crop planting method plays a major role in the successful production of NT vegetables (Table 1), the production system utilized for NT vegetables has shown to have a definite influence on the resulting productivity of the crop.

2.1 Bare soil systems

A bare soil NT system will improve weed control compared to CT (Moore et al. 1994, Walters et al. 2008), since the use of NT avoids bringing new weed seeds closer to the soil surface where they can germinate and increase weed densities (Derken et al. 1993). The presence of surface residue and lack of soil disturbance in NT were likely contributing factors that significantly influenced weed control even in the absence of herbicides (Walters et al. 2008). However, most growers will include cover crops when using NT for various reasons.

2.2 Cover crop based systems

Cover crops reduce the amount of soil erosion that is likely to occur compared to a bare soil system, as well as reducing water evaporation from the soil and increasing infiltration, generally resulting in greater soil moisture than bare soil systems. Cover crops will also provide additional weed control compared to that achieved with only a bare soil system. Furthermore, cover crops provide many other benefits that improve soil characteristics. A cover crop is any living ground cover that is planted into or after the primary crop and is commonly killed before the next crop is planted. The primary benefit of cover crops is reduction of water runoff and soil erosion, which ultimately results in improved soil productivity. Griffith et al. (1986) indicated that the use of cover crop residues for NT planting protects the soil surface from erosion by absorbing the impact of raindrops, thus reducing soil particle detachment and decreasing the acceleration of runoff; additionally, increased water infiltration and reduced soil water evaporation under NT generally increases plant-available water and subsequent crop yield potential. The presence of 1 to 2 $Mg{\cdot}ha^{-1}$ of crop residues on the soil surface on sloping lands can reduce water runoff and soil erosion losses by 40 to 80% compared to bare soil (Meyer et al. 1970).

Cover crop residues are often used as part of a weed management program in vegetable cropping systems (Leather 1983, Masiunas 1998, Putnam 1986). Cover crop mulch systems modify the microenvironment, which has an impact on weed populations and vegetable crop yields (Masiunas 1998). Although the combination of NT and cover-cropping practices have certain advantages over traditional tillage (e.g., plowing and disking of soil), there has been limited research in vegetable crops, particularly when used in conjunction with chemical weed control (Rapp et al. 2004). Cover crops are often integrated into NT production systems for their weed suppressive ability and numerous researchers have attempted to use cover crops as a method for weed control in vegetable crop production.

2.2.1 Types of cover crops used in NT systems

Two types of cover cropping systems are typically used in vegetable production, winter and summer annuals. The majority of efforts have focused on fall seeded cereal grains (especially winter rye) or perennial legumes, such as clovers (*Trifolium* spp.) or vetches (*Vicia* spp.), although some research has been conducted on the use of summer annuals for use as living mulches. Winter annual cover crops have been successfully incorporated into NT production systems and are the most widely used type of cover crop in NT systems. Summer annuals are rarely used in vegetable production, although they can provide several advantages during the summer months including weed suppression, increasing nitrogen levels in soil for subsequent crops, preventing the leaching of soil nitrogen, improving soil physical properties, and adding organic matter to the soil.

Winter annual cover crops include many cereal grains such as barley (*Hordeum vulgare* L.), wheat, and winter rye, as well as legumes like clovers and hairy vetch (*Vicia villosa* Roth.). Cereals like winter rye or wheat are the most popular cover crops, since that are relatively easy to establish and fast growing, and the seed is readily available and relatively inexpensive. In contrast, legumes do not cover the soil as quickly, although they do improve soil nitrogen levels that can be used by the following crop. An ideal cover crop should prevent erosion, suppress weeds, scavenge excess nutrients, and add organic matter back into the soil. Small grain cover crop residues suppress weeds by modifying light, temperature, moisture, and the chemical environment of germinating weeds (Putnam 1986,

Teasdale et al. 1991). Small-seeded annual weed species, such as redroot pigweed and common waterhemp (*Amaranthus rudis* Sauer), require light for germination and residues on the soil surface will prevent germination (Teasdale et al. 2004); however, as straw mulch residues decompose, more light reaches the soil surface, resulting in germination of these small seeded annual weeds and greater weed pressures later in the growing season. Thus, small grain cover crops can contribute to weed control in NT systems but herbicides or other weed control tactics are generally necessary to obtain optimal weed control and crop yield.

Several different cereal grains are often used as cover crops for vegetable production (Masiunas 1998). Barley is a fast growing, cool season, annual grain crop that can be used as a cover crop for vegetable plantings. This plant can provide nonchemical weed suppression, as it will often shade and smother weeds, or provide competition for soil moisture and nutrients. In addition, barley has an allelopathic effect on weed germination. Although winter wheat is typically grown as a cash grain crop, it can provide similar benefits as other cereal crops. It has shown to work well in NT systems for weed control and will generally provide similar results as winter rye (Walters & Young 2010). Winter rye is one of the best winter hardy cover crops as it overwinters well and produces considerable biomass (Putnam 1986, Weston 1990). It is also effective at capturing nutrients from soils and provides a persistent weed suppressive mulch during the summer, although it can remove soil moisture and immobilize nitrogen. However, some type of additional weed control is generally required with winter rye residue to provide season-long suppression of weeds (Masiunas et al. 1995, Teasdale 1993).

Many types of legumes can also be used in NT vegetable production systems. Although there are many types of clovers that can be included as cover crops, white clover (*Trifolium repens* L.) has been widely used in various vegetable crops. They are widely adapted perennial nitrogen producers that protect soils from erosion and suppress weeds. White clovers can be used in living mulch systems, as they can be broadcast over vegetables in late spring to establish under the primary vegetable crop (Infante and Morse 1996). This plant tends to grow slowly while shaded, and then grows more rapidly when it receives more light. Hairy vetch is widely used as a winter annual legume cover crop. It will consistently produce high biomass to suppress weeds with high nitrogen content and can be easily killed in the spring by herbicides, mowing, or rolling (Teasdale 1999). Teasdale (1993) also observed that light reduction may be more important than allelopathy or physical impedance for weed suppression by hairy vetch residues. However, it only captures limited amounts of nutrients from soils during the fall and winter and will only suppress weeds for a limited amount of time due to rapid decomposition.

Cover crop research has focused primarily on winter annual crops, although summer annual cover crops have potential applications in many regions. There are several summer annual cover crops that can be used including buckwheat (*Fagopyrum esculentum* Moench), sorghum-sudangrass [*Sorghum X drummondii* (Steudel) Millsp. & Chase], and soybean [*Glycine max* (L.) Merr.] (Creamer & Baldwin 1999). These plants when grown at close spacings will completely shade the soil surface which will suppress and outcompete weed growth. Buckwheat can be grown during the summer months and will effectively suppress weed growth and recycle nutrients. Sorghum-sudangrass is a warm-season annual grass that grows well under hot, dry conditions and will produce high amounts of biomass; it is also very effective at suppressing weeds, reducing soil erosion, and recycling nutrients. Soybean produces an erect, bushy plant that establishes quickly, improves nitrogen fertility in the soil, and suppresses weeds.

2.2.2 Effects of intercropping cover crops for NT systems

Intercropping of different cover crop species can overcome some of the problems associated with the use of a single cover crop species. The intercrop mixture of winter rye and hairy vetch is widely used for vegetable production as it produces more biomass, does a better job at protecting the soil, and provides better weed suppression than either species grown alone (Mangan et al. 1995, Schonbeck et al. 1993). Winter rye intercropped with hairy vetch provided 50% fewer weed seedlings compared to hairy vetch alone (Burgos & Talbert, 1996). Many research studies have indicated that cover cropping systems using winter rye in combination with hairy vetch can significantly suppress weeds in NT vegetable production systems and in some instances, can eliminate the need for additional weed control. However, in most situations, additional weed control measures will need to be implemented as the use of cover crops alone will generally provide insufficient total-season weed control.

2.2.3 Management of cover crops in NT systems

There are two basic approaches that are generally used in managing cover crops (Paine & Harrison 1993). In mulch residue systems, the cover crop growth is killed in some manner before planting the vegetable crop, whereas in living mulch (LM) systems, the companion crop grows at the same time as the vegetable crop. The cover crop management system used in NT really depends on the particular vegetable crop that is being grown, as certain crops are more amenable to residue mulch systems while others can be managed in living mulch systems.

Cover crops to be left on the soil surface for NT production must be killed, either with herbicides or in a mechanical manner (Creamer & Baldwin 1999). Fall planted cover crops, such as wheat or winter rye, are often herbicide-killed in the spring. Nonselective contact herbicides, such as glyphosate or paraquat dichloride, are often used to desiccate cover crops as well as other perennial and immature annual weeds growing in the field; and, these herbicides should be used within two weeks of seeding or transplanting the vegetable crop to ensure complete vegetative kill, otherwise another application will have to be made to kill the weeds that are present. Although many cover crops are herbicide killed, there are several methods for mechanically killing cover crops including undercutting, mowing and rolling. A rolling stalk chopper or similar device can also be used to roll down cover crops. Flail mowing and rolling can effectively kill mature winter rye, hairy vetch, crimson clover, wheat, and mixtures of winter rye and hairy vetch (Dabney et al. 1991).

Living mulches are cover crops planted either before or with the primary crop and are maintained as a living ground cover throughout the growing season of the crop. Living mulches grow alongside or within a vegetable crop and can significantly reduce weed populations, but they can often be difficult to manage in vegetable cropping systems because they compete with the crop. The use of LMs can minimize erosion, decrease soil temperatures, improve the rate of water infiltration, improve soil structure, enhance soil microbial activity and increase crop yield (Hartwig & Ammon 2002). Since LMs can compete for moisture and nutrients, they are not recommended for low-growing, shallow-rooted, or drought-susceptible vegetable crops. Various grasses, legumes, and *Brassica* species have been used as living mulches for NT vegetable production. Those LMs that are seeded and established shortly before the vegetable crop is planted tend to provide less competition to the crop and less weed suppression than those established several months prior to planting of the vegetable crop (Masuinas 1998). The success of a LM in reducing weed populations depend on its ability to rapidly establish a ground cover and smother weeds without competing with the vegetable crop (Putnam 1990). Living mulches that have been used in vegetable production include perennial ryegrass (*Lolium perenne* L.), creeping red fescue (*Festuca rubra* L. subsp. *commutate*), ladino clover (*Trifolium repens* L.), red clover (*Trifolium pratense* L.), and sorghum [*Sorghum bicolor* (L.) Moench]. Perennial ryegrass is low growing and will usually not grow higher than the vegetable crop which allows the vegetable crop to photosynthesize at somewhat a normal rate. Although clovers improve nitrogen fertility of soils, most clovers are low growing and produce allelochemicals (Harborne 1987) that suppress weed populations. However, LMs, such as perennial ryegrass, sorghum, and clovers compete with vegetable crops for light, moisture and nutrients (Shennan 1992). Bottenberg et al. (1997) and Masuinas et al. (1997) indicated that perennial ryegrass provided a better LM than did red clover in cabbage production; perennial ryegrass did not grow above the cabbage foliage, but the growth of red clover was able to overtop the cabbage which restricted light from reaching the crop canopy causing yield reductions.

The application of low herbicide rates may reduce the competitiveness of LMs, allowing their use in vegetable production. However, Walters & Young (2008) found that a NT herbicide suppressed winter rye LM system provided excessive amounts of zucchini squash stunting which significantly reduced yield. Nicholson & Wien (1983) also suggested that light competition between a white clover LM and cabbage probably reduced crop yields due to the shading of lower cabbage leaves by excessive clover growth. In contrast, Infante & Morse (1996) indicated that legumes intentionally seeded (or interseeded) into a standing crop can be effectively established in NT after transplanting broccoli to suppress weeds without reducing crop yield. Although interseeded cover crops often suppress weeds, they also generally result in vegetable crop yield reductions compared to other more conventional weed management practices. These reductions in crop yields are often due to the direct competition between the cover crop and the main crop. Although Masiunas et al. (1997) found that a hairy vetch LM established before transplanting cabbage would suppress weeds, it would eventually grow above the cabbage canopy later in the growing season and reduce yields. The delayed seeding of cover crops has been an effective means of minimizing yield losses in many crops including broccoli (Brainard & Bellinder 2004). Furthermore, Brainard et al. (2004) found that hairy vetch seeded at 20 days after transplanting cabbage might provide an alternative for weed control in this crop since it: 1) provides significant biomass for soil improvement; 2) does not reduce cabbage yields; and 3) provides some weed suppression.

Living mulches are more difficult to manage than conventional cropping systems and are not suitable in all situations. The interspecific competition for light, water, and nutrients between the LM and vegetable crop can limit the use of the system (Fisher & Burrill 1993; Galloway & Weston 1996). The careful selection of less vigorous genotypes is essential for the success of living mulches for vegetable production systems (Nicholson & Wien 1983). Furthermore, the competitiveness of LMs can be reduced by using strip tillage systems, mowing, or using reduced rates of herbicides (Hoyt et al. 1994, Paine & Harrison 1993).

2.2.4 Effects of allelochemicals produced by cover crops for NT systems

Cover crop mulches often release allelochemicals that aid in suppressing weed populations. Those cover crops that contain a high level of allelochemicals are well-suited for mulch residue mediated weed suppression. The soil surface coverage provided by cover crops mulches often correlate with weed suppression (Teasdale et al. 1991). The amount of soil surface coverage is important since mulches block the light stimulus that is required for the germination of many small seeded weed species (Barnes & Putnam 1983, Moore et al. 1994, Teasdale 1993). When cover crops, such as winter rye, are incorporated into the soil, the resulting weed control is often significantly reduced (Walters & Young 2010, Walters et al. 2008), which is most likely due to several factors including less soil surface coverage by mulch residues, bringing new weed seed to the soil surface, quicker decomposition of incorporated residues, and lower levels of allelochemicals in the weed seed germination zone (Masiunas 1998). Generally, once cover crop residues are incorporated into the soil, allelochemicals quickly decompose and are leached away from the upper soil levels where weed seeds germinate (Dias 1991). In contrast, when cover crop residues remain on the soil surface, weed growth is suppressed for a longer period of time since allelochemicals degrade slower (Masiunas 1998).

The residues of many cover crops release allelochemicals that inhibit weed seed germination and growth (Creamer et al. 1996, Mwaja et al. 1995, Ohno et al. 2000, Weston 1996). Wheat residues contain ferulic acid (4-hydroxy-3-methoxycinnamic acid) which has been shown to inhibit the germination and root growth of many important weeds including large crabgrass (*Digitaria sanguinalis* (L.) Scop.), pitted morningglory (*Ipomoea lacunosa* L.), common ragweed (*Ambrosia artemisiifolia* L.), and prickly sida (*Sida spinosa* L.) (Hicks et al. 1989, Liebl & Worsham 1983). Furthermore, in barley, the alkaloid gramine has been shown to inhibit weed growth (Harborne 1987). Sorghum residues have been shown to contain the phenolic compounds, p-coumaric, m-hydroxybenzoic, and protocatechuic acids which can inhibit weed seed germination and seedling growth (Lehle & Putnam 1982, Panasiuk et al. 1986, Weston et al. 1989). Oil seed rape (*Brassica napus* L.) releases glucosinolate breakdown products, including isothiocyanates, oxazolidinethiones, ionic thiocyanate and organic cyanides (Brown & Morra, 1996, Haramoto & Gallandt 2004).

Compared to other cover crops, the effect of allelochemicals on weed suppression has been extensively studied in winter rye. This cover crop is especially important since it has been widely documented to suppress the density of weeds in NT production systems (Barnes & Putnam 1983, Teasdale et al. 1991, Weston 1990, Zasada et al. 1997). Putnam & DeFrank (1983) reported that winter rye reduced the emergence of common ragweed by 43%, green foxtail [*Setaria viridis* (L.) Beauv.] by 80%, redroot pigweed by 95% and common purslane (*Portulaca olearacea* L.) by 100%. Shilling et al. (1985) indicated that winter rye residues used in a NT system reduced the biomass of common lambsquarters by 99%, redroot pigweed by 96% and common ragweed by 92% compared to a non-mulched tilled control. Barnes & Putnam (1983) found that winter rye provides better weed control if allelochemicals are actively produced in roots and released into the soil; and, once the winter rye plant dies, most weed control is achieved through the decaying mulch on the soil surface simply providing a physical barrier to weed germination and growth. Furthermore, Yenish et al. (1995) reported that the duration of weed suppression by a winter rye cover crop more closely follows the disappearance of allelochemicals from residues than the disappearance of the residue itself. The decomposing winter rye residues on the soil surface produce a wide range of allelochemicals including phenylacetic acid, 4-phenylbutric acid, 2,4-dihydroxy-1,4-benzoxazin-3-one (DIBOA), and 2-benzoxazoline (BOA). Both DIBOA and BOA have been shown to inhibit germination and seedling growth of several grass and broadleaf weed species (Barnes & Putnam 1983, Chase et al. 1991, Creamer et al. 1996, Shilling et al. 1985). The amounts of DIBOA and BOA found in rye often correlates with the inhibition of weed growth (Mwaja et al. 1995, Yenish et al. 1995). Furthermore, Chase et al. (1991) indicated that large-seeded weed species or those species that have deeper seed placement in the soil profile were less affected by allelochemicals produced by winter rye, which was most likely due to higher concentrations of allelochemicals near the soil surface where small-seeded species typically germinate. The decline in DIBOA concentrations as winter rye matures, and the fact that many winter rye cultivars mature at different rates, may partially explain the discrepancies observed in previous studies from weed suppression from winter rye (Reberg-Horton et al. 2005).

Although the allelopathic potential of hairy vetch and other legume cover crops have been documented, weed suppression by legumes is generally less compared to grass cover crops. Hairy vetch residues contain many different compounds including alcohols, aldehydes, furans, and monoterpenes which have the ability to suppress weed growth (Bradow & Connick 1990). Hairy vetch residues decompose more rapidly than those of winter rye (Mohler & Teasdale 1993, Schonbeck et al. 1993) and surface coverage by hairy vetch residues has been shown to be more important for weed control than allelochemical release (Curran et al. 1994, Teasdale et al. 1991). Therefore, weed control by hairy vetch residues is lower compared to that obtained by winter rye residues.

3. Importance of herbicides for No-Tillage vegetables

Some type of herbicide application is generally required to optimize weed control and maximize vegetable productivity in NT systems, since cover crop residues can generally be expected to provide early-season weed suppression (for about the first 4 to 6 weeks) but not full-season weed control. In NT squash (Walters et al. 2004, 2005) and cucumber (Walters et al. 2007) production systems, the use of winter rye enhanced weed control even when a standard herbicide program was used. In NT pumpkin production, redroot pigweed and common waterhemp control was achieved early in the growing season with only winter rye residues but control was not sustained throughout the growing season and required some other method of weed management, such as herbicides, to optimize control (Walters et al. 2008). As stated earlier, broadleaf weed control is a major problem in NT vegetable production systems because there are few herbicides available in most vegetable crops to control these weeds, and the number of available herbicides is dependent on the specific vegetable grown. Although there are several PRE herbicides available for many different vegetable crops, the emergence of weeds after soil residual herbicides have lost their activity often leads to excessive weed populations later in the production season. In comparison, between-row tillage is often used in CT to reduce weed populations that PRE herbicides fail to control.

3.1 Preemergence herbicides for No-Tillage vegetables

Preemergence herbicides are widely used in NT vegetable production to control weeds. These are applied to the soil surface prior to crop or weed emergence, and when transplants are used, they are generally applied to weed-free soil before transplanting. Preemergence herbicides should be effective for at least 30 days, since this would prevent establishment of early germinating weeds from providing excessive competition to vegetable plants. Stall (2001) indicated that weeds emerging during the first four weeks of cucurbit crop establishment will suppress crop yield, while those emerging after this time period will generally not reduce yields.

There are only a few labeled PRE herbicides for use in NT production systems for cole crop, cucurbit and solanaceous vegetables, including bensulide, clomazone, ethalfluralin, halosulfuron-methyl, napropamide, oxyfluorfen, rimsulfuron, and S-metolachlor (Table 2). Bensulide is applied either preplant or PRE in many vegetable crops, especially many of those within the cole crop, cucurbit, and solanaceous vegetable groups. However, the delay of irrigation or rainfall to activate this herbicide in the soil by more than 36 hours may result in poor weed control. Although this herbicide is registered for control of many different annual grasses and a few broadleaf weeds, it tends to provide only marginal weed control even at low weed infestation levels. Clomazone has been registered for a number of vegetable crops for many years, and is a valuable herbicide that is critical for weed control programs in several crops, including cabbage and some cucurbits. Although clomazone is an excellent PRE grass herbicide, it only controls a limited number of broadleaf weed species. Ethalfluralin is a PRE herbicide that provides control of a few broadleaf and grass weeds in cucurbit vegetable crops. However, clomazone and ethalfluralin are typically sold as a pre-mix that is used in cucurbit vegetable crops. Prior to the labeling of halosulfuron-methyl and S-metolachlor for pumpkin, this clomazone and ethalfluralin pre-mix was widely used by most growers since it provided control of many different weed species. The bleaching of crop and weed leaves will often be observed due to the clomazone component in the mixture. The clomazone and ethalfluralin pre-mix will provide PRE control of annual grasses and many broadleaf weeds including common lambsquarters, various pigweed

Herbicide	Vegetable crop	Application	Weeds controlled
Bensulide	Br, Ca, Cu, Pu, Sq	PRE	Broadleaves, grasses
Clethodim	Br, Ca, Cu, Pu, Sq, To	POST	Grasses
Clomazone	Ca, Cu, Sq	PRE	Broadleaves, grasses
Clomazone + Ethalfluralin	Pu, Sq	PRE	Broadleaves, grasses
Ethalfluralin	Pu, Sq	PRE	Broadleaves, grasses
Halosulfuron-methyl	Pu, Cu, To	PRE, POST	Broadleaves, nutsedges
Metribuzin	To	PRE, POST	Broadleaves, grasses, nutsedges
Napropamide	Br, Ca, To	PRE	Broadleaves, grasses
Oxyfluorfen	Br, Ca	PRE	Broadleaves
Rimsulfuron	To	PRE	Broadleaves, grasses
S-metolachlor	Pu, To	PRE	Broadleaves, grasses, nutsedges
Sethoxydim	Br, Ca, Cu, Pu, Sq, To	POST	Grasses

Br = Broccoli, Ca = Cabbage, Cu = Cucumber, Pu = Pumpkin, Sq = Squash, To = Tomato. PRE is preemergence to weed emergence and POST is postemergence to vegetable crop. The herbicides, bensulide, clomazone, ethalfluralin, metribuzin, napropamide, oxyfluorfen, rimsulfuron, and S-metolachlor provide control of only a limited number of broadleaf or grass weed species. The control of specific weed species by halosulfuron-methyl or metribuzin depends on whether applied PRE or POST. Nutsedges are *Cyperus* spp.

Table 2. Selected herbicides for use in no-tillage production systems for various vegetable crops.

species, common purslane, velvetleaf (*Abutilon theophrasti* Medik), common ragweed, and Pennsylvania smartweed *(Polygonum pensylvanicum* L.). Halosulfuron-methyl is a herbicide that can be used both PRE and POST in cucurbit vegetable crops and several different fruiting vegetables including tomato. Although this herbicide will provide PRE control of many different broadleaf weeds, it provides no grass weed control and only suppresses yellow (*Cyperus esculentus* L.) and purple nutsedge (*Cyperus rotundus* L.). Heavy rains following PRE applications of halosulfuron-methyl can often lead to severe crop injury. Napropamide can be applied PRE to either direct seeded or transplanted cole crops, such as broccoli and cabbage, and solanaceous fruiting vegetables including pepper (*Capsicum annuum* L.) and tomato. Although napropamide does not control established weeds, it will provide PRE control of numerous annual broadleaf and grass weeds. Oxyfluorfen can be used PRE in both broccoli and cabbage crops as a pre-transplant treatment to provide control of carpetweed (*Mollugo verticillata* L.), redroot pigweed, common purslane, and Pennsylvania smartweed. Pre-transplant applications of oxyfluorfen may result in early leaf cupping or crinkling crop injury and is more severe if crop leaves directly contact treated soil, although crops will rapidly outgrow this injury. However, more severe crop injury will result if transplants are under some type of stress. It is important to note that oxyfluorfen should not be applied to soil if an acetanilide herbicide such as S-metolachlor has been applied to the field during the current growing season as severe crop injury may occur. Rimsulfuron can be used PRE in tomato for the control of a wide variety of broadleaf and grass weeds, although it will only provide partial control of weeds such as crabgrass (*Digitaria* spp.), common cocklebur (*Xanthium strumarium* L.), common lambsquarters, common ragweed, velvetleaf, and black (*Solanum nigrum* L.) and hairy nightshade [*Solanum villosum* (L.) Mill.]. Preemergence applications may not provide adequate control of weeds > 2.5 cm in height or weeds that have an established root system prior to the activation of rimsulfuron. S-metolachlor can be applied PRE in NT pumpkins and tomatoes for control of numerous grasses and broadleaf weeds, as well as yellow nutsedge. S-metolachlor will not control emerged weeds and must be applied to a weed-free soil or in tank mixtures with products that provide POST control of weeds present at the time of application; tank-mixtures of S-metolachlor with a contact herbicide (glyphosate or paraquat dichloride) can be used to provide POST control of many weeds, with later germinating weeds controlled by S-metolachlor.

It is important to note that moisture is essential for activation of these PRE herbicides once they have been applied. Within 5 to 7 days after application, about 12 to 25 mm of rainfall or sprinkler irrigation is needed to activate most PRE herbicides used in vegetable production. If adequate moisture is not provided during this time period then weed control provided by PRE herbicides will be drastically reduced. For those herbicides that can be used PRE and POST, if moisture cannot be managed via rainfall or sprinkler irrigation, allowing weeds to emerge and then applying POST will most likely result in better weed control.

3.2 Postemergence herbicides for No-Tillage vegetables

The use of effective POST herbicides for control of annual grass and broadleaf weeds is important to achieve success in NT vegetable production systems. Hoyt et al. (1994) indicated that the lack of effective POST herbicides is a problem for NT vegetable production systems. Although grass weeds can be effectively controlled in most vegetable crops with minimal effort, many difficult to control broadleaf weeds are not sufficiently controlled due to the lack of effective herbicides that can be sprayed POST. Postemergence grass control can usually be obtained in most vegetables with the use of clethodim or sethoxydim. Both of these herbicides are widely used selective POST herbicides that control annual and perennial grass weeds in broadleaf vegetable crops. Although several vegetable crops including cabbage, cucumber, pumpkin, and tomato have registered POST broadleaf weed herbicides which have been successfully used in NT systems, many vegetables crops including squash lack labeled effective POST broadleaf herbicides. There are a limited number of POST herbicides labeled for broadleaf weed control in NT cole crop, cucurbit and solanaceous vegetables, including halosulfuron-methyl, metribuzin, and rimsulfuron. Although halosulfuron-methyl is often applied PRE, POST applications can be made to established plants of cucumber, pumpkin and tomato for control of yellow nutsedge, redroot pigweed, velvetleaf, common ragweed, and many other broadleaf weeds. For optimum control of yellow and purple nutsedge, sequential applications on areas where this weed has emerged or re-grown may be required. However, there is the potential for crop stunting and a slight maturity delay with the use of halosulfuron-methyl when used POST. Many times, POST applications will often extensively slow the growth of cucurbit vines. To reduce injury, it can also be used as a directed POST application to row middles for many different vegetable crops. If halosulfuron-methyl is applied POST on drought stressed weeds, its activity will most likely be reduced and the resulting control is often inadequate. The carryover from halosulfuron-methyl is 0 to 36 months depending on the next crop that is grown and this should be considered when this herbicide is used. Metribuzin is often applied POST in established tomatoes and provides effective control of many different broadleaf weeds and a few grass weeds. Repeated POST applications of metribuzin are often required to provide optimal control of several weeds such as jimsonweed (*Datura stramonium* L.), common ragweed, and velvetleaf. If applications are made to tomato growing under stressful conditions, crop injury or delayed maturity may result. Rimsulfuron can be used POST in tomato to control a wide range of broadleaf and grass weeds. Weed control is best achieved when POST applications of rimsulfuron are made to actively growing weeds that are less than 2.5 cm in height. Applications should be made after tomato plants reach at least the cotyledon stage. Similar to metribuzin, if applications are made to tomato growing under stressful conditions, temporary crop chlorosis may occur, but symptoms normally disappear within 2 weeks. To optimize weed control in tomato, PRE and then POST or sequential POST applications of rimsulfuron can be made.

4. No-Tillage vegetable cropping systems

4.1 Brassicas

4.1.1 Cabbage

Several researchers found that cabbage yields in NT were similar to that of CT (Hoyt et al. 1996, Morse 1995, Morse & Seward 1986), although Knavel and Herron (1981) indicated that spring cabbage yields were reduced in NT when compared to CT. Wilhot et al. (1990) related cabbage yield reductions in NT to poor plant establishment/impeded crop growth more than to the effects of a NT production system. Furthermore, similar to what has been observed for other vegetable crops, when weed control methods and PRE herbicides were used in NT cabbage, yields were similar to those of CT (Bellinder et al. 1984).

Weed control has been the limiting factor for implementation of NT cabbage production systems. Masiunas et al. (1997) indicated that the cropping system utilized affected both broadleaf and grass weed densities in NT cabbage production. Although winter rye mulch is often utilized in NT cabbage production, Morse & Seward (1986) indicated that hairy vetch and Austrian winter pea [*Pisum sativum* spp. *arvense* (L.) Poir.] were better mulch covers than winter rye for NT cabbage production, which was most likely due to the nitrogen released by the two legumes through mineralization of the plant residues. Furthermore, Schonbeck et al. (1993) indicated that hairy vetch produced greater cabbage yields than winter rye, which was most likely due to the immobilization of soil nitrogen due to the high carbon to nitrogen ratio in winter rye. In contrast, Masiunas et al. (1997) found that the use of fall-seeded winter rye was the most promising mulch system in NT cabbage for weed suppression; and, the weed suppression obtained from the winter rye NT system was similar to that obtained from CT using trifluralin applied PRE. Winter rye mulch suppressed broadleaf weed emergence for 6 weeks compared to CT (Masiunas et al. 1997).

Living mulches also show some potential for suppressing weeds in cabbage and other *Brassicas*. The use of LMs for all or part of a *Brassica* crop growing season is becoming of interest to growers to extend weed control for a more sustainable weed management system. The integration of cover crops by interseeding into an established vegetable crop may serve to provide a more effective way to manage weeds. Castello (1994) found that broccoli head size and weights in a NT living mulch system using white clover was similar to that produced in CT. In contrast, Brandsaeter et al. (1998) found that clover interseeded in cabbage provided some late-season weed suppression, but this alternative weed management strategy tended to reduce cabbage yields.

4.1.2 Broccoli

Broccoli is a crop that can be easily produced in a NT production system. Abdul-Baki et al. (1997) found that fall-produced broccoli yields were similar between NT and CT production systems when surface residues from a killed summer cover crop provided sufficient soil coverage in the NT system. Furthermore, Morse (1995) indicated that yields of broccoli grown in NT increased by about 10% compared to CT. Broccoli transplant establishment in NT was found to be similar or better compared to CT, which directly related to the high yields observed in NT (Infante and Morse, 1996). Lastly, Morse (2000) indicated that broccoli yields increased in a NT cover crop mulch system compared to a NT bare soil system.

Production systems for NT broccoli can often be successful without using herbicides, when appropriate high-residue cover crops are effectively killed by flail mowing or rolling and broccoli transplants are properly established and maintained in these evenly distributed cover crop mulches (Morse 1999b). Morse (2001) indicated that the use of a winter rye and hairy vetch mixture that was rolled in the spring was the best combination evaluated for production of NT summer broccoli. The use of forage soybean or foxtail millet (*Setaria italica* L.P. Beauv) mulch alone or in combination provided NT yields that were similar to CT, with applied herbicides having little influence on broccoli productivity (Abdul-Baki et al. 1997). Although NT broccoli yield is inversely correlated with the amount of weed biomass produced (Morse 2001), the use of herbicides can be reduced when large, vigorous transplants of broccoli are set in narrow double-rows in persistent, heavily mulched NT production systems, since this will result in significant amounts of weed suppression.

Many different cover crops and herbicides have been utilized to improve NT broccoli production. Broccoli produced larger heads and higher yields in a NT system utilizing a combination of a legume (e.g., hairy vetch) with winter rye than with winter rye alone or no cover crop (Mangan et al. 1995). Although cover crops, such as winter rye or hairy vetch, integrated into NT *Brassica* crop production systems improve weed control, the use of herbicides is still required to provide a more effective weed management system. Hoyt et al. (1996) indicated that the use of oxyfluorfen PRE prior to tranplanting significantly improves weed control and increases the success of using NT for cabbage production. The application of pretransplant herbicides, such as metolachlor or oxyfluorfen, generally reduces weed biomass in NT broccoli production (Abdul-Baki et al. 1997). Although there are many PRE or pre-transplant herbicides available for use in NT broccoli and cabbage, the lack of POST herbicides for broadleaf weed control still remains a major hindrance to the adoption of NT practices, since the lack of late-season broadleaf weed control will affect both yield and harvest efficiency.

4.2 Cucurbits

4.2.1 Cucumber

Similar to many other vegetable crops, cucumbers are generally managed with CT practices, such as plowing and repeated cultivations (Lonsbary et al. 2004). Weston (1990) found that cucumber, similar to most other cucurbits, was easy to establish in NT culture. Ogutu & Caldwell (1999) found that the use of cucumber transplants provided more biomass accumulation at 3 weeks after planting in NT resulting in higher early yields due to earlier flowering and fruit set than those that were direct seeded. Furthermore, although pickling cucumber leaf number, leaf area index and vine growth were reduced by NT, no reduction in total yield was observed compared to CT; and, the reduced vegetative growth in NT may actually be an advantage for the mechanical harvesting of this crop (Lonsbary et al. 2004). Adequate weed control in NT cucumber production systems must be achieved in some manner before this system will be widely used for this crop (Walters et al. 2007). The most consistent establishment of cucumber plants in NT occurred in winter wheat or rye residues, which provided a substantial level of weed suppression for at least 60 days following herbicide application to cover crops (Weston 1990). Walters et al. (2007) indicated that a winter rye cover crop alone would provide some but not sufficient, season-long redroot pigweed and smooth crabgrass [*Digitaria ischaemum* (Schreb. *ex* Schweig.) Schreb. *ex* Muhl.] control for cucumber grown in NT. Although broadleaf weed control is improved, herbicide-killed winter rye will not sufficiently suppress many difficult-to-control broadleaf weeds (depending on seasonal growing conditions) in NT cucumber production even if used with the standard PRE herbicide combination of clomazone + ethalfluralin + halosulfuron.

Weaver (1984) indicated that if cucumber plants are kept weed-free for the first four weeks after planting, yields would be similar to those kept weed free for the entire growing season. Thus, an appropriate PRE herbicide would appear to need a residual period of 24 to 36 days, as this would prevent establishment of early germinating weeds which provide excessive competition to young cucumber seedlings (Friesen 1978). Although cucumber will provide some weed suppressive ability once it forms a vine across the soil surface, other weed control measures are generally necessary to achieve adequate weed control.

The use of clomazone and ethalfluralin does not provide consistent satisfactory weed control in NT cucumber culture (Ogutu & Caldwell 1999, Walters et al. 2007). The PRE herbicide mixture of clomazone + ethalfluralin + halosulfuron provides both broadleaf and grass weed control and high cucumber yields in a NT production system when used in combination with a winter rye cover crop (Walters et al. 2007). However, many-difficult to control broadleaf weeds are not adequately controlled by these herbicides when used PRE in a cover crop residue NT system, which often provides various problems for cucumber growers including yield suppression and reduced harvest efficiency. An advantage for cucumber compared to squash is that halosulfuron can also be used POST to suppress many difficult-to-control broadleaf weeds.

4.2.2 Squash

Many squash growers are interested in NT production because of the ecological and potential economic benefits provided by this type of production system. Growers tend to apply PRE herbicides regardless of whether CT or NT is used, with squash seeded before herbicide treatment or transplanted in herbicide treated soil. However, the emergence of weeds after soil residual herbicides have dissipated leads to excessive weed problems during fruit harvest (Walters et al. 2005). Harvesters cannot locate squash fruit as easily on plants that are shaded by weeds compared to those growing in a weed-free field, and this contributes to reduced harvest efficiency and yield loss. Several studies have indicated that squash grown in NT have similar yields to those grown in CT (Knavel & Herron 1986, NeSmith et al. 1994, Walters & Kindhart 2002, Walters et al. 2005), although yields were only comparable if weeds were adequately controlled. Since a major limitation to NT squash production is weed control, improved weed management practices must be developed before NT systems in this crop will be readily adopted.

Although clomazone + ethalfluralin is the PRE herbicide mixture most often utilized by squash growers, it often provides poor control of certain broadleaf weeds, such as the various species of *Amaranthus*. Walters et al. (2004, 2005) found that a PRE application of clomazone + ethalfluralin resulted in the best overall weed control without having a detrimental effect on zucchini squash yields in NT. Although applying clomazone + ethalfluralin PRE to winter rye residues in NT squash production improved redroot pigweed control compared with no herbicide, the level of control was generally not adequate (< 85% control) by 8 weeks after planting (Walters et al. 2005). Clomazone + ethalfluralin did not provide sufficient season-long weed control, which especially caused problems in locating squash fruit during hand-harvesting. Walters et al. (2004) indicated that although the herbicide clomazone and the no-herbicide produced high early-season squash yields in NT culture, the productivity in these treatments declined as weed pressures increased due to limited weed control. The PRE herbicide combinations of clomazone + ethalfluralin and clomazone + imazamox provided the best overall weed control without having detrimental effects on squash yields in a NT system (Walters et al. 2005).

Living mulch systems have been shown to generally result in crop yield reductions compared to more traditional weed control methods (Liebman & Staver 2001, Teasdale 1998, Wiles et al. 1989). Walters & Young (2008) found that a NT winter rye living mulch system provided excessive amounts of zucchini squash stunting which significantly reduced yields. Furthermore, as a living mulch in NT squash production, winter rye resulted in 80 and 82% control of redroot pigweed and smooth crabgrass about 8 weeks after transplanting, respectively, in the absence of herbicides compared to the no herbicide bare soil system (Walters & Young 2008).

Few herbicides are labeled for squash production and none will consistently provide season-long weed control (Walters et al. 2004), since most, except the POST grass herbicides, are only labeled for PRE applications. Herbicides available for use in squash include bensulide (PRE), clethodim (POST), clomazone (PRE), ethalfluralin (PRE), and sethoxydim (POST) (Table 2). Although the PRE combination of clomazone + ethalfluralin is widely used in squash production, the weed control provided by this herbicide combination is generally inadequate. Due to the limited number of herbicides and inadequate weed control of those herbicides available for use in summer squash, registration of additional herbicides or the development of alternative methods of weed control is needed to allow for the widespread use of NT in this crop (Walters et al. 2004).

4.2.3 Pumpkin

Several studies have all indicated that NT and CT produce comparable pumpkin yields when sufficient weed control is achieved in NT production systems (Galloway & Weston 1996, Rapp et al. 2004, Walters et al. 2008). Pumpkin vegetation will provide some soil shading and weed suppression once vines form across the soil surface, but other weed control measures are generally necessary to achieve adequate weed control in NT pumpkin production. The use of herbicides and cover crops often play an important role in the management of weeds in NT pumpkin production.

The use of effective herbicides in combination with cover crops integrated into NT planting systems may provide a feasible option for pumpkin growers trying to enhance weed control. Although Harrelson et al. (2007) indicated that all cover crop residues evaluated, which included winter wheat, winter rye, perennial ryegrass, triticale (×*Triticosecale rimpaui* Wittm.), barley, oats (*Avena sativa* L.) and crimson clover (*Trifolium incarnatum* L.), produced acceptable NT pumpkin yields and fruit size, small grain cover crops, such as winter wheat or winter rye, are generally used to suppress weed densities in NT pumpkin production systems (Morse et al. 2001, Walters et al. 2008). The presence of surface residue and lack of soil disturbance in NT pumpkin production were likely contributing factors that significantly influenced weed control even in the absence of herbicides (Walters et al. 2008). In NT pumpkin production systems, sparse or unevenly distributed cover crop residues often result in fields having high weed densities that lead to low pumpkin yields and poor fruit quality (Morse et al. 2001). Several studies have indicated that although broadleaf weed control is improved, herbicide-killed winter rye will not effectively suppress many broadleaf weeds in NT pumpkin production even if used in conjunction with a standard herbicide program (Rapp et al. 2004, Walters & Young 2010, Walters et al. 2008). Weed densities in pumpkin vary with environmental conditions, tillage strategy, and amount of cover crop residue (Rapp et al. 2004). Although crop residues provided by herbicide-killed winter wheat or winter rye will improve grass and broadleaf weed control, the densities of many difficult to control broadleaf weeds (e.g., many *Amaranthus* spp.) in NT pumpkin production at harvest remain similar to those produced in bare soil. Winter rye or winter wheat cover crop residues alone will provide some, but insufficient weed control for pumpkins grown in NT (Walters et al. 2008, Walters & Young 2010).

The production of high-residue, evenly distributed mulches over the soil surface can enhance weed suppression in NT pumpkins, which can often reduce or even eliminate the need for PRE herbicides (Morse et al. 2001). In growing seasons with high weed pressures, winter rye residues without herbicide application were effective in suppressing weed populations in pumpkins for only about 6 to 7 weeks after planting (Rapp et al. 2004). Walters et al. (2008) found that redroot pigweed and common waterhemp control in NT pumpkins was achieved early in the growing season with only winter rye residues, but control was not sustained throughout the growing season. Pumpkin productivity in NT production systems was highly correlated with giant foxtail (*Setaria faberi* Herrm.), common cocklebur, redroot pigweed, and total weed control, with correlations indicating that pumpkin yields increased with greater weed control (Walters et al. 2008). Furthermore, pumpkin fruit number and weight, as well as average fruit size were correlated with both early- and late-season control of all weed species ($0.47 \geq r \leq 0.86$, $P \leq 0.01$; Walters et al. 2008).

The lack of effective herbicides has hindered the adoption of NT pumpkin production. Weed control is essential to obtain the highest possible pumpkin yields in NT production systems and tank mixtures of various herbicides are generally necessary to maximize weed control (Brown & Masiunas 2002, Kammler et al. 2008). Although weeds are a major problem in pumpkin NT production systems, there are a limited number of registered herbicides available for weed control. The majority of herbicides registered for pumpkins are used PRE and provide limited control of broadleaf weeds and nutsedge (Grey et al. 2000, Brown & Masiunas 2002) and are often ineffective when weather conditions are not ideal for activation. Several registered herbicides including clomazone + ethalfluralin, halosulfuron-methyl and S-metolachlor have made NT more successful, since cultivation is not an option in this type of production system. Walters et al. (2008) found that PRE use of clomazone + ethalfluralin or clomazone + ethalfluralin with halosulfuron-methyl tended to improve weed control in a NT, winter rye residue production system. Galloway & Weston (1996) found that ethalfluralin applied alone provided only short term weed suppression with control observed for only 4 to 5 weeks after application. Rapp et al. (2004) found that PRE application of ethalfluralin and halosulfuron, provided effective weed control in a NT winter rye residue production system.

Walters & Young (2008) indicated that although cover crops, such as winter wheat or winter rye, can be integrated into NT pumpkin production systems along with labeled herbicides to improve weed control, improvement in weed management systems beyond current practices and available herbicides is still necessary to maximize pumpkins yields. In NT pumpkin production systems, the potential yield reduction from herbicide injury does not outweigh the yield gains that are provided by reliable and effective weed control (Rapp et al. 2004, Walters et al. 2008). Thus, the judicious use of herbicides is an important part of any effective weed management program for NT pumpkin.

4.3 Tomato

The few studies on NT tomato production systems have provided conflicting results. Although Beste (1973) indicated that yields of direct seeded processing tomatoes grown in NT were similar to those grown in CT, Doss et al. (1981) reported that marketable staked tomato yields decreased in NT compared to CT. However, Shelby et al. (1988) indicated that the use of NT is a feasible alternative to CT for tomato production, since staked tomato yields in CT were generally comparable to yields obtained in NT. Furthermore, staked tomatoes in a NT hairy vetch mulch system yielded higher than CT tomatoes (Abul-Baki & Teasdale 1993). The production of fresh-market tomatoes in NT hairy vetch residue has been successful in providing high economic returns, especially in regards to reducing herbicide and nitrogen inputs (Teasdale 1999).

Fall-seeded winter rye can be used as a weed management tool in NT tomato production, as tomato yields using winter rye residues were comparable to treatments without winter rye, provided that weed control was sufficient (Smeda & Weller 1996). Masiunas et al. (1995) indicated that winter rye residues in reduced tillage cropping systems can provide weed control and tomato yields similar to those in CT systems that have had a pre-plant incorporated (PPI) soil application of trifluralin and metribuzin. Furthermore, the non-selective herbicide, glyphosate, that was used to kill the winter rye, also resulted in eliminating any existing winter annual weeds, which tended to result in similar or higher tomato yields compared to those obtained by just mechanical mowing of the winter rye. However, in most field situations, additional POST weed management would be necessary to maintain control of weeds through the critical weed-free period (about 6 weeks after transplanting) for tomato, although in some instances, winter rye residues can suppress weeds for up to 60 days after transplanting (Masiunas et al. 1995).

The availability of several POST herbicides for both broadleaf and grass weed control in tomato (Table 2) provides a greater overall potential to achieve optimum weed control compared to many other vegetable crops. Although many PRE or pre-transplant herbicides are available for use in tomato, the availability of POST herbicides are important to control later emerging weeds that can affect both yields and harvest efficiency. In NT tomato, redroot pigweed and morningglory control at 68 and 93 DAT was adequate with POST applications of metribuzin (Shelby et al. 1988). Sequential POST metribuzin applications followed by a POST grass herbicide in NT tomato provided adequate broadleaf and grass weed control that resulted in high marketable yields. These sequential POST applications in NT tomato production mimic the use of cultivation in CT tomato to provide effective late season weed control.

5. Summary and conclusions

The effectiveness of NT production systems depends on the vegetable crop grown, the crop establishment method, establishment of high residue mulch from a cover crop on the soil surface, and available PRE and POST herbicides for the vegetable crop. NT systems seem to work better in those vegetables that: 1) have vines that rapidly spread across the soil surface (e.g., cucumber and pumpkin) which suppress weed growth; 2) provide rapid canopy closure (e.g., broccoli planted on narrow rows) to prevent weed growth; or 3) have several labeled PRE and POST herbicides that will provide both early-and late-season broadleaf and grass weed control. Often times, vegetable crops that are transplanted into NT produce greater yields than those that are direct seeded. For example, cabbage yield reductions in NT were related more to poor plant establishment than to the effects of a NT production system (Wilhot et al. 1990). In most vegetable production systems, sparse or unevenly distributed cover crop residues on the soil surface often result in fields having high weed densities that lead to low yields and often poor fruit quality. In NT systems, high-residue cover crop mulches can suppress weed growth and often reduce or even eliminate the need for applied herbicides (Morse 2001). The inclusion of cover crops in vegetable production systems will not only improve weed control (Teasdale 1999), but will also conserve soil moisture, increase soil organic matter content, and provide other soil conservation advantages (Johnson & Hoyt 1999). Additionally, alternative tillage systems, such as NT, can further extend vegetable production into regions that are highly erodible.

Although cover crops contribute to weed control in NT production systems, herbicides or other weed control tactics are generally required to obtain optimal weed control and crop yield. The use of labeled PRE and POST herbicides is important to achieve optimum weed control in NT vegetable production. The use of PRE herbicides is important for early-season weed control in vegetable crops; and, although there are several PRE herbicides available for many different vegetable crops, the emergence of weeds after soil residual herbicides have dissipated often leads to excessive weed populations later in the production season. There are only a few POST herbicides labeled for broadleaf weed control for NT vegetable crops. This lack of effective POST herbicides is a major limiting factor in NT vegetable production systems (Hoyt et al. 1994), as the effective control of annual grasses and broadleaf weeds by POST herbicides is important to achieve success in this type of production system. Although grass weeds can be effectively controlled in most vegetable crops with PRE and POST herbicides, many difficult to control broadleaf weeds are often not sufficiently controlled due to the lack of effective herbicides that can be applied POST.

Since high weed populations are generally observed in NT production systems, weed control is essential to obtain the highest possible vegetable yields in this type of production system. The use of cultural practices that promote better vegetable crop establishment, more rapid plant growth and canopy closure will also result in improved weed suppression and high crop yields. Although NT systems utilizing cover crops are becoming more common for many vegetable crops, the major limitation for widespread grower limitation of NT practices is weed management. Thus, fields that have weed problems should be avoided or weed densities should be reduced in some manner prior to using NT production practices; and, if necessary, herbicides may have to be used during the production of the cover crop to minimize weed populations before transplanting (Morse 1999a). Cover crop mulch residues contribute to weed control in integrated weed management systems for NT crop production, but a major issue for growers is that they require more intensive management than CT systems. Although results from most studies have indicated that improvement in weed management systems beyond current practices and available herbicides is still required to maximize vegetable productivity in NT production systems, cover crop residues integrated in NT vegetable production systems along with the judicious use of herbicides can potentially suppress weeds in NT vegetable production and provide yields similar to CT.

6. References

Abdul-Baki, A.A. & Teasdale, J.R. (1993). A no-tillage tomato production system using hairy vetch and subterranean clover mulches. *HortScience* 28, 106-109.

Abdul-Baki, A.A.; Morse, R.D.; Devine, T.E. & Teasdale, J.R. (1997). Broccoli production in forage soybean and foxtail millet cover crop mulches. *HortScience* 32, 836-839.

Barnes, J.P. & Putnam, A.R. (1983). Rye residues contribute weed suppression in no-tillage cropping systems. *J. Chem. Ecol.* 9, 1045-1057.

Bellinder, R.R.; Hines, T.E. & Wilson, H.P. (1984). Herbicide programs and tillage systems for cabbage. *Proc. Northeastern Weed Sci. Soc.* 38, 191-194.

Beste, C.E. (1973). Evaluation of herbicides in no-till planted cucumbers, tomatoes, and lima beans. *NE Weed Sci. Soc. Proc.* 27, 232-239.

Blevins, R.L.; Cook, D.; Phillips, S.H. & Phillips, R.E. (1971). Influence of no-tillage on soil moisture. *Agron. J.* 63, 593-596.

Blum, U.L.; King, T.; Gerig, M.; Lehman, M. & Wosham, A.D. (1997). Effects of clover and small grain cover crops and tillage techniques on seedling emergence of some dicotyledonous weed species. *Amer. J. Alternative Agr.* 12, 146-161.

Bottenberg, H.; Masiunas, J.; Eastman, C. & Eastburn, D. (1997). Yield and quality constraints of cabbage planted in rye mulch. *Biol. Agric. Hort.* 14, 323-342.

Bradow, J.M. & Connick, W.J., Jr. (1990). Volatile seed germination inhibitors from plant residues. *J. Chem. Ecol.* 16, 654-666.

Brainard, D.C. & Bellinder, R.R. (2004). Weed suppression in a broccoli winter rye intercropping system. *Weed Sci.* 52, 281-290.

Brainard, D.C.; Bellinder, R.R. & Miller, A.J. (2004). Cultivation and interseeding for weed control in transplanted cabbage. *Weed Tech.* 18, 704-710.

Brandsaeter, L.O.; Netland, J. & Meadow, R. (1998). Yields, weeds, pests and soil nitrogen in a white cabbage-living mulch system. *Biol. Agric. Hortic.* 16, 291-309.

Brown, P.D. & Morra, J.M. (1996). Hydrolysis products of glucosinolates in *Brassica napus* tissues as inhibitors of seed germination. *Plant Soil* 181, 307-316.

Brown, D. & Masiunas, J. (2002). Evaluation of herbicides for pumpkin (*Cucurbita* spp.). *Weed Tech.* 16, 282-292.

Burgos, N.R. & Talbert, R.E. (1996). Weed control and sweet corn (*Zea mays* var. *rugosa*) response in a no-till system with cover crops. *Weed Sci.* 44, 355-361.

Castello, M.J. (1994). Broccoli growth, yield and level of applied infestations in leguminous living mulches. *Biol. Agr. & Hort.* 10, 207-222.

Chase, W.R.; Nair, M.G. & Putnam, A.R. (1991). 2,2'-oxo-1,1'-azobenzene: selective toxicity of rye (*Secale cereale* L.) allelochemicals to weed and crop species: II. *Chem. Ecol.* 17, 9-19.

Creamer, N.G. & Baldwin, K.R. (1999). Summer cover crops. Horticultural information leaflet 37. N.C. Coop. Ext. Serv., Raleigh.

Creamer, N.G.; Bennett, M.A.; Strinner, B.R.; Cardina, J. & Regnier, E.E. (1996). Mechanisms of weed suppression in cover crop-based production systems. *HortScience* 31, 410-413.

Curran, W.S.; Hoffman, L.D. & Werner, E.L. (1994). The influence of a hairy vetch (*Vicia villosa*) cover crop on weed control and corn (*Zea mays*) growth and yield. *Weed Tech.* 8, 777-784.

Dabney, S.M.; Buehring, N.W. & Reginelli, D.B. (1991). Mechanical control of legume cover crops. In: *Cover crops for clean water*, Hargrove, W.L. (Ed.), Soil and Water Cons. Soc., Ankeny, Iowa, USA, 146-147.

Derken, D.A.; Lafond, G.P.; Thomas, A.G.; Loeppky, H.A. & Swanton. C.J. (1993). Impact of Agronomic practices on weed communities: Tillage systems. *Weed Sci.* 41, 409-417.

Dias, L.S. (1991). Allelopathic activity of decomposing straw of wheat and oat and associated soil on some crop species. *Soil and Tillage Res.* 21, 113-120.

Doss, B.D.; Turner, J.L. & Evans, C.E. (1981). Influence of tillage, nitrogen, and rye cover crop on growth and yield of tomatoes. *J. Amer. Soc. Hort. Sci.* 106, 95-97.

Fisher, A. & Burrill, L. (1993). Managing interference in a sweet corn-white clover living mulch system. *Amer. J. Alter. Agric.* 8, 51-56.

Friesen, G.H. (1978). Weed interference in pickling cucumbers (*Cucumis sativus*). *Weed Sci.* 26, 626-628.

Galloway, B.A. & Weston, L.A. (1996). Influence of cover crop and herbicide treatment on weed control and yield in no-till sweet corn (*Zea mays* L.) and pumpkin (*Cucurbita maxima* Duch.). *Weed Tech.* 10, 341-346.

Grey, T.L.; Bridges, D.C. & NeSmith, D.S. (2000). Tolerance of cucurbits to the herbicides clomazone, ethalfluralin, and pendimethalin. I. summer squash. *HortScience* 35, 632-636.

Griffith, D.R; Manning, J.V. & Box, J.E. (1986). Soil and moisture management with reduced tillage. In: *No-tillage and surface-tillage agriculture*, Sprague, M.A. & G.B. Triplett (Eds.), Wiley, New York, 19-55.

Hebblethwaite, J. (1997). Survey-environmental friendly farming eclipses intensive tillage. *Partners/CTIC* 15(6), 1-4, 13-14.

Haramoto, E.R. & Gallandt, E.R. (2004). *Brassica* cover cropping for weed management: a review. *Renew Agric. Food Syst.* 19, 187-198.

Harborne, J.B. (1987). Chemical signals in the ecosystem. *Ann. Bot.* 60 (4 Suppl.), 39-57.

Harrelson, E.R.; Hoyt, G.D.; Havlin, J.L. & Monks, D.W. (2007). Effect of winter cover crop residue on no-till pumpkin yield. *HortScience* 42, 1568-1574.

Hartwig, N.L. & Ammon, H.U. (2002). Cover crops and living mulches. *Weed Sci.* 50, 688-699.

Hicks, S.K.; Wendt, C.W.; Gannaway, J.R. & Baker, R.B. (1989). Allelopathic effects of wheat straw on cotton germination, emergence, and yields. *Crop Sci.* 29, 1057-1061.

Hoyt, G.D. & Monks, D.W. (1996). Weed management in strip-tilled Irish and sweet potato production. *HortTechnology* 6, 238-241.

Hoyt, G.D.; Monks, D.W. & Monaco, T.J. (1994). Conservation tillage for vegetable production. *HortTechnology* 4, 129-135.

Hoyt, G.D.; Bonanno, A.R. & Parker, G.C. (1996). Influence of herbicides and tillage on weed control, yield, and quality of cabbage (*Brassica oleraceae* L. var. *capitata*). *Weed Tech.* 10, 50-54.

Infante, M.L. & Morse, R.D. (1996). Integration of no tillage and overseeded legume living mulches for transplanted broccoli production. *HortScience* 31, 376-380.

Johnson, A.M. & Hoyt, G.D. (1999). Changes to the soil environment under conservation tillage. *HortTechnology* 9, 380-393.

Kammler, K.J.; Walters, S.A. & Young, B.G. (2008). Halosulfuron tank-mixtures and adjuvants for weed control in pumpkin production. *HortScience* 43, 1823-1825.

Knavel, D.E. & Herron, J.W. (1981). Influence of tillage system, plant spacing, and N on head weight, yield, and nutrient concentration of spring cabbage. *J. Amer. Soc. Hort. Soc.* 106, 540-545.

Knavel, D.E. & Herron, J.W. (1986). Response of vegetable crops to nitrogen rates in tillage systems with and without vetch and ryegrass. *J. Amer. Soc. Hort. Sci.* 111, 502-507.

Kruidhof H.M.; Bastiaans, L. & Kropff, M.J. (2009). Cover crop residue management for optimizing weed control. *Plant Soil* 318, 169-184.

Leather, G.R. (1983). Weed control using allelopathic crop plants. *J. Chem. Ecol.* 9, 983-989.

Lehle, F.R. & Putnam, A.R. (1982). Quantification of allelopathic potential of sorghum residues: novel indexing of Richard's function fitted to cumulative cress seed germination curves. *Plant Physiol.* 62, 1212-1216.

Liebl, R.A. & Worsham, A.D. (1983). Inhibition of pitted morningglory (*Ipomoea lacunosa* L.) and certain other weed species by phytotoxic components of wheat (*Triticum aestivum* L.) straw. *J. Chem. Ecol.* 9, 1027-1042.

Liebman, M. & Staver, C.P. (2001). Crop diversification for weed management, In: M. Liebman, C.L. Mohler & C.P. Staver (Eds.), *Ecological management of agricultural weeds*, Cambridge University Press, New York, USA, 322-374.

Lonsbary, S.K.; O'Sullivan, J. & Swanton, C.J. (2004). Reduced tillage alternatives for machine-harvested cucumbers. *HortScience* 39, 991-995.

Mangan, F.; DeGregorio, R.; Schonbeck, M.; Herbert, S.; Guillard, K.; Hazzard, R.; Sideman, E. & Litchfield, G. (1995). Cover cropping systems for Brassicas in the northeastern United States: 2. Weed, insect and slug incidence. *J. Sustainable Agr.* 5(3), 15-37.

Masiunas, J.B. (1998). Production of vegetables using cover crop and living mulches – a review. *J. Veg. Crop Prod.* 4(1), 11-31.

Masiunas, J.B.; Weston, L.A. & Weller, S.C. (1995). The impact of rye cover crops on weed populations in a tomato cropping system. *Weed Sci.* 43, 318-323.

Masuinas, J.B.; Eastburn, D.M.; Mwaja, V.N. & Eastman, C.E. (1997). The impact of living and cover crop mulch systems on pests and yields of snap beans and cabbage. *J. Sustainable Agr.* 9(2/3), 61-89.

Meyer, L.D.; Wischmeier, W.H. & Foster, G.R. (1970). Mulch rates required for erosion control on steep slopes. *Soil. Sci. Soc. Am. Proc.* 34:928-931.

Mohler, C.L. & Teasdale, J.R. (1993). Response of weed emergence to rate of *Vicia villosa* Roth and *Secale cereale* L. residue. *Weed Res.* 33, 487-500.

Moore, M.J.; Gillespie, T.J. & Swanton, C.J. (1994). Effect of cover crop mulch on weed emergence, weed biomass, and soybean (*Glycine max*) development. *Weed Tech.* 8, 512-518.

Morse, R.D. (1995). In-row soil loosening improves plant survival and yield of no-till cabbage and broccoli. In: *Proc. Fourth Nat. Symp. on Stand Establishment,* K.J. Bradford & T.K. Hartz (Eds.), Monterey, California, USA, 23-26 June 1993, 113-116.

Morse, R.D. (1999a). No-till vegetable production-its time is now. *HortTechnology* 9, 373-379.

Morse, R.D. (1999b). Cultural weed management methods for high-residue/no-till production of transplanted broccoli (*Brassica oleraceae* L. gp. *italica). Acta Hort. (ISHS)* 504, 121-128.

Morse, R.D. (2000). High-residue, no-till systems for production of organic broccoli. In: *Proc. 23rd Annual Southern Conservation Tillage Conference for Sustainable Agriculture,* P.A. Bollich (Ed.), Monroe, Louisiana, USA, 19-21 June, 48-51.

Morse, R.D. (2001). No-herbicide, no-till summer broccoli-quantity of rye and hairy vetch mulch on weed suppression and crop yield. In: *Proc. 24th Annual Southern Conservation Tillage Conference for Sustainable Agriculture,* J.H. Stiegler (Ed.), Oklahoma City, Oklahoma, USA, 9-11 July, 85-94.

Morse, R.D. & Seward, D.L. (1986). No-tillage production of broccoli and cabbage. *Applied Agric. Res.* 1, 96-99.

Morse, R.; Elkner, T. & Groff, S. (2001). No-till pumpkin production principles and practices. Pennsylvania Marketing and Research Program, College Station, 16 p.

Mwaja, V.N.; Masiunas, J.B. & Weston, L.A. (1995). The effect of fertility on biomass, phytoxicity, and allelochemical content of cereal rye. *J. Chem. Ecol.* 21, 81-96.

NeSmith, D.S.; Hoogenboom, G. & McCracken, D.V. (1994). Summer Squash Production Using Conservation Tillage. *HortScience* 29(1), 28-30.

Nicholson, A.G. & Wien, H.C. (1983). Screening of turfgrass and clovers for use as living mulches in sweet corn and cabbage. *J. Amer. Soc. Hort. Sci.* 108, 1071-1076.

Ogutu, M.O. & Caldwell, J.S. (1999). Stand differences in no-till and plasticulture direct seeded and transplanted cucumbers (*Cucumis sativus* L.). *Acta Hort. (ISHS)* 504, 129-134.

Ohno, T.; Doolan, K.; Zibilske, L.M.; Liebman, M.; Gallandt, E.R. & Berube, C. (2000). Phytotoxic effects of red clover amended soils on wild mustard seedling growth. *Agric. Ecosyst. Environ.* 78, 187-192.

Paine, L.K. & Harrison, H. (1993). The historical roots of living mulch and related practices. *HortTechnology* 3, 137-143.

Panasuik O.; Bills, D.D. & Leather, G.R. (1986). Allelopathic influence of *Sorghum bicolor* on weeds during germination and early development of seedlings. *J. Chem. Ecol.* 12, 1533-1548.

Putnam, A.R. (1986). Allelopathy: Can it be managed to benefit horticulture? *HortScience* 21, 411-413.

Putnam, A.R. (1990). Vegetable weed control with minimal herbicide inputs. *HortScience* 25, 165-169.

Putnam, A.R. & DeFrank, J. (1983). Use of phytotoxic plant residues for selective weed control. *Crop Prot.* 2, 173-181.

Putnam, A.R.; DeFrank, J. & Barnes, J.P. (1983). Exploitation of allelopathy for weed control in annual and perennial cropping systems. *J. Chem. Ecol.* 9, 1001-1010.

Rapp, H.S.; Bellinder, R.R.; Wien, H.C. & Vermeylen, F.M. (2004). Reduced tillage, rye residues, and herbicides influence weed suppression and yield of pumpkins. *Weed Tech.* 18, 953-961.

Reberg-Horton, S.C.; Burton, J.D.; Danehower, D.A.; Ma, G.; Monks, D.W.; Murphy, J.P.; Ranells, N.N.; Williamson, J.D. & Creamer, N.G. (2005). Changes over time in the allelochemical content of ten cultivars of rye (*Secale cereale* L.). *J. Chem. Ecol.* 31, 179-193.

Schonbeck, M.S.; Herbert, S.; DeGregorio, R.; Mangan, F.; Guillard, K.; Sideman, E.; Herbst, J. & Jaye, R. (1993). Cover cropping system in the northeastern United States: 1. Cover crop and vegetable yields, nutrients and soil conditions. *J. Sustainable Agric.* 3(3/4), 105-132.

Shelby, P.P., Jr.; Coffey, D.L.; Rhodes, G. N., Jr. & Jeffery L.S. (1988). Tomato production and weed control in no-tillage versus conventional tillage. *J. Amer. Soc. Hort. Sci.* 113, 675-678.

Shennan, C. (1992). Cover crops, nitrogen cycling, and soil properties in semi-irrigated vegetable production systems. *HortScience* 27, 749-754.

Shilling, D.G.; Liebl, R.A. & Worsham, A.D. (1985). Rye (*Secale cereale* L.) and wheat (*Triticum aestivum* L.) mulch: the suppression of certain broadleaved weeds and the isolation and identification of phytotoxins. In: *The chemistry of allelopathy: biochemical interactions among plants*, Thompson, A.C. (ed.), Amer. Chem. Soc., Washington, D.C., 257-271.

Smeda, R.J. & Weller, S.C. (1996). Potential of rye (*Secale cereale*) for weed management in transplant tomatoes (*Lycopersicon esculentum*). *Weed Sci.* 44, 596-602.

Stall, W.M. (2001). Weed control in cucurbit crops (muskmelon, cucumber, squash, and watermelon. In: *Weed management in Florida fruits and vegetables*, Stall, W.M. (Ed.), Florida Coop. Ext. Ser. HS-90, Univ. Florida, Gainesville, Florida, USA, 23-27.

Teasdale, J.R. (1993). Interaction of light, soil moisture, and temperature with weed suppression by hairy vetch residue. *Weed Sci.* 41, 46-52.

Teasdale, J.R. (1998). Cover crops, smother plants and weed management. In: *Integrated weed and soil management*, Hatfield, J.L., D.D. Buhler & B.A. Stewart (eds.), Ann Arbor Press, Chelsea, Michigan, USA, 247-270.

Teasdale, J.R. (1999). Contribution of cover crop mulches to weed management. In: *Proc. 1999 New England Vegetable and Berry Growers Conference and Trade Show*, Sturbridge, Massachusetts, USA, 347-350.

Teasdale, J.R.; Beste, C.E. & Potts, W.E. (1991). Response of weeds to tillage and cover crop residue. *Weed Sci.* 39, 195-199.

Teasdale, J.R.; Abdul-Baki, A.A.; Mill, D.J. & Thorpe, K.W. (2004). Enhanced pest management with cover crop mulches. *Acta Hort. (ISHS)* 638, 135-140.

Wallace, R.W. & Bellinder, R.R. (1992). Alternative tillage and herbicide options for successful weed control in vegetables. *HortScience* 27, 745-749.

Walters, S.A. & Kindhart, J.D. (2002). Reduced tillage practices for summer squash production in southern Illinois. *HortTechnology* 12, 11-14.

Walters, S.A. & Young, B.G. (2008). Utility of winter rye living mulch for weed management in zucchini squash production. *Weed Tech.* 22, 724-728.

Walters, S.A. & Young, B.G. (2010). Effect of herbicide and cover crop on weed control in no-tillage jack-o-lantern pumpkin (*Cucurbita pepo* L.) production. *Crop Prot.* 29, 30-33.

Walters, S.A.; Young, B.G. & Kindhart, J.D. (2004). Weed control in no-tillage squash production. *Acta Hort.* 638, 201-207.

Walters, S.A.; Nolte, S.A. & Young, B.G. (2005). Influence of winter rye and pre-emergence herbicide combinations in no-tillage zucchini squash production. *HortTechnology* 15, 238-243.

Walters, S.A.; Young, B.G. & Nolte, S.A. (2007). Cover crop and pre-emergence herbicide combinations in no-tillage fresh market cucumber production. *J. Sustain. Agr.* 30(3), 5-19.

Walters, S.A.; Young, B.G. & Krausz, R.F. (2008). Influence of tillage, cover crop, and preemergence herbicides on weed control and pumpkin yield. *Inter. J. Veg. Sci.* 14(2), 148-161.

Weaver, S.E. (1984). Critical period of weed competition in three vegetable crops in relation to management practices. *Weed Res.* 24, 317-325.

Weston, L.A. (1990). Cover crop and herbicide influence on row crop seedling establishment in no-tillage culture. *Weed Sci.* 38, 166-171.

Weston, L.A. (1996). Utilization of allelopathy for weed management in agroecosystems. *Agron. J.* 88, 860-866.

Weston, L.A.; Harmon, R. & Mueller, S. (1989). Allelopathic potential of sorghum-sudan grass hybrid (Sudex). *J. Chem. Ecol.* 13, 403-421.

Wilhot, J.H.; Morse, R.D. & Vaughan, D.H. (1990). Strip tillage production of summer cabbage using high residue levels. *Appl. Agric. Res.* 5, 338-342.

Wiles, L.J.; William, R.D. & Crabtree, G.D. (1989). Analyzing competition between a living mulch and a vegetable crop in an interplanting system. *J. Amer. Soc. Hort. Sci.* 114, 1029-1034.

Yenish, J.P.; Worsham, D. & Chilton, W.S. (1995). Disappearance of DIBOA-glucoside, DIBOA, and BOA from rye (*Secale cereale* L.) cover crop residue. *Weed Sci.* 43, 18-20.

Zasada, I.A.; Linker, H.M. & Coble, H.D. (1997). Initial weed densities affect no-tillage weed management with a rye (*Secale cereale*) cover crop. *Weed Tech.* 11, 473-477.

3

Weed Control and the Use of Herbicides in Sesame Production

W. James Grichar, Peter A. Dotray and D. Ray Langham
Texas AgriLife Research, 3507 Hwy 59E, Beeville, TX 78102; Texas AgriLife Research, 1102 E FM 1294, Lubbock, TX 79403; Sesaco Corporation, San Antonio, TX 78217 USA

1. Introduction

Sesame (*Sesamum indicum* L.) is one of the oldest crops known to humans. There are archeological remnants of sesame dating to 5,500 BC in the Harappa Valley in the Indian subcontinent (Bedigian & Harlan, 1986). Assyrian tablets from 4,300 BC in a British museum described how the gods ate bread and drank sesame wine together before battles to restore order to the universe (Weiss, 1971). Most people remember the words "Open sesame" from Ali Baba and the 40 Thieves to open a cave full of riches. It is similar to the sesame capsules because their opening produced great riches. Sesame was a major oilseed in the ancient world because of its ease of extraction, great stability, and drought resistance. In India today, almost as in the olden days, a farmer can take his crop to an expeller that consists of grinding mortar and pestle stones driven by a bullock. He can place the oil in a vessel, take it back to his home and have cooking oil for a year without the oil going rancid (S.S. Rajan, personal communication).

Sesame is a survivor crop. It has been planted for over 7,500 years in Asia and Africa in very poor growing conditions. In parts of Thailand, farmers broadcast the seed and came back at the end of the season and see which plant had won - the sesame or the weeds (W. Wongyai, personal communication). More often than not, the sesame won. Sesame cultivars in those areas were tall, had very long internodes, and grew above the weeds. In Rajasthan, India, sesame is the last crop that can be grown adjacent to the deserts under extreme dry conditions. In several droughts in the U.S., sesame was the only crop that survived without irrigation (Langham & Wiemers, 2002).

Although this chapter draws from research from many countries, the emphasis is on herbicides in the U.S., the only country where sesame is completely mechanized and where herbicides are critical for economic production. Sesame was introduced to the U.S. from Africa and was called beni/benne/benni. Betts (1999) quotes letters from Thomas Jefferson that document his trials with sesame between 1808 and 1824. Jefferson stated that sesame "…is among the most valuable acquisitions our country has ever made. … I do not believe before that there existed so perfect a substitute for olive oil." He talks about the rule of thumb that still exists today - that sesame will do well where cotton (*Gossypium hirsutum* L.) does well. Sesame was produced in Texas on a limited scale during the 1950's and early 1960's, first in northeast Texas and later shifting to the High Plains, where consistent yield increases resulted from irrigation and more favorable climate conditions (Brigham & Young,

1982). The sesame was cut with a binder, hand shocked, and manually fed into a combine when dry. Due to a change in guest worker laws in the mid 1960's, the hand labor from Mexico became unavailable, and the sesame crop disappeared (Langham & Wiemers, 2002). Sesame returned to Texas in 1987 with varieties that did not require binding and shocking. The sesame could be swathed into a windrow, allowed to dry, and picked up with a pick-up attachment on a combine. Since that time, new varieties have been developed that can be left standing in the field to dry down, and then combined directly. Today, sesame has spread from Texas to parts of Oklahoma and southern Kansas.

One of the more difficult problems in planting sesame is that the seeds are small and need to be placed at precise depths and densities (Langham, 2007). Seeds cannot be planted too deep that the cotyledons never reach the surface, and yet they cannot be planted too shallow that the moisture around the seed is lost to evaporation. Once the cotyledons emerge, which are small compared to other crops, the sesame plants do not grow very fast. This slow growth and development is compounded by its drought resistance because sesame will partition a large portion of photosynthetic resources to create more root mass to penetrate the soil to find moisture. In the first 30 days, sesame plants reach about 28 cm in height; however, sesame will double to 60 cm in the next 11 days, triple to 90 cm in the following 8 days, and quadruple to 120 cm in the following 9 days. Depending on row spacing and phenotype, mechanization of sesame requires careful weed management for the first 30 to 60 days after planting (Langham, 2007).

The presence of weeds can negatively influence sesame yields. Kropff and Spitters (1991) reported that the major factor influencing sesame yield in a competitive situation is the ratio between the relative leaf area of the weed and the crop at the time of crop canopy closure. The effects of weeds on sesame establishment and growth have been well-documented. Balyan (1993), Gurnah (1974), Singh et al., (1992), and Upadhyay (1985) reported weed-induced reductions of sesame yield up to 65% and a need for a critical weed-free period up to 50 days after planting. Under weedy conditions, Eagleton et al., (1987) recorded a weed biomass six times that of sesame 48 days after planting and Bennett (1993) reported a weed biomass 1.3 fold that of sesame 42 days after planting. Without an herbicide, hundreds of hectares have been disced under in the U.S. due to excessive weed pressure.

Mechanically harvested non-dehiscent varieties present another problem that is not present in manual harvest, which comprises 99% of all sesame harvested in the world (Langham & Wiemers, 2002). If there are weeds in manual harvest, only the sesame plants are cut and placed in the shocks. However, in mechanical harvest, sesame and weeds are cut together. In Venezuela, a binder cuts the sesame and weeds together while they are still green, which is not a problem because the weeds dry down at the same time as the sesame. The only concern is that a high population of weeds may delay the combining because weeds may envelop plants and trap moisture or thicker stem weeds such as pigweed (*Amaranthus* spp.) will take longer to dry down. In direct combining, the weeds can be a big problem because they are normally green and add moisture to the combine bin. There are many cases where the sesame seeds are dry and weed seeds are not. Thick weed stems can add moisture, but the major problem is weed seeds. Since it is logistically difficult to scalp off the weed seeds at harvest, moisture from the weeds will transfer to sesame seeds. Sesame is 50% oil and needs to be harvested at 6% moisture or below in order to be transported by trucks and stored in silos. High moisture under these conditions can lead to heating and ruining of the seed. A second concern is that mechanically harvested sesame moves through a series of augers from the combine screen, to the combine bin, to the grain buggy, to the truck, to the silo, to the cleaning equipment, and within the cleaning process. Moist sesame can be damaged by this movement forming free fatty acids and leading to spoiling (Langham & Wiemers, 2002).

The small size of the sesame seed is similar to the size of many weed seeds (Langham, 2008; Langham et al., 2010). When sesame is used for oil, weed seeds within the sesame samples are not as critical unless they are toxic. However, a large percentage of sesame is used in the edible markets that require 99.99% purity. There are seeds such as johnsongrass [*Sorghum halepense* (L.) Pers] and other grass seeds that would seemingly be easy to remove because of their size and shape; however, these seeds go through the round holes of the cleaning screens end first and are difficult to separate in gravity tables because they have a similar specific gravity to sesame (Langham, 2008). In decortication of the seed for bakery products and tahini, the seed from lanceleaf sage (*Salvia reflexa* Hornem.) can cause a unique problem. When the lanceleaf sage seed has been hydrated, the seed surface formed a gelatinous substance that can cause all the sesame seeds around it to stick and form balls. Kochia [*Kochia scoparia* (L.) Schrad], buffalobur (*Solanum rostratum* Dunal), Russian thistle (*Salsola iberica* (Sennen & Pau) Botsch. ex Czerep.), tickseed also known as bugseed (*Corispermum hyssopifolium Nutt.*), and several species of grass seeds are other weeds that are difficult to separate from sesame. Any weed seed that is in a sesame sample in a large percentage is difficult to separate out, no matter the size and specific gravity, without having to slow down the processing or reprocessing. In Japan, purity needs to be 100% since processors have to pay claims to customers that find anything other than pure sesame seeds (author's personal observation).

Broadleaf weeds such as morningglory (*Ipomoea* spp.) and smellmelon (*Cucumis melo* L.) affect sesame growth and development. These weeds come up in flushes after a rainfall or irrigation event and after sesame canopy formation (Grichar et al., 2001a; Grichar et al., 2009). They can continue growing under weak light conditions, climb the sesame plants to the top of the canopy, and when they reach the light, greatly expand their infestation. As soon as they reach light, their leaf size increases dramatically. In high populations, these climbing weeds form a mat on top of the sesame and cause problems at harvest because it is difficult to separate adjacent rows of sesame (Langham et al., 2010). Many farmers go into these areas and treat with a herbicide such as glyphosate and sacrifice the sesame to keep the weeds from producing seeds and increasing the problem in future years. In general, annual plants are more susceptible to glyphosate than are perennial plants containing well-established underground propagules (Akin and Shaw, 2004). This difference in susceptibility is primarily due to the ratio of herbicide-intercepting foliage compared with the number of active sinks that need to be inhibited for plant death to occur (Franz et al., 1997). Successful control of perennial weeds with foliar-applied herbicides depends on the rapid absorption and basipetal translocation of the biologically active compound (e.g., glyphosate) into the underground storage organs in sufficient quantities to kill the entire plant before metabolism can degrade the compound (Sprankle et al., 1975).

Sesame is mainly grown in countries where abundant and inexpensive labor is available (Schrodter & Rawson, 1984). However, the trend in agriculture around the world is towards mechanization. Sesame has disappeared in Japan and parts of Mexico as the sesame growing areas mechanized. In Korea, sesame hectares have continually decreased since 1987 as the labor migrates to the cities (C. Kang, personal communication). With weak seedling vigor, limited competitive ability, and a lack of cheap labor, the use of herbicides are essential for commercial mechanized sesame production.

There has been considerable progress in mechanizing the crop by the development of non-dehiscent capsules (Langham & Wiemers, 2002) that hold the seed until combining and release the seed within the combine with minimum threshing. In addition, the growth habit of phenotypes has been changed to more readily feed into combines. The one area of best management practices for sesame that is still in development is the use of herbicides. Several agronomic practices have reduced the need for herbicides in dry areas. Preplant weed control followed by cultivation between rows has helped reduce weeds until the crop has reached a sufficient height to form a canopy. In areas that have early-season rainfall, herbicides use is essential. In addition, the trend towards minimum and no-till practices will require both preemergence and postemergence herbicides. A preemergence herbicide is applied to the soil before emergence of the specified weed or crop, whereas a postemergence herbicide is applied after emergence of the specified weed or crop (Senseman, 2007). There are two types of postemergence: over-the-top and directed application. In the latter, the herbicide is applied with a hooded sprayer with herbicide being applied between the rows and to the bottom of the crop - normally the lower 5 to 10 cm. In many areas where glyphosate tolerant crops are readily grown, there are no longer hoe crews available to manually clean the fields.

In mechanical harvest, there is an additional window of weed control that is important. The major form of weed control after the first 30 to 50 days of planting is the formation of the sesame canopy which blocks out light. At about 60 days after planting, current sesame varieties begin losing the leaves under the canopy where there is no light. As the plants mature, they self-defoliate and leaves are shed by about 100 days after planting. Without a harvest aid, it takes 40 to 50 days from the time that the plants lose all their leaves until the sesame is dry enough to combine. The leaves are a major part of the sun-blocking canopy, and as the weight of the leaves are lost, the branches become more erect, which allows more sunlight to penetrate the canopy. With autumn rains there may be new flushes of weeds, particularly fast-growing annual grasses. These late emerging weeds can be controlled in four ways: 1) applying postemergence-directed herbicides that have soil residual properties; 2) use of narrower row spacing; 3) planting the rows north/south so that there is light to the ground only at mid-day; and 4) using harvest aids to shorten the sesame drying period and which also kill and dry weeds.

Bennett (1993) found that alternating of grass and broadleaf crops in Queensland, Australia, helped in reducing weed populations since broadleaf weeds were more easily controlled in the grass crops and the grass weeds were more easily controlled in the sesame crop. However, this method is not as effective in the U.S. In many areas where either corn (*Zea mays* L.) or sorghum (*Sorghum bicolor* L.) was grown in the previous growing season, broadleaf weeds appeared late in the season and often were not controlled until after they have produced seed. Many of these problems result from the inability to disc the weeds mechanically due to lack of soil moisture. In growing wheat (*Triticum aestivum* L.) in the winter and spring prior to sesame, there are two problems: (1) there can be residues from broadleaf herbicides applied to the wheat in the spring that are toxic to sesame, and (2) there are many broadleaf weeds that will not germinate until the warm summer temperatures. In all areas, there are winter weeds that will not germinate until the sesame plants lose their leaves.

Planting sesame in fields with low weed pressure and knowledge of herbicide carryover from the previous crop to sesame are important in reducing weed competition and possible sesame injury. To date, the primary means of controlling weeds has been with cultivation. However, cultivation cannot reliably control weeds within the seed row that emerge while the sesame is emerging. Since sesame grows slowly in the first three to four weeks, many growers have waited three to four weeks to cultivate. Sesame roots follow moisture and with rain or irrigation in the first few weeks after planting, the roots may grow laterally and stay near the surface. Cultivating too close to the plant can cut roots and plants will wilt quickly and possibly die. In times of a dry season, roots grow more vertically allowing for closer cultivation. The cultivation process can throw soil up on the base of the plant covering any small weed after sesame plants are 10 to 15 cm in height. When a tractor is used for cultivation, sesame can be cultivated when it is slightly taller than the tractor axle, but it should be done in the afternoon when the plants are less turgid. Flower petals may fall, but the young capsules are rarely knocked off by the tractor (Langham et al., 2010). Breaking or creasing the main stem damages the sesame and prevents the plant from developing.

In many sesame growing areas, the trend has been to move to no-till practices excluding the use of cultivation. Most varieties used in the U.S. were developed for use on row spacing of 50 to 100 cm and were not suitable for narrow row spacing primarily because of the large leaves creating too much competition between the sesame plants. Most of the sesame grown in North and South America was bred from varieties developed by D.G. Langham in Venezuela in the 1940-50s. Without herbicides and insecticides, he found that large leaves canopied faster and outgrew many of the insects (B. Mazzani, personal communication). The current breeding programs have created potential varieties with smaller leaves that will allow for row spacing as close as 15 cm apart. With this narrow row spacing, the canopy can develop and close within 30 days of planting, which can be about the time that some preemergence herbicides are no longer effective. There is the potential to develop varieties which develop closure in 21 days, but there will always be a trade-off between too much inter-row sesame competition and rapid canopy.

Although the main thrust of this paper has been the controlling of weeds in sesame, there is always a concern as to whether sesame will become a weed in other crops used in rotation with sesame. There are many herbicides used in other crops that will prevent sesame from germinating. To date, only postemergence applications of glyphosate have consistently controlled sesame from the juvenile stage on through maturity. However, prometryn, flumioxazin, imazapic, trifloxysulfuron, mesotrione, flumetsulam, and foransulam have been effective in controlling sesame in some studies (Grichar et al., 2001a; Grichar et al., 2009). Many postemergence herbicides used in other crops will delay sesame maturity enough for the crop to canopy over-the-top of the sesame.

Until the advent of Roundup Ready® cotton, there was concern that sesame could become a problem weed in a cotton rotation. Under normal planting conditions, cotton germinates about 5 degrees cooler than sesame and has a faster growth rate in the first 30 days than sesame. Cotton planted during a normal planting window rarely will have sesame as a weed. The problem with volunteer sesame has primarily been in areas where cotton planting has been delayed due to environmental conditions or for integrated pest management. When there has been a volunteer sesame issue, most of the cotton herbicides will damage sesame, but will rarely kill it. As long as the cotton stand is good, the cotton will outgrow and canopy the sesame, but with a low cotton population, sesame would persist. However, sesame was never a problem in the harvest of the cotton.

Volunteer sesame could be a problem in groundnut (*Arachis hypogaea* L.), but with peanut herbicides such as imazapic or imazethapyr, volunteer sesame is no longer an issue. Volunteer sesame was never an issue in monocot crops such as corn, sorghum, and small grains because there are many good broadleaf herbicides that can control sesame. Theoretically, sesame could be a weed in many vegetable crops, but with a wide range of herbicides approved for those crops and the usual presence of manual labor, volunteer sesame has not been a problem in any vegetable crop to date.

2. Herbicides, weed control, and sesame tolerance

Several herbicides provide excellent control of weeds with minimal to no damage to sesame. However, in evaluating herbicides, there have been conflicting results, and it is difficult to sort out why some herbicides work in one area and do not work in another. Also, in some cases, at the same location, the herbicides effectively control weeds and little sesame injury is noted in one year; however, the opposite may be true the following year.

With most herbicides, herbicide dose, formulation, soil texture, pH, moisture, method of incorporation, and temperature before and after application are all a factors affecting herbicide persistence (Smith, 1989). Since soil organic matter, temperature, and aeration are more favorable for microbial activity in the topsoil than in the subsoil, degradation rates may decrease if a herbicide is leached into the subsoil (Smith, 1989). Soil pH can affect degradation directly if the stability of the herbicide is dependent upon acidity or alkalinity, and indirectly via its effects on the absorption of the herbicide to the soil (Smith, 1989). Increased rates of non-biological reactions and biological processes are favored by increasing temperature, herbicide degradation rates should increase also. Adequate moisture is also essential for microbiological activity (Smith, 1989). Martin (1995) reported that rainfall amounts during germination and establishment can markedly affect herbicide phytotoxicity to sesame, a possible factor in the reported erratic behavior of many herbicides. Many herbicides will delay sesame maturity while a few herbicides will completely kill the sesame. In many of the studies mentioned below, it will be seen with some herbicides that even with severe stand reduction, sesame yields are good because the plants can compensate for open space by putting out branches with capsules.

In some herbicide studies in the U.S. where multiple varieties were used, there have been differences in varietal susceptibility. Some of the clues have not been followed up because the moving baselines of new varieties has been fast, and the emphasis has always been placed on the use of the most recent released variety to use in herbicide evaluations. More work needs to be done in this area; particularly to determine whether a specific genotype may have more tolerance to a particular herbicide.

A review of sesame herbicide information from 21 countries has shown that there are approximately 16 herbicides that are used or have the potential to be used in commercial sesame production somewhere in the world (Langham et al., 2007). Some of these products are not available in the U.S. or have been discontinued. Table 1 shows the active ingredients of these 16 current herbicides that show the greatest potential for weed control in sesame production. The table does not contain herbicides such as flumioxazin that is used commercially in other parts of the world, but have resulted in considerable sesame injury in the U.S. (Grichar et al., 2001a; Grichar & Dotray, 2007).

Just as important as knowing the potential use of herbicides, it is important to note herbicides that have resulted in severe sesame injury or have had mixed results. In some cases, another application method of a herbicide in Table 2 can be toxic, e. g., glyphosate postemergence over-the-top.

Use	Preemergence	Postemergence	Postemergence-directed
Commercial	Alachlor Diuron Fluchloralin Fluometuron Glyphosate Linuron Metobromuron *S*-metolachlor Pendimethalin Trifluralin	Clethodim Diuron Fluazifop-P-butyl Sethoxydim Haloxyfop	Diuron Glyphosate (only between rows or wiper application)
Potential	Acetochlor Diuron + linuron *S*-metolachlor + diuron *S*-metolachlor + linuron	Pendimethalin *S*-metolachlor Alachlor Acetochlor	Diuron + linuron Linuron Diuron Prometryn

Table 1. Current and potential herbicides for use in sesame.

There are many preemergence herbicides that have been successfully used in sesame growing regions worldwide. These would include: alachlor, diuron, fluchloralin, fluometuron, linuron, metobromuron plus metolachlor, metolachlor, pendimethalin, and trifluralin. In the U.S., the main herbicides are *S*-metolachlor, diuron, linuron, and alachlor. Fluchloralin and metobromuron are not available in the U.S. Glyphosate is often applied with the preemergence herbicide to control emerged weeds. Herbicides act differently under certain environmental conditions which include variability in soil texture, organic matter, temperature, pH, humidity, rainfall timing and intensity, and under different methods and timing of application (Grichar et al., 2001a; 2001b). Pendimethalin and trifluralin are particularly difficult to use with results ranging from exceptional weed control with no damage to the sesame to little or no sesame stand (Grichar & Dotray, 2007). Poor sesame stands with the use of pendimethalin or trifluralin have resulted from incorporating either of the herbicides too deep. Since sesame is planted shallow, it is difficult to properly incorporate the dinitroaniline herbicides effectively and not have the herbicides come in contact with the sesame seed or roots (Grichar & Dotray, 2007).

It is important to realize that the many preemergence herbicides reduce sesame populations, but in mechanized sesame growing, this reduction is not noticed because of the cultural practices. One of the most difficult aspects of growing sesame is getting an uniform stand. The seeds are very small as compared to other field crops such as corn, soybean, cotton, wheat, and peanuts. One of the trends in mechanized agriculture is to singulate the larger seeded crops to attain the optimum plant population. Singulation has not worked in sesame because the seeds need adjacent seeds to help emerge out of the soil. Even with seed that has over 95% germination, rarely do more than 60% of the seeds emerge (Langham et al., 2010). In addition, there are many variations in soil type and row configurations within the sesame growing areas. In order to compensate for poor land preparation, the seeding rate is increased. Sesame varieties have been selected to compensate in high populations by self-thinning and in low populations by branching (Langham 2007). Various studies have shown that the yields are comparable between the untreated check and herbicide treatments that have some stand reduction (unpublished data).

Active ingredient	Preemergence	Postemergence over-the-top	Postemergence directed	Harvest Aid
2,4-DB	Toxic	Toxic		
Acetochlor	Potential	Potential		
Acifluorfen		Toxic	Potential	
Alachlor	Commercial	Potential		
Allidochlor (CDAA)	Mixed results			
Ametryn	Toxic			
Amiprophosmethyl	Toxic			
Asulam	Semi-selective			
Atrazine	Toxic	Toxic		
Benefin	Toxic			
Benfuresate	Toxic			
Bensulide	Selective			
Bentazon		Toxic		
Bifenox		Toxic		
Bromoxynil		Toxic		
Carbuthioate	Semi-toxic			
Carfentrazone			Semi-toxic	Not effective
Chloramben	Mixed results			
Chlorimuron		Toxic		
Chloroxuron	Toxic			
Chlorpropham (CIPC)	Mixed results			
Chlorsulfuron	Mixed results			
Chorthal-dimethyl	Semi-toxic			
Clethodim		Commercial		
Clomazone	Toxic			
Clopyralid	Semi-selective	Toxic		
Cloransulam	Toxic	Toxic		
Dicamba		Toxic		
Dichlobenil	Toxic			
Dichlormate	Semi-selective			
Diclosulam	Toxic	Toxic		
Diethatyl	Semi-selective			
Diethylacetanilide	Semi-selective			
Diflufenican	Semi-toxic	Toxic		
Diflufenzopyr		Toxic		
Dimethenamid	Mixed results			
Dinitramine	Toxic			
Dinoseb	Toxic			
Diphenamid	Selective	Selective		
Diquat				Effective
Diuron	Commercial	Commercial	Potential	
DSMA		Semi-toxic		

Active ingredient	Preemergence	Postemergence over-the-top	Postemergence directed	Harvest Aid
Endothall	Toxic	Toxic		
EPTC	Mixed results			
Ethalfluralin	Mixed results			
Fenoxaprop	Inconclusive			
Fluazifop-P-butyl		Commercial		
Fluchloralin	Commercial			
Flufenacet	Toxic			
Flumetsulam	Toxic	Toxic		
Flumioxazin	Toxic	Semi-toxic	Mixed results	
Fluometuron	Commercial	Semi-selective		
Fluorodifen	Toxic			
Fomesafen		Mixed results		
Glufosinate-ammonium			Mixed results	Effective
Glyphosate	Commercial	Toxic	Mixed results	Effective
Haloxyfop		Selective		
Imazapic	Toxic	Toxic		
Imazethapyr	Semi-selective	Toxic		
Isopropalin	Toxic			
Lactofen		Toxic	Semi-toxic	
Linuron	Commercial	Toxic	Potential	
Mesotrione		Toxic		
Methabenthiazuron		Semi-selective		
Methazole	Mixed results			
Metobromuron	Commercial			
Metolachlor	Commercial			
Metribuzin	Toxic			
Metsulfuron	Mixed results			
Monolinuron	Mixed results			
Monuron	Selective			
MSMA		Semi-toxic		
Napropamide	Mixed results			
Naptalam (NPA)	Toxic	Toxic		
Nicosulfuron	Mixed results	Toxic		
Nitralin	Mixed results			
Nitrofen	Toxic			
Norea	Mixed results			
Norflurazon	Toxic			
Oxadiazon	Semi-selective	Semi-toxic		
Oxasulfuron		Toxic		
Oxyfluorfen	Semi-selective	Toxic		
Paraquat		Toxic	Semi-toxic	Effective
Pebulate	Semi-selective			

Active ingredient	Preemergence	Postemergence over-the-top	Postemergence directed	Harvest Aid
Pendimethalin	Commercial	Potential		
Perfluidone	Selective			
Phenmediphan		Toxic		
Piraflufen ethyl		Semi-selective		
Proatryne	Selective			
Profluralin	Selective			
Prometryn	Toxic	Toxic	Potential	
Pronamide	Toxic			
Propachlor	Selective			
Propanil	Semi-selective			
Propazine	Mixed results	Toxic	Selective	
Prosulfuron	Mixed results	Toxic		
Pyraflufen ethyl		Semi-toxic	Selective	Not effective
Pyridate		Mixed results		
Pyrithiobac	Toxic	Toxic	Toxic	
Rimsulfuron	Selective	Toxic		
Sesone	Mixed results			
Sethoxydim		Commercial		
Simazine	Toxic			
S-metolachlor	Commercial	Potential		
Sufentrazone	Toxic			
Sulfonamide	Mixed results			
Thiobencarb	Toxic			
Triasulfuron	Mixed results			
Trifloxysulfuron	Mixed results	Toxic	Toxic	
Trifluralin	Commercial	Mixed results		
Vernolate	Toxic			

[a]In the evaluation the following categories of effectiveness are used:
Commercial: used commercially in at least one country
Potential: potential to use commercially
Selective to sesame: does not damage sesame
Semi- selective to sesame: some damage to sesame, but helps
Mixed results with some showing some selectivity and others showing toxicity
Toxic: substantial reduction of production
Semi- toxic: enough reduction that probably cannot be used
Effective as a harvest aid
Not effective as a harvest aid

Table 2. Summary of herbicides that have been evaluated for weed control and sesame tolerance[a].

Until 2000, little or no research has been done on the use of postemergence herbicides in sesame (Grichar et al., 2001b). Most of the herbicide work has been at crop establishment. From initial work done in the U.S. in Arizona, several postemergence herbicides have done a very good job controlling grasses and not damaging the sesame. Grass herbicides, fluazifop-P-butyl, haloxyfop, and sethoxydim have been used successfully in many parts of the world. More recently, clethodim has proven equally good controlling both annual and perennial grasses (particularly johnsongrass) and not damaging sesame (Grichar et al., 2001b). There is a label in the U.S. for clethodim (Select Max®) use in sesame which allows spraying in all phases except flowering (Langham et al., 2010). Concerns have been raised on the use of clethodim after extensive glyphosate applications and improper clean-out of spray tanks. Sesame capsule inhibition has been noted when glyphosate carryover has been noted in spray tanks that have been used to apply clethodim. The cleaning and removal of any glyphosate residues in spray tanks after each herbicide use is essential to prevent herbicide carryover.

To date there is no postemergence over-the-top broadleaf herbicide that will control the weeds without damaging the sesame (Grichar et al., 2001b). There are products such as alachlor and metolachlor that cause minimum injury to sesame when applied postemergence, will not control emerged weeds, but will provide some soil residual activity (Grichar et al., 2001a; Grichar et al., 2009). In the case of herbicides such as diuron, sesame will recover, but the farmer will notice stunting and leaf necrosis on sesame leaves for about 10 days after herbicide application (Grichar et al., 2009). In some sesame herbicide research, severe sesame plant stunting and leaf necrosis has resulted in good weed control and produced higher yields than the untreated check because of the loss of production to weeds in the untreated check (Grichar et al., 2009). A controversial use of herbicides is what is known as a "rescue treatment", which is using a herbicide that will injure the sesame, but will bring weeds under some control and allow the sesame to be harvested at an economic return. As an example, a farmer used clopyralid on a portion of a field that was being overwhelmed by common cocklebur (*Xanthium strumarium* L.). Where he did not spray, he lost the crop; however, where he sprayed there was damage to the sesame with control of the cocklebur and he harvested about 660 kg/ha. However, many sesame growers in the U.S. are not tolerant of any type of sesame herbicide injury even knowing that the sesame will recover.

Starting in 2003, research has been conducted using postemergence-directed herbicides with and without the use of hooded sprayers. This work is very encouraging; however, there are many cropping patterns that preclude the use of hooded sprayers. There is a label for glyphosate (Roundup Max®) that allows wiper applicators or hooded sprayers to be used between sesame rows (Langham et al., 2010). While this does not provide effective weed control in the sesame seed row, it helps with vining weeds such as morningglory species (*Ipomoea* spp.) and smellmelon (*Cucumis melo* L.) that spread across the rows. While morningglory is becoming increasingly tolerant of glyphosate, glyphosate will slow the growth of morningglory and reduce the damage to the sesame from this weed. In the case of *Amaranthus,* which quickly can become taller than the sesame, wiper applicators using glyphosate have been very successful, particularly in areas with high relative humidity, as long as the glyphosate does not drip on the sesame. Initial work with spraying glyphosate on the sesame stem showed little injury; however, in subsequent studies, there have been instances of severe damage. In further observations, when the sesame was under moisture stress, there was little damage, but when the plants were in a rapid growth phase following rainfall or irrigation.

One of the major problems in using postemergence-directed herbicides has been the timing of the application and the height of the spray application on the sesame stem as related to the height of the plant. Recent work has shown that there are differences in applying herbicides at

5 cm above the surface versus 15 cm; differences in applying 4 weeks after planting versus 6 weeks; and differences is the heights of the plants in different locations in a field. In waiting for the sesame to get tall enough to spray a postemergence-directed herbicide, weeds also become tall and herbicides may not control taller weeds (Langham et al., 2010).

In reviewing research using postemergence herbicides, it is sometimes difficult to understand exactly at what stage of growth the herbicide was applied (Langham et al. 2007). Many of the documents will cite the number of days after planting or the height of the plants. However, there are many differences in the cultivars of the world in terms of number of days in each stage and in the heights of the plants in each stage as shown in Table 3.

In order to standardize terminology, a phenology chart has been developed to specify the beginning and end points of the stages (Langham, 2007). Table 4 summarizes sesame phenology.

	Days from planting		Phase length	
Phase	Range	Mean	Range	Mean
Vegetative	29-59	42	29-59	42
Reproductive	56-116	89	16-70	47
Ripening	77-140	108	(14)[b]-54	11
Drying	102-181	150	11-57	38

[a] Based on sesame germplasm from Sesaco Corporation (Langham 2007)

[b] In some cultivars, there are dry capsules above green leaves while the upper portion of the plant is still flowering creating a negative range.

Table 3. Range and mean of number of days in phases for sesame germplasm.[a]

Stage/Phase	End point of stage	DAP[a]	No. weeks
Vegetative			
Germination	Emergence	0-5	1-
Seedling	3rd pair true leaf length=2nd	6-25	3-
Juvenile	First buds	26-37	1+
Pre-reproductive	50% open flowers	38-44	1-
Reproductive			
Early bloom	5 node pair of capsules	45-52	1
Mid bloom	Branches/minor plants stop flowering	53-81	4
Late bloom	90% of plants with no open flowers	82-90	1+
Ripening	Physiological maturity	91-106	2+
Drying			
Full maturity	All seed mature	107-112	1-
Initial drydown	1st dry capsules	113-126	2
Late drydown	Full drydown	127-146	3

[a] DAP, days after planting. These numbers are based on S26 (Sesaco Corp.) in 2004 near Uvalde, TX under irrigation.

Table 4. Phases and/or stages of sesame.

Future work on sesame herbicides should specify the stage of the sesame. Recent application timing work has shown that some herbicides are phytotoxic in the seedling stage, are neutral in the juvenile stage, and reduce yield in the pre-reproductive through mid bloom stages. Additional work is needed to verify these initial findings as to the exact neutral stages, but there is enough data to know that plant stage at application is critical.

A second problem in reviewing the literature is that some of the work has not been carried through to completion of the sesame crop (Langham et al. 2007). Sesame has a remarkable ability to compensate. Recent work has compared the stunting/damage ratings of some contact-based herbicides and showed that the amount of damage to sesame was reduced over time and the yields of stunted/damaged materials was comparable to the weed-free checks. Sesame injury ratings should only be done by researchers familiar with sesame. Sesame yields are related to the number of capsules and the seed weight per capsule per square meter. There have been herbicide treatments that apparently damage the sesame, i.e., the yellow splotching of leaves by such herbicides as diuron, but the number and weight of the capsules were not affected. In some cases, the herbicide delayed flowering, but the plants flowered longer. A reduction in plant height may not affect yield.

Below is a discussion of the most promising and effective herbicides for use across the sesame growing areas of the world. A discussion of research in various sesame growing areas is also included.

3. Alachlor

Alachlor, a chloroacetamide herbicide, has been widely used in corn, groundnut, snap bean (*Phaseolus vulgaris* L.), and soybean for preemergence annual grass and broadleaf weed control (Wilson et al., 1988). Bijanzadeh and Ghadiri (2006) reported that alachlor alone controlled redroot pigweed (*Amaranthus retroflexus* L.) 68 to 72% in one year and at least 92% in another, but the efficacy of atrazine plus alachlor increased when tank-mixed together.

Alachlor is the most widely used sesame herbicide in the world. However, little work has been done in the U.S. because the Environmental Protection Agency (EPA) has indicated that additional uses of alachlor would not be approved due to groundwater concerns. Commercial preemergence uses of alachlor include the following: in Thailand, a field guide recommends alachlor at 1.2 to 1.5 L/ha in case of labor shortage (Anonymous, 1997). In Honduras, a grower guide states that alachlor proved to be very effective in the control of weeds in sesame (Anonymous, 2002). In Mexico, a grower guide for Michoacan recommends the use alachlor as a preemergence alone or in combination with linuron and diuron. In all instances, 250 to 300L of water was used as carrier volume (Anonymous, 2007a). In El Salvador, a growers guide recommends 2.8 L/ha of alachlor (Anonymous, 2007b).

Research on alachlor use in sesame dates back 40 years. In Bulgaria, Lyubenov and Kostadinov (1970) conducted experiments with sesame sown on Chernozem Smolnitsa soil. Alachlor applied preemergence at 4 kg/ha effectively controlled weeds and increased sesame seed yields and seed oil content. In Ethiopia, Moore (1973a; 1973b) found that alachlor between 1.6 and 2.9 kg/ha was the safest of the herbicides to be tested, provided high yields, but residual activity was poor. In California, in studies with alachlor applied preplant incorporated under furrow irrigation in multiple locations and years, alachlor provided excellent weed control of *Amaranthus* spp., wild mustard (*Brassica kaber* L.) and various grasses with minimal injury to the sesame, but little control of volunteer barley (*Hordeum vulgare* L.) and marginal control of other broadleaf weeds was noted (St. Andre, unpublished data). In the only experiment carried to maturity, sesame yield following alachlor at 2.25 kg/ha was 1,051 kg/ha, while yields from the weedy and weed-free control were 259 and 1,075 kg/ha, respectively (St Andre, unpublished data). In India, Subramanian and Sankaran (1977, 1981) conducted experiments over 4 seasons in both summer and winter crops to study the efficiency of alachlor. They found that alachlor at 1.75 kg/ha controlled desert horse purslane (*Trianthema portulacastrum* L.) and purple nutsedge (*Cyperus rotundus* L.) and provided the maximum net income and the highest return per rupee invested in weed control. Graph et al., (1985) showed preemergence treatments with 1.0 to 2.0 kg/ha of alachlor did not injure sesame but caused damage when applied with a preplant incorporated trifluralin treatment. In Australia, Schrodter and Rawson (1984) evaluated alachlor in 3 experiments over two years. They concluded that alachlor was the safest herbicide with yield, population, and vigor similar to the weed-free control. The overall conclusion was that alachlor at 2.25 kg/ha applied preemergence was the most acceptable herbicide treatment for sesame. In Ethiopia, work with various herbicides indicated that the greatest yields were obtained with alachlor at 2.9 kg/ha (Anonymous, 1973). Kim et al., (1986) conducted field trials in the sesame-producing uplands of Korea, to study herbicide efficacy and phytotoxicity in crops grown under polyethylene film. Alachlor at 1.5 L/ha produced sesame yields equivalent to that obtained with manual weed control. In Venezuela, Pineda et al., (1988) tried alachlor as a preemergence herbicide and found it was comparable to the untreated check with respect to sesame yield. In India, Bansode and Shelke (1991) assessed six weed control treatments (an unweeded control, hand-weeding plus hoeing 3 weeks after sowing), and alachlor at 0.75 or 1.5 L/ha applied preemergence in field trials during the kharif of 1988 with sesame cv. Punjab-1 and T-85. Alachlor applied preemergence to cv. Punjab-1 combined with hand-weeding plus hoeing provided greatest yields (689 kg) compared to hand-weeding plus hoeing alone (583 kg) and all other treatments.

In a Venezuela grower guide, Caraballo et al., (1986) found that alachlor applied at 6, 5, 4, and 3 L/ha yielded 1,008, 833, 848, and 810 kg/ha, respectively, compared to the weedy check yield of 536 kg/ha and weed-free check yield of 1,042 kg/ha. In India, Dungaral et al., (2003) conducted a field experiment during the kharif seasons of 1997 and 1998 to evaluate the relative efficacy of alachlor applied alone or in combination with one hoeing at four weeks after sowing to control weeds in sesame cv. TC 25. On average, season-long weed competition caused 50% reduction in seed yield. Among the herbicides, the preemergence application of alachlor at 2.0 kg/ha combined with one hoeing at 4 weeks after sowing registered the greatest weed control efficiency, which enhanced yield attributes leading to greater seed yield (530 kg/ha) and net return (Rs. 4275/ha). In India, Anil and Thakur (2005) concluded that the greatest sesame yields were obtained with alachlor at 1.5 kg/ha alone or in combination with hand weeding. In the U.S., B. Sadler (2007, personal communication) has grown sesame for bird hunting for the past twenty years. He has used both alachlor and metolachlor in alternating years in order to control yellow (*Cyperus esculentus* L.) and suppress purple nutsedge.

4. Metolachlor and *S*-metolachlor

Metolachlor or *S*-metolachlor are commonly used in various crops for control of small-seeded broadleaf weeds, some annual grasses, and yellow nutsedge (Grichar et al., 1996). *S*-metolachlor will control small-seeded annual grasses, but does provide inconsistent control of large-seeded annual grasses (Grichar et al., 2004a; 2004b). Many growers have reported peanut stunting when soil applications of metolachlor have been followed by rain (Grichar et al., 1996). Grichar et al., (1996) reported that postemergence applications of metolachlor followed by irrigation within 24 h could be effective for yellow nutsedge control and reduce the chance of peanut injury from soil applications. Combinations of factors, such as herbicide dose, moisture conditions at planting, soil organic matter, and pH may affect peanut injury by chloroacetamide herbicides such as *S*-metolachlor (Cardina & Swann, 1988; Wehtje et al., 1988; Osborne et al., 1995; Mueller et al., 1999). Cardina and Swann (1988) reported that metolachlor often delayed peanut emergence and reduced peanut growth when irrigation followed planting; however, yield loss was observed only when metolachlor was applied at a 3X rate.

In many areas of the world, metolachlor has shown similar results as alachlor on sesame and is being used more frequently because it requires less active ingredient per hectare to achieve similar results. Commercial preemergence uses of metolachlor include the following: in Thailand, a field guide recommends metolachlor at 1.2 to 1.25 L/ha in case of labor shortage (Anonymous, 1997). In Australia, grower guides in the Northern Territories (Bennett 1998) and in South Burnett (Sapin et al., 2000) recommend the use of metolachlor. In El Salvador, a grower guide recommends 1.4 L/ha of metolachlor (Anonymous, 2007c). Reportely, alachlor has been used in commercial fields in Mexico, Venezuela, Brazil, Nigeria, Ethiopia, Nicaragua, Guatemala, and Argentina.

In preemergence experiments in the U.S., metolachlor treatments (2.2 kg/ha) had a slight reduction in sesame vigor, provided good broadleaf control initially, but allowed broadleaf weeds to germinate later (St Andre, unpublished data). Metolachlor applied preemergence at 2.1 kg/ha was one of the best overall treatments, but preplant incorporation of metolachlor affected early vigor and stunted the sesame (D. Howell, unpublished data). In Egypt, metolachlor alone at 1.2 and 1.8 kg/ha and a premix of metolachlor and metobromuron (Galex®) was tested. The premix provided good broadleaf weed control while both metolachlor alone and the premix provided good annual grass control (Hussein et al., 1983). In Nicaragua, metolachlor at 1.1 and 2.2 kg/ha provided good grass control, did not injure the sesame, and doubled the yield from that of the untreated check (Soto-Soto & Silva-Vasquez, 1987). In Ethiopia, metolachlor at 1.7 kg/ha provided good grass and broadleaf control, which resulted in a significant yield increase (Zewdie, 1994). In Australia, Martin (1995) reported that metolachlor adequately controlled weeds but caused unacceptable crop injury. Despite that report, farmers use metolachlor for commercial sesame fields (M. Bennett, L. Serafin, and P. O'Shanesy, personal communication). Metolachlor at 0.6, 1.1, 2.2, and 3.4 kg/ha resulted in variable sesame plant populations, had no effect on sesame plant height, resulted in consistent weed control, and provided greater yields than the untreated plots (Grichar et al., 2001a).

In later work at a south Texas location, *S*-metolachlor caused no sesame stand reduction or injury; however, at the Lubbock location, stand reduction and injury was noted in one of two years (Grichar et al., 2009). Also, sesame stand reductions have been observed in Oklahoma where *S*-metolachlor was applied followed by irrigation, but there was no problem when planted into moisture (C. Medlin & C. Godsey, personal communication). However, in 2009, there was no stand reduction (J. Armstrong, personal communication). In Argentina, metolachlor at 0.8 kg/ha provided good control of *Amaranthus quitensis* but marginal control of *Raphanus sativus*. *S*-metolachlor did provide similar yields to the weedy and weed-free checks (L. Lanfranconi, unpublished data). *S*-metolachlor alone in comparison with diuron, linuron, and a premix of linuron and diuron has produced sesame yield similar to the weed-free check at several location in Texas (Table 5). When *S*-metolachlor was applied postemergence over-the-top of sesame at the juvenile stage, minimal stunting and yield differences were observed (authors' personal observations).

		Lorenzo		Uvalde [a]	Average yield vs. untreated
Treatment	Rate	2007	2008	2008	
	Kg ai/ha	Kg/ha			Percent
Untreated		1,224	763	1,233	100
Diuron	0.6	1,417	835	1,127	106
Diuron	1.2	1,350	751	1,211	102
Diuron	2.4	1,215	742	1,105	95
Linuron	0.6	1,280	829	1,199	103
Linuron	1.2	1,278	879	1,300	108
Linuron	2.4	1,267	625	1,289	97
S-metolachlor	0.7	1,168	773	1,233	99
S-metolachlor	1.4	1,138	834	1,161	99
S-metolachlor	2.8	1,185	888	1,237	105
Linuron + diuron	0.3 + 0.3	1,203	790	1,199	100
Linuron + diuron	0.6 + 0.6	1,327	767	1,121	100
Linuron + diuron	1.1 + 1.1	1,374	781	1,239	105
LSD (0.05)		NS	140	NS	

[a] No yield taken in 2007 due to glyphosate drift from adjacent sorghum fields.

Table 5. Sesame yield response to preemergence herbicides at two locations in Texas.

In the U.S., *S*-metolachlor has been used under temporary labels from the EPA on over 50,000 hectares of sesame and provided good weed control with no evidence of a reduction in yield. Hundreds of hectares that have not had *S*-metolachlor applied have been plowed under because of severe weed pressure. *Amaranthus spp.* and small-seeded grasses are the most damaging weeds to sesame in the U.S. and *S*-metolachlor provides excellent control of these weeds when applied immediately after planting of sesame (Grichar et al., 2001a; 2009).

5. Diuron

Diuron is systemic urea herbicide that inhibits photosynthesis and has been used to control various weeds in cotton (Culpepper et al., 2004). Reddy et al., (2007) reported that ragweed parthnium (*Parthenium hysterophorus* L.) was highly sensitive to pigment inhibitors and photosynthetic inhibitors such as diuron compared to herbicides with other modes of action. Commercial preemergence uses of diuron in sesame include the following: in Mexico, a grower guide recommends a diuron mixture with alachlor at 0.5 kg plus 1.0 kg/ha, respectively, for commercial fields (Anonymous, 2007a). Diuron, when used preemergence, controls many broadleaf weeds that cannot be controlled by *S*-metolachlor. Also, diuron has the potential to be used postemergence as a rescue treatment when broadleaf weeds are growing profusely.

In preemergence experiments in Venezuela, diuron at 0.6 and 1.2 kg/ha reduced sesame yield, but yield would have been much lower without weed control (Mazzani, 1957). In one year in the U.S., diuron at 0.8 and 1.7 kg/ha resulted in adequate weed control without apparent crop injury; however, in another year, there was stand reduction and chlorosis (Culp & McWhorter, 1959). Further work in Venezuela showed that diuron provided reasonable control of weeds with no significant reduction in yield (Montilla, 1964; Mazzani,

1966). In Sri Lanka, diuron at 0.6 and 0.8 kg/ha effectively controlled weeds with no significant reduction in yield (Appendurai, 1967); however in Ethiopia, diuron caused serious crop damage in both irrigated and rainfed trials (Moore, 1974). In Egypt, diuron at 1.0 kg/ha was tested alone and in tank mixtures with pendimethalin. The herbicide combinations controlled both grass and broadleaf weeds and resulted in greater yields (Ibrahim et al., 1988). In contrast, Viera et al., (1998) reported that diuron mixtures (0.8, 1.0, and 1.3 kg/ha) with pendimethalin and alachlor caused greater phytotoxicity with the greatest dose. However, there was no difference in the height of the first fruiting branch, the number of capsules per plant, and the yield between the different herbicide treatments and manual weeding (Viera et al., 1998). In Brazil, diuron at 1.0 kg/ha enhanced seed production (Beltrao et al., 1991). In later work by Grichar et al., (2009), they reported that diuron at 1.12 kg/ha reduced sesame stands and caused sesame injury in one year in the Texas High Plains area; however, in south Texas no adverse effects with diuron were seen in the two years. Sesame yield from plots treated with diuron have not been different from the weed-free check (Table 5).

Diuron also has a potential for use as a postemergence both over- the-top and directed. In post-directed studies, the authors have found that diuron controls emerged weeds with minimal damage to the sesame. In Venezuela a grower guide (Avila, 1999) recommends that when the plants are about 30 cm tall [juvenile stage], diuron should be used at 1.5 L/ha. They reported that diuron controlled the weeds with minimum damage to the sesame. In Venezuela, Caraballo (1986) did a timing study using diuron at 1.5 L/ha with 0.5 L of surfactant. Applications at 15, 22, 29, 36, 43, and 50 days after planting resulted in yields of 947, 896, 817, 911, 762, and 770 kg/ha, respectively, compared to 557 kg/ha in the weedy check and 1117 kg/ha in the hand-weeded check. Diuron controlled 94% of the broadleaves and 89% of the grasses.

In recent work with postemergence applications of diuron, an application at the late juvenile stage has shown a discoloration of the leaves and some height reduction, but yields have been comparable to the weed-free check (authors' personal observations). Recent timing studies have shown that time of application is critical. Minimal sesame damage has been found when diuron has been applied in the late juvenile stage; however, earlier applications in the seedling stage severely damaged the sesame and applications during sesame flowering reduced yield (authors' personal observations).

6. Linuron

Linuron, a substituted urea herbicide, has been used extensively in cotton and carrot (*Daucus carota* L.) as a preemergence or postemergence herbicide since the 1960's for control of annual broadleaf weeds such as pigweed spp., common ragweed (*Ambrosia artemisiifolia* L.), common groundsel (*Senecio vulgaris* L.), and common purslane (*Portulaca oleracea* L.) (Bell et al., 2000; Bellinder et al., 1997; Saint-Louis et al., 2005). Linuron can be used in combination with 2,4-DB to control sicklepod (*Cassia obtusifolia* L.) in soybean; however, a height differential must be established between soybean and sicklepod to cover the weeds without contacting more than the lower 25 to 30% of the soybean plant to reduced herbicide injury (Shaw & Coats, 1988).

Commercial preemergence uses of linuron in sesame include its use in Mexico. A grower guide recommends a linuron mixture with alachlor at 0.5 kg plus 1.5 L/ha, respectively, for commercial fields (Anonymous, 2007a).

In studies using linuron applied preemergence, Santelmann et al. (1963) found slight phytotoxicity and a reduction in sesame yield with linuron at 2.24 kg/ha. In Bulgaria, Lyubenov and Kostadinov (1970) found preemergence application of mixtures of 3 kg/ha of linuron and 3 kg/ha of alachlor gave effective control of weeds and increased seed yields and seed oil content. In Egypt, Hussein et al. (1983) found linuron at 1.8 kg/ha increased the seed yield 60% as compared to the weedy check. Seed oil content was not affected. In Australia, Schrodter and Rawson (1984) evaluated linuron in 3 experiments over two years. They concluded that linuron increased the yield over the weedy check, but did not provide yields comparable to alachlor and the weed-free control. In Nicaragua, Soto-Soto & Silva-Vasquez (1987) conducted trials in two sites. They concluded that linuron provided the best control of broadleaf weeds and metolachlor provided the best control of grasses; neither damaged the sesame; and recommended rates of 2.1 L/ha for both herbicides. In Argentina, linuron applied preemergence at 1.5 and 2.0 L/ha controlled weeds and provided yields close to the check (L. Lanfranconi, unpublished data). In the U.S., linuron at 0.6 to 2.4 kg/ha has produced yields as good as the weed-free check (Table 5).

Linuron in combination with glyphosate as a postemergence-directed spray has caused severe sesame injury. In more recent work, however, linuron alone did not injure sesame and controlled problem weeds such as morningglory and smellmelon. Linuron can complement the use of metolachlor by providing additional broadleaf weed control.

7. Fluometuron

Fluometuron controls many annual dicotyledon weeds, however, it does not completely control some of the more troublesome weeds found in crops such as cotton (Burke & Wilcut, 2004). These troublesome weeds that fluometuron does not completely control include *Amaranthus* spp., *Ipomoea* spp., prickly sida (*Sida spinosa* L.), and sicklepod (*Senna obtusifolia* L.) (Buchanan, 1992; Crowley et al., 1979; Culpepper & York, 1997). Fluometuron applied postemergence may injure cotton and delay maturity (Guthrie & York, 1989). Guthrie and York (1989) stated that growers may resort to this type of application when an insufficient height differential between the crop and weeds prohibits postemergence-directed herbicide applications. Commercial preemergence uses of fluometuron in sesame include Costa Rica, where a grower guide (Anonymous, 2007d) recommends using an application of fluometuron at 2 kg/ha.

In preemergence experiments in India, fluometuron did not perform as well as alachlor or dichlormate (Subramanian & Sankaran, 1977). In Bulgaria, fluometuron at 1.0 kg/ha applied 2 days after sowing controlled annual broadleaf weeds. The quality and fat content of sesame seeds were not affected (Georgiev, 1980). In India, Subrananian & Sankaran (1981) found that fluometuron at 0.25 to 1.75 kg/ha did not perform as well as alachlor. In the U.S., fluometuron rates of 0.3 and 1.1 kg/ha had no effect on sesame height or population, provided good weed control, and had comparable yields to the check in south Texas (Grichar et al., 2001a). Later, Grichar et al., (2009) reported that fluometuron at 1.12 kg/ha in the High Plains region of Texas reduced sesame stand and caused injury in one of two years while no stand reduction or injury was noted at the south Texas location.

Recent work has evaluated fluometuron for use as a postemergence herbicide. There is some damage to the sesame when applied postemergence; however, it may be a good rescue herbicide. The sesame damage parallels diuron in that there is less damage in the late juvenile stage.

8. Prometryn

Prometryn has been widely used as a residual soil-applied and postemergence-directed herbicide in cotton grown west of the Mississippi River in the U.S. (Byrd, 2000) and controls many annual grasses and broadleaf weeds (Corbett et al., 2002; Burke & Wilcut, 2004). Prometryn is the only registered herbicide in the U.S. that provides excellent broad-spectrum control of weeds such as little mallow (*Malva parviflora* L.), shepherdspurse (*Capsella bursa-pastoris* L.), common purslane (*Portulaca oleracea* L.), and burning nettle (*Urtica urens* L.) in celery (*Apium graveolus* L.) (Daugovish et al., 2007). Currently, there is no commercial use of prometryn in sesame.

In preemergence experiments in Ethiopia, prometryn at 1.0 kg/ha was used safely on irrigated sesame while prometryn at 1.85 kg/ha resulted in less than 10% sesame injury. In a similar trial under natural rainfall, prometryn at 2.2 kg/ha completely eliminated the crop (Anonymous, 1973). In other studies in Ethiopia under irrigated conditions, prometryn applied preemergence at 3.2 kg/ha provided excellent weed control with negligible crop damage. However, under rain-fed conditions, prometryn at 0.8 kg/ha caused 100% sesame mortality (Moore, 1974). In Egypt, prometryn at 1.9 kg/ha caused sesame injury compared to pendimethalin alone and in tank mixtures with linuron and diuron (Ibrahim et al., 1988). In the U.S., preemergence applications of prometryn at 0.5 kg/ha caused no sesame injury, but prometryn applied preplant incorporated almost completely eliminated the sesame (D. Howell, unpublished data). Prometryn at 0.6 and 1.1 kg/ha resulted in lower sesame populations, lowered plant height, and also significantly reduced yields compared with metolachlor at 1.1 or 2.2 kg/ha (Grichar et al., 2001a). However, in a later study in south Texas and the High Plains of Texas, prometryn injured sesame but yields were not reduced from that of *S*-metolachlor (Grichar et al., 2009).

Farmer experience with prometryn is instructive in how a simple change in planting procedures can change results dramatically. In Arizona in the early 1980s, prometryn applied preplant incorporated provided excellent weed control with no apparent damage to sesame. In one year, suddenly there were very poor stands. In analyzing the situation, the farming practices had been to use a double disc opener to open the soil followed by press wheels after the seed was planted to close the gap. Farmers had decided that at times the press wheels did not close the trench resulting in moisture loss and poor germination. A few farmers put a chain on the back of each planter unit to drag soil back over the trench. These were the farmers that were not getting a stand. Basically, the double disc openers were pushing the soil layer with the prometryn to the sides, and allowed the sesame to germinate through the prometryn-treated herbicide zone. With the new farming practice, the soil with prometryn was brought back over the seed, and the sesame would not germinate. One simple farming practice changed success to failure.

In the past few years in Texas, under some conditions, prometryn applied preemergence has caused severe injury while in other instances little or no sesame injury has been noted. Another concern with prometryn is the effect on the sesame when the herbicide is applied to the previous crop. In West Texas, it is common to have localized hail storms that destroy cotton fields. In many cases, it is too late to replant cotton, but early enough to plant sesame. Sesame has followed thousands of hectares of failed cotton treated with prometryn with little or no injury to the sesame.

In studies with prometryn applied postemergence over-the-top or postemergence-directed, no injury has been noted with prometryn applied postemergence-directed but severe sesame injury has been found when applied over-the-top (author's personal observations). Excellent weed control, especially morningglory spp., has been noted with the postemergence-directed applications.

9. Clethodim, Fluazifop-P-butyl, Sethoxydim and Haloxyfop

Large seed grasses such as Texas millet [*Urochloa texana* (Buckl.) R. Webster] and rhizome johnsongrass can be a serious problem in sesame fields and are not controlled with current preemergence herbicides; therefore, postemergence grass herbicides are an absolute necessity. Postemergence control of annual grasses can be obtained with several herbicides (Grichar, 1991 a,b; Prostko et al., 2001). Grichar (1991a) found that sethoxydim applied early postemergence provided more effective annual grass control than late postemergence applications. However, grass size did not affect control with clethodim. Sethoxydim was reported to provide poor Texas millet control under less than ideal moisture conditions (Grichar, 1991a). Grichar (1991a) speculated that reduced moisture conditions resulted in less uptake and translocation of the herbicide within the plant (Chernicky et al., 1984; Fawcett et al., 1987). Clethodim controls annual grasses at lower use rates than sethoxydim. Prostko et al. (2001) noted 90% Texas millet, southern crabgrass (*Digitaria ciliaris* L.), or crowfootgrass [*Dactyloctenium aegyptium* (L.) Willd.] control occurred when clethodim followed an application of a soil applied dinitroaniline herbicide.

Commercial postemergence uses of grass herbicides in sesame include the following: in a Venezuela grower guide (Avila, 1999) stated that for grasses, fluazifop-P-butyl at 200 to 400 ml/ha worked well. In Australia, a grower guide (Sapin et al., 2000) stated that sesame was susceptible but tolerated fluazifop-P-butyl, haloxyfop, and sethoxydim. In Nigeria, a survey of agricultural crops (Anonymous, 2004a) suggested that fluazifop-P-butyl was used by two sesame farmer cooperatives. In the U.S., a producer guide (Langham et al., 2010) states that there is a label for clethodim but it should not be used during the flowering phase. Clethodim use during flowering will prevent capsule formation for 1 to 10 node pairs. Reportedly, fluazifop-P-butyl is commercially used in Mexico, Guatemala, Brazil, Paraguay, and Argentina while sethoxydim is used in Australia, and clethodim is used in the U.S.

In studies in Somalia, Malik and Muhammed-Ramzan (1992) showed that fluazifop-P-butyl at 3.7 L/ha and hand weeding provided effective control of grasses. In the U.S., Grichar et al., (2001b) reported that fluazifop-P-butyl and sethoxydim increased sesame yield over the untreated check and this was attributed to the control of Texas millet and southern crabgrass. In field studies in south Texas, clethodim has had no effect on sesame; however, when used by growers there has been considerable sesame injury under certain conditions. This injury has manifested itself as an inhibition of capsule formation. Field experiments which tried to replicate farmer results by using every permutation of 1x, 2x, and 3x rates of clethodim and 0, 1x, 2x, and 3x rates of crop oil showed no damage to sesame. However, in 2005, minute traces of glyphosate were added to clethodim and the farmer results were replicated. It was hypothesized that glyphosate residues had contaminated the clethodim in commercial air and ground sprayers. Further analysis of the effects of glyphosate drift from adjacent fields showed the same symptoms of yellowing of the growing tip, poor growth, and lack of formation of capsules for a period of time. Additional testing of timing of clethodim applications has shown that certain conditions, the application of clethodim during flowering will result in no capsule formation for 0 to 10 node pairs. However, there is not yellowing of the growing tip and plant growth is normal. Minimal damage to sesame (in the form of lack of capsule formation) has been noted with fluazifop-P-butyl and sethoxydim.

10. Glyphosate and/or Glufosinate-ammonium

Glyphosate is one of the safest and most frequently used herbicides in the world (Tao et al., 2007). It is a non-selective herbicide that controls many weed species. Products containing glyphosate are registered in more than 130 countries and are approved for weed control in more than 100 crops (Fernandez-Cornejo & McBride, 2000). Use of glyphosate increased dramatically with the introduction of glyphosate resistant crops. Crops that are glyphosate resistant allow glyphosate to be used as a selective herbicide and have offered additional options for weed control and have brought tremendous economic and agronomic benefits to growers around the world (Prostko et al., 2003; Thomas et al., 2006). A weakness of glyphosate is poor morningglory control (Corbett et al., 2004; Jordan et al., 1997).
Another development from the use of glyphosate-tolerant (Roundup Ready®) crops has been, in some instances, reduced weed pressure or increased weed species shifts when a Roundup Ready® crop has been grown for a number of years (Culpepper et al., 2000; Hilgenfeld et al., 2004; Marshall et al., 2000). In these areas, glyphosate has been so effective at controlling weeds that farmers are not concerned with a build-up of weed seed in the soil. Also, the overuse of this herbicide has resulted in an increase in glyphosate resistance to several weed species including *Amaranthus* spp., Italian ryegrass (*Lolium multiflorum* L.), and marestail or horseweed (*Conyza canadensis L. Cronquist*) (Bradshaw et al., 1997; Culpepper et al., 2006; Feng et al., 2004; Koger and Reedy, 2005; Mueller et al., 2003; Peterson, 1999).
Glyphosate is cleared in the U.S. for use in sesame as a burndown, with wiper applicators, and/or hooded sprayers in row middles (Langham et al., 2010). For burndown use, glyphosate should be applied before, during, or just after planting but before the sesame seedlings emerge. There have been no reports of glyphosate damage with the exception of late application where the seedlings have cracked the ground and exposed the plant to direct contact with the glyphosate. In the commercial use of glyphosate between the row middles, many weedy fields have been cleaned of weeds with no damage to the sesame. Wiper applications have been successful in controlling *Amaranthus* spp. as long as there is a height differential between the weeds and sesame with the weeds taller than the sesame. The wipers need to be adjusted throughout the field and careless low wipers that touch the sesame will either kill or severely damage the sesame.
Glyphosate applied postemergence over-the-top to sesame will result in plant death. In commercial fields, aerial recognition mistakes have taken airplanes spraying cotton over sesame fields and have killed sesame. Glyphosate drift from spraying adjacent fields have led to kill or yellowing of the sesame and a lack of capsule formation for one to three weeks depending of the amount of drift (Langham et al., 2010). When capsule formation does somewhat recover, the capsules will be smaller and will have less seeds and seed weight.
In postemergence-directed studies, glyphosate applied up to the 15-cm stem height resulted in 28% stunting; however, when applied to the 5-cm sesame stem height, stunting was no greater than 15% (Grichar, unpublished data). Glyphosate plus diuron stunted sesame 10% when applied up to 15 cm; however, no other herbicides stunted sesame more than 4% when applied to sesame at either height. In 2007, sesame stunting was greater than 2006 and more herbicides caused stunting. Stunting was more severe with glyphosate, glufosinate-ammonium, pyrithiobac, and trifloxysulfuron when applied 15 cm in sesame height compared with applications made 5 cm in height. Only glufosinate-ammonium, pyraflufen-ethyl, diuron plus linuron (Layby Pro®), and linuron at 5 cm caused less than 10% sesame stunting (Grichar, unpublished data). However, in subsequent years, all combinations using

glyphosate severely damaged the sesame. Further analysis showed a correlation between stage and condition of the sesame. The older sesame was less susceptible, but the clearest correlation was the amount of stress. When there was severe drought stress, there was less damage than when the plants were rapidly growing after recent irrigations or rains.
Glufosinate-ammonium is a nonselective postemergence herbicide like glyphosate that may have potential for use in sesame in many of the same ways. It controls a wide range of weed species and is especially effective on morningglory that can be difficult to control with glyphosate (Askew et al., 1997; Corbett et al., 2004; Hydrick and Shaw, 1995). Glufosinate-ammonium inhibits the synthesis of glutamine from glutamine and ammonia by inhibiting the activity of glutamine synthesis (Coetzer et al., 2002). This causes accumulation of ammonium and inhibition of photosynthesis (Sauer et al., 1987; Wild & Manderscheid, 1984). Glufosinate-ammonium is degraded rapidly by soil microorganisms (Wauchope et al., 1992). It controls several grasses and broadleaf weeds including *Amaranthus* spp. (Coetzer et al., 2002). Use of glufosinate-ammonium was limited to burndown treatments on noncrop areas and in no-till plantings; however, advances in genetic transformation of plants have facilitated the development of glufosinate-ammonium resistant crops such as corn, cotton, and soybean (Bradley et al., 2000; Coetzer et al., 2002; Wilson et al., 2007).
Postemergence-directed use of glufosinate-ammonium has produced similar results to glyphosate with mixed results ranging from no damage to severe damage. Work with glyphosate and glufosinate-ammonium as postemergence-directed sprays has been abandoned because these herbicides can severely damage the sesame and do not provide any late preemergence activity.

11. Trifluralin and Pendimethalin

The dinitroaniline herbicides, such as trifluralin and pendimethalin, are used to reduce weed populations and aid in the establishment and production of many crops including groundnut, soybean, and grain sorghum (Dotray et al., 2004; Grichar & Colburn, 1993; Grichar et al., 2005a, b; Grichar, 2006). The dinitroaniline herbicides provide excellent control of annual grasses (Buchanan et al., 1982; Chamblee et al., 1982; Wilcut et al., 1995) and are the only soil-applied herbicides registered for use in peanut that will provide full-season control of Texas millet (Wilcut et al., 1987a,b; Wilcut et al., 1995).
Uptake of dinitroaniline herbicides is primarily through roots and emerging shoots (Ashton & Crafts, 1981; Appleby & Valverde, 1989). Parker (1966) showed that trifluralin was more inhibitory to *Sorghum bicolor* when absorbed through roots than emerging shoots. It is possible that the dinitroaniline herbicides will be concentrated in the extreme upper portions of the soil profile and weed seed may be able to germinate below the zone where dinitroaniline herbicides are located (Johnson et al., 2002). In this case, emerging shoots pass through treated soil, whereas developing roots would be below the herbicide treated soil. The dinitroaniline herbicides have very low water solubility and are subject to losses due to photodecomposition and volatilization (Weber, 1990). Therefore, incorporation soon after herbicide application is important for effective weed control.
The effectiveness of soil-applied herbicides is dependent upon several factors, including movement of the herbicide into the soil either through water provided by rainfall or irrigation, or by mechanical incorporation (Prostko et al., 2001; Ross & Lembi, 1999). Chenault et al. (1992) reported that pendimethalin or trifluralin provided greater than 78% control of barnyardgrass (*Echinochloa crus-galli* (L.) P. Beauv) depending on incorporation

method. Tolerance to the dinitroaniline herbicides has been evaluated extensively in many crops. These herbicides injure susceptible plants by binding to β-tublin molecules, which ultimately leads to an inhibition of cell mitosis (Appleby & Valverde, 1989). Information on absorption and translocation within plants is less clearly defined; however, direct entry into plant tissue is considered limited, and unless the dinitroaniline herbicide enters meristematic tissues, the herbicide will have little effect on plant growth (Miller et al., 2003).

Previous research by Grichar et al. (2001a; 2009) reported sesame injury following the use of dinitroaniline herbicides applied preplant incorporated using various incorporation methods. Grichar et al. (2001a) reported that ethalfluralin, pendimethalin, and trifluralin reduced sesame stand numbers when compared with the untreated check. In that study the dinitroaniline herbicides were incorporated 2.5 cm deep with a tractor-driven power tiller. In another study, Grichar et al. (2009) reported that a spring-tooth harrow, with the lack of the ability to adjust incorporation depth, caused similar problems. However, the rolling cultivator mixing wheels, which were set to a depth of less than 2.5 cm, resulted in excellent sesame stands. Therefore, only a shallow incorporation of the dinitroaniline herbicides must be done when used in sesame to ensure a good stand. They concluded that it was best if the dinitroaniline herbicides were applied preemergence. Of the dinitroaniline herbicides, only pendimethalin formulated as Prowl $H_20^{®}$ can be applied preemergence (Anonymous 2004b); however, annual grass control following pendimethalin applied preemergence is often poor (Byrd & York, 1987; Culpepper, 1996).

Commercial uses of trifluralin in sesame include: in Honduras, a grower guide (Anonymous, 2002) states that use of trifluralin applied preemergence has proved to be very efficient in the control of weeds in sesame while in Costa Rica, a grower guide (Anonymous, 2007d) recommends using a preemergence application of trifluralin at 2.0 L/ha.

Martin and Crawford (1963) and Martin (1964) reported that trifluralin at 1.1 to 1.8 kg/ha was effective and non-toxic; however, trifluralin at 2.8 kg/ha killed sesame. In Venezuela, Montilla (1964) tried trifluralin at 1, 2, 3 L/ha, and the sesame did not germinate. In Ethiopia, Moore (1974) reported that trifluralin applied preplant incorporated at 0.75 and 1.4 kg/ha provided the greatest yields in sesame. Hussien et al. (1983) reported that trifluralin at rates greater than 0.84 kg/ha was harmful to sesame. However, it controlled annual grasses and increased the yield over the weedy check by 45%. Schrodter and Rawson (1984) reported that pendimethalin at 1.5 and 3.0 kg/ha and trifluralin at 0.84 kg/ha reduced sesame plant populations. Plant selectivity by herbicide placement is influenced greatly by the movement of the herbicide in soils (Ennis, 1964). If the dinitroaniline herbicides move, they may come in contact with the absorptive sites of sesame and cause sesame injury (Grichar et al., 2001a). In India, Shukla (1984) found that pendimethalin was toxic to sesame. In Israel, Graph et al. (1985) reported that preplant incorporation of trifluralin at 0.125 to 0.188 kg/ha was selective to sesame when the crop was sown on relatively warm soil, but early sowing resulted in inhibited root growth, retardation, and crop damage. In Korea, Kim et al. (1986) found that pendimethalin provided effective weed control using 1.27 kg/ha, but caused crop damage and yield reductions. In Egypt, Ibrahim et al. (1988) found that the best weed control and significantly greater seed yields and seed and yield components resulted from treatment with pendimethalin alone or in tank mixtures with linuron or diuron. In Somalia, Malik and Muhammed-Ramzak (1992) reported that pendimethalin at 3.7 L/ha provided the greatest weed control and significantly higher yield over the weedy check with no phytotoxic effects on sesame. Grichar et al. (2001a) reported yield increases over the untreated check with pendimethalin and trifluralin. They concluded that lack of yield differences among herbicide treatments which injured or reduced sesame stands could be attributed to the ability of the sesame plant to compensate for reduced stands. Sesame can produce excellent yields with only six to ten plants/m of row (author's personal observation). The rate of a dinitroaniline herbicide can affect sesame stand establishment. The one-half rate of ethalfluralin, pendimethalin EC, and trifluralin or the 3/4X dose of pendimethalin (Prowl H_2O®) resulted in greater stand counts than the 1 to 2X rate of these herbicides when incorporated with rolling cultivator mixing wheels (Grichar & Dotray, 2007).

12. Harvest aids

There are four reasons to use a harvest aid: (1) accelerate the drying to allow earlier harvest in better weather conditions (less rainfall, higher temperatures, and longer daylength), (2) create a uniform field where there are differences in moisture (lower areas with more moisture generally mature later) or late germination (seed that is planted in dry soil can germinate as much as 6 weeks later), (3) control weeds to dry down the entire field (green weeds can delay harvest or increase moisture in the combine bin), and (4) stop regrowth after a late rain (sesame can revert from the drying phase to the reproductive phase). Initial results show that use of a dessicant before the sesame drying phase can reduce the yields by as much as 10%. Initial results show that paraquat and diquat dry down the crop faster than any other harvest aid, but they will not kill weeds or stop regrowth. Glyphosate and glufosinate-ammonium take longer to dry down the crop, but do a more thorough job in a uniform drying down.

13. Conclusions

Herbicides are available that can help control weeds during the production of sesame. Control of weeds is the most important part of sesame production. There are millions of non-mechanized and hundreds of thousands of mechanized hectares of sesame grown every year with good economic return and minimum loss to weeds. However, improved weed control systems will contribute to increased net returns of the crop. The strategy that is being considered is to use a preemergence herbicide that has residual control and will provide effective soil residual control for approximately 4 to 6 weeks followed by a postemergence herbicide that will control small weeds and possibly provide residual control of weeds that have not germinated.

In all of the testing, there are few herbicides that do not affect sesame under some conditions; however, it is clear that in weedy conditions, sesame cannot produce economical yields. Therefore, some damage must be acceptable and with this minimal damage to the sesame, many herbicides have produced excellent economic yields. In the 1920s, Iowa farmers used to say that they plant 3 kernels of corn, "One for the worm, one for the crow, and one for me." Perhaps, in this century sesame farmers will need to plant extra sesame seed, "Some for the herbicide, and most for me."

14. References

Akin, D. & Shaw, D. (2004). In-season applications of glyphosate for control of redvine (*Brunnichia ovata*) in glyphosate-resistant soybean. Weed Technol. 18, 325-331.

Anil, K. & Thakur, K. S. (2005). Effect of sowing method and weed control practices on production potential of sesamum (*Sesamum indicum*)-based intercropping system under rainfed condition, Indian Journal of Weed Science 37, 133-134.

Anonymous. (1973). Report 1972-3, Institute of Agricultural Research. Ethiopia, p. 168.

Anonymous. (1997). A Guide Book for Field Crops in Thailand. Field Crops research Institute, Dept. of Agriculture. Bangkok, Thailand, p.166.

Anonymous. (2002). Estudio de prefactibilidad del ajonjoli. Secretaria de agricultura y ganaderia. Honduras. www.sic.gob.hn/portal/agro/infoagro/Ajonjoli, accessed 1 Apr 2007.

Anonymous. (2004a). Survey report of ten selected agro raw materials in Nigeria, Federal Ministry of Science and Technology, Abuja. http://www.rmrdc.gov.ng/Surveyreport2005/Beniseed.pdf, accessed 14 Apr. 2007.

Anonymous, (2004b). H_20: The power of water. BASF Corp. Research Triangle Park, NC. 4p.

Anonymous. (2007a). Guia Tecnica para producer ajonjoli de temporal en Michoacan, Instituto Nacional de Investigaciones Forestales, agricolas, y pecunarias, http://www.sagarpa.gob.mx/dlg/michoacan/inifap/ajonjoli_tdap.pdf, accessed 29 Mar 2007.

Anonymous. (2007b). Guia tecnica para el cultivo del "ajonjoli", http://72.14.209.104/search?q=cache:sQUIg6rKNzAJ:www.agronegocios.gob.sv/comoproducir/guias/ajonjoli.pdf+ajonjoli+el+salvador+herbicida&hl=en&ct=clnk&cd=1&gl=us, accessed 13 Apr 2007.

Anonymous. (2007c). Guia tecnica para el cultivo del "ajonjoli", http:// www.agronegocios.gob.sv/comoproducir/guias/ajonjoli.pdf+ajonjoli+el+salvador +herbicida&hl=en&ct=clnk&cd=1&gl=us, accessed 13 Apr 2007.

Anonymous. (2007d). Ajonjoli (*Sesamum indicum* L.), extracted from Aspectos tecnicos de 45 cultivos agricolas de Costa Rica, MAG. http://www.mag.go.cr/bibioteca_virtual_ciencia/tec_ajonjoli.pdf, accessed 13 Apr 2007.

Appendurai, R. (1967). Weed control in field crop cultivation in Ceylon.-P.A.N.S. Sect C 13(2), 123-6.

Appleby, A. & Valverde, B. (1989). Behavior of dinitroaniline herbicides in plants. Weed Technol. 3, 198-206.

Ashton, F. & Crafts, A. (1981). Dinitroanilines, In: Mode of Action of Herbicides. JohnWiley & Sons, Inc., New York. pp. 201-223.

Askew, S.; Shaw, D. & Arnold, J. (1997). Weed control in Liberty-Link soybean. Proc. South. Weed Sci. Soc. 50, 59.

Avila, J. (1999). Cultivo de ajonjoli (*Sesamum indicum* L.), Maracay, Ven., Fondo Nacional de Investigaciones Agropecuarias, Centro do Investigaciones Agropecuarias del Estado Portuguesa, 83 p.

Balyan, R. (1993). Integrated weed management in oilseed crops in India. Proc. India Sump., Indian Soc. Weed Sci. 1, 317-323.

Bansode, B. & Shelke, D. (1991). Integrated weed management in sesame. J. Maharashtra Agric. Univ. 16, 275.

Bedigian, D. & Harlan, J. (1986). Evidence for cultivation of sesame in the ancient world. Econ. Bot. 40(2), 137–154.

Bell, C.; Boutwell, B., Ogbuchiekwe, E. & McGiffen, M. (2000). Weed control in carrots: the efficacy and economic value of linuron. HortSci. 35, 1089-1091.

Bellinder, R.; Kirkwyland, J. & Wallace, R. (1997). Carrot (*Daucus carota*) and weed response to linuron and metribuzin applied at different crop stages. Weed Technol. 11, 235-240.

Beltrao, N.; Vieira, D., Nobrega, L. & Santos, J. (1991). Effects of fertilizers, cultivar and weed control in sesame, Pesquisa Agropecuaria Brasilieira 26(5), 605-611.

Bennett, M. (1993). Sesame Research Report 1991-92. Wet season, Katherine. Northern Territory, Australian Dept. of Primary Industries and Fisheries, Technical Bulletin No. 215.

Bennett, M. (1998). Sesame seed. The new rural industries, a handbook for farmers and investors. www.ridc.gov.au/pub/handbook/sesame.html, accessed 3 Apr 2007.

Betts, E. (1999). Thomas Jefferson's Garden Book (1766–1824). Thomas Jefferson Memorial Foundation, Inc., Charlottesville, Virginia

Bijanzadeh, E. & Ghadiri, H. (2006). Effect of separate and combined treatments of herbicides on weed contol and corn (*Zea mays*) yield. Weed Technol. 20, 640-645.

Bradley, P.; Johnson, W., Hart, S., Buesinger, M. & Massey, R. (2000). Economics of weed management in glufosinate-resistant corn (*Zea mays* L.) Weed Technol. 14, 495-501.

Bradshaw, L; Padgette, S., Kimball, S. & Wells, B. (1997). Perspectives on glyphosate resistance . Weed Technol. 11,189-198.

Brigham, R. & Young, J. (1982). Sesame performance tests, Lubbock, 1979. Texas Agr. Exp. Sta. Prog. Rep. 3688.

Buchanan, G. (1992). Trends in weed control methods. In Weeds of Cotton: Characterization and Control, McWhorter, C. & Abernathy, J. (Eds.), pp. 47-72, The Cotton Foundation, Memphis, TN

Buchanan, G.; Murray, D. & Hauser, E. (1982). Weeds and their control in peanuts. In Peanut Science and Technology. Pattee, H. & Young, C. (Eds.), pp. 206-249, Amer. Peanut Res. Educ. Soc., Inc., Yoakum, TX.

Burke, I. & Wilcut, J. (2004). Weed management in cotton with CGA-362622, fluometuron, and pyrithiobac. Weed Technol. 18, 268-276.

Byrd, Jr., J. (2000). Report of the 1999 Cotton Weed Loss Committee. 2000 Proc. Beltwide Cotton Conf. 2, 1455-1458.

Byrd, Jr., J. & York, A. (1987). Annual grass control in cotton (*Gossypium hirsutum*) with fluazifop, sethoxydim, and selected dinitroaniline herbicides. Weed Sci. 35, 388-394.

Caraballo de S. L.; Flores, J.O & Luna, J.A. (1986). Evaluación de cuatro dosis y dos fechas de aplicación del herbicida alacloro en ajonjolí (Sesamum indicum L.) en IVJornadas de biología y combate de malezas. Maturín.

Cardina, J. & Swann, C. (1988). Metolachlor effects on peanut growth and development. Peanut Sci. 15, 57-60.

Chamblee, R.; Thompson, Jr.,W. & Bunn, T. (1982). Management of broadleaf signalgrass (*Brachiaria platyphylla*) in peanuts (*Arachis hypogaea*). Weed Sci. 30, 40-44.

Chenault, E.; Bean, B. & Salisbury, C. (1992). Influence of incorporation methods on pigweed (*Amaranthus palmeri* S. Wats) and barnyardgrass [*Echinochloa crus-galli* (L.) Beauv] control with trifluralin and pendimethalin in sorghum (*Sorghum bicolor* L. Moench). Proc. South. Weed Sci. Soc. 45, 57.

Chernicky, J.; Gossett, B. & Murphy, T. (1984). Factors influencing control of annual grasses with sethoxydim or Ro-13-8895. Weed Sci. 32, 174-177.

Coetzer, E.; Al-Khatib, K. & Peterson, D. (2002). Glufosinate efficacy on *Amaranthus* species in glufosinate-resistant soybeans (*Glycine max*). Weed Technol. 16, 326-331.

Corbett, J.; Askew, S., Porterfield, D. & Wilcut, J. (2002). Bromoxynil, prometryn, pyrithiobac, and MSMA weed management systems for bromoxynil-resistant cotton (*Gossypium hirsutum*). Weed Technol. 16, 712-718.

Corbett, J.; Askew, S., Thomas, W. &Wilcut, J. (2004). Weed efficacy evaluations for bromoxynil, glufosinate, glyphosate, pyrithiobac, and sulfosate. Weed Technol. 18, 443-453.

Crowley, R.; Teem, D., Buchanan, G. & Hoveland, C. (1979). Responses of *Ipomoea* spp. and *Cassia* spp. to preemergence herbicides. Weed Sci. 27, 531-535.

Culp, T. & McWhorter, G. (1959). Annual report of cooperative industrial crops and weed investigations - 1959, Crops Research Division, ARS, USDA, Stoneville, MS. 75 p.

Culpepper, A. (1996). Sicklepod (*Senna obtusifolia*) management in sulfonylurea-tolerant soybean (*Glycine max*) and weed management systems in no-tillage bromoxynil-tolerant cotton (*Gossypium hirsutum*). M. S. Thesis. North Carolina State Univ., Raleigh, NC, 138pp.

Culpepper, A. & York, A. (1997). Weed management in no-tillage bromoxynil-tolerant cotton (*Gossypium hirsutum*). Weed Technol. 11, 335-345.

Culpepper, A.; York, A., Batts, R. & Jennings, K. (2000). Weed management in glufosinate- and glyphosate-resistant soybean (*Glycine max*). Weed Technol. 14, 77-88.

Culpepper, A.; Flanders, J., York, A. & Webster, T. (2004). Tropical spiderwort (*Commelina benghalensis*) control in glyphosate-resistant cotton. Weed Technol. 18, 432-436.

Culpepper, A.; Grey, T., Vencill, W, Kichker, J., Webster, T., Brown, S., York, A., Davis, J., & Hanna, W. (2006). Glyphosate-resistant Palmer amaranth (*Amaranthus palmeri*) confirmed in Georgia. Weed Sci. 54,620-626.

Daugovish, O.; Fennimore, S. & Smith, R. (2007). Herbicide evaluation for fresh market celery. Weed Technol. 21,719-723.

Dotray, P.; Keeling, J., Grichar, W., Prostko, E., & Lemon, R. (2004). Peanut response to ethalfluralin, pendimethalin, and trifluralin preplant incorporated. Peanut Sci. 30, 34-37.

Dungaral, H.; Chaplot, P. & Nagda, B. (2003). Integrated weed management in sesame (*Sesamum indicum* L.). Indian J. Weed Sci. 35, 236-238.

Eagleton, G.; Sandover, S. & Dickson, M. (1987). Research Report: Sesame Seed 1982-1986. Dept. Agric. Kununurra, West Australia.

Ennis, Jr., W. (1964). Selected toxicity in herbicides. Weed Res. 4, 93-104.

Fawcett, J.; Harvey, R., Arnold, W., Bauman, T., Eberlein, C., Kells, J., Moshier, L., Slife, F. & Wilson, R. (1987). Influence of environment on corn (*Zea mays*) tolerance to sethoxydim. Weed Sci. 35, 568-575.

Feng, P.; Tran, M., Chiu, T., Sammons, R., Heck, G. & CaJacob, C. (2004). Investigation into glyphosate-resistant horseweed (*Conyza canadensis*): retention, uptake, translocation, and metabolism. Weed Sci. 52, 498-505.

Fernandez-Cornejo, J. & McBride, W. (2000). Genetically Engineered Crops for Pest Management in U.S. Agriculture: Farm Level Effects. Agriculture Economic Rep. #786 (AER-786). Washington, DC: Economic Research Service, United States

Department of Agriculture: Web page: http://www.ers.usda.gov/publications. Accessed: April 16, 2008.

Franz, J.; Mao, M. & Sikorski, J. (1997). Glyphosate -A Unique Global Herbicide. Washington, DC: American Chemical Society. Pp. 143-186.

Georgiev, S. (1980). Effectiveness of some herbicides on the control of weeds on sesame Fields. Rasteniev dni Nauki. 15(7), 70-76.

Graph, S.; Kleifeld, Y., Herzlinger, G., Blumenfeld, T. & Bargutti, A. (1985). Weed control in sesame. Phytoparasitica. 13, 259.

Grichar, W. (1991a) Control of Texas panicum (*Panicum texanum*) and southern crabgrass (*Digitaria ciliaris*) in peanuts (*Arachis hypogaea*) with postemergence herbicides. Peanut Sci. 18, 6-9.

Grichar, W. (1991b) Sethoxydim and broadleaf herbicide interaction effects on annual grass control in peanuts (*Arachis hypogaea*). Weed Technol. 5, 321-324.

Grichar, W. & Colburn, A. (1993). Effect of dinitroaniline herbicides upon yield and grade of five runner cultivars. Peanut Sci. 20, 126-128.

Grichar, W.; Colburn, A. & Baughman, P. (1996). Yellow nutsedge (*Cyperus esculentus*) control in peanut (*Arachis hypogaea*) as influenced by method of metolachlor application. Weed Technol. 10, 278-281.

Grichar, W.; Sestak, D., Brewer, K., Besler, B., Stichler, C., & Smith, D. (2001a). Sesame (*Sesamum indicum* L.) tolerance and weed control with soil-applied herbicides. Crop Protect. 20, 389-394.

Grichar, W.; Sestak, D., Brewer, K., Besler, B., Stichler, C. & Smith, D. (2001b). Sesame (*Sesamum indicum* L.) tolerance with various postemergence herbicides. Crop Protect. 20, 685-689.

Grichar, W.; Besler, B., Dotray, P., Johnson, W. III, & Prostko, E. (2004a). Interaction of flumioxazin with dimethenamid or metolachlor in peanut (*Arachis hypogaea*). Peanut Sci. 31, 12-16.

Grichar, W.; Lemon, R., Dotray, P. & Besler, B. (2004b). Control of problem weeds and net returns with herbicide programs in peanut (*Arachis hypogaea*). In: Weed Biology and Management, Inderjit (Ed). pp.485-515, Kluwer Academic Publ. Dordrecht, The Netherlands.

Grichar, W.; Besler, B. & Brewer, K. (2005a). Weed control and grain sorghum (*Sorghum bicolor*) response to postemergence applications of atrazine, pendimethalin, and trifluralin. Weed Technol. 19, 999-1003.

Grichar, W.; Besler, B., Lemon, R. & Brewer, K. (2005b). Weed management and net returns using soil-applied and postemergence herbicide programs in peanut (*Arachis hypogaea* L.). Peanut Sci. 32, 25-31.

Grichar, W. (2006). Using soil-applied herbicides in glyphosate-resistant soybeans along the Texas Gulf Coast. Weed Technol. 20, 633-639.

Grichar, W. & Dotray, P. (2007). Weed control and sesame (*Sesamum indicum* L.) response to preplant incorporated herbicides and method of incorporation. Crop Protect. 26, 1826-1830.

Grichar, W.; Dotray, P. & Langham, D. (2009). Sesame (*Sesamum indicum* L.) response to preemergence herbicides. Crop Protect. 28, 928-933.

Gurnah, A. (1974). Critical weed competition periods in annual crops. Proceedings of the Fifth East African Weed Control Conf., Nairobi, Kenya, pp. 89-98.

Guthrie, D. & York, A. (1989). Cotton (*Gossypium hirsutum*) development and yield following fluometuron postemergence applied. Weed Technol. 3, 501-504.

Hilgenfeld, K; Martin, A., Mortensen, D. & Mason, S. (2004). Weed management in a glyphosate resistant soybean system: Weed species shifts. Weed Technol. 18, 284-291.

Hussein, M.; El-Hatlab, A., Abd-El Raouf, M., Shaban, S. & El-Deek, M. (1983). Effect of soil herbicides on weeds, yield and quality of sesame, *Sesamum indicum* L. Z. Acker. Pflanzenbou 152, 173-185.

Hydrick, D. & Shaw, D. (1995). Non-selective and selective herbicide combinations in stale seedbed soybean (*Glycine max*). Weed Technol. 9, 158-165.

Ibrahim, A.; El Wekil, H. Yehia, Z. & Shaban, S. (1988). Effect of some weed control treatments on sesame (*Sesamum indicum* L.) and associated weeds. Journal of Agronomy and Crop Science 160, 319-324.

Johnson, W.,III; Prostko, E. & Mullinix, B., Jr., (2002). Texas panicum (*Panicum texanum*) control in strip-tillage peanut (*Arachis hypogaea*). Peanut Sci. 29, 141-145.

Jordan, D.; York, A., Griffin, J. Clay, P., Vidrine, P. & Reynolds, D. (1997). Influence of application variables on efficacy of glyphosate. Weed Technol. 11, 354-362.

Kim, J.; Kang, C., Oh, S. & Lee, H. (1986). Studies on the weed control and phytotoxicity of herbicides under polyethylene film mulching in the sesame upland. Research of the Rural Development Admin., Plant Environment Mycology and Farm Products Ultilization. Korea Republic 28, 72-78.

Koger, C. & Reddy, K. (2005). Role of absorption and translocation in the mechanism of glyphosate resistance in horseweed (*Conyza canadensis*). Weed Sci. 53, 84-89.

Kropff, M.& Spitters, C. (1991). A simple model of crop loss by weed competition from early observations on relative leaf area of weeds. Weed Res. 31, 97-105.

Langham, D. & Wiemers, T. (2002). Progress in mechanizing sesame in the U.S. through breeding, In: Trends in New Crops and New Uses, J. Janick and A. Whipkey (Eds.), pp. 157-173, ASHS Press, Alexandria, VA.

Langham, D.(2007). Phenology of sesame. In: Issues in New Crops and New Uses, J. Janick and A. Whipkey (Eds.), pp. 144-182, ASHS Press, Alexandria, VA.

Langham, D; Grichar, W. & Dotray, P. (2007). Review of herbicide research on sesame (*Sesamum indicum* L.). Version 1. www.sesamegrowers.org, accessed Aug, 2010.

Langham, D.(2008). Growth and Development of Sesame. Sesaco Corp. 42 p.

Langham D.; Riney, J., Smith, G., Wiemers, T., Pepper, D. & Speed, T. (2010). Sesame Producers Guide. Sesaco Corp. 33 p.

Lyubenov, Y. & Kostadinov, K. (1970). Possibilities for chemical weed control in sesame (*Sesamum indicum*). Preliminary Project, Rastitelna Zashchita. 4, 31-35.

Malik, N. & Muhammed-Ramzan. (1992). Effect of herbicides and cultural operations on weed control and grain yield sesame, Journal of Agricultural Research Lahore 30, 147-152.

Marshall, M; Al-Khatib, K. & Maddox, L. (2000). Weed community shifts associated with continuous glyphosate applications in corn and soybean rotation. Proc. West. Soc. Weed Soc. 53, 22.

Martin, J. &. Crawford, J.H. (1963). Sesame and Castorbean research in South Carolina, Res. Ser. 41, S.C. Agri. Exp. Sta., Clemson, S.C.

Martin, J. (1964). Sesame and Castorbean Research in South Carolina.-Res.Ser.59 South Carolina Agric. Expt. Stn .,Clemson, South Carolina, U.S.A.

Martin, C. (1995). Development of an effective weed control system for sesame in the Northern Territory. Proceedings of Sesame Workshop, Darwin and Katherine, New Territory, pp. 121-127.

Mazzani, B. (1957). Mejoramiento del ajonjoli en Venezuela, Ministerio de Agricultura y Cria, Maracay, Venez., p. 127.

Mazzani, B. (1966). El cultivo del ajonjoli.-Agronomia 3, 1-10.

Miller, A.; Bellinder, R., Bradley, B., Rauch, B., Goffinet, M., & Welser, M. (2003). Cabbage (*Brassica oleracea*) response to pendimethalin applied postemergence. Weed Technol. 17, 256-260.

Montilla, D. (1964). Resultados de Pruebas de Algunos Herbicidas en el Cultivo de Ajonjoli en Turen, Agronomia Tropical, Venez. 14,3.

Moore, J. (1973a). Sesame herbicide trial-setit Humera. Progress Report for the Period April 1972 to March 1973, Ethiopia Institute of Agricultural Research Station. Sesame herbicide screening trial Melka Werer. Pp. 175-177.

Moore, J. (1973b). Sesame herbicide screening trial-Melka Werer, Progress Report for the period April 1972 to March 1973, Ethiopia Institute of Agricultural Research Station. Sesame herbicide screening trial Melka Werer. Pp.168-174.

Moore, J. (1974). Evaluation of herbicides in irrigated and raingrown sesame in the lowlands of Ethiopia. Proceedings of the Fifth East African Weed Control Conference, Nairobi, Kenya, pp.108-130.

Mueller, T.; Shaw, D. & Witt, W. (1999). Relative dissipation of acetochlor, alachlor, metolachlor, and SAN 582 from three surface soils. Weed Technol. 13, 341-346.

Mueller, T; Massey, J., Hayes, R., Main, C.& Stewart, C. (2003). Shikimate accumulation in both glyphosate- sensitive and glyphosate-resistant horseweed (*Conyza canadensis*). J. Agric. Food Chem. 51, 680-684.

Osborne, B.; Shaw, D. & Ratliff, R. (1995). Response of selected soybean (*Glycine max*) cultivars to dimethenamid and metolachlor in hydroponic conditions. Weed Technol. 9, 178-181.

Pineda P.; Juan B. & Jesus M. Avila M. (1988). Alternativas para el control de *Macrophomina phaseolina* y *Fusarium oxysporum* pategenos del ajonjoli *(Sesamum indicum L,* Agronomía Tropical. 38,79-84.

Parker, C. (1966). The importance of shoot entry in the action of herbicides applied to the soil. Weeds. 14, 117-121.

Peterson, D. (1999). The impact of herbicide-resistant weeds on Kansas agriculture. Weed Technol. 13, 632-635.

Prostko, E.; Johnson, III, W. & Mullinix, Jr., B. (2001). Annual grass control with preplant incorporated and preemergence applications of ethalfluralin and pendimethalin in peanut (*Arachis hypogaea*). Weed Technol. 15, 36-41.

Prostko, E.; Norsworthy, J. & Raymer, P. (2003). Soybean (*Glycine max*) response to glyphosate, iflubenzuron, and boron combinations. Weed Technol. 17, 186-189.

Reddy, K.; Bryson, C., & Burke, I. (2007). Ragweed parthenium (*Parthenium hysterophorus*) control with preemeergence and postemergence herbicides. Weed Technol. 21, 982-986.

Ross, M. & Lembi, C. (1999). Herbicide incorporation techniques and equipment. In: Applied Weed Science. 2nd ed, Stewart, C., Stagman, J., Carnis, M., (Eds.), Pp. 371-375, Upper Saddle River, NJ: Prentice-Hall.

Saint-Louis, S.; DiTommaso, A., & Watson, A. (2005). A common ragweed (*Ambrosia artemisiifolia*) biotype in southwestern Quebec resistant to linuron. Weed Technol. 19, 737-743.

Santelmann, P.; Elder, W. & Murlock, R. (1963). The effect of several pre-emergence herbicides on guar, cowpeas, mungbeans, and sesame. Proc. South. Weed Sci. Soc. 16, 83.

Sapin, V.;. Mills, G, Schmidt, D., & O'Shanesy, P. (2000). Growing sesame in South Burnett. Department of Primary Industries, Queensland Government, www.dpi.qld.gov.au/fieldcrops/2888.html, accessed Nov 2003.

Sauer, H.; Wild, A. & Ruhle, W. (1987). The effect of phosphinothricin (glufosinate) on photosynthesis II. The causes of inhibition of photosynthesis . Z. Naturforsch. 42C, 270-278.

Schrodter, G. & Rawson, J. (1984). Herbicide evaluation studies in sesame. Aust. Weeds. 3, 47-49.

Senseman, S. 2007 (editor). Herbicides Handbook. 9th ed. Lawrence, KS: Weed Science Society of America. 458 p.

Shaw, D. & Coats, G. (1988). Herbicides and cultivation for sicklepod, *Cassia obtusifolia,* control in soybeans, *Glycine max*. Weed Technol. 2, 187-190.

Shukla, V. 1984. Chemical weed control in sesame, Pesticides 18, 13-14.

Singh, D.; Dagar, J. & Gangwar, B. (1992). Infestation by weeds and their management in oilseed crops - a review. Agric. Rev. 13, 163-175.

Smith, A. (1989). Degradation, fate, and persistence of phenoxyalkanoic acid herbicides in soil. Rev. Weed Sci. 4, 1-24.

Soto-Soto, E. & Silva-Vasquez, J. (1987). Trials of pre-emergence in the cultivation of sesame (*Sesamum indicum* L.), Informe de Las Labores de la Seccion de Agronomica Centro Experimental del Algodon, pp. 131-145.

Sprankle, P.; Meggitt, W. &Pinner, D. (1975). Absorption, action, and translocation of glyphosate. Weed Sci. 23, 235-240.

Subramanian, A. & Sankaran, S. (1977). Studies on the relative efficiency of preemergence herbicides in *Sesamum* under graded levels of nitrogen. Program-and-Abstracts of Papers, Weed Science Conference and Workshop in India (Paper 63), 37-38.

Subramanian, A. & Sankaran, S. (1981). Chemical weed control of common purslane (*Trianthema portulacastrum* L.), Abstracts of Papers, Annual Conference of Indian Society of Weed Science, undated 26.

Tao, B.; Zhou, J., Messersmith, C. & Nalewaja, J. D. (2007). Efficacy of glyphosate plus bentazon or quizalofop on glyphosate-resistant canola or corn. Weed Technol. 21, 97-101.

Thomas W.; Britton, T., Clewis, S., Askew, S. & Wilcut, J. (2006). Glyphosate-resistant cotton (*Gossypium hirsutum*) response and weed management with trifloxysulfuron, glyphosate, prometryn, and MSMA. Weed Technol. 20, 6-13.

Upadhyay, U. (1985). Weed management in oilseed crops. In Oilseed Production Constraints and Opportunities. Srivastava, H., Bhaskaran, S., Vatsya, B., Menon, K.K.G. (Eds.), pp. 491-499, New Delhi, Oxford & IBH Publishing Co.

Viera, D.; Beltrao, N., Nobrega, L., Azevedo, D. & Oliveira, J. (1998). Chemical control of weeds in sesame crop, Communicado Technico Embrapa Algodao (72): 8 pp.

Wauchope, R.; Butler, T., Hornsby, A., Augustijn-Beckers, R. & Burt, J. (1992). TheSCS/CES pesticide properties database for environmental decision-making. Rev. Environ. Contam. Toxicol. 123, 1.

Weber, J. (1990). Behavior of dinitroaniline herbicides in soils. Weed Technol. 4, 394-406.

Wehtje, G.; Wilcut, J., Hicks, T. & McGuire, J. (1988). Relative tolerance of peanuts to alachlor and metolachlor. Peanut Sci. 15, 53-56.

Weiss, E. (1971). Castor, sesame, and safflower. Leonard Hill Books, London. p. 311–525.

Wilcut, J.; Wehtje, G. & Walker, R. (1987a). Economics of weed control in peanuts (*Arachis hypogaea*) with herbicides and cultivations. Weed Sci. 35, 711-715.

Wilcut, J.; Wehtje, G. & Patterson, M. (1987b). Economic assessment of weed control systems for peanuts (*Arachis hypogaea*). Weed Sci. 35, 433-437.

Wilcut, J.; York, A., Grichar, W. & Wehtje, G. (1995). The biology and management of weeds in peanut (*Arachis hypogaea*), In Advances in Peanut Science, Pattee, H. & Stalker, H. (Eds.).pp. 207-244. Amer. Peanut Res. Educ. Soc., Stillwater, OK.

Wild, A. & Manderscheid, R. (1984). The effect of phosphinothricin on the assimilation of ammonia in plants. Z. Naturfosch. 39C, 500-504.

Wilson, H.; Hines, T., Hatzios, K. & Doub, J. (1988). Efficacy comparisons of alachlor and metolachlor formulations in the field. Weed Technol. 2, 24-27.

Wilson, D.; York, A. & Jordan, D. (2007). Effect of row spacing on weed management in glufosinate-resistant cotton. Weed Technol. 21, 489-495.

Zewdie, K. (1994). Effect of pre- and post-emergence herbicides on weed control and yield of sesame under irrigation., Proceedings of the Ethiopian Weed Science Society, Addis Abeba (Ethiopia), 15-16 Dec 1994.

4

Defining Interactions of Herbicides with Other Agrochemicals Applied to Peanut

[1]David L. Jordan, Professor
[2]Gurinderbir S. Chahal, Graduate Research Assistant
[3]Sarah H. Lancaster, Former Graduate Research Assistant
[4]Joshua B. Beam, Former Graduate Research Assistant
[5]Alan C. York, William Neal Reynolds Professor Emeritus
[1]Department of Crop Science, North Carolina State University, Box 7620, Raleigh, NC 27695-7620.
[2]Department of Crop Science, North Carolina State University, Box 7620, Raleigh, NC 27695-7620.
[3]Department of Crop Science, North Carolina State University, Box 7620, Raleigh, NC 27695-7620. Current position: Assistant Professor, Department of Plant and Soil Sciences, Oklahoma State University, 368 Agricultural Hall, Stillwater, OK 74078-6028.
[4]Department of Crop Science, North Carolina State University Box 7620, Raleigh, NC 27695-7620. Current position: North Carolina Business Manager, Sunny Ridge Farm Inc., 1900 5th Street Northwest, Winter Haven, FL 33881-2106.
[5]Department of Crop Science, North Carolina State University Box 7620, Raleigh, NC 27695-7620. United States of America

1. Introduction

Management strategies to protect peanut (*Arachis hypogaea* L.) from pest damage require multiple applications of fungicides, herbicides, and insecticides. Additionally, micronutrients and plant growth regulators are often applied to improve nutrient balance and to manage peanut growth and development. Over fifty active ingredients can be used to manage pests in peanut, often with more than one formulated product commercially available. Timing of application of pesticides, micronutrients, and plant growth regulators often coincide during the growing season, and co-application of these agrochemicals is desirable if pesticide, micronutrient, and plant growth regulator performance and peanut tolerance are not compromised. In addition to potential interactions related to physiological effects on plants and other organisms, application variables such as commercial formulation, adjuvant, water quality, and environmental stress can affect agrochemical compatibility.

Physical compatibility, in particular formation of precipitates in spray tanks and equipment, is a concern for farmers when co-applying agrochemicals. Defining potential interactions among these agrochemicals is important in developing appropriate weed management programs and implementing integrated pest management strategies for peanut.
Considerable research has been conducted during the last four decades to define interactions among agrochemicals (Barrett, 1993; Green, 1989; Green and Bailey, 1987; Hatzios and Penner, 1985; Putnam and Penner, 1974). Most of these reviews are focused on interactions of herbicides in mixture with other herbicides, fungicides, insecticides, nematicides, and adjuvants, in general, but not for a particular crop. Some of these reviews summarized the mechanisms responsible for the interactions of herbicides with other agrochemicals and the statistical methodology for characterization of agrochemical combinations (Barrett, 1993; Green, 1989; Hatzios and Penner, 1985; Jianhua *et al.*, 1995). Since these reviews were published, many of new agrochemicals have received registration for different crops and for other uses. Defining interactions of these new agrochemicals is important when developing pest management strategies for a cropping system. This chapter reviews some of the interactions discussed in the earlier work, but also elucidates the interactions and/or compatibility of herbicides with other agrochemicals used in peanut production systems.

2. Peanut production systems

Mechanized production systems utilize a wide range of agrochemicals to manage peanut growth and development and minimize the impact of pests on peanut yield and quality (Lynch and Mack, 1995; Sherwood *et al.*, 1995; Wilcut *et al.*, 1995). Pests that can potentially impact peanut are diverse (Table 1). Yield loss from weed interference or from damage caused by insects, diseases, and nematodes can be substantial in peanut if pest control strategies are not implemented in a timely manner.
Monocotyledonous weeds, including annual and perennial grasses and sedges, as well dicotyledonous weeds, are prevalent in peanut production systems in the United States (Webster, 2009; Wilcut *et al.*, 1995). Comprehensive herbicide programs, in combination with appropriate cultural practices, are employed to manage weeds and minimize interference and subsequent yield loss (Wilcut *et al.*, 1987a 1987b 1990 1995). Herbicides are often applied in mixtures either prior to planting (preplant incorporated in conventional tillage or preplant to emerged weeds in reduced tillage), immediately following planting (preemergence), or after peanut and weeds have emerged (postemergence) (Burke *et al.*, 2004; Clewis *et al.*, 2007; Richburg *et al.*, 1995 1996; Wilcut *et al.*, 1994a 1994b 1995).
Agrochemicals with efficacy against insects and plant parasitic nematodes are often applied in the seed furrow at planting and include organophosphate and carbamate insecticides (Brecke *et al.*, 1996; Drake *et al.*, 2009; Funderburk *et al.*, 1998; Minton *et al.*, 1990; Minton and Morgan, 1974; Riley *et al.*, 1997). In-furrow insecticides also reduce incidence of tomato spotted wilt of peanut (caused by tomato spotted wilt virus, a *Tospovirus* vectored by several species of thrips) (Brown *et al.*, 2003; Hurt *et al.*, 2003). Pyrethroid insecticides are often applied to peanut foliage to control beet armyworm, corn earworm, fall armyworm, potato leaf hopper, and two-spotted spider mites. Chlorpyrifos can be applied at pegging, 45 to 70 days after peanut emergence, to control lesser cornstalk borer (Mack *et al.*, 1989 1991) and southern corn rootworm (Brandenburg and Herbert, 1991; Chapin and Thomas, 1993).

Depending on environmental and edaphic conditions and a range of agronomic and pest management practices, application of insecticides may be needed throughout the growing season to protect peanut from pest damage.

Disease, caused by viruses, bacteria, or fungi, can reduce peanut yield considerably when not controlled (Sherwood *et al.*, 1995). Fungicides are applied routinely to peanut to control foliar-borne diseases, including early leaf spot, late leaf spot, and web blotch (Brenneman *et al.*, 1994; Culbreath *et al.*, 2008; Shew and Waliyar, 2005). Fungicides are also applied to control the soil-borne disease stem rot and Sclerotinia blight (Brenneman *et al.*, 1994; Culbreath *et al.*, 2008; Smith *et al.*, 1992). Although variation is noted among geographical regions, years, and environmental conditions, during a typical growing season fungicides are applied either singly or in combination beginning approximately 45 days after peanut emergence and continuing throughout the remainder of the growing season, which can approach 135 or more days (Sherwood *et al.*, 1995; Smith and Littrell, 1980). Fungicide programs to control early and late leaf spot and stem rot often include bi-weekly sprays during this period. Fungicides applied to control these diseases provide protection for a period of two weeks under most environmental conditions (Shew and Waliyar, 2005). The soil fumigant metam sodium is often applied to peanut to control *Cylindrocladium* black rot (Cline and Beute, 1986). A period of at least two weeks between fumigation and peanut planting is required to allow the fumigant to dissipate, making weed control prior to planting challenging under some environmental conditions, especially excessive rainfall, that allow weeds to emerge between fumigation and planting operations (Van Gundy and McKenry, 1977).

The micronutrients boron and manganese are applied routinely to optimize peanut growth and development and, in the case of boron, to ensure proper kernel development (Gascho and Davis, 1995; Harris and Brolman, 1966; Powell *et al.*, 1996). Because peanut is often grown on coarse-textured soils, boron can be deficient due to leaching. Single, and in some cases, multiple applications of boron-containing foliar solutions are applied 45 to 70 days after peanut emergence (Gascho and Davis, 1995). Manganese deficiency occurs frequently in peanut because of liming to achieve a target soil pH above 6.0. Correcting a manganese deficiency is achieved by foliar applications when visible symptoms become apparent, although some growers apply manganese irrespective of plant symptomology (Powell *et al.*, 1996).

Excessive vine growth of peanut can reduce row visibility at digging and vine inversion (Mitchem *et al.*, 1996). Prohexadione calcium is currently the only plant growth regulator applied to manage vine growth in order to facilitate efficient digging. Prohexadione calcium inhibits gibberellin biosynthesis in responsive plants (Grossman *et al.*, 1994) and is applied when 50% of vines from adjacent peanut rows have met, and an application is repeated 2 to 3 weeks later (Mitchem *et al.*, 1996). This timing of application is generally 70 to 90 days after peanut emergence (Mitchem *et al.*, 1996). In addition to prohexadione calcium, a wide range of products are available at the distributor level that contain micronutrient combinations, synthetic plant growth regulators, and other ingredients perceived to have value. Many of these products are not applied routinely to peanut.

3. Agrochemicals used in peanut

A diversity of pesticide active ingredients is available for peanut (Table 2) (Brandenburg, 2010; Jordan, 2010; Shew, 2010). Currently, 19 herbicide active ingredients, 16 insecticide active ingredients, and 20 fungicide active ingredients representing the major modes of action can be applied during the peanut growing season. Three fumigants, two micronutrients, and one plant growth regulator are registered for use in peanut. Within herbicide, insecticide, fungicide, and fumigant categories, a range of formulated products are available for most active ingredients. These products are often manufactured and sold through distributor networks and are numerous. Additionally, spray adjuvants are recommended with some, but not all, agrochemicals to increase performance and compatibility.

Presence of biotic and abiotic stresses often occur simultaneously during the peanut growing season, and timing of application for many agrochemicals overlap (Table 3). Practitioners prefer limiting the number of trips across fields in order to increase efficiency of managing peanut. This approach is preferable because of convenience, savings in time, reduced application costs, and freeing labor for other operations. Additionally, applying multiple pesticides with different modes of action is an important resistance management strategy for pests (Brandenburg, 2010; Jordan, 2010; Shew, 2010). This approach is feasible as long as adverse interactions, primary increased crop injury or decreased pest control, do not occur. Defining interactions among agrochemicals is important in assisting growers and their advisors as they make decisions on co-application of these products.

4. Herbicide – Herbicide interactions

A considerable amount of research has been conducted to define interactions among herbicides used in peanut. Herbicides applied in combination either preplant incorporated or preemergence generally increase the spectrum of weed control or the length of residual weed control (Wilcut *et al.*, 1987b 1995). For example, pendimethalin is often applied in combination with alachlor, dimethenamid-*P*, metolachlor, or *S*-metolachlor to improve early season weed control (Bridges *et al.*, 1984; Wehtje *et al.*, 1988; Wilcut *et al.*, 1994b 1995; Wilcut and Swann, 1990). Alachlor, dimethenamid-*P*, metolachlor, or *S*-metolachlor can be applied with diclosulam, flumioxazin, or imazethapyr preemergence to enhance weed control with a single application (Clewis *et al.*, 2007; Grichar *et al.*, 1992, 1996, 2000, 2008; Scott *et al.*, 2001). Combinations of preplant incorporated or preemergence herbicides currently registered for use in peanut have not been shown to increase peanut injury over either herbicide component applied alone (Wilcut *et al.*, 1995). However, several herbicides that are no longer registered for peanut increased peanut injury when co-applied as compared to the herbicides applied alone (Wilcut *et al.*, 1995).

In reduced tillage systems, herbicides are needed to control winter weeds and summer annual weeds that have emerged prior to planting peanut. These herbicide applications include glyphosate, paraquat, or 2,4-D alone or in combinations with other herbicides. Combinations of glyphosate and 2,4-D broaden the spectrum of weed control compared with each herbicide applied alone (Flint and Barrett, 1989a). However, in some instances, 2,4-D can negatively affect efficacy of glyphosate, but this interaction is typically noted only on grass weeds (Flint and Barrett, 1989b). Efficacy of paraquat is generally not negatively affected by 2,4-DB (Wehtje *et al.*, 1992a). Glyphosate and paraquat can also be applied with herbicides that provide residual weed control. This approach is designed to control emerged weeds and provide residual weed control prior to and following planting (Wilcut *et al.*, 1995).

Paraquat is often applied at peanut emergence or up to 28 days after peanut emergence (Carley *et al.*, 2009; Wilcut *et al.*, 1990). Other non-residual herbicides such as bentazon or acifluorfen plus bentazon as well as residual herbicides such as alachlor, diclosulam, dimethenamid-*P*, imazethapyr, metolachlor, or *S*-metolachlor are applied postemergence to broaden the spectrum of control (Askew *et al.*, 1999; Bailey *et al.*, 1999; Grey *et al.*, 2002; Grichar and Colburn, 1996). Injury associated with paraquat can be reduced by co-application with bentazon (Jordan *et al.*, 2003b; Wehtje *et al.*, 1992b). However, the chloroacetamide herbicides alachlor, dimethenamid-*P*, metolachor, or *S*-metolachlor applied with paraquat can increase peanut injury (Jordan *et al.*, 2003b). Diclosulam and imazethapyr did not affect injury potential from paraquat (Jordan *et al.*, 2003b). Weed control with these herbicide combinations generally increases depending on the weed species and size of the weed (Wilcut *et al.*, 1995). For example, bentazon and imazethapyr co-applied can increase control of emerged common cocklebur and yellow nutsedge, while control of annual grasses by paraquat can be reduced when paraquat is co-applied with bentazon (Wehtje *et al.*, 1992b; Wilcut *et al.*, 1994b). Residual control by chloroacetamide herbicides, diclosulam, and imazethapyr was not affected by paraquat applied alone or with bentazon (Grichar *et al.*, 2000; Wilcut *et al.*, 1995).

Co-application of postemergence herbicides with efficacy against dicotyledonous weeds and sedges generally increases control of weeds or broadens the spectrum of control compared with components of the mixture applied alone (Green, 1989; Hatzios and Penner, 1985; Jianhua *et al.*, 1995; Wilcut *et al.*, 1995). In contrast, efficacy of clethodim and sethoxydim, often referred to as graminicides, can be reduced when applied in mixture with herbicides that control dicotyledonous weeds and sedges (Culpepper *et al.*, 1998 1999; Grichar, 1991; Jordan and York, 1989; Minton *et al.*, 1989; Mueller *et al.*, 1989; Myers and Coble, 1992; Vidrine *et al.*, 1995). The interaction of bentazon and sethoxydim is one of the most notable examples of reduced graminicide efficacy caused by a herbicide that controls dicotyledonous plants and sedges (Rhodes and Coble, 1984a 1984b; Wanamarta and Penner, 1989; Wanamarta *et al.*, 1989). Annual and perennial grass control by sethoxydim is reduced by bentazon through reduced absorption of sethoxydim into grasses (Rhodes and Coble, 1984b; Wanamarta and Penner, 1989; Wanamarta *et al.*, 1989). The mechanism of reduced control is associated with physical interactions of the herbicides in the spray solution prior to reaching the target weed (Penner, 1989; Thelen *et al.*, 1995). Acifluorfen and imazethapyr also can reduce efficacy of clethodim and sethoxydim (Burke and Wilcut, 2003; Grichar, 1991; Lassiter and Coble, 1987; Myers and Coble, 1992). In contrast to reduced grass control when these herbicides are co-applied, control of dicotyledonous plants and sedges is not reduced by clethodim and sethoxydim (Dotray *et al.*, 1993; Holshouser and Coble, 1990; Isaacs *et al.*, 2003). Efficacy of clethodim can also be reduced by acifluorfen, acifluorfen plus bentazon, bentazon, imazethapyr, imazapic, lactofen, and 2,4-DB (Grichar *et al.*, 2002; Myers and Coble, 1992; York *et al.*, 1993). The magnitude of reduced efficacy can be minimized or eliminated by applying the herbicides sequentially, increasing the graminicide rate, or applying more efficacious adjuvants (Burke *et al.*, 2004; Jordan, 1995; Myers and Coble, 1992; Wanamarta and Penner, 1989; Wanamarta *et al.*, 1989). Grass species, plant size, and plant stress also can affect the magnitude of negative interactions (Green, 1989; Hatterman-Valenti *et al.*, 2006). York and Wilcut (1995) reported that bentazon reduced control of yellow and purple nutsedge by imazethapyr.

Chloroacetamide herbicides can be applied postemergence without injuring peanut (Grichar *et al.*, 1996, 2008; Jordan *et al.*, 2003b). While these herbicides provide residual control of grasses and some dicotyledonous and sedge weeds, they do not control weeds that have emerged (Foy and Witt, 1997; Grichar *et al.*, 2000; Richburg *et al.*, 1995). These herbicides can be applied with herbicides that have efficacy against emerged weeds. Dimethenamid-*P* and *S*-metolachlor did not reduce grass control by the graminicides clethodim or sethoxydim or the dicotyledonous and sedge herbicides acifluorfen, acifluorfen plus bentazon, or imazapic (Grichar *et al.*, 2000; Wilcut *et al.*, 1995). However, visible injury caused by acifluorfen increased when acifluorfen was applied with chloroacetamide herbicides (Jordan *et al.*, 2003b). Johnson *et al.* (1993) reported that injury from postemergence application of paraquat was not increased when following several chloroacetamide herbicides applied at planting, in contrast with injury observed when the herbicides were co-applied.

5. Herbicide – Insecticide interactions

Timing of application of herbicides and insecticides overlap during much of the growth cycle of peanut (Table 3). As with other crops, potential interactions between herbicides and insecticides applied in the seed furrow to control thrips and suppress plant parasitic nematodes can occur (Hauser *et al.*, 1976 1981). Acephate and aldicarb applied in the seed furrow at planting did not affect injury potential of peanut following postemergence application of acifluorfen plus bentazon or bentazon; however, the insecticide phorate applied in the seed furrow enhanced visible injury associated with bentazon, although this injury was generally transient (Swann and Herbert, 1999). Although interactions of nicosulfuron (Bailey and Kapusta, 1994; Morton *et al.*, 1994; Rahman and James, 1993) and pyrithiobac-sodium (Allen and Snipes, 1995) increased injury in corn (*Zea mays* L.) and cotton (*Gossypium hirsutum* L.), respectively. However, chlorpyrifos applied at planting did not affect peanut response to diclosulam, *S*-metolachlor, or flumioxazin applied preemergence or acifluorfen, acifluorfen plus bentazon, imazapic, or paraquat plus bentazon applied postemergence (Jordan *et al.*, 2008). Efficacy of graminicides can be affected by insecticides applied to peanut. Carbaryl and dimethoate applied postemergence in combination with sethoxydim reduced annual grass control; no adverse effect was noted when acephate was mixed with sethoxydim (Byrd and York, 1988). Pyrethroid insecticides did not affect efficacy of postemergence herbicides (Allen and Snipes, 1995).

6. Herbicide – Fungicide interactions

Similar to herbicides and insecticides, timing of application of postemergence herbicides and fungicides to control foliar and soil-borne diseases overlap considerably during the peanut growing season (Table 3). Fungicides are applied beginning approximately 45 days after peanut emergence and can be applied until a few weeks prior to digging and vine inversion. Efficacy of clethodim and sethoxydim can be reduced by co-application with copper-containing fungicides or azoxystrobin, chlorothalonil, and pyraclostrobin (Jordan *et al.*, 2003a; Lancaster *et al.*, 2005a 2008). Fluazinam and tebuconazole did not reduce grass control compared with graminicides applied alone (Jordan *et al.*, 2003a; Lancaster *et al.*, 2005a 2005b). Efficacy of herbicides that control dicotyledonous and sedge weeds is not generally affected by fungicides (Jordan *et al.*, 2003a). As was noted for interactions of herbicides, weed species and size and plant stress can affect the magnitude of interactions between herbicides and fungicides (Jordan *et al.*, 2003a).

Although not used in peanut, efficacy of glyphosate was not affected by azoxystrobin, pyraclostrobin, or tebuconazole (Grichar and Prostko, 2009). Weed control by metribuzin, rimsulfuron, and thifensulfuron-methyl applied to tomato (*Lycopersicon esculentum* Mill.) was not affected by azoxystrobin or pyraclostrobin (Robinson and Nurse, 2008). However, pyraclostrobin increased tomato injury from thifensulfuron-methyl when co-applied (Robinson and Nurse, 2008). Chlorothalonil increased persistence of metolachlor in soil although cyproconazole, flutriafol, and tebuconazole did not affect dissipation of metolachlor (White *et al.*, 2009).

7. Herbicide – Micronutrient interactions

Boron and manganese are the primary micronutrients applied to peanut. Occasionally, these can affect herbicide performance. For example, efficacy of clethodim and imazethapyr was reduced by micronutrients for some, but not all, weeds evaluated (Jordan *et al.*, 2006; Lancaster *et al.*, 2005b).

8. Herbicide – Plant growth regulator interactions

Prohexadione calcium is the primary plant growth regulator available for use in peanut. Efficacy of the herbicides acifluorfen, acifluorfen plus bentazon, bentazon, imazethapyr, imazapic, lactofen, and 2,4-DB was not affected by prohexadione calcium (Beam *et al.*, 2002). However, other plant growth regulator products developed by agrochemical distributor chains are numerous and have not been evaluated sufficiently to make recommendations on compatibility with herbicides.

9. Co-application of multiple components

The previous discussion focused on co-applications that have only two components. However, there is considerable interest in compatibility of three or more pesticides, micronutrients, adjuvants, or plant growth regulators and their impact on pest control and crop management. With respect to weed control, efficacy of clethodim, sethoxydim, and 2,4-DB were compared when these herbicides were applied alone or with fungicides and insecticides (Jordan *et al.*, 2003a; Lancaster *et al.*, 2005a 2005b). Although results often supported previous findings with components from two groups of pesticides, no clear relationships were established with respect to combinations of three or more pesticides (Lancaster *et al.*, 2005a 2005b). More recently, research is being conducted to compare herbicide efficacy with mixtures containing various levels of fungicide, insecticide, micronutrient, or adjuvant (Chahal *et al.*, 2009a 2009b).

10. Herbicide effects on other agrochemicals

The focus of this review has been the impact of agrochemicals used in peanut on herbicide efficacy. However, defining the impact of herbicides on insect and disease control and response of peanut to micronutrient and plant growth regulator applications is important. Preliminary research has shown that the herbicides clethodim and 2,4-DB do not affect performance of chlorothalonil, pyraclostrobin, or the prepackage combination of prothioconazole plus tebuconazole (Chahal *et al.*, 2009a 2009b). Paraquat and 2,4-DB did not affect chlorothalonil efficacy (Choate *et al.*, 1998). Katan and Eshel (1973) discussed possible mechanisms of interactions among herbicides and pathogens. With respect to peanut, Baysinger *et al.* (1999) reported that sporulation of early leaf spot was reduced by acifluorfen and lactofen. In-vitro, chlorothalonil efficacy against early blight (*Alerternaria solani*) was reduced by metribuzin while susceptibility of *Pseudo cercospera herpotrichoides* to cyproconazole increased following exposure to dicamba, bromoxynil, or ioxynil (Kataria and Gisi, 1990; Levesque and Rahe, 1992).

The influence of the postemergence herbicides clethodim, imazapic, lactofen, and 2,4-DB on boron and manganese absorption into peanut leaves was evaluated, and results suggested that while herbicides could affect accumulation of boron and manganese in leaf tissue, the adjuvant associated with these herbicides may have been the primary factor in influencing absorption (Jordan *et al.*, 2006 2009a).

11. Application variables that can affect interactions

A wide range of application variables can affect interactions of herbicides with other agrochemicals. Adjuvant selection, herbicide rate, commercial formulation, active ingredient, length of time between applications of components, spray volume, water quality, weed species, and environmental conditions can affect interactions of agrochemicals. For example, the negative effect of bentazon was reduced by including ammonium sulfate and other more efficacious adjuvants with clethodim and sethoxydim (Jordan, 1995: Jordan and York, 1989; Penner, 1989; Wanamarta and Penner, 1989; York *et al.*, 1990). Applying a higher rate of the herbicide that may be adversely affected can compensate for interactions (Chernicky and Slife, 1986; Rhodes and Coble, 1984a 1984b). Differential response to clethodim was noted when applied with different formulations of chlorothalonil (Jordan *et al.*, 2003a). Increasing the interval between applications of components of the mixture can overcome negative interactions, especially herbicide-herbicide interactions (Green, 1989; Grichar and Boswell, 1987; Putnam and Ries, 1967). Applying graminicides in higher spray volumes can exacerbate the negative influence of herbicides and fungicides on weed control by graminicides (Buhler and Burnside, 1984; Jordan *et al.*, 2003a; Kells and Wanamarta, 1987). Water quality, in particular presence of cations that can form complexes in the spray solution, can influence the propensity of herbicides to perform poorly in pesticide combinations (Buhler and Burnside, 1983a 1983b; Hatzios and Penner, 1985; Sandberg *et al.*, 1978; Stahlman and Phillips, 1979; Thelen *et al.*, 1995; Whisenant and Bovey, 1993; Wills and McWhorter, 1985).

Environmental conditions that affect plant response to agrochemicals can influence the magnitude of interactions. Negative effects of interactions associated with efficacy of systemic herbicides, especially graminicides, are exacerbated when grasses are stressed and physiological processes that reduce absorption and translocation occur (Burke *et al.*, 2004; Burke and Wilcut, 2003; Green, 1989; Wanamarta and Penner, 1989; Wanamarta *et al.*, 1989).

12. Challenges of defining interactions and making recommendations

Major challenges in recommending practices associated with interactions of herbicides with other agrochemicals include the diversity of products available, the diversity of weeds present in the field that can vary in response to herbicides, differences in water quality within and across production regions, and adjuvant recommendations associated with herbicides versus other agrochemicals. Compounding these complex variables is the unpredictable response often observed due to environmental conditions. Additionally, new active ingredients are being marketed that have not been evaluated for possible interactions, and as patents expire, some of these active ingredients are formulated differently from the product receiving the initial registration. The extensive number of possible combinations, especially when multiple components are considered, is challenging from a research standpoint when considering the logistics of research trials needed to address possible interactions. Although attempts are made to establish research trials in a manner similar to practitioner operations, techniques associated with maintaining spray solutions and applying materials with small-plot equipment differ from commercial applications and adds to the challenges in research.

One reasonable criticism of the current approach to defining interactions, which is dictated by the number of combinations and the logistics of experimentation, is the lack of defining the impact of interactions on the larger context of the production system. For example, when a reduction in weed control by a fungicide occurs, how detrimental to peanut yield and economic value is this reduction when considering alternatives to prevent the interaction from occurring? Also, dose response curves are often used to define interactions of pesticides, and while these can be more informative than selection of a single rate of the components in the mixture, the number of treatment combinations required for this approach is often not feasible because of resource constraints.

13. Future research

A considerable knowledge base has been developed to define interactions of herbicides with other agrochemicals with respect to weed control in peanut. However, the effect of herbicides on performance of fungicides and insecticides is limited but no less important than defining impacts on herbicide efficacy. As new active ingredients and new formulations of active ingredients become available, additional research will be needed to define interactions among these agrochemicals. Although interactions of herbicide-herbicide combinations have been defined broadly and in some cases in detail, research elucidating the mechanism of reduced control associated with co-application of fungicides, insecticides, or plant growth regulator and micronutrients is limited (Duke *et al.*, 2007). Finally, determining the impact of interactions in the overall production system would be beneficial.

Pest	Latin bionomical and authority
Weeds[a]	
Bermudagrass	*Cynodon dactylon* (L.) Pers.
Bristly starbur	*Acanthospermum hispidum* DC.
Broadleaf signalgrass	*Urochloa platyphylla* (Nash) R.D. Webster
Common cocklebur	*Xanthium strumarium* L.
Common lambsquarters	*Chenopodium album* L.
Common ragweed	*Ambrosia artemisiifolia* L.
Crabgrass spp.	*Digitaria* spp.
Crowfootgrass	*Dactyloctenium aegyptium* (L.) Willd.
Eclipta	*Eclipta prostrata* L.
Florida beggarweed	*Desmodium tortuosum* (Sw.) DC.
Florida pusley	*Richardia scabra* L.
Goosegrass	*Eleusine indica* (L.) Gaertn.
Hairy indigo	*Indigofera hirsuta* Harvey
Jimsonweed	*Datura stramonium* L.
Johnsongrass	*Sorghum halepense* (L.) Pers.
Morningglory spp.	*Ipomoea* spp.
Nutsedge spp.	*Cyperus* spp.
Palmer amaranth	*Amaranthus palmeri* S. Wats.
Pigweed spp.	*Amaranthus* spp.
Prickly sida	*Sida spinosa* L.
Sicklepod	*Senna obtusifolia* (L.) H.S. Irwin & Barneby
Spurge spp.	*Chamaesyce* spp.
Texas millet	*Urochloa texana* (Buckl.) R. Webster
Tropic croton	*Croton glandulosus* var. septentrionalis Muell.-Arg.
Insects[b]	
Beet armyworm	*Spodoptera exigua* Hübner
Corn earworm	*Heliothis zea* Boddie
Lesser cornstalk borer	*Elasmopalpus lignosellus* Zeller

Pest	Latin bionomical and authority
Southern corn rootworm	*Diabrotica undecimpunctata* Howardi
Thrips	*Frankliniella* spp.
Diseases[c]	
Aspergillus crown rot	*Aspergillus niger*
Botrytis	*Botrytis cinerea* Pers.
Cylindrocladium black rot	*Cylindrocladium parasiticum* Crous, Wingfield, and Alfenas
Early leaf spot	*Cercospora arachidicola* Hori
Late leaf spot	*Cercosporidium personatum* Berk. & Curtis
Pythium	*Pythium* spp.
Rhizoctonia limb rot	*Rhizoctonia solani* Kuhn
Sclerotinia blight	*Sclerotinia minor* Jagger
Spotted wilt	Tomato spotted wilt, caused by a *Tospovirus*
Stem rot	*Sclerotium rolfsii* Sacc.
Web blotch	*Phoma arachidicola* Marasas, Pauer, and Boerema
Plant parasitic nematodes[c]	
Lesion nematode	*Pratylenchus brachyurus*
Northern root knot	*Meloidogyne hapla*
Peanut root knot	*Meloidogyne arenaria*
Ring	*Criconemella ornata*
Sting	*Belonolaimus longicaudatus*
Two-spotted spider mite[b]	*Tetranychus urticae* Koch

[a]Webster, T. M. 2009. Weed survey - southern states. Proc. South. Weed Sci. Soc. 62:509-524.
[b]Brandenburg, R. L. 2010. Peanut insect management. Pages 85-102 *in* 2010 Peanut Information. North Carolina Cooperative Extension Service Publication AG-331.
[c]Shew, B. B. 2010. Peanut disease management. Pages 103-130 *in* 2010 Peanut Information. North Carolina Cooperative Extension Service Publication AG-331.

Table 1. Common and scientific names for major pests found in peanut in the United States.

Herbicides	Insecticides	Fungicides	Fumigants	Plant growth regulators
Acifluorfen	Acephate	Azoxystrobin	Dichloropropene plus chloropicrin	Prohexadione calcium
Alachlor	Aldicarb	Basic copper sulfate	Metam sodium	
Bentazon	*Bacillus thuringiensis*	Boscalid	1,3 dichloropropene	
Carfentrazone	Carbaryl	Chlorothalonil		
Clethodim	Chlorpyrifos	Dodine		
Diclosulam	Disulfoton	Fludioxomil		
Dimethenamid -*P*	Esfenvalerate	Fluoxastrobin		
Ethafluralin	Fenpropathrin	Flutolanil		
Flumioxazin	Gamma-cyhalothrin	Mancozeb		
Glyphosate	Indoxacarb	Mancozeb and copper hydroxide		
Imazapic	Lambda-cyhalothrin	Mefenoxam		
Imazethapyr	Malathion	Metconazole		
Lactofen	Methomyl	Pentachloronitrobenzene (PCNB)		
Metolachlor	Phorate	Propiconazole		
Paraquat	Propargite	Prothioconazole		
Pendimethalin	Spinosad	Pyraclostrobin		
Sethoxydim		Sulfur		
S-metolachlor		Tebuconazole		
Sulfentrazone		Thiophanate methyl		
2,4-DB		Trifloxystrobin		

[a]Brandenburg, R. L. 2010. Peanut insect management. Pages 85-102 *in* 2010 Peanut Information. North Carolina Cooperative Extension Service Publication AG-331.
[b]Jordan, D. L. 2010. Weed management in peanuts. Pages 55-83 *in* 2010 Peanut Information. North Carolina Cooperative Extension Service Publication AG-331.
[c]Shew, B. B. 2010. Peanut disease management. Pages 103-130 *in* 2010 Peanut Information., Peanut Information 2010. North Carolina Cooperative Extension Service AG-331.

Table 2. Pesticide active ingredients registered for use in peanut in the United States during 2010.[a,b,c]

	April	May	June	July	August	September
Weeds						
Dicotyledonous						
Monocotyledonous						
Insects						
Beet armyworm						
Corn earworm						
Lesser cornstalk borer						
Southern corn rootworm						
Tobacco thrips						
Diseases						
Aspergillus crown rot						
Botrytis						
Cylindrocladium black rot						
Cercospora spp.						
Pythium						
Rhizoctonia limb rot						
Sclerotinia blight						
Spotted wilt						
Stem rot						
Web blotch						
Plant parasitic nematodes						
Two-spotted spider mite						
Nutrient deficiency						
Boron						
Manganese						
Peanut vine management						
Prohexadione calcium						

Table 3. Biotic and abiotic stresses and approximate timing of management that can occur in peanut during the growing season in the United States.

14. References

Allen, R. L. and C. E. Snipes. 1995. Interactions of foliar insecticides applied with pyrithiobac. Weed Technol. 9:512-517.

Askew, S. D., J. W. Wilcut, and J. R. Cranmer. 1999. Weed management in peanut *(Arachis hypogaea)* with flumioxazin preemergence. Weed Technol. 13:594-598.

Bailey, J. A. and G. Kapusta. 1994. Soil insecticide and placement influence corn (*Zea mays*) tolerance to nicosulfuron. Weed Technol. 8:598-606.

Bailey, W. A., J. W. Wilcut, D. L. Jordan, C. W. Swann, and V. B. Langston. 1999. Response of peanut *(Arachis hypogaea)* and selected weeds to diclosulam. Weed Technol. 13:771-776.

Barret, M. 1993. Interactions of herbicides and other agrochemicals in plants: Interactions in mixtures with other herbicides and with safeners, fungicides, insecticides, and nematicides. Pages 113-132 *in* J. Altman (ed.) Pesticide Interactions in Crop Production: Beneficial and Deleterious Effects. CRC press Boca Raton, FL.

Baysinger, J. A., H. A. Melouk, and D. S. Murray. 1999. Acifluorfen, acifluorfen plus 2,4-DB, and lactofen decreased the sporulation of early leaf spot lesions. Peanut Sci. 26:4-8.

Beam, J. B., D. L. Jordan, A. C. York, J. E. Bailey, T. G. Isleib, and T. E. McKemie. 2002. Interaction of prohexadione calcium with agrichemicals applied to peanut (*Arachis hypogaea* L). Peanut Sci. 29:29-35.

Brandenburg, R. L. 2010. Peanut insect management. Pages 85-102 *in* 2010 Peanut Information. North Carolina Cooperative Extension Service Publication AG-331.

Brandenburg, R. L. and D. A. Herbert, Jr. 1991. Effect of timing on prophylactic treatments for southern corn rootworm (Coleoptera:Chrysomelidae) in peanut. J. Econ. Entomol. 84: 1894-1898.

Brecke, B. J., J. E. Funderburk, I. D. Teare, and D. W. Gorbet. 1996. Interactions of early-season herbicide injury, tobacco thrips injury, and cultivar on peanut. Agron. J. 88:14–18.

Brenneman, T. B., H. R. Sumner, L. R. Chandler, J. M. Hammond, and A. K. Culbreath. 1994. Effect of application techniques on performance of propiconazole for peanut disease control. Peanut Sci. 21:134-138.

Bridges, D. C., R. H. Walker, J. A. McGuire, and N. R. Martin. 1984. Efficiency of chemical and mechanical methods for controlling weeds in peanuts (*Arachis hypogaea*). Weed Sci. 32:584-591.

Brown, S., J. Todd, A. Culbreath, J. Baldwin, J. Beasley, B. Kemerait, and E. Prostko. 2003. Minimizing spotted wilt of peanut. University of Georgia Cooperative Extension Service Bulletin 1165.

Buhler, D. D. and O. C. Burnside. 1983a. Effect of spray components on glyphosate toxicity to annual grasses. Weed Sci. 31:124-130.

Buhler, D. D. and O. C. Burnside. 1983b. Effect of water quality, carrier volume, and acid on glyphosate phytotoxicity. Weed Sci. 31:163-169.

Buhler, D. D. and O. C. Burnside. 1984. Effect of application factors on the phytotoxicity of fluazifop-butyl, haloxyfop-methyl, and sethoxydim. Weed Sci. 32:574-583.

Burke, I. C., A. J. Price, J. W. Wilcut, D. L. Jordan, A. S. Culpepper, and J. T. Ducar. 2004. Annual grass control in peanut (*Arachis hypogaea*) with clethodim and imazapic. Weed Technol. 18:88-92.

Burke, I. C. and J. W. Wilcut. 2003. Physiological basis for antagonism of clethodim by imazapic on goosegrass (*Eleusine indica* (L.) Gaertn.). Pestic. Biochem. Physiol. 76:37-45.

Byrd, J. D., Jr. and A. C. York. 1988. Interactions of carbaryl and dimethoate with sethoxydim. Weed Technol. 2:433-436.

Carley, D. S., D. L. Jordan, R. L. Brandenburg, and L. C. Dharmasri. 2009. Factors influencing response of Virginia market type peanut (*Arachis hypogaea*) to paraquat under weed-free conditions. Peanut Sci. 36:180-189.

Chahal, G. B. S., D. L. Jordan, J. Burton, S. B. Clewis, R. L. Brandenburg, and B. B. Shew. 2009a. Weed control in peanut with tank mixtures containing five components. Proc. South. Weed Sci. Soc. 62:124.

Chahal, G. B. S., D. L. Jordan, B. B. Shew, R. L. Brandenburg, J. D. Burton, and D. Danehower. 2009b. Weed and disease control in peanut as influenced by co-application of agrichemicals. Proc. Am. Peanut Res. and Education Soc. 41:51.

Chapin, J. W. and J. S. Thomas. 1993. Effects of chlorpyrifos on pod damage, disease incidence, and yield in two peanut fungicide programs. Peanut Sci. 20:102-106.

Chernicky, J. P. and F. W. Slife. 1986. Effects of sublethal concentrations of bentazon, fluazifop, haloxyfop, and sethoxydim on corn (*Zea mays*). Weed Sci. 34:171-174.

Choate, J., G. Wehtje, and K. L. Bowen. 1998. Interaction of paraquat-based weed control with chlorothalonil-based disease control in peanut. J. Production Agriculture. 11:191-195.

Clewis, S. B., W. J. Everman, D. L. Jordan, and J. W. Wilcut. 2007. Weed management in North Carolina peanuts (*Arachis hypogaea*) with *S*-metolachlor, diclosulam, flumioxazin, and sulfentrazone systems. Weed Technol. 21:629-635.

Cline, W. O. and M. K. Beute. 1986. Effect of metam sodium, peanut genotype and inoculum density on incidence of Cylindrocladium black rot. Peanut Sci. 13:41-45.

Culbreath, A. K., R. C. Kemerait, Jr., and T. B. Brenneman. 2008. Management of leaf spot diseases of peanut with prothioconazole applied alone or in combination with tebuconazole or trifloxystrobin. Peanut Sci. 35:149-158.

Culpepper, A. S., A. C. York, and C. Brownie. 1999. Influence of bromoxynil on annual grass control by graminicides. Weed Sci. 47:123-128.

Culpepper, A. S., A. C. York, K. M. Jennings, and R. B. Butts. 1998. Interaction of bromoxynil and postemergence graminicides in large crabgrass (*Digitaria sanguinalis*). Weed Technol. 12:554-559.

Dotray, P. A., L. C. Marshall, W. B. Parker, D. L. Wyse, D. A. Somers, and B. G. Gengenbach. 1993. Herbicide tolerance and weed control in sethoxydim-tolerant corn (*Zea mays*). Weed Sci. 41:213-217.

Drake, W. L., D. L. Jordan, R. L. Brandenburg, B. R. Lassiter, P. D. Johnson, and B. M. Royals. 2009. Peanut cultivar response to damage from tobacco thrips and paraquat. Agron. J. 101:1388-1393.

Duke, S. O., D. E. Wedge, A. L. Cerdeira, and M. B. Matallo. 2007. Herbicide Effects on Plant Disease. Pest Manage. Sci. 18:36-40.

Flint, J. L. and M. B. Barrett. 1989a. Effects of glyphosate combinations with 2,4-D or dicamba on field bindweed (*Convolvulus arvensis*). Weed Sci. 37:12-18.

Flint, J. L. and M. B. Barrett. 1989b. Antagonism of glyphosate toxicity to johnsongrass (*Sorghum halepense*) by 2,4-D and dicamba. Weed Sci. 37:700-705.

Foy, C. L. and H. L. Witt. 1997. SAN 582, alachlor, and metolachlor control triazine-resistant (TR) smooth pigweed *(Amaranthus hybridus)* in no-till corn *(Zea mays).* Weed Technol. 11:623-625.

Funderburk, J. E., D. W. Gorbet, I. D. Tearre, and J. Stavisky. 1998. Thrips injury can reduce peanut yield and quality under conditions of multiple stress. Agron. J. 90:563–566.

Gascho, G. J. and J. G. Davis. 1995. Soil fertility and plant nutrition. Pages 383-418 *in* H. E. Pattee and H. T. Stalker (eds.) Advances in Peanut Science. Am. Peanut Res. and Education Soc., Stillwater, OK.

Green, M. J. 1989. Herbicide antagonism at the whole plant level. Weed Technol. 3:217-226.

Green, J. M. and S. P. Bailey. 1987. Herbicide interactions with herbicides and other agricultural chemicals. Pages 37-61 *in* C. G. McWhorter and M. R. Gebhardt (eds.) Methods of Applying Herbicides., Weed Sci. Soc. Am., Champaign, IL.

Grey, T. L., D. C. Bridges, E. F. Eastin, and G. E. MacDonald. 2002. Influence of flumioxazin rate and herbicide combinations on weed control in peanut *(Arachis hypogaea* L.). Peanut Sci. 29:24-29.

Grichar, W. J. 1991. Sethoxydim and broadleaf herbicide interaction effects on annual grass control in peanuts *(Arachis hypogaea).* Weed Technol. 5:321-324.

Grichar, W. J., B. A. Besler, K. D. Brewer, and T. A. Baughman. 2002. Grass control in peanut (*Arachis hypogaea*) with clethodim and selected broadleaf herbicide combinations. Peanut Sci. 29:85-88.

Grichar, W. J. and T. E. Boswell. 1987. Herbicide combinations in peanut (*Arachis hypogaea*). Weed Technol. 1:290-293.

Grichar, J. E. and A. E. Colburn. 1996. Flumioxazin for weed control in Texas peanut (*Arachis hypogaea* L.). Peanut Sci. 23:30-36.

Grichar, W. J., A. E. Colburn, and P. A. Baumann. 1996. Yellow nutsedge *(Cyperus esculentus)* control in peanut *(Arachis hypogaea)* as influenced by method of metolachlor application. Weed Technol. 10:278-281.

Grichar, W. J., P. A. Dotray, and T. A. Baughman. 2008. Yellow nutsedge (*Cyperus esculentus*) control and peanut tolerance to *S*-metolachlor and diclosulam combinations. Weed Technol. 22:442-447.

Grichar, W. J., R. G. Lemon, D. C. Sestak, and K. D. Brewer. 2000. Comparison of metolachlor and dimethenamid for yellow nutsedge (*Cyperus esculentus* L.) control and peanut (*Arachis hypogaea* L.) injury. Peanut Sci. 27:26-30.

Grichar, W. J., P. R. Nester, and A. E. Colburn. 1992. Nutsedge (*Cyperus* spp.) control in peanuts (*Arachis hypogaea*) with imazethapyr. Weed Technol. 6:396-400.

Grichar, J. L. and E. P. Prostko. 2009. Effect of glyphosate and fungicide combinations on weed control in soybeans. Crop Protection. 28:619-622.

Grossman, K., K. S. Koenig, and J. Kwiatkowski. 1994. Phytohormonal changes in intact shoots of wheat and oilseed rape treated with the acylcyclohexanedione growth retardant prohexadione calcium. Physiol. Plant. 90:139-143.

Harris, H. C. and J. B. Brolman. 1966. Comparison of calcium and boron deficiencies of peanut. I. Physiological and yield differences. Agron. J. 58:575-578.

Hatterman-Valenti, H. M., A. Pitty, and M. D. K. Owen. 2006. Effect of environment on giant foxtail (*Setaria faberi*) leaf wax and fluazifop-P absorption. Weed Sci. 54:607–614.

Hatzios, K. K. and D. Penner. 1985. Interaction of herbicides with other agricultural chemicals in higher plants. Rev. Weed Sci. 1:1-64.

Hauser, E. W., G. A. Buchanan, W. J. Ethredge, M. D. Jellum, and S. R. Ceci. 1976. Interactions among peanut cultivars, herbicide sequences, and a systemic insecticide. Peanut Sci. 3:56-62.

Hauser, E. W., G. A. Buchanan, J. E. Harvey, W. L. Currey, D. W. Gorbet, and N. A. Minton. 1981. Pesticide interactions with peanut cultivars. Peanut Sci. 8:142-144.

Holshouser, D. L. and H. D. Coble. 1990. Compatibility of sethoxydim with five postemergence broadleaf herbicides. Weed Technol. 4:128-133.

Hurt, C., R. Brandenburg, D. Jordan, B. Shew, T. Isleib, M. Linker, A. Herbert, P. Phipps, C. Swann, and W. Mozingo. 2003. Managing tomato spotted wilt virus in peanuts in North Carolina and Virginia. North Carolina Cooperative Extension Service Publication AG-638.

Isaacs, M. A., H. P. Wilson, and J. E. Toler. 2003. Combinations of sethoxydim with postemergence broadleaf herbicides in sethoxydim-resistant corn (*Zea mays*). Weed Technol. 17:224-228.

Jianhua, Z., A. S. Hamill, and E. W. Susan. 1995. Antagonism and synergism between herbicides: Trends from previous studies. Weed Technol. 9:86-90.

Johnson, W. C., Jr., D. L. Colvin, and B. G. Mullinix, Jr. 1993. Comparative response of three peanut cultivars to multiple herbicide applications. Peanut Sci. 20:17-20

Jordan, D. L. 1995. Influence of adjuvants on the antagonism of graminicides by broadleaf herbicides. Weed Technol. 9:741-747.

Jordan, D. L. 2010. Weed management in peanuts. Pages 55-83 *in* 2010 Peanut Information. North Carolina Cooperative Extension Service Publication AG-331.

Jordan, D. L. 2010. Peanut production practices. Pages 25-53 *in* 2010 Peanut information. North Carolina Cooperative Extension Service Publication AG-331.

Jordan, D. L., R. L. Brandenburg, P. D. Johnson, B. M. Royals, and B. Watson. 2008. Interactions of chlorpyrifos and herbicides applied to peanut (*Arachis hypogaea* L.). Peanut Sci. 35:32-37.

Jordan, D. L., A. S. Culpepper, W. J. Grichar, J. T. Ducar, B. J. Brecke, and A. C. York. 2003a. Weed control with combinations of selected fungicides and herbicides applied postemergence to peanut (*Arachis hypogaea* L.). Peanut Sci. 30:1-7.

Jordan, D. L., S. H. Lancaster, J. E. Lanier, P. D. Johnson, J. B. Beam, and A. C. York. 2009a. Influence of application variables on efficacy of manganese-containing fertilizers applied to peanut. Proc. Am. Peanut Res. and Education Soc. 41:82-83.

Jordan, D. L., S. H. Lancaster, J. E. Lanier, P. D. Johnson, J. B. Beam, A. C. York, R. L. Brandenburg, F. R. Walls, S. Casteel, and C. Hudak. 2006. Influence of application variables on efficacy of boron-containing fertilizers applied to peanut (*Arachis hypogaea* L.). Peanut Sci. 33:104-111.

Jordan, D. L., J. F. Spears, and J. W. Wilcut. 2003b. Tolerance of peanut (*Arachis hypogaea* L.) to herbicides applied postemergence. Peanut Sci. 30:8-13.

Jordan, D. L. and A. C. York. 1989. Effects of ammonium fertilizers and BCH 81508 S on antagonism with sethoxydim plus bentazon mixtures. Weed Technol. 3:450-454.

Katan, J. and Y. Eshel. 1973. Interactions between herbicides and plant pathogens. Residue Rev. 45:145-177.

Kataria, H. R. and U. Gisi. 1990. Interactions of fungicide-herbicide combinations against plant pathogens and weeds. Crop Protection. 9:403-409.

Kells, J. J. and G. Wanamarta. 1987. Effect of adjuvant and spray volume on quackgrass *(Agropyron repens)* control with selective postemergence herbicides. Weed Technol. 1:129-132.

Lancaster, S. H., D. L. Jordan, and P. D. Johnson. 2008. Influence of graminicides formulation on compatibility with other pesticides. Weed Technol. 22:580-583.

Lancaster, S. H., D. L. Jordan, A. C. York, I. C. Burke, F. T. Corbin, Y. S. Sheldon, J. W. Wilcut, and D. W. Monks. 2005a. Influence of selected fungicides on efficacy of clethodim and sethoxydim. Weed Technol. 19:397-403.

Lancaster, S. H., D. L. Jordan, A. C. York, J. W. Wilcut, D. W. Monks, and R. L. Brandenburg. 2005b. Interactions of clethodim and sethoxydim with selected agrichemicals applied to peanut. Weed Technol. 19:456-461.

Lassiter, R. B. and H. D. Coble. 1987. Carrier volume effects on the antagonism of sethoxydim by bentazon. Weed Sci. 35:541-546.

Levesque, C. A. and J. E. Rahe. 1992. Herbicide interactions with fungal root pathogens, with special reference to glyphosate. Annual Review Phytopathology. 30:579-602.

Lynch, R. E. and T. P. Mack. 1995. Biological and biotechnical advances for insect management in peanut. Pages 95-159 *in* H. E. Pattee and H. T. Stalker (eds.) Advances in Peanut Science. Am. Peanut Res. and Education Soc., Stillwater, OK.

Mack, T. P., J. E. Funderburk, R. E. Lynch, E. G. Braxton, and C. B. Backman. 1989. Efficacy of chlorpyrifos in soil to lesser cornstalk borer (Lepidoptera: Pyralidae). J. Econ. Entomol. 82: 1224-1229.

Mack, T. P., J. E. Funderburk, and M. G. Miller. 1991. Efficacy of selected granular insecticides in soil in 'Florunner' peanut fields to larvae of lesser cornstalk borer (Lepidoptera: Pyralidae). J. Econ. Entomol. 84: 1899-1904.

Minton, B. W., M. E. Kurtz, and D. R. Shaw. 1989. Barnyardgrass (*Echinochloa crus-galli*) control with grass and broadleaf weed herbicide combinations. Weed Sci. 37:223-227.

Minton, N. A., A. S. Csinos, and L. W. Morgan. 1990. Relationship between tillage, nematicide, fungicide, and insecticide treatments on pests and yield of peanut double-cropped with wheat. Plant Disease 74:1025-1029.

Minton , N. A. and L. W. Morgan. 1974. Evaluation of systemic and nonsystemic pesticides for insect and nematode control in peanuts. Peanut Sci. 1:91-98.

Mitchem, W. E., A. C. York, and R. B. Batts. 1996. Peanut response to prohexadione calcium, a new plant growth regulator. Peanut Sci. 23:1-9.

Morton, C. A., R. G. Harvey, J. L. Wedberg, J. J. Kells, D. A. Landis, and W. E. Lueschen. 1994. Influence of corn rootworm insecticide on the response of field corn (*Zea mays*) to nicosulfuron. Weed Technol. 8:289-295.

Mueller, T. C., W. W. Witt, and M. Barrett. 1989. Antagonism of johnsongrass *(Sorghum halepense)* control with fenoxaprop, haloxyfop, and sethoxydim by 2,4-D. Weed Technol. 3:86-89.

Myers, P. F. and H. D. Coble. 1992. Antagonism of graminicide activity on annual grass species by imazethapyr. Weed Technol. 6:333-338.

Penner, D. 1989. The impact of adjuvants on herbicide antagonism. Weed Technol. 3:227-231.

Powell, N. L., C. W. Swann, and D. C. Martens. 1996. Foliar fertilization of Virginia-type peanut with MnEDTA- crop grade, pod yield, and value. Peanut Sci. 23:98-103.

Putnam, A. R. and D. Penner. 1974. Pesticide interactions in higher plants. Residue Rev. 50:73-110.

Putnam, A. R. and S. K. Ries. 1967. The synergistic action of herbicide combinations containing paraquat on *Agropyron repens* (L.) Beauv. Weed Res. 7:191-199.

Rahman, A. and T. K. James. 1993. Enhanced activity of nicosulfuron in combination with soil-applied insecticides in corn (*Zea mays*). Weed Technol. 7:824-829.

Rhodes, G. N., Jr. and H. D. Coble. 1984a. Influence of application variables on antagonism between sethoxydim and bentazon. Weed Sci. 32: 436-441.

Rhodes, G. N., Jr. and H. D. Coble. 1984b. Influence of bentazon on absorption and translocation of sethoxydim in goosegrass. Weed Sci. 32:595-597.

Richburg III, J. S., J. W. Wilcut, D. L. Colvin, and G. R. Wiley. 1996. Weed management in southeastern peanut (*Arachis hypogaea*) with AC 263,222. Weed Technol. 10:145–152.

Richburg III, J. S., J. W. Wilcut, and E. F. Eastin. 1995. Weed management in peanut (*Arachis hypogaea*) with imazethapyr and metolachlor. Weed Technol. 9:807-812.

Riley, D. G., G. K. Douce, and R. M. McPherson. 1997. Summary of losses from insect damage and costs of control in Georgia 1996. Georgia Agricultural Experiment Station, University of Georgia Special Publication No. 91.

Robinson, D. E. and R. Nurse. 2008. Effect of herbicide-fungicide tank-mix combinations on weed control and tomato tolerance. Acta Horticulturae (ISHS) Pages 129-134.

Sandberg, C. L., W. F. Meggitt, and D. Penner. 1978. Effect of diluent volume and calcium on glyphosate phytotoxicity. Weed Sci. 26:476-479.

Scott, G. H., S. D. Askew, and J. W. Wilcut. 2001. Economic evaluation of diclosulam and flumioxazin systems in peanut (*Arachis hypogaea*). Weed Technol. 15:360-364.

Sherwood, J. L., M. K. Beute, D. W. Dickson, V. J. Elliot, R. S. Nelson, C. H. Opperman, and B. B. Shew. 1995. Biological and biotechnical control in *Arachis* diseases. Pages 160-206 *in* H. E. Pattee and H. T. Stalker (eds.) Advances in Peanut Science. Am. Peanut Res. and Education Soc., Stillwater, OK.

Shew, B. B. 2010. Peanut disease management. Pages 103-130 *in* 2010 Peanut Information. North Carolina Cooperative Extension Service Publication AG-331.

Shew, B. B. and F. Waliyar. 2005. Peanut Diseases: Ecology and Control. Encyclopedia of Pest Management 1:1-5. (DOI: 10.1081/E-EPM-120041160).

Smith, D. H. and R. H. Littrell. 1980. Management of peanut foliar diseases with fungicides. Plant Disease. 64:356-361.

Smith, F. D., P. M. Phipps, and R. J. Stipes. 1992. Fluazinam: A new fungicide for control of Sclerotinia blight and other soilborne pathogens of peanut. Peanut Sci. 19:115-120.

Stahlman, P. W. and W. M. Phillips. 1979. Effects of water quality and spray volume on glyphosate phytotoxicity. Weed Sci. 27:38-41.

Swann, C. W. and D. A. Herbert, Jr. 1999. Influence of in-furrow thrips insecticides on response of peanut to bentazon and acifluorfen. Proc. South. Weed Sci. Soc.. 52:67.

Thelen, K. D., E. P. Jackson, and D. Penner. 1995. Characterizing the sethoxydim-bentazon interaction with proton nuclear magnetic resonance spectrometry. Weed Sci. 43:337-341.

Van Gundy, S. D. and M. V. McKenry. 1977. Action of Nematicides. Pages 263-283 *in* J. G. Horsfall, J. G. And E. B. Cowling (eds.) Plant Disease: An Advanced Treatise, Vol. 1. Academic Press, NY.

Vidrine, P. R., D. B. Reynolds, and D. C. Blouin. 1995. Grass control in soybean (*Glycine max*) with graminicides applied alone and in mixtures. Weed Technol. 9:68-72.

Wanamarta, G. and D. Penner. 1989. Identification of efficacious adjuvants for sethoxydim and bentazon. Weed Technol. 3:60-66.

Wanamarta, G., D. Penner, and J. J. Kells. 1989. The basis of bentazon antagonism on sethoxydim absorption and activity. Weed Sci. 37:400-404.

Webster, T. M. 2009. Weed survey - southern states. Proc. South. Weed Sci. Soc. 62:509-524.

Wehtje, G., J. W. Wilcut, T. V. Hicks, and J. McGuire. 1988. Relative tolerance of peanuts to alachlor and metolachlor. Peanut Sci. 15:53–56.

Wehtje, G., J. W. Wilcut, and J. A. McGuire. 1992a. Paraquat behavior as influenced by 2, 4-DB in peanut (*Arachis hypogaea*) and selected weeds. Peanut Sci. 19:51-55.

Wehtje, G., J. W. Wilcut, and J. A. McGuire. 1992b. Influence of bentazon on the phytotoxicity of paraquat to peanuts *(Arachis hypogaea)* and associated weeds. Weed Sci. 40:90-95.

Whisenant, S. G. and R.W. Bovey. 1993. Water quality effects on stability and phytotoxicity of picloram and clopyralid. J. Range Manage. 46: 254-256.

White, P. M., T. L. Porter, and A. K. Culbreath. 2009. Fungicide dissipation and impact on metolachlor-aerobic soil degradation and soil microbial dynamics. Science of Total Environment. 408:1393-1403. (doi:10.1016/j.scitotenv.20.9.11.012).

Wilcut, J. W., J. S. Richburg III, E. F. Eastin, G. R Wiley, F. R Walls, Jr., and S. Newell. 1994a. Imazethapyr and paraquat systems for weed management in peanut *(Arachis hypogaea)*. Weed Sci. 42:601-607.

Wilcut, J. W. and C. W. Swann. 1990. Timing of paraquat applications for weed control in Virginia-type peanut (*Arachis hypogaea*). Weed Sci. 38:558–562.

Wilcut, J. W., A. C. York, W. J. Grichar, and G. R. Wehtje. 1995. The biology and management of weeds in peanut *(Arachis hypogaea)*. Pages 207-244 *in* H. E. Pattee and H. T. Stalker (eds.) Advances in Peanut Science. Am. Peanut Res. and Education Soc., Stillwater, OK.

Wilcut, J. W., G. R. Wehtje, D. L. Colvin, and M. G. Patterson. 1987a. Economic assessment of herbicide systems for minimum-tillage peanut and conventional tillage peanut. Peanut Sci. 14:83–86.

Wilcut, J. W., G. R. Wehtje, and T. V. Hicks. 1990. Evaluation of herbicide systems in minimum and conventional tillage peanut (*Arachis hypogaea*). Weed Sci. 38:243–248.

Wilcut, J. W., G. R. Wehtje, and M. G. Patterson. 1987b. Economic assessment of weed control systems for peanuts (*Arachis hypogaea*). Weed Sci. 35:433–437.

Wilcut, J. W., A. C. York, and G. R. Wehtje. 1994b. The control and interaction of weeds in peanut (*Arachis hypogaea*). Rev. Weed Sci. 6:177–205.

Wills, G. D. and C. G. McWhorter. 1985. Effect of inorganic salts on the toxicity and translocation of glyphosate and MSMA in purple nutsedge (*Cyperus rotundus*). Weed Sci. 33:755–761.

York, A. C., D. L. Jordan, and J. W. Wilcut. 1990. Effects of $(NH_4)_2SO_4$ and BCH 81508 S on efficacy of sethoxydim. Weed Technol. 4:76-80.

York, A. C. and J. W. Wilcut. 1995. Interaction of bentazon and imazethapyr applied postemergence to nutsedge *(Cyperus* spp.). Peanut Sci. 22:150-154.

York, A. C., J. W. Wilcut, and W. J. Grichar. 1993. Interactions of 2,4-DB with postemergence herbicides. Peanut Sci. 20:57-61.

5

Computational Biology, Protein Engineering, and Biosensor Technology: a Close Cooperation for Herbicides Monitoring

Giuseppina Rea[1], Fabio Polticelli[2], Amina Antonacci[1], Maya Lambreva[1], Sandro Pastorelli[1], Viviana Scognamiglio[1], Veranika Zobnina[2] and Maria Teresa Giardi[1]

[1]*Institute of Crystallography, National Research Council, Monterotondo Scalo, Rome,*
[2]*Department of Biology, University Roma Tre, Rome,*
[1,2]*Italy*

1. Introduction

Application of herbicides has led to a marked increase in the productivity and preservation of agricultural products, as a result of which, cultural techniques for weed control, such as altering soil pH, salinity, fertility levels or mechanical approaches, have been abandoned. These compounds are also used extensively in industrial sites, roadsides, ditch banks, irrigation canals, fence lines, recreational areas, lawns, railroad embankments, and power line rights-of-way, to remove undesirable plants that might cause damage, present fire hazards, or impede work crews. They also reduce costs of mowing procedures. However, due to the toxic effect, their control is carried out by a system of national registration which limits the manufacture and/or sale of pesticide products to those who have been approved (Montesinos 2003).

In this context, herbicides were classified into families based on their chemical similarity or, as proposed by the global Herbicide Resistance Action Committee (HRAC) group, according to their target sites and modes of action (Table 1). Standards and regulations for the classifications, labelling, and packaging of pesticides were first set up by the EUROPEAN ECONOMIC COMMUNITY (EEC) Council Directive 67/548/ in 1967.

At present the issue regarding herbicides is quite intricate, because according to the Food and Agriculture Organization (FAO), their exclusion would lead to a strong reduction in farming production; however, several toxic effects on biological systems associated with their use were proved by epidemiological and experimental studies (Waller et al., 2010; Roberts et al., 2010; Frazier 2007). After the first cases of animals poisoned by heavy utilization of herbicides, the monitoring of these compounds to avoid accumulation in the human body were strongly intensified. In particular in 1963, the World Health Organization (WHO) and FAO created the Codex Alimentarius Commission, which joined 173 signatories from the European Community (EC) countries in order to control the tolerable limits of pollutants in food. Twenty years later, the EC established a legal framework for the regulation of pesticides in all member countries. The Commission is responsible for the registration of pesticides actively used in all European countries. This authority is granted

Mode of action	Herbicide classes					
Amino acid synthesis inhibitors	Amino acid derivatives	Imidazolinones	Sulfonylureas	Sulfonamides	Thiopyrimidines	Glycines
Cell membrane disrupters	Diphenylethers	Bipyridiliums				
Growth regulators	Benzoics	Phenoxys	Pyridines			
Respiration inhibitors	Organic Arsenicals					
Photosynthesis inhibitors	Triazines	Ureas	Nitriles	Miscellaneous	Phenolic compounds	Diazines
Lipid biosynthesis inhibitors	Cyclo hexanediones	Arylphenoxy propanoates				
Root growth inhibitors	Dinitroanilines					
Shoot growth inhibitors	Substituted amides	Carbamothioates				
Pigment synthesis inhibitors	Isoxazoles	Isoxazolidinone	Pyridazinones			

Table 1. Herbicide classification by mode of action.

by the Council of the European Community under Council Directive 91/414/EEC, adopted in 1991 and effective as of 1993.

Despite European policies to reduce the use of herbicides, EU statistics data for the period 1992-2003 showed that the annual consumption had not decreased (Eurostat Statistical Book, 2007). Hence, due to the occurrence of several toxic effects induced by herbicides, severe restrictions were adopted to safeguard particularly children, whose immature liver enzymes system is unable to detoxify these compounds. Concrete examples are the EU Baby Food Directives 2003/13/EC and 2003/14/EC, that fixed the maximum acceptable daily intake of pesticide residues in foods for infants and young children to a level lower than 10 µg/kg.

Later in 2004, the production and use of persistent organic pollutants was forbidden by the United Nations Environmental Protection Programme (UNEP). Subsequently, the EC Regulation No. 396/2005 of the European Parliament and of the Council on Maximum Residue Levels (MRLs) of pesticides in products of plant and animal origin defined a new fully harmonized set of rules for pesticide residues, which became effective in 2008. Recently, the new maximum residue levels of pesticides in food and feed of plant and animal origin was defined in the Regulation 2008/149/CE. In particular, for those herbicides most commonly found in surface and ground waters, allowed concentrations are 0.1 µg/L and 0.5 µg/L, for a single pollutant and total pollutants, respectively.

Nowadays the Directive 2009/128/EC of the European Parliament and of the Council is adopted to achieve the sustainable use of pesticides. Member States should monitor the use of plant protection products containing active substances of particular concern and establish timetables and targets for their use, in particular when it is an appropriate means to achieve risk reduction targets.

The United States (US) organization and legislation concerning herbicides sale and distribution are quite different from the above mentioned EC directives. The Environmental Protection Agency (EPA) is the agency primarily responsible for safety review and legal registration, regulating pesticides in the US.

In 1996, US Congress unanimously passed a landmark pesticide food safety law, called the Food Quality Protection Act (FQPA), which takes the protection of children into special consideration, and asked the EPA to conduct an Endocrine Disruptor Screening Program (EDSP) to monitor the effect of pesticides on the endocrine systems of living organisms. Globally, different policies were undertaken regarding pesticides use. For example in 2003, while the use of atrazine was banned by the EC, EPA studies affirmed that all triazine herbicides were without any harmful effect on the US population, infants or children (Sass & Colangelo 2006). In 2006, the EPA also initiated a new program called "registration review" to re-evaluate all pesticides. The program's aim was to review the active ingredient of each pesticide every 15 years to ensure that all pesticide products in the marketplace could still be used safely. This process, called re-registration, considers the human health and ecological effects of pesticides and results in actions to reduce risks. The Agency completed more than 99% of tolerance reassessments by the end of 2006. The EPA issued the first test orders for pesticides concerning their potential effects on the endocrine system on October 29, 2009.

2. Relevance of herbicide detection for environmental and health claims

Herbicides are potent contaminants of ground and surface water as they are transported far away from the point of application via runoff and, as a result, contaminate otherwise pristine habitats, even in remote areas where they are not used (Readman et al., 1993; Relyea 2005; Haynes et al., 2000). In addition to their persistence, mobility, and widespread contamination of water, some herbicides brought about considerable ecological damage such as the disruption of predator-prey relationships, posing a threat to the survival of major ecosystems and a loss of biodiversity (Schneider 2009).

Another important aspect is herbicide resistance, an inherited ability of a plant to survive and reproduce following exposure to a dose of herbicide that would normally be lethal to the wild type. Herbicide resistance may occur naturally in plants as a result of random and rare mutations, or may be induced by genetic engineering (Shewry et al., 2008). Herbicide exploitation kills susceptible plants, allowing the herbicide-resistant plants to survive and reproduce without competition. The continuous use of herbicides allows the reproduction of resistant plants which then become dominant in the environment (Al-Ahmad et al., 2005; Murphy & Lemerle, 2006). In addition, increasing problems with herbicide- resistant weed populations have increasingly occurred in countries with intensive agriculture cropping systems (Green & Owen, 2010; Vila-Aiub et al., 2008). Changes in farming methods such as crop rotation, manipulation of planting time, hand weeding and application of herbicides with different target sites, are effective practices preventing resistance development.

Regrettably, herbicides also negatively affect human health. According to the World Health Organization between 1 and 25 million people suffer of herbicide poisoning each year. It is estimated that as many as 20,000 people in the US will develop cancer every year due to herbicides residues in their food, but the number is much higher when we include the enormous number of people suffering from herbicide poisoning symptoms unknowingly.

In this context, from a public health viewpoint, the prevention of diseases represents a priority and several investigations have been performed to reveal the presence of herbicides in water and food products (Lorenzin 2007; Sondhia 2010; Fussell et al., 2002; Cesnik et al., 2006).

In the literature many significant effects associated with exposure to herbicides have been documented; in particular, herbicides can cause short-term adverse health effects, called acute effects, as well as chronic adverse effects that can occur months or years after exposure. In addition, these effects are not necessarily exclusively caused by exposure to herbicides or other organic contaminants, but may be associated with a combination of environmental compounds which can have a synergistic effect with organic pollutants (Witte et al., 1995).

Several studies conducted on farmer populations, or on people particularly exposed to herbicides found high rates of eyes stinging, rashes, blisters, blindness, nausea, dizziness, asthma, diarrhoea and even death, as examples of acute health effects (Senthilselvan et al., 1992). On the other hand, examples of chronic effects include cancer, birth defects, reproductive damage, neurological and developmental toxicity, immunotoxicity, and disruption of the endocrine system (Kristensen et al., 1997; Fukuyama et al., 2009; Schreinemachers 2010; Turner et al., 2010; Ochoa-Acuña et al., 2009; Tanner et al., 2009).

Regarding neurological effects, herbicides can be potent neurotoxins. When people are exposed to neurotoxins they may feel dizzy, lightheaded, confused and may have reduced coordination and ability to think, as short-term effects. Long term exposure can result in reduced intelligent quotient, learning disability and permanent brain damage in people, especially children, who live in areas with high levels of herbicide contamination in water and food. Recent studies in several countries with a high use of herbicides indicate that there is an increasing incidence of carcinogenic effects, especially for children (Zahm & Ward, 1998). National and international trends indicate that cancer rates have increased, including lymphocytic leukemia, childhood brain cancer, neuroblastoma, non-Hodgkin's lymphoma, testicular cancer, ovarian cancer and all cancers combined (Ries et al., 1998; Gurney et al., 1996; DeVesa et al., 1995).

Several herbicides are hormone disrupting chemicals, interfering with hormone biosynthesis, metabolism and resulting in a deviation from normal homeostatic control or reproduction. These compounds can cause physical birth defects, hormonal effects on the developing foetus or affect a child's functional capacities (Weselak et al., 2008). In addition, hormone disruptors are linked to many health problems including reproductive cancers (Fan et al., 2007). Twenty-four pesticides still on the market, including the herbicide atrazine, are known to be endocrine-disrupters, which are able to increase rates of endometriosis, hypospadias, undescended testicles and consequently testicular cancer, precocious puberty in girls, reduced sperm counts and fertility problems.

Given the latest statistics on the pathological effects caused by herbicides and the analytical problems of inadequate detection levels, as well as the insufficient quality control in many laboratories, the monitoring data are frequently a poor indication of the level of pollution in the environment (Guzzella et al., 2006). Key herbicides are included in the monitoring schedule of most countries, however the cost of analysis and the necessity to sample at critical times of the year (linked to periods of pesticide use) often preclude development of an extensive data set. In addition, several limits exist concerning the analyses related to the inadequate facilities, impure reagents, and financial constraints (Brena et al., 2005).

3. Description of analytical methods and biosensors for herbicide detection

This "unhealthy" scenario requires the challenging development of sensitive analytical control systems to reveal the presence of herbicides and protect humans and ecosystems. Properly assembled biosensors can satisfy these requirements, also providing reliability and flexibility of the assays (Giardi & Pace, 2005).

A major difficulty in estimating environmental quality related to herbicides contamination is due to seasonal change of field application and the extremely low levels of the maximum admissible concentrations set by the EC. Nowadays, the data on herbicides pollution are still quite scarce since monitoring data are based on a few investigations carried out with methods able to detect relatively high concentrations of herbicides. Currently, in parallel to traditional analytical methods, novel detection systems have been already developed based on biosensor technology which provides rapid, inexpensive and reliable tools for herbicide monitoring and screening analyses, to answer the concern on this issue.

In this context, chromatographic techniques, such as gas chromatography (GC) and high-performance liquid chromatography (HPLC) with UV and/or mass spectrometry (MS) detection, surely represent the most trustworthy and common techniques used to monitor the presence of herbicides. The classical analytical techniques are unlikely to provide adequate sensitivity, while advanced instrumental methods are highly sensitive, but generally expensive, require skilled operators and are not easily amenable to on-site field testing. In addition, it must be emphasized that herbicides can greatly differ in chemical structure and chromatographic behaviour, so it is still impossible to apply a unique method to discriminate all of the different compounds that could be present in a real-world sample. Herbicides usually represent a very small fraction of the whole sample under investigation, so pre-treatments such as clean up and/or pre-concentration steps are required to make their identification possible. As a consequence, the qualitative and quantitative analysis of herbicide residues is time consuming and involves high costs.

Because of the large numbers of samples to be measured, the necessity of expensive equipment, organic solvents and laborious sample preparation, the development of a fast, automated and inexpensive test is of great interest. These concerns encouraged researchers to seek out alternative methods providing the desired analytical information.

The utility and high specificity of immunochemical technology for the detection of organic molecules has been well established in various research applications. In a study by Gascon and co-workers (1997), an ELISA analysis was developed for the detection of atrazine with a very low limit of detection, taking into account the European regulations for water and food. Chouhan and co-workers (2010) also reported a sensitive chemiluminescence (CL)-based immunoassay technique based on dipstick and flow injection analytical formats for the detection of atrazine.

In recent years, the development of new advanced methodologies for rapid, inexpensive and in field environmental monitoring based on biosensing devices represents a promising research challenge. Biosensors, following IUPAC definition (Thevenot et al., 1999), are devices able to provide quantitative analytical information exploiting a biological recognition element linked to a transduction system. They consist of three parts, as depicted in Figure 1. The first element is the biomediator (a biomimic or biologically derived material e.g. tissue, microorganisms, organelles, cell receptors, enzymes, antibodies, nucleic acids, and biological sensitive elements created with genetic engineering), the second element is the transducer or the detector (physicochemical, optical, piezoelectric, electrochemical, etc.) that transforms the signal resulting from the analyte's interaction with the biological element into another signal that can be measured and quantified; the third element is the associated electronics or signal processor, responsible for the output of the results in a user-friendly way (Cavalcanti et al., 2007). Biosensors require immobilization of biological elements on the surface of the sensor (metal, polymer, glass, etc.) using physical or chemical techniques. According to a recent report on the biosensor market, titled "Biosensors in Medical Diagnostics: A Global Strategic Business Report" published by Global Industry Analysts Inc., the global market for biosensors and other bioelectronics has grown from $6.1 billion in 2004 to $8.2 billion in 2009, at an AAGR (average annual growth rate) of about 6.3%, and it is projected to increase still further in 2011 (RB-159R Biosensors and Bioelectronics report).

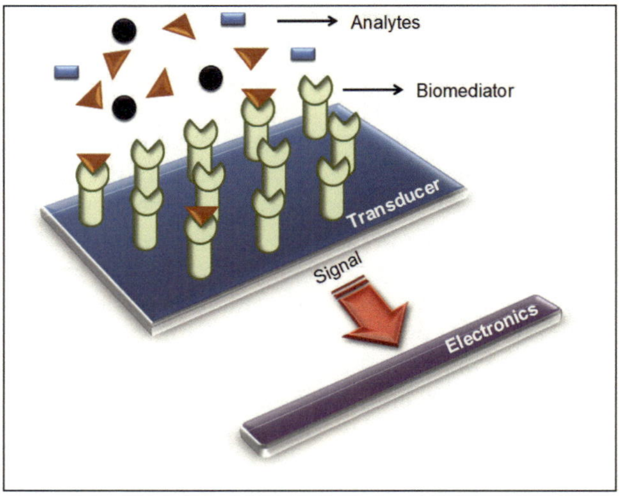

Fig. 1. Schematic representation of typical biosensor's components and activity mode.

A significant number of biosensors were designed, implemented and tested by several research groups with the aim to develop suitable methodologies with the best features in terms of versatility, stability, life-time and long-term activities, sensitivity and selectivity, detection limits, linear range, reproducibility and low cost.

In 1993 McArdle and co-workers described an amperometric biosensor incorporating the enzyme tyrosinase for the detection of the inhibiting herbicide atrazine. A similar electrochemical biosensor was developed by Mazzei and co-workers (1995), based on a plant tissue bioelectrode. In 2002 Shao and co-workers developed a cyanobacterial-based biosensor able to detect herbicides and other environmental pollutants. A magnetic nanoparticle-based biosensor incorporating alkaline phosphatase enzyme was proposed by Loh and co-workers in 2008 and applied to the determination of the herbicide 2,4-dichlorophenoxyacetic acid (2,4-D).

Studies in the framework of the development of biosensors for the detection of environmental pollutants exploit photosynthetic microorganisms or parts of them, such as thylakoids. Photosynthetic systems are naturally-occurring anisotropic supramolecular arrangements of proteins and small molecules that are able to harvest light energy and funnel it towards building up biomass and releasing oxygen. For these purposes, photosynthetic organisms are equipped with multi-enzymatic complexes embedded in thylakoidal or free membranes known as photosystems. The hierarchical organization of these pigment-protein complexes is at the basis of their unique efficiency. Functional and structural knowledge of photosynthetic systems has been steadily increasing, and as a result, fundamental and applied research have made it possible to integrate biological photosystems or their functional sub-structures into artificial assemblies in order to get them to carry out their tasks in a controlled environment for specific applications. Biosensors and nanobiosensors for the detection of herbicides fall into this category.

In detail, Photosystem II (PSII) is the multienzymatic chlorophyll-protein complex located in the thylakoid membrane of algae, cyanobacteria and higher plants (Figure 2). PSII is an integral part of the electron transport chain that catalyses photosynthetic primary charge separation. This protein complex consists of over 25 polypeptides, which make up the light-harvesting chlorophyll protein complex, the reaction centre and the water-splitting system, also called the oxygen evolving complex. The scaffold of the PSII reaction centre (RC) is formed by two protein subunits, D1 and D2, each composed of five transmembrane α-helices (named from A to E) with their N- and C-termini exposed to the stromal and

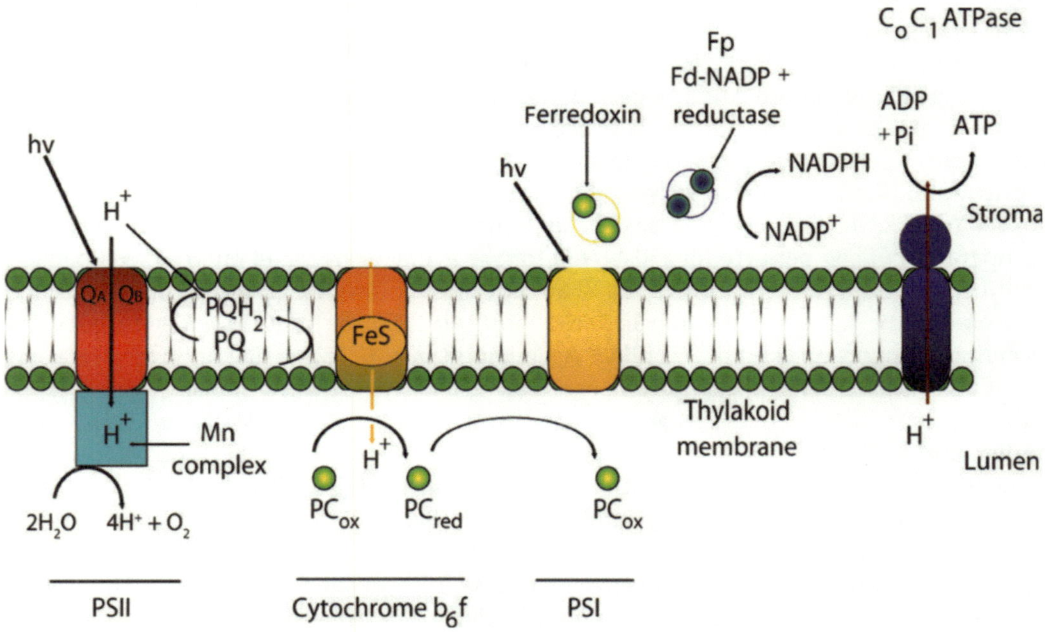

Fig. 2. Model of the photosynthetic membrane of plants showing the electron transport components (cross-sectional view). The complete membrane forms a vesicle. The pathways of electrons are shown by solid arrows. Photosystem II (PSII), photosystem I (PSI), and cytochrome b_6f complexes, all involved in the electron transfer chain, are also shown (Buonasera et al., 2010).

luminal sides, respectively. All the photosynthetic redox active components are located within the D1-D2 heterodimer, including the plastoquinones Q_A and Q_B (PQs) (Ferreira et al., 2004; Loll et al., 2005 a).

4. Atrazine: a case study, exploitation and mode of action

Most herbicides introduced in agriculture during the last 40 years impede chloroplast functions. The inhibitors of the photosynthetic reactions (urea/triazines and phenolic herbicides) comprise more than half of all commercially available herbicides of the 70's. Among them, atrazine has been a cornerstone for high-yield crop production, mostly in corn, sorghum and sugar cane farms for the last 50 years (http://www.atrazine.com). Atrazine, 2-chloro-4-(ethylamine)-6-(isopropylamine)-s-triazine, is a symmetric triazine. It is a soil-applied herbicide, which is actively absorbed by the roots, transported via the apoplastic system and accumulated mainly in leaf margins and tips (Hilton et al., 1976; Bouldin et al., 2006). Inside the cells, it readily penetrates and accumulates into the chloroplasts probably until an equilibrium concentration is attained between the chloroplast and the cytoplasm (Shimabukuro & Swanson, 1969). Atrazine is selective and highly effective in the pre- and post-emergence control of a broad range of yield-robbing weeds; it is safe to the crop and fits in a variety of farming systems; it is the most widely used herbicide (except in EU since 2004) in conservation tillage systems, preventing soil erosion. Studies on weed control in corn plantations published in the North Central Weed Science Society Research Report showed that despite the availability of many new herbicidally active ingredients, corn yields with atrazine continue to be higher than with non-atrazine treatments (Fawcett 2008). This analysis assembled the data accumulated over a period of 20 years, from 1986 to 2005, including 236 relevant studies with a total of 5871 qualifying treatments. For the entire 20-years period, the average yield with atrazine was 5.1% higher than without atrazine. In addition to all the advantages mentioned above, its low price also contributes to the widespread exploitation. As an example, according to the National Agricultural Statistics Service survey, the average US atrazine rate used in 2005 was 1.13 lb/acr (or 1.27 kg/ha) and the cost of atrazine per acre was only $2.46 (or $6.08/ha) compared to $12.34 per acre (or $30.49/ha) for the average cost of 14 alternative broadleaf control herbicides in corn cultivation (Fawcett 2008). Considering the income from the increased corn yields and the low herbicide cost, for 2005 the study estimated a total US benefit for farmers of about $1.39 billion. A similar estimation of atrazine utilization rate and its significance in cornfield production, based on analyses of US weed management systems, can be found in numerous research papers (Swanton et al., 2007; Williams et al., 2009; Williams et al., 2010).

Atrazine was registered for use in the United States in 1959 (US EPA, 1994). Because of its high water solubility and intensive utilization, it is the most common herbicide found in rivers, streams and groundwater at concentrations that very often exceed the MRL imposed by several European directives, even when used appropriately (Cox 2001). This set in motion the process for a regulatory ban in 2004 which became effective one year later in all EU countries (Ackerman 2007). Nonetheless, atrazine is still present in over 20% of the 3000 sites analysed in Italy (Paris et al., 2009).

However, in 2006 the EPA completed its re-registration eligibility process and still allowed atrazine use. Lately, the EPA has started broad re-evaluation of human health and ecological risk assessments associated with atrazine (US EPA, 2009). Nonetheless, atrazine remains one of the most widely used herbicides in North America and is used in more than 60 countries around the world - in Africa, North and South America, Asia and the Middle East.

The initial investigations on the possible mode of action of the photosynthetic herbicides (urea/triazines- and phenol-type), and in particular atrazine, performed intensively during the 50-60's of the last century, showed that they inhibit the Hill reaction in isolated chloroplasts (Moreland et al., 1959; Moreland & Hill, 1962). It appeared that the triazines and urea-type herbicides had a similar mode of action, but it was still uncertain at that time. Based on the herbicide-induced stimulation of the chlorophyll *a* fluorescence signal, Duysens and Sweers (1963) postulated that the herbicide inhibits the re-oxidation of primary quinone acceptor of PSII - Q_A^-. Subsequently, it was shown that the photosynthetic herbicides block the PSII electron transport immediately after the Q_A and the interruption of the electron transport occurs as a consequence of the herbicide binding to, at that time, an unknown protein in the chloroplast thylakoid membranes (Forbush & Kok, 1968; Pfister & Arntzen, 1979). The discovery of the first triazine-resistant plant (*Senecio vulgaris*) brought to light evidence that the resistance is due to alteration in the primary target, but not in the uptake, translocation or degradation metabolism of the herbicide (Pfister et al., 1979). The related protein was identified by means of photoactive herbicide derivatives, which under UV irradiation covalently bind the target protein. In this way, it was shown that azidotriazin binds to a 32 kDa protein – the D1 protein of PSII (Pfister et al., 1981). In the beginning of the '80s it was believed that the photosynthetic herbicides inhibit the PSII electron transport allosterically (reviewed by Van Rensen 1982; Renger 1986). Tischer and Strotmann (1977) and later on also other groups (Oettmeier & Masson, 1980; Haworth & Steinback, 1987; Oettmeier et al., 1987; Giardi et al., 1988) demonstrated competitive binding between different herbicides. Vermaas et al. (1983) found that the binding affinity of atrazine sharply decreases when the Q_B site is occupied by azidoquinone. The herbicide affinity to D1 depends on the redox state of Q_B; it is high when the Q_B is in the oxidized state and weakly bound to the Q_B-pocket, and low when Q_B is in the semi-reduced state and tightly linked to D1 (Lavergne 1982). These findings suggested a competitive binding between the herbicide and the PSII electron acceptor Q_B.

Nowadays it is widely accepted that herbicides such as diuron and atrazine block the electron transport between the primary (Q_A) and the secondary (Q_B) quinones of PSII by competitive substitution of plastoquinone in the Q_B-site of the D1 protein. An initial model of the herbicide-binding region and the orientation of the molecule in the binding site was developed by Trebst (1986, 1987). Later, a number of additional models were developed that described in more details the topology of the Q_B region and the possible herbicide/D1 interactions (Xiong et al., 1996; Lancaster & Michel, 1999; Kern & Renger, 2007). More recently an additional quinone binding site, named the Q_C site, has been identified, shedding light on the possible mechanism of quinol-quinone exchange which would involve a quinone entry tunnel and an additional/alternative quinol exit tunnel (Guskov et al., 2009). The above mentioned models are based on molecular modelling and X-ray diffraction studies of the structure of the RC of the bacterium *Rhodopseudomonas viridis* and of the PSII of cyanobacterium *Thermosynechococcus elongatus*, which are homologs of eukaryotic PSII. The Q_B site is formed between the D and E trans-membrane helices of D1 protein, from Gly207 down to Phe274 aminoacid residues, and the plastoquinone forms directly hydrogen bonds with Ser264, His215 and Phe265 (Kern & Renger, 2007). Lancaster and Michel (1999) performed a detailed study on the atrazine interaction with the bacterial RC in

Rhodopseudomonas viridis. According to their model, the atrazine molecule is bound to the RC protein directly by three hydrogen bonds (Ile, Ser and Tyr residues) and indirectly via water molecules by four other hydrogen bonds. They emphasized the role of the orientation of the triazine molecule inside the RC binding site for the magnitude of the herbicide toxicity. The atrazine and the plastoquinone bind to overlapping, though not completely identical, regions localized in the common domain, but both target the serine aminoacid residue (Lancaster & Michel, 1999). Additional information about the aminoacids involved in the herbicide/Q_B pocket interactions was obtained from analyses of herbicide resistant or sensitive mutants. All 77 identified mutations in the PSII reaction centre D1 protein in cyanobacteria, algae and higher plants conferring herbicide resistance or sensitivity are localized in the region encompassing Phe211-Leu275 (Oettmeier 1999). Among them, 27 mutations consisted of Ser264 replacement, a finding in agreement with the well accepted concept that this aminoacid is the principal binding target of triazines and urea type herbicides. Other frequently cited aminoacid substitutions resulting in changes in the atrazine/D1 interaction are localized on Ala at positions 250 and 251, 8 and 9 mutations, respectively (Oettmeier 1999; Johanningmeier et al., 2000). The increasing number of data coming from atrazine resistant/sensitive mutants and the advances in the computation of structure design and modelling could significantly expand our knowledge on herbicide and PSII reaction centre interactions (Rea et al, 2009).

5. Relevance of bioinformatic tools. Molecular docking, binding energy calculation and molecular dynamics

In silico studies of macromolecular systems are becoming increasingly useful and reliable with the improvement of our knowledge of their physico-chemical properties and with the availability of more powerful hardware resources.

Nowadays, structural bioinformatics tools coupled to modern molecular biological techniques allow the tailoring of macromolecules as high affinity receptors for organic compounds of biomedical/environmental/industrial relevance to be used as biosensing devices for these compounds. In this framework, in the absence of high-resolution crystal structures, molecular docking techniques allow prediction of the binding site and mode of action of a given molecule to a macromolecule, a first step towards the rational redesign of the macromolecule to build an efficient biosensor. Several docking packages can also calculate the ligand-macromolecule interaction energy allowing the *in silico* evaluation of a macromolecular system affinity for a given ligand, and the effect of point mutations on this parameter. Energy minimization and molecular dynamics simulations constitute complementary and, to a certain degree, alternative methods to evaluate the affinity of a macromolecule for a ligand. In fact, energy minimization techniques represent a fast method to refine the putative complexes obtained by molecular docking and to predict the ligand–macromolecule interaction energy. An example of such an application is recent work carried out in our lab in which a large number of mutants of the PSII proteins D1 and D2 were generated *in silico* and the atrazine binding affinity of the mutant proteins was calculated by a combination of molecular docking and energy minimization techniques, to predict mutations able to increase PSII affinity for atrazine (Rea et al., 2009). The validation of the computational approach through comparison with available experimental data confirmed the efficacy of this approach. In addition, the production and characterization of one of the predicted mutants confirmed an increase in atrazine affinity for the D1-D2 heterodimer of one order of magnitude, with evident benefits for PSII-based atrazine biosensing applications (Rea et al., 2009).

A different approach is taken by molecular dynamics (MD) simulations. MD simulations are a powerful tool to study the evolution of a protein conformation in response to various stimuli, such as substrate/ligand binding, site directed mutagenesis, etc.. In the case of protein inhibitors, as is the case with herbicides, MD simulations provide essential information regarding the relevant interactions established by the inhibitor with the protein moiety and can guide the design of novel, more powerful inhibitors. However, when a given protein system is used in biosensing applications the reverse strategy can also benefit from molecular dynamics studies. In this case, rather than using simulations to design better inhibitors, MD simulations can provide an atomic level view of the protein-inhibitor interactions and guide the design of site directed mutants aimed at improving the affinity of the protein system for the inhibitor in order to improve its "fitness" for biosensing applications.

In this framework, MD simulations of PSII in the presence of herbicide inhibitors help to define the details of the molecular interactions stabilizing the herbicide-PSII complex and to pinpoint possible binding pocket modifications that could lead to an increased binding affinity.

6. Molecular dynamics simulations of PSII-atrazine complex. Relevance for biosensor applications of PSII

As already detailed above, atrazine is known to bind in the eukaryotic D1 protein region encompassing residues Phe211-Leu275, that partially overlap the Q_B binding pocket (Giardi et al., 1988; Oettmeier 1999). Analysis of mutations conferring herbicide resistance or sensitivity indicated that Ala 250, Ala251 and Ser264 are located close to the atrazine binding pocket and probably directly interact with it (see above, Oettmeier 1999; Johanningmeier et al., 2000). Using this information, in the previous study cited above we obtained a molecular view of the atrazine-D1 interactions using a combination of molecular docking and energy minimization techniques (Rea et al., 2009). As shown in Figure 3, the model we obtained is in good agreement with the available mutational data in which atrazine is predicted to bind in a pocket made up by residues 211-218 on one side, residues 251-255 on top and residues 264-275 on the opposite side, the same pocket hosting the Q_B ring in the available crystal structure of *Thermosynechococcus elongatus* PSII (Loll et al., 2005 b). In the attempt to verify the reliability of this model and to refine the structure of the atrazine-PSII complex, we recently undertook MD simulations of the whole PSII macromolecular complex embedded in a lipid bilayer in the presence of atrazine. The *Thermosynechococcus elongatus* PSII crystal structure contains two macromolecular complexes in the asymmetric unit. However, it has been recently demonstrated that the *in vivo* functional PSII unit is composed of a single macromolecular complex (Takahashi et al., 2009). Thus MD simulations were carried out on the "monomeric" PSII macromolecular complex. Methodological details can be found in Table 2.

The atrazine molecule was placed inside the Q_B binding pocket in a such way that the following residues were found within 4.0 Å distance from the molecule: His215, Leu218, His252, Phe255, Gly256, Ser264, Phe265, Leu271 and Phe274 (Rea et al., 2009). A hydrogen bond between the nitrogen atom of the ethylamino moiety (N5) of atrazine and the backbone amide group of Phe265 was possible for such a conformation.

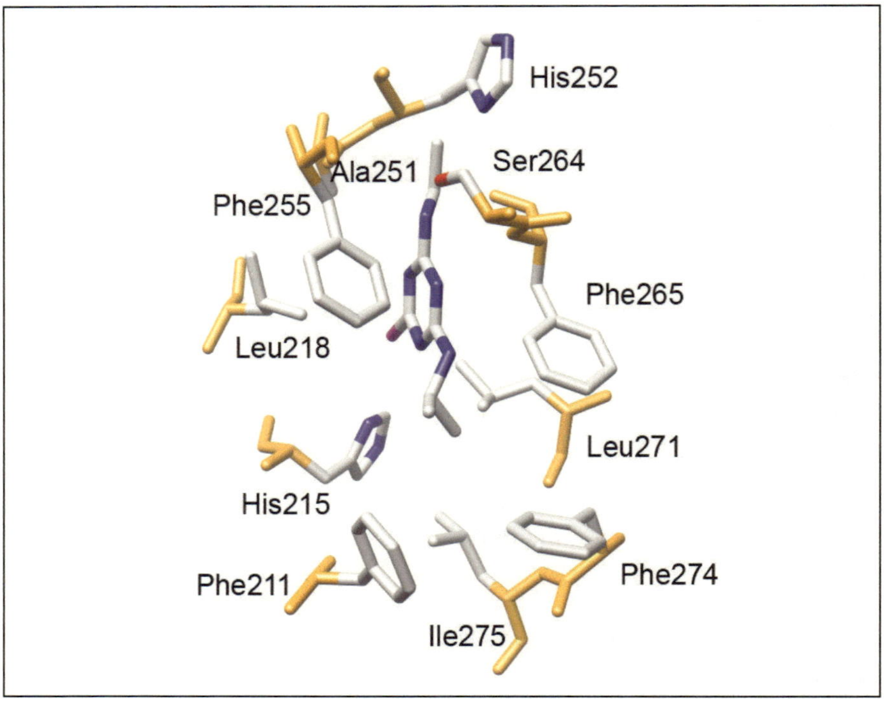

Fig. 3. Schematic view of the atrazine binding pocket within the PSII D1 protein obtained by docking simulations and energy minimization techniques (for details see Rea et al., 2009).

Starting structure	*T. elongatus* PSII (3.0 Å resolution, PDB entry 2AXT)
MD simulations package	GROMACS v. 4.0.7 (Van der Spoel et al., 2005)
Protein subunits force field	AMBER 99SB (Wang et al., 2000)
Cofactors force field	General AMBER force field (Wang et al., 2004)
Cofactors charges calculation method	GAUSSIAN 03 package (Frisch et al., 2004)
Charge fitting and parametrization	Antechamber (Wang et al., 2006)
Membrane bilayer	Pre-equilibrated DOPC bilayer model (Siu et al., 2008)
MD simulations total time	10 ns
Time step	2 fs
Temperature	300 K

Table 2. Methodological details of MD simulations of the PSII-atrazine complex.

Already during the energy minimization run, additional hydrogen bonds between the aromatic ring nitrogen (N2) of atrazine and the amide group of Phe265, and between the ethylamino group (N5) of atrazine and the hydroxyl group oxygen of Ser264 were formed as well. Overall, only small changes of the atrazine position with respect to the position of the amino acids forming the active site of the starting conformation took place at this stage.

During the MD simulation run, atrazine changed substantially its position within the Q_B binding pocket (Figure 4).

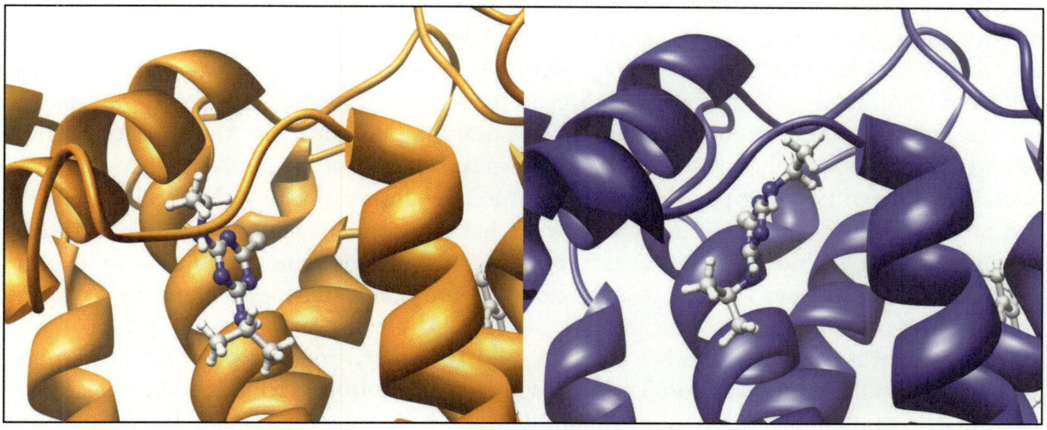

Fig. 4. Schematic representation of the atrazine-PSII complex initial structure (orange) and final structure (blue) after 10 ns MD simulations.

In particular after 10 ns MD simulations atrazine changes its orientation and binds in a deeper position inside the Q_B pocket. In detail, only 5 amino acids (His215, Leu218, Phe255, Phe265 and Phe274) out of the 9 within 4 Å distance from atrazine in the initial conformation, remained in the vicinity of atrazine at the end of the MD run. However, additional residues were found closer to the atrazine molecule (Phe211, Met214, Tyr246, Ile248, Ala251, Asn266, Asn267, Ser268, Leu271) (Figure 5).

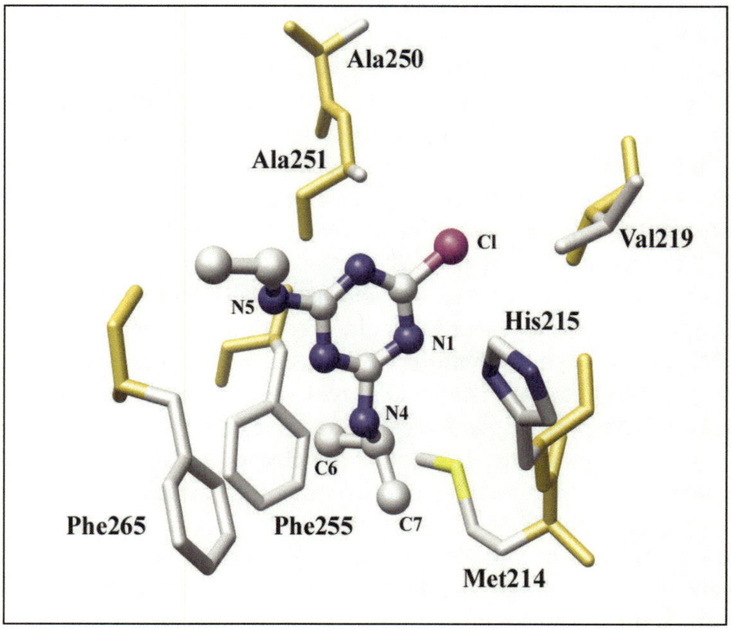

Fig. 5. Detailed representation of the chemical environment of the atrazine binding pocket after the MD simulations run.

Analysis of the atrazine binding mode at the end of the simulation trajectory surprisingly reveals only few specific interactions with residues of the D1 protein which stabilize the bound molecule. In particular, hydrophobic interactions are observed between the C6 methyl group of atrazine and Phe255 phenyl ring, between the C7 methyl group of atrazine and the Met214 sidechain, while a single strong hydrogen bond is established between the N1 atom of atrazine and the His215 Nδ proton. On the other end, apparently no binding partner stabilizes other groups on the atrazine molecule. In particular the chloride atom is bound in an energetically unfavorable position in the vicinity of the aliphatic residue Val219 and no hydrogen bonding partner is observed for the atrazine N4 and N5 protons. In two out of three cases, rational design of site-directed mutants are likely to increase atrazine affinity for the Q_B binding niche. In fact, the atrazine chloride atom could be stabilized by mutation of Val219 into a polar residue that can provide a hydrogen bond donor to the chloride atom.
In the same way, the atrazine N5 proton could find a hydrogen bond partner through replacement of the aromatic Phe265 residue with a polar hydrogen bond acceptor.
Indeed, the results discussed above indicate that MD simulations can be an effective strategy for the design of an improved herbicide binding pocket in PSII and experiments are being carried out to evaluate the reliability of this approach.

7. Relevance of genetic engineering to improve sensitivity and selectivity of biomediators for biosensor development

The combination of computational analyses and molecular biology tools makes possible the realization of more stable, sensitive, selective and specific biomediators for the creation of effective biosensors. The improvement of these parameters is of outstanding relevance for biosensor reliability, and strongly attracts the interest of commercial companies accelerating the acceptance of this technology.
Nowadays, genetic engineering allows the modification of specific nucleotide sequences of an organism genome to obtain proteins with novel improved properties, and innovative biotechnological approaches make it possible to integrate these systems, or their functional sub-structures, into artificial assemblies for specific applications such as environmental monitoring. Several biomediators have been already developed exploiting molecular biological techniques to produce enzymes and/or protein with improved features in the detection of specific analytes (Wang et al., 2009).
In the context of the photosynthesis-based biosensors, activities in different research areas allowed the design and development of engineered photosynthetic microorganisms with improved sensitivity and stability features to be used as bio-recognition elements for the detection of environmental contaminants. Different approaches, such as space research and physical elicitations, have been applied to select microorganisms with improved tolerance to extreme environmental conditions. The newly selected organisms generated for biosensor purposes were able to maintain a stable photosynthetic efficiency and an increased oxygen evolution capacity (Rea et al., 2008).
In particular, we carried out modifications of the D1 reaction centre proteins, as they play a crucial role in electron tunnelling-mediated charge separation and transmembranal electric field generation, acting principally on reduction, release and migrations of (plasto) quinones. Random mutagenesis targeted to the D1-encoding *psb*A gene was exploited as a directed evolution strategy to produce a huge mutant library of chlamydomonas carrying

novel D1 proteins with different aminoacidic composition. In addition, thanks to the support of bioinformatics studies, site-directed mutagenesis was also exploited to generate specific point mutations in the D1 protein, in order to modify the properties of the PQ/atrazine binding affinity.

Chlamydomonas D1 random and site-directed mutants were produced by particle gun bombardments of the chloroplast genome (Przibilla et al., 1991). The Del1 chlamydomonas strain was used as a recipient host for the mutant's generation (Preiss et al., 2001). This strain has a defined deletion in the chloroplast-encoded *psb*A gene and is unable to grow photoautotrophically, as it cannot produce a functional D1 protein. Acetate is needed as carbon source as minimal media do not support its growth. Minimal media were used to select photosynthetically active colonies generated after the integration of the *psb*A variant produced both by random and site-directed PCR (Dauvillee et al., 2004). Selected mutants were then characterised by analysing their photosynthetic performance and the sensitivity and/or resistance to different classes of herbicides assessed (Tibuzzi et al., 2007; Rea et al., 2009; Giardi et al., 2009; Scognamiglio et al., 2009). After the characterization, the best performing mutants were immobilized on screen-printed electrodes and integrated in amperometric or potentiometric circuits. Both electrochemical and optical devices were arranged in multi-arrayed setups.

8. Biosensors already developed

Although a variety of whole-cell-based bacterial sensors have been applied in environmental assays for pollutant monitoring, generally they display a poor response to herbicides (Table 3).

If the electron transfer from the reaction centre to the quinone pool is blocked, such as during the binding of the photosynthetically active pesticides, these parameters change dramatically and can be monitored by electro-optical analysis in a pesticide concentration dependent manner (Figure 6). In this context, an optical biosensor based on the green photosynthetic alga *Chlamydomonas reinhardtii* described by Tibuzzi and coworkers (2007) was employed to monitor several classes of herbicides, such as atrazine, diuron, ioxynil, terbuthylazine, prometryn and linuron, in a low concentration range (10^{-8}-10^{-10} M) (Table 3). In particular, a miniaturized optical biosensor instrument was designed and produced for multiarray fluorescence measurements of several biomediators in series, with applications in environmental monitoring and agrofood analysis. In the work by Rea and coworkers (2009), a computational study was performed to design and construct a set of mutant strains from the green photosynthetic alga *C. reinhardtii*, with higher sensitivity towards several classes and subclasses of herbicides (Table 3).

In this context, an *in silico* study was performed to predict mutations within the D1-D2 heterodimer which improve its specificity, sensitivity, and binding affinity for atrazine. In detail, taking advantage of the high sequence homology observed between *Thermosynecococcus elongatus* D1 and D2 proteins and the corresponding proteins from *C. reinhardtii* (87% and 89% amino acid sequence identity, respectively), the three-dimensional structure of the latter proteins was homology modelled. On the basis of this model, a series of D1 and D2 mutants were generated *in silico* and the atrazine affinity of wild type and mutant proteins was predicted by binding energy calculations to identify mutations able to increase PSII affinity for atrazine.

DEVELOPED BIOSENSORS	ADVANTAGES	DISADVANTAGES	REFERENCE
Whole-cell-based bacterial biosensors Cyanobacterial PSII-based biosensors	- simple and rapid pre-treatment steps - real samples/complex matrix analyses - ability to recognize different classes of chemicals	- intrinsic instability, short half-life and specificity, poor response	Weitz et al., 2001 Merz et al., 1996
Optical biosensor based on various microalgae or chloroplast and thylakoids membranes	- different recognition elements for various classes of pesticides, insecticides and organophosphorus compounds	- high limits of detection in a concentration range from 10^{-8} to 10^{-3} M or 10^{-9} to 10^{-5} M	Marty et al., 1995 Naessens et al., 2000 Euzet et al., 2005 Giardi et al., 2005 Breton et al., 2006
Optical multiarray biosensor which employ several mutant strains from *C. reinhardtii*	- high specificity towards classes and subclasses of herbicides - low limits of detection in a concentration range from 10^{-10} to 10^{-8} M	- low specificity towards specific target analytes	Tibuzzi et al., 2007 Rea et al., 2009 Giardi et al., 2009 Scognamiglio et al., 2009
Biosensing platform with different biomediators and double detection systems (optical and amperometric)	- high specificity towards classes of herbicides - low limits of detection in a concentration range from 10^{-10} to 10^{-8} M - high stability of immobilisation biological recognition elements	- low specificity towards specific target analytes	Buonasera et al., 2010
Amperometric biosensors based on mutant strains from *C. reinhardtii*	- real samples/complex matrix analyses	- low specificity towards specific target analytes	Giardi et al., 2005
Amperometric biosensors based on thylakoid from *Spinacia oleracea* and *Senecio vulgaris*	- real samples/complex matrix analyses	- low specificity towards specific target analytes	Touloupakis et al., 2005

Table 3. Main features of developed biosensors.

New advances in the same context were achieved in amplifying the range of recognition elements and measurement of a significant number of different classes of environmental pollutants. These advances occurred through the development of a biosensing system which uses sets of mutant organisms with different affinities towards pesticides. A library of functional mutations in the unicellular green alga *C. reinhardtii* for preparing biomediators was presented by Giardi and coworkers (2009). Exploiting bioinformatics to design new mutant strains resulted in the construction of microorganisms which showed different limits of detection for diazines, triazines and urea herbicides, underlined the high potential of bioinformatics and molecular biology in the design of desired biological material suitable for biosensor use.

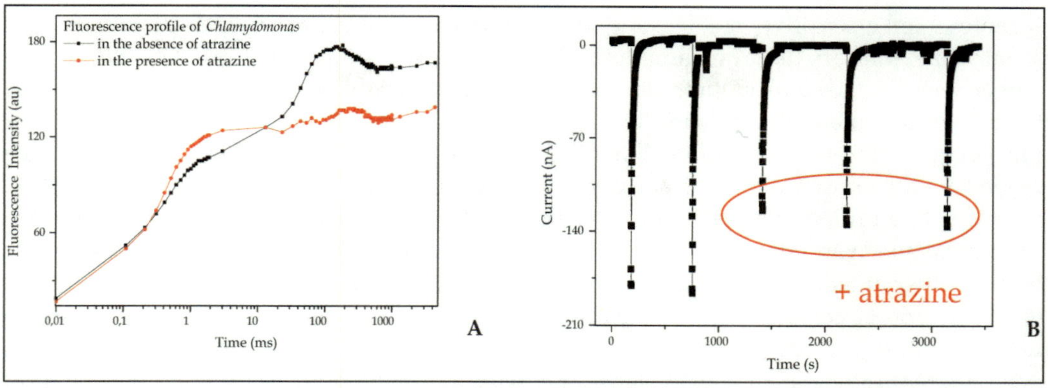

Fig. 6. (**A**) Fluorescence profile of *C. reinhardtii* PSII in the absence and in the presence of atrazine. The binding of the herbicide is related to the main parameters of the profile in a concentration dependent manner. (**B**) Current profiles of *C. reinhardtii* PSII in the absence and in the presence of atrazine (10^{-7} M). The binding of the herbicide decreases the current signal in a concentration dependent manner.

A multi-biomediator fluorescence biosensor based on a new versatile portable instrument was assembled by Scognamiglio and coworkers (2009). The biosensor instrument was composed of a 24 cell array configuration able to host different mutant strains for the detection of a variety of herbicide classes such as triazines, diazines and ureas (Figure 7).

As we can observe from the described advances in biosensor technology, the main features of a successful biosensor are characterised by the interchangeable recognition elements, which provide the versatility to measure large numbers of analytes. In Buonasera and coworkers (2010), a biosensing platform was constructed to provide an analytical tool applicable to the daily pre-screening of a broad spectrum of samples. The platform combined the most used transduction systems for biosensors, amperometric and optical systems, and used genetically modified microorganisms as versatile biomediators, allowing detection of different subclasses of herbicides.

Fig. 7. Biosensing platform set-up consisting of 24 cells able to host an array of several biomediators.

It represented a sensitive, reliable, and low-cost system able to detect water pollutants such as atrazine, diuron, linuron, and terbuthylazine down to 10^{-8}-10^{-10} M. Combining the amperometric and optical detection systems, the platform was able to determine the toxicological potential of samples, through the determination of the biomediator physiological activity inhibition. Fluorescence modification and current reduction were related to the concentration of herbicide and quantified by a dedicated data acquisition software. In addition, the opportunity to use a wide range of biological materials made the platform a good candidate for the development of a biosensor with required features.

Several other practical aspects seem to be important for the development of biosensors. One of these aspects considers the variability of real samples, whose composition is usually unknown and can vary widely from sample to sample. Suitable biosensors have to demonstrate reliability in field tests followed by validation by standard analytical methodologies. In Giardi and coworkers (2005) a fluorescence multi-biosensor was reported based on the thylakoids activity from different microorganisms used for the determination of several pollutants on real samples from the Tiber river, the Aqua Marcia, the Valle del Sorbo, and the Po river, tested contemporaneously by gas chromatography-mass spectrometry. Similar results were found with both methods. In Touloupakis and coworkers (2005) an amperometric multibiosensor using various photosynthetic preparations as biosensing elements for the detection of herbicides and pollutants on real samples was described. The photosynthetic thylakoid from *Spinacia oleracea*, *Senecio vulgaris* and its mutant resistant to atrazine were immobilized on the surface of screen printed electrodes composed of a graphite-working electrode and Ag/AgCl reference electrode deposited on a polymeric substrate (Figure 8). The presence of pollutants was revealed as the effect on PSII due to the sum of various herbicides, mainly triazines (0.210 μg/l) and phenolic compounds (0.041 μg/l). The performance of a biosensor is also related to the stability, the operating lifetime and the reusability of the biodevice, which is of critical importance and can affect the success of its use. In biosensor production, the biological material is usually immobilised, entrapped or cross-linked so as to produce an intimate connection or communication between the biomediator and the transducer. Many techniques have been

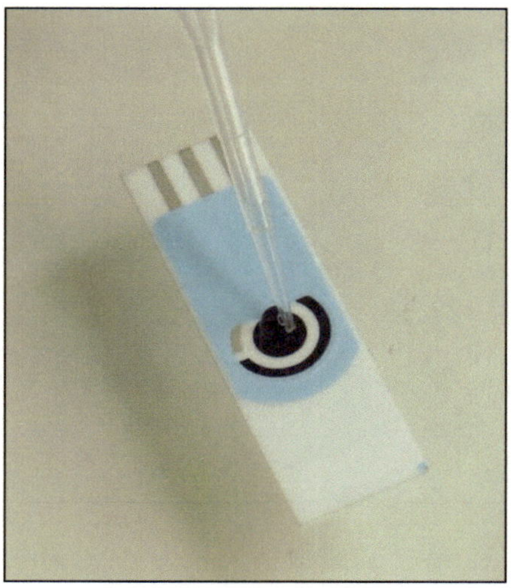

Fig. 8. Screen printed electrode used for biomediator immobilization.

introduced to overcome manufacturing problems associated with stability. In Buonasera and coworkers (2010) different immobilisation procedures, following physical and chemical approaches, were described as being suitable for several biological recognition elements and to be applied in amperometric and/or fluorescence measurements, some of these biosensors are already commercially available (see www.biosensor.it) (Table 4).

LEAKING	IMMOBILIZATION METHOD	RESIDUAL ACTIVITY %
High	Filter paper disk	56
High	Alumina filter disk	26
High	Glass microfiber filter	49
High	DEAE cellulose	50
Medium	Nitrocellulose	90
Medium	Agar	45
Medium	Agarose	33
Low	Carrageenan	30
Low	Alginate	20
Low	Gelatin	50
High	Lyophilisation	70
Low	Glutaraldehyde	70
Low	Magnetic-beads polymer	60
Low	Binding on gold films	35
Low	Collagen	30
Low	Bovine serum albumine (BSA-GA)	70
Low	Cross-linking on Gold	20
Low	Cross-linking on TiO_2	55
Low	Polyacrylamid	41
Low	Polyurethane	28
Low	Photocrosslinkable resin	10
Low	Vinyl	15
Low	Poly(vinylalcohol)	38
Low	Tyrylpyridinium groups	45
Low	Thiophen polymer	57
	error less than 10%	

Table 4. Immobilization methods applied to several biomediators. The stability of each procedure was evaluated measuring the residual activity and leaking before and after immobilization.

9. Conclusion and future perspectives

The exploitation of herbicides for weed control is vital to increase the yields and productivity in agriculture. Without the use of herbicides, it would have been impossible to fully mechanize the production of cotton, sugar beets, grains, potatoes, and corn. As a consequence, given the harmful economic implications of poor harvesting, herbicide production is the principal driver of the farming industry. However, the continuous and massive application of these compounds can negatively affect human health and ecosystems. These consequences result in an increased demand for risk assessment and prompt the regulatory agencies to update legislation aimed at controlling environmental contaminations. In this scenario, the development of analytical devices able to detect the low levels of herbicide contaminants defined by the EU directives, and to distinguish among different classes of compounds, is essential. Several instruments which partially satisfying these requirements have been already developed.

Future activity should be focused on the development of new types of bio-sensing elements for building up a platform of modular biosensors which can be easily adopted for the simultaneous detection of several herbicides. We are currently manufacturing an array of novel whole-cell biosensors based on the activity of engineered photosynthesis enzymes with improved sensitivity and stability features, and ectopic expressed fluorescent proteins as sensing elements. The new complex biosensor array will be based on both optical and electronic transduction for multi-parameter detection. It will be able to monitor the herbicide levels and to diagnose their biological impact. This improvement should provide the impetus for the technological transfer from laboratory devices to in-field operation systems. The new devices will lead to a tremendous breakthrough in the detection of contaminants and quality control in risk assessment sectors by providing a rapid broad spectrum screening tool.

10. Acknowledgements

This study was granted by the EU FP7-SME-2008-1 projects: *SENSBIOSYN* n° 232522 to GR; *BEEP-C-EN* n° 232082 to MTG; *MULTIBIOPLAT* EUROTRANS-BIO-2007-34 to GR; CASPUR High-Performance-Computing Grants 2009-10 to FP.

11. References

Ackerman, F. (2007). The Economics of Atrazine. *International Journal of Occupational and Environmental Health*, 13: 441–449.

Al-Ahmad, H.; Galili, S. & Gressel, J. (2005). Poor competitive fitness of transgenically mitigated tobacco in competition with the wild type in a replacement series. *Planta*, 222(2): 372-385.

Bouldin, J.L.; Farris, J.L.; Moore, M.T.; Smith, S. Jr. & Cooper, C.M. (2006). Hydroponic uptake of atrazine and lambda-cyhalothrin in Juncus effusus and Ludwigia peploides. *Chemosphere*, 65: 1049-1057.

Brena, B.M.; Arellano, L.; Rufo, C.; Last, M.S.; Montaño, J.; Egaña Cerni, E.; Gonzalez-Sapienza, G. & Last, J.A. (2005). ELISA as an affordable methodology for monitoring groundwater contamination by pesticides in low-income countries. *Environmental Science & Technology*, 39(11): 3896-3903.

Breton, F.; Euzet P.; Piletsky, S.A.; Giardi, M.T. & Rouillon, R. (2006). Integration of photosynthetic biosensor with molecularly imprinted polymer-based solid phase extraction cartridge. *Analytica Chimica Acta*, 569: 50-57.

Buonasera, K.; Pezzotti, G.; Scognamiglio, V.; Tibuzzi, A. & Giardi, M.T. (2010). New platform of biosensors for prescreening of pesticide residues to support laboratory analyses. *Journal of Agricultural and Food Chemistry*, 58(10): 5982-5990.

Cavalcanti, A.; Shirinzadeh, B.; Freitas, R.A. & Kretly, L.C. Jr. (2007). Medical nanorobot architecture based on nanobioelectronics. *Recent Patents on Nanotechnology*, 1(1): 1-10.

Cesnik, H.B.; Gregorcic, A.; Bolta, S.V. & Kmecl, V. (2006). Monitoring of pesticide residues in apples, lettuce and potato of the Slovene origin, 2001-04. *Food Additives and Contaminants*, 23(2): 164-73.

Chouhan, R.S.; Rana, K.V.; Suri, C.R.; Thampi, R.K. & Thakur, M.S. (2010). Trace-level detection of atrazine using immuno-chemiluminescence: dipstick and automated flow injection analyses formats. *Journal of AOAC International*, 93(1): 28-35.

Cox, C. (2001). Atrazine: Environmental contamination and ecological effects. *Journal of Pesticide Reform*, 21: 12-20.

Dauvillee, D.; Hilbig, L.; Preiss, S. & Johanningmeier, U. (2004). Minimal extent of sequence homology required for homologous recombination at the psbA locus in *Chlamydomonas reinhardtii* chloroplasts using PCR-generated DNA Fragments. *Photosynthesis Research*, 79(2): 219-224.

DeVesa, S.S.; Blot W.J.; Stone, B.J.; Miller, B.A.; Tarone, R.E. & Fraumeni, J.F. Jr. (1995). Recent cancer trends in the United States. *Journal of the National Cancer Institute*, 87(3): 175-182.

Duysens, L.N.M. & Sweers, H.E. (1963). Mechanism of two photochemical reactions in algae as studied by means of fluorescence. In: *Studies on Microalgae and Photosynthetic Bacteria*, J. Ashida (Ed), 353-372, Tokyo University Press, Tokyo.

Environmental Protection Agency. (2009). Atrazine Science Reevaluation: Potential Health Impacts. EPA-HQ-OPP-2009-0759-0003 Accessed: October 27, 2009. www.epa.gov/pesticides/reregistration/atrazine/atrazine_update.htm.

EUROPEAN ECONOMIC COMMUNITY (EEC) 15 July 1991. Council Directive 91/414/ on the placing of plant protection products on the market. Official Journal of the European Communities, p. 1-290.

EUROPEAN ECONOMIC COMMUNITY (EEC) 27 June 1967. Council Directive 67/548/ on the approximation of laws, regulations and administrative provisions relating to the classification, packaging and labelling of dangerous substances. Official Journal of the European Communities, p. 196-198.

Eurostat statistical books. The use of plant protection products in the European Union. Data 1992–2003. [http://epp.euro stat.ec.europa.eu/cache/ITY_OFFPUB/KS-76-06-669/EN/KS-76-06- 669-EN.PDF].

EC REGULATION N. 2009/128/EC OF THE EUROPEAN PARLIAMENT AND OF THE COUNCIL 21 October 2009. Official Journal of the European Union. L 309/71-86.

EC REGULATION N. 149/2008 OF THE EUROPEAN COUNCIL 29 January 2008. Official Journal of the European Union. L 58/1-16.

EC REGULATION N. 396/2005 OF THE EUROPEAN PARLIAMENT AND OF THE COUNCIL 23 February 2005. Official Journal of the European Union. L 70/1-16.

Euzet, P.; Giardi, M.T. & Rouillon, R. (2005). A crosslinked matrix of thylakoids coupled to the fluorescence transducer in order to detect herbicides. *Analytica Chimica Acta*, 539: 263-269.

Fan, W.Q.; Yanase, T.; Morinaga, H.; Gondo, S.; Okabe, T.; Nomura, M.; Komatsu, T.; Morohashi, K.I.; Hayes, T.B.; Takayanagi, R. & Nawata, H. (2007). Atrazine-induced aromatase expression is SF-1 dependent: implications for endocrine disruption in wildlife and reproductive cancers in humans. *Environmental Health Perspectives*, 115: 720–727.

Fawcett. (2008). Twenty years of university corn yield data: with and without atrazine, proceedings north central weed science society. North Central Weed Science Society.

Ferreira, K.N.; Iverson, T.M.; Maghlaoui, K.; Barber, J. & Iwata, S. (2004). Architecture of the photosynthetic oxygen-evolving center. *Science*, 303: 1831-1838.

Forbush, B. & Kok, B. (1968). Reaction between primary and secondary electron acceptors of photosystem II of photosynthesis. *Biochimica et Biophysica Acta*, 162: 243-253.

Frazier, L.M. (2007). Reproductive disorders associated with pesticide exposure. *Journal of Agromedicine*, 12(1): 27-37.

Frisch, M. J. et al., (2004). Gaussian 03, Revision C.02 , Gaussian, Inc., Wallingford CT.

Fukuyama, T.; Tajima, Y.; Ueda, H.; Hayashi, K.; Shutoh, Y.; Harada, T. & Kosaka, T. (2009). Allergic reaction induced by dermal and/or respiratory exposure to low-dose phenoxyacetic acid, organophosphorus, and carbamate pesticides. *Toxicology*, 261(3): 152-161.

Fussell, R.J.; Jackson Addie, K.; Reynolds, S.L. & Wilson, M.F. (2002). Assessment of the stability of pesticides during cryogenic sample processing. 1. Apples. *Journal of Agricultural and Food Chemistry*, 50(3): 441-448.

Gascon, J.; Oubina, A. & Barcelò, D. (1997). Detection of endocrine-disrupting pesticides by enzyme-linked immunosorbent assay (ELISA): application to atrazine. *Trends in Analytical Chemistry*, 16(10): 544-562.

Giardi, M.T.; Scognamiglio, V.; Rea, G.; Rodio, G.; Antonacci, A.; Lambreva, M.; Pezzotti, G. & Johanningmeier, U. (2009). Optical biosensors for environmental monitoring based on computational and biotechnological tools for engineering the photosynthetic D1 protein of *Chlamydomonas reinhardtii*. *Biosensor and Bioelectronics*, 25(2): 294-300.

Giardi, M.T.; Guzzella, L.; Euzet, P.; Rouillon, R. & Esposito, D. (2005). Detection of herbicide subclasses by an optical multibiosensor based on an array of photosystem II mutants. *Environmental Science & Technology*, 39: 5378-5384.

Giardi, M.T. & Pace, E. (2005). Photosynthetic proteins for technological applications. *Trends in Biotechnology*, 23(5): 257-263.

Giardi, M.T.; Marder, J.B. & Barber, J. (1988). Herbicide binding to the isolated Photosystem II reaction centre. *Biochimica et Biophysica Acta*, 934: 64-71.

Green, J.M. & Owen, M.D. (2010). Herbicide-Resistant Crops: Utilities and Limitations for Herbicide-Resistant Weed Management (dagger). *Journal of Agricultural and Food Chemistry*, DOI: 10.1021/jf101286h.

Gurney, J.G.; Davis, S.; Severson, R.K.; Fang, J.Y.; Ross, J.A. & Robison, L.L. (1996). Trends in cancer incidence among children in the US. *Cancer*, 78: 532-541.

Guskov, A.; Kern, J.; Gabdulkhakov, A.; Broser, M.; Zouni, A. & Saenger1, W. (2009). Cyanobacterial photosystem II at 2.9-A° resolution and the role of quinones, lipids, channels and chloride. *Nature Structural & Molecular Biology*, 16(3): 334-342.

Guzzella, L.; Pozzoni, F. & Giuliano, G. (2006). Herbicide contamination of surficial groundwater in Northern Italy. *Environmental Pollution*, 142(2): 344-353.

Haworth, P. & Steinback, K.E. (1987). Interaction of herbicides and quinone with the QB-protein of the diuron-resistant *Chlamidomonas reinhardtii* mutant Dr2. *Plant Physiology*, 83: 1027-1031.

Haynes, D.; Müller, J. & Carter, S. (2000). Pesticide and herbicide residues in sediments and seagrasses from the great barrier reef world heritage area and queensland coast. *Marine Pollution Bulletin*, 41(7-12): 279-287.

Hilton, H.W.; Nomura, N.S.; Kameda, S.S. & Yauger, W.L. (1976). Some patterns of herbicide and growth regulator intake, persistence, and distribution in sugarcane. *Archives of Environmental Contamination and Toxicology*, 4: 385-394.

Johanningmeier, U.; Sopp, G.; Brauner, M.; Altenfeld, U.; Orawski, G. & Oettmeier, W. (2000). Herbicide resistance and supersensitivity in Ala250 or Ala251 mutants of the D1 protein in *Chlamydomonas reinhardtii*. *Pesticide Biochemestry Physiology*, 66: 9-19.

Kern, J. & Renger, G. (2007). Photosystem II: Structure and mechanism of the water:plastoquinone oxidoreductase. *Photosynthesis Research*, 94: 183-202.

Kristensen, P.; Irgens, L.M.; Andersen, A.; Bye, A.S. & Sundheim, L. (1997). Birth defects among offspring of Norwegian farmers, 1967-1991. *Epidemiology*, 8(5): 537-544.

Lancaster, C.R.D. & Michel, H. (1999). Refined crystal structure of reaction centres from *Rhoropseudomonas viridis* in complexes with the herbicide atrazine and two chiral atrazine derivatives also lead to a new model of the bound carotenoid. *Journal of Molecular Biology*, 286: 883-898.

Lavergne, G. (1982). Mode of action of 3-(3,4 dichlorophenyl)-1,1 dimethylurea. Evidence that the inhibitor competes with plastoquinone for binding to a common site on the acceptor side of Photosystem II. *Biochimica et Biophysica Acta*, 682: 345-353.

Loh, K-S.; Lee, Y.H.; Musa, A.; Salmah, A.A. & Zamri I. (2008). Use of Fe3O4 nanoparticles for enhancement of biosensor response to the herbicide 2,4-dichlorophenoxyacetic acid. *Sensors*, 8: 5775-5791.

Loll, B.; Kern, J.; Saenger, W.; Zouni, A. & Biesiadka, J. (2005 a). Towards complete cofactor arrangement in the 3.0 A resolution structure of photosystem II . *Nature*, 438: 1040-1044.

Loll, B.; Kern, J.; Zouni, A.; Saenger, W.; Biesiadka, J. & Irrgang, K.D. (2005 b). The antenna system of photosystem II from *Thermosynechococcus elongatus* at 3.2 Å resolution. *Photosynthesis Research*, 86: 175-184.

Lorenzin, M. (2007). Pesticide residues in Italian Ready-Meals and dietary intake estimation. *Journal of Environmental Science and Health. Part. B, Pesticides, food contaminants, and agricultural wastes*, 42(7): 823-833.

Mage, D.T.; Allen, R.H.; Gondy, G.; Smith, W.; Barr, D.B. & Needham, L.L. (2004). Estimating pesticide dose from urinary pesticide concentration data by creatinine correction in the Third National Health and Nutrition Examination Survey

(NHANES-III). *Journal of Exposure Analysis and Environmental Epidemiology*, 14: 457-465.

Marty, J.L.; Garcia, D. & Rouillon, R. (1995). Biosensors: potential in pesticide detection. *Trends in Analytical Chemistry*, 14(7): 329-333.

Mazzei, F.; Botrè, F.; Lorenti, G.; Simonetti, G.; Porcelli, F.; Scibona, G. & Botrè, C. (1995). Plant tissue electrode for the determination of atrazine. *Analytica Chimica Acta*, 316(1): 79-82.

McArdle, F.A. & Persaud, K.C. (1993). Development of an enzyme-based biosensor for atrazine detection. *Analyst*, 118: 419-423.

Merz, D.; Geyer M.; Moss, D.A. & Ache H.J. (1996). Chlorophyll fluorescence biosensor for the detection of herbicides. *Fresenius' Journal of Analytical Chemistry*, 354: 299–305.

Montesinos, E. (2003). Development, registration and commercialization of microbial pesticides for plant protection. *International Microbiology*, 6(4): 245-252.

Moreland, .D.E. & Hill, K.L. (1962). Interference of herbicides with the Hill reaction of isolated chloroplasts. *Weeds*, 10: 229-236.

Moreland, D.E.; Gentner, J.L.; Hilton, J.L. & Hill, K.L. (1959). Studies on the mechanism of herbicidal action of 2-chloro-4,6-bis(ethylamino)-s-triazine. *Plant Physiology*, 43: 432-435.

Murphy, C.E. & Lemerle, D. (2006). Continuous cropping systems and weed selection. *Euphytica*, 148: 61–73.

Naessens, M.; Leclerc, J.C. & Tran-Minh C. (2000). Fiber optic biosensor using *Chlorella vulgaris* for determination of toxic compounds. *Ecotoxicology and Environmental Safety*, 46: 181-185.

Ochoa-Acuña, H.; Frankenberger, J.; Hahn, L. & Carbajo, C. (2009). Drinking-water herbicide exposure in Indiana and prevalence of small-for-gestational-age and preterm delivery. *Environmental Health Perspectives*, 117: 1619–1624.

Oettmeier, W. (1999). Herbicide resistance and supersensitivity in photosystem II. *Cellular and Molecular Life Sciences*: 55: 1255-1277.

Oettmeier, W.; Kude, C. & Soll, H-J. (1987). Phenolic herbicides and their methylethers: binding characteristics and inhibition of photosynthetic electron transport and photophosphorylation. *Pesticide Biochemistry Physiology*, 27: 50-60.

Oettmeier, W. & Masson, K. (1980). Synthesis and thylakoid membrane binding of the radioactively labeled herbicide dinoseb. *Pesticide Biochemistry Physiology*, 14: 86-97.

Paris, P.; De santis, T.; Esposito, D.; Giuliani, R. & Pace, E. (2009). Atti del XVII convegno S.I.R.F.I. Protezione dei corpi idrici superficiali dall'inquinamento da agro-farmaci. Vidotto, F. ; Ferrero, A. (Eds.), 17-35, Bologna, Italy, (ISBN 978-88-95616- 49802-5).

Pfister, K.; Steinback, K.K.; Gardner, G. & Arntzen, C.J. (1981). Photoaffinity labelling of a herbicide receptor in chloroplast membranes. *Proceedings of the National Academy of Sciences of the United States of America*, 78: 981-985.

Pfister, K. & Arntzen, C.J. (1979). The mode of action of Photosystem II-specific inhibitors in herbicide-resistant weed biotypes. *Zeitschrift für Naturforschung. C, Journal of Biosciences*, 34: 996-1009.

Pfister, K.; Radosevich, S.R. & Arntzen, C.J. (1979). Modification of herbicide binding to photosystem II in two biotypes of *Sencio vulgaris* L. *Plant Physiology*, 78: 981-985.

Preiss, S.; Schrader, S. & Johanningmeier, U. (2001). Rapid, ATP-dependent degradation of a truncated D1 protein in the chloroplast. *European Journal of Biochemistry*, 268(16): 4562-4569.

Przibilla, E.; Heiss, S.; Johanningmeier, U. & Trebst, A. (1991). Site-specific mutagenesis of the D1 subunit of photosystem II in wild-type chlamydomonas. *The Plant Cell*, 3: 169-174.

RB-159R Biosensors and Bioelectronics report released by Business Communications Co., Inc. (bccresearch.com)

Rea, G.; Antonacci, A.; Lambreva, M.; Margonelli, A.; Ambrosi, C. & Giardi M.T. The NUTRA-SNACKS project: basic research and biotechnological programs on nutraceutics. In: Bio-Farms for Nutraceuticals: Functional Food and Safety Control by Biosensors. Editors: M.T. Giardi, G. Rea & B. Berra. Landes Bioscience, Springer Publishers, Church St. Georgetown USA. *In press*.

Rea, G.; Polticelli, F.; Antonacci, A.; Scognamiglio, V.; Katiyar, P.; Kulkarni, S.A.; Johanningmeier, U. & Giardi, M.T. (2009). Structure-based design of novel *Chlamydomonas reinhardtii* D1–D2 photosynthetic proteins for herbicide monitoring. *Protein Science*, 18(10): 2139-2151.

Rea, G.; Serafini, A.; Margonelli, A.; Faraloni, C.; Zanini, A.; Bertalan, I.; Johanningmeier, U. & Giardi. M.T. (2008). Ionizing radiation impacts photochemical quantum yield and oxygen evolution activity of photosystem II in photosynthetic microorganisms. *International Journal of Radiation Biology*, 84 (11): 867-877.

Readman, J.W.; Albanis, T.A.; Barcelo, D.; Galassi, S.; Tronczynski, J. & Gabrielides, G.P. (1993). Herbicide contamination of Mediterranean estuarine waters: results from a MED POL pilot survey. *Marine Pollution Bulletin*, 26 (11): 613-619.

Relyea, R.A. (2005). The impact of insecticides and herbicides on the biodiversity and productivity of aquatic communities. *Ecological Applications*, 15(2): 618-627.

Renger, G. (1986). Herbicide interaction with Photosystem II: recent developments. *Physiologie Végétale*, 24: 509-521.

Ries, L.A.G.; Kosary, C.L.; Hankey, B.F.; Miller, B.A.; Clegg, L.; & Mariotto, A.; et al. (1998). SEER Cancer Statistics Review, 1973–1995. NIH Publ no 98-2789. Bethesda, MD: National Cancer Institute.

Roberts, D.M.; Buckley, N.A.; Mohamed, F.; Eddleston, M.; Goldstein, D.A.; Mehrsheikh, A.; Bleeke, M.S. & Dawson, A.H. (2010). A prospective observational study of the clinical toxicology of glyphosate-containing herbicides in adults with acute selfpoisoning. *Clinical Toxicology (Philadelphia, Pa.)*, 48(2): 129–136.

Sass, J.B. & Colangelo, A. (2006). European Union bans atrazine, while the United States negotiates continued use. *International Journal of Occupational and Environmental Health*, 12(3): 260-267.

Shewry, P.R.;Jones, H.D. & Halford, N.G. (2008). Plant biotechnology: transgenic crops. *Advances in Biochemical Engineering/Biotechnology*; 111: 149-186.

Schneider, M.I.; Sanchez, N.; Pineda, S.; Chi, H. & Ronco, A. (2009). Impact of glyphosate on the development, fertility and demography of *Chrysoperla externa* (Neuroptera: Chrysopidae): ecological approach. *Chemosphere*; 76(10): 1451-1455.

Schreinemachers, D.M. (2010). Perturbation of lipids and glucose metabolism associated with previous 2,4-D exposure: a cross-sectional study of NHANES III data, 1988-1994. *Environmental Health*, 9: 11.

Scognamiglio, V.; Raffi, D.; Lambreva, M.; Rea, G.; Tibuzzi, A.; Pezzotti, G.; Johanningmeier, U. & Giardi, M.T. (2009). *Chlamydomonas reinhardtii* genetic variants as probes for fluorescence sensing system in detection of pollutants. *Analytical and Bioanalytical Chemistry*, 394(4): 1081-1087.

Senthilselvan, A.; McDuffie, H.H. & Dosman, J.A. (1992). Association of asthma with use of pesticides. Results of a cross-sectional survey of farmers. *The American Review of Respiratory Disease*, 146(4): 884-887.

Shao, C. Y.; Howe, C. J.; Porter, A. J. R. & Glover, L. A. (2002). Novel cyanobacterial biosensor for detection of herbicides. *Applied and Environmental Microbiology*, 68(10): 5026-5033.

Shimabukuro, R.H. & Swanson, H.R. (1969). Atrazine metabolism, selectivity, and mode of action. *Journal of Agriculture and Food Chemistry*, 17: 199-204.

Siu, S.W.I.; Vacha, R.; Jungwirth P. & Böckmann, R. A. (2008). Biomolecular simulations of membranes: physical properties from different force fields. *The Journal of Chemical Physics*, 128: 125103.

Sondhia, S. (2010). Persistence and bioaccumulation of oxyfluorfen residues in onion. *Environmental Monitoring and Assessment*, 162: 163–168.

Swanton, C. J.; Gulden, R. H. & Chandler, K. (2007). A rationale for atrazine stewardship in corn. *Weed Science*, 55: 75–81.

Takahashi, T.; Inoue-Kashino, N.; Ozawa, S.; Takahashi, Y.; Kashino, Y. & Satoh, K. (2009). Photosystem II complex in vivo is a monomer. *The Journal of Biological Chemistry*; 284: 15598-15606.

Tanner, C.M.; Ross, G.W.; Jewell, S.A.; Hauser, R.A.; Jankovic, J.; Factor, S.A.; Bressman, S.; Deligtisch, A.; Marras, C.; Lyons, K.E.; Bhudhikanok, G.S.; Roucoux, D.F.; Meng, C.; Abbott, R.D. & Langston, J.W. (2009). Occupation and risk of parkinsonism: a multicenter case-control study. *Archives of Neurology*, 66(9): 1106-1113.

Thevenot, D.R.; Tóth, K.; Durst, R.A. & Wilson G.S. (1999). Electrochemical biosensors: recommended definitions and classification. Pure and applied chemistry. *Chimie Pure et Appliquée Links*, 71: 2333-2348.

Tibuzzi, A.; Rea, G.; Pezzotti, G.; Esposito, D.; Johanningmeier, U. & Giardi, M.T. (2007). A new miniaturized multiarray biosensor system for fluorescence detection. *Journal of Physics: Condensed Matter*, 19: 395006-395018.

Tischer, W. & Strotmann, H. (1977). Relationship between inhibitor binding by chloroplasts and inhibition of photosynthetic electron transport. *Biochimica et Biophysica Acta*, 460: 113-125.

Touloupakis, E.; Giannoudi, L.; Piletsky, S.A.; Guzzella, L.; Pozzoni, F. & Giardi, M.T. (2005). A multi-biosensor based on immobilized Photosystem II on screen-printed electrodes for the detection of herbicides in river water. *Biosensor & Bioelectronics*, 20(10): 1984-1992.

Trebst, A. (1987). The three-dimenstional structure of the herbicide binding niche on the reaction center polypeptides of photosystem II. *Zeitschrift für Naturforschung. C, Journal of Biosciences*, 42c: 742-750.

Trebst, A. (1986). The topology of the plastoquinone and herbicide binding peptides of Photosystem II in the thylakoid membrane. *Zeitschrift für Naturforschung. C, Journal of Biosciences*, 41c: 240-245.

Turner, M.C.; Wigle, D.T. & Krewski, D. (2010). Residential pesticides and childhood leukemia: a systematic review and meta-analysis. *Environmental Health Perspectives*, 118: 33–41.

US Environmental Protection Agency (US EPA). (1994). Atrazine, simazine and cyanazine, notice of initiation of special review. *Federal Register*, 59: 60412–60443.

Van der Spoel, D.; Lindahl, E.; Hess, B.; Groenhof, G.; Mark, A.E. & Berendsen, H.J.C. (2005). GROMACS: fast, flexible and free. *Journal of Computational Chemistry*, 26: 1701–1718.

Van Rensen, J.J.S. (1982). Molecular mechanisms of herbicide action near photosystem II. *Physiol Plantarum*, 54: 515-521.

Vermaas, W.F.J.; Arntzen, C.J.; Gu, L.Q. & Uy, C.A. (1983). Interactions of herbicides and azidoquinones at a photosystem II binding site in the thylakoid membrane. *Biochimica et Biophysica Acta*, 723: 266-275.

Vila-Aiub, M.M.; Vidal, R.A.; Balbi, M.C.; Gundel, P.E.; Trucco, F. & Ghersa, C.M. (2008). Glyphosate-resistant weeds of South American cropping systems: an overview. *Pest Management Science*, 64(4): 366-371.

Waller, S.A.; Paul, K.; Peterson, S.E. & Hitti, J.E. (2010). Agricultural-related chemical exposures, season of conception, and risk of gastroschisis in Washington State. *American Journal of Obstetrics and Gynecology*, 202: 241.

Wang, H.; Nakata, E. & Hamachi I. (2009). Recent progress in strategies for the creation of protein-based fluorescent biosensors. *Chembiochem : a European Journal of Chemical Biology*, 10(16): 2560-2577.

Wang, J.; Wolf, R.M.; Caldwell, J.W.; Kollamn, P.A. & Case, D.A. (2004). Development and testing of a general Amber force field. *Journal of Computational Chemistry*, 25: 1157-1174.

Wang, J.; Wang, W.; Kollman P. A. & Case, D. A. (2006). Automatic atom type and bond type perception in molecular mechanical calculations. *Journal of Molecular Graphics and Modelling*, 25: 247260-247280.

Wang, J.; Cieplak, P. & Kollman P.A. (2000). How well does a restrained electrostatic potential (RESP) model perform in calculating conformational energies of organic and biological molecules?. *Journal of Computational Chemistry*, 21: 1049-1074.

Weitz, H.J.; Ritchie J.M.; Bailey, D.A.; Horsburgh A.M.; Killham K. & Glover, L.A. (2001). Construction of a modified mini-Tn5 luxCDABE transposon for the development of bacterial biosensors for ecotoxicity testing. *FEMS Microbiology Letters*, 197: 159–165.

Weselak, M.; Arbuckle, T.E.; Wigle, D.T.; Walker, M.C. & Krewski, D. (2008). Pre- and post-conception pesticide exposure and the risk of birth defects in an Ontario farm population. *Reproductive Toxicology (Elmsford, N.Y.)*, 25(4): 472-480.

Williams, M.M.; Boerboom, C.M. & Rabaey, T.L. (2010). Significance of atrazine in sweet corn weed management systems. *Weed Technology*, 24: 139-142.

Williams, M.M.; Davis, A.S.; Rabaey, T.L. & Boerboom, C.M. (2009). Linkages among agronomic, environmental, and weed management characteristics in North American sweet corn production. *Field Crops Research*, 113: 161–169.

Witte, I.; Jacobi, H. & Juhl-Strauss, U. (1995). Correlation of synergistic cytotoxic effects of environmental chemicals in human fibroblasts with their lipophilicity. *Chemosphere*, 31(9): 4041-4049.

Xiong, J.; Govindjee, J. & Subramaniam, S. (1996). Modeling of the D1/D2 proteins and cofactors of the Photosystem II reaction center: Implications to herbicide and bicarbonate binding. *Protein Science*, 5: 2054-2073.

Zahm, S. & Ward, M. (1998). Pesticides and Childhood Cancer. *Environmental Health Perspectives*, 106(3): 893-908.

6

Statistical Based Real-Time Selective Herbicide Weed Classifier

Irshad Ahmad[1] and Abdul Muhamin Naeem[2]
[1]College of Computer and Information Sciences, Al Jouf University
[2]NetSol Technologies
[1]Saudi Arabia
[2]Pakistan

1. Introduction

Weeds are "Any plant growing in the wrong place at the wrong time and doing more harm than good". Weeds compete with the crop for water, light, nutrients and space, and therefore reduce crop yields and also affect the efficient use of machinery. A lot of methods are used for weed control. Mechanical cultivation is commonly practiced in many vegetable crops to remove weeds, aerate soil, and improve irrigation efficiency, but this technique cannot selectively remove weeds from the field. The most popular used method for weed control is to use agricultural chemicals (herbicides and fertilizer products). In fact, the success of agriculture is attributable to the effective used of chemicals.

2. Weed control

Weed control is a critical farm operation and can significantly affect crop yield. Herbicides have vital importance in weed control and high crop yield however these have potential to produce harmful effects [1]. Herbicides are applied to whole field uniformly without considering the weed density. Weeds are often patchy rather than even or randomly distributed in the crop fields [2]. Total variable costs in 2002 for U.K were within a range of £1,720/ha and £1,870/ha for main crop potatoes, of which herbicides accounted for between 3% and 4% of costs, fungicides accounted for about 8% of variable costs and nematicides accounted for about 14%-16% of variable costs. United States farmers applied about $16 billion of herbicides in 2005 (The Value of Herbicides in U.S. Crop Production: 2005 Update, Crop Life Foundation), in 1965 pesticide use was $474.1 million for the United States. By 1970 the use of pesticides doubled to $960 million for the United States and between 1975 and 1999 pesticide use grew 383% for the United States (Agribusiness and Applied Economics Report No. 456), representing a significant portion of the variable costs of agricultural production. Obviously, if a more sophisticated chemical delivery system is develop which applied chemicals where weeds existed and abstained where there are no weeds, chemical usage would be reduced and chemicals would be more effectively placed. These practices would result in lower environmental loading and increased profitability in the agricultural production sector. Selectively spraying, spot spraying, or intermittent spraying are different names which are attached to this herbicide application method.

(a)

(b)

Fig. 1. (a) Automated Weed Sprayer Arm (b) Control Panel
(images are courtesy of HARDI Australia Pty Ltd)

The amount of herbicides in a control patch sprayer has been potentially reduced when real-time weed sensing is used. Patch spraying using remote sensing and machine vision are successful systems [3].

Weed Features: A verity of visual characteristics that have been used in plant identification can be divided into three categories: Spectral Reflectance, Morphology and texture.

The photosensor-based plant detection systems [4], [5] can detect all the green plants and spray only the plants. A machine-vision guided precision band sprayer for small-plant foliar spraying [6] demonstrated a target deposition efficiency of 2.6 to 3.6 times that of a conventional sprayer, and the non-target deposition was reduced by 72% to 99%.

Certain accurate methods for weed detection have been developed, which included wavelet transformation to discriminate between crop and weed in perspective agronomic images [7] and spectral reflectance of plants with artificial neural networks [8]. Other researchers have investigated texture features [7] or biological morphology such as leaf shape recognition [6]. So in real time for the identification and classification of crop rows in images, a lot of fast methods have been implemented [9]; some of them are based on Hough transform [10], Fourier transform [13], Kalman filtering [11] and linear regression [12]. Consequently, there are various vision systems available on autonomous weed control robots for mechanical weed removal.

3. Statistical weed classifier

Statistical classification is a supervised machine learning procedure in which entities are placed into cluster based on quantitative information on one or more characteristics inherent in the items and based on a training set of previously labeled items.
Figure. 2 shows the Flow Chart of a Real-Time Specific Weed Recognition System which were developed to accomplish the broad and narrow weed classification. The algorithm was based on a variance of an image taken from the grayscale image which is obtained from the color image after pre-processing to detect the target area in the fields.

a. Image Pre-processing

Color images were taken from the field. Three arrays were defined to store Red, Green and Blue colors of RGB image in their respective arrays. Then the corresponding pixels from these three arrays were converted in to a single gray scale pixel using the formula

$$\text{GrayPixel}=0.299\text{Red} +0.587\text{Green} + 0.1\ 14\text{Blue} \tag{1}$$

The gray levels are from 0 to 255. To distinguish weeds from background objects in a grayscale image, a grayscale segmentation image-processing step is conducted where objects are classified into one of two classes (weeds and background) by their grayscale difference. Reference [14], indicated that weeds in field images must be carefully segmented; otherwise the feature extraction will yield unreliable results from analyzing soil and weeds. To identify weeds and classify them into one of two classes (broad and narrow) feature extraction are developed.

b. Classification of Images using Statistical Population

Variance and Sample Variance Statistical approach is used to describe the texture of an image. Variance is of particular importance in texture description of plants. After converting the color image into grayscale and segmentation step, the variance is then calculated. Variance for a 2D image from population data can be calculated as

$$\delta^2 = \frac{\sum_{i=0}^{M}\sum_{j=0}^{N}(x_{ij}-\mu)^2}{M*N} \tag{2}$$

Where

$$\mu = \frac{\sum_{i=0}^{M}\sum_{j=0}^{N}x_{ij}}{M*N} \tag{3}$$

M represents the total number of rows and N represents the total number of columns in the image. Variance of a 2D image from a sample data can be calculated using a formula

$$S^2 = \frac{\sum_{i=0}^{m}\sum_{j=0}^{n}(x_{ij} - \overline{x})^2}{m * n} \tag{4}$$

Where

$$\overline{x} = \frac{\sum_{i=0}^{m}\sum_{j=0}^{n} x_{ij}}{m * n} \tag{5}$$

After calculating the variance of an image, the variance is compared with the thresholds TI and T2 to classify the weed into broad, narrow, and little weed as

If S2 < TI, then there is Little Weed in the processed Image

Else if TI < S2 < T2, then it is Narrow Weed

Else if S2 > T2, then it is Broad weed

TI and T2 are set after a series of experiments done on the images.

Figure. 3 show the classification images of broad and narrow weeds, which are taken in the field. These images are processed by using Statistical Population Variance and Sample Variance of an image. The algorithm gave 100% accuracy to detect the presence or absence of weed cover.

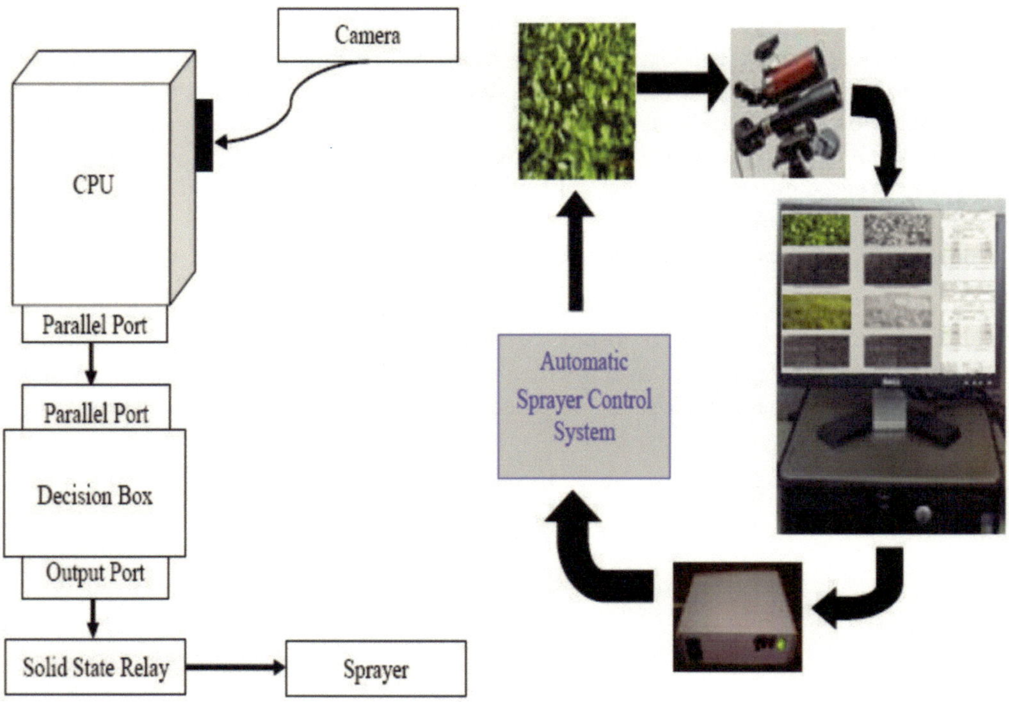

Fig. 2. Flow Chart of Sprayer System

For areas where weeds are detected, results show 98 percent classification accuracy over 140 sample images with 70 samples from each class as shown in Table 1. The population variance and the sample variance of an image are calculated. Different samples were taken.

Random Samples of Intensities taken per image	Results found correct %			Time for Calculating Variance/Image (millisecond)
	Broad	Narrow	Little Weed	
Population (76800)	98%	98%	100%	20
4800	97%	98%	100%	2.5
1200	96%	98%	100%	0.625
300	96%	98%	100%	0.3125

Table 1. Results of the weeds in fig 3 using population variance and sample variance for different samples

The time taken for calculating Population Variance and Sample Variance is given in Table 1. Sample Variance is calculated much faster than Population Variance while retaining the same accuracy for weed detection. The result of taking the

Population and Samples were found the same. Less number of samples is good for high processing speed in real time environment.

4. References

[1] Sunil, K.M., P.R. Weckler and R.K. Taylor, 2007. Effective Spatial Resolution for Weed Detection. 2007 *ASABE Annual International Meeting Sponsored by ASABE Minneapolis Convention Center Minneapolis*, Minnesota 17 - 20 June, 2007.

[2] Wane N, Zhang. E.F, Sun Y. and D.E. Peterson, 2001, "Design of an Optical weed ditreibution for improved post emergence control decision", *Weed Science*, 40,546-553.

[3] Siddiqi, M.H., S.B.T. Sulaiman, I. Faye and I. Ahmad, A Real Time Specific Weed Discrimination System Using Multi-Level Wavelet Decomposition, *Int. J. Agric. Biol., Vol. 11, No. 5, 2009*

[4] Shearer, S. A. and P. T. Jones. 1991. Selective application of post-emergence herbicides using photoelectrics. *Transactions of the Transactions of American Society of Association Executives* 34(4):1661-1666.

[5] J. E.Hanks 1996. Smart sprayer selects weeds for elimination. Agricultural Research. 44(4): 15. [4] Giles, D. K. and D. C. Slaughter. 1997. Precision band sprayer with machine-vision guidance and adjustable yaw nozzles. *Transactions of the American Society of Association Executives* 40(1):29-36.

[6] Manh, A.G., G. Rabatel, L. Assemat and M.J. Aldon, 2001. Weed leaf image segmentation by deformable templates. *J. Agric. Eng. Res.*,80: 139–146

[7] Meyer, G., T. Metha, M. Kocher, D. Mortensen and A. Samal, 1998. Textural imaging and discriminate analysis for distinguishing weeds for spot spraying. *Trans. ASAE,* 41: 1189–1197

[8] Fontaine, V. and T.G. Crowe, 2006. Development of line-detection algorithms for local positioning in densely seeded crops. *Canadian Biosyst. Eng.*, 48: 19–29

[9] Moshou, D., E. Vrindts, D.B. Ketelaere, D.J. Baerdemaeker and H. Ramon, 2001. A neural network based plant classifier. Comput. Electron. Agric., 31: 5–16.

[10] Leemans, V. and M.F. Destain, 2006. Application of the Hough Transform for seed row location using machine vision. Biosyst. Eng., 94: 325- 336

[11] Hague, T. and N.D. Tillet, 2001. A band pass filter-based approach to crop row location and tracking. Mechatronics, 11: 1–12

[12] Sogaard, H.T. and H.J. Olsen, 2003. Determination of crop rows by image analysis without segmentation. Comput. Electron. Agric., 38: 141–158

[13] Vioix, J.B., J.P. Douzals, F. Truchetet, L. Assemat and J.P. Guillemin, 2002. Spatial and spectral method for weeds detection and localization. EURASIP JASP, 7: 679–685

[14] D. M. Woebbecke, G. E. Meyer, K. Von Bargen and D. A. Mortensen, "Shape features for identifying weeds using image analysis," Transactions of the ASAE, vol. 38, no.1, pp. 271-281, 1995.

7

Variable Rate Herbicide Application Using GPS and Generating a Digital Management Map

Majid Rashidi and Davood Mohammadzamani
Department of Agricultural Machinery, Faculty of Agriculture, Islamic Azad University, Takestan Branch, Iran

1. Introduction

This chapter covers developing a precision method of variable rate application (VRA) for application of cyanazine pre-emergence herbicide which eventuates to save considerable pre-emergence herbicide, reduces its adverse effects on the environment and agricultural products, and increases crop yield. For this purpose a digital management map is generated using the global positioning system (GPS). A field of about 6500 m^2 is selected for the grid soil sampling. After that local and Universal Transverse Mercator (UTM) coordinates of the field are determined using total station surveying equipments and four static GPS receivers. Data processing is then accomplished using a personal computer equipped with surveying software. Some soil characteristics such as soil texture and soil organic matter content are also determined by soil sampling and analyzing the soil samples. Five interpolation methods are then used to determine the make-up at other points of the grid. By using Cross Validation method for evaluation of these interpolators and considering manufacture recommendations for cyanazine herbicide application based on soil texture and soil organic matter content, management zones with different herbicide application rates are determined and eventually a digital management map is generated. For implementation of the generated digital management map, a direct injection system is designed and constructed. This system is based on GPS data for positioning of sprayer, comparing the GPS data with digital management map data, measuring of speed, and finally injection of active ingredient inside carrier fluid using solenoid injectors proportionate to any management zone on the digital management map. Using the generated digital management map and equipments of VRA, optimized rate of required herbicide for the selected field is determined. Finally, total required herbicide with VRA is compared with uniform rate application for the entire selected field.

2. What is cyanazine?

Cyanazine is a synthetic chemical that is widely used as a pre-emergence herbicide to control broad-leaf weeds and grasses in agricultural crops. This chemical is in the s-triazine family of herbicides. Some common trade names for cyanazine include Bladex and Fortrol. Cyanazine is also available commercially premixed with another s-triazine, atrazine.

3. The history of cyanazine

Cyanazine was first registered for use as an herbicide by Shell Chemical Company in 1971. In the U.S., over 90% of its use in agriculture is to control weeds in corn fields. Its highest use is in corn-growing states of the Midwest. It is used primarily as a pre-emergent herbicide on corn. It is usually applied once during the growing season to control weeds before the corn-seedlings emerge from the soil. It is also used to control weeds in sorghum, cotton, barley, wheat, oil rape seed, sugar cane, potatoes, and in forestry.

4. The usage of cyanazine

Cyanazine ranked as the 5th most used herbicide in U.S. agriculture in 1990-93, with an estimated 32 million pounds of active ingredient (AI) used per year. Cyanazine was third in herbicide usage in New York State (NYS), with 650 thousand pounds of AI used annually during the same time period.

5. The current regularity status of cyanazine

Cyanazine, along with the s-triazine herbicides atrazine and simazine, was placed under Special Review by the U.S. Environmental Protection Agency (EPA) in 1994. Cyanazine was placed under Special Review because of concerns raised about its cancer-causing potential in experimental animals and possible risks to humans exposed to this herbicide. On August 2, 1995, Du Pont Chemical Co., then the primary manufacturer and registrant, voluntarily proposed to phase out its production of cyanazine and to stop production for use in the U.S. by December 31, 1999. Sale and use of existing stocks of cyanazine will be prohibited after September 30, 2002. The EPA sets the maximum levels of cyanazine allowed in public drinking water supplies. The maximum contaminant level (MCL) for cyanazine has been set at no more than 1 microgram per liter of drinking water (one microgram is one-millionth of a gram). The EPA also sets the limits on the maximum levels of cyanazine residues allowed in food for human consumption, and in animal feed. These maximum levels are called tolerances. The Food and Drug Administration (FDA) and the U.S. Department of Agriculture (USDA) are the federal agencies responsible for monitoring the residues of cyanazine in domestic and imported foods. Foods that exceed the tolerances can be seized or destroyed by local or federal government officials.

6. Who might be exposed to cyanazine?

People possibly exposed to cyanazine include:

- Agricultural workers who have mixed, handled or applied cyanazine, or herbicide mixtures containing cyanazine
- Family members that had lived on farms that have used cyanazine
- People who have been involved in cyanazine manufacture, or in preparing commercial mixtures of herbicides that contain cyanazine
- People who have handled or laundered clothing contaminated with cyanazine
- People who have consumed cyanazine-contaminated water
- People who have consumed foods with residues of cyanazine and its breakdown products

7. How can save herbicides and reduce their adverse effects?

Sprayer controllers have been developed by agricultural equipment vendors to minimize variation of applied rates of chemicals within fields. The control systems that allow these devices to compensate for changes in vehicle speeds now also provide the potential to apply variable rates of herbicides according to preplanned maps. The types of sprayer systems and controllers capable of variable rate control are discussed here, along with their advantages and disadvantages. Communications between task computers used to store maps and these sprayer controllers are also discussed.

8. Variable rate technology equipments for weed control

Perhaps you apply pre-emergence herbicides for which recommended rates are based on soil texture and soil organic matter content. Furthermore you recognize large variability in soil texture and soil organic matter content within your field units. If so, variable rates may improve overall herbicide performance and reduce costs while reducing its adverse impacts on the environment and agricultural products. Perhaps your farming operation has grown to the point that you are no longer completely familiar with all of the fields and local weed pressure areas within them. Perhaps you have other operators for your application equipment who are even less familiar with those fields than you are. Any of these may be reasons to consider the application of chemicals from a map-based or real-time sprayer system.

Most of us have performed a form of variable rate application with a traditional sprayer. By traditional, we are referring to a system in which the chemical is tank-mixed with a carrier (generally water), and the nozzles and pressure regulating valve are calibrated to provide a desired volumetric application of chemical solution at a certain forward speed. Any change in the boom pressure or vehicle travel speed from that of the calibration results in an application rate different from the desired rates. We have all used this to our advantage at times. For example, when observing an area of heavy weed infestation you might manually increase the pressure or reduce speed, thereby applying a higher (and somewhat unknown) rate of herbicide. Some precision application technologies rely on the use of a map of planned application rates, coupled with a global positioning system (GPS) receiver, to determine the appropriate herbicide rate for a given area in the field. Moreover, you can apply sensor based (real-time) approach to reach this ideal.

If you have begun adopting some precision farming technologies, then you might have a yield monitor and a GPS receiver. Since the GPS receiver is necessary for map-based application of agricultural inputs you already may have one of the big items on hand. Two other components are required to conduct VRA of herbicides. First, some form of "Task Computer" will be required to provide a signal indicating the current target rate for the current location. Second, a system for physically changing the application rate to match the current target rate will be required. Let's examine the technologies available for this part of the overall system first.

There are a number of different types of control systems on the market today that are adaptable to precision application. For the purposes of this discussion we will lump them into three categories. The first is total flow-based control of a tank mixture. The second is chemical injection based control, and the third is chemical injection control with carrier control. Incidentally, all of these systems evolved out of the desire to automatically match application rates to variations in ground speed. This eliminates much of the errors in application that could occur if ground speeds change from the calibrated setup. These systems are effective at reducing this error. With the application rate managed by an electronic system, the ability to apply variable rates is a logical next step. This requires that the target application rate, or set point, be changeable according to the rate established for that location.

8.1 Flow-based control system

The flow based control of a tank mixture is the simplest of the three types discussed here. These systems combine a flow meter, a ground speed sensor, and a controllable valve (servo valve or proportional solenoid valve), with an electronic controller to apply the desired rate of the tank mixture. A microprocessor in the console uses information regarding sprayer width and desired liters per hectare to calculate the appropriate flow rate for the current ground speed. The servo valve is then opened or closed until the flow meter measurement matches the calculated flow rate. If a communication link can be established between this controller and a "map system", a VRA can be made. An illustration of the components comprising such system is shown in Figure 1.

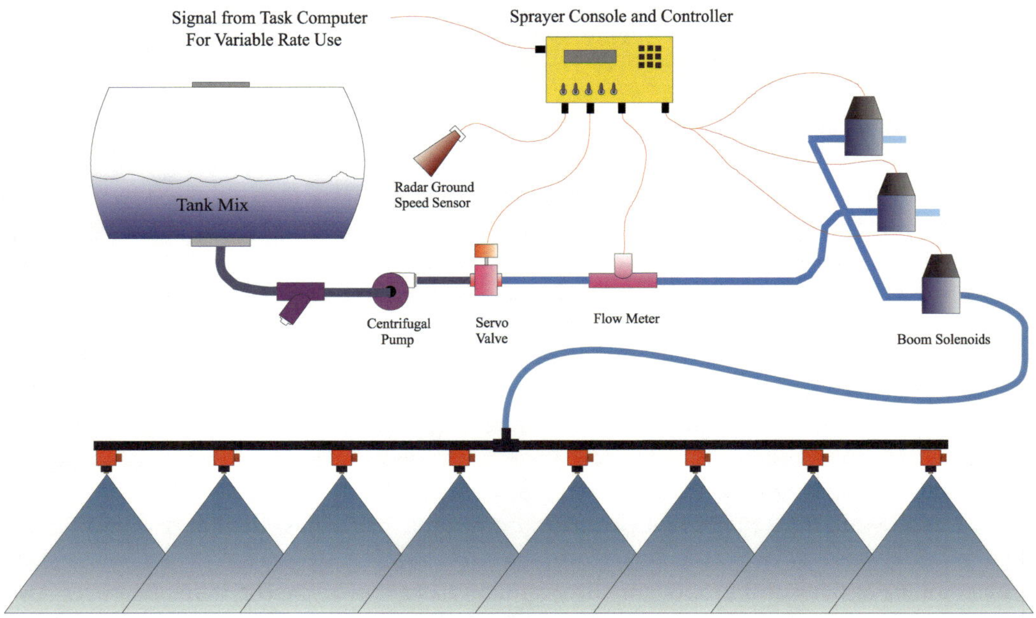

Fig. 1. A flow-based control system (adapted from Humburg [7]).

Common alternatives for varying the total flow are:

- Varying the system pressure through (a) direct pressure regulation, (b) by-pass pressure control, (c) eccentricity of the pump's rotor and (d) pulse width modulated nozzles
- Varying the nozzle diameter

The first approach and its technical solutions are limited by the square root relationship (Equation 1) between pressure P_i and flow through a nozzle orifice Q_d so that doubling the flow rate requires a four-fold increase in pressure. Therefore, the range of operating pressures is relatively narrow. The coefficient k is an experimentally determined coefficient which depends on the type and size of the nozzle and liquid used.

$$Q_d = \frac{\sqrt{P_i}}{k} \tag{1}$$

Another limiting factor is the pressure range over which conventional pressure nozzles will provide a defined spray quality and volume distribution pattern (turn-down ratio). This means that the range of application rates that can be applied with a given size of conventional nozzle by changing the liquid pressure is limited to ± 25% of the nominal output (1.25:1). As the pressure drops below a specified level, the spray pattern becomes distorted and application uniformity is sacrificed. When nozzles are operated above the recommended pressure range, too many small droplets are generated. Because of these two limitations of the application rate range, traditional sprayers are not suitable for site-specific control strategies.

The second approach to controlling the sprayer output with a wider range of dose rates consists of using a twin-fluid nozzle with a dose rate range of 3:1 or a variable flow (swirl-type) nozzle with a range of 4:1. Variable-duration, pulsed spray emission technology was developed for flow rate control with traditional spray nozzles. This is a relatively new variable rate application technology that is referred to as 'pulse width modulation' (PWM). It utilizes an electronically actuated solenoid valve coupled directly to the sprayer nozzle. An advantage of this technology over pressure-based systems is that the usable range of application rates available through one type of nozzle is greatly increased. Utilizing a duty cycle range (pulse width) of 10 to 100 % and the use of PWM nozzles would result in a flow control range of 10:1. To obtain this kind of flow control with a pressure-based system, the system pressure would have to vary 100:1. This is clearly out of the workable range for sprayer nozzles. Not only a wide range of flow control can be obtained using a pulse width modulated sprayer system, it can also be changed relatively quickly. The nozzle valves' capability of changing the flow 10:1 has been given as less than one second.

Another approach to achieving a high turn ratio with common sprayers has been developed recently. These systems involve the use of multiple nozzles in each nozzle location along a boom with the ability to pneumatically switch between output orifices and to adjust nozzle pressures. By using different combinations of orifice sizes and pressures it is possible to achieve a turn ratio of approximately 10:1. In this case the application rate ranges from 50 to 500 L ha^{-1}.

A further approach, i.e. a reflectance-based system uses nozzles fitted with solenoid valves that open briefly to apply spray when the nozzles pass over green vegetation. This system is commercially available along with another system, which is based on the same principle.

The following prototype is an example of a machine-vision system guided sprayer. This system was developed and tested by Tian et al. (1999). To create an intelligent sensing and spraying system, a real-time machine vision sensing system was integrated with an automatic herbicide sprayer (Figure 2). Multiple video images were used to cover the target area. For greater accuracy each individual spray nozzle was controlled separately. Instead of trying to identify each individual plant in the field, weed infestation zones were detected.

A "triple-tank" system for variable application of three different chemical solutions was also developed by the Institute of Agronomy in Bonn in cooperation with the Kverneland Group. This system was set up with three parallel nozzle supply lines; solenoid valves are used to

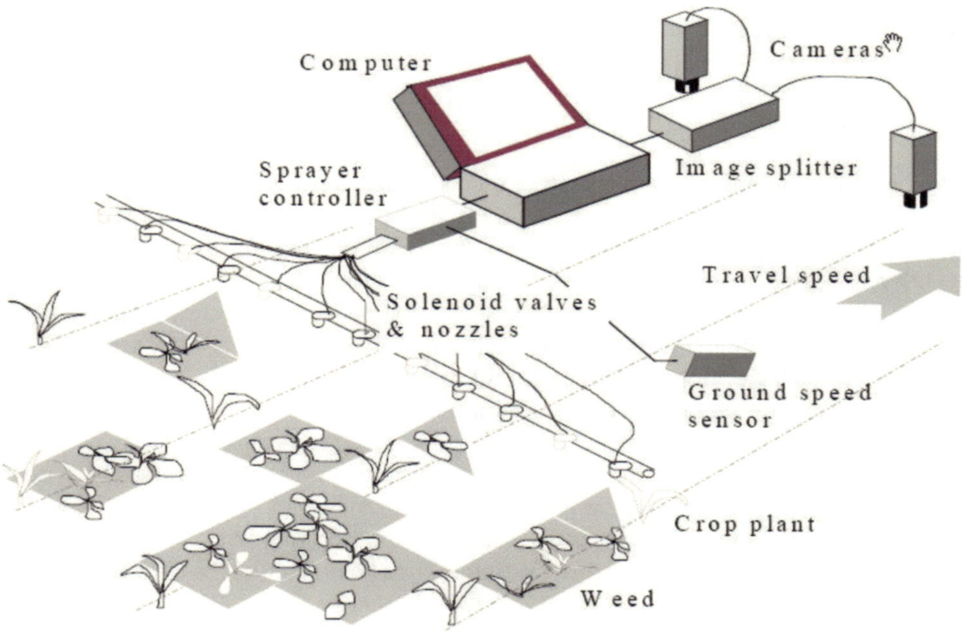

Fig. 2. A machine-vision system guided sprayer (adapted from Tian et al. [24]).

switch between boom sections as instructed by the sprayer's control unit. Each of the lines is connected to a tank with a spray mixture with an appropriate chemical concentration.

Examples of commercial systems with flow-based control capability include Micro-Trak's 9000 series controller, Mid-Tech's 6100 series, Raven Industries SCS 440 or higher, and Dickey John's Land Manager and PCS systems. These systems have the advantage of being reasonably simple. They are also able to make rate changes across the boom as quickly as the control system can respond to a new rate command, which is generally quite fast. As with any technology flow-based controllers also have limitations. The flow sensor and servo valve control the flow of tank mixture by allowing greater or lesser pressure to be delivered to the spray nozzles. This can result in large changes in droplet size in the spray, and potential problems with drift. Some systems will warn you when the commanded flow rate is outside the best operating range for your nozzles. You can adjust the vehicle speed to get the flow rate back into an acceptable range. Also, an operator may have to deal with leftover mixture and is exposed to the chemical during the mixing process. If you want a relatively simple system and can live with these limitations, this one should meet your needs while giving you the capability of VRA of herbicides.

8.2 Direct chemical injection system

An alternative approach to chemical application and control uses direct injection of the chemical into a carrier fluid such as water. These systems utilize the controller and a chemical metering pump to manage the rate of chemical injection rather than the flow rate of a tank mixture. The flow rate of the carrier (water) is usually constant (occasionally variable), and the injection rate is varied to accommodate changes in ground speed or changes in the commanded rate based on maps or sensors. Again, if the controller has been designed, or modified, to accept an external command, the system can be used to do VRA. The components of a system are shown in Figure 3.

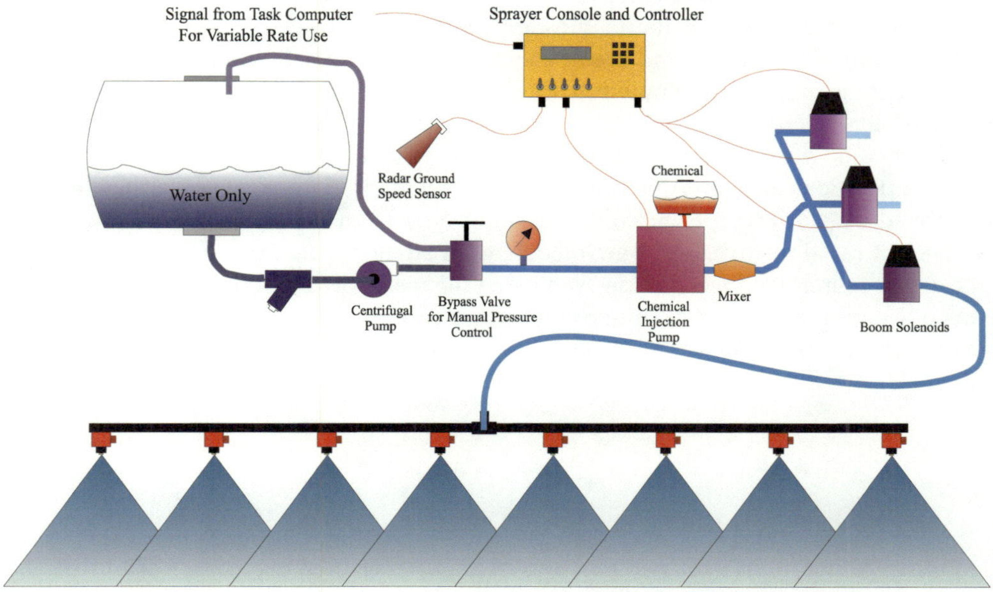

Fig. 3. A direct chemical injection system (adapted from Humburg [7]).

Chemical injection eliminates leftover tank mixture and reduces chemical exposure risk. An additional advantage of this system is that the constant flow of carrier can be adjusted to operate the boom nozzles to provide droplets with a desirable size and distribution. The principle disadvantage for variable rate control is the long transport delay between the chemical injection pump and the discharge nozzles at the ends of the boom. The volume of this plumbing must be applied before the new rate reaches the nozzles. This can cause large delays in the rate change and "Christmas Tree" patterns of application as the new concentration of chemical works its way out through the boom. For example, a simulation of a farmer-owned broadcast sprayer conducted at South Dakota State University indicated that nearly 30.5 m of forward travel would occur before a newly commanded rate would find its way to the end nozzles of that sprayer. These limitations have lead to systems that use both carrier and injection control. Raven Industries, Micro-Trak, and Dickey John all have injection pump systems. All would also recommend that for VRA they be used in conjunction with carrier control as described below.

8.3 Direct chemical injection system with carrier control

Chemical injection with carrier control requires that the control system change both the chemical injection rate and the water carrier rate to respond to speed or application rate changes. One control loop manages the injection pump while a second controller operates a servo valve to provide a matching flow of water. A perfect system of this type would deliver a mixture of constant concentration just as if it were coming from a premixed tank. The system can have many of the advantages of both of the earlier systems. Direct injection of chemical means that there is no leftover mixture to worry about, and the operator is not exposed to chemicals in the process of tank mixing. Changeover from one rate to another occurs as quickly as both chemical and carrier controllers can make the change, which is generally very fast. The components comprising such system are shown in Figure 4. Disadvantages include a more complex system with higher initial cost, and the problem of pushing varying amounts of liquid through the spray nozzles as rates change, with the resulting changes in droplet and spray characteristics. Available systems that fit into this category include, but are not limited to, the Raven SCS 700 series, the Mid-Tech TASC 6300 system, or the Micro-Trak TNi1740. If you do a lot of spraying and wish to avoid the hazards of tank mixing, these systems will give you a great deal of control over your spraying operations and offer the capability of applying variable rates of herbicides from a pre-planned map. A few specific control systems have been mentioned here. However, this is an area of rapid change, and new models with advanced features debut regularly. It is suggested to search the World Wide Web using the manufacturer's name as a keyword as a means of locating product descriptions and specifications. Most systems will fit into one of the categories described here.

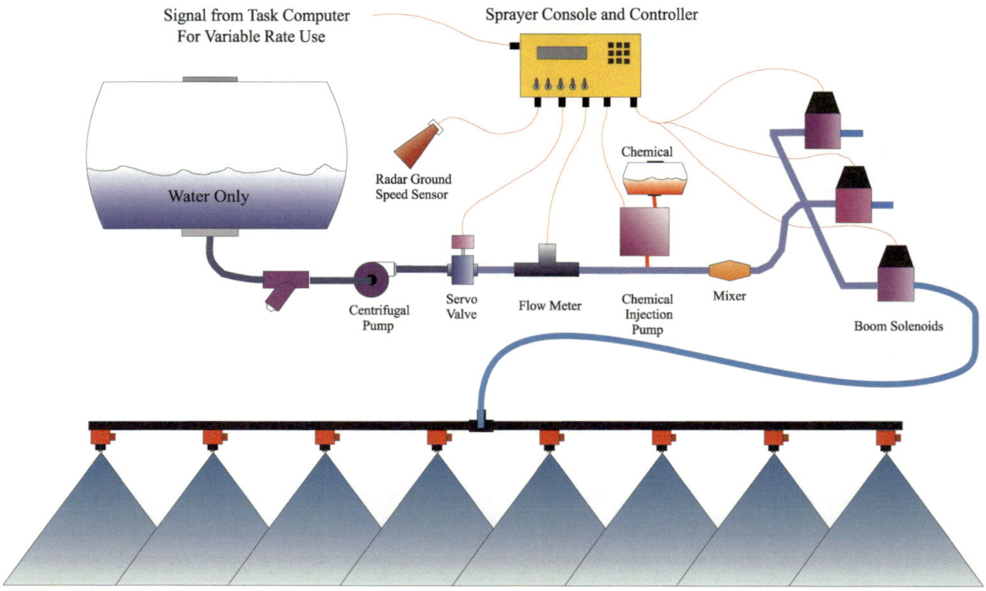

Fig. 4. A direct chemical injection system with carrier control (adapted from Humburg [7]).

There is a range of possible solutions for the technical realization of the direct injection method. In considering the suitability of these solutions, it is first necessary to determine the requirements. In an ideal case, sprayers with a direct injection system should cover the whole operating range of common field sprayers currently available on the market. The most important factors and requirements can be divided into two groups. The first includes requirements which are relevant to the on-line approach to site specific herbicide application. The second group includes requirements which are related to injection metering systems only. The basic requirements are all listed below.

- Requirements for on-line site-specific application:

1. Application rate of the carrier
2. Application rate of the chemical
3. Minimum total response time of the application system

4. Forward speed
5. Position of weed detection device (sensor)
6. High spatial resolution of sprayer
7. Uniformity of mixture concentration across a working width (lateral distribution)
8. Application of several different herbicide/additive products according to weed population

- Requirements for injection metering system:

1. Fast change of dose rates according to changes in operating parameters - minimum response time of injection system
2. Accurate metering of herbicides across the range of dose rates found in practice (flow rate of carrier/chemical)
3. Optimal number and position of injection points
4. Dimensioning of the injection system in accordance with the required nozzle/system pressure
5. Ability to deliver and inject a wide range of herbicides with varying physical properties
6. Good miscibility and solubility of herbicides with carrier (homogeneity of mixture)
7. No or, if applicable, few herbicide / spray residues
8. Easy rinsing of chemical supply lines
9. Easy and safe handling of concentrate tanks
10. Capability of being fitted to most existing sprayers
11. Robust construction of the system and use of durable materials

8.4 Putting it all together

The discussion so far has centered on how the different controller and plumbing systems achieve a given rate of application. The other part of implementing variable rate, or site-specific, weed control concerns how we store and communicate commanded rates to these sprayer systems. In simple terms, this requires a "task computer" and a communications link. The task computer holds the map of rates that you have planned. This map would most likely have been developed on your desktop computer with a mapping program. That program must save the application map in a form understandable to your task computer. Note that the task computer could actually be a conventional notebook computer running the desktop software, but the industry is moving towards more rugged devices with fewer moving parts. Examples of these include Raven's AMS198 and John Deere's Green Star system. The Ag Leader PF3000 system combines this task computer concept into the yield monitor console so that the unit can serve both purposes. Other systems are undoubtedly available, these are only examples. The map is typically loaded into the task computer on a PCMCIA card that uses no moving parts. Current practice also includes connecting the GPS receiver to the computer. The software running on the task computer then determines the current rate command based on the coordinates it receives from the GPS receiver and sends the rate for that management zone to the sprayer controller.

How the chemical rate information is passed from the task computer to the sprayer is another issue. Current practice in most cases is to use the RS 232 Serial Interface to connect the task computer to the sprayer controller. This standard interface is able to send strings of characters and numbers from a task computer that the receiving device can use if they are in exactly the right format. A properly formatted message might begin, for example, with a specific character to signify a chemical rate and be followed by a specific number of digits that represent the actual rate to the controller. These messages are currently specific to each controller manufacturer. Raven, Micro-Trak, and Dickey John allow direct connection of an RS 232 cable for this purpose. Mid-Tech uses a Data Link communications managing module between the task computer and their sprayer controller. In each case it is necessary for your task computer software to be fully aware of the format of the rate message required by the device with which it is communicating. Companies generally make this format available to anyone who needs it, including mapping software developers. If your mapping program has "drivers" for your brand of sprayer system, communication between the software and sprayer should not be a problem. "Drivers" are small computer files or programs that tell your software the specific ways to deliver information to another specific device. If drivers are not available, it will require more work and some understanding of your software and serial communications to make the two devices function together. This communications link is usually used in both directions as the sprayer controller sends the current measured application rate back to the task computer which records this information as a part of a map record.

Whatever your level of technology usage today, it is valuable to stay informed with regard to the changes occurring in production agriculture. Not all new technologies offer clear and large economic benefits to all producers. However, being familiar with the technology will allow you to decide which pieces of the precision puzzle may be used to help you survive and thrive in a competitive world.

9. Generating a digital management map using GPS

9.1 Selected field

A field about 6500 m^2 at the Research Site of Qazvin Province Agricultural and Natural Resources Research Center in south-west of Qazvin province is selected to generate a digital management map.

9.2 Surveying

For surveying, four benchmarks are delineated on the selected field using the 30 × 20 × 20 cm concrete blocks. The settlement location of these blocks is arbitrary so that these blocks are used later as locations for settlement of total station surveying equipments and four static GPS receivers. Using total station surveying equipments, local coordinates of four benchmarks are determined so that coordinate (1000, 1000, 100) is allocated to B_4 benchmark and then relative coordinates of three other benchmarks, i.e. B_1, B_2 and B_3 are determined concern to benchmark B_4 (Figure 5).

The local coordinates and the contours of the selected field are obtained by settling total station surveying equipments on benchmark B_4 and settling reflector on various locations of the selected field. Then, the preliminary local map of the selected field is generated using the LAND software. By means of the LAND software a 42-cell grid is also created and laid out on the selected field (Figure 5). Each cell of the grid is 148 m^2. As the coordinates obtained by the total station surveying equipments are local and can not be used in the precision faming, these coordinates are converted to Universal Transverse Mercator (UTM) coordinates. For this purpose four static GPS receivers with 5 mm accuracy are used for positioning of the four benchmarks. A static GPS receiver used in this study is shown in Figure 6.

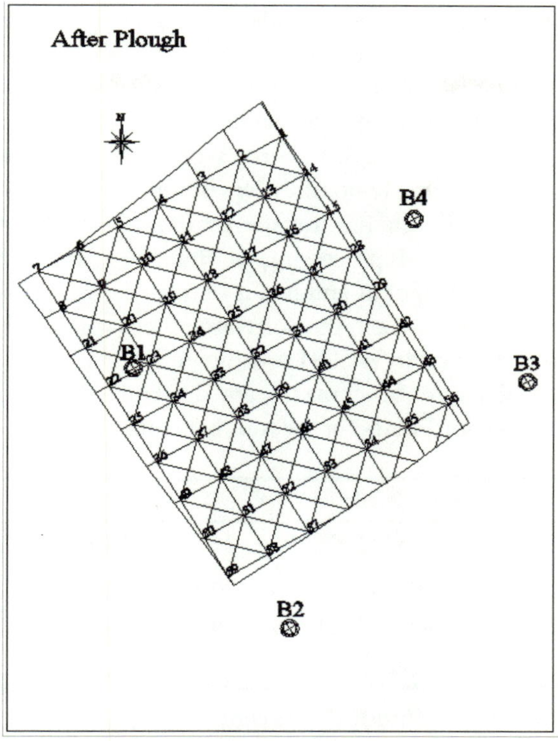

Fig. 5. Field grid and position of the four benchmarks.

Fig. 6. A static GPS receiver.

The static GPS receivers are installed on tripods and their heights are measured manually. Observation of satellites is last almost four hours so that more position data and consequently more accuracy are obtained. The number of data received by the static GPS receivers is 14676, 14934, 15024 and 2991 for the B_1, B_2, B_3 and B_4 benchmarks, respectively. The number of data received by the static GPS receiver installed on the B_4 benchmark is less than those of other benchmarks due to possibly less observation of the GPS satellites. For the purpose of processing, the GPS data are transferred to a personal computer using the HC LOADER software. Handling and processing of the GPS data is performed using the COMPASS software. First the height of antenna is defined for the software and this work is performed for the four antennas. Then, the software automatically processes the GPS data and data processing is performed in WGS84 coordinate system. After processing longitude, latitude and altitude of the four benchmarks are determined. Table 1 shows longitude, latitude and altitude (UTM coordinates) of the four benchmarks.

Benchmark	Longitude	Latitude	Altitude (m)
B_1	49:54:37.28 E	36:05:00.39 N	1292.329
B_2	49:54:39.26 E	36:14:57.56 N	1288.264
B_3	49:54:42.16 E	36:15:00.22 N	1287.967
B_4	49:54:40.79 E	36:16:02.14 N	1293.663

Table 1. Longitude, latitude and altitude (UTM coordinates) of the four benchmarks

The LAND software is used again to convert local coordinates to UTM coordinates. In this stage UTM coordinates of all grid points are obtained by defining UTM coordinates of the four benchmarks in the LAND software and obtaining vector of position transfer. The three-dimensional contour map generated for true perception of the selected field is shown in Figure 7.

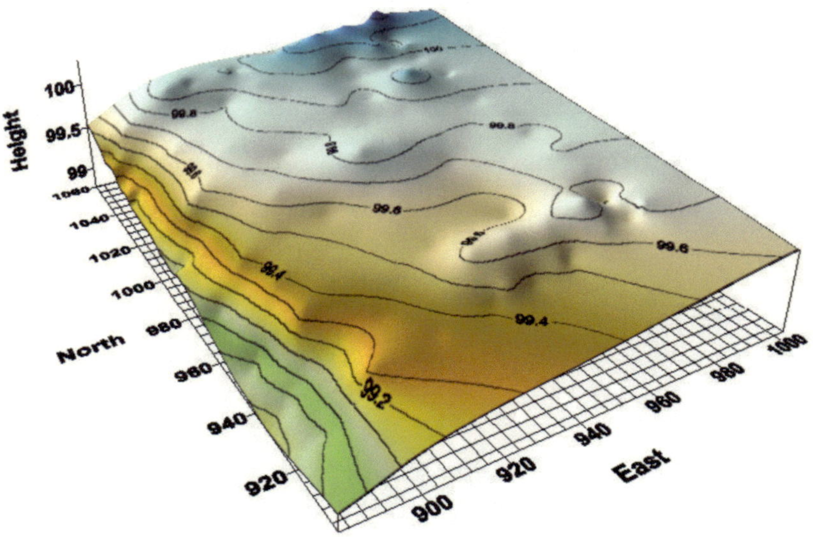

Fig. 7. Three-dimensional contour map of the selected field.

9.3 Soil sampling

In order to generate digital management map for VRA of cyanazine pre-emergence herbicide, soil texture and soil organic matter content are determined in the center of all cells of the grid which is laid out on the selected field. All soil samples are collected by bulking augured core (internal diameter 7.5 cm) from the 0-30 cm soil layer. Soil depth of 30 cm is the average depth for expansion of roots, i.e. active crop root zone. After collection, soil samples are placed in airtight polyethylene bags and transported back to the Soil and Water Laboratory. Finally, texture and organic matter content of all soil samples are determined as described by Soil Survey Manual. The laboratory test results indicate that the minimum, maximum and range of organic matter content of the soil samples are 0.43%, 1.25% and 0.82% (by weight), respectively. In addition, the mean and standard deviation of organic matter content of soil samples are 0.86% and 0.18%, respectively. Also, texture of soil samples vary between loam, sandy loam and loamy sand.

9.4 Conformity of UTM position layer with herbicide application rate layer

After obtaining test results of soil samples, soil texture and soil organic matter content in the center of each cell of the grid are assigned to UTM position of the center of each cell. In order to extend soil texture and soil organic matter content of center of each cell to other grid points, five interpolations methods, i.e. Inverse Distance to a Power, Kriging, Minimum Curvature, Moving Average and Radial Basis Function can be used. By using Cross Validation method for evaluation of interpolators, it is demonstrated that Minimum Curvature method is the best interpolation method for estimating grid points where sampling has not been done.

9.5 Digital management map

Digital management map for VRA of cyanazine pre-emergence herbicide can be generated based on the manufacture recommendations for application rate for different soil textures and soil organic matter contents (Table 2). It can be seen from Table 2 that application rate increases with increasing soil organic matter content and as soil texture varies from sand and sandy loam to clay loam and clay.

Considering manufacture recommendations (Table 2) and soil test results which indicate soil organic matter content ranges from 0.43% to 1.25%, and soil texture varies between loam, sandy loam and loamy sand, four management zones with four different herbicide application rates as 1.4, 1.7, 2.9 and 3.5 L ha^{-1} are determined, and eventually digital management map for VRA of cyanazine pre-emergence herbicide is generated as two-dimensional and three-dimensional maps indicating four distinct zones corresponding to the different soil conditions, and consequently different herbicide application rates (Figure 8).

10. Implementation of the digital management map

For implementation of the generated digital management map, a direct chemical injection system was designed and constructed. This system was based on GPS data for positioning of the sprayer, comparing the GPS data with digital management map data, measuring of velocity and finally injection of active ingredient inside carrier fluid using solenoid injectors proportionate to any management zone on the digital management map. The most important factor for evaluation of the developed direct chemical injection system was response (delay) time. This time was defined the period from the instant the injection begins until the chemical concentration reaches 95 % of the equilibrium rate. The results showed that response time depends significantly on carrier fluid pressure and injection position of active ingredient inside carrier fluid. A schematic illustration of the developed system is shown in Figure 9.

Soil texture	**Soil organic matter content (%)**					
	< 1.0	**1.0**	**2.0**	**3.0**	**4.0**	**≥ 5.0**
Sand	0.60	0.75	1.25	1.50	1.75	2.00
Sandy Loam	0.75	1.25	1.50	1.75	2.00	2.25
Loam, Silty Loam, Silt	1.25	1.50	1.75	2.00	2.25	2.50
Sandy Clay Loam, Clay Loam, Silty Clay Loam	1.50	1.75	2.00	2.25	2.50	2.75
Sandy Clay, Silty Clay, Clay	1.75	2.00	2.25	2.50	2.75	3.00
Peat or muck	Not recommended					

Table 2. Recommended application rate (L ha^{-1}) of cyanazine pre-emergence herbicide based on soil texture and soil organic matter content range

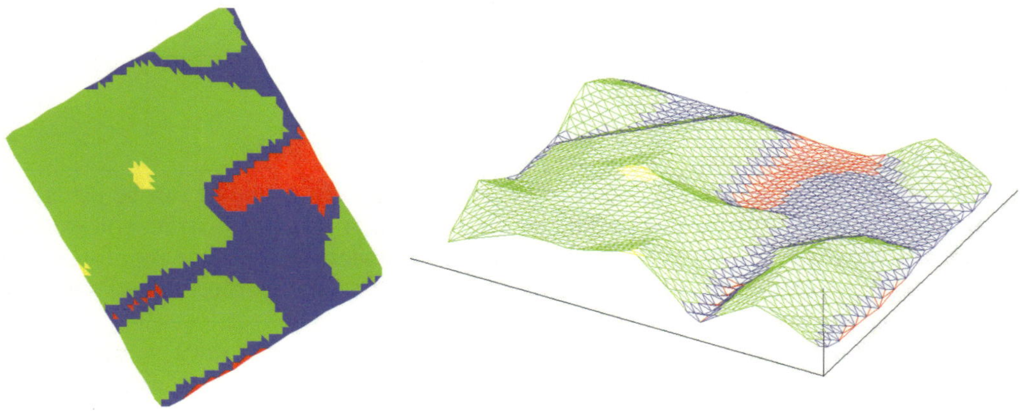

(I) (II)

Color	**Soil organic matter content range (%)**	**Area (m^2)**	**Area ratio (%)**	**Herbicide application rate (L ha^{-1})**
■	1.25-1.55	408	6.40	1.4
■	1.56-1.85	1616	25.1	1.7
■	1.86-3.35	4374	67.9	2.9
■	3.36-3.65	40	0.60	3.5

Fig. 8. Two-dimensional (I) and three-dimensional (II) digital management map of the selected field for VRA of cyanazine pre-emergence herbicide

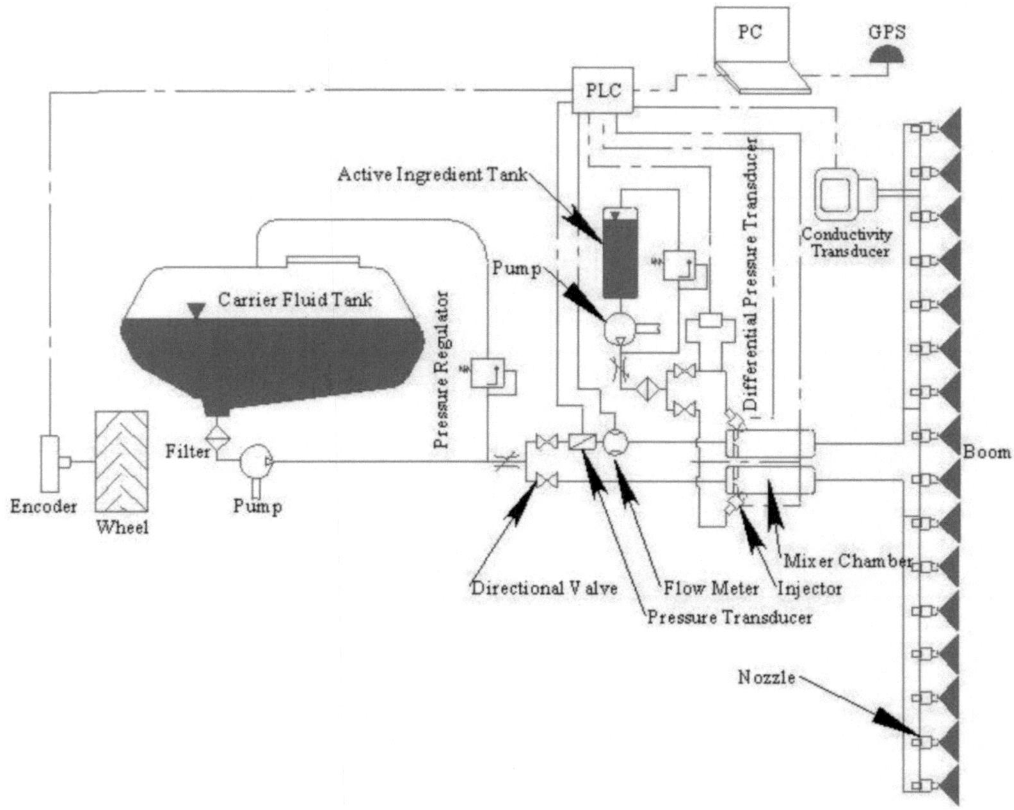

Fig. 9. Schematic illustration of the developed direct chemical injection system

11. Comparison between VRA and uniform rate application

As shown in two-dimensional and three-dimensional digital management maps (Figure 8) 6.4, 25.1, 67.9 and 0.6% of the selected field need application rates as 1.4, 1.7, 2.9 and 3.5 L ha^{-1}, respectively. Based on the generated digital management map for VRA, total required herbicide for the entire selected field is determined to be 1.6 L. If herbicide application is based on the digital management map and VRA instead of 2.9 L ha^{-1} which is the herbicide application rate for 67.9% of the selected field, herbicide application can be decreased up to 13%. Also, herbicide application can be done economically, and suppressing of weed growth in all management zones will be successful and without further adverse effects on the environment and agricultural crops.

If herbicide application rate of 1.4 and 1.7 L ha^{-1} is considered as herbicide application rate of the entire selected field, herbicide application can be decreased as 44.2% and 32.2%, respectively (Table 3). However, suppressing of weed growth in some management zones may be unsuccessful. Conversely, if herbicide application rate of 2.9 and 3.5 L ha^{-1} is considered as herbicide application rate of the entire selected field, herbicide application can be increased as 15.7% and 39.6%, respectively (Table 3). In this situation suppressing of weed growth in all management zones can be successful, but additional herbicide application will have adverse effects on the environment and agricultural crops.

At present, many farmers apply more herbicide than the manufacture recommendations for herbicide application rate in order to reach secure results for suppressing of weed growth. But using digital management map for VRA, herbicide application can be done economically, and suppressing of weed growth will be successful and without further adverse effects on the environment and agricultural crops.

Management zone No.	Area (ha)	Area ratio (%)	Herbicide application rate (L ha^{-1})	Needful herbicide (L)	A	B	C
1	0.0408	6.40	1.4	0.057	0.901	0.713	44.2 Decrease
2	0.1616	25.1	1.7	0.275	1.095	0.520	32.2 Decrease
3	0.4374	67.9	2.9	1.269	1.867	-0.253	15.7 Increase
4	0.0040	0.60	3.5	0.014	2.253	-0.639	39.6 Increase

A: Required herbicide for the entire selected field based on uniform rate application of each management zone (L)
B: Difference between column A and required herbicide for the entire selected field based on variable rate application i.e. 1.6 L (L)
C: Increase or decrease of required herbicide for the entire selected field based on column B and 1.6 L (%)

Table 3. Comparison between VRA and uniform rate application

12. References

[1] Abbaspour-Gilandeh, Y., R. Alimardani, A. Khalilian, A. Keyhani and S.H. Sadati, 2006. Energy requirement of site-specific and conventional tillage as affected by tractor speed and soil parameters. *Int. J. Agric. Biol.*, 8: 499-503.

[2] Al-Gaadi, K.A. and P.D. Ayers, 1999. Integrating GIS and GPS into a spatially variable rate herbicide application system. *Trans. ASAE*, 15: 255-262.

[3] Bauer, W.D. and M. Schefcik, 1994. Using differential GPS to improve crop yields. *GPS World* (February): 38-41.

[4] Blumhorst, M.R., J.B. Weber and L.R. Swain, 1990. Efficacy of selected herbicides as influenced by soil properties. *Weed Technol.*, 4: 279-283.

[5] Carrara, M., A. Comparetti, P. Febo, S. Orlando, 2004. Spatially variable rate herbicide application on durum wheat in Sicily. *Biosystems Engineering*, 87: 387-392.

[6] Corwin, D.L. and R.E. Plant, 2005. Applications of apparent soil electrical conductivity in precision agriculture. *Computers and Electronics in Agriculture*, 46: 1-10.

[7] Humburg, D. 2003. Site-specific management guidelines: variable rate equipment-technology for weed control (SSMG-7), Department of Agricultural and Biosystems Engineering, South Dakota State University, Brooking, SD 57007.

[8] Humburg, D. 2003. Site-specific management guidelines: Standardization and Precision Agriculture--'The Promised Land' (SSMG-8), Department of Agricultural and Biosystems Engineering, South Dakota State University, Brooking, SD 57007.

[9] Fisher, K.B., G.J. Shropshire, C.L. Peterson and E.A. Dowding, 1993. A spatially variable liquid fertilizer applicator for wheat. *ASAE Paper* No. 93-1074. St. Joseph, Mich.: ASAE.

[10] Jafari, A., S.S. Mohtasebi, H.E. Jahromi and M. Omid, 2006. Weed detection in sugar beet fields using machine vision. *International Journal of Agriculture & Biology*, 8: 602-605.

[11] Loghavi, M. 2004. *The Precision Farming Guide for Agriculturists*, Agricultural Extension, Education and Research Organization, Iran.

[12] Mohammadzamani, D. and M. Rashidi, 2009. Generating a digital management map for variable rate herbicide application using the global positioning system. *Am.-Eurasian J. Sustain. Agric.*, 3(1): 101-106.

[13] Mohammadzamani, D., M. Rashidi and B.G. Khabbaz, 2009. Generating digital management map for herbicide application in VRA spraying by using GPS. In: Proceedings of Biennial Conference of the Australian Society for Engineering in Agriculture (SEAg), 13-16 September 2009, Brisbane, QLD, Australia.

[14] Mohammadzamani, D., S. Minaei, R. Alimardani, M. Almassi, M. Rashidi and H. Norouzpour, 2009. Variable rate herbicide application using the global positioning system for generating a digital management map. *Int. J. Agric. Biol.*, 11: 178-182.

[15] Nishiwaki, K., K. Amaha and R. Otani, 2004. Development of positioning system for precision sprayer. In: Proceedings of Automation Technology for Off-Road Equipment Conference, 7-8 October 2004, Kyoto, Japan. *ASAE Publication* No.: 701P1004.

[16] Ollila, D.G., J.A. Schumacher and D.P. Frochlich, 1990. Integrating field grid sense system with direct injection technology. *ASAE Paper* No. 90-1628. St. Joseph, Mich.: ASAE.

[17] Qiu, W., G.A. Watkins, C.J. Sobolik and S.A. Shearer, 1994a. A feasibility study of direct injection variable-rate herbicide application. *Trans. ASAE*. 41: 291-299.

[18] Qiu, W., S.A. Shearer and G.A. Watkins, 1994b. Modeling of variable-rate herbicide application using GIS. *ASAE Paper* No. 94-3522. St. Joseph, Mich.: ASAE.

[19] Rastgar, M. 2000. *Herbicides and Control Methods*, 2nd Edition, Academic Publishing Center, Tehran, Iran.

[20] Sirjacobs, D., B. Hanquet, F. Lebeau and M.F. Destain, 2002. On-line soil mechanical resistance mapping and correlation with soil physical properties for precision agriculture. *Soil & Tillage Res.*, 64: 231-242.

[21] Snedeker, S.M., 2004. Breast cancer and environmental risk factors; Herbicides and breast cancer risk, an evaluation of cyanazine. College of Veterinary Medicine, Cornell University, Box 31, Ithaca, NY 14853-5601.

[22] Soil Survey Staff. 1996. Soil survey laboratory methods manual. Soil Survey Investigations Rep. 42. Version 3.0. U.S. Gov. Print. Washington, DC.

[23] Sudduth, K.A., N.R. Kitchen, W.J. Wiebold, W.D. Batchelor, G.A. Bollero, D.G. Bullock, D.E. Clay, H.L. Palm, F.J. Pierce, R.T. Schuler and K.D. Thelen, 2005. Relating apparent electrical conductivity to soil properties across the north-central USA. Computers and Electronics in Agriculture, 46: 263-283.

[24] Tian, L., Reid, J. F. and Hummel, J. 1999. Development of a precision sprayer for site-specific weed management. *Trans. ASAE* 42: 893-900.
[25] Vondricka, J. 2007. Study on the Response Time of Direct Injection Systems for Variable Rate Application of Herbicides. Ph.D. Thesis. University of Bonn.
[26] Weber, J.B., M.R. Tucker and R.A. Isaac, 1987. Making herbicide rate recommendations based on soil tests. *Weed Technol.*, 1: 41-45.
[27] West, C.P., A.P. Mallarino, W.F. Weider and D.B. Marx, 1989. Spatial variability of soil chemical properties in grazed pastures. *Soil Sci. Soc. America*, 53: 784-789.

8

Soil Electrical Conductivity as One Possible Tool for Predicting of Cirsium Arvense Infestation Occurence

Milan Kroulik[1], Atonin Slejska[2],
Dana Kokoskova[2] and Veronika Venclova[1]
[1]Czech University of Life Sciences, Kamycka 129, Prague 6 – Suchdol, 165 21
[2]Crop Research Institute in Prague, Drnovska 507, Praha 6 – Ruzyne, 161 06
Czech Republic

1. Introduction

From a practical point of view it is necessary to use an exact low cost and time consuming approximation to obtain information about an actual weed infestation. In this study the intensity of *Cirsium arvense* (L.) SCOP infestation was monitored in a 12 ha experimental field, where malting barley (2002) and winter wheat (2003) were grown.

The sampling points for *C. arvense* infestation were established in a square raster with one 18 m long side of one raster unit.. *Cirsium arvense* occurrence was manually counted. During the data collection at the sampling points, the number of *C. arvense* plants which were situated outside of sampling points was counted as well. Each *C. arvense* patch was localized by GPS and saved as digital coordinates as well. On the basis of the field survey and the *C. arvense* infestation monitoring, two data sets were collected. The first data set contains the *C. arvense* densities in the raster, and the second data set offers information about *C. arvense* occurrence in patches. The soil variability was described by means of soil electrical conductivity (EC_a) measurement in the year 2003. The number of *C. arvense* plants from the raster and values of EC_a were included into the evaluation.

Acquired data were evaluated in GIS using a geostatistical procedure. Correlation analysis brought the following results: relatively high correlation R = 0.64 in the year 2002 and R = 0.91 in the year 2003 were found between EC_a and *C. arvense* infestation. The results indicated a statistically significant correlation at a 99% confidence level, it means a close dependence between *C. arvense* infestation and EC_a was observed. This fact proved our presumption about *C. arvense* response to soil properties. According to our results it can be stated that higher values of EC_a are be observed at places where higher density of *C. arvense* is present.

The aim of this research is: (1) the evaluation of the *C. arvense* infestation and spatial distribution, (2) the herbicide effect, (3) the field heterogeneity by means of the soil electrical conductivity (EC_a) measurement and (4) the comparison of relations between EC_a and *C. arvense* infestation.

Several studies proved that many weeds, including grass weeds, are spread non-uniformly within a field. Furthermore, the weed patches are relatively stable within a season and among seasons (Hamouz et al. 2002, 2004; Godwin & Miller 2003; Krohmann et al. 2002; Werner & Garbe 1998). On the other hand, Soukup et al. (2003) noticed biological differences of crops in crop rotation and specific characteristics of general year-to-year crop rotation which may have an important influence on annual changes in weed infestation. Moreover, weed distribution proved to be heterogeneous during several years and with different crops (Gerhards et al. 2000). Targeted protection could be done only on the basis of every year and more than once repeated diagnostics. Further, weed seeds in soil seed bank are dispersed during soil cultivation and harvest (Godwin & Miller 2003). Weed patches are often extended in the direction of machinery movement (Hamouz et al. 2004). These facts have to be taken into account for the site-specific weed management.

A considerable sum of money was spent on monitoring weed infestation. In general, the number of weeds is hold down on a sustainable threshold value of plants per square meter. The economical threshold varies within fields as well as within different weed species (Scotford & Miller 2005). For example, it is reported (SAC 2001) that low populations (10 plants m^{-2}) of volunteer barley in oilseed rape can reduce yield by 5 %, whereas populations of broad leaved weeds (excluding cleavers) can be up to 200 plants m^{-2} without any significant effect on crop yield. In accordance with the level of malignancy it is generally possible to find areas in a field, where the chemical weed management could be omitted or reduced (Hamouz et al. 2000). As a matter of routine the same dose of herbicide is applied to hold the weeds under the economic thresholds, despite the fact that the weeds infestation varies (Hamouz et al. 2000). Of course, there is a chance for herbicide savings by using site specific herbicide application. The results of Gerhards et al. (1997) demonstrated that the site specific weed management was technically feasible but further investigations are needed to verify and evaluate site specific weed control methods.

The site-specific herbicide application presumes, that the herbicide spray can be omitted at certain areas within the field with no or low weed infestation. The area with higher infestation necessary to be treated should be sprayed with the dose adjusted precisely according to the weed infestation level (Sökefeld et al. 2000; Gerhards & Oebel 2006). Soukup (2000) noticed possible savings of herbicides in the range of 30 to 50 %, which would have significant economical and ecological benefits. But there are some difficulties linked with this idea. First of all, the weed detection in the required time interval for the treatment is very difficult as well as the setting of a precise dose for the optimal treatment. Furthermore, it is inexpedient to carry herbicides which are not needed in the field. The type of weed detection and the proper detection time is the crucial factor for the whole weed detection system (on-line or off-line), especially for the acceptable delay time of the direct injection system (Sökefeld et al. 2004).

As far as the information of weed infestation is concerned, from a practical point of view, it is important to use an exact real approximation with low time and low cost consumption. One way of mapping is manual weed detection connected with GPS. However, walking over the whole field on foot is very time consuming and expensive and not possible for larger fields. According to Soukup (2000), manual classification of weed infestation with the raster 50 x 36 m in a big field takes time from 0.5 to 2.5 hours per one hectare. Manual classification is impractical in this regard. Utilization of tractors, harvesters or off-road cars is preferable to do good weed mapping.

An alternative method is checking just the areas of interest. These areas could be chosen according to vegetations indexes obtained by remote sensing or by radiometers mounted onto the machines (Godwin & Miller 2003). Hamouz (2008) describes an algorithm for detection of *Cirsium arvense* in cereals using an aircraft with high resolution multispectral camera. He calculated and tested the classification accuracy of various vegetation indices including NDVI. The best correlation coefficient and also the highest classification accuracy was reached using DVI index.

Spectral vegetations indexes, calculated as a ratio of particular wave lengths, were described and used in many studies. These indexes are further compared with other characteristics of investigated environment like coverage or weed infestation (Scotford & Miller, 2005). Remote investigation, alternatively for sampling plots establishment, is also used for example to determine plant nutritious conditions, yield, soil characteristics, organic matter, and weed infestation, (Cox 2002; Zhang 2002; Selige 2003). The advantage of satellite pictures, as opposed to sensors and sampling points, is in providing information about the whole area and detailed overview of spatial variability. Sampling points taking considerably limits quality of spatial variability description, because of its difficulty and high labour consumption (Basso et al. 2003).

Oebel and Gerhards (2005, 2006) tested variable herbicide application in cereals and also in sugar beet, maize and winter rape. They used a manual and automatic real time mapping system, application maps and economic thresholds.

During the automatic weed classification 69 % of weed plants were recognized in sugar beet and 72 % in maize. These systems require costly and sophisticated technical and software equipments.

2. Materials and methods

An intensity of *Cirsium arvense* (L.) SCOP populations was monitored in a 12 ha experimental field at Prague-Ruzyně district, Czech Republic (50°05´N, 14 °18´E), where malting barley (2002) and winter wheat (2003) were grown. The field was tilled by conventional ploughing technology. The soil type at the study site is *Orthic Luvisol,* according to the FAO classification, with different share of skeleton. The soil texture is non-homogenous with different textures of loam and sandy loam. The altitude of the field is about 340 m above sea level, the average rainfall is 450 mm per year and the average temperature 7.8°C. The experimental field slopes from north to south and it merges to a plane in lower part. In the same direction on the left hand side from the axial fall line a shallow ground wave is present.

The sampling points were established in the 18 x 18 m raster (Figure 1). *Cirsium arvense* occurrence was manually counted at 0.25 m^2 squares along the raster in April. The number of plants per 0.25 m^2 was converted to the number of plants per 1 m^2. During the data collection at the sampling points, the number of *C. arvense* plants which were situated outside of sampling points was counted as well. Each *C. arvense* patch was localized by GPS and registered as well.

The soil variability was described by means of soil electrical conductivity measurement (Figure 2).

Soil conductivity was measured by using the contact method. Output signal from the sensor was simultaneously recorded with the GPS position signal. The records were written into the data set in 5 second intervals. Data were processed using geostatistical methods. In order to process data accurately and to eliminate measuring errors, several modifications on the initial EC_a values were performed prior to statistical processing and evaluation. The majority of errors, when measuring EC_a, occurred when the machine started a new line.

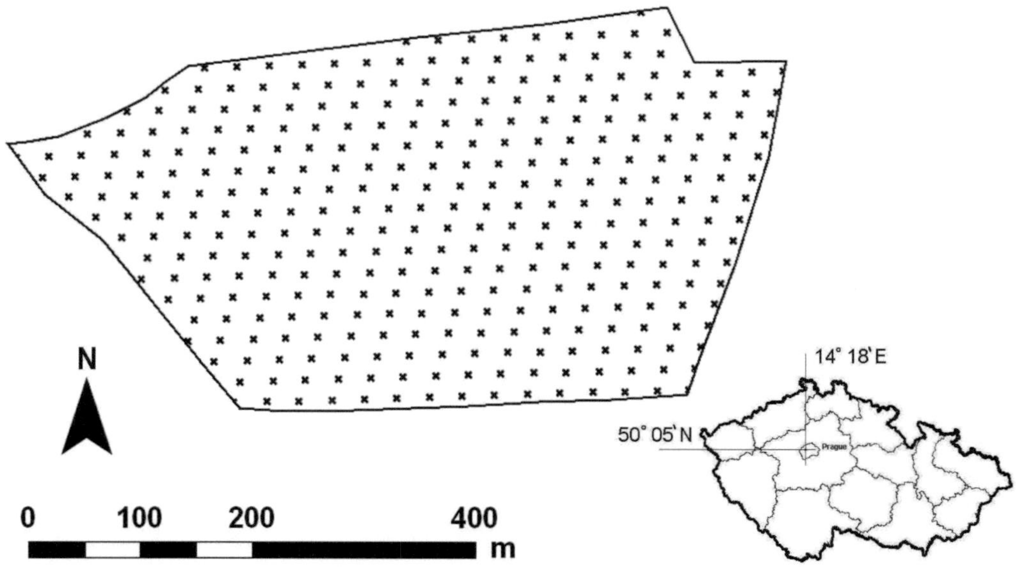

Fig. 1. Map of sampling points in the experimental field.

Fig. 2. Tractor drawn device for soil conductivity measurement

Thus, values that did not describe precisely the factor measured were removed from the initial data set (for example errors occurring during measurement interruptions on headlands and turning points of a vehicle). These values were eliminated by trimming the marginal points recorded. Values larger then the double value of the average were also excluded from the initial data set. The time series were smoothened during the subsequent modification. The values of EC_a usually show oscillations from the curve. A simple running average method was applied to smooth the time series of all measurements. The following formula was used:

$$\hat{Y}_t = \frac{1}{3}(Y_{t-1} + Y_t + Y_{t+1}) \quad (1)$$

where *Y* are original values at time t.
The number of *C. arvense* plants from the raster and values of EC_a were included into the evaluation. Spatial dependence of sampled values was described by variogram parameters. Experimental variograms were calculated and fitted by models. Variogram parameters such as Nugget (C_0), Sill (C_0+C) and Range (A_0) were calculated. The spatial relation itself is expressed as a portion of the nugget (C_0) in the sill value (C_0+C). The infestation map was completed with patches of *C. arvense* consequently without geostatistical analysis. Spatial interpolation of values was carried out by *Ordinary Kriging* interpolation method. Validity of interpolation method was confirmed by Cross-Validation method. Estimated values were collected after this process. The errors between the measured data and the estimates were analysed. The goodness-of-prediction statistic was used as the criterion for checking and comparing the map accuracies (G) (Kravchenko 2003).
The obtained EC_a map and the *C. arvense* infestation map were transferred to the raster after the interpolation process. These two rasters showed a relatively close interval of values (similar values) which is clear from the map as different types of colors. These raster maps were merged into a final file in the next step. EC_a values from the localized patches were recorded as well. All these procedures were made in ArcGIS 9.2. This procedure was necessary to apply because the data of particular measurements were not possible to record at the same measurement point of the field. It means that the comparison of that two measured data sets is not possible to achieve without this procedure. Thus, it was possible to apply statistical evaluation procedure for the data prepared in this way in the next step. The procedure in data evaluation was the calculation of a correlation coefficient which showed Pearson product moment correlation between a pair of values.
According to the field survey a herbicide application map was created in the year 2002. A uniform dose of herbicides was applied in the sprayer mode on - off. The sprayer was operated manually. The real points of beginning and end of spraying were registered during the spray job. The herbicides Dicopur M750 (active ingredient *MCPA*) and Banvel 480S (active ingredient *dicamba*) were applied. Four weeks after herbicide application the observation of herbicide effect was carried out by a field survey. The number of *C. arvese* plants in the raster and on patches was counted again Herbicide treatment was done uniformly throughout the whole field also in the year 2003.
Satellite pictures were taken by QuickBird satellite in August in the year 2002 with a graphics resolution of 2.8 m. The satellite pictures were taken after main crop harvest. A normalized difference vegetation index (NDVI) map was created. The following software was used for data processing: MS Office XP, ArcGIS 9.1 and GS + 5.1.1. The picture was distributed by QuickBird©Digital GlobeTM, Distribution Eurimage/ ARCDATA PRAHA, s.r.o.

3. Results

On the basis of the field survey and the C. *arvense* infestation monitoring, two data sets were collected. The first data set contains the C. *arvense* densities in the raster, and the second data set offers information about the occurrence of C. *arvense* in patches. Descriptive statistics of data shows Table 1 for the year 2002 a Table 2 for the year 2003.

Variable / Property	Raster	Patches centers
Mean value	5.65	20.42
Median	0.00	16.00
Standard deviation	11.80	19.34
Skew	3.53	2.29
Minimum	0.00	4.00
Maximum	84.00	124.00

Table 1. Descriptive statistics of data set (number of Cirsium arvense per m^2) (2002).

Variable / Property	Raster	Patches centers
Mean value	14.86	11.13
Median	8.00	12.00
Standard deviation	20.91	6.49
Skew	1.941	3.38
Minimum	0.00	4.00
Maximum	116.00	80.00

Table 2. Descriptive statistics of data set (number of Cirsium arvense per m^2) (2003).

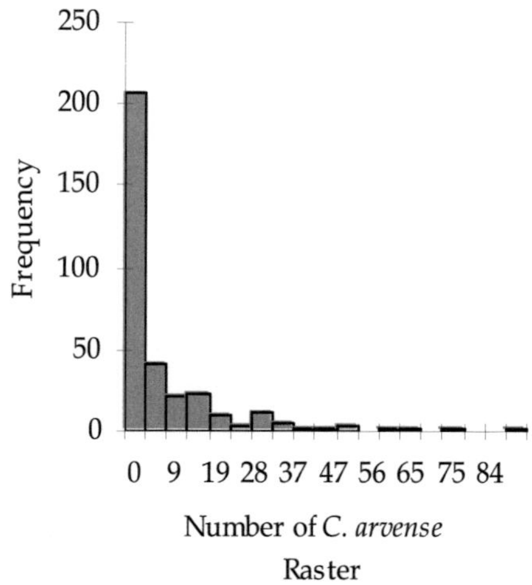

Fig. 3. Histogram of values from sampling raster (number of Cirsium arvense per m^2) (2002).

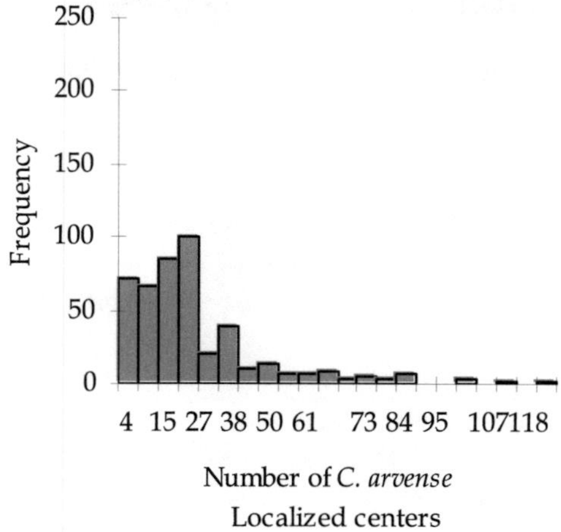

Fig. 4. Histogram of values from random localized patches (number of Cirsium arvense per m2) (2002).

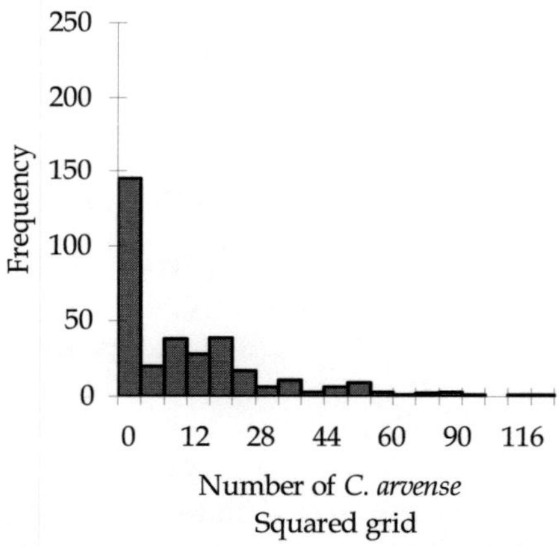

Fig. 5. Histogram of values from sampling raster (number of Cirsium arvense per m2) (2003).

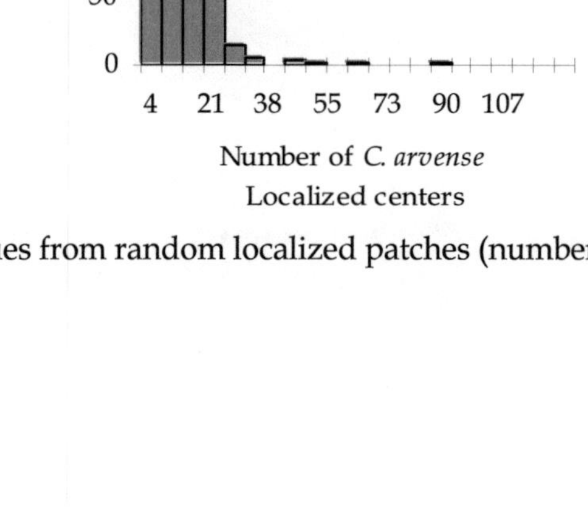

Fig. 6. Histogram of values from random localized patches (number of Cirsium arvense per m2) (2003).

Each data sets display left side asymmetry according to values of the skew. High skew value of basic data set from the raster is caused by a considerable high number of points, where no weed plants were present. Less skew value of second data set shows normal distribution approximation. Figures 3 and 4 show histograms of values from the raster and the random localized patches for the year 2002. Figures 5 and 6 show histograms of values from the raster and the random localized patches for the year 2003.
During the modeling of variogram for number of *C. arvense* plants, it was not possible to define exactly the variogram structure. It is evident that the value of sill is equal to nugget (Table 3).

Variable / Property	Number of *C. arvense* Raster (2002)	Number of *C. arvense* Raster (2003)
Nugget C_0	140	429
Sill C_0+C	140	429
Range A_0 (m)	-	-
R^2	0.38	0.12
RSS	3061	17323
C_0/C_0+C (%)	100.00	100.00
Model	Pure nugget	Pure nugget

Table 3. Parameters of model variogram, (number of Cirsium arvense per m2).

The measurement errors as well as the variability character can influence the values of the nugget (Heisel et al. 1999; Ilsemann et al. 2001; Lopez-Granados et al. 2002). The higher value of ratio calculated from the formula where nugget is divided by sill, the lower spatial dependencies are observed. Pure nugget was observed in this case. According to Lopez-Granados et al. (2002) and Cambardella & Karlen (1999), the ratio C_0/C_0+C higher than 75 % represents spatial independent data. According to the nugget value it was proved that the spatial dependence of *C. arvense* infestation was under the value of the distance between two adjacent sampling points and it could be concluded that the distance between points was too big.

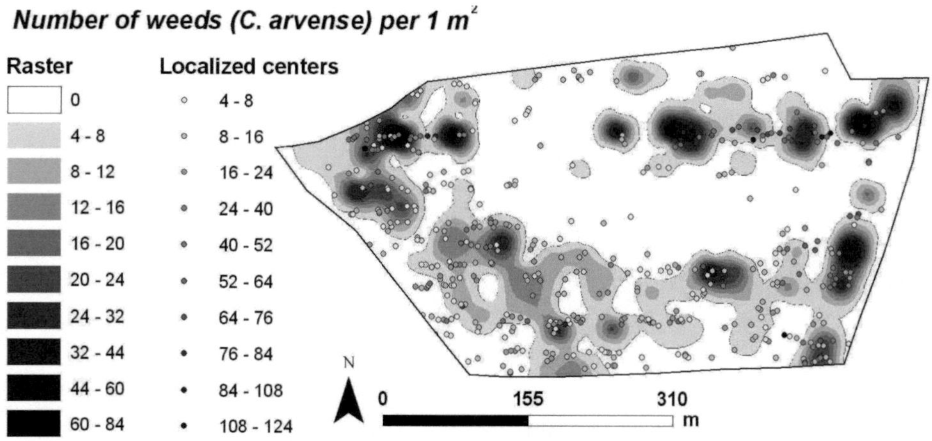

Fig. 7. Cirsium arvense infestation map (2002).

Despite the fact that the variograms structure was not suitable for the interpolation, the Kriging method was used. An exponential models without nugget was used to describe the spatial distribution of *C. arvense*. The variability of the *C. arvense* distribution is evident (Figure 7 and 10). The *C. arvense* infestation showed significant variability and spatial distribution was found at two areas in the field. The infestation is characterized by cumulating the weeds (south, west and northeast). The middle part of the field except of a few patches was not infested with *C. arvense*.

On the basis of the mentioned results, the variable herbicide application against *C. arvense* was carried out. According to the measured data the actual consumption of herbicide was: 1 l ha^{-1} of Dicopur M750, 0.2 l ha^{-1} of Banvel 480S and 210 l ha^{-1} of water. The spray was applied approximately onto 73.8 % of the total field area, which represents 8.86 ha.

Descriptive statistics of the *C. arvense* infestation recorded about four weeks after herbicides application is shown in Table 4. The results of application efficiency are shown in Figure 8. Repeated occurrence of *C. arvense* was observed at the places with the highest *C. arvense* concentration before the herbicides were applied.

Variable / Property	Raster	Patches centers
Mean value	2.21	12.32
Median	0.00	8.00
Standard deviation	4.76	9.61
Skew	3.25	2.73
Minimum	0.00	4.00
Maximum	40.00	60.00

Table 4. Descriptive statistics of data set (four weeks after herbicide application) (number of Cirsium arvense per m^2) (2002).

The NDVI index evaluation was taken as additional information in order to complete and precise the research (Figure 9). In this case lighter colour represents higher NDVI index and vice versa. Results of this map showed an exact demarcation of the areas where herbicides were applied and the area without herbicide treatment. Weeds undergrowth was visible at the places without herbicide treatment.

We can also derive from Figures 7 and 10 that the spatial distribution of *C. arvense* plants was presumably partly dependent on soil properties and generally on site specific conditions.

The descriptive statistics of EC_a data set is shown in Table 5.

Variable / Property	EC_a
Mean value	24.63
Median	24.30
Standard deviation	6.57
Skew	0.27
Minimum	7.49
Maximum	45.18

Table 5. Descriptive statistics of soil electric conductivity [ECa (mS m^{-1})] data set.

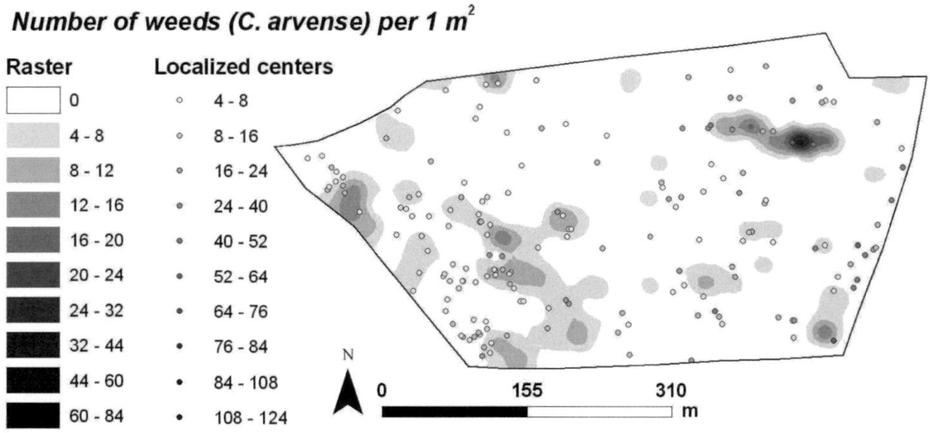

Fig. 8. Application efficiency map about four weeks after herbicide application.

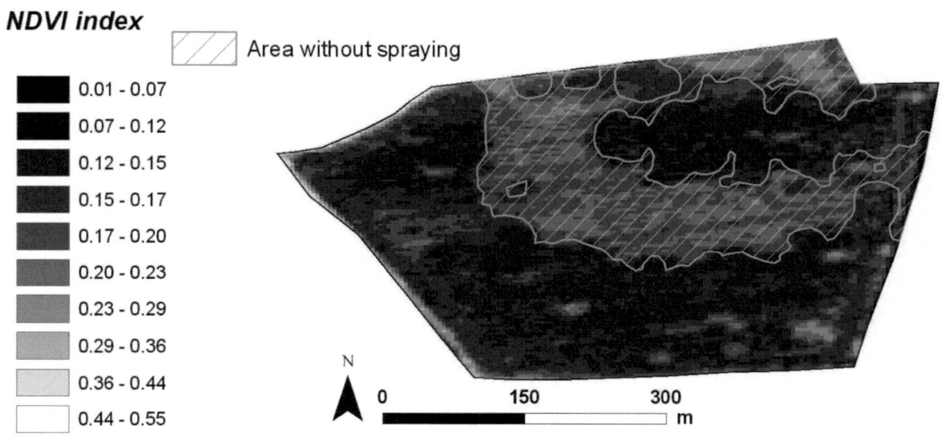

Fig. 9. Normalized difference vegetation index (NDVI) map.

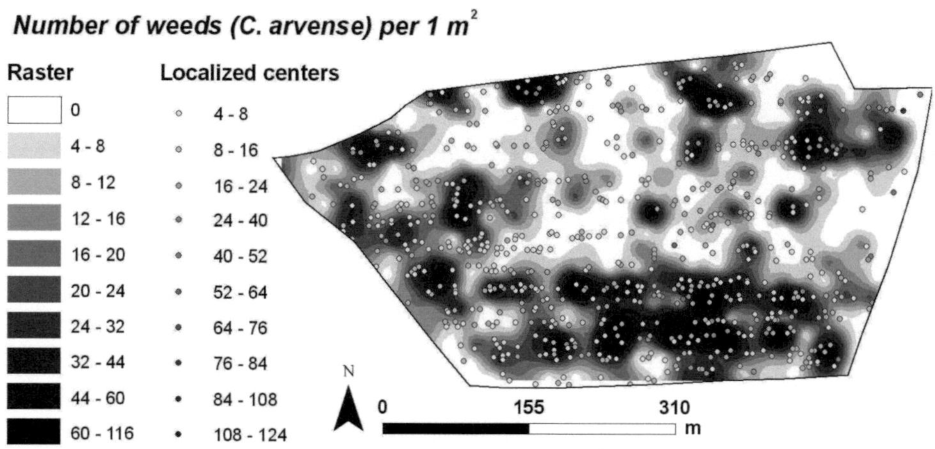

Fig. 10. Cirsium arvense infestation map (2003).

The range of values expressed as the maximum and minimum as well as variation coefficient illustrates the variability of the individual data sets. Asymmetry from the normal distribution is expressed as a coefficient of asymmetry. According to Lopez-Granados (2002), the normality condition is met, if the interval of inclination lies between -2 and 1. Low inclination values prove that data show a normal distribution.

Figure 11 shows the histograms of EC_a values. The picture proves normal distribution of EC_a values.

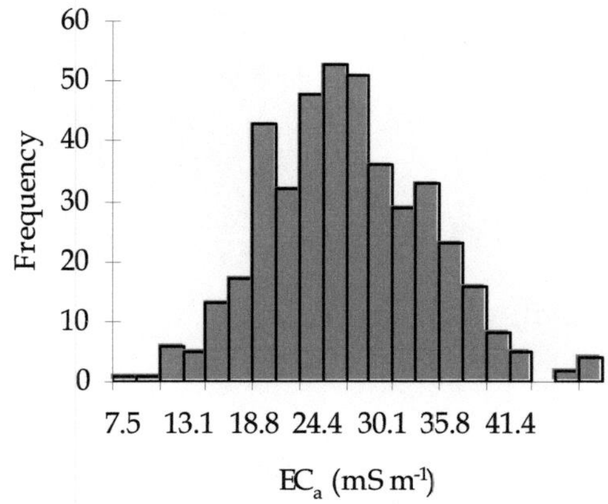

Fig. 11. Histogram of soil electric conductivity [ECa (mS m^{-1})] values.

The exponential model of the variogram of EC_a values with nugget was chosen. Parameters of model variogram were taken off (Table 6).

Variable / Property	EC_a
Nugget C_0	3.91
Sill C_0+C	34.22
Range A_0 (m)	39.50
R^2	0.96
RSS	14.70
C_0/C_0+C (%)	11.42
Model	Exponential

Table 6. Parameters of model variogram, [soil electric conductivity ECa (mS m^{-1}) values].

Spatial relation of 11.42 % according to ratio C_0/C_0+C was observed in this case. According to variogram parameters, a spatial interpolation of the EC_a was done. The EC_a map is shown in Figure 12. *G*-value of 82.9 % was observed. It is evident from Figure 12 that the variability of tested data set is relatively high. Higher values of measured factors were indicated in the south and south-eastern part of the tested field. In relation to slope of the field, the colour scheme of the picture matches the segmentation of the terrain.

Correlation analysis brought the following results: relatively high correlation R = 0.64 in the year 2002 and R = 0.91 in the year 2003 were found between EC_a and *C. arvense* infestation level. The results indicated a statistically significant correlation at a 99% confidence level. According to the correlation coefficient value a close dependence between *C. arvense* infestation and EC_a was observed.

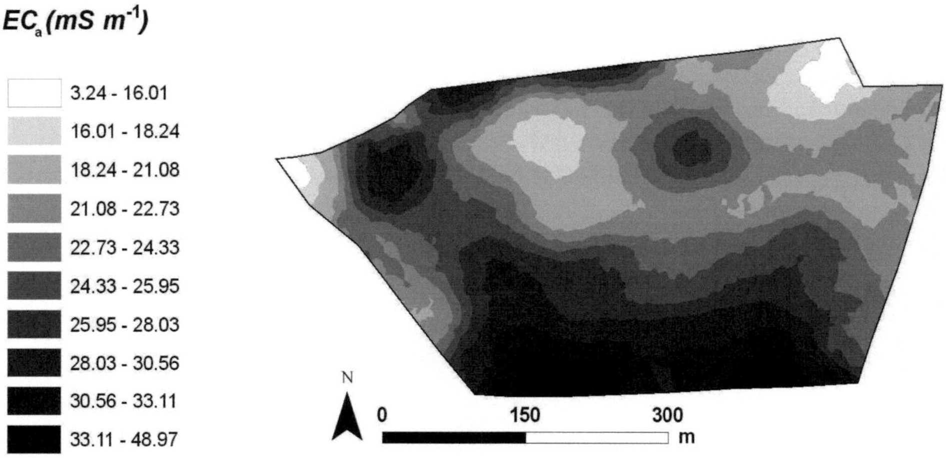

Fig. 12. Soil electric conductivity (ECa) map.

4. Discusion

Generally, indirect measuring methods will play an important role in precision farming system development. High density of the data measured, time accessibility, low cost and undemanding measurements are the crucial points for the possible application of precision farming tools into a common practice. Data from the soil sensors could be widely used in precision agriculture system. Especially soil conductivity measurement is very advantageous tool for soil properties description (Zhang et al. 2002; Godwin & Miller 2003). Soil electrical conductivity is especially affected by soil humidity and soil grain contents (Godwin & Miller 2003). Despite the amount of various sensors used in precision agriculture system, soil EC_a measurement represents a simple and cheap instrument for soil variability determination in the field.

When excluding climatic conditions influence, the formation of plants cover (arable crops, weeds, wild plants) is influenced mainly by soil conditions (Soukup et al. 2003). The agronomical practices have an important role as well. Further discussed results of many research projects do not explain unambiguously the dependence of the weed infestation on the soil properties. Dunker & Nordmeyer (2000) reported certain correlations between weed occurrence and soil properties in their results. Kurstjens & Perdok (2000) show in their study, that relationships between weed control and good crop growth may not only depend on weed and crop characteristics but also on soil conditions and tractor-implement settings.

On the other hand Medlin et al. (2001) pointed out that prediction of weed infestations with environmental properties was specific for each field, year, and species. However, it was evident from the weed infestation maps that the distribution of *C. arvense* infestation was not only random. It was proved that the *C. arvense* plants distribution could be also affected by different soil properties. It was also noticeable from the same map that the patches which underlie beyond the raster are concentrated at the places with previously noticed *C. arvense* appearance. *C. arvense* infestation was not extended in the direction of movement by a soil cultivator. Similar results reported Hamouz et al. (2004). Donald (1994) on the basis of literature review describes, that Canada thistle (*C. arvense)* shoots density varies across patches and often decreases near patch borders, but not as a uniform trend. Canada thistle shoot biomass exhibited a bell-shape distribution across a 35-m-wide path in Colorado. Also in our case we recognised circular shape of infestation patches.

In our research, the variability of *C. arvense* plants distribution was observed, but the evaluation according to the raster did not bring the exact result. This fact was proved by *G* parameter derived from the evaluation of prediction quality. The goodness-of-prediction fit was observed in this case (G = 0.06 % (2002) and G = - 8.13 % (2003)), monitoring in a raster was not sufficient for the description of real weed infestation conditions and its intensity. Negative and close to zero *G* values indicate that the field average predicts the values at unsampled locations as accurately (or even better) than the raster sampling estimates (Kravchenko 2003). Donald (1994) in his work described spatial dependencies for weeds spreading. However he used a raster with 1.8 m long side which would not be possible for our experiments. Donald (1994) also used a so called weed infestation average value in the experimental method which thus tended to use a uniform herbicides application. But this was in contravention of field observation.

The map of EC_a represents reliable output data which can relatively exactly describe the spatial variability. Division of spatial relations into classes can be found e.g. in Lopez-Granados et al. (2002) and Cambardella & Karlen (1999) work. Magnitude of the spatial relation is expressed as a ratio of the nugget divided by the total sill of the variogram. If this ratio is $\leq$ 25 % the observed relation is strong spatial relation. Positive *G* values indicate that the map obtained by interpolating data from the raster samples is more accurate than the field average. Jaynes et al. (1994) stated that it is possible to control variable herbicide application on the basis of soil conductivity measurement. According to the correlation coefficient value a close dependence between *C. arvense* infestation and EC_a was observed (significant dependence on the 99% confidence level was observed). This fact proved our presumption about *C. arvense* response to soil properties.

On the other hand it is necessary to say that lots of other factors may significantly influence the final EC_a measurement results. Concerning *C. arvense* infestation in the field in relation to EC_a data measured, it could not be assumed precisely, that higher values of EC_a will be observed explicitly where higher density of *C. arvense* is present. EC_a measurement does not totally substitute further soil survey. However, it may provide important data for the decision making processes, when applying the precision agriculture principle. The results indicate sufficient density of sampling points and suitably chosen evaluation methods. Map of EC_a represents a valuable outcome.

5. Acknowledgements

This research was supported by the Ministry of Education, Youth and Sports of the Czech Republic, Project No. MSM 604 6070905, by the Ministry of Agriculture, Project No. MZE 0002700604 and by the Project FP7 SP1 Cooperation Grant No. 211386

6. References

Basso, B.; Sartori, L.; Bertocco, M.; Oliviero G. (2003). Evaluation of variable depth tillage: economic aspects and simulation of long term effects on soil organic matter and soil physical properties, Stafford J.V., Werner A. (eds.): *4th European Conference on Precision Agriculture,* pp. 61-67, ISBN 9076998213 Berlin, Germany, June 2003. Wageningen Academic Publisher, Wageningen

Cambardella, C.A.; Karlen, D.L. (1999). Spatial analysis of soil fertility parameters. *Precision Agric.*, Vol. 1, No. 1, 5-14, ISSN 1385-2256

Cox, S. (2002). Information technology: The global key to precision agriculture and sustainability. *Comput. Electron. Agric.*, Vol. 36, No. 2-3, 93-111, ISSN 0168 -1699

Donald W.W. (1994). Geostatistics for mapping weeds, with a Canada Thistle (*Cirsium arvense*) patch as a case study. *Weed. Sci.*, Vol. 42, No. 4, 648-657, ISSN 0043-1745

Dunker, M.; Nordmeyer, H. (2000). Reasons for the distribution of weed species on arable fields - field and greenhouse experiments concerning the influence of soil properties. J. Plant Dis. Prot., Vol. 17, 55-62, ISSN1861-3829

Gerhards, R., Sökefeld, M.; Schulze-Lohne, K.; Mortensen, D.A.; Kühbauch, W. (1997). Site specific weed control in winter wheat. *J. Agron. Crop Sci.*, Vol. 178, No. 4, 219-225, ISSN 0931-2250

Gerhards, R., Sökefeld, M.; Timmermann, C.; Krohmann, P.; Kühbauch, W. (2000). Precision weed control - more than just saving herbicides. *J. Plant Dis.* Prot., Vol. 17, 179-186, ISSN1861-3829

Gerhards, R.; Oebel, H. (2006). Practical experiences with a system for site-specific weed control in arable crops using real-time image analysis and GPS-controlled patch spraying. *Weed Res.*, Vol. 46, No. 3, 185-193, ISSN 0043-1737

Godwin, R.J.; Miller, P.C.H. (2003). A Review of the Technologies for Mapping within-field Variability. *Biosyst. Eng.*, Vol. 84, No. 4, 393-407, ISSN 1537-5110

Lopez-Granados, F.; Jurado-Expósito, M.; Atenciano, S.; Garcia-Ferrer, A.; Sánchez de la Orden, M.; Garcia-Torres, L. (2002). Spatial variability of agricultural soil parameters in southern Spain. *Plant Soil,* Vol. 246, No. 1, 97-105, ISSN 0032-079X

Hamouz, P.; Soukup, J.; Tyšer, L. (2000). Weed distribution and spatial stability in arable field. Vanek V., Kolar. L., Pavlikova D., Tlustos P. (eds.): *Reasonable Use of Fertilizers focused on Precision Farming,* pp. 101-105, ISBN 80-213-0691-2, Prague, Czech Republic, November 2000, CULS, Prague.

Hamouz, P.; Soukup, J.; Gerhards, R. (2002). Weed distribution and weed spatial stability on the field in the crop sequence. *J. Plant Dis. Prot.,* Vol. 18, 269-275, ISSN1861 -3829

Hamouz, P.; Soukup, J.; Holec, J.; Jursik, M. (2004). Field-scale variability of weediness on arable land. Plant Soil Environ., Vol. 50, No. 3, 134-140, ISSN1214-1178

Hamouz, P.; Nováková, K.; Soukup, J.; Holec, J. (2008). Detection of *Cirsium arvense* L. in winter wheat using a multispectral imaging system. *J. Plant Dis. Prot.*, Vol. 21, 163-166, ISSN1861-3829

Heisel, T.; Ersbøll, A.K.; Andreasen, Ch. (1999). Weed mapping with Co-Kriging using soil properties. *Precision Agric.*, Vol. 1, No. 1, 39-52, , ISSN 1385-2256

Ilsemann, J.; Simone, G.; Bachmann, J. (2001). How many soil samples are necessary to obtain a reliable estimate of mean nitrate concentrations in an agricultural field? *J. Plant Nutr. Soil Sci.*, Vol. 164, No. 5, 585-590, ISSN1436-8730

Jaynes, D.B.; Novak J.M.; Mooran T.B.; Cambardella C.A. (1994). Estimating Herbicide partitio coefficients from electromagnetic induction measurements. *J. Environ. Qual.*, Vol. 24, No. 1, 26-41, ISSN 0047-2425

Kurstjens, D.A.G.; Perdok, U.D. (2000). The selective soil coveringmechanism of weed harrows on sandy soil. *Soil Tillage Res.*, Vol. 55, No. 3-4, 193-206, ISSN 0167-1987

Kravchenko, A.N. (2003). Influence of Spatial Structure on Accuracy of Interpolation Methods. *Soil Sci. Soc. Am. J.*, Vol. 67, No. 5, 1564-1571, ISSN 0361-5995

Krohmann, P.; Timmermann, C.; Gerhards, R.; Kühbauch, W. (2002). Reasons for the persistence of weed populations. *J. Plant Dis. Prot.* , Vol. 18, 261-268, ISSN 1861-3829

Medlin, C.R.; Shaw, D.R.; Cox, M.S.; Gerard, P.D.; Abshire, M.J.; Wardlaw, M.C. (2001). Using soil parameters to predict weed infestations in soybean. *Weed Res.*, Vol. 49, No. 3, 367-374, ISSN 0043-1737

Oebel, H.; Gerhards, R. (2005). Site-specific weed control using digital image analysis and georeferenced application maps: On-farm experiences. Stafford J.V. (eds.): *Precision Agriculture '05*, pp. 131-138. ISBN: 978-90-76998-69-5, Uppsala, Sweden, Jun 2005, Wageningen Academic Publisher, Wageningen.

Oebel, H.; Gerhards, R. (2006). Camera-guided weed control - a technology for practice. *J. Plant Dis. Prot.*, Vol. 20, 181-187, ISSN1861-3829

SAC (2001). Weed management in winter oilseed rape. *SAC Technical Note*, Edinburgh, UK.

Scotford, I.M.; Miller, P.C.H. (2005). Application of spectral reflectance techniques in northern European cereal production: A review. *Biosyst. Eng.*, Vol. 90, No. 3, 235-250, ISSN: 1537-5110

Selige, T.; Nätscher, L.; Schmidhalter, U. (2003). Spatial detection of topsoil properties using hyperspectral sensing. Stafford J.V., Werner A. (ed.): *4th European Conference on Precision Agriculture*, pp. 633-638. ISBN 9076998213 Berlin, Germany, June 2003. Wageningen Academic Publisher, Wageningen

Sökefeld, M.; Gerhards, R.; Kühbauch, W. (2000). Site-specific weed control - from weed recording to herbicide application. *J. Plant Dis. Prot.*, Vol. 17, 227-233, ISSN1861-3829

Sökefeld, M.; Hloben, P.; Lammers, P.S. (2004). Possibilities and limits of direct injection of herbicides for site-specific weed control. *J. Plant Dis. Prot.*, Vol. 19, 431-437, ISSN1861-3829

Soukup, J. (2000). Weed control in precision farming. Vanek V., Kolar L., Pavlikova D., Tlustos P. (eds.): *Reasonable Use of Fertilizers focused on Precision Farming*, pp. 56-61, ISBN 80-213-0691-2, Prague, Czech Republic, November 2000, CULS, Prague.

Soukup, J.; Hamouz, P.; Holec, J. (2003). Weed control in precision farming. In: Vanek V., Balik J., Pavlikova D., Tlustos P. (eds.): *Reasonable Use of Fertilizers dedicated to Sustainable Development and Precision Farming*, pp. 92-97, ISBN 80-213-1083-9, Prague, Czech Republic, November 2003, CULS, Prague.

Werner, B.; Garbe, V. (1998). Bedeutung der Unkrautverteilung im Winterraps für eine gezielte Bekämpfung nach Schadensschwellen. *Z. Pflanzenk. Pflanzen., Sonderheft.*, Vol. 17, 279-288, ISSN 0938-9938

Zhang, N.; Wang, M.; Wang, N. (2002). Precision Agriculture - a worldwide overview. *Comput. Electron. Agric.*, Vol. 36, No. 2-3, 113-132, ISSN 0168-1699

9

Herbicides in the Soil Environment: Linkage Between Bioavailability and Microbial Ecology

M. Celina Zabaloy[1], Graciela P. Zanini[1], Virginia Bianchinotti[1], Marisa A. Gomez[1] and Jay L. Garland[2]
[1]Universidad Nacional del Sur
[2]Dynamac Corp.
[1]Argentina
[2]United States

1. Introduction

Modern agriculture relies heavily on herbicides for the control of weeds in crops and pastures to maximize yields and economical benefits to sustain an increasing world population. The introduction of herbicide-resistant traits in several crops, such as glyphosate-resistant (GR) soybean, maize and canola, has further increased herbicide consumption worldwide (Cerdeira & Duke, 2006). United States consumed roughly 200 million kg in 2001, with glyphosate representing 20 % of the total. Glyphosate is, undoubtedly, the most used herbicide worldwide (Woodburn, 2000). In Argentina, where GR soybean accounts for almost 90 % of planted soybean, it was estimated that 160 million l of glyphosate were used with this crop in 2004, representing 37 % of the total herbicide consumed in agriculture (Altieri & Pengue, 2006; Pengue, 2004).

The environmental fate of herbicides is a matter of recent concern given that only a small fraction of the chemicals reach the target organisms (Pimentel, 1995), leading to potential impacts of residual herbicides in soil and water have on human, animal and crop health. Bunce (1993) wrote in 1993 "It is useful to keep in mind the concept that a pollutant is a substance in the wrong place, at the wrong time, or in the wrong amount". While herbicides are very important to agriculture, under certain circumstances they may act as pollutants that can deteriorate soils, ground waters and surface waters. While most herbicides are not intentionally applied onto soil, they can enter the soil environment from 1) direct interception of spray by the soil surface during early season or post-harvest applications, 2) runoff of the herbicide from vegetation and 3) leaching from dead plant material. The herbicide concentration may vary from a few µg to mg per kg soil, as most of the applied chemical is retained within the top 5 cm of soil. This chapter will present aspects of the behavior of herbicides in soils, focusing on soil retention and microbial degradation as main factors controlling persistence. The potential impact of herbicides on non-target soil microbes, including their processes and interactions, will be also discussed.

Adsorption to soil is of critical importance for the regulation of herbicide persistence and mobility throughout the environment because sorption processes control the amount of herbicide present in the soil solution. These processes are dependent on several factors related to soil characteristics such as mineral composition, organic matter content, soil solution chemistry, and chemical characteristics of the herbicide. Soil-bound herbicide or residues are temporarily inactivated, which prevents harmful effects on soil biota but also makes them less bioavailable for microbial degradation because most microbes may not be able to utilize herbicides in the sorbed state (Ainsworth et al., 1993). Soil biochemical and biological processes are critical for ecosystems functioning, as microbes have key roles in organic matter transformations, nutrient cycling and degradation of organic pollutants, including pesticides (Beck et al., 2005). Biological degradation mediated by microbial enzymes is the main route for pesticides detoxification in soils (van Eerd et al., 2003). Most isolated herbicide-degrading microorganisms belong to bacterial species, but fungi are also well-known for their capacity to degrade complex substrates, and may be more important than present isolation approaches have suggested (Smith & Collins, 2007). Differential toxicity of herbicides to soil microorganisms may alter community structure, including potential increases in plant or animal pathogens. Herbicides may also cause changes in microbial community function and concomitant impacts on soil health and ecosystem processes. Even though functions may appear unaltered, due to species redundancy in soil, the extinction of resistant species may compromise the continuity of such processes.

The enormous variety of herbicides commercially available today makes it impossible to review all of them. Thus, this work will focus on some of the herbicides most used in the (semiarid) Pampa region of Argentina and worldwide (i.e., glyphosate, 2,4 dichlorophenoxyacetic acid, metsulfuron-methyl), based on our own research data.

2. Factors influencing the fate of herbicides in soil

2.1 Physicochemical interactions with soil

Soil is one of the main regulators of herbicide mobility in the environment. Many chemical and biological processes that determine the retention or transport of herbicides take place on the soil surface. These processes include adsorption phenomena, chemical degradation, and biological degradation. While all these processes are interrelated, occurring in parallel (Cheng, 1990), it is important to first understand adsorption since it regulates the bioavailability of herbicides in the environment, i.e. the ability to be used by microorganisms and thus be biodegraded (Laor et al., 1996; Boesten, 1993; Martins & Mermoud, 1998). Adsorption determines the quantity of herbicide that is retained on the soil surface and therefore is one of the primary processes that affect the transport of these compounds in soils. Thus to relate bioavailability and microbial ecology it is helpful to understand this primary process. Soils are complex assemblies of solids, liquids, and gases. A typical mineral soil contains 50% solid material (45% mineral and 5 % organic matter) and 50% pore space. The mineral particles in the soil are distributed into three sizes: sand, silt, and clay. Between the solid components of soil is space forming pores that plays a major role in movement of water, solutes and air. The adsorption processes depend fundamentally on the composition and properties of the solid component as well as the physicochemical characteristics of the herbicide. The solid component is formed mainly by primary and secondary minerals and by organic matter. These materials provide the specific sites for herbicide adsorption. Their properties and behavior have been treated extensively (Dixon & Weed, 1990; Greenland & Hayes, 1978).

Important characteristics of herbicides include: structure of the compound (including functional groups), water solubility, vapor pressure, octanol-water partitioning constant (K*ow*), and acidity. Table 1 shows the structures and some physicochemical properties of glyphosate, 2,4-D and metsulfuron-methyl.

The distribution of an herbicide in the soil depends on partitioning between the soil solution and the solid phase (Figure 1). The chemical is partitioned between the soil solution and the solid phase. The proper term for this process is adsorption equilibrium, which can be written to describe the interaction between any herbicide and any soil component as follows:

$$\mathrm{S+H(aq)} \rightleftharpoons SH \tag{1}$$

Where S represents a surface site of soil, H(aq) the herbicide in soil solution and SH the herbicide attached to the surface site. Surface sites where the herbicide can be adsorbed are numerous and varied in soils. These sites are provided by soil minerals (clays, Fe and Mn oxides, etc) as well as by organic matter. Equation (1) gives an idea of the general process involving adsorption, but it does not specify the mechanism by which it occurs, which are varied in the complex soil system (formation of surface complexes, electrostatic interactions, hydrophobic interactions, ion exchange, etc.).

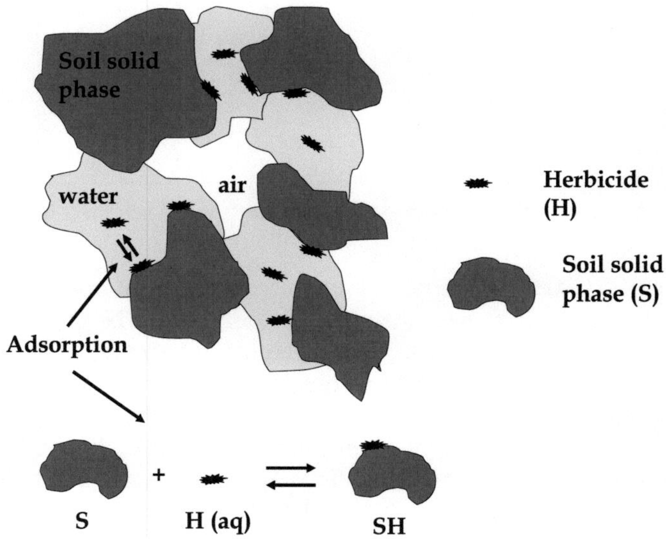

Fig. 1. Distribution of an herbicide in soil

Defining the bioavailability of an herbicide requires an understanding of the strength of its 1) interaction with a particular soil and 2) concentration of herbicide in the soil solution. This can be known by using adsorption isotherms. An adsorption isotherm shows the relationship between the herbicides concentration in the soil solution (C, correspond to H(aq) in Equation (1) and the amount adsorbed (q, correspond to SH in Equation (1)) at constant temperature and after equilibrium was reached (Stumm, 1992). As an example, Figure 2 shows adsorption isotherms of the herbicide metsulfuron methyl (MM) on different soils of the semiarid pampean region of Argentina (Zanini et al., 2009). Although isotherms with 30 different soils were measured in that study, the figure presents the results for three selected soils. These soils are characterized by having rather similar specific surface areas (SSA) and clay contents (% clay), but rather different total organic carbon (TOC) content. The physical and chemical characteristics of the 30 soils are shown in Table 2.

Herbicide	Chemical Structure	pKa	Water Solubility	Log K_{ow}	Vapour Pressure (Pa)
Metsulfuron Methyl (MM)	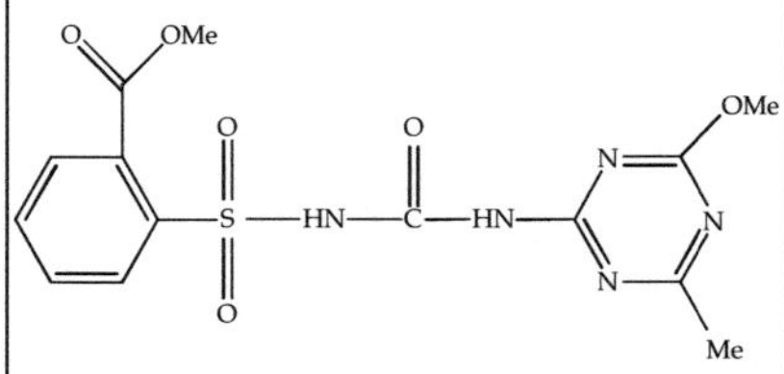	3.3	548 mg L^{-1} (pH 5) 2.79 g L^{-1} (pH 7) 213 g L^{-1} (pH9, 20°C)	1.8 (pH 5) 0.018 (pH 7) 0.0002 (pH9, 25°C)	1.1×10^{-10}
Glyphosate (Gly)	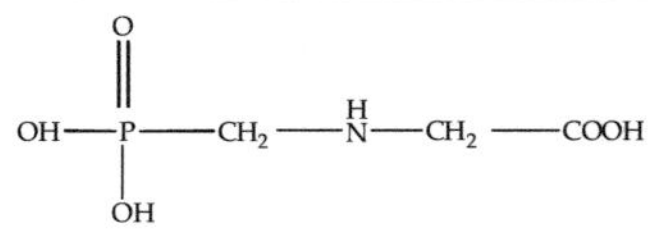	2.3 5.3 10.9	1.157 wt% in water at 25°C	-4.1	Negligible
2,4-(Dichlorophenoxy) acetic acid (2,4-D)	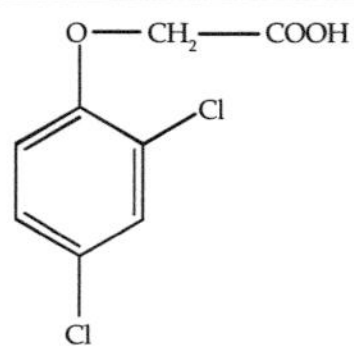	3.2	4.8 mg L^{-1} (20°C-nonionised, est.)	3.2	1×10^{-5} (20°C)

Table 1. Structure and some physicochemical properties of the selected herbicides (Roberts, 1998).

As stressed by Sparks (2003), isotherms are only descriptions of macroscopic data and do not definitively prove a reaction mechanism. Mechanisms must be gleaned from molecular investigations, e.g. the use of spectroscopic techniques. However, the fit of experimental data with theoretical and/or empirical equations for adsorption isotherms is very useful in determining some parameters that provide information on the strength of soil-herbicide interaction, which will give an idea of the bioavailability of the herbicide in a particular soil. There are several adsorption isotherms equations applied to soils and sediments (Haws et al., 2006; Hinz, 2001). In this chapter, only the simplest and most widely applied equations are discussed.

2.1.1 Linear equation

The linear, or partitioning equation is expressed as (Pateiro-Moure et al., 2009; Cooke et al., 2004):

$$q = K_d C \tag{2}$$

where K_d is the partition coefficient and q and C as defined above. The parameter K_d provides a measure of the ratio of the amount of material adsorbed to the amount in solution. The higher the value of K_d, the greater the affinity of the herbicide for the surface, resulting in lower bioavailability. The problem with the application of this equation is that linear behavior of the system in the range of concentrations of interest must be proved. If experimental data do not show a linear response in all the concentration range, the use of K_d values obtained from linear regression will cause over- or underestimation of the true behavior in the non-linear ranges. Calculating K_d with only a pair of values (C, q) may not be very useful to evaluate bioavailability across a range of environmentally relevant concentrations. It is recommended to perform an adsorption isotherm in the range of concentrations of interest, to test for linearity. Since adsorption of hydrophobic organic pollutants has been shown to be well correlated with the organic carbon content of soil and relatively independent of other soil properties, K_d is sometimes expressed on the basis of TOC (Laor et al., 1996):

$$K_OC = \frac{K_d}{0.01TOC} \quad (3)$$

where TOC is expressed in % units.

Most experimental data do not respond to the linear equation; the most common models that describe non-linear adsorption isotherms are the Freundlich equation and the Langmuir equation.

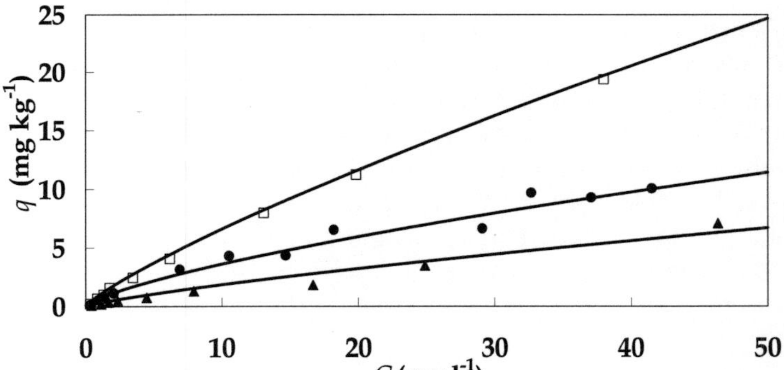

Fig. 2. Binding isotherms at pH=6 for samples with different TOC%. 4.02% (empty squares), 2.18% (filled circles), 0.98% (filled triangles). From Zanini et al. (2009).

2.1.2 Freundlich equation

The Freundlich equation is perhaps the most widely applied model in environmental soil chemistry to describe nonlinear sorption behavior (Valverde-García et al., 1998; Kibe et al., 2000). It is an empirical adsorption model (Stumm, 1992; Sparks, 1986) and it can be written as:

$$q = K_f C^{1/n} \quad (4)$$

Where, K_f is the distribution coefficient and n is a correction factor. The lines of Figure 2 have been drawn according to the Freundlich equation. The fitting parameters are present in Table 2 and discussed below. It is important to note that when n = 1 Equation (4) becomes Equation (2) and $K_f = K_d$. In addition, when C is equal to unity the distribution coefficient gives the amount adsorbed at that concentration.

While researchers have often used the K_f and 1/n parameters to make conclusions concerning mechanisms of adsorption, and have interpreted multiple slopes from

Freundlich isotherms as evidence of different binding sites, such interpretations are speculative (Sparks, 2003). This is especially true in very complex and heterogeneous systems such as those formed by soil particles. For these systems, fitting of experimental data with isotherm equations should only be used for comparative purposes and to give some interpretation of the shape of the isotherms. Comparison of parameters should be performed with caution. It is necessary to be sure that the K_f values present the same units. The best way to avoid mistakes is to compare different sets of experimental data made under the same conditions and with isotherms performed in the same units of concentration. As in the case of the linear equation, parameters derived from the Freundlich equation should not be used to predict for behavior outside of the range of experimental data.

2.1.3 Langmuir equation

This model has been employed in many fields to describe sorption on colloidal surfaces (Zanini et al., 2006; Xi et al., 2010). The Langmuir adsorption equation can be written as:

$$q = \frac{K_L C_b}{1 + K_L C} \tag{5}$$

Where K_L is a constant related to the binding strength, b is the maximum amount of herbicide that can be adsorbed (monolayer coverage) and q and C were defined previously. This equation has several assumptions that Langmuir (1918) made in its development. Most of these assumptions are not valid for the heterogeneous surface found in soils. However, many researchers used this model to describe adsorption on soils (Gimsing et al., 2007; Ketelsen & Meyer-Windel, 1999). As with K_f above, K_L is useful for comparative purposes but they do not provide information on reaction mechanisms. Some researchers fit the experimental data with both Langmuir and Freundlich equations to compare methodological approaches (Campbell & Davies, 1995; Martínez-Villegas et al., 2004).

2.2 Isotherm parameters and soil properties

In order to understand the bioavailability of an herbicide it is important to know the factors that affect its adsorption on soil. A good approach is to perform adsorption isotherms under different experimental conditions, and then relate the parameters of the isotherm to the soil properties. This will be demonstrated for a series of data on MM adsorption on the 30 different soils of the semiarid pampean region of Argentina listed in Table 2. As indicated above, the Freundlich equation was applied to this set of data. Table 2 shows the parameters K_f and n for all soils. All these soils are subject to similar farming practices (no till and production of the same kind of crops), thus the quality of the soil organic matter is expected to be similar, and the adsorptive differences among soils should be mainly given by differences in TOC. In most soils 1/n is lower than 1 and thus their isotherms are L shaped (Hunter, 2002) (Table 2). This kind of shape was also found by Pusino et al. (2003) for the adsorption of primisulfuron on soils, suggesting that the affinity of surface sites for MM is decreasing as the surface is becoming populated with MM. It may also suggest a decrease in vacant adsorption sites as MM concentration increases.

It is necessary to be careful when K_f values are compared. If the values of 1/n for the different soils are equal or similar, K_f values can be directly compared, and large K_f mean a strong herbicide-soil interaction. However, if the values of 1/n are rather different the comparison is not straightforward.

Values of K_f and 1/n were used to calculate the adsorption of MM at different equilibrium concentration for the 30 analyzed soils. From these calculations, plots relating adsorbed amounts with a given soil property can be constructed. For example, Figure 3 a shows the adsorbed amount at equilibrium concentration of 10 mg l^{-1} as a function of TOC. A positive and significant relationship between q and TOC is observed in the Figure. Although not shown here, this positive and significant relationship was found for all the studied concentrations (10, 20, 30 and 40 mg l^{-1}). The results show that TOC is a very important factor that affects MM adsorption in the entire range of MM concentrations investigated. This is known for other herbicides (Kah & Brown, 2006; Weber et al., 2002). However, Cramer et al. (1993) found no clear relationship between adsortion of metsulfuron methyl and soil organic matter in Colorado soils, and the adsorbed amount showed only a weak correlation with organic matter content.

Soils	TOC %	Sand %	Clay %	SSA m2 g^{-1}	pH	K_f	$1/n$	R^2
1	0.98	53.7	28.4	8.3	7.50	0.16 (0.01)a	0.95 (0.03)a	0.99
2	1.28	64.1	25.3	3.4	5.94	0.24 (0.04)	0.91 (0.05)	0.99
3	1.29	51.9	38.9	12.0	6.51	0.22 (0.03)	0.98 (0.04)	0.99
4	1.40	48.5	33.9	5.2	6.05	0.54 (0.08)	0.74 (0.05)	0.95
5	1.43	56.5	28.2	4.5	6.90	0.24 (0.09)	1.02 (0.10)	0.96
6	1.44	57.4	33.4	4.9	6.05	0.61 (0.06)	0.70 (0.03)	0.98
7	1.58	54.5	33.4	5.4	6.51	0.22 (0.08)	0.86 (0.11)	0.94
8	1.76	44.7	41.8	7.3	6.19	0.78 (0.12)	0.55 (0.05)	0.97
9	1.82	45.8	39.8	8.4	6.55	0.36 (0.07)	0.86 (0.05)	0.98
10	1.88	43.7	44.2	7.9	6.47	0.74 (0.16)	0.68 (0.07)	0.96
11	1.91	50.3	41.8	4.6	6.04	0.96 (0.29)	0.40 (0.09)	0.86
12	1.94	42.7	39.4	4.6	6.46	0.54 (0.08)	0.74 (0.05)	0.97
13	2.06	47.2	39.2	6.1	6.77	0.20 (0.07)	1.02 (0.10)	0.97
14	2.07	44.0	38.3	5.8	6.30	0.53 (0.13)	0.88 (0.08)	0.98
15	2.10	46.4	38.9	4.8	6.51	0.31 (0.07)	0.85 (0.07)	0.97
16	2.18	52.7	28.2	4.6	6.90	0.78 (0.07)	0.69 (0.07)	0.98
17	2.46	52.2	35.9	3.7	6.14	0.76 (0.27)	0.77 (0.10)	0.94
18	2.50	42.5	36.7	10.6	6.58	0.60 (0.12)	0.98 (0.06)	0.99
19	2.56	30.7	49.7	7.7	6.59	1.18 (0.12)	0.55 (0.06)	0.93
20	2.59	32.6	44.0	5.6	6.08	0.91 (0.12)	0.69 (0.04)	0.91
21	2.75	36.1	40.9	5.6	5.80	0.84 (0.21)	0.65 (0.08)	0.97
22	2.88	43.0	35.4	5.2	6.44	0.56 (0.10)	0.92 (0.06)	0.98
23	2.93	31.2	48.6	5.2	7.80	0.36 (0.11)	0.80 (0.09)	0.96
24	3.07	47.2	33.7	4.3	6.36	1.07 (0.25)	0.67 (0.07)	0.96
25	3.10	45.3	33.9	4.4	7.10	0.61 (0.14)	0.69 (0.07)	0.98
26	3.34	50.4	30.6	9.0	6.51	0.85 (0.16)	0.89 (0.06)	0.98
27	3.91	46.5	31.2	5.5	6.10	1.10 (0.32)	0.79 (0.09)	0.99
28	4.02	52.0	28.6	6.1	6.65	0.96 (0.05)	0.83 (0.02)	0.99
29	4.62	61.4	25.3	4.0	5.86	0.80 (0.15)	0.93 (0.06)	0.99
30	4.85	44.3	32.8	6.0	8.03	1.35 (0.22)	0.77 (0.05)	0.98

Values within brackets correspond to standard error.

Table 2. Selected physical and chemical properties of the studied soils.

In order to investigate the effects of soil inorganic compounds on the adsorption of MM, the adsorbed amount was also plotted as a function of clay percent (Figure 3 b). There is no significant correlation indicating that inorganic compounds are not important on adsorption. The lack of interaction with inorganic compounds is not always the case for the adsorption of sulfonylureas. Pusino et al. (2003), for example, reported that inorganic solids such as amorphous Fe oxides and Al^{3+} and Fe^{3+} exchanged montmorillonites were active in adsorbing primsulfuron. The absence of important amounts of Fe oxides and smectites exchanged with trivalent cations in the studied soils (Blanco & Stoops, 1993) might explain the weak effect that inorganic components have on the adsorption of MM.

The above discussion highlights the variable behavior of MM among soils. These differences may result from variation in the properties of the inorganic compound, organic matter, or other soil properties such as pH. Another important factor to take into account is pH, especially if the herbicide has acid or basic groups. Figure 4 shows the adsorption isotherms of MM on soil at pH 4, 6 and 8. MM adsorption decreases as the pH increases, in agreement with the general trend observed for sulfonylureas (Hay, 1990). This figure shows that changes in pH can affect the adsorption of MM. This behavior is usually explained in terms of charge development at the surface of soil particles and speciation of the herbicide in aqueous solutions as a function of pH (Berglöf et al., 2003). Since the surface charge of soil particles becomes more negative as the pH increases, the adsorption of the negatively charged MM species becomes less favored by increasing pH as a consequence of electrostatic repulsion. In addition, although the adsorption of the neutral MM species should not be affected by electrostatics, its concentration decreases with increasing pH, also causing less favorable adsorption with higher pH.

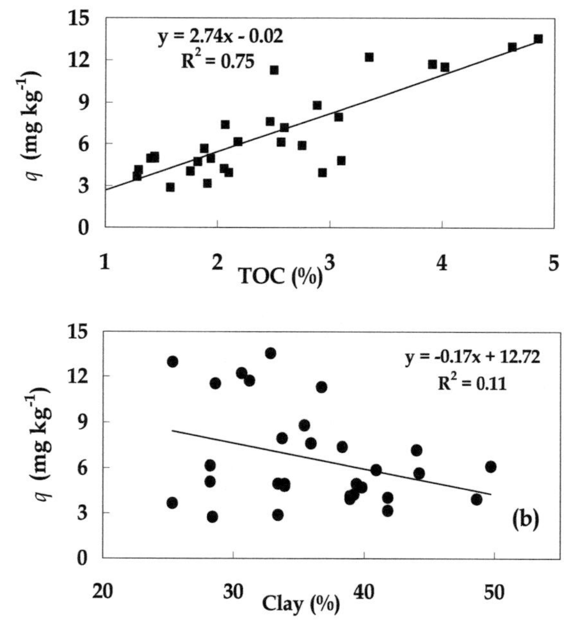

Fig. 3. (a) q (at 20 mg l^{-1} equilibrium concentration) as a function of TOC % for data at pH=6.(b) q (at 20 mg l^{-1} equilibrium concentration) as a function of the clay content for data at pH=6.

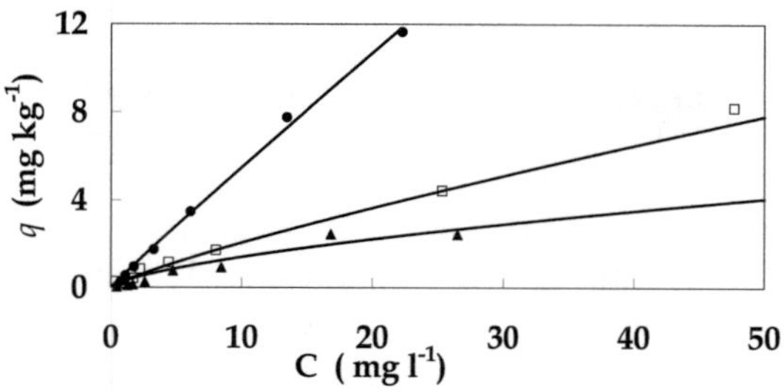

Fig. 4. Binding isotherms at (filled circle) pH=4, (empty square) pH=6, (filled triangle) pH=8. Lines have been drawn according to the Freundlich isotherm. From Zanini et al. (2009).

Some of the parameters obtained from adsorption isotherms are useful for an indirect estimation of the mobility of herbicides in soils. This can be obtained from the groundwater ubiquity score, GUS (Gustafson, 1989) defined as:

$$GUS = \log t_{1/2}(4 - \log K_{OC}) \quad (6)$$

where GUS is a dimensionless index, $t_{1/2}$ is the herbicide half-life in soil and K_{OC} was defined previously (Equation 3). According to Oliveira Jr. et al. (2001) herbicides with GUS < 1.8 are ranked as non-leachers, those with GUS > 2.8 are leachers, whereas those with 1.8 < GUS < 2.8 are considered transitional. It must be remarked that $t_{1/2}$ values should be measured for the specific soil under study because it may change greatly from soil to soil (Juhler et al., 2008; Bedmar et al., 2006).

In summary, it can be stated that adsorption depends on the physicochemical characteristics of the herbicide and the particular soil properties. In order to understand the processes that affect the bioavailability of an herbicide it is necessary to perform adsorption isotherms with the soils under study. Since adsorption and bioavailability change greatly from one soil to another, literature data can help to understand the general behavior of an herbicide, but they cannot give specific information about the behavior on a particular soil. Another important conclusion is that the mobility of an herbicide cannot be assessed only by knowing its physicochemical data. It is also very important to consider the presence of microorganisms in the soil system, as they can significantly affect the value of $t_{1/2}$ in Equation 6.

2.3 Spatial distribution of microbial populations

As discussed earlier, soil is characterized by the heterogeneity in physicochemical and structural characteristics that provide many different micro-habitats for microbial life. The distribution of soil microorganisms varies both horizontally and vertically and from the micro site (few millimeters) to the regional (kilometers) scale. The abundance and diversity of most soil organisms are highest in the top 0-10 cm of soil and decline with depth in parallel with organic matter contents, the source of energy, nutrients and carbon (C) for the vast majority of soil microorganisms. Most microorganisms are located surrounding the water layer attached to soil particles within micro-aggregates. Pallud et al. (2004) estimated that soil bacteria occupy only 0.1 mm^3 of the 500 mm^3 of pores in 1 cm^3 of soil. Moreover, different physiological groups present at the same density in soil might have significantly different microscale spatial distributions (e.g., forming clumps of cells in a few "hot spots" versus evenly spread out across soil particle surfaces).

Biodegradation of herbicides constitutes a clear example of the importance of understanding the spatial distribution of soil microorganisms. Spatial variability in both glyphosate mineralization and general soil microbial characteristics, was observed even across small areas (decimeter scale) within a single field in two Norwegian sandy loam soils (Stenrød et al., 2006), reflecting the importance of soil physicochemical parameters controlled by surface topography. Similarly, Vieuble-Gonod et al. (2005) reported that potential for 2,4-D mineralization was heterogeneous from field to microhabitats. High mineralization potential was not distributed randomly in the soil, but rather as systematic hot spots organized at centimeter scales (Vieuble-Gonod et al., 2005). Most pesticides may represent an occasional source of C and nutrients for soil microorganisms, so they can be completely dissipated from the soil environment by either a single microbial species or the joint action of a microbial consortium. In the later case, the degradation pathway in soil involves the cooperative activity of several strains that possess enzymes that catalyze different degradation steps. This cooperation is only possible with intimate contact between microbial cells and their substrates, as metabolites resulting from one step of the pathway may act as substrate for a different strain. Moreover, the bacterial distribution at the microscale may facilitate spreading of degradative genes located in plasmids or transposons (McGowan et al., 1998; DiGiovanni et al., 1996). Pallud et al. (2004) found that for low abundances of 2,4 D degraders, there was strong spatial isolation within the degraders populations, with less than 2 cells per colonized patch. 2,4-D amendment caused an increase in degrader abundance and concurrent spreading of degraders, reducing the distance between colonized patches, although the number of cells per patch remained low (< 28). They argued that the spatial spreading of bacteria was an ecological strategy that increased the probability of encountering the substrate (2,4-D), and proposed that this was achieved either through active cell movement (chemotaxis) or degradative plasmids transfer to indigenous microbial populations (Pallud et al., 2004).

The zone of soil directly influenced by the presence of plant roots, known as rhizosphere, is of particular importance. Plant roots act on microbes essentially through the input of a wide variety of organic compounds (e.g., sugars, amino acids, cellulose, proteins, phenolic acids), known as rhizodeposition, and by providing a surface for attachment, creating an environment that can greatly differ from the surrounding bulk soil. As a consequence, higher microbial biomass and activities are found in the rizosphere as opposed to bulk soil. Several studies have reported that rhizosphere enhances biodegradation of chlorophenoxyacetic acids (Shaw & Burns, 2004; 2005; Merini et al., 2007), metsulfuron-methyl (Ghani & Wardle, 2001) and atrazine (Piutti et al., 2002). Biodegradation pathways and strategies will be discussed in the following section.

2.4 Biodegradation: co-metabolism vs. growth-linked metabolism

Biodegradation is the enzyme-mediated transformation of a xenobiotic by living microbial cells. In soil systems, biodegradation is a fundamental attenuation process for pesticides and is controlled by biotic factors (i.e. microbial activity) and a number of physicochemical processes such as sorption and desorption, diffusion, and dissolution (Chen et al., 2009). Pesticide degradation by microorganisms that are capable of using the chemical as a source of C and energy for growth, is called mineralization. This metabolic strategy results in the complete dissipation of the chemical and its conversion to CO_2, water and inorganic elements. In this case, the biomass of the degrading population increases at the expense of the substrate. The rate of change in herbicide concentration in the medium follows the dynamic of the expanding microbial population, i.e., as the herbicide concentration decreases in the solution, growth of the microbial population reaches a plateau at a high cell density. Conversely, the partial transformation of an herbicide by microorganisms that gains no C or nutrients and energy, is called co-metabolism. In most cases, co-metabolism of herbicides involves microbial growth at the expense of a co-substrate that provides C and energy, but the pesticide in itself does not support microbial proliferation. The biomass of the herbicide degrading microbial population and the concomitant rate of herbicide degradation is not affected by the herbicide concentration in solution. Even though an herbicide may be partially transformed by co-metabolism, intermediate metabolites may be completely degraded by other microorganisms in soil.

2.4.1 Metabolic pathway of 2,4–D

One of the most studied herbicide degradation pathways is that of 2,4–D, which can be readily used as a C and energy source by environmental microorganisms. Numerous 2,4–D degrading bacteria have been isolated and characterized (Tiedje et al., 1969; Don & Pemberton, 1981; Kamagata et al., 1997; Muller et al., 2001). Most of these strains are members of genera belonging to the β and γ subdivisions of the class *Proteobacteria* and were isolated from 2,4–D treated environments (Kamagata et al., 1997; Lee et al., 2005). These β and γ subgroups carry *tfd* genes homologous to the canonical genes found in *Cupriavidus necator* JMP134 (Lerch et al., 2007), the model organism for 2,4–D degradation studies (Figure 5). These genes are located on conjugative plasmids like pJP4 which carries *tfdABCDEF* (Don & Pemberton, 1981). On this plasmid, *tfdA* encodes a 2,4-D/α-ketoglutarate dioxygenase, which transforms 2,4–D into 2,4-dichlorophenol (DCP), while *tfdB* encodes a dichlorophenol hydroxylase that transforms DCP in 3,5-dichlorocatechol. The *tfdCDEF* operon encodes enzymes involved in the *ortho*-cleavage of the aromatic ring and subsequent reactions (Fukumori & Hausinger, 1993a;b; Vallaeys et al., 1996). Zabaloy et al. (unpublished results) recovered several *Cupriavidus*-like isolates from an agricultural soil in Argentina, able to grow with up to 1.1 mM herbicide as sole C source with complete primary degradation in < 72 h. These isolates harbored *tfdA* and *tfdB* genes similar to the canonical degradation genes described by Vallaeys et al. (1996), as determined by PCR and restriction fragment length polymorphism (RFLP). Recently, isolation of 2,4–D degrading bacteria from pristine environments has unveiled the existence of other degradative genes, namely the *cadRABKC* operon, which are responsible for 2,4–D catabolism in slow-growing *Proteobacteria* (Kitagawa et al., 2002).

Considerably less information is available regarding 2,4–D degradation by fungi. Donnelly et al. (1993) reported that the basidiomycete *Phanerochaete chrysosporium* was able to degrade 2,4–D when provided with external nitrogen sources. Vroumsia et al. (2005) screened the ability of ninety strains of filamentous fungi to degrade the herbicide in liquid media, finding that 2,4–D was less accessible to degradation than its metabolite DCP, although both compounds were inefficiently used. The kinetics studies performed on the most efficient strains revealed a one-day lag phase before 2,4–D degradation and no lag phase for DCP.

2,4-D application to agricultural soils triggers specific degradation pathways in existing degrading bacterial populations (Baelum et al., 2006; 2008). Degradation of 2,4–D in soil initially occurs at low rates, as the specific degrading population increases in size. During that 1-3 day period, which corresponds to the lag phase of degraders, 2,4–D degradation probably proceeds by co-metabolism in soil (Vieublé-Gonod et al., 2006; Lerch et al., 2009). Zabaloy et al., 2010 examined aerobic degradation of 2,4-D in soil microcosms treated with environmentally-relevant level (ERL, 5 mg kg^{-1} soil) and high level (HL, 50 mg kg^{-1} soil) of 2,4-D after 3 and 14 days of incubation, using the BD Oxygen Biosensor System (BDOBS) . The use of 2,4-D as sole source of C and energy (50 mg l^{-1}) was initially retarded (> 40 h at day 3) and was maximal 2 weeks after treatment (Figure 6). They argued that

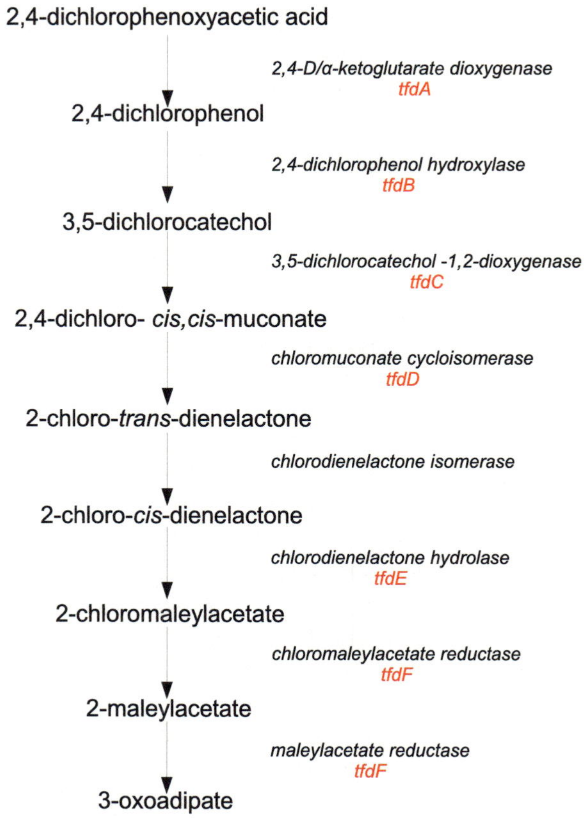

Fig. 5. Metabolic pathway of 2,4-D/α-ketoglutarate dioxygenase acid in *Cupriavidus necator* JMP134. Adapted from Caspi et al. (2010).

between day 3 and 14 a shift in dominant degrader populations might have occurred in the HL, as opposed to the ERL microcosms, and the increase in use of 2,4-D was reflecting the activity of specific degraders that were favoured by the high dose.

2.4.2 Metabolic pathways of glyphosate

Microbial degradation of glyphosate has been extensively explored and several degrading bacteria, belonging to Arthrobacteriaceae, Bacillaceae, Rhodobacteriaceae, Alcaligenaceae, Pseudomonaceae, Enterobacteriaceae and Rizhobiaceae, have been isolated and characterized (Kononova & Nesmeyanova, 2002). Most degrading isolates posses the capability to use glyphosate as a source of P, once extracellular inorganic phosphate becomes limiting in the environment (McGrath et al., 1997). Bacterial degradation of glyphosate follows either of two metabolic pathways. For example, a C–P lyase catalyzes the breakdown of the C–P bond in *Pseudomonas* sp. PG2982, releasing inorganic phosphate (Pi) and sarcosine, which is subsequentially mineralized to CO_2 and NH4+ (Kishore & Jacob, 1987). Glyphosate dehydrogenase catalyzes the conversion of glyphosate to aminomethylphosphonic acid (AMPA) and glioxylate as primary metabolites in *Geobacillus caldoxylosilyticus* T20, and a C–P lyase releases Pi from AMPA afterwards (Obojska et al., 2002). Although bacteria are considered as the main biological degrader of glyphosate in soils, fungi also may be important (Singh &Walker, 2006). Krzyko-Lupicka & Orlik (1997)

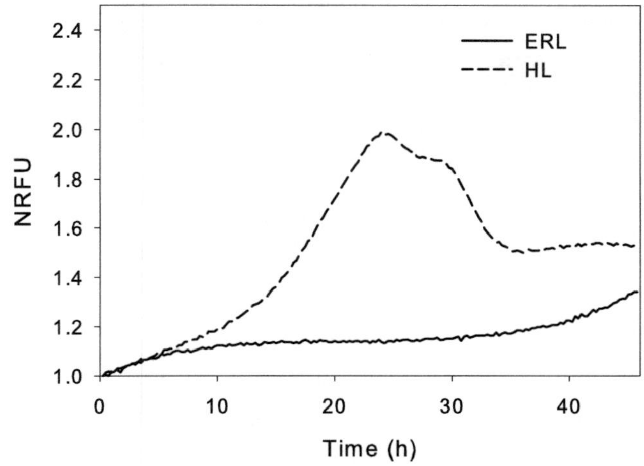

Fig. 6. Fluorescence curve (i.e., oxygen consumption) in BDODS plate with 2,4-D as sole source of C and energy, in 2,4-D-acclimated soil slurries after 14 days of incubation. Microcosms received 5 and 50 mg kg^{-1} soil of 2,4-D (ERL and HL). NRFU= normalized relative fluorescence units. From Zabaloy et al., 2010.

reported that the diversity of species isolated from soil diminished in media containing glyphosate, with a predominance of strains of *Mucor, Fusarium* and *Trichoderma.* Almost all the tolerant species isolated grew well when glyphosate was used as the unique source of P, but only a few were able to grow on it as the sole C source. Glyphosate also served as a nitrogen source for a *Penicillium chyrsogenum* strain isolated from soil (Lipok et al., 2003) and as C source for indigenous yeasts isolated from treated and untreated soils (Romero et al., 2004). Two yeast species were identified as active biodegraders (*Yarrowia lipolitica* and *Candida krusei*) but failed to uptake the herbicide when phosphate was present, suggesting that glyphosate was important for C and P nutrition in the yeasts (Romero et al., 2004).

Glyphosate degradation in soil is predominantly a co-metabolic process, as it is not used as a C and energy source by the vast majority of microorganisms (Forlani et al., 1999; Singh & Walker, 2006). As opposed to degradation in pure-culture experiments, glyphosate degradation in soil occurs without a lag phase, further suggesting a co-metabolic process as enzymes were present in soil before the application of the herbicide (Borggaard & Gimsing, 2008). Moreover, no adaptation to glyphosate degradation has been observed in soils with long histories of herbicide treatment (Gimsing et al., 2004). Lancaster et al. (2009) reported that the total amount of $^{14}CO_2$ evolved from glyphosate was reduced with repeated herbicide applications compared to a single application, which proved that biodegradation was not enhanced (i.e., no evidence of accelerated degradation) and was probably the result of a co-metabolic process. It is not well-known which metabolic pathway prevails in soils (Borggaard & Gimsing, 2008) although most isolated bacteria (Liu et al., 1991; Kishore & Jacob, 1987) and fungi (Sailaja & Satyaprasad, 2006) possess the sarcosine pathway. Even though AMPA is the main metabolite detected in soil, this could be attributed to the rapid mineralization of sarcosine and the persitance of AMPA in the environment.

2.4.3 Biodegradation of metsulfuron-methyl

Many studies have focused on the environmental fate and behavior of metsulfuron-methyl, but research on microbial degradation is still scarce. Zanardini et al. (2002) isolated a

Pseudomonas strain that degraded metsulfuron through co-metabolism and Boschin et al. (2003) studied the degradation pathway in the common soil fungus *Aspergillus niger*. Yu et al. (2005) isolated a *Curvularia* sp. capable of using the herbicide as a sole source of C and energy and studied several features of herbicide degradation in pure culture and in soils. Vázquez et al. (2008) isolated several filamentous fungi able to grow with metsulfuron as sole source of C and energy. Only *Penicillium* and *Trichoderma* strains were able to complete their life cycle in metsulfuron-containing medium. *Trichoderma* strains showed the best capacity to grow using the herbicide and were selected to perform tolerance assays; all could grow from spores in minimal medium containing metsulfuron and two showed heavy sporulation with increasing concentrations (up 1×10^{-2} ppm) (Vázquez et al., 2009). He et al. (2007) inoculated wheat rhizosphere with a highly effective metsulfuron-degrading *Penicillium* sp., previously isolated from treated soil, reporting that the inoculation enhanced the degradation of the target herbicide. Regardless of the pure-culture experiments that showed microbial degradation of metsulfuron methyl, limited mineralization of this herbicide in soil has been reported (Pons & Barriuso, 1998; Andersen et al., 2001). Ghani & Wardle (2001) studied ^{14}C-labeled metsulfuron-methyl in soil-plant microcosms and reported that 42% of the applied metsulfuron was respired or incorporated into microbes in the planted treatment while 36% was used in the unplanted system by day 131. They argued that greater utilization of metsulfuron in the planted microcosm would have been influenced largely by a greater microbial biomass in that treatment. Despite the positive rhizosphere effect on herbicide mineralization, more than 50% was still present in soil even four months after an application at recommended rates.

3. Approaches to link bioavailability and biodegradation

Bioavailability is influenced by a variety of factors, including physical characteristics of the sorbent (e.g., particle shapes, sizes, and internal porosities), chemical properties of the sorbates and sorbents, and biological factors (e.g., microbial density and degradative capacity). Generally, sorbed compounds are assumed to be less accessible to attached or suspended microorganisms, which preferentially or exclusively utilize herbicides in the aqueous phase. In this view, herbicide is available for biodegradation only after desorption, followed by diffusion into solution. The sorbed fraction remains protected from microbial attack as a result of: 1) physical sequestration of the herbicide in the organo-chemical matrix, 2) chemical stabilization in the sorbent surface, and/or 3) reduction of aqueous-phase concentrations to levels that do not sustain microbial growth (Ainsworth et al., 1993; Lerch et al., 2009). However, some other investigations revealed that silica-sorbed 2,4–D (Park et al., 2001) and soil-sorbed atrazine (Park et al., 2003) can be directly utilized by degrading bacteria. Park et al. (2001) proposed two plausible explanations for the observed enhanced bioavailability of silica-sorbed herbicide: 1) attached biomass is able to access adjacent elevated concentrations of herbicide before complete dilution in the liquid phase; 2) attached cells are capable of higher metabolic rates. Similarly, soil organic matter is implicated in sorption processes and therefore, affects the availability of herbicides to degrading microbes (Benoit et al., 1999). This section will briefly present recent research approaches that have been successfully used to link bioavailability and degradation of herbicides.

Benoit et al. (1999) studied the degradation of ^{14}C ring-labeled 2,4–D and two chlorophenols, adsorbed on different organic materials (wood chips, straw, lignin, humic acids) and aluminum oxide in soil incubations. They observed that mineralization of these compounds, when incubated in direct contact with soil, varied greatly according to the nature of the sorbent, but was generally higher in more humified organic matter (humic acid) than in less transformed organic matter (wood, lignin and straw). However, separation of chemical-sorbent associations from soil during incubation in polyamide bags with non-decomposed and composted straw showed higher mineralization levels than in direct touch with soil for all compounds, despite slower mineralization rates. They proposed that straw-associated microorganisms actively degraded 2,4–D and chlorophenols. Schnürer et al. (2006) tested the effect of surface sorption on the bioavailability of glyphosate by adding goethite to an organic soil, and using respirometric and attenuated total reflectance Fourier transform (ATR-FTIR) spectroscopy approaches. Addition of goethite reduced the negative effects of glyphosate on microbial respiration, as surface sorption reduced toxic effects of the herbicide or its metabolites. However, ATR-FTIR data showed that sorbed herbicide was bioavailable and was degraded despite the reduction in soil respiration in the presence of glyphosate. Hermosín et al. (2006) evaluated the bioavailability of organoclay-based formulations of 2,4–D for bacterial degradation in pure culture and leaching potential in soil columns. They observed that the rate of mineralization of ^{14}C-2,4–D from the organoclay complexes was related to the rate of release from the complexes, suggesting that desorption into the aqueous phase was the limiting step for biodegradation. The organoclay-based formulations reduced the leaching losses of 2,4–D in soil columns, and the amount of herbicide leached was considerably less than the amounts of 2,4–D mineralized. They concluded that these formulations slowly released 2,4-D, reducing risk of leaching losses in soil, while maintaining accessibility for bacterial degradation. Sørensen et al. (2006) studied sorption and biodegradation of ^{14}C-labeled glyphosate and the phenoxyacid herbicide 4-chloro-2-methylphenoxy-acetic acid (MCPA) in soil and subsurface samples from a sandy agricultural site and a clay rich till in Denmark. These authors observed that MCPA sorbed to a minor extent and was mineralized rapidly in most samples, except in the deepest layers at both sites, and no relation was found between sorption and mineralization for this herbicide. Interestingly, the highest extent of mineralization of MCPA occurred in the top soil which coincided with the largest sorption and lowest desorption. Conversely, samples which showed higher sorption and low desorption exhibited no or reduced mineralization of glyphosate indicating a limited glyphosate bioavailability. Lerch et al. (2009) assessed the link between bioavailability and mineralization of ^{13}C-2,4–D over a 6-month study, by using stable isotope probing (SIP) coupled to fatty acid methyl ester (FAME) to study degrader populations. These authors reported that the proportion of readily available (water extracted) as well as potentially available (solvent extracted) herbicide residues decreased rapidly to less than 1.2 % of the initial amount added to soil at day 8, which corresponded to the period of highest biodegradation activity. From day 8 onwards, labelled C-2,4–D was present in the form of non-extractable residues (NER), which nontheless were biodegradable. The ^{13}C-2,4–D enriched FAME profiles during this period of incubation were similar to those of the populations degrading 2,4–D when it was still available. They proposed that either the degradation of NER was due to the activity of the same specific degraders involved in degradation of available 2,4–D, limited by desorption of ^{13}C-2,4–D in the soil solution, or that specific degraders were present in a dormant state and/or their fatty acids were being recycled by cells from the same taxonomical group. Overall, these studies show that while adsorption-desorption phenomenons affect bioavailability, there is evidence that sorbed herbicides may be accessible for biodegradation.

The above discussion has practical implications, when predicting the potential *in situ* biodegradation of a certain herbicide. In the field, additional factors other than presence of potential degraders and bioavailability must be considered, including the 1) presence of other contaminants that can compete for adsorption sites and for access to microbial enzymes (Haws et al., 2006); 2) availability of nutrients and co-factors necessary for degraders growth and activity; 3) intrinsic environmental factors (e.g., temperature, oxygen concentration, surface charge, water availability, pH, etc.). The presence and nature of crop residues should also be considered, as they can have great impact on the bioavailability of herbicides in the agroecosystem, helping reduce hazardous effects on soil and water resources (Benoit et al., 1999).

4. How to assess the impact of herbicide exposure on soil microbial communities

Although the desorption of an herbicide from soil particles into the aqueous phase facilitates its biodegradation, the bioavailable fraction is also a potential risk for non-degrading microbial populations. Microbial-mediated processes in soils are of critical importance to ecosystem functions, including transformation of organic matter, nutrient release and degradation of xenobiotics. Therefore, an active soil microbial population is considered a key component of good soil quality (Parkin et al., 1996; Pell et al., 2006). Several biological parameters have been used to assess soil quality and health as affected by agricultural practices (Anderson, 2003; Benedetti & Dilly, 2006). Microbes are expected to be more effective indicators than physical and chemical parameters as they are able to respond immediately to environmental changes (Nannipieri et al., 2002).

The effects of pesticides on the microbiota can be assessed at the whole community-level (e.g., respiration, enzyme activities, biomass, total bacteria counts, etc.) or at sub-community level (i.e., specific physiological or phylogenetic groups). The use of molecular tools has greatly improved the ability to detect pesticide-induced changes, as they allow better resolution of the microbial community structure. The recommended approach for assessing the effects of pesticides on microbial communities is the simultaneous measurement of multiple ecological, structural and functional end points in soil microcosms or terrestrial model ecosystems, rather than reliance on a single assay (Nannipieri et al., 2002; Burrows & Edwards, 2004; Joergensen & Emmerling, 2006). It should be noted that there is little value in assessing the effects of unrealistically high herbicide concentrations in agricultural soils, as there is no reason to expect that those levels would be reached under normal agricultural use. This section is not intended to be an extensive literature review, but rather show the considerable variation in response among soil microbial communities and the diversity of parameters available to assess potential negative impacts of herbicides on the microbiota.

4.1 Microbial respiration

Besides being a generally accepted measure of total soil microbial activity, respiration has been used as a sensitive indicator of pesticide and heavy metal toxicity (e.g. Anderson (2003); Yao et al. (2006)). Zabaloy & Gómez (2008) observed that metsulfuron methyl at 100 $\mu g\ kg^{-1}$ soil depressed cumulative respiration (measured as evolved CO_2 at the end of the 6 weeks incubation) in a Typic Haplustoll [TH] soil while it had no effect in a Petrocalcic Paleustoll [PP] soil, even at a dose of 10 $mg\ kg^{-1}$ soil. Similar results have been reported by

Dinelli et al. (1998) and Accinelli et al. (2002) in soils amended with low doses of sulfonylurea (triasulfuron, primisulfuron methyl and rimsulfuron). Zabaloy & Gómez (2008) proposed that the lower tolerance of the microbial community of TH soil was the result of low adsorption and degradation of herbicide due to higher pH in TH soil (7.4) compared to PP (6.1) (Figure 7). Phytotoxic effects of metsulfuron have been reported in soils with high pH (Walker et al., 1989). Higher degradation of metsulfuron methyl in acidic soils compared to alkaline soils is due to the combined actions of chemical hydrolysis and microorganisms (Pons & Barriuso, 1998; Andersen et al., 2001). No mineralization of either metsulfuron methyl or tribenuron methyl was observed in soils of pH > 8, unless the compounds have been pre-hydrolyzed (Andersen et al., 2001). Several studies reported that the effects of glyphosate and 2,4-D on microbial respiration at low rates, equivalent to agronomic doses, are negligible (e.g. Wardle & Parkinson (1990); Busse et al. (2001); Zabaloy & Gómez (2008)).

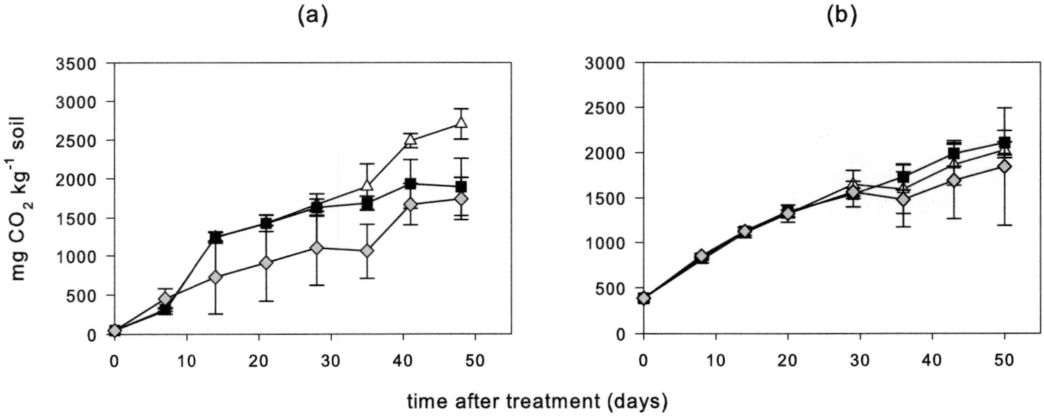

Fig. 7. Effect of two rates of metsulfuron methyl on cumulative CO_2 evolution of Typic Haplustoll (a) and Petrocalcic Paleustoll (b) soils. Symbols: (filled squares) 0.01 mg a.i. kg^{-1} air-dried soil; (gray diamonds) 0.1 mg a.i. kg^{-1} air-dried soil; (empty triangle) control (distilled water). Error bars indicate standard deviation. Error bars not shown were smaller than the symbols. From Zabaloy & Gómez (2008)

4.2 Enzyme activities

Many studies have shown that enzyme activities are sensitive enough to detect the effects of soil pollutants, including heavy metals (Avidano et al., 2005), insecticides (Yao et al., 2006) and herbicides (Sannino & Gianfreda, 2001). Dehydrogenases exist as an integral part of intact cells and represent the oxidative activities of soil microbes, whereas fluorescein diacetate (FDA) hydrolysis can be catalyzed by intracellular and extracellular lipases, esterases and proteases produced by microorganisms (Shaw & Burns, 2006). Both are well-established methods to measure the microbial mineralizing capacity in soil and are suitable to assess broad-spectrum biological activity in the short-term (Nannipieri et al., 2002).
Zabaloy et al. (2008) reported that metsulfuron-methyl and 2,4-D had transient, relatively small (<25% change from control) effects on soil enzyme activities within two weeks after herbicide addition. While both herbicides induced an early reduction in FDA, 2,4-D also stimulated DHA in the different soils analyzed. In contrast, glyphosate caused a significant reduction (50 %) of intracellular dehydrogenase activity, suggesting a strong influence on bacterial metabolism (Zabaloy et al., 2008). Metsulfuron-methyl at comparable doses

inhibited urease, amylase and protease activities in loamy sand and clay loam soils (Ismail et al., 1998). There is general agreement on the lack of significant effects of agricultural rates of metsulfuron (Dinelli et al., 1998; Accinelli et al., 2002) and 2,4-D (Frioni, 1981;Wardle & Parkinson, 1990) on different enzymes. Variable effects of glyphosate and glufosinate on soil enzymatic activities have been reported. In general, literature reports mainly stimulatory effects of glyphosate on enzyme activities for doses within a range of 2-200 mg a.i. kg^{-1} soil (Sannino & Gianfreda, 2001; Accinelli et al., 2002; Araújo et al., 2003; Lupwayi et al., 2007).

4.3 Microbial biomass and abundance

The number and biomass of microorganisms are basic properties of ecological studies, and which can be related to parameters describing microbial activity and soil health (Bölter et al., 2006). Substrate-induce respiration is a commonly-used, sensitive parameter for the observation of pollutant impacts on soil microorganisms (Brohon et al., 2001). Under standardized conditions, the metabolism of glucose added in excess is limited by the amount of active aerobic microbes in soil. Initially, there is no microbial growth and the respiratory response is proportional to glucose-responsive microbial biomass already present in soil (Höper, 2006). The glucose-responsive and more active part of the microbial community, determined by the SIR biomass, is more sensitive to pollutants than the total microbial biomass, as measured biochemically (Höper, 2006; Chander et al., 2001; Zabaloy et al., 2008).
The number of physiological groups of bacteria has also proved to be useful to measure structural changes in soil due to several anthropogenic factors. Glyphosate is an organophosphonate that can be used as a source of P, C or N by either gram-positive or gram-negative bacteria (van Eerd et al., 2003). Accordingly, increases in bacterial abundance and biomass (Zabaloy et al., 2008) and fungal counts (Araújo et al., 2003; Ratcliff et al., 2006) after glyphosate doses comparable to field rates have been observed. Supporting the hypothesis of a bacterial role in glyphosate dissipation, Gimsing et al. (2004) found a high correlation between glyphosate mineralization rates and *Pseudomonas* spp. counts in five different Danish soils. Moreover, two soils with high glyphosate mineralization rates also showed high CFU counts (Gimsing et al., 2004). Conversely, low rates of 2,4-D (< 10 mg kg^{-1} soil) have no effects on heterotrophic bacteria counts (Ka et al., 1995; Merini et al., 2007; Zabaloy et al., 2010). The abundance of cellulose degraders and *Azotobater* were reported to decrease with 2,4-D treatment (Frioni, 1981), although the dose used was several times higher than the expected concentration in soil after a field rate application.

4.4 Microbial community structure

Community structure could be defined as the abundance and proportion of distinct phylogenetic and functional groups. Functional groups are defined by the substrates used for energy metabolism (Pankhurst et al., 1996). Community-level end points may not be sensitive enough to detect minor shifts in microbial community structure, due to the inherent functional redundancy that is recognized to exist in soil microbial communities. The disappearance of a certain member of the microbial community as a result of herbicide (or other pollutant) exposure may eliminate key ecosystem functions and/or impair the ability of the microbial community to respond to other environmental perturbations (i.e., reducing resilience). Physiological, biochemical or genetic profiling methods give insight of such potential shifts at the subcommunity level. Popular methods include community-level physiological profiles

(CLPP), phospholipids fatty acid analysis (PLFA), and various DNA fingerprint techniques (e.g. denaturing or thermal gradient gel electrophoresis [DGGE/TGGE], terminal restriction fragment length polymorphism [T-RFLP]). These and other methods have been summarized in the excellent reviews by Torsvik et al. (1996), Preston-Mafham et al. (2002), Lynch et al. (2004), Kirk et al. (2004), Ogram et al. (2007) and Garland et al. (2007).

Unintended consequences of herbicide applications may be the reduction of sensitive populations and/or stimulation of a certain microbial group with or without detriment to co-existing microbial populations that may compete for available resources. Several investigations that used culture-independent methods reported only slight, short-lived effects of field levels of glyphosate (Weaver et al., 2007; Accinelli et al., 2007) and 2,4-D (Chinalia & Killham, 2006; Macur et al., 2007; Vieublé-Gonod et al., 2006) on microbial communities. No major changes in community structure, assessed by CLPP and PLFA, occurred with application of field rate concentrations of glyphosate in soils from two pine plantations in California (Ratcliff et al., 2006). Both higher abundance of PLFA biomarkers of gram-negative bacteria (Weaver et al., 2007; Lancaster et al., 2009) and fungal to bacterial biomass ratios (Powell et al., 2009) have been reported in glyphosate-treated soils. In a recent study, Zabaloy et al. (2009) reported minor effects of glyphosate on sole C sources utilization with BDOBS. However, the number of 16S ribosomal gene copies, as determined by quantitative PCR (qPCR), increased in a glyphosate-treated soil relative to the control soil, although T-RFLP analysis did not show consistent selective enrichment for specific bacteria species (i.e., no specific phylotype dominated in glyphosate-treated microcosms) (Zabaloy et al., 2009). Due to the enormous diversity of soil microbial communities, more relevant results could be obtained by targeting specific functional groups that are more likely to be directly affected by the herbicide or indirectly by herbicide-induced changes in the soil environment. Interestingly, no effects of glyphosate on denitrifying bacteria nor rhizosphere fungal abundances (qPCR) or communities composition (T-RFLP) have been reported (Hart et al., 2009). Glyphosate was reported to inhibit growth of mycorrhizal fungi and could favor the growth of less desirable fungal species, like soil-borne pathogens (Johal & Huber, 2009). Krzysko-Lupicka & Sudol (2008) observed a bias towards the selection of autochtonous *Fusarium* strains after treatment with the herbicide. This could be related with changes in microbial populations that alter the equilibrium and ultimately lead to diminishing biodiversity, as the postulated decrease in the (pseudomonad) antagonists of fungal pathogens observed by Kremer & Means (2009) in long-term field studies.

The most noticeable effect of 2,4-D on community structure is the enrichment of degrading populations that use this compound as a source of C and energy. Zabaloy et al., 2010 reported a persistent 2,4-D degrading population able to use the herbicide as C and energy source in an agricultural soil where herbicide applications had ceased 2 years before the study. The number of degraders increased immediately after treatment of soil microcosms with 2,4- D and remained high until the end of the incubation, while culturable aerobic heterotrophic bacteria counts were not affected by the herbicide (Figure 8). The addition of succinate (S) as an alternative source of C to soil microcosms did not stimulate degrader population, which confirmed that 2,4-D degradation in this soil was mainly a metabolic process performed by specific degraders. Similar results have been obtained by a number of researchers that used a range of herbicide concentrations in different agricultural soils (e.g. Ka et al. (1995); Merini et al. (2007); Macur et al. (2007); Lerch et al. (2009). One practical implication of the proliferation of soil microbes able to degrade some herbicides, such as foliar-applied chlorophenoxy acids, is that this phenomenon guarantees self-cleaning of herbicide-impacted agricultural soils, reducing the risk of contamination.

4.5 Pollution-induced community tolerance

The pollution-induced community tolerance (PICT) concept is based on the assumption that long-term exposure of a community to a given toxicant will lead to a higher tolerance for this pollutant (Blanck et al., 1988; Blanck, 2002). PICT is tested by collecting intact communities from polluted and reference sites and exposing these communities to contaminants under controlled conditions. Increased community tolerance resulting from the elimination of sensitive species and addition of tolerant species is considered strong evidence that changes were caused by the pollutant. A fundamental step in the PICT measurements is the selection of an ecologically relevant parameter as endpoint that reflects the toxic effects at the community level (Blanck, 2002).

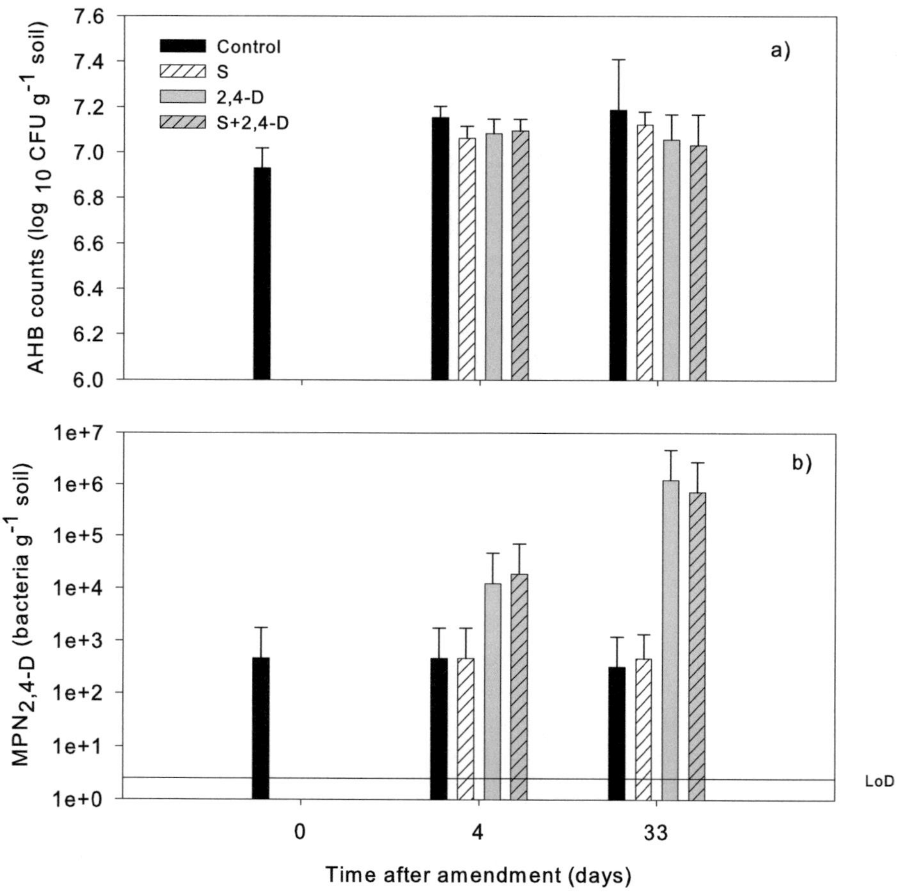

Fig. 8. Effect of combined amendments of 2,4-D and succinate (S) on aerobic heterotrophic bacteria (AHB) counts (a) and most probable number of 2,4-D degraders ($MPN_{2,4-D}$) for soil microcosms sampled after 0, 4 and 33 days of incubation. AHB data are given as means ±S.E (n=3). $MPN_{2,4-D}$ data are represented as median of three replicates and 95% confidence intervals. LoD, limit of detection. From Zabaloy et al., 2010.

Microbial activity may be affected by soil characteristics as well as other environmental factors other than contamination. However, increased tolerance to a specific contaminant is less sensitive to variation in physicochemical variables, and more likely a direct result of contaminant exposure (Siciliano & Roy, 1999; Gong et al., 2000). The PICT approach has been used to study effects of chemicals on microbial communities with various methods such as Biolog™ plates (Schmitt et al., 2004), respirometer (Gong et al., 2000) and methane oxidation assay (Seghers et al., 2003). Zabaloy et al., 2010 used BDOBS to assay mineralization of coumaric acid as an indication of PICT to 2,4-D in an agricultural and a forest soil. This study revealed that past field exposure of the agricultural soil to 2,4-D was enough to develop resistant microbial populations, while the herbicide exerted a more severe inhibitory effect on coumaric acid use in the pristine forest soil (Figure 9). In a similar study, Seghers et al. (2003) reported that long-term use of atrazine and metolachlor selected towards a methanotrophic community more tolerant to the methane oxidation inhibitor 2,4-D in an agricultural soil.

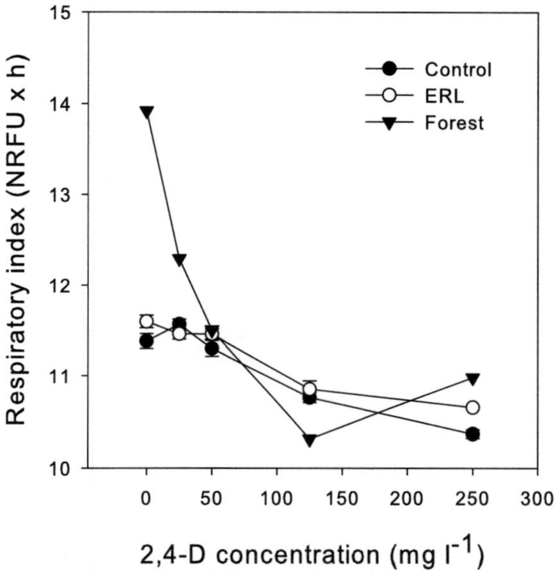

Fig. 9. Respiratory index with coumaric acid as C source, in agricultural soil treated with 5 mg kg^{-1} de 2,4-D (ERL) or untreated (control) and forest soil, exposed to increasing doses of 2,4-D (25-250 mg l^{-1}) in BDOBS. Values represent means ±S.E (n=3); for forest soil is the average of two samples. From Zabaloy et al., 2010

5. Conclusion

Although desorption has been considered a pre-requisite for biodegradation of soil-bound herbicides, there is increasing evidence that sorbed compounds may still be degraded by attached cells. However, there is still considerable work ahead for researchers to understand the mechanisms and populations intervening in these processes. Integrative approaches are essential to study physicochemical and biological factors that affect sorption, bioavailability and biodegradation of herbicides in soil. Development of new molecular methods coupling function and structure may improve our understanding of the role of microbial populations in herbicides degradation and how these compounds affect non-degrading members of the microbial community. Overall, a number of studies have shown that the herbicides 2,4–D, metsulfuron methyl and glyphosate at recommended rates have only transient impacts on soil microbial communities, being glyphosate the one with larger effects, while metsulfuron methyl may be toxic under certain soil conditions (e.g. high pH).

6. References

Accinelli, C., Koskinen, W. C., Becker, J. M. & Sadowsky, M. J. (2007). Environmental fate of two sulfonamide antimicrobial agents in soil, *Journal of Agricultural and Food Chemistry* 55: 2677-2682.

Accinelli, C., Screpanti, C., Dinelli, G. & Vicari, A. (2002). Short-time effects of pure and formulated herbicides on soil microbial activity and biomass, *International Journal of Environmental and Analytical Chemistry* 82: 519–527.

Ainsworth, C. C., Frederickson, J. K. & Smith, S. C. (1993). Effect of sorption on the degradation of aromatic acids and bases, *in* D. M. Linn, T. H. Carski, M. L. Brusseau & F. H. Chang (eds), *Sorption and Degradation of Pesticides and Organic Chemicals in Soil*, Soil Science Society of America, Madison, WI, pp. 125–144.

Altieri, M. & Pengue,W. (2006). GM soybean: Latin America's new coloniser, *Seedling* pp. 13– 17.

Andersen, S. M., Hertz, P. B., Holst, T., Bossi, R. & Jacobsen, C. (2001). Mineralisation studies of $^{14}CO_2$-labeled metsulfuron-methyl, tibenuron-methyl, chlorsulfuron and thifensulfuron-methyl in one Danish soil and groundwater sediment profile, *Chemosphere* 45: 775–782.

Anderson, T.-H. (2003). Microbial eco-physiological indicators to assess soil quality, *Agriculture, Ecosystems and Environment* 98: 285–293.

Araújo, A. S. F., Monteiro, R. T. R.&Abarkeli, R. B. (2003). Effect of glyphosate on the microbial activity of two Brazilian soils, *Chemosphere* 52: 799–804.

Avidano, L., Gamalero, E., Cossa, G. P. & Carraro, E. (2005). Characterization of soil health in an Italian polluted site by using microorganisms as bioindicators, *Applied Soil Ecology* 30: 21–33.

Baelum, J., Henriksen, T., Hansen, H. C. B. & Jacobsen, C. S. (2006). Degradation of 4-chloro-2-methylphenoxyacetic acid in top- and subsoil is quantitatively linked to the class *III tfdA* gene, *Applied and Environmental Microbiology* 72: 1476–1486.

Baelum, J., Nicolaisen, M. H., Holben, W. E., Strobel, B. W., Sørensen, J., & Jacobsen, C. S. (2008). Direct analysis of tfdA gene expression by indigenous bacteria in phenoxy acid amended agricultural soil, *The ISME Journal* 2: 677–687.

Beck, L., Römbke, J., Breure, A. M. & Mulder, C. (2005). Considerations for the use of soil ecological classification and assessment concepts in soil protection, *Ecotoxicology and Environmental Safety* 62: 189–200.

Bedmar, F., Perdigón, J. A. & Monterubbianesi, M. G. (2006). Residual phytotoxicity and persistence of chlorimuron and metsulfuron in soils of Argentina, *Journal of Environmental Biology* 27: 175–179.

Benedetti, A. & Dilly, O. (2006). Introduction, *in* J. Bloem, D. H. Hopkins & A. Benedetti (eds), *Microbiological Methods for Assessing Soil Quality*, CABI Publishing, Wallingford, UK, pp. 3–14.

Benoit, P., Barriuso, E. & Soulas, G. (1999). Degradation of 2,4-D, 2,4-dichlorophenol and 4-chlorophenol in soil after sorption on humified and nonhumified organic matter, *Journal of Environmental Quality* 28: 1127–1135.

Berglöf, T., Koskinen, W. C., Duffy, M. C., Norberg, K. A. & Kylin, H. (2003). Metsulfuron methyl sorption-desorption in field-moist soils, *Journal of Agricultural and Food Chemistry* 51: 3598–3603.

Blanck, H. (2002). A critical review of procedures and approaches used for assessing Pollution- Induced Community Tolerance (PICT) in biotic communities, *Human and Ecological Risk Assessment* 8: 1003–1034.

Blanck, H., Wängberg, S.-A. & Molander, S. (1988). Pollution-induced community tolerance-a new ecotoxicological tool, *in* J. J. Cairns & J. R. Pratt (eds), *Functional testing of aquatic biota for estimating hazards of chemicals*, American Society for Testing and Materials, Philadelphia, PA, USA, pp. 219–230.

Blanco, M. & Stoops, G. (1993). Genesis of pedons with discontinuous argillic horizons in the Holocene loess mantle of the southern Pampean landscape, Argentina, *Journal of South American Earth Science* 47: 353–359.

Boesten, J. J. T. I. (1993). Bioavailability of organic chemicals in soil related to their concentration in the liquid phase: a review, *The Science Total Environment* 134 Supplement 1: 397–407.

Bölter, M., Bloem, J., Meiners, K. & Möller, R. (2006). Enumeration and biovolume determination of microbial cells, *in* J. Bloem, D. H. Hopkins & A. Benedetti (eds), *Microbiological Methods for Assessing Soil Quality*, CABI Publishing, Wallingford, UK, pp. 93–107.

Borggaard, O. K. & Gimsing, L. (2008). Fate of glyphosate in soil and the possibility of leaching to ground and surface waters: a review, *Pest Management Science* 64: 441–456.

Boschin, G., Agostina, A., Arnoldi, A., Marotta, E., Zanardini, E., Negri, M., Valle, A. & Sorlini, C. (2003). Biodegradation of chlorsulfuron and metsulfuron-methyl by *Aspergillus niger* in laboratory conditions, *Journal of Environmental Science and Health. Part B* 38: 737–746.

Brohon, B., Delolme, C. & Gourdon, R. (2001). Complementarity of bioassays and microbial activity measurments for the evaluation of hydrocarbon-contaminated soils quality, *Soil Biology and Biochemistry* 33: 883–891.

Bunce, N. J. (1993). *Introduction to Environmental Chemistry*,Wuerz Publishing Ltd.,Winnipeg, Canada.

Burrows, L. A. & Edwards, C. A. (2004). The use of integrated soil microcosms to assess the impact of carbendazim on soil ecosystems, *Ecotoxicology* 13: 143–161.

Busse, M. D., Ratcliff, A. W., Shestak, C. J. & Powers, R. F. (2001). Glyphosate toxicity and the effects of long-term vegetation control on soil microbial communities, *Soil Biology and Biochemistry* 33: 1777–1789.

Campbell, L. & Davies, B. (1995). Soil sorption of cesium modeled by the Langmuir and Freundlich isotherm equations, *Applied Geochemistry* 10: 715–723.

Caspi, R., Altman, T., Dale, J. M., Dreher, K., Fulcher, C. A., Gilham, F., Kaipa, P., Karthikeyan, A. S. Kothari, A., Krummenacker, M., Latendresse, M., Mueller, L. A., Paley, S. M., Popescu, L., Pujar, A., Shearer, A., Zhang, P. & Karp, P. (2010). The MetaCyc Database of metabolic pathways and enzymes and the BioCyc collection of pathway/Genome Databases, *Nucleic Acids Research* .

Cerdeira, A. L. & Duke, S. O. (2006). The current status and environmental impacts of glyphosate-resistant crops: a review, *Journal of Environmental Quality* 35: 1633–1658.

Chander, K., Dyckmans, J., Hoeper, H., Joergensen, R. G. & Raubuch, M. (2001). Long-term effects on soil microbial properties of heavy metals from industrial exhaust deposition, *Journal of Plant Nutrition and Soil Science* 164: 657–663.

Chen, H., He, X., Rong, X., Chen, W., Cai, P., Liang, W., Li, S. & Huang, Q. (2009). Adsorption and biodegradation of carbaryl on montmorillonite, kaolinite and goethite, *Applied Clay Science* 46: 102–108.

Cheng, H. H. (1990). *Pesticides in the Soil Environment: Processes, Impacts, and Modeling*, Soil Science Society of America, Inc, Madison, Wisconsin, USA.

Chinalia, F. A. & Killham, K. S. (2006). 2,4-Dichlorophenoxyacetic acid (2,4-D) biodegradation in river sediments of Northeast Scotland and its effect on the microbial communities (PLFA and DGGE), *Chemosphere* 64: 1675–1683.

Cooke, C., Shaw, G. & Collins, C. D. (2004). Determination of solid-liquid partition coefficients (kd) for the herbicides isoproturon and trifluralin in five UK agricultural soils, *Environmental Pollution* 132: 541–552.

Cramer, B. K., Westra, P. & Zimdahl, R. L. (1993). Sorption, dissipation, and leaching of metsulfuron in Colorado soils, *Weed Science* 47: 353–359.

DiGiovanni, G. D., Neilson, J. W., Pepper, I. L. & Sinclair, N. A. (1996). Gene transfer of *Alcaligenes eutrophus* JMP134 plasmid pJP4 to indigenous soil recipients, *Applied and Environmental Microbiology* 62: 2521–2526.

Dinelli, G., Vicari, A. & Accinelli, C. (1998). Degradation and side effects of three sulfonylurea herbicides in soil, *Journal of Environmental Quality* 27: 1459–1464.

Dixon, J. B. & Weed, S. B. (1990). *Minerals in Soil Environments*, Soil Science Society of America, Inc.

Don, R. H. & Pemberton, J. M. (1981). Properties of six pesticide degradation plasmids isolated from *Alcaligenes paradoxus* and *Alcaligenes eutrophus*, *Journal of Bacteriology* 145: 681- 686.

Donnelly, P., Entry, J. & Crawford, D. (1993). Degradation of atrazine and 2,4-dichlorophenoxyacetic acid by mycorrhizal fungi at three nitrogen concentrations in vitro, *Applied and Environmental Microbiology* 59: 2642–2647.

Forlani, G., Mangiangalli, A., Nielsen, E. & Suardi, C. M. (1999). Degradation of the phosphonate herbicide glyphosate in soil: evidence for a possible involvement of unculturable microorganisms, *Soil Biology and Biochemistry* 31: 991–997.

Frioni, L. (1981). Efecto de atrazina, linurón y 2,4-D amina sobre algunas propiedades biológicas de un suelo. II Ensayo de laboratorio, *Revista Argentina de Microbiología* pp. 9–16.

Fukumori, F. & Hausinger, R. P. (1993a). *Alcaligenes eutrophus* JMP134 "2,4-dichlorophenoxyacetate monooxygenase" is an α-ketoglutarate-dependent dioxygenase, *Journal of Bacteriology* 175: 2083–2086.

Fukumori, F. & Hausinger, R. P. (1993b). Purification and characterization of 2,4-dichlorophenoxyacetate/α-ketoglutarate dioxygenase, *Journal of Biological Chemistry* 32: 24311–24317.

Garland, J. L., Campbell, C. D. & Mills, A. L. (2007). Physiological profiling of microbial populations, *in* C. J. Hurst, R. L. Crawford, J. L. Garland, D. A. Lipson, A. L. Mills & L. D. Stezenbach (eds), *Manual of Environmental Microbiology*, 3 edn, ASM Press, Washington DC, USA, pp. 126–138.

Ghani, A. &Wardle, D. (2001). Fate of ^{14}C from glucose and the herbicide metsulfuron-methyl in a plant-soil microcosm system, *Soil Biology and Biochemistry* 33: 777–785.

Gimsing, A. L., Borggaard, O. K., Jacobsen, O. S., Aamand, J. & Sørensen, J. (2004). Chemical and microbiological soil characteristics controlling glyphosate mineralisation in Danish surface soils, *Applied Soil Ecology* 27: 233–242.

Gimsing, A., Szilas, C. & Borggaard, O. (2007). Sorption of glyphosate and phosphate by variable-charge tropical soils from Tanzania, *Geoderma* 138: 127–132.

Gong, P., Gasparrini, P., Rho, D., Hawari, J., Thiboutot, S., Ampleman, G. & Sunahara, G. I. (2000). An *in Situ* respirometric technique to measure pollution-induced microbial community tolerance in soils contaminated with 2,4,6-trinitrotoluene, *Ecotoxicology and Environmental Safety* 47: 96–103.

Greenland, D. & Hayes, M. (1978). *The Chemistry of Soil Constituents*, JohnWiley Sons, London, Great Britain.

Gustafson, D. I. (1989). Groundwater ubiquity score: a simple method for assessing pesticide leachability, *Environmental Toxicology and Chemistry* 8: 339–357.

Hart, M. M., Powell, J. R., Gulden, R. H., Dunefield, K. E., Pauls, K. P., Swanton, C. J., Klironomos, J. N., Antunes, P. M., Koch, A. M. & Trevors, J. T. (2009). Separating the effect of crop from herbicide on soil microbial communities in glyphosate-resistant corn, *Pedobiologia* 52: 253–262.

Haws, N. W., Ball, W. P. & Bouwer, E. J. (2006). Modeling and interpreting bioavailability of organic contaminant mixtures in subsurface environments, *Journal of Contaminant Hydrology* 82: 255–292.

Hay, J. V. (1990). Chemistry of sulfonylurea herbicides, *Pesticide Science* 29: 247–261.

He, Y., Shen, D., Hu, L. & Zhu, Y. (2007). Study on metsulfuron-methyl degradation in simulated wheat (*Triticum aestivum* L.) rhizospheric soil with *Penicillium* sp. inoculation, *Water, Air, and Soil Pollution*, 179: 297–307.

Hermosín, M. C., Celis, R., Fanceda, G., Carrizosa, M., Ortega-Calvo, J. & Cornejo, J. (2006). Bioavailability of the herbicide 2,4–D formulated with organoclays, *Soil Biology and Biochemistry* 38:2117-2124

Hinz, C. (2001). Description of sorption data with isotherm equations, *Geoderma* 99: 225–243.

Höper, H. (2006). Substrate-induced respiration, *in* J. Bloem, D. H. Hopkins & A. Benedetti (eds), *Microbiological Methods for Assessing Soil Quality*, CABI Publishing, Wallingford, UK, pp. 84–92.

Hunter, R. (2002). *Foundations of Colloid Science*, Oxford University Press, Great Britain.

Ismail, B. S., Yapp, K. F. & Omar, O. (1998). Effects of metsulfuron-methyl on amylase, urease, and protease activities in two soils, *Australian Journal of Soil Research* 36: 449–456.

Joergensen, R. G. & Emmerling, C. (2006). Methods for evaluating human impact on soil microorganisms based on their activity, biomass, and diversity in agricultural soils, *Journal of Plant Nutrition and Soil Science* 169: 295–309.

Johal, G. & Huber, D. (2009). Glyphosate effects on diseases of plants, *European Journal of Agronomy* 31: 144–152.

Juhler, R. K., Henriksen, T. H., Ernestsen, V., Vinther, F. P. & Rosenberg, P. (2008). Impact of basic soil parameters on pesticides disappearance investigated by multivariate partial least square regression and statistics, *Journal of Environmental Quality* 37: 1719–1732.

Ka, J. O., Burauel, P., Bronson, J. A., Holben, W. & Tiedje, J. M. (1995). DNA probe analysis of microbial community selected in field by long-term 2,4-D-treated field soils, *Soil Science Society of America Journal* 59: 1581–1587.

Kah, J. M. & Brown, C. D. (2006). Adsorption of ionisable pesticides in soils, *Reviews of Environmental Contamination Toxicology* 188: 149–217.

Kamagata, Y., Fulthorpe, R., Tamura, K., Takami, H., Forney, L. & Tiedje, J. (1997). Pristine environments harbor a new group of oligotrophic 2,4- dichlorophenoxyacetic acid degrading bacteria, *Applied and Environmental Microbiology* 63(6): 2266–2272.

Ketelsen, H. & Meyer-Windel, S. (1999). Adsorption of brilliant blue FCF by soils, *Geoderma* 90: 131–145.

Kibe, K., Takahashi, M., Kameya, T. & Urano, K. (2000). Adsorption equilibriums of principal herbicides on paddy soils in Japan, *The Science of the Total Environment* 263: 115–125.

Kirk, J. L., Beaudette, L. A., Hart, M., Moutoglis, P., Klironomos, J. N., Lee, H. & Trevors, J. T. (2004). Methods of studying soil microbial diversity, *Journal of Microbiological Methods* 58: 169–188.

Kishore, G. M. & Jacob, G. S. (1987). Degradation of glyphosate by *Pseudomonas* sp. PG2982 via a sarcosine intermediate, *Journal of Biological Chemistry* 262: 12164–12168.

Kitagawa, W., Takami, S., Miyauchi, K., Masai, E., Kamagata, Y., Tiedje, J. M. & Fukuda, M. (2002). Novel 2,4-dichlorophenoxyacetic acid degradation genes from oligotrophic *Bradyrhizobium* sp. strain HW13 isolated from a pristine environment, *Journal of Bacteriology* 184: 509–518.

Kononova, S. V. & Nesmeyanova, M. A. (2002). Phosphonates and their degradation by microorganisms, *Biochemistry (Moscow)* 67(2): 184–195.

Kremer, R. & Means, N. (2009). Glyphosate and glyphosate-resistant crop interactions with rhizosphere microorganisms, *European Journal of Agronomy* 31: 153–161.

Krzyśko-Lupicka, T. & Orlik, A. (1997). The use of glyphosate as the sole source of phosphorus or carbon for the selection of soil-borne fungal strains capable to degrade this herbicide, *Chemosphere* 34: 2601–2605.

Krzysko-Lupicka, T. & Sudol, T. (2008). Interactions between glyphosate and autochthonous soil fungi surviving in aqueous solution of glyphosate, *Chemosphere* 71: 1386–1391.

Lancaster, S. H., Hollister, E. B., Senseman, S. A. & Gentry, T. J. (2009). Effects of repeated glyphosate applications on soil microbial community composition and the mineralization of glyphosate, *Pest Management Science* 66: 59–64.

Langmuir, I. (1918). The adsorption of gases on plane surface of glass, mica and platinum, *Journal of the American Chemical Society* 40(9): 1361–1403.

Laor, Y., Strom, P.&Farmer,W. J. (1996). The effect of sorption on phenanthrene bioavailability, *Journal of Biotechnology* 51: 227–234.

Lee, T., Kurata, S., Nakatsu, C. & Kamagata, Y. (2005). Molecular analysis of bacterial community based on 16S rDNA and functional genes in activated sludge enriched with 2,4-dichlorophenoxyacetic dcid (2,4-D) under different cultural conditions, *Microbial Ecology* 49.

Lerch, T., Dignac, M.-F., Barriuso, E., Bardoux, G. & Mariotti, A. (2007). Tracing 2,4-D metabolism in *Cupriavidus necator* JMP134 with ^{13}C-labelling technique and fatty acid profiling, *Journal of Microbiological Methods* 71: 162–174.

Lerch, T. Z., Dignac, M.-F., Nunan, N., Bardoux, G., Barriuso, E. & Mariotti, A. (2009). Dynamics of soil microbial populations involved in 2,4-D biodegradation revealed by FAME-based Stable Isotope Probing, *Soil Biology and Biochemistry* 41: 77–85.

Lipok, J., Dombrovska, L., Wieczorek, P. & Kafarski, P. (2003). The ability of fungi isolated from stored carrot seeds to degrade organophosphonate herbicides, *Pesticide in Air, Plant, Soil and Water System, Proceedings of the XII Symposium Pesticide Chemistry*.

Liu, C. M., McLean, P. A., Sookdeo, C. C. & Cannon, F. C. (1991). Degradation of the herbicide glyphosate by memebers of the family *Rhizobiaceae*, *Applied and Environmental Microbiology* 57: 1799–1804.

Lupwayi, N. Z., Hanson, K. G., Harker, K. N., abd R. E. Blackshaw, G. W. C., O'Donovan, J. T., Johnson, E. N., Gan, Y., Irvine, R. B. & Monreal, M. A. (2007). Soil microbial biomass, functional diversity and enzyme activity in glyphosate-resistant wheat-canola rotations under low-disturbance direct seeding and conventional tillage, *Soil Biology and Biochemistry* 39: 1418–1427.

Lynch, J. M., Benedetti, A., Insam, H., Nuti, M. P., Smalla, K., Torsvik, V. & Nannipieri, P. (2004). Microbial diversity in soil: ecological theories, the contribution of molecular techniques and the impact of transgenic plants and transgenic microorganisms, *Biology and Fertility of Soils* 40: 363–385.

Macur, R. E., Wheeler, J. T., Burr, M. D. & Inskeep, W. P. (2007). Impacts of 2,4-D application on soil microbial community and on populations associated with 2,4-D degradation, *Microbiological Research* 162: 37–45.

Martínez-Villegas, N., Flores-Vélez, L. M. & Domínguez, O. (2004). Sorption of lead in soil as a function of pH: a study case in Mexico, *Chemosphere* 57: 1537–1542.

Martins, J. M. & Mermoud, A. (1998). Sorption and degradation of four nitroaromatic herbicides in mono and multi-solute saturate/unsaturated soil batch sylstems, *The Science of the Total Environment* 33: 187–210.

McGowan, C., Fullthorpe, R. R.,Wrigth, A. M. & Tiedje, J. M. (1998). Evidence for interspecies gene transfer in the evolution of 2,4-dichlorophenoxyacetic acid degraders, *Applied and Environmental Microbiology* 64: 4089–4092.

McGrath, J. W., Ternan, N. G. & Quinn, J. P. (1997). Utilization of organophosphonates by environmental microorganisms, *Letters in Applied Microbiology* 24: 69–73.

Merini, L. J., Cuadrado, V., Flocco, C. G. & Giulietti, A. M. (2007). Dissipation of 2,4-D in soils of the Humid Pampa region, Argentina: a microcosm study, *Chemosphere* 68: 259–265.

Muller, R. H., Kleinsteuber, S. & Babel, W. (2001). Physiological and genetic characteristics of two bacterial strains utilizing phenoxypropionate and phenoxyacetate herbicides, *Microbiological Research* 156: 121–131.

Nannipieri, P., Kandeler, E. & Ruggiero, P. (2002). Enzyme activities and microbiological and biochemical processes in soil, *in* R. G. Burns & R. P. Dick (eds), *Enzymes in the Environment: Activity, Ecology and Applications*, Marcel Dekker Inc., chapter 1, pp. 1–33.

Obojska, A., Ternan, N. G., Lejczak, B., Kafarski, P. & McMullan, G. (2002). Organophosphonate utilization by the termophile *Geobacillus caldoxylosilyticus* T20, *Applied and Environmental Microbiology* 68: 2081–2084.

Ogram, A., Castro, H. & Chauhan, A. (2007). Methods of soil microbial community analysis, *in* C. J. Hurst, R. L. Crawford, J. L. Garland, D. A. Lipson, A. L. Mills & L. D. Stetzenbach (eds), *Manual of Environmental Microbiology*, 3 edn, ASM Press, Washington D.C., chapter 53, pp. 652–662.

Oliveira Jr., R. S., Koskinen, W. C. & Ferreira, F. A. (2001). Sorption and leaching potential of herbicides on Brazilian soils, *Weed Research* 41: 97110.

Pallud, C., Dechesne, A., Gaudet, J. P., Debouzie, D. & Grundmann, G. L. (2004). Modification of spatial distribution of 2,4-dichlorophenoyacetic acid degrader microhabitats during growth in soil columns, *Applied and Environmental Microbiology* 70: 2709–2716.

Pankhurst, C. E., Ophel-Keller, K., Doube, B. M. & Gupta, V. V. S. R. (1996). Biodiversity of soil microbial communities in agricultural systems, *Biodiversity Conservation* 5: 197–209.

Park, J.-H., Feng, Y., Ji, P., Voice, T. C. & Boyd, S. A. (2003). Assessment of bioavailability of soil-sorbed atrazine, *Applied and Environmental Microbiology* 69: 3288–3298.

Park, J. H., Kay, D., Zhao, X. D., Boyd, S. A. & Voice, T. C. (2001). Kinetic modelling of bioavailability for sorbed-phase 2,4-dichlorophenoxyacetic acid, *Journal of Environmental Quality* 30.

Parkin, T. B., Doran, J. W. & Vizcaíno-Franco, E. (1996). Field and laboratory tests of soil respiration, *in* J. W. Doran & A. J. Jones (eds), *Methods for Assessing Soil Quality*, no. 49 edn, ASA, CSSA, and SSSA, Madison, WI, pp. 231–245.

Pateiro-Moure, M., Pérez-Novo, C., Arias-Estévez, M., Rial-Otero, R. & Simal-Gándara, J. (2009). Effect of organic matter and iron oxides on quaternary herbicide sorption-desorption in vineyard-devoted soils, *Journal of Colloid and Interface Science* 333: 431–438.

Pell, M. J., Stenström, J. & Granhall, U. (2006). Soil respiration, *in* J. Bloem, D. H. Hopkins & A. Benedetti (eds), *Microbiological Methods for Assessing Soil Quality*, CABI Publishing, Wallingford, UK, pp. 117–126.

Pengue, W. A. (2004). Transgenic crops in Argentina and its hidden costs, *in* E. Ortega & S. Ulgiati (eds), *Proceedings of IV Biennial International Workshop Advances in Energy Studies*, Campinas, Brazil, pp. 91-101.

Pimentel, D. (1995). Amounts of pesticides reaching target pests: Environmental impacts and ethics, *Journal of Agricultural and Environmental Ethics* 8: 17–29.

Piutti, S., Hallet, S., Rouseaux, S., Philippot, L., Soulas, G. & Martin-Laurent, F. (2002). Accelerated mineralisation of atrazine in maize rhizosphere soil, *Biology and Fertility of Soils* 36: 434–441.

Pons, N. & Barriuso, E. (1998). Fate of metsulfuron-methyl in soils in relation to pedo-climatic conditions, *Pesticide Science* 53: 311–323.

Powell, J. R., Levy-Booth, D. J., Gulden, R. H., Asbil, W. L., Campbell, R. G., Dunfield, K. E., Hamill, A. S., Hart, M. M., Lerat, S., Nurse, R. E., Pauls, K. P., Sikkema, P. H., Swanton, C. J., Trevors, J. T. & Klironomos, J. N. (2009). Effects of genetically modified, herbicide-tolerant crops and their management on soil food web properties and crop litter decomposition, *Journal of Applied Ecology* 46: 388–396.

Preston-Mafham, J., Boddy, L. & Randerson, P. F. (2002). Analysis of microbial community functional diversity using sole-carbon-source utilisation profiles: a critique, *FEMS Microbiology Ecology.* 42: 1–14.

Pusino, A., Pinna, V. & Gessa, C. (2003). Azimsulfuron sorption-desorption on soil, *Journal of Agricultural and Food Chemistry* 52: 3462–3466.

Ratcliff, A.W., Busse, M. D. & Shestak, C. J. (2006). Changes in microbial community structure following herbicide (glyphosate) additions to forest soils, *Applied Soil Ecology* 34: 114- 124.

Roberts, T. R. (ed.) (1998). *Metabolic Pathways of Agrochemicals: Part 1. Herbicides and Plant Regulators*, Royal Society of Chemistry.

Romero, M., Reinoso, E., Kiernan, A. M. & Crdoba, S. (2004). Biodegradation of glyphosate by wild yeasts, *Revista Mexicana de Micología* 19: 45–50.

Sailaja, K. & Satyaprasad, K. (2006). Degradation of glyphosate in soil and its effect on fungal population, *Journal of Environmental Science and Engineering* 48: 189–190.

Sannino, F. & Gianfreda, L. (2001). Pesticides influence on soil enzymatic activities, *Chemosphere* 45: 417–425.

Schmitt, H., van Beelen, P., Tolls, J. & Van Leeuwen, C. L. (2004). Pollution-induced community tolerance of soil microbial communities caused by the antibiotic sulfachloropyridazine, *Environmental Science and Technology* 38: 1148–1153.

Schnürer, Y., Persson, P., Nilsson, M., Nordgren, A. & Giesler, R. (2006). Effects of sorption on microbial degradation of glyphosate, *Environmental Science and Technology* 40: 4145- 4150.

Seghers, D., Bulcke, R., Reheul, D., Siciliano, S., Top, E. & Verstraete, W. (2003). Pollution induced community tolerance (PICT) and analysis of 16S rRNA genes to evaluate the long-term effects of herbicides on methanotrophic communities in soil, *European Journal of Soil Science* 54: 679–684.

Shaw, L. J. & Burns, R. G. (2004). Enhanced mineralization of [U-^{14}C]2,4-dichlorophenoxyacetic acid in soil from the rhizosphere of *Trifolium pratense*, *Applied and Environmental Microbiology* 70(8): 4766–4774.

Shaw, L. J. & Burns, R. G. (2005). Rhizodeposits of *Trifolium pratense* and *Lolium perenne*: their comparative effects on 2,4-D mineralization in two contrasting soils, *Soil Biology and Biochemistry* 37: 995–1002.

Shaw, L. J. & Burns, R. G. (2006). Enzyme activity profiles and soil quality, *in* J. Bloem, D. H. Hopkins & A. Benedetti (eds), *Microbiological Methods for Assessing Soil Quality*, CABI Publishing, Wallingford, UK, pp. 158–171.

Siciliano, S. D. & Roy, R. (1999). The role of soil microbial tests in ecological risk assessment: Differentiating between exposure and effects, *Human and Ecological Risk Assessment* 5: 671–682.

Singh, B. K. & Walker, A. (2006). Microbial degradation of organophosphorous compounds, *FEMS Microbiological Review* 30: 428–471.

Smith, J. L. & Collins, H. P. (2007). Management of organisms and their processes in soils, *in* E. A. Paul (ed.), *Soil Microbiology, Ecology and Biochemistry*, Academic Press, Oxford, UK, pp. 471–502.

Sørensen, S. R., Schulz, A., Jacobsen, O. S. & Aamand, J. (2006). Sorption, desorption and mineralisation of the herbicides glyphosate and MCPA in samples from two Danish soil and subsurface profiles, *Environmental Pollution* 141: 184–194.

Sparks, D. (1986). *Soil Physical Chemistry*, CRC Press, Inc., Boca Raton, Florida, USA. Sparks, D. (2003). *Environmental Soil Chemistry*, Academic Press, California, USA.

Stenrød, M., Charnay, M.-P., Benoit, P. & Eklo, O. M. (2006). Spatial variability of glyphosate mineralization and soil microbial characteristics in two Norwegian sandy loam soils as affected by surface topographical features, *Soil Biology and Biochemistry* 38: 962- 971.

Stumm, W. (1992). *Chemistry of the Solid-Water Interface*, John Wiley Sons, Inc., New York, USA.

Tiedje, J. M., Duxbury, J. M., Alexander, M. & Dawson, J. E. (1969). 2,4-D metabolism: pathway of degradation of chlorocatechols by *Arthrobacter* sp., *Journal of Agricultural and Food Chemistry* 17: 1021–1026.

Torsvik, V., Sørheim, R. & Goksøyr, J. (1996). Total bacterial diversity in soil and sediment communities a review, *Journal of Industrial Microbiology*. 17: 170–178.

Vallaeys, T., Fullthorpe, R. R., Wrigth, A. M. & Soulas, G. (1996). The metabolic pathway of 2,4-dichlorophenoxyacetic acid degradation involves different families of *tfdA* and *tfdB* genes according to PCR-RFLP analysis, *FEMS Microbiology Ecology*. 20: 163–172.

Valverde-García, A., Gonzalez-Pradas, E., Villafranca-Sanchez, M., Rey-Bueno, F. D. & García- Rodriguez, A. (1998). Adsorption of tirad and dimethoate on Almeria soils, *Soil Science Society American Journal* 52: 1571–1574.

van Eerd, L. L., Hoagland, R. E., Zablotowicz, R. M. & Hall, J. C. (2003). Pesticide metabolism in plants and microorganisms, *Weed Science* 51: 472–495.

Vázquez, B., Curvetto, N. & Bianchinotti, M. (2008). Hongos degradadores de metsulfurón metilo presentes en suelos agrícolas de Argentina. Aislamiento y evaluación, *Proceedings of VI CLAM Congreso Latinoamericano de Micología*, p. 308.

Vázquez, M., Bianchinotti, M. & Curvetto, N. (2009). Resistencia de aislamientos de *Trichoderma harzianum* a altas concentraciones de metsulfurón metilo, *Boletín de la Sociedad Argentina de Botánica*, Vol. 44 (Supl.), pp. 130–131.

Vieublé-Gonod, L., Chadoeuf, J. & Chenu, C. (2005). Spatial distribution of microbial 2,4-dichlorophenoxy acetic acid mineralization from field to microhabitat scales, *Soil Science Society of America Journal* 70(1): 64–71.

Vieublé-Gonod, L., Martin-Laurent, F. & Chenu, C. (2006). 2,4-D impact on bacterial communities, and the activity and genetic potential of 2,4-D degrading communities in soil, *FEMS Microbiology Ecology* 58: 529–537.

Vroumsia, T., Steiman, R., Seigle-Murandi, F., Benoit-Guyod, J. & Groupe pour l'Étude du Devenir des Xénobiotiques dans l'Environnement (GEDEXE), G. (2005). Fungal bioconversion of 2,4-dichlorophenoxyacetic acid (2,4-D) and 2,4-dichlorophenol (2,4-DCP), *Chemosphere* 60: 1471–1480.

Walker, A., Cotterill, E. G. & Welch, S. J. (1989). Adsorption and degradation of chlorsulfuron and metsulfuron-methyl in soils from different depths, *Weed Research* 29: 28–37.

Wardle, D. A. & Parkinson, D. (1990). Effects of three herbicides on soil microbial activity and biomass, *Plant and Soil* 122: 21–28.

Weaver, M. A., Krutz, L. J., Zablotowicz, R. M. & Reddy, K. N. (2007). Effects of glyphosate on soil microbial communities and its mineralization in a Mississippi soil, *Pest Management Science* 63: 388–393.

Weber, J. B., Wilkerson, G. G. & Reinhardt, C. F. (2002). Calculating pesticide sorption coefficients (K_d) using selected soil properties, *Chemosphere* 55: 157–166.

Woodburn, A. T. (2000). Glyphosate: production, pricing, and use worldwide, *Pest Management Science* 56: 309–312.

Xi, Y., Mallavarapu, M. & Naidu, R. (2010). Adsorption of the herbicide 2,4-D on organopalygorskite, *Applied Clay Science* 49: 255–261.

Yao, X., Min, H., Lü, Z. & Yuan, H. (2006). Influence of acetamiprid on soil enzymatic activities and respiration, *European Journal Soil Biology* 42: 120–126.

Yu, Y., Wang, X., Luo, Y., Yang, J., Yu, J. & Fan, D. (2005). Fungal degradation of metsulfuron methyl in pure cultures and soil, *Chemosphere* 60: 440–466.

Zabaloy, M. C., Garland, J. L. & Gómez, M. A. (2008). An integrated approach to evaluate the impacts of the herbicides glyphosate, 2,4-D and metsulfuron-methyl on soil microbial communities in the Pampas region, Argentina, *Applied Soil Ecology* 40: 1–12.

Zabaloy, M. C., Garland, J. L. & Gómez, M. A. (2009). Assessment of microbial community function and structure in soil microcosms exposed to glyphosate, *Proceedings of the VI Congress of the Argentine Society of General Microbiology- SAMIGE*, Carlos Paz, Argentina, p. 40. URL: www.samige.org.ar/2009.php

Zabaloy, M. C., Garland, J. L. & Gómez, M. A. (2010). Assessment of the impact of 2,4-dichlorophenoxyacetic acid (2,4-D) on indigenous herbicide-degrading bacteria and microbial community function in an agricultural soil, *Applied Soil Ecology* 46: 240-246.

Zabaloy, M. C. & Gómez, M. A. (2008). Microbial respiration in soils of the Argentine Pampas after metsulfuron-methyl, 2,4-D and glyphosate treatments, *Communications in Soil Science and Plant Analysis* 39: 370–385.

Zanardini, E., Arnoldi, A., Boschin, G., Agostina, A., Negri, M. & Sorlini, C. (2002). Degradation pathways of chlorsulfuron and metsulfuron-methyl by a *Pseudomonas fluorescens* strain, *Annals of Microbiology* 52: 25–37.

Zanini, G., Avena, M., Fiol, S. & Arce, F. (2006). Effects of pH and electrolyte concentration on the binding between a humic acid and an oxazine dye, *Chemosphere* 63: 430–439.

Zanini, G. P., Maneiro, C., Waiman, C., Galantini, J. A. & Rossel, R. A. (2009). Adsorption of metsulfuron methyl on soils under no-till system in semiarid Pampean Region, Argentina, *Geoderma* 149: 110–115.

10

Application of Mutated Acetolactate Synthase Genes to Herbicide Resistance and Plant Improvement

Masanori Shimizu[1], Kiyoshi Kawai[2], Koichiro Kaku[2], Tsutomu Shimizu[2] and Hirokazu Kobayashi[3]
[1]*School of Health Promotional Science, Hamamatsu University, 1230 Miyakoda, Hamamatsu 431-2102*
[2]*Life Science Research Institute, Kumiai Chemical Industry Co., Ltd., Kakegawa 439-0031*
[3]*Laboratory of Plant Molecular Improvement and Global COE Program (MEXT), Graduate School of Nutritional and Environmental Sciences, University of Shizuoka, 52-1 Yada, Suruga, Shizuoka 422-8526*
Japan

1. Introduction

Herbicides have been used to enhance the productivity of plants including crops by killing the weeds which compete with the growth of cultivated plants. They have also been utilized as a tool to improve plants by means of genetic engineering, whereby transformed plants containing genes for enzymes which impart tolerance to herbicides are selected. These genes are designated as "selectable markers" and are utilized for the production of genetically-modified (GM) plants. Selectable markers used for the selection of plants in which genes of interest are successfully integrated into the genome of host plants include genes that impart tolerance to antibiotics or herbicides, the majority of which are derived from bacteria: genes for neomycin phosphotransferase II (*nptII*) from Tn5 in *Escherichia coli*, 5-enoylpyruvate shikimate-3-phosphate synthase (*epsps*) from *Agrobacterium* sp. CP4, phosphinothricine acetyltransferase (*pat, bar* for bialaphos resistance) from *Streptomyces viridochromogenes*, and aminoglycoside-3″-adenyltransferase (*aadA*) for spectinomycin resistance from *Shigella flexneri* (Hare and Chua, 2002). The safety of genes used for antibiotic resistance is questionable given the possibility that these genes might be transferred into pathogenic bacteria that may be converted to antibiotic-resistant strains. A search for appropriate selectable markers from plants is therefore desirable.

The impact on the environment is another important factor that must be considered, and efforts need to be made to minimize so-called "genetic pollution" or detrimental effects on the ecosystem. The transfer of foreign genes into other non-transgenic plants is most reliably performed via pollen. Apprehension associated with this process is dispelled when considering the transformation of plastids such as chloroplasts. Since genes in plastids of most plant species are inherited maternally, they represent genes that are not transferred into other plants via pollen. Therefore, the development of methodologies based on the genetic manipulation of plastid genomes, in addition to that of nuclear genomes where the engineering has already been established, is necessary for ecological safety. Our efforts have focused on the use of acetolactate synthase (ALS, EC 2.2.1.6), also referred to as acetohydroxyacid synthase (AHAS), an enzyme involved in the biosynthesis of branched-chain amino acids in chloroplasts in higher plants. This enzyme is uniquely suited for use in biotechnology and basic research.

2. ALS and ALS-inhibiting herbicides

ALS is a common enzyme that catalyzes the first step of the biosynthetic pathway of the branched-chain amino acids valine, leucine and isoleucine. ALS is the primary target site for at least five structurally distinct classes of herbicides including sulfonylureas (SUs), imidazolinones (IMs), triazolopyrimidine sulfonamides (TPs), pyrimidinylsalicylates (pyrimidinylcarboxylates, PCs), and sulfonylaminocarbonyl-triazolinones (Figure 1, see the chapter written by Sato et al.) (Shimizu et al., 2002). ALS-inhibiting herbicides are widely used in agriculture given their high weed control efficacy, high crop-weed selectivity, low use rates and low levels of mammalian toxicity (Sharner & Singh 1997).

Plant ALS genes encoding the catalytic (large) subunits were first isolated from Arabidopsis and tobacco utilizing the yeast ALS gene as a heterologous hybridization probe (Mazur et al., 1987). Since then, some plant ALS genes encoding catalytic subunits have been cloned and characterized (Bernasconi et al., 1995; Fang et al., 1992; Grula et al., 1995; Rutledge et al., 1991). The plant ALS regulatory (small) subunit has been shown to enhance the catalytic activity of the large subunit and to confer sensitivity to feedback inhibition by branched-chain amino acids (Lee & Duggleby 2001). Plants and cultured plant cells resistant to SU- and IM-type ALS-inhibiting herbicides have been generated using conventional mutation breeding methods and *in vitro* cell selection (Hart et al., 1993; Newhouse et al., 1991; Rajasekaran et al., 1996). ALS genes encoding catalytic subunits have been cloned from some

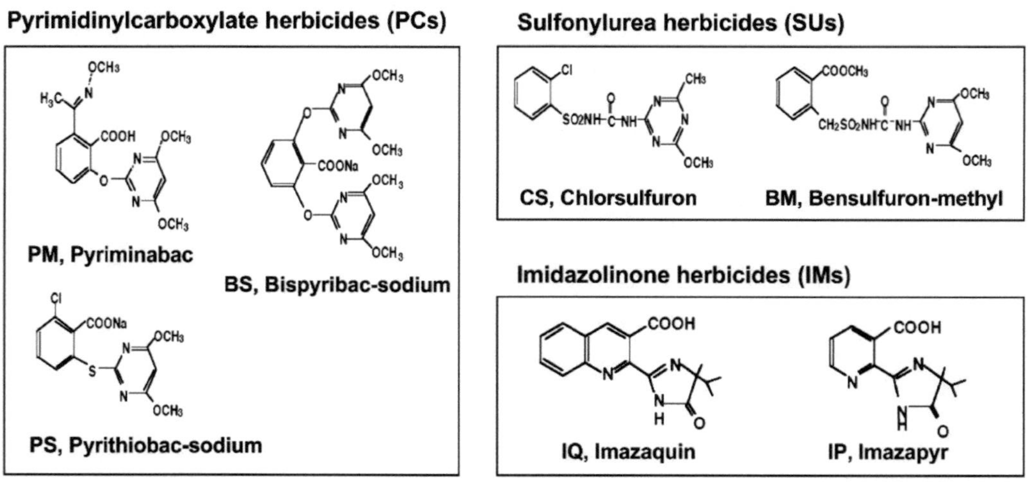

Fig. 1. Herbicides that inhibit ALS activity. These herbicides can be classified into the three classes PC, SU and IM. BS, PS and PM belong to the PC class of herbicides, CS and BM to the SU class of herbicides, and IQ and IP to the IM class of herbicides.

of these plants, and their sequences were found to possess single or double mutations. These mutated ALS (mALS) genes have been revealed to confer resistance to ALS-inhibiting herbicides.

3. ALS mutations interfering with herbicide actions

Herbicide-resistant ALS genes are useful not only for the generation of transgenic plants that express resistance to the corresponding herbicide, but also for introducing foreign traits into plants as selectable markers. We have isolated a double-mutated ALS gene from rice cells (*OsALS-W548L/S627I*) (Figure 2), which confers a high level of resistance to the PC-type ALS-inhibiting herbicide bispyribac-sodium (BS) (Figure 1), and demonstrated that it could be used as a selectable marker for generating transgenic rice plants (Kawai et al., 2007a; Kawai et al., 2007b). We also found that the single amino acid substitution S627I in the ALS gene (*OsALS-S627I*) imparts high levels of resistance to the PCs pyrithiobac-sodium (PS) and pyriminobac (PM). It was postulated that these mALS genes coupled with the PC-type ALS-inhibiting herbicides might be promising selectable markers for various plant species. Indeed, it has been shown that *OsALS-W548L/S627I* works as an effective selectable marker gene for the transformation of wheat (Ogawa et al., 2008) and soybean (Tougou et al., 2009).

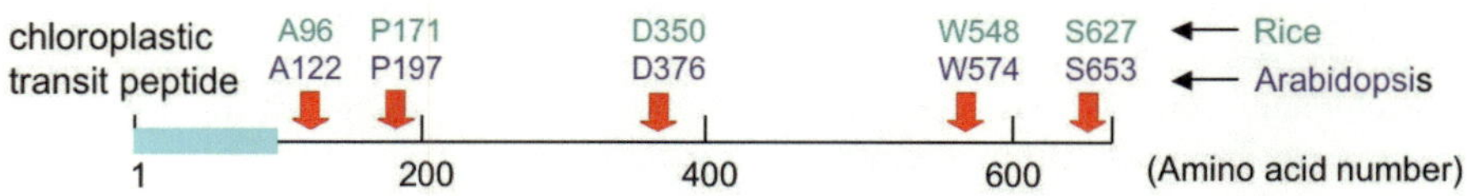

Fig. 2. Schematic representation of amino acid mutations conferring resistance to ALS-inhibiting herbicides. Amino acid residue numbers shown under the peptide are those of rice ALS.

Recombinant Arabidopsis mALSs, AtALS-W574L/S653I and -S653I, were expressed in *Escherichia coli* cells. These recombinant mALSs exhibited resistance to PCs, and showed similar sensitivity against herbicides to rice recombinant ALSs with the corresponding mutations (Table 1) (Kawai et al., 2008). We have shown that these Arabidopsis ALS genes can also be utilized as selectable markers for the genetic transformation of Arabidopsis (Kawai et al., 2010). It has been revealed that selection by PCs can clearly distinguish resistant seedlings from non-resistant seedlings of Arabidopsis at very low concentrations of herbicide compared with kanamycin selection (Figure 3). The concentrations of PCs employed for the selection were about 1000-fold lower than that of kanamycin. We performed *in vivo* ALS assays in an effort to determine whether the Arabidopsis seedlings, selected by resistance to PCs, were indeed transformants. Although the *in vivo* ALS assay was originally developed for the analysis of ALS-resistant weeds (Gerwick et al., 1993), we reasoned that the procedure could be applied for the evaluation of transformants. PC-type ALS-inhibiting herbicide-resistant seedlings showing *in vivo* ALS activity were further analyzed to verify integration of the T-DNA region within the genome by PCR. Results suggested that the assay could reliably be used to evaluate transformation. The advantage of using mALS genes over other selectable markers is that the *in vivo* ALS assay confirms both integration of the mALS gene and expression of the corresponding protein in the selected plants. Furthermore, the *in vivo* ALS assay allows for a larger number of samples to be easily tested at relatively lower costs compared to PCR-based screening methods. Differences in levels of acetoin accumulation were observed between the independent transgenic lines. This observation may reflect copy number differences or differential expression of ALS due to positional effects in the Arabidopsis genome. If so, transgenic plants expressing a high or desirable level of the gene of interest may be identified at an early stage of transformation.

Herbicide [b]	RS ratio [a]			
	AtALS-S653I	OsALS-S627I	AtALS-W574L/S653I	OsALS-W548L/S627I
CS	4.2	2.4	4,400	200
BM	43	73	>9,100	>14,000
IQ	21	6.8	>56	>45
IP	>14	>10	>14	>10
BS	83	41	>17,000	>16,000
PS	350	200	>2,900	>9,100
PM	3,300	2,500	>8,300	>13,000

[a] RS ratios for mutated ALSs were obtained by calculating the ratio of the I_{50} value for each mutated ALS to the I_{50} value for the wild-type.

[b] SUs: CS, chlorsulfuron; BM, bensulfuron-methyl; IMs: IQ, imazaquin; IP, imazapyr; PCs: PM, pyriminobac; PS, pyrithiobac-sodium; BS, bispyribac-sodium.

Table 1. Degree of resistance of recombinant mALSs to ALS-inhibiting herbicides

Fig. 3. Comparison of screening seedlings of Arabidopsis transgenic with *AtALS-W574L/S653I* in pMLH7133 binary vector (Kawai et al., 2010) encoding *nptII* by BS or kanamycin. A, 0.045 ppm BS; B, 50 ppm kanamycin.

4. Species-specific properties of ALS mutations

As mentioned above, the degrees of resistance of Arabidopsis and rice recombinant ALS proteins with identical mutations to PCs are very similar. However, the sensitivity of transgenic Arabidopsis to PCs indicated that the degree of resistance to PCs of transformants expressing Arabidopsis mALSs was greater than those of transformants expressing rice mALSs (Figure 4). It is known that plant ALSs have a signal peptide that is required for translocation of the protein into the chloroplast (Duggleby & Pang, 2000), although the exact size of the signal peptide remains to be determined experimentally. Several reports have indicated that the size of the signal peptide ranges between 70 and 85 amino acids (Chang & Duggleby, 1997; Rutledge et al., 1991; Wiersma et al., 1990). If the cleavage site of the signal peptide is assumed to be at position 85, then the sequence homology of rice and Arabidopsis ALS is only 23%. Therefore, signal peptide processing and transport of the protein into the chloroplast may be involved in limiting rice ALS enzyme activity in Arabidopsis. We also considered another potential reason for the observed difference in ALS activity. It has been shown that Arabidopsis ALS is composed of four catalytic subunits and four regulatory subunits (Lee & Duggleby, 2001; McCourt et al., 2006). Thus, ALS derived from transformants expressing rice ALS will presumably be chimeric, *i.e.*, composed of both rice and Arabidopsis catalytic subunits. As a result, the enzyme activity may be reduced compared with that of ALS composed of only Arabidopsis enzyme. The full-length amino acid sequences of ALSs derived from monocotyledonous and dicotyledonous plants were clearly divided into two distinct clusters in the phylogenetic

Fig. 4. Comparison of sensitivity to BS of Arabidopsis wild-type and T3 transformants. Plants, planted in pots (9-cm diameter), were sprayed with 6.3 to 100 μg mL^{-1} (approximately 14 to 220 μM) BS with an application dose of 6.3 g to 100 g ha^{-1}. The photograph was taken 2 weeks after spraying. A, wild-type; B, *OsALS-W548I/S627I* 30-8; C, *AtALS-W574L/S653I* 1-3.

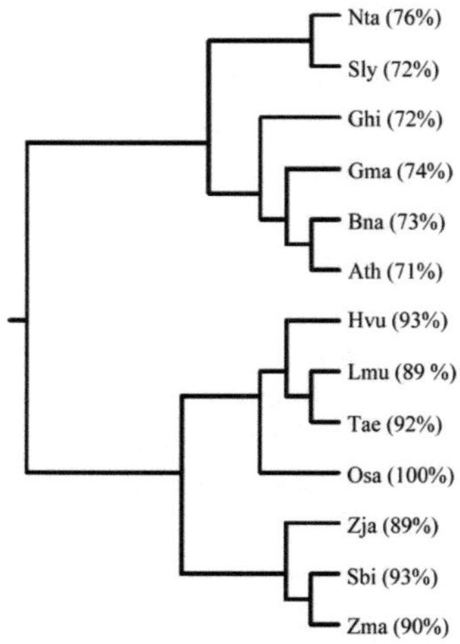

Fig. 5. Phylogenetic analysis with full-length amino acid sequences of ALSs. The NJ-tree was constructed using the ClustalW program (http://align.genome.jp/). The percentage indicates the amino acid homology of each plant with rice (Osa). Amino acid sequences of corn (Zma), wheat (Tae) (lacking the N-terminal region), barley (Hvu) (lacking the N-terminal region), Italian ryegrass (Lmu), Arabidopsis (Ath), tobacco (Nta), cotton (Ghi) and rapeseed (Bna) were obtained from the GenBank database. Putative amino acid sequences of ALS from sorghum (Sbi) (lacking the N-terminal region) and soybean (Gma) were identified through a BLAST search of the JGI database (http://genome.jgi-psf.org/) and that from tomato (Sly) was identified through a BLAST search of the Tomato SBM (http://www.kazusa.or.jp/tomato/). The nucleotide sequences of ALS of Japanese lawn grass (Zja) have been determined by genome walking using a DNA Walking SpeedUp Kit (Seegeen, Inc., Korea, unpublished data). Lack of the N-terminal region slightly increased the amino acid homology of sorghum, wheat and barley with rice compared to the complete ALS protein sequences since the homology in this region was very low.

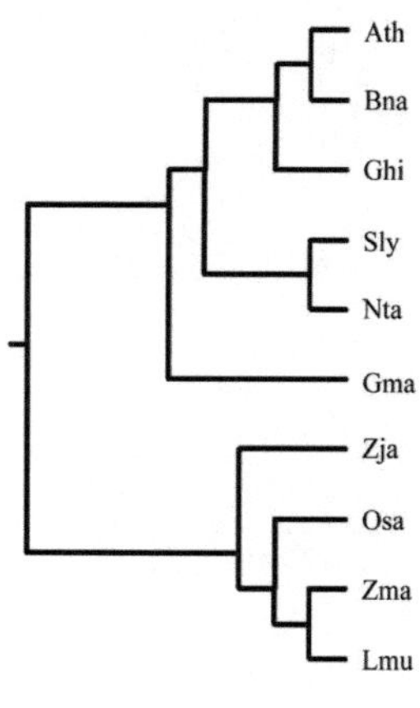

Fig. 6. Phylogenetic analysis with amino acid sequences of the putative signal peptide region of ALSs. Seventy amino acid residues from the first methionine of complete ALS protein sequences were used for the analysis.

tree, each cluster being highly conserved (Figure 5). The putative signal peptide amino acid sequences of ALSs were also divided into two clusters in the phylogenetic tree (Figure 6). These findings suggest it would be best to use rice and Arabidopsis mALS genes for generating monocotyledonous and dicotyledonous transgenic plants, respectively. Given differences in the sensitivity to PCs and in the expression level of induced mutant ALSs between plant species, the preparation of various combinations of mALS genes and PCswould be an effective strategy in applying this selection system to a broad range of plant species.

We artificially generated other types of mALS genes of Arabidopsis, which yielded recombinant ALS proteins with A122V, P197S, W574L, S653N and P197H/R198S mutations. Recombinant ALS proteins from these genes were prepared as glutathione S-transferase-fused proteins, and the sensitivity of the proteins to ALS-inhibiting herbicides were examined. It was found that the level of resistance of these recombinant ALS proteins to ALS-inhibiting herbicides varied for the compounds tested (Table 2), while mALS-P197S, W574L and P197H/R198S proteins showed similar sensitivity to herbicides to that of rice ALS proteins with the corresponding mutations (Kawai et al., 2008). These results indicated that some Arabidopsis-mALS genes are useful as selectable marker genes for the genetic transformation of plants when used together with ALS-inhibiting herbicides to which mALSs express high resistance.

Herbicide [b]	RS ratio [a]				
	A122V	P197S	W574L	S653N	P197H/R198S
CS	10	300	>8,300	4.3	2,400
BM	190	9,100	>9,100	72	>9,100
IQ	>55	25	>56	>55	1.8
IP	>14	3.7	>14	>14	>14
BS	20	5.3	2,800	53	80
PS	1	56	>2,900	68	2
PM	140	13	6,500	700	13

[a] RS ratios for mutated ALSs were obtained by calculating the ratio of the I_{50} value for each mutated ALS to the I_{50} value for the wild-type.
[b] SUs: CS, chlorsulfuron; BM, bensulfuron-methyl; IMs: IQ, imazaquin; IP, imazapyr; PCs: PM, pyriminobac; PS, pyrithiobac-sodium; BS, bispyribac-sodium.

Table 2. Degree of resistance of recombinant Arabidopsis mALSs to ALS-inhibiting herbicides

5. Nuclear gene-targeting as an ultimate clean technology

More than two decades have passed since the basic technology of plant transformation was established (Herrera-Estrella et al, 1983). Although a variety of GM crops have been cultivated world-wide, they have not been fully accepted by consumers, especially in Japan and Europe, leading to the conclusion that the technology may not satisfy consumers' desires. Selectable-marker genes in addition to the genes of interest necessary to improve plant growth and resistance are always required for plant transformation. With respect to selectable markers, the use of antibiotic-resistant genes may not obviate the possibility of generating antibiotic-resistant bacteria in the intestines of cattle. Most imported GM crops are generated with selectable-marker genes derived from microorganisms and promoters taken from plant pathogens such as cauliflower mosaic virus and *Agrobacterium* spp., and are therefore less accepted by consumers and the general public. In an effort to overcome these problems, strategies for excising selectable-marker genes from transgenic plants have been developed (Hare & Chua, 2002). Eventually, strategies employing a combination of selectable markers originating from plants and the use of herbicides harmless to the environment and humans are worth investigating to shorten the time it takes to create GM crops compared to technologies employing selectable-marker excision. Another approach to allay consumer anxiety is the employment of plant-derived DNA sequences including selectable markers without their excision. When DNA sequences are introduced into plant species from which they are derived, the transformed plants are designated as "intragenic" (Nielsen, 2003). Such an approach has been successfully performed with a mALS gene in Arabidopsis (Ahmad et al., 2009). Furthermore, if we can replace an internal wild-type gene with its point-mutated gene by gene-targeting, the resultant plants are completely equivalent to those generated by conventional breeding or mutagenesis. One gene for ALS has been reported to be present in the genome of Arabidopsis (Endo et al., 2006) or rice (Endo et al., 2007), and replacement of the internal wild-type gene with a mutated species which confers herbicide resistance on those plants has been successfully demonstrated.

6. Plastid transformation

Plastid transformation was first achieved in 1988 for the unicellular alga *Chlamydomonas reinhardtii* by Boynton *et al.* (Boynton et al., 1988), and was followed in 1990 by the transformation of tobacco by Maliga *et al* (Svab et al., 1990). Genetic engineering approaches utilizing chloroplasts possess a number of attractive advantages compared with nuclear transformation, and include: (i) a high level of transgene expression (Daniell et al., 2002), (ii) delivery of multiple genes in a single transformation event (Daniell & Dhingra, 2002), (iii) the absence of gene silencing (DeCosa et al., 2001), (iv) the absence of position effects due to site-specific transgene integration (Daniell et al., 2004), and (v) the absence of pleiotropic effects given localization of the transgene products inside the chloroplast (Daniell et al., 2001). These advantages have lead to trials of chloroplast transformation in many plants such as Arabidopsis (Sikdar et al., 1998), potato (Sidorov et al., 1999), rice (Khan & Maliga, 1999), tomato (Ruf et al., 2001), *Lesquerella fendleri* (Skarjinskaia et al., 2003), oilseed rape (Hou et al., 2003), carrot (Kumar S et al., 2004a), cotton (Kumar S et al., 2004b), soybean (Dufourmantel et al., 2004), lettuce (Lelivelt et al., 2005), and cabbage (Liu et al., 2007). The successful recovery of genetically-stable transplastomic plants is dependent on the ability to selectively amplify the plastid genomes, which are quite low in copy number, following delivery of the genes by particle bombardment. The key factor affecting transformation efficiency is the choice of selectable marker. There are two types of plastid selectable marker genes: 'primary selectable markers' to be used for direct selection (*aadA*, *nptII* and *aphA-6* for aminoglycoside phosphotransferase), and 'secondary selectable markers' (*bar* and *epsps*) that are not suitable for direct selection when only a few copies of plastid DNA (ptDNA) have settled down, but will allow selection when many copies of ptDNA are integrated (Maliga, 2004; Lutz et al., 2007). The 'primary selective markers' are of bacterial origin and confer resistance to an antibiotic: *aadA* to spectinomycin and streptomycin (Svab and Maliga, 1993;

Zoubenko et al., 1994) and *neo* (Carrer et al., 1993) and *aphA-6* (Huang et al., 2002) to kanamycin. The 'secondary selective markers', *bar* and *epsps*, are also derived from bacteria and confer resistance to herbicides such as phosphinothricin (Lutz et al., 2001) and glyphosate (Ye et al., 2003), respectively. To date, introduction of mALSs to the chloroplast genome has not been attempted perhaps because ALS was thought to be unsuitable as a selectable marker, as in the case of *bar* or *epspe* (Cao et al., 1992, Ye et al., 2003). We therefore attempted to utilize mALSs in chloroplast engineering strategies.

7. Clean gene transformation technology for chloroplast engineering

Transformation technologies of nuclear genomes have been developed to eliminate antibiotic marker genes, an approach referred to as nuclear genome-clean gene transformation technology (CGTT) (Yoder et al., 1994). Over the past several years, consumer and environmental organizations have expressed ethical and biosafety concerns about the use of antibiotic- and herbicide-resistance genes derived from microorganisms (Miki & McHugh, 2004). This concept has also been applied to chloroplast genetic engineering, in which antibiotic-resistant genes such as *aadA* were eliminated from the chloroplast genome, resulting in marker-free transplastomic plants or replacement with plant-derived marker genes. To date, four strategies have been developed for the generation of marker-free transplastomic plants: (i) homology-based excision via direct repeated regions (Iamtham & Day, 2000); (ii) cotransformation–segregation (Carrer & Maliga, 1995); (iii) transient co-integration of marker genes (Klaus et al., 2004), and (iv) excision by phage site-specific recombinases (Corneille et al., 2001). On the other hand, two marker genes applied to tobacco have been reported to be derived from plants: genes for betaine aldehyde dehydrogenase (BADH) from spinach (*Spinacia oleracea*) (Daniell et al., 2001) and feedback-insensitive anthranilate synthase α-subunit (ASA2) from tobacco (Barone et al., 2009). However, use of the *BADH* gene has not been consistently reproduced (Maliga, 2004; Whitney & Sharwood, 2008).

Use of such technologies prevents the transfer of antibiotic-resistant genes to surrounding weeds and microorganisms in soil, and to bacteria in animal guts after oral intake. Integration of foreign genes into the plastid genome strengthens gene containment since plastids are inherited maternally in many crop plants, avoiding the pollen-mediated spread of transgenes (Maliga, 1993; Daniell et al., 1998; Scott & Wilkinson, 1999). Homologous recombination in plastids allows for accurate gene targeting into a well-characterized genome and elimination of bacterial vector sequences (Svab et al., 1990). High levels of gene expression have resulted from an increasing number of foreign genes being located in plastids (McBride et al, 1995; Staub et al., 2000; Kanamoto et al., 2006). Notwithstanding all of the advantages associated with marker-free technology in chloroplast transformation, there remains one outstanding problem that should be resolved in the field. GM and non-GM plants must be clearly and easily distinguished from one another. Although PCR-based methods are the most convenient for ascertaining contamination in bulk samples, they are unsuitable for checking a single seed or plant in terms of efficiency and use of resources. The use of herbicides was proposed as an appropriate method to solve this problem, although the generation of herbicide-resistant plants must be considered. Some reports have indicated that herbicides which inhibit ALS or acetyl-CoA carboxylase, such as glyphosate and others, accelerated the generation rate of weeds and crops tolerant to the herbicide (Preston & Powles, 2002; Shimizu et al., 2002; Tranel & Wright, 2002; Tranel et al., 2007,

Heap, 2010). Employing a rotation supply of some herbicides was proposed as a countermeasure against the occurrence of herbicide-resistant weeds (Gressel, 1984).

8. Mutated ALS genes as plastid sustainable markers

The employment of some plant-origin genes for herbicide tolerance has solved the problem and allayed the public's anxiety. We have focused on the use of mALS genes as sustainable markers. It is well known that ALS imparts herbicide tolerance by mutation at several amino acid residues (Figure 2). Herbicide-tolerant plants have been reported for rice, tobacco and Arabidopsis (Chang et al., 1998; Tan et al., 2005; Shimizu et al., 2002; Kawai et al., 2007b; Okuzaki et al., 2007). Several mutated species of Arabidopsis ALS have been expressed in *Escherichia coli*, and their sensitivity to inhibitors was examined (Tables 1 and 2) (Kawai et al., 2008). These results showed that P197S, W574L, S653I, P197H/R198S and W574L/S653I were resistant to SUs, IMs, SUs, IMs, and all three types of herbicides, respectively. However, little is known about the effect of herbicides on the growth of plants with mALSs. It has recently been reported that mutation of W548L/S627I and G95A in rice ALS imparts tolerance to all three types of herbicides and pyrimidinylcarboxylate herbicides, respectively (Kawai et al., 2007b; Okuzaki et al., 2007). We examined whether introduction of mALSs into the chloroplast genome can be applied to a strategy involving the rotation supply of different herbicides by characterizing the transplastomic lines with respect to: (i) the influence of hyper-expression of mALSs on plant growth, (ii) feedback regulation by the regulatory subunit *in vivo*, (iii) the dependency of herbicide resistance on each mutation similarly observed *in vitro* (Tables 1 and 2) (Kawai et al., 2008; Okuzaki et al., 2007), and (iv) the availability of multiple combinations of different mutations and herbicides. We have reported on the introduction of some mALS genes into the chloroplast genome and examined the sensitivity of transformants to ALS-inhibiting herbicides. The results indicated that mALS genes are useful as sustainable markers, which function to exclude non-transformed crops while maintaining transformed plants. These markers have shown selectable tolerance to different types of herbicide. We have proposed that the rotation supply of different herbicides can be effective when used with transgenic plants harboring mALS genes (Shimizu et al., 2008).

The chloroplast transformation vectors pLD201- mALS (Figure 7A), possessing the *aadA* and *mALS* (transit peptide truncated) genes inserted between tobacco sequences *rbcL* for the large subunit of ribulose-1,5-bisphosphate carboxylase/oxygensae and *accD* for homologous recombination, were introduced by particle bombardment (Figure 7A). The integration of *mALS* into the chloroplast genome in regenerated tobacco plants was confirmed by PCR using the 5 primer sets shown in Figure 7A. Tobacco chloroplast transformation was performed using pLD-201-mALS harboring *G121A, A122V, P197S, P197S/S653I* or *W574L/S653I*. G121A in Arabidopsis ALS corresponds to G95A in rice (Okuzaki et al., 2007). The resultant transplastomic plants were maintained on hormone-free Murashige and Skoog (MS) medium (Figure 7B). It is concluded that the chloroplast genome in these transgenic plants were almost transplastomic (Figure 7C).

We investigated the involvement of the regulatory subunit in ALS activity. The regulatory subunit plays a role in feedback regulation by Val, Ile and Leu and in general enzyme activity (Lee & Duggleby, 2001). The determination of ALS activity in leaves, where regulatory subunit molecules are present, has been performed in the presence of 1,1-cyclopropanedicarboxylic acid, which blocks acetolactate metabolism, resulting in no

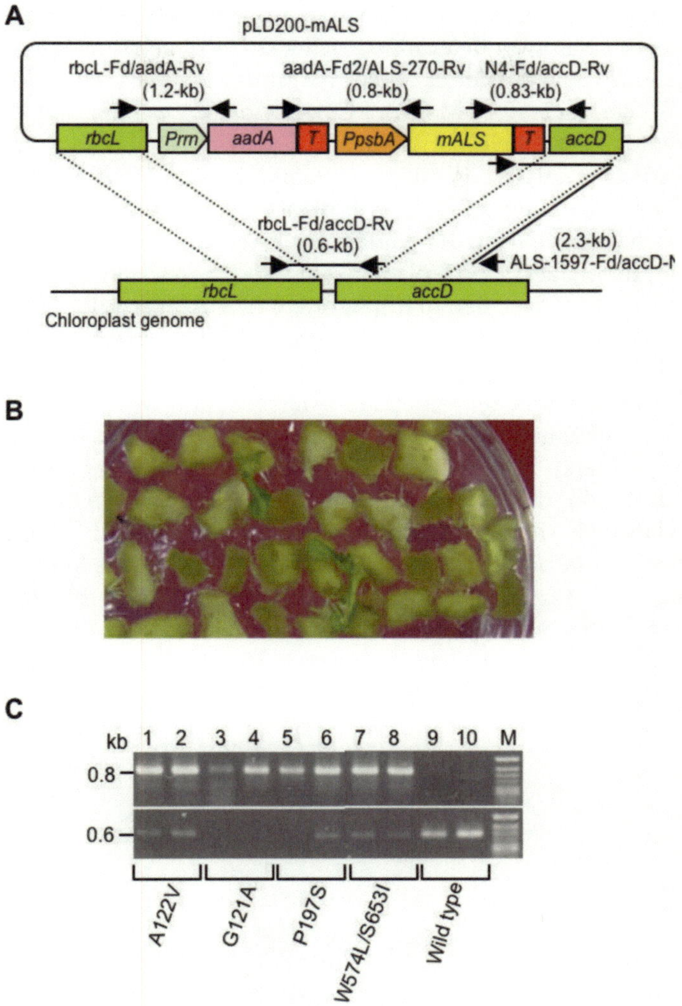

Fig. 7. Transformation of tobacco chloroplast with *aadA* and mALS. A, Structure of the chloroplast transformation vector. *Prrn-aadA* and *PpsbA-mALS* show the transgenes introduced into the chloroplast genome. The region located between *rbcL* and *accD* may be integrated into the chloroplast genome by homologous recombination. PCR was performed to confirm transplastomic integration using the primer sets shown, and the expected sizes of the PCR products are shown in parentheses. B. Following bombardment, leaf slices were grown on RMOP (Shimizu et al., 2008) containing 0.5 mg L^{-1} spectinomycin. A spectinomycin plate after 6 weeks is shown. Regenerated plants represent candidate transformants. C. Population of the transformed chloroplast genome. PCR analysis of different lines of each chloroplast transformant harboring A122V (lanes 1 and 2), G121A (lanes 3 and 4), P197S (lanes 5 and 6), P197S/S653I (lanes 7 and 8), W574L/S653I (lanes 9 and 10), and wild-type (lanes 11 and 12). PCR reactions were performed using 25 cycles. The 0.8-kb product represents part of the transgene introduced into the chloroplast genome, and the 0.6-kb product is derived from endogenous chloroplast genome without a transgene insert.

feedback regulation. The activity of native ALS from wild-type tobacco in the absence of ALS-inhibiting herbicides was determined using a colorimetric assay, and yielded a red color in the samples. The red color changed to a transparent or pale yellow color following the addition of SU herbicide (0.1 μM BM), PC herbicide (0.1 μM PS), and IM herbicide (5 μM IP), indicating that these herbicides inhibited ALS activity. This assay was employed for the evaluation of mALS activity in transplastomic plants (*G121A*, *A122V*, *P197S* and *W574L/S653I*). The ALS activity of *G121A* plants was strongly resistant to PS, weakly resistant to BM and sensitive to IP (Figure 8), whereas *A122V* plants were particularly resistant to IP (Figure 8), and *P197S* plants were strongly resistant to BM, showed medium resistance to PS, and were sensitive to IP (Figure 8). The ALS activity of *W574L/S653I* plants was strongly resistant to PS, BM and IP (Figure 8). The selectable tolerance of plants transplastomic with *G121A*, *A122V* and *W574L/S653I* (Figure 9) were similar to those obtained when using the same recombinant mALSs that only expressed the catalytic subunit in *E. coli*, to which endogenous *E. coli* regulatory subunits, it was concluded, were not associated (Tables 1 and 2) (Kawai et al., 2008; Okuzaki et al., 2007). Therefore, the regulatory subunits do not affect the sensitivity of these mALSs to herbicides in transplastomic plants. On the other hand, the behavior of mALS P197S differed from that of the aforementioned mutations. The novel tolerance of *P197S* plants to PC and SU herbicides was demonstrated with regard to mALS activity in response to herbicides in leaves (Figure 8), whereas mALS P197S expressed in *E. coli* was resistant to SU but not to PC herbicides (Table2) (Kawai et al., 2008). This result suggests that the regulatory subunit contributes towards imparting mALS P197S with resistance to PC herbicides.

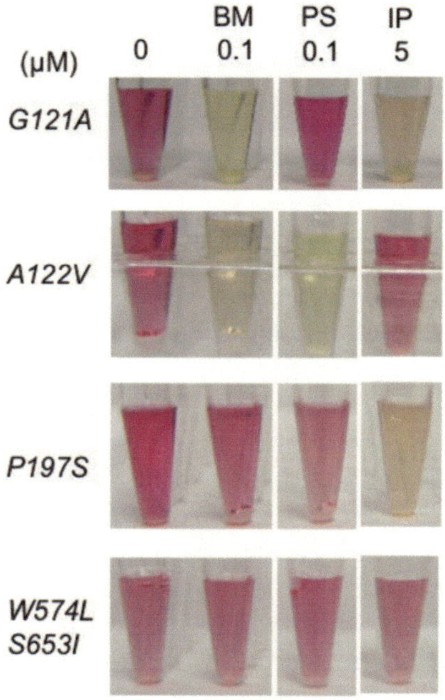

Fig. 8. Inhibition of ALS activity with herbicides in plants transplastomic with *mALSs*. ALS activity in tobacco transplastomic with *mALSs* was colorimetrically examined. ALS activity of tobacco, wild-type and plants transformed with *G121A, A122V, P197S, P197S/S653I* or *W574L/S653I* was determined in the presence of 0.1 μM BM, 0.1 μM PS or 5 μM IP.

In an effort to investigate the influence of feedback regulation caused by hyper-expression of the ALS gene, transplastomic plants were grown on medium containing herbicide. We analyzed herbicide resistance in transplastomic plants harboring four different mALSs. *W574L/S653I*-plants showed synergistic tolerance, similar to that observed when the corresponding *mALS* gene was introduced into the nuclear genome of rice (Kawai et al., 2007b). The tolerance of *P197S*-plants to PC and SU herbicides was also demonstrated during plant growth (Figure 9). Additionally, two other transplastomic plants (*G121A* and *A122V*) showed sensitivity to herbicides with respect to the activity in leaves (Figure 8) and during plant growth (Figure 9). Our results provide evidence to suggest that the sensitivity of mALSs to herbicides in plants is not affected by feedback regulation. The highly-expressed mALS molecules may not be fully active due to the resultant stoichiometrically insufficient number of regulatory subunits (Lee & Duggleby, 2001). Therefore, the ALS activity of transplastomic plants was almost equivalent to that of wild-type plants in the absence of herbicide.

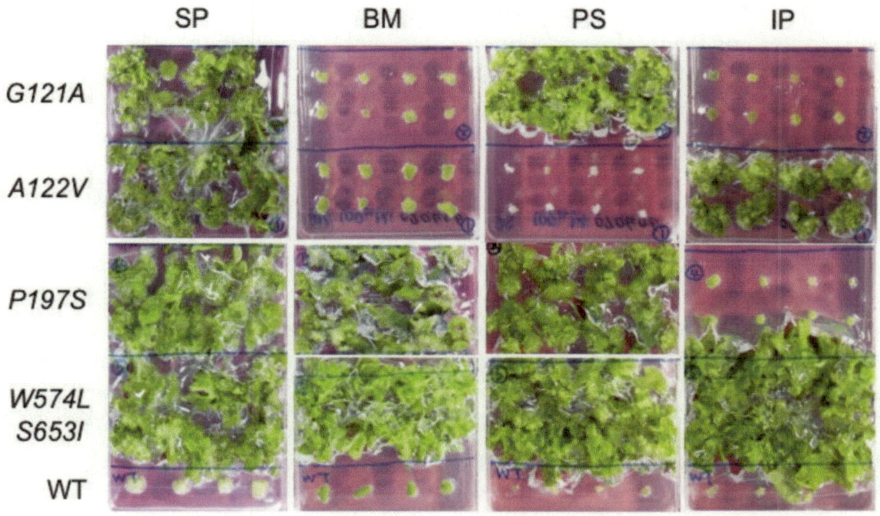

Fig. 9. Regeneration of transplastomic plants on medium containing ALS-inhibiting herbicides. Plants transgenic with A122V, G121V, P197S, P197S/S653I or W574L/S653I were regenerated on RMOP medium (Shimizu et al., 2008) containing 0.5 g L^{-1} spectinomycin (SP), 0.1 μM BM, 0.1 μM PS or 1 μM IP.

In an effort to confirm that the herbicide-related traits of the transplastomic plants were inherited by the next generation, T1 seeds, the self-pollinated progeny of the transplastomic lines, were planted on medium containing the corresponding herbicide or spectinomycin. Both seed types were able to grow on MS medium (Figure 10). Although wild-type plants were sensitive to IP and SP, all *A122V* seeds were uniformly resistant to SP and IP (Figure 10). This study revealed that transplastomic plants with *mALS*s grow normally on MS medium without significant differences compared to wild-type plants, indicating that hyper-expression of mALSs does not influence plant growth. These transplastomic plants containing mALS were able to grow in the presence of the corresponding herbicide, indicating that mALSs are useful as sustainable markers in the field, and lending support to proposals that involve the rotation of three or more combinations of herbicide and transplastomic plants. The advanced technology described here would allow for the efficient and controlled management of weeds resistant to ALS-inhibiting herbicides.

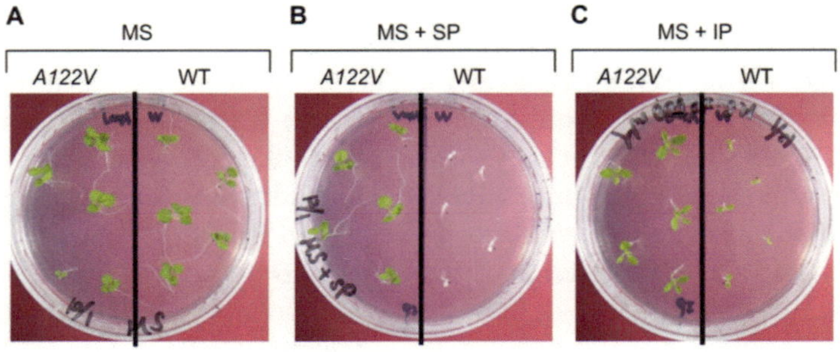

Fig. 10. Inheritance of herbicide tolerance in the seed progeny of chloroplast transgenic plant. The T_1 seeds transplastomic with *PpsbA-A122V* (the left in all panels, A122V) and wild-type (the right in all panels, WT) were germinated on MS medium alone (panel A, MS), or medium containing 0.5 mg L^{-1} spectinomycin (panel B, MS+SP) or 1 μM IP (panel C, MS+IP).

9. Application of mALSs integrated into plastid genomes

Herbicide-resistant weeds have been reported in many countries (Tranel & Wright, 2002; Tranel et al., 2007, Heap, 2010) including weeds resistant to ALS-inhibiting herbicides. New technology is required to assist in the management of weeds resistant to these herbicides. We propose a strategy involving herbicide rotation to overcome the aforementioned problem. To this end, we have developed transplastomic plants that possess tolerance to PC, IM and SU/PC. We identified three types of ALS mutations that conferred specific resistance to the three classes of herbicides used with the transplastomic plants and showed that *G121A*, *A122V* and *P197S* plants were resistant to PC, IM, and SU/PC herbicides, respectively (Figure 9). Use of these transplastomic markers in crop plants could allow for the implementation of a new strategy based on the rotation of three or more combinations of herbicides. The advanced technology described in this review provides the basis for the efficient and strict management of weeds resistant to ALS-inhibiting herbicides. Investigations concerning herbicide resistance have been performed using chloroplast transformation. For example, the petunia *epsps* gene was introduced into the tobacco chloroplast genome and resulted in transplastomic plants resistant to glyphosate (Daniell et al., 1998). Similarly, the *bar* gene for phosphinothricin resistance was used to investigate the resulting plant phenotype (Lutz et al., 2001). Since this gene is derived from microorganisms and not plants, it is less suitable for use in CGTT-based approaches. However, epsps is worthy of consideration in strategies involving herbicide rotation schemes as described above since epsps is present in higher plants. The glyphosate and ALS-inhibiting herbicides are thought to be nontoxic to living organisms, except plants and microorganisms (Peterson & Shama, 2005). Plant-derived epsps might be useful as an additional tool for use in a herbicide rotation system for the management of herbicide-resistant weeds. We have tried to adapt *mALSs* for use as selectable markers in chloroplast transformation but have not succeeded to date. As with *epsps* and *bar* (Cao et al., 1992, Ye et al., 2003), mALSs might be unsuitable for use as selectable markers. The technology described here may be employed in CGTT-based applications in association with *aadA* elimination following transformation.

10. Conclusion

A number of genes have been employed for the generation of genetically-modified crops possessing tolerance to herbicides in an effort promote crop growth and discourage the growth of competing plants such as weeds. Herbicide-resistant genes are also invaluable for use as selectable markers in the genetic transformation of plants. The majority of herbicide-resistant genes are derived from soil bacteria such as *Agrobacterium* and *Streptomyces*, organisms which have never been utilized as ingredients in products for human consumption. With respect to the use of plant-derived genes for herbicide tolerance, attention may be paid in order to facilitate public awareness and acceptance of the technologies involved. These genes are also useful in strategies involving intragenic transformation through homologous recombination to generate plants free from any exogenous DNA fragments. Our research efforts have focused on ALS. Use of this gene has several advantages including: (i) a single locus is present in Arabidopsis and rice, thus allowing for the straightforward implementation of gene targeting strategies, (ii) multiple classes of herbicides which interfere with different domains of ALS molecules are available, thereby providing the opportunity to generate plants with selected tolerance so as to reduce the occurrence of herbicide-resistant weeds in programs employing the rotation supply of different herbicides, and (iii) availability as a sustainable marker in chloroplast transformation in addition to a selectable marker for nuclear transformation. We have introduced the mutations G121A, A122V, P197S, P197H, R198S, W574L, S653I and others into Arabidopsis ALS and delivered these genes into nuclear and chloroplast genomes of plants. Use of these nuclear and transplastomic markers in crop plants would facilitate the implementation of a new strategy based on the rotation of multiple combinations of herbicides and mALSs to prevent the generation of herbicide-resistant weeds. Furthermore, the use of mALSs in gene-targeting for nuclear transformation and homologous recombination in plastid engineering would bring us closer to our goal of an ultimate clean technology, and allow for the production of GM plants in which only the ALS gene is mutated without integration of any other external DNA sequences.

11. References

Adachi, T., Takase, H. & Tomizawa, K. (2007). Introduction of a 50 kbp DNA Fragment into the Plastid Genome. *Biosci. Biotechnol. Biochem.*, 71, 2266-2273

Ahmad, A., Kaji, I., Murakami, Y., Funato, N., Ogawa, T., Shimizu, M., Niwa, Y. & Kobayashi, H. (2009). Transformation of *Arabidopsis* with plant-derived DNA sequences necessary for selecting transformants and driving an objective gene. *Biosci. Biotechnol. Biochem.*, 73, 936-938

Barone, P., Zhang, X.H. & Widholm, J.M. (2009). Tobacco plastid transformation using the feedback-insensitive anthranilate synthase [alpha]-subunit of tobacco (ASA2) as a new selectable marker. *J. Exp. Bot.*, 60, 3195-3202

Boynton, J.E., Gillham, N.W., Harris, E.H., Hosler, J.P., Johnson, A.M., Jones, A.R., McBride K.E., Svab, Z., Schaaf, D.J., Hogan, P.S., Stalker, D.M. & Maliga, P. (1995).

Amplification of a chimeric *Bacillus* gene in chloroplast leads to an extraordinary level of insecticidal protein in tobacco. *Nat. Biotechnol.*, 13, 362–365

Brixey, P.J., Guda, C. & Daniell, H.(1997). The chloroplast *psbA* promoter is more efficient in *Escherichia coli* than the T7 promoter for hyperexpression of a foreign protein. *Biotechnology Letter*, 19, 395-399

Cao, J., Dun, X. & McElroy, D. (1992). Regeneration of herbicide resistant transgenic rice plants following microprojectile-mediated transformation of suspension culture cells. *Plant Cell Rep.*, 11, 586-591

Carrer, H., Hockenberry, T.N., Svab, Z. & Maliga, P. (1993). Kanamycin resistance as a selectable marker for plastid transformation in tobacco. *Mol. Gen. Genet.*, 241, 49–56

Carrer, H. & Maliga, P. (1995). Targeted insertion of foreign genes into the tobacco plastid genome without physical linkage to the selectable marker gene. *Nat. Biotechnol.*, 13, 791–794

Chang, A.K. & Duggleby, R.G. (1998). Herbicide-resistant forms of *Arabidopsis thaliana* acetohydroxyacid synthase, characterization of the catalytic properties and sensitivity to inhibitors of four defined mutants. *Biochem. J.*, 333, 765-777

Corneille, S., Lutz, K., Svab Z. & Maliga, P. (2001). Efficient elimination of selectable marker genes from the plastid genome by the CRE-lox site-specific recombination system. *Plant J.*, 72, 171–178

Daniell, H., Carmona-Sanchez, O. & Burns, B. (2004). Chloroplast derived antibodies, biopharmaceuticals and edible vaccines. In: *Molecular Farming*, R. Fischer & S. Schillberg, (Eds.), 113–133, Wiley-VCH-Verlag, Weinheim, Germany

Daniell, H & Dhingra, A. (2002). Multigene engineering, dawn of an exciting new era in biotechnology. *Curr. Opin. Biotechnol.*, 13, 136–141

Daniell, H., Dhingra, A. &, Allison, L. (2002). Chloroplast transformation from basic molecular biology to biotechnology. *Rev. Plant Physiol. Biochem.*, 1, 1–20

Daniell, H., Datta, R., Varma, S., Gray, S. & Lee, S.B. (1998). Containment of herbicide resistance through genetic engineering of the chloroplast genome. *Nat. Biotechnol.*, 16, 345–348

Daniell, H., Lee, S.B., Panchal, T. & Wiebe, P.O. (2001). Expression of cholera toxin B subunit gene and assembly as functional oligomers in transgenic tobacco chloroplasts. *J. Mol. Biol.*, 311, 1001–1009

Daniell, H., Muthukumar, B. & Lee, S.B. (2001). Marker free transgenic plants: engineering the chloroplast genome without the use of antibiotic selection. *Curr. Genet.*, 39, 109–116

DeCosa, B., Moar, W., Lee, S.B., Miller, M. & Daniell, H. (2001). Overexpression of the Bt cry2Aa2 operon in chloroplasts leads to formation of insecticidal crystals. *Nat. Biotechnol.*, 19, 71-74

Dufourmantel, N., Pelissier, B., Garçon, F., Peltier, G., Ferullo, J.M. & Tissot, G. (2004). Generation of fertile transplastomic soybean. *Plant Mol. Biol*, 55, 479–489

Endo, M., Osakabe, K., Ichikawa, H. & Toki, S. (2006). Molecular characterization of true and ectopic gene targeting events at the acetolactate synthase gene in *Arabidopsis*. *Plant Cell Physiol.*, 47, 372-379

Endo, M., Osakabe, K., Ono, K., Handa, H., Shimizu, T. & Toki S. (2007). Molecular breeding of a novel herbicide-tolerant rice by gene targeting. *Plant J.*, 52, 157-166

Fang, L.Y., Gross, P.R., Chen, C.H. & Lillis, M. (1992). Sequence of two acetohydroxyacid synthase genes from *Zea mays*. *Plant Mol. Biol.*,18, 1185–1187

Gerwick, B.C., Mireles, L.C. & Eilers, R.J. (1993). Rapid diagnosis of ALS/AHAS resistant weeds. *Weed Technol.* 7, 519–524

Gressel J (1984). Evolution of herbicide-resistant weeds. *Ciba Found Symp.*, 102, 73-93

Grula, J.W., Hudspeth, R.L., Hobbs, S.L. & Anderson, D.M. (1995). Organization, inheritance and expression of acetohydroxyacid synthase genes in the cotton allotetraploid *Gossypium hirsutum*. *Plant Mol. Biol.*, 28, 837–846

Hare, P.D. & Chua, N.-H. (2002). Excision of selectable marker genes from transgenic plants. *Nat. Biotechnol.*, 20, 575-580

Hart, S.E., Saunders, J.W. & Penner, D. (1993). Semidominant nature of monogenic sulfonylurea herbicide resistance in sugar beet (*Beta vulgaris*). *Weed Sci.*, 41, 317–324

Heap, I. (2010). The International Survey of Herbicide Resistant Weeds. Online Internet. August 02, Available www.weedscience.com

Herrera-Estrella, L., Block, M.D., Messens, E., Hernalsteens, J.P., Montagu, M.V. & Schell. J. (1983). Chimeric genes as dominant selectable markers in plant cells. *EMBO J.*, 2, 987-995

Hershey, H.P., Schwartz, L.J., Gale, J.P. & Abell, L.M. (1999). Cloning and functional expression of the small subunit of acetolactate synthase from Nicotiana plumbaginifolia. *Plant Mol. Biol.*, 40, 795–806.

Hou, B.-K., Zhou, Y.-H., Wan, L.-H., Zhang, Z.-L., Shen, G.-F., Chen, Z-H & Hu, Z.-M. (2003). Chloroplast transformation in oilseed rape. *Trans. Res.*, 12, 111-114

Huang, F.C., Klaus, S.M.J., Herz, S., Zuo, Z., Koop, H.U. & Golds, T.J. (2002). Efficient plastid transformation in tobacco using the aphA-6 gene and kanamycin selection. *Mol. Genet. Genomics*, 268, 19–27

Iamtham, S. & Day, A (2000). Removal of antibiotic resistance genes from transgenic tobacco plastids. *Nat. Biotechnol.*, 18, 1172–1176

Kanamoto, H., Yamashita, A., Asao, H., Okumura, S., Takase, H., Hattori, M., Yokota, A. & Tomizawa, K. (2006). Efficient and stable transformation of *Lactuca sativa* L. cv. Cisco (lettuce) plastids. *Transgenic. Res.*, 15, 205-217

Kawai, K., Kaku, K., Izawa, N., Fukuda, A., Tanaka, Y. & Shimizu, T. (2007a). Functional analysis of transgenic rice plants expressing a novel mutated ALS gene of rice. *J. Pestic. Sci.*, 32, 385–392

Kawai, K., Kaku, K., Izawa, N., Shimizu, M., Kobayashi, H. & Shimizu, T. (2008). Herbicide sensitivities of mutated enzymes expressed from artificially generated genes of acetolactate synthase. *J. Pestic. Sci.*, 32, 128–137

Kawai, K., Kaku, K., Izawa, N., Shimizu, T., Fukuda, A. & Tanaka, Y. (2007b). A novel mutant acetolactate synthase gene from rice cells, which confers resistance to ALS-inhibiting herbicides. *J. Pestic. Sci.*, 32, 89–98

Kawai, K., Kaku, K., Izawa, N., Shimizu, M., Kobayashi, H. & Shimizu, T. (2010). Transformation of Arabidopsis by mutated acetolactate synthase gene from rice and *Arabidopsis* that confer specific resistance to pyrimidinylcarboxylate-type ALS inhibitors. *Plant Biotech.*, 27, 75-84

Khan, M.S. & Maliga, P. (1999). Fluorescent antibiotic resistance marker for tracking plastid transformation in higher plants. *Nat. Biotechnol.*, 17, 910–915

Klaus, S.M.J., Huang, F.C., Golds, T.J. & Koop, H.-U. (2004). Generation of marker-free plastid transformants using a transiently cointegrated selection gene. *Nat. Biotechnol.*, 22, 225–229

Kumar, S., Dhingra, A. & Daniell, H. (2004a). Plastid expressed betaine aldehyde dehydrogenase gene in carrot cultured cells, roots and leaves confers enhanced salt tolerance. *Plant Physiol.*, 136, 2843–2854

Kumar, S., Dhingra, A. & Daniell, H. (2004b). Stable transformation of the cotton plastid genome and maternal inheritance of transgenes. *Plant Mol. Biol.*, 56, 203–216.

Lee, Y.T. & Duggleby, R.G. (2001). Identification of the regulatory subunit of *Arabidopsis thaliana* acetohydroxyacid synthase and reconstitution with its catalytic subunit. *Biochemistry*, 40, 6836–6844

Lelivelt, C.L., McCabe, M.S., Newell, C.A., Desnoo, C.B., van Dun, K.M., Birch-Machin, I., Gray, J.C., Mills, K.H. & Nugent, J.M. (2005). Stable plastid transformation in lettuce (Lactuca sativa L.). *Plant Mol. Biol.*, 58, 763-774

Liu, C.W., Lin, C.C., Chen, J.J. & Tseng, M.J. (2007). Stable chloroplast transformation in cabbage (*Brassica oleracea* L. var. capitata L.) by particle bombardment. *Plant Cell Rep.*, 26, 1733-1744

Lutz, K.A., Azhagiri, A.K., Tungsuchat-Huang, T. & Maliga, P. (2007). A guide to choosing vectors for transformation of the plastid genome of higher plants. *Plant Physiol.*, 145, 1201–1210

Lutz, K.A., Knapp, J.E. & Maliga, P. (2001). Expression of *bar* in the Plastid Genome Confers Herbicide Resistance. *Plant Physiol.*, 125, 1585–1590

Maliga P (1993). Towards plastid transformation in flowering plants. *Trends Biotechnol.*, 11, 101–107

Maliga, P. (2004). Plastid transformation in higher plants. *Annu. Rev. Plant Biol.*, 55, 289–313

McCourt, J.A., Pang, S.S., King-Scott, J., Guddat, L.W. & Duggleby, R.G. (2006). Herbicide-binding sites revealed in the structure of plant acetohydroxyacid synthase. *Proc. Natl. Acad. Sci. USA*, 103, 569–573

Mazur, B.J., Chui, C.F. & Smith, J.K. (1987). Isolation and characterization of plant genes coding for acetolactate synthase, the target enzyme for two classes of herbicides. *Plant Physiol.*, 85, 1110–1117

Miki, B. & McHugh, S. (2004). Selectable marker genes in transgenic plants: applications, alternatives and biosafety. *J. Biotechnol.*, 107, 193–232

Newhouse, K., Singh, B., Shaner, D. & Stidham, M. (1991). Mutations in corn (*Zea mays L.*) conferring resistance to imidazolinone herbicides. *Theor. Appl. Genet.*, 83, 65–70

Nielsen, K.M. (2003). Transgenic organisms--time for conceptual diversification? *Nat. Biotechnol.*, 21, 227-228

Ogawa, T., Kawahigashi, H., Toki, S. & Handa, H. (2008). Efficient transformation of wheat by using a mutated rice acetolactate synthase gene as a selectable marker. *Plant Cell Rep.*, 27, 1325–1331

Okuzaki, A., Shimizu, T., Kaku, K., Kawai, K. & Toriyama, K. (2007). A novel mutated acetolactate synthase gene conferring specific resistance to pyrimidinyl carboxy herbicides in rice. *Plant Mol. Biol.*, 64, 219–224

Ott, K.H., Kwagh, J.G., Stockton, G.W., Sidorov, V. & Kakefuda, G. (1996). Rational molecular design and genetic engineering of herbicide resistant crops by structure modeling and site-directed mutagenesis of acetohydroxyacid synthase. *J. Mol. Biol.*, 263, 359-368

Peterson, R.K. & Shama, L.M. (2005). A comparative risk assessment of genetically engineered, mutagenic, and conventional wheat production systems. *Transgenic. Res.*, 14, 859-875

Preston, C. & Powles, S.B. (2002). Evolution of herbicide resistance in weeds, initial frequency of target site-based resistance to acetolactate synthase-inhibiting herbicides in Lolium rigidum. *Heredity*, 88, 8-13

Randolph-Anderson, B.L., Robertson, D., Klein, T.M., Shark, K.B. & Sanford, J.C. (1988). Chloroplast transformation in *Chlamydomonas* with high velocity microprojectiles. *Science*, 240, 1534–1538

Rajasekaran, K., Grula, J.W. & Anderson, D.M. (1996). Selection and characterization of mutant cotton (*Gossypium hirsutum L.*) cell lines resistant to sulfonylurea and imidazolinone herbicides. *Plant Sci.* 119, 115–124

Ruf, S., Hermann, M., Berger, I.J., Carrer, H. & Ralph, B (2001). Stable genetic transformation of tomato plastids and expression of a foreign protein in fruit. *Nat. biotechnol.*, 19, 870-875

Rutledge, R.G., Quellet, T., Hattori, J. & Miki, B.L. (1991). Molecular characterization and genetic origin of the *Brassica napus* acetohydroxyacid synthase multigene family. *Mol. Gen. Genet.* 229, 31–40

Schlittler, M., Carroll, J.A., Spatola, L., Ward, D., Ye, G. & Russell, D.A. (2000). High-yield production of a human therapeutic protein in tobacco chloroplasts. *Nat. Biotechnol.*, 18, 333–338

Scott,. S.E. & Wilkinson, M.J. (1999). Low probability of chloroplast movement from oilseed rape (*Brassica napus*) into wild *Brassica rapa*. *Nat. Biotechnol.*, 17, 390–392

Sharner, D.L. & Singh, B.K. (1997). Acetohydroxyacid synthase inhibitors. In: Roe RM, Burton JD, Kuhr RJ (eds) *Herbicide activity: Toxicology, Biochemistry and Molecular Biology*. IOS press, Amsterdam, pp 69–110

Shimizu, M., Goto, M., Hanai, M., Shimizu, T., Izawa, N., Kanamoto, H., Tomizawa, K., Yokota, A. & Kobayashi, H. (2008). Selectable tolerance to herbicides by mutated acetolactate synthase genes integrated into the chloroplast genome of tobacco. *Plant physiol.*, 147, 1976-1983

Shimizu, T., Nakayama, I., Nagayama, K., Miyazawa, T. & Nezu, Y. (2002). ALS inhibitors. In: *Herbicide Classes in Development: Mode of Action, Targets, Genetic Engineering, Chemistry*, P. Böger, K. Wakabayashi &K. Hirai, (Eds.), 1–14, Springer-Verlag, Berlin, Germany

Sidorov, V.A., Kasten, D., Pang, S.Z., Hajdukiewicz, P.T., Staub, J.M. & Nehra, N.S. (1999). Stable chloroplast transformation in potato: use of green fluorescent protein as a plastid marker. *Plant J.*, 19, 209–216

Sikdar. S., Serino, G., Chaudhuri, S. & Maliga, P. (1998). Plastid transformation in Arabidopsis thaliana. *Plant Cell Rep.*, 18, 20–24

Skarjinskaia, M., Svab, Z. & Maliga, P. (2003). Plastid transformation in *Lesquerella fendleri*, an oilseed Brassicacea. *Transgenic Res.*, 12, 115–122

Staub, J.M., Garcia, B., Graves, J., Hajdukiewicz, P.T., Hunter, P., Nehra, N., Paradkar, V., Svab, Z., Hajdukiewicz, P. & Maliga, P. (1990). Stable transformation of plastids in higher plants. *Proc. Natl. Acad Sci. USA*, 87, 8526–8530

Svab, Z., & Maliga, P. (1993). High-frequency plastid transformation in tobacco by selection for a chimeric aadA gene. *Proc. Natl. Acad Sci. USA,* 90, 913–917

Tan, S., Evans, R.R., Dahmer, M.L., Singh, B.K. & Shaner, D.L. (2005). Imidazolinone-tolerant crops: history, current status and future. *Pest Manag. Sci.,* 61, 246–257

Tougou, M., Yamagishi, N., Furutani, N., Kaku, K., Shimizu, T., Takahata, Y., Sakai, J., Kanematsu, S. & Hidaka, S. (2009). The application of the mutated acetolactate synthase gene from rice as the selectable marker gene in the production of transgenic soybeans. *Plant Cell Rep.,* 28, 769–776

Tranel, P.J., Wright, T.R. & Heap, I.M. (2007). ALS mutations from herbicide-resistant weeds. http://www.weedscience.org/mutations/MutDisplay.aspx

Tranel, P.J. & Wright, T.R. (2002). Resistance of weeds to ALS-inhibiting herbicides, What have we learned? *Weed Science,* 50, 700-712

Whitney, S.M. & Sharwood, R.E. (2008). Construction of a tobacco master line to improve rubisco engineering in chloroplasts. *J. Exp. Bot.,* 59, 1909–1921

Ye, G.-N., Colburn, S.M., Xu, C.W., Hajdukiewicz. P.T.J. & Staub, J.M. (2003). Persistence of Unselected Transgenic DNA during a Plastid Transformation and Segregation Approach to Herbicide Resistance. *Plant Physiol.,* 133, 402-410

Yoder, J.I. & Goldsbrough, A.P. (1994). Transformation systems for generating marker-free transgenic plants. *Nat. Biotechnol.,* 12, 263–267

Zhang, X.-H., Brotherton, J.E., Widholm, J.M. & Portis, A.R. Jr. (2001). Targeting a nuclear anthranilate synthase alpha-subunit gene to the tobacco plastid genome results in enhanced tryptophan biosynthesis. Return of a gene to its pre-endosymbiotic origin. *Plant Physiol.,* 127, 131-141

Zoubenko, O.V., Allison, L.A., Svab, Z. & Maliga, P. (1994). Efficient targeting of foreign genes into the tobacco plastid genome. *Nucleic Acids Res.,* 22, 3819–3824

11

Transgenic Tall Fescue and Maize with Resistance to ALS-Inhibiting Herbicides

Hiroko Sato[1], Tadashi Takamizo[1], Junko Horita[2], Kiyoshi Kawai[2], Koichiro Kaku[2] and Tsutomu Shimizu[2]
[1]*National Institute of Livestock and Grassland Science*
[2]*Life Science Research Institute, Kumiai Chemical Industry Co.*
Japan

1. Introduction

Transgenic crops such as maize (*Zea mays* L.), soybeans (*Glycine max* L. Merr.), canola (*Brassica napus* L.) and cotton (*Gossypium hirsutum* L.) have been widely used in the field. In 2009, transgenic crops were cultivated in approximately 134 million hectares in 25 countries, mainly USA, Brazil, Argentina, India, Canada and China (http://www.isaaa.org). The adoption of transgenic crops has steadily increased since 1996 because of their many benefits for farmers.
Herbicide resistance is one of the most important agronomic traits conferred onto transgenic crops. Herbicide-resistant crops comprise 62 percent of all transgenic crops (http://www.isaaa.org), and are produced by the introduction of herbicide-resistant genes using genetic transformation. Herbicide resistance can be used as an efficient tool to allow easier weed management. It facilitates control of weed species and contributes to reducing costs, labor, and the waste of chemical spray. Herbicide resistance can also facilitate the selection of transgenic cells from non-transgenic cells as a selectable marker in genetic transformation.
Acetolactate synthase (ALS)-inhibiting herbicides are widely used around the world. ALS-inhibiting herbicide-resistant weeds were first found in kochia (*Kochia scoparia* L. Shrad) (Primiani et al., 1990) and prickly lettuce (*Lactuca serriola* L.) (Mallory-Smith et al., 1990). Subsequently, plants and cultured cells resistant to ALS-inhibiting herbicides have been generated using both conventional mutation breeding and somatic cell selection. Since then, the *ALS* genes have been cloned and characterized. In most cases, resistance to ALS-inhibiting herbicides has been found to be conferred by single or double amino-acid mutations at a particular position in ALS. Mutated *ALS* genes can be used not only for the generation of herbicide-resistant crops, but also as selectable markers.
We are now producing transgenic tall fescue (*Festuca arundinacea* Schreb.) and maize that are resistant to ALS-inhibiting herbicides using novel mutated *ALS* genes. This chapter focuses on mutated *ALS* genes and their application to the production of herbicide-resistant crops and selection of transgenic cells as selectable markers.

1.1 ALS-inhibiting herbicides

ALS (EC 2.2.1.6; also referred to as acetohydroxyacid synthase, AHAS) is the first common enzyme in the biosynthetic pathway leading to the branched-chain amino acids, isoleucine, leucine and valine (Fig. 1). It is a highly conserved enzyme in higher plants. ALS moves to the chloroplast with the use of a transit peptide. Although ALS functions in plastids, ALS is a dominant and nuclear gene, and thus follows normal Mendelian inheritance.

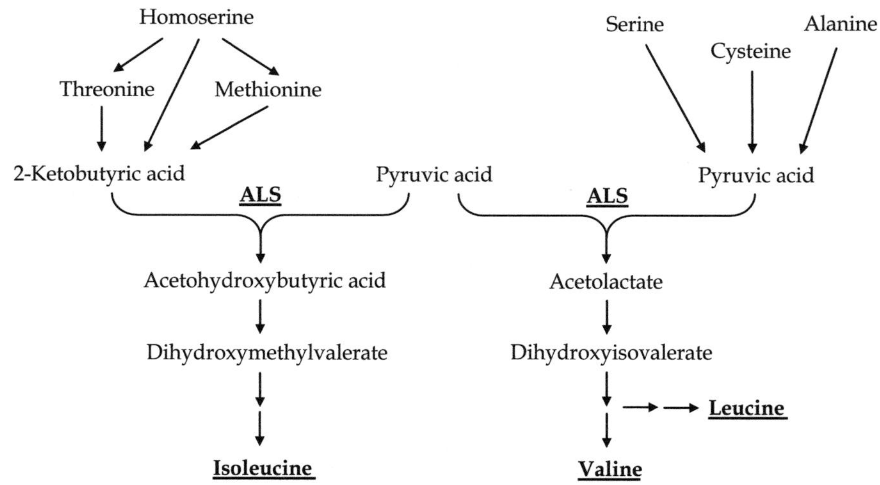

Fig. 1. The biosynthesis pathway of branched-chain amino acids. ALS-inhibiting herbicides inhibit ALS.

ALS is the target enzyme of at least five structurally distinct classes of herbicides; pyrimidinylcarboxylates (PCs), sulfonylureas (SUs), imidazolinones (IMs), triazolopyrimidine sulfonamides and sulfonylaminocarbonyltriazolinones (Shimizu et al., 2002). These herbicides all bind to ALS, but not all at the same attachment points. ALS-inhibiting herbicides are widely used around the world and account for about 17.5 % of the total global herbicide market (Green, 2007). There are more than 50 commercial herbicides from these five classes of herbicides used for selective weed control. Some representative ALS-inhibiting herbicides are shown in Fig. 2. These herbicides control an immense variety of grass and broadleaf weeds.

When we spray plants with these herbicides, plants cannot biosynthesize essential amino acids due to the inhibition of ALS and they come to die. ALS-inhibiting herbicides are used for controlling weed species at relatively low application rates and have both foliar and soil residual activity. Furthermore, ALS does not exist in mammals; thus, ALS-inhibiting herbicides are thought to be less toxic to mammals.

1.2 Mutated *ALS* genes confer resistance to ALS-inhibiting herbicides

Resistance to ALS-inhibiting herbicides in plants has in most cases been conferred by either single or double-mutant amino-acid substitutions at a particular position in ALS. Different types of mutation have been found to confer resistance to different classes of herbicide (Table 1).

The most commonly encountered mutations involve the residues of alanine at position 96 (A96), proline at position 171 (P171), tryptophane at position 548 (W548) and serine at position 627 (S627); Mutations were described using the rice numbering system. The mutation of the residue of tryptophane 548 substituted with leucine (W548L) was first isolated together with the mutation of the residue of proline at position 171 substituted with alanine (P171A) in tobacco (*Nicotiana tabacum* L.) by selection using SU (Lee et al., 1988). Subsequently, the mutation has been found in maize (Bernasconi et al., 1995) and canola

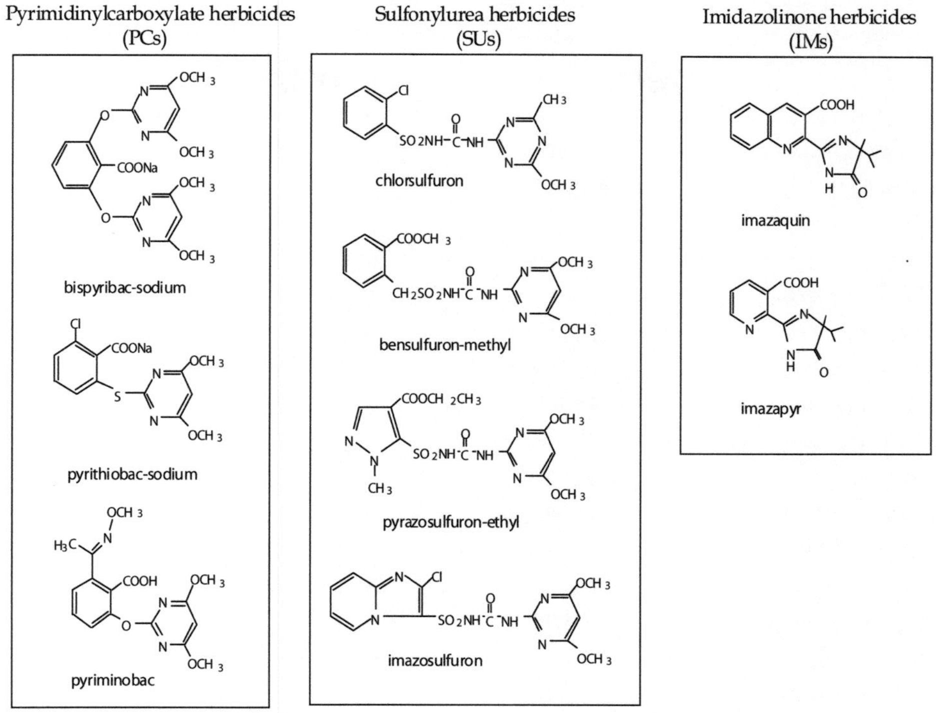

Fig. 2. ALS-inhibiting herbicides.

(Hattori et al., 1995). This mutation confers resistance to different classes of ALS-inhibiting herbicides, SUs and IMs. At this position, other amino-acid substitutions, W548C and W548S, have been identified in cotton (Rajasekaran et al., 1996).

A mutation of S627 was first found in IM-resistant *Arabidopsis thaliana* (Haughn & Somerville, 1990). In contrast to W548L, this mutation of S627N confers resistance to IM, but not to SU. The mutation at this position leading to S627A, S627N, S627T and S627F has been analyzed in *Arabidopsis* by site-directed mutagenesis (Lee et al. 1999).

Double mutations such as P171A/W548L in tobacco (Lee et al., 1988), P171S/S627N in *Arabidopsis* (Hattori et al., 1992) and A96T/P171S in sugar beets (*Beta vulgaris* L.) (Wright et al., 1998) have been reported.

Some amino acid substitutions conferring herbicide resistance are well conserved in plant ALS (Tan et al., 2005). We can artificially develop an herbicide-resistant *ALS* gene using this information even if mutations that confer herbicide resistance have not been characterized in the target plant. When we produce transgenic plants, it is desirable to use transgenes derived from host plant DNA as much as possible. This will be applicable to the production of cisgenic plants with public acceptance (Schouten et al., 2006).

1.3 A two-point mutated rice (*Oryza sativa* L.) *ALS* gene conferred resistance to a PC herbicide

Double mutations have been found in the rice *ALS* gene through cell culture using bispyribac-sodium (BS), a PC herbicide (Kawai et al., 2007b). Before isolation of the mutated *ALS* gene, no paper had reported a mutated *ALS* gene as conferring resistance to PC herbicides. The mutations were selected from BS-resistant calli produced spontaneously by somaclonal variation during tissue culture.

Plant species	Mutation[a)]	Method[b)]	Selection agent
Zea mays	A96T	CMB	Imidazolinone
Beta vulgaris	A96T	SCS	Imidazolinone
Arabidopsis thaliana	A96V	SDM	
Arabidopsis thaliana	M98E	SDM	
Arabidopsis thaliana	M98I	SDM	
Arabidopsis thaliana	M98H	SDM	
Arabidopsis thaliana	P171S	CMB	Sulfonylurea
Nicotiana tabacum	P171Q	SCS	Sulfonylurea
Nicotiana tabacum	P171A	SCS	Sulfonylurea
Nicotiana tabacum	P171S	SCS	Sulfonylurea
Beta vulgaris	P171S	SCS	Sulfonylurea
Brassica napus	P171S	SDM	
Arabidopsis thaliana	P171 deletion	SDM	
Arabidopsis thaliana	R173A	SDM	
Arabidopsis thaliana	R173E	SDM	
Arabidopsis thaliana	F180R	SDM	
Zea mays	W548L	CMB	Imidazolinone
Nicotiana tabacum	W548L	SCS	Sulfonylurea
Brassica napus	W548L	SCS	Sulfonylurea
Oryza sativa	W548L	SCS	Pyrimidinylcarboxylate
Gossypium hirsutum	W548S	SCS	Sulfonylurea
Gossypium hirsutum	W548C	SCS	Sulfonylurea
Arabidopsis thaliana	W548L	SDM	
Nicotiana tabacum	W548F	SDM	
Arabidopsis thaliana	W548S	SDM	
Arabidopsis thaliana	W548 deletion	SDM	
Zea mays	S627D	CMB	Imidazolinone
Arabidopsis thaliana	S627N	SCS/SDM	Imidazolinone
Zea mays	S627N	SCS	Imidazolinone
Oryza sativa	S627I	SCS	Pyrimidinylcarboxylate
Arabidopsis thaliana	S627T	SDM	
Arabidopsis thaliana	S627F	SDM	
Arabidopsis thaliana	S627 deletion	SDM	
Oryza sativa	G95A	SCS	Pyrimidinylcarboxylate

a) Mutations were described using the rice numbering system. Amino acids are described by one letters. A=alanine; C=cysteine; D=aspartic acid; E=glutamic acid; F=phenylalanine; G=glycine; H=histidine; I=isoleucine; L=leucine; M=methionine; N=asparagine; P=proline; Q=glutamine; R=arginine; S=serine; T=threonine; V=valine; W=tryptophane. b) Mutated ALSs were obtained through conventional mutation breeding (CMB), somatic cell selection (SCS) or site-directed mutagenesis (SDM).

Table 1. Mutations in ALS conferring resistance to ALS-inhibiting herbicides (Adapted from Kawai et al., 2007b).

The mutations involved W548L and the residue of serine at position 627 being substituted with isoleucine (S627I). These mutations are a new combination of spontaneous mutations with a novel substitution at the S627 position. The resistance to BS was extremely high as compared with those to SUs and IMs. The single mutations of W548L and S627I in ALS conferred resistance to BS, and the degree of resistance was higher in W548L than in S627I (Fig. 3A). The resistance to BS among these single-mutated ALSs was shown to be lower than that of the double-mutated ALS (Fig. 3A). The W548L mutation also conferred resistance to chlorsulfuron (CS), a SU herbicide, while the S627I mutation conferred no obvious resistance (Fig. 3B). A comparison of the degree of resistance to CS between the W548L single mutation and the W548L/S627I double mutation revealed that they shared the same degree of resistance to CS (Fig. 3B.) Therefore, when it was introduced into an *ALS* gene carrying the W548L mutation, the S627I mutation was shown to drastically enhance BS resistance in particular.

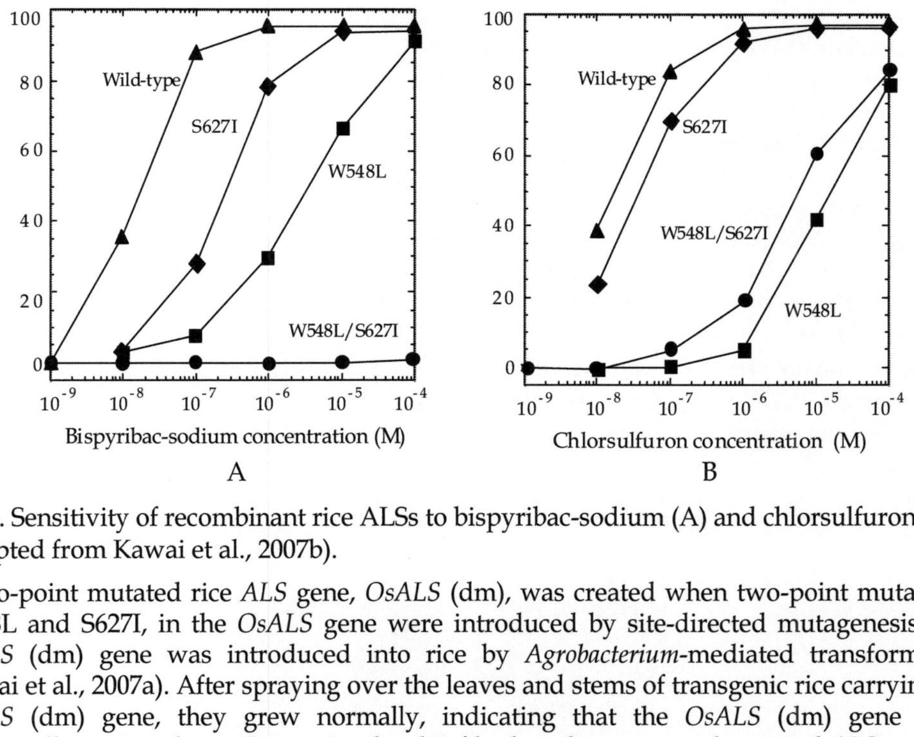

Fig. 3. Sensitivity of recombinant rice ALSs to bispyribac-sodium (A) and chlorsulfuron (B) (Adapted from Kawai et al., 2007b).

A two-point mutated rice *ALS* gene, *OsALS* (dm), was created when two-point mutations, W548L and S627I, in the *OsALS* gene were introduced by site-directed mutagenesis. The *OsALS* (dm) gene was introduced into rice by *Agrobacterium*-mediated transformation (Kawai et al., 2007a). After spraying over the leaves and stems of transgenic rice carrying the *OsALS* (dm) gene, they grew normally, indicating that the *OsALS* (dm) gene acted functionally in rice plants. Expression levels of both endogenous and mutated *ALS* genes of transgenic rice plants were correlated with the resistance of transgenic plants to BS. The BS resistance of transgenic plants was stably inherited to the progeny in a Mendelian manner.

2. Production of transgenic herbicide-resistant tall fescue for turf

Tall fescue is a major cool-season perennial grass species. It is an outcrossing, open-pollinated, and highly self-imcompatible grass species; therefore, generally genetic improvement takes a long time. Genetic transformation can help to overcome the problem and facilitate grass improvement. Because of its agronomic importance, many *Agrobacterium*-mediated transformation systems have been developed in tall fescue (Bettany et al., 2003; Dong et al., 2005; Wang et al., 2005; Gao et al., 2008). Tall fescue is widely used not only as forage in pastures, but also as turf for lawns, golf courses, athletic fields, roadsides, and other places. Applying herbicide to tall fescue in pasture or meadow is unrealistic because of the cost and safety, but its application in turf is promising. In turfgrass, weed management is very important, and herbicide resistance can be used as an efficient tool to allow easier maintenance.

We introduced the *OsALS* (dm) gene into turf-type tall fescue to confer herbicide resistance using *Agrobacterium*-mediated transformation (Sato et al., 2009). *Agrobacterium tumefaciens* strain EHA105 carries the binary vector pMLH7133-*OsALS* (dm) consisting of the *OsALS* (dm) gene and hygromycin phosphotransferase gene (*hpt*) under the control of the enhanced cauliflower mosaic virus (CaMV) 35S promoter (Kawai et al., 2007a). Embryogenic calli were induced from shoot tips of the turf-type tall fescue cultivar Tomahawk germinated *in vitro*. Infected calli were selected by incubation with hygromycin. Hygromycin-resistant calli were regenerated, transferred to soil and grown in a greenhouse.

Introduction of the *OsALS* (dm) and *hpt* genes was confirmed by PCR analysis. The PCR products amplified by the *OsALS* (dm) primers from both regenerated and wild-type plants were equivalent in size to a fragment amplified from the binary vector pMLH7133-*OsALS* (dm). Because the *ALS* genes are well conserved in plants, the PCR from the wild-type plant would be amplified from the endogenous tall fescue *ALS* gene (*FaALS*). In the *OsALS* (dm) gene, two new *Mfe*I sites are produced at the mutation sites (Osakabe et al., 2005), and thus the primers were designed to cover one *Mfe*I site to distinguish the *OsALS* (dm) from the *FaALS* gene. After the PCR products were digested with *Mfe*I, the regenerated plants and pMLH7133-*OsALS* (dm) yielded two fragments, whereas the wild-type plant yielded a single fragment. The copy number of integrated genes was estimated by Southern blot analysis and ranged from one to five.

Transgenic plants were sprayed on the leaves with a commercial ALS-inhibiting herbicide containing BS. Wild-type plants were confirmed to die completely after herbicide treatment (Fig. 4). On the other hand, transgenic plants were unaffected by the treatment and showed resistance to the herbicide (Fig. 4).

Fig. 4. Herbicide application to wild-type (left) and transgenic plants (right). The picture was taken 45 days after herbicide treatment.

ALS activity in the transgenic plants under the herbicide treatment was analyzed by colorimetric enzymatic assay (Osakabe et al., 2005) with some minor modifications. This assay is able to estimate ALS activity in plant tissues with or without herbicide treatment based on a comparison of acetoin accumulation (Gerwick et al., 1993). Red or pink coloration indicates a high accumulation of acetoin produced by the ALS activity, and yellow or brown

indicates a low accumulation of acetoin. When the leaf tissues were incubated without BS, both wild-type and transgenic plants produced pink coloration (Fig. 5A). When incubated with BS, only transgenic plants produced pink coloration while the wild-type plants produced a brown color (Fig. 5A).
When ALS activity with BS was measured by a spectrophotometer, the ALS activity in transgenic plants was almost equivalent to that in wild-type plants without BS and showed higher activity than in wild-type plants (Fig. 5B). In the assay without BS, the ALS activity tended to be higher in transgenic plants than in wild-type plants because OsALS (dm) protein would be produced in addition to the endogenous FaALS protein. The transgenic plants showed lower ALS activity with BS than without BS, probably because the FaALS protein was inhibited by BS treatment. These results indicated that the transgenic plants actively produced OsALS (dm) protein under herbicide treatment.

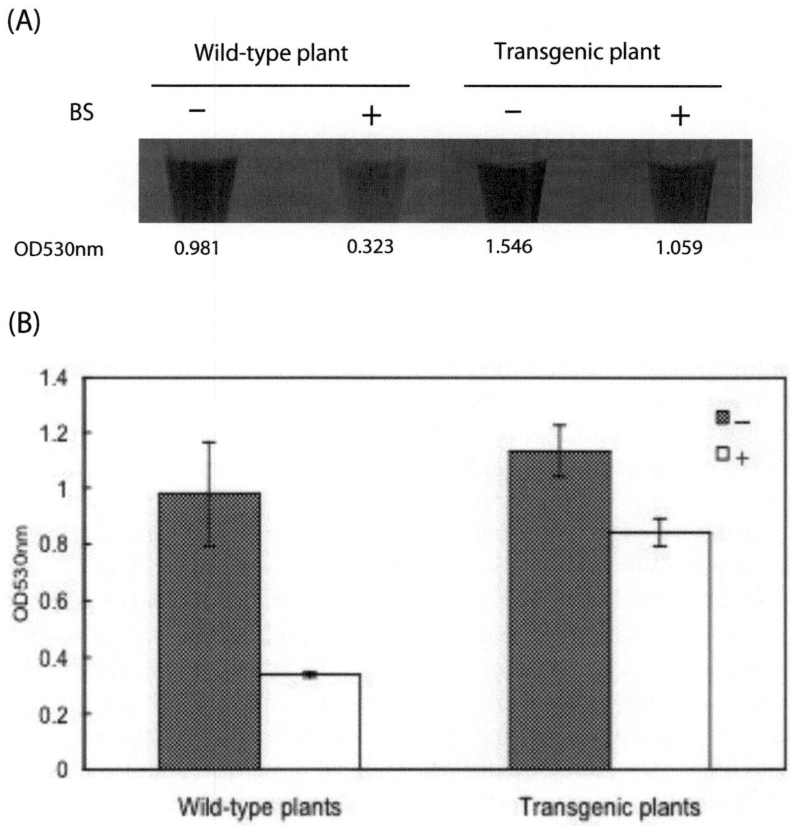

Fig. 5. Colorimetric enzymatic assay in leaves of wild-type and transgenic plants resistant to herbicide. The leaf tissues were incubated with (+) or without (-) BS. (A) Comparison of acetoin accumulation. (B) Measurement of ALS activity. Error bars represent the SE for wild-type plants (n=3) and transgenic plants (n=9) (Adapted from Sato et al., 2009).

Although the transgenic plants were confirmed to show herbicide resistance in the greenhouse, they should be further examined to ensure that herbicide resistance is stable under field conditions. However, since tall fescue is an open-pollinated and anemophilous grass, it is possible that transgenes could be dispersed into the environment through pollen.

Lee (1996) discussed two environmental risks associated with transgenic turfgrass. The first risk is the possibility that transgenes will be spread by crossing transgenic plants with weed species. The second is the chance that transgenic plants will themselves become weeds. Tall fescue produces large amounts of pollen-containing allergenic proteins that cause hay fever in susceptible people. In tall fescue, plants with cytoplasmic male sterility have been developed to limit grass pollen allergy (Fujimori, 2002). To minimize the risk of dispersal of transgenic pollen in the field, we are crossing such cytoplasmic male-sterile plants with our transgenic plants.

3. Production of transgenic maize using a mutated *ALS* gene derived from host maize DNA

Four commercial hybrid maize varieties resistant to ALS-inhibiting herbicides (IM and SU) were developed by a somatic cell selection method with a B73 x A188 callus tissue culture (Tan et al., 2005) with a single mutation (W574L, S653N and T155A). A double mutation (P197A and W574L) that showed enhanced resistance to ALS-inhibiting herbicides was later discovered and called a highly herbicide-resistant ALS (HRA).

The first transgenic maize resistant to ALS-inhibiting herbicides was produced by Pioneer Hi-Bred with this *HRA* gene. The transgenic maize not only has resistance to ALS-inhibiting herbicides but also to glyphosate introduced by Bacillus-derived 5-enolpyruvylskimate-3-phosphate synthase (EPSPS) in the same vector. As for regulatory elements, an *ALS* gene-derived promoter with three copies of a CaMV 35S enhancer and potato (*Solanum tuberosum* L.) protease inhibitor II-derived terminator was used. Nicosulfuron and rimsulfuron were used to check the resistance to ALS-inhibiting herbicides of the transgenic maize. This event has already been named DP-098140, OECD UI: DP-098140-6 (https://bch.cbd.int/database/record-v4.shtml?documentid=48466) and has been approved in several countries.

It is important to pay attention to the production of consumer-acceptability of transgenic crops in certain countries, including Japan. In order to produce transgenic maize plants carrying only host-derived genes which are more acceptable, we isolated a maize *ALS* gene inducing both a promoter and terminator region from a Japanese inbred line, and then introduced the same mutations as in rice but at different positions (W542L and S621I) to confer ALS-inhibiting herbicide resistance. Its resistance to BS was also confirmed by analyzing the enzymatic activities. This mutated *ALS* gene was again introduced to maize by an improved *Agrobacterium*-mediated transformation method (Ishida et al., 2007). Japanese maize inbred lines were at first screened for their tissue culture response, and Mi29 (Ikegaya et al., 1999) was adopted for its high *in vitro* regenerative ability. Immature embryos of Mi29 at 7-10 days after fertilization were infected with *Agrobacterium* containing either the standard binary vector or a super-efficient one and cultured on selection medium with 0.1 or 0.5 microM BS after a one-week co-cultivation period. Transgenic calli resistant to BS were transferred to regeneration medium, and regenerated shoots were further transferred to rooting medium. The overall transformation frequency was 5-30% depending upon the stage and quality of the immature embryo. Transgenic BS-resistant maize grew to maturity and set seeds. T_1 progenies were obtained by crossing the transgenic maize with wild type. The inheritance of the transgene was confirmed by PCR analysis and BS application to their progenies. The progenies showed the segregation ratio (resistant: susceptible 1:1) expected for a single locus (Fig. 6).

Fig. 6. Segregation of resistance to ALS-inhibiting herbicide containing BS in progenies of wild-type x transgenic BS-resistant maize. 100-fold diluted commercial herbicide was sprayed one week after germination. The picture was taken after another week.

4. The use of mutated *ALS* genes as selective markers

Selectable markers facilitate the selection of transgenic cells from non-transgenic cells in genetic transformation. Without them, the transgenic cells that integrate transgenes stably would be lost in non-transgenic cells, which would grow well in the absence of a selection agent. The most widely used selectable markers are antibiotic-resistant genes such as the *hpt* gene and the *nptII* gene encoding neomycin phosphotransferase. HPT has a low likelihood of inducing toxicity and allergenicity (Zhuo et al., 2009; Lu et al., 2007), and NPTII was determined to be nontoxic for human or animal consumption (Nap et al., 1992). Herbicide-resistant genes are also used as selectable markers. The *bar* gene encodes the enzyme phosphinothricin acetyltransferase (PAT) and confers resistance to phosphinothricin, glufosinate or bialaphos herbicides. PAT is specific and does not possess food toxins or allergens in human food and animal feed (Hérouet et al., 2005). The lack of toxicity or allergenicity of EPSPS has also been proved (Hammond et al., 2004).

However, some consumers are opposed to the use of these selectable markers because they are derived from bacterial or fungal DNA. Therefore, the use as selectable markers of mutated *ALS* genes, derived from plant DNA, has recently increased. Mutated *ALS* genes derived from *Arabidopsis* have been reported to be useful as selectable markers in various plants, such as tobacco (Gabard et al., 1989), rice (Li et al., 1992), potato (Anderson et al., 2003), oilseed mustard (*Brassica juncea*) (Ray et al., 2004) and maize (Zhang et al. 2005). Mutated Os*ALS* genes have been demonstrated to be useful as selectable markers in rice (Osakabe et al., 2005; Okuzaki et al., 2007). Using these mutated *OsALS* genes, transgenic plants have been produced in various plants such as rice (Osakabe et al., 2005; Okuzaki et al., 2007), soybeans (Tougou et al., 2009), tall fescue (in preparation) and wheat (*Triticum aestivum* L.) (Ogawa et al., 2008).

Some studies have suggested that homology-dependent gene silencing is associated with the presence of either multiple copies of homologous transgenes and promoters (Matzke & Matzke, 1995) or a transgene and a homologous endogenous gene (Meyer, 1995). In general, constitutive promoters, such as the CaMV 35S promoter, rice actin 1 promoter and maize ubiquitin promoter, are used to drive selectable markers. In our transgenic tall fescue, multiple integrated transgenes were observed, and ALS activity was insufficient to confer herbicide resistance in susceptible plants (Sato et al., 2009). The CaMV 35S promoter was used for two genes (*OsALS (dm)*, *hpt*) in the same binary vector; therefore, the chances of gene silencing may have increased by overuse of the CaMV 35S promoter. Okuzaki et al.

(2007) reported that some transgenic rice calli with multiple copies of a mutated *ALS* gene driven by the maize ubiquitin promoter did not regenerate, whereas transgenic calli with only one or two transgenes did. On the other hand, no relationship between herbicide resistance and copy number was apparent in wheat transformation using the rice ALS promoter (Ogawa et al., 2008). It was assumed that the rice ALS promoter is not a strong one and is expressed in a tissue-specific manner (Osakabe et al., 2005). Strong expression in all tissues by constitutive promoters tends to cause deleterious effects, and the use of the endogenous ALS promoter would be preferable for more stable expression.

5. Conclusion

In this chapter, we introduced mutated *ALS* genes and their application to the production of herbicide-resistant crops and selection of transgenic cells. Our transgenic tall fescue and maize were confirmed to show ALS-inhibiting herbicide resistance in the greenhouse. So far, many transgenic herbicide-resistant crops have been developed. Though relatively new, their contribution to production-based agriculture has been significant. In future, we expect our herbicide-resistant crops to allow easier weed management.

Although herbicides are effective weed management tools, a cultivation system that depends on the application of a single type of herbicide with the same site of action would tend to increase the frequency of emergence of herbicide-resistant weed species or group of herbicides. Herbicide-resistant weeds evolve through random mutation events. In particular, there are more weed species that are resistant to ALS-inhibiting herbicides than to any other herbicides (Tranel & Wright, 2002) because resistance to ALS-inhibiting herbicides is conferred by single or double mutations.

Recently, the adoption of stacked cultivars in which multiple transgenic traits were introduced has been produced in maize, soybeans and other crops. The use of a combination of several herbicides with other mechanisms and plants resistant to those herbicides is useful to inhibit and delay effectively the generation of herbicide-resistant weed species.

6. References

Andersson, M., Trifonova, A., Andersson, A. B., Johansson, M., Bülow, L. & Hofvander, P. (2003). A novel selection system for potato transformation using a mutated AHAS gene. Plant Cell Rep. 22, 261-267.

Bernasconi, P., Woodworth, A. R., Rosen, B. A., Subramanian, M. V. & Siehl, D. L. (1995). A naturally occuring point mutation confers broad range tolerance to herbicides that target acetolactate synthase. J. Biol. Chem. 270, 17381-17385.

Bettany, A. J. E., Dalton, S. J., Timms, E., Manderyck, B., Dhanoa M. S. & Morris, P. (2003). *Agrobacterium tumefaciens*-mediated transformation of *Festuca arundinacea* (Schreb.) and *Lolium multiflorum* (Lam.). Plant Cell Rep. 21, 437-444.

Dong, S. & Qu, R. (2005). High efficiency transformation of tall fescue with *Agrobacterium tumefaciens*. Plant Sci. 168, 1453-1458.

Fujimori, M. (2002). Breeding male sterile tall fescue. Japan Livestock Technol. 562, 15-18 (in Japanese).

Gabard, J. M., Charest P. J., Iyer V. N. & Miki, B. L. (1989). Cross-resistance to short residual sulfonylurea herbicides in transgenic tobacco plants. Plant Physiol. 91, 574-580.

Gao, C., Long, D., Lenk, I. & Nielsen, K. K. (2008). Comparative analysis of transgenic tall fescue (*Festuca arundinacea* Schreb.) plants obtained by *Agrobacterium*-mediated transformation and particle bombardment. Plant Cell Rep. 27, 1601-1609.

Gerwick, B. C., Mireles, L. C. & Eilers, R. J. (1993). Rapid diagnosis of ALS/AHAS-resistant weeds. Weed Technol. 7, 519-524.

Green, J. M. (2007). Review of glyphosate and ALS-inhibiting herbicide crop resistance and resistant weed management. Weed Technol. 21, 547-558.

Hammond, B., Dudek, R., Lemen, J. & Nemeth, M. (2004). Results of a 13 week safety assurance study with rats fed grain from glyphosate tolerant corn. Feed Chem. Toxicol. 42, 1003-1014.

Hattori, J., Brown, D., Mourad, G., Labbé, H., Ouellet, T., Sunohara, G., Rutledge, R., King, J. & Miki, B. (1995). An acetohydroxy acid synthase mutant reveals a single site involved in multiple herbicide resistance. Mol. Gen. Genet. 246, 419-425.

Hattori, J., Rutledge, R., Labbé, H., Brown, D., Sunohara, G. & Miki, B. (1992). Multiple resistance to sulfonylureas and imidazolinones conferred by an acetohydroxyacid synthase gene with separate mutations for selective resistance. Mol. Gen. Genet. 232, 167-173.

Haughn, G. W. & Somerville, C. R. (1990). A mutation causing imidazolinone resistance maps to the *Csr1* locus of *Arabidopsis thaliana*. Plant Physiol. 92, 1081-1085.

Hérouet, C., Esdaile, D. J., Mallyon, B. A., Debruyne, E., Schulz A., Currier, T., Hendrickx, D., van der Kils, R. J. & Rouan, D. (2005). Safety evaluation of the phosphinothricin acetyltransferase proteins encoded by the *pat* and *bar* sequences that confer tolerance to glufosinate-ammonium herbicide in transgenic plants. Reg. Toxic. Pharm. 41, 134-149.

Ikegaya, F., Koinuma, K., Ito, E., Inoue, Y., Nozaki, K., Fujita, K. & Mochizuki, N. (1999). Development and characteristics of new inbred line 'Mi29' for silage maize. Bulletin of Kyusyu National Agricultural Experiment Station 35, 71-83 (in Japanese with English summary).

Ishida, Y., Hiei, Y. & Komari, T. (2007). *Agrobacterium*-mediated transformation of maize. Nature protocols 2, 71614-1621.

Kawai, K., Kaku, K., Izawa, N., Fukuda, A., Tanaka, Y. & Shimizu, T. (2007a). Functional analysis of transgenic rice plants expressing a novel mutated ALS gene of rice. J. Pestic. Sci. 32, 385-392.

Kawai, K., Kaku, K., Izawa, N., Shimizu, T., Fukuda, A. & Tanaka, Y. (2007b). A novel mutant acetolactate synthase gene from rice cells, which confers resistance to ALS-inhibiting herbicides. J. Pestic. Sci. 32, 89-98.

Lee, K. Y., Townsend, J., Tepperman, J., Black, M., Chui, C. F., Mazur, B., Dunsmuir, P. & Bedbrook, J. (1988). The molecular basis sulfonylurea herbicide resistance in tobacco. EMBO J. 7, 1241-1248.

Lee, L. (1996). Turfgrass biotechnology. Plant Sci. 115, 1-8.

Lee, Y. T., Chang, A. K. & Duggleby, R. G. (1999). Effect of mutagenesis at serine 653 of *Arabidopsis thaliana* acetohydroxyacid synthase on the sensitivity to imidazolinone and sulfonylurea herbicides. FEBS Lett. 452, 341-345.

Li, Z., Hayashimoto, A. & Murai, N. (1992). A sulfonylurea herbicide resistance gene from *Arabidopsis thaliana* as a new selectable marker for production of fertile transgenic rice plants. Plant Physiol. 100, 662-668.

Lu, Y., Xu, W., Kang, A., Luo, Y., Guo, F., Yang, R., Zhang, J & Huang, K. (2007). Prokaryotic expression and allergenicity assessment of hygromycin B phosphotransferase protein derived from genetically modified plants. J. Food Sci. 72, M228-M232.

Mallory-Smith, C. A., Thill, D. C. & Dial, M. J. (1990). Identification of sulfonylurea herbicide-resistant prickly lettuce (*Lactuca serriola*). Weed Technol. 4, 163-168.

Matzke, M. A. & Matzke A. J. M. (1995). How and why do plants inactivate homologous (trans)-genes? Plant Physiol. 107, 679-685.

Meyer, P. (1995). Understanding and controlling transgene expression. Trends Biotechnol. 13, 332-337.

Nap, J. P.,Bijvoet, J. & Willem, J. (1992). Biosafety of kanamycin-resistant transgenic plants. Transg. Res. 1, 239-249.

Ogawa, T., Kawahigashi, H., Toki, S. & Handa, H. (2008). Efficient transformation of wheat by using a mutated rice acetolactate syhthase gene as a selectable marker. Plant Cell Rep. 27, 1325-1331.

Okuzaki, A., Shimizu, T., Kaku, K., Kawai, K. & Toriyama, K. (2007). A novel mutated acetolactate synthase gene conferring specific resistance to pyrimidinyl carboxy herbicides in rice. Plant Mol. Biol. 64, 219-224.

Osakabe, K., Endo, M., Kawai, K., Nishizawa, Y., Ono, K., Abe, K., Ishikawa, Y., Nakamura, H., Ichikawa, H., Nishimura, S., Shimizu, T. & Toki, S. (2005). The mutant form of acetolactate synthase genomic DNA from rice is an efficient selectable marker for genetic transformation. Mol. Breed. 16, 313-320.

Primiani, M., Cotterman, M. J. C. & Saari, L. L. (1990). Resistance of kochia (*kochia scoparia*) to sulfonylurea and imidazolinone herbicides. Weed Technol. 4, 169-172.

Rajasekaran, K., Grula, J. W. & Anderson, D. M. (1996). Selection and characterization of mutant cotton (*Gossypium hirsutum* L.) cell lines resistant to sulfonylurea and imidazolinone herbicides. Plant Sci. 119, 115-124.

Ray, K., Jagannath, A., Gangwani, S. A., Burma, P. K. & Pental, D. (2004). Mutant *acetolactate synthase* gene is an efficient in vitro selectable marker for the genetic transformation of *Brassica juncea* (oilseed mustard). J. Plant Physiol. 161, 1079-1083.

Sato, H., Shimizu, T., Kawai, K., Kaku, K. & Takamizo, T. (2009). Conferred resistance to an acetolactate synthase-inhibiting herbicide in transgenic tall fescue (*Festuca arundinacea* Schreb.). HortSci. 44, 1254-1257.

Schouten, H. J., Krens, F. A. & Jacobsen, E. (2006). Do cisgenic plants warrant less stringent oversight? Nature Biotech. 24, 753.

Shimizu, T., Nakayama, I., Nagayama, K., Miyazawa, T. & Nezu, Y. (2002). Acetolactate synthase inhibitors, in Böger, P., Wakabayashi, K. & Hirai, K. of Editors (ed.), *Herbicide classes in development*, Springer-Verlag, Berlin, Germany, pp. 1-41.

Tan, S., Evans, R. R., Dahmer, M. L., Singh, B. K. & Shaner, D. L. (2005). Imidazolinone-tolerant crops: history, current status and future. Pest Manage. Sci. 61, 246-257.

Tarnel, P. J. & Wright, T. R. (2002). Resistance of weeds to ALS-inhibiting herbicides: what have we learned? Weed Sci. 50, 700-712.

Tougou, M., Yamagishi, N., Furutani, N., Kaku, K., Shimizu, T., Takahata, Y., Sakai, J., Kanematsu, S. & Hidaka, S. (2009). The application of the mutated acetolactate synthase gene from rice as the selectable marker gene in the production of transgenic soybeans. Plant Cell Rep. 28, 769-776.

Wang, Z. Y. & Ge, Y. (2005). *Agrobacterium*-mediated high efficiency transformation of tall fescue (*Festuca arundinacea*). J. Plant Physiol. 162, 103-113.

Wright, T. R., Bascomb, N. F., Sturner, S. F. & Penner, D. (1998). Biochemical mechanism and molecular basis for ALS-inhibiting herbicide resistance in sugarbeet (*Beta vulgaris*) somatic cell selections. Weed Sci. 46, 13-23.

Zhang, Y., Yin, X., Yang, A., Li, G. & Zhang, J. (2005). Stability of inheritance of transgenes in maize (*Zea mays* L.) lines produced using different transformation methods. Euphytica. 144, 11-22.

Zhuo, Q., Piao, J. H., Tian, Y., Xu, J. & Yang, X. G. (2009) Large-scale purification and acute toxicity of Hygromycin B phosphotransferase. Biomedical Environ. Sci. 22, 22-27.

12

Pollen Mediated Gene Flow in GM Crops: The Use of Herbicides as Markers for Detection. The Case of Wheat

Iñigo Loureiro, Concepción Escorial, Inés Santín and Cristina Chueca
Instituto Nacional de Investigación y Tecnología Agraria y Alimentaria (INIA)
Spain

1. Introduction

ISAAA has estimated that genetically modified (GM) crops, mainly soybean, maize, cotton and canola, are cultivated worldwide in an area that has increased from 1.7 million hectares in 1996 to 134 million hectares in 2009, of which more than 80% have an herbicide-tolerant trait (ISAAA 2010). This work reviews the agricultural and environmental concerns about the likelihood for gene flow from GM wheat (*Triticum aestivum* L.). Wheat is the world's most important crop species, grown on over 210 million hectares. There are no GM wheat varieties commercially available but transgenic wheat varieties are being successfully developed and field-tested. That makes wheat in the pipeline of genetically engineered crops to be cultivated. Although wheat is predominantly a self-pollinating crop, pollen from one plant can travel via wind to other receptive plant, being outcrossing between wheat cultivars possible at variable rates. Coexistence problems in wheat could thus arise if no measures are taken before releasing and marketing any transgenic cultivar, as has occurred with other GM crops such as oilseed rape or maize, where measures were implemented after commercial transgenic introduction. Besides this, wild *Aegilops* species like *Ae. geniculata* Roth., *Ae. cylindrica* Host., *Ae. biuncialis* Vis. or *Ae. triuncialis* L. can form natural interspecific hybrids with wheat where they grow in sympatry. These natural hybrids are highly sterile, although seeds may occasionally be found. Data presented aim to contribute to the determination of the extent of this phenomena. These data are necessary to manage the possible impact of transgenic wheat hybrids before the transgenic crop can be grown under field conditions. Herbicide-tolerant wheat parental varieties can be used to obtain resistant progeny detectable by herbicide selection, providing a high approach to the potential occurrence of intra and interspecific pollen mediated gene flow.

2. Herbicide resistance as a marker for gene flow

In spite of the knowledge of GM herbicide tolerant wheat cultivars, whose use is limited by availability and regulatory constraints, in the experiments presented in this book chapter we have used non GM wheat cultivars possessing homozygous dominant genes for herbicide response. Chlorotoluron and difenzoquat tolerant wheat cultivars were used to obtain hybrid-resistant progenies detectable by herbicide selection.

The herbicide chlorotoluron is a commercially available selective phenylurea that is widely used for broad-leaf and annual grass weed control in winter cereals. The genetic control of tolerance to chlorotoluron in bread wheat is determined by a major single dominant gene, *Su1*, located on the short arm of chromosome 6B (Krugman et al., 1997). This herbicide is selective in winter wheat crops although there are wheat cultivars susceptible to chlorotoluron (Sixto et al., 1995; Bozorgipour & Snape, 1997). Wheat wild relatives as *Aegilops* spp. are also susceptible to it. In the presence of herbicide selection pressure, herbicide resistance allows for the detection of hybrids between resistant wheat cultivars and susceptible ones and between *Aegilops* spp. and resistant wheats. In our studies, we have used chlorotoluron tolerant wheat cultivars as Castan or Deganit.
The herbicide difenzoquat is a mitosis inhibitor used for the post-emergence control of wild *Avena* spp. in winter cereals. *Aegilops* species are susceptible to this herbicide. Chinese Spring (CS) is a wheat cultivar possessing herbicide resistance alleles endowing resistance that can be used to obtain hybrid resistant progeny. The genetic control of tolerance to difenzoquat in bread wheat is determined by a major single gene (Busch et al., 1989).
During our work we have conducted two types of assays which have enabled us to identify resistant hybrids: growing plants with herbicides in hydroponic assays and herbicide spraying assays.

3. Pollen dispersal in wheat

Wheat pollen dispersal is not a new issue in agriculture. The varietal purity of the seed has always played a fundamental role in the development, yield and final quality of crops. It has long been known that pollen contamination not only takes place in cross-pollinated crops, it is also possible in self-pollinating crops when different varieties of the same crop are cultivated and sufficient separation distance is not maintained (Sanchez-Monge, 1955).
One of the most effective methods for preventing pollen contamination between crossable genotypes is the use of isolation distances. The isolation distance required will depend on flower characteristics, compatibility with neighboring crops, pollen quantity and viability, mode of pollen dissemination and environmental conditions, which are of the upmost importance. Not all genotypes show the same ability in crosses. Wheat cultivars could show differences in the factors included in their reproductive biology; the flowering period of a wheat plant takes around 8 days. During these days each flower is open from 8 to 60 minutes. Wheat produces a low number of pollen grains (10,000 per anther) only the 5 to 7% of the pollen drops on the stigma, the great majority is dispersed by wind (de Vries, 1971). The period of pollen viability is low, never above three hours (D'Souza, 1970). Pollen viability declines, with time and exposure to environmental stresses. From a hybridization rate of 86% obtained with fresh pollen maintained at 15° C (at RH 65 ± 5%), hybridization was only 12% after one hour at 25°C , while no seeds were found at 30°C. At 15°C seed set declined 14 % and 23% at 20°C (Loureiro et al., 2007). Receptivity of stigma and flower opening were also environmental and genetically dependent (de Vries, 1971). Under our circumstances, in a year with favourable conditions (77% RH and 20 ± 2°C), a maximum seed set of 78% was obtained for Pavon x CS wheat cultivars hand crosses. These values were of 39% in a less favorable year.

4. Outcrossing in wheat. The problem of coexistence

Wheat is a self-pollinating crop but outcrossing is possible between cultivars at variable rates that are related with populations, genotypes and environmental conditions (Jain, 1975).

The main studies on pollen dispersal in wheat appear in two stages. In the 1960s and beginning of the 1970s managing pollen drift was a major concern within the context of commercial production of hybrid wheat, where achieving high levels of genetic purity and satisfactory seed set on male sterile plants were essential (Pickett 1993). In recent years, pollen dispersal in wheat has again received considerable attention, within the context of the legislation applied to cultivars issued from the advances in biotechnology. Transgenic wheat varieties are being successfully developed and field-tested, primarily as glyphosate-tolerant wheat (Blackshaw & Harker, 2002; Zhou et al., 2003), and there is extensive research on a wide range of GM wheat traits (e.g. Fusarium resistance, drought resistance); probably in the next few years certified cultivars of transgenic wheat shall be commercially available.

There is concern that once transgenic wheat is released for commercial production, there will be a potential pollen flow from GM wheat to non GM-wheat (van Acker et al., 2003). As a consequence the product could not fulfil all the requirements of some international markets and farmers could lose the ability of choose between conventional, organic or GM-based crop productions, in compliance with the relevant EU legislation on labelling and/or purity standards. EU regulations framework establishes a 0.9% labelling threshold for the adventitious presence of GM material in non-GM products. Thus, problems could appear in wheat if no measures are taken prior to the release and commercialisation of any transgenic cultivars to establish the basis that allows the coexistence of all type of wheat with the GM wheat.

Outcrossing studies between *T. aestivum* cultivars have been conducted by different authors in the absence of any pollen competition on male sterile receptor plants. In this sense emasculate plants provide information on the upper levels of outcrossing under specific conditions and help in evaluating safety distances that avoid outcrossing and potential pollen-mediated gene-flow. Outcrossing rates in these studies are very different among experiments in terms of frequency of hybrid seed set and maximum seed set distance (from 12 to 73% at distances near to the pollen source, from 0.3 to 9 % at around 10 m distance) (Khan et al., 1973; de Vries 1974). In a three-year study we assessed the maximum potential outcrossing under field conditions between the wheat cultivars Pavon (receptor) and Chinese Spring (3 x 3 m source donor). Bread wheat can also coexist in the field with the second major cultivated wheat species, the durum wheat tetraploid *Triticum turgidum* L. (tetraploid, AABB) that is closely related to bread wheat which bulk of production is concentrated in the Middle East, North America and the Mediterranean region. For this reason durum wheat *T. turgidum* L. var. *durum* cultivar Nita was also included in the study. Outcrossing was measured by seed set on emasculated recipient plants. Frequencies of seed set at 0 m distance were 45% (37-56%) for *T. aestivum* cultivars and 18% (5-30%) with *T. turgidum* (Loureiro et al., 2007). Under semiarid conditions of this assay, viable pollen was found at 14 m from the source, the maximum distance analyzed, with a distance of 8 m at which cross-pollination decreases below 1%. There is a strong positive correlation between outcrossing and the amount of pollen in air, for this reason hybridisation at distances close from the pollen source are similar to maximum hybridisation when emasculated plants were used as receptors. However as the distance from the pollen source increases the pollen concentration rapidly decline, 90% of the pollen in wheat remains within 6 meters from its source (Jensen, 1968; Loureiro, 2005). A mean seed set of 45% at 0 m decrease to 10% at 2 m (Figure 1). At 10 m seed set was of 1% in agreement with data of Stopkopf & Rai (1972) ; de Vries (1974) and Zhao et al. (2000) and slightly higher than data of Lu et al. (2002). Other authors have found a slower decrease on seed set in relation to distance from the pollen source (Johnson et al., 1967; Bitzer & Patterson 1967; Khan et al., 1973). An exponential predictive curve (Figure 1) provides the upper level of the magnitude of this event (Loureiro et al., 2007). In these circumstances, 5 m would be required to avoid adventitious GM presence above the 0.9% marked by the European legislation. This isolation could be higher downwind with 7 m required to meet the threshold.

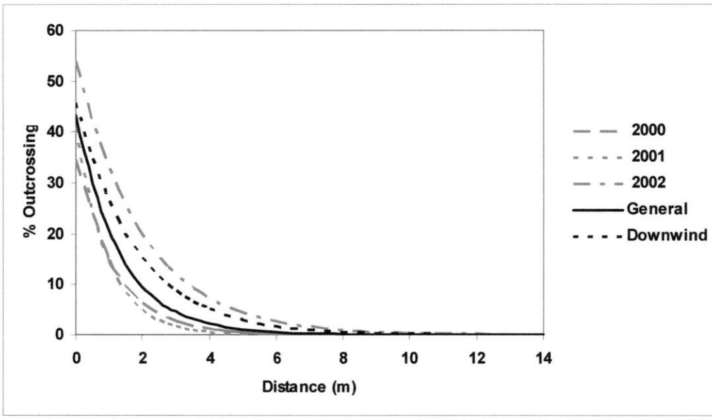

Fig. 1. Mean seed set related to distance under no pollen competition in field assay.

The outcrossing between wheat cultivars have been also assessed natural conditions of pollen competition. Experiments were carried out in the year 2005 at "La Canaleja (Instituto Nacional de Investigación y Tecnología Agraria y Alimentaria, INIA) and at "El Encín" (Instituto Madrileño de Investigación y Desarrollo Rural y Agrario, IMIDRA) experimental stations, Madrid, Spain. The layout of the experiment was such that it permitted observations on the extent of natural crossing of a wheat pollen donor with different recipient cultivars and in different directions and distances. The experimental field design consisted in a 50 x 50 m central square plot sown with a *T. aestivum* chlorotoluron tolerant cultivar pollen donor (Castan in "El Encin" and Deganit in "La Canaleja") at field density and chlorotoluron susceptible receptors (Altria and Recital) placed in the four sides of the pollen source at distances of 0, 1, 3, 5, 10, 20, 40, 80 and 100 m. In any case the mean of outcrossing reached 2% at 0 m distance. This value was always below 5% downwind even in close proximity (Loureiro et al., 2005). Outcrossing was detected at the very low level of 0.07% at 100 m from the source.

These outcrossing rates are in the range of published frequencies averaging 1%, but that can vary between 0 to 6.7% at distances below 1 m (Griffin, 1987; Hucl, 1996; Zhao et al., 2000; Hucl & Matus-Cadiz, 2001; Loureiro et al., 2005), although hybrid seed set is also possible at greater distances.

5. Hybridization with wild relatives

Genes could also be transferred from GM crops to wild relatives through interspecific hybridization. Prior to the commercialization of GM crops the research on the natural hybridization between crops and related wild species was very limited. Most of the research was done with the purpose of breeding and with the aim of transferring desirable traits between species, with crops always used as female parent in intergeneric and interspecific crosses. But the picture is quite different and numerous crops are known to have wild relatives that can hybridize with them somewhere in the world. Gene flow between cultivated species and their weedy and wild relatives has been documented in species such as oilseed rape (*Brassica napus* L.) (Jørgensen & Andersen, 1994), maize (*Zea mays* L.) (Doebley, 1990), sorghum (*Sorghum halepense* (L.) Pers) (Arriola & Ellstrand, 1996), sunflower (*Helianthus annuus* L.) (Arias & Rieseberg, 1994) and sugarbeet (*Beta vulgaris* L.) (Bartsch & Pohl-Orf, 1996). Hybridization with wild relatives has been a real issue implicated in the evolution of some of the most aggressive weeds. In order to prevent the diffusion of a character that could provide adaptative advantages, thus making weed and wild species more invasive (Darmency, 1994), it is important to understand the potential for gene flow and transgene introgression from cultivated wheat into other species, mainly their wild relatives.

Any future market launch and use of genetically modified wheat must be undertaken with extreme care, since a number of closely related species, primarily of the genus *Aegilops*, share their habitat with wheat and some natural hybrids between *Aegilops* spp. and wheat have been documented in field borders (van Slageren, 1994). Hybridization of herbicide-resistant genetically modified wheat with populations of free living relatives could make these plants increasingly difficult to control, especially if they are already recognized as agricultural weeds and if they acquire resistance to widely used herbicides (Darmency, 1994). The transfer of herbicide resistance genes from wheat to *Aegilops cylindrica* Host., a noxious weed in the wheat producing areas of the western United States, has been detected in the field and created problems for its control (Seefeldt et al., 1998; Wang et al., 2001; Gandhi et al., 2006). Other wild *Aegilops* species like *Ae. geniculata* Roth., *Ae. biuncialis* Vis. and *Ae. triuncialis* L. also form natural intergeneric hybrids with bread wheat where they grow in sympatry and with overlapping flowering times (van Slageren, 1994; Loureiro et al., 2006; Zaharieva & Monneveux, 2006), a phenomenon underlining the close genetic links of the two genera. Hybrids between *Ae. geniculata* and *Ae. triuncialis* and wheat have been found in several countries of Europe, mainly in Spain and France, while *Ae. biuncialis*-wheat natural hybrids have been described in Lebanon (van Slageren, 1994). These natural hybrids are highly sterile, although seeds may occasionally be found in *Ae. geniculata* hybrids (van Slageren, 1994; Loureiro et al., 2008).

In order to study the extent of natural hybridization, we collected spikes from one *Ae. geniculata* population that was spread extensively along a wheat field (in close proximity, Fig. 2A) where one natural hybrid has been previously detected (Fig. 2 B). A total of 3200 seeds were collected and grown in the greenhouse. Six hybrid individuals were identified from 3158 germinated seedlings, so the spontaneous hybridization rate was of 0.19% (Loureiro et al., 2006). This natural hybridization rate was similar to the 0.24% and 0.39% obtained in the assays carried under simulated field conditions explained below (Loureiro et al., 2007). Our semiarid field conditions, with frequent high temperatures and low relative humidity during the flowering periods, negatively affect to the viability and dispersal of the wheat pollen (Waines and Hegde, 2003; Loureiro, 2005). Therefore, rates of crop-wild hybridization may be higher under environmental conditions that are more favorable to hybridization.

An useful herbicide resistance screening test has been conducted to detect the potential occurrence of gene flow from *T. aestivum* to *Aegilops* using herbicide tolerant wheat cultivars as pollen donors. *Aegilops* spp. seeds are sown at appropriate depths in 1 L plastic pots (10 cm diameter, 10 seeds per pot) containing soil and sand in a 1:1 (V/V) mixture. Plants were treated at the three leaf stage with a commercially formulated herbicide at the amount of

Fig. 2. A) An extensive stand of *Ae. geniculata* with some *Ae. triuncialis* in a roadside near Zamora, Castilla-León, Spain. B) Spikes of a natural hybrid plant between *Ae. geniculata* and *T. aestivum* on the edge of wheat field. Hybrids were identified in the field by their intermediate spike morphology.

herbicide recommended in the field. In the case of Chinese Spring used as parental in crosses, the spraying was done with difenzoquat (Superaven, 330 g a.i. kg^{-1}, Cyanamid Ibérica, S.A.) at 3 kg a.i. ha^{-1}. For Castan and Deganit, plants were sprayed 1 day after planting with a commercial formulation of chlorotoluron (Oracle, 500 g a.i. L^{-1}, DuPont Ibérica, S.A.) at 2 kg a.i. ha^{-1}.

The damage produced by the herbicide to the growth of the susceptible plants was apparent 21 days after treatment. The response to the herbicides was evaluated visually 30 days after treatment. Herbicide applications were made using a Research Track Spray Cabinet (Devries Manufacturing, Hollandale, MN, USA) equipped with a Teejet 8002-E flat fan nozzle calibrated to spray 176 L ha^{-1} at 130 kPa. After spraying, the pots can be placed in the glasshouse or in a growing chamber and watered as required. Temperature was maintained at 24/16 ± 2°C (day/night temperature).

We can see in the Figure 3A that the herbicide killed the *Ae. geniculata* plants 30 days after treatment, while the Deganit tolerant wheat cultivar and the F_1 hybrid plants survived the treatments. Figure 3B shows the response to difenzoquat, with the CS tolerant wheat cultivar and the hybrids between this cultivar and *Ae. biuncialis* surviving the herbicide treatment while the *Ae. biuncialis* plants are dead. The results indicated that the bioassay was adequate for detecting hybrids. This kind of bioassay will be useful for the identification of hybrids in *Aegilops* wild populations growing near fields sown with wheat carrying a dominant trait for resistance to herbicides and in the quantification of the rate of hybridization.

These bioassays using herbicides as markers for hybrid detection were used to evaluate the hybridization between cultivated wheat and two *Aegilops* wild relatives during two seasons in simulated field conditions under Central Spain conditions (Loureiro et al., 2007). Ten 1 m x 1 m pollinator experimental plots sowed with *T. aestivum* cv Deganit at field density (400 seeds m^{-2}) were established per *Aegilops* spp. for each of two consecutive years of experimentation. Two to 3 days before anthesis one pot of *Aegilops* spp. was placed inside each pollinator plot. The wheat flowering period was monitored each year. Spikes from *Aegilops* plants were collected at maturity separately from each individual. Progeny from

Fig. 3. A) Response to the herbicide chlorotoluron (2 kg a.i. ha^{-1}) 30 days after treatment of *Triticum aestivum* cv Deganit (left), *Ae. geniculata* (right) and their F_1 hybrids. B) Response to difenzoquat (3 kg a.i. ha^{-1}) 21 days after treatment of *T. aestivum* cv Chinese Spring (left), *Ae. biuncialis* (right) and their F_1 hybrids. The herbicide application allows the identification of the hybrids.

Fig. 4. A) *Aegilops-Triticum* hybrid detection by herbicide screening in the greenhouse. B) Herbicide resistant hybrid between *Ae. geniculata* and *T. aestivum* cv Castan wheat identified by screening with the chlorotoluron applied at 2 kg a.i. ha^{-1}.

each *Ae. geniculata* and *Ae. biuncialis* plant was screened separately to check for resistance to chlorotoluron in the greenhouse (Fig. 4A). Percentage of hybridization was estimated as a ratio of survivor chlorotoluron-resistant hybrids to the total number of *Aegilops* seeds sprayed. Figure 4B shows a chlorotluron resistant hybrid between *Ae. geniculata* and Castan. The spike morphology of interspecific hybrids, intermediate between wheat and *Aegilops*, was similar to that of those obtained previously by hand-crossing under greenhouse conditions and allowed for their identification. The different ploidy levels of *T. aestivum* (2n = 42) and the two *Aegilops* spp. (2n = 28) also enabled us to confirm the hybrid status of all surviving individuals on the basis of their chromosome number in root meristems (2n = 35). The estimated hybridization rates using the data from both years were similar in both species and averaged 0.34% for *Ae. biuncialis* and 0.31% for *Ae. geniculata.* Assuming these hybridization rates and that the average seed production per plant is of 58.8 and 80.2

seeds/plant for *Ae. biuncialis* and *Ae. geniculata*, respectively, in a hypothetical field population of 100 plants growing in wheat close proximity in 1 year, the next year we would find around 17 *Ae. biuncialis* x wheat and 24 *Ae. geniculata* x wheat hybrids that could germinate or remain viable in the soil for more than 1 year. This study was carried out under experimental conditions where the factors that influenced cross-pollination as experimental plot layout or flowering synchrony, were optimized to promote hybridization. Thus, the results provided are a better indication of the maximum potential for hybridization under field conditions than of actual hybridization in agronomic settings, although it can vary within and probably among wild *Aegilops* populations and wheat varieties (Farooq et al., 1989; Hedge & Waines, 2004).

Hybridization frequency is only a component of the rate of interspecific gene flow; the ability of the hybrids to reproduce and survive in nature for the first generations is another limiting factor in terms of introgression. Fertile progenies of an *Ae. geniculata* x wheat hybrid were described as early as early as 1838 in the South of France (van Slageren, 1994). After a few years of cultivation, seed producing fertile plants that increasingly looked like wheat were obtained. The fact that hybrids between wheat and *Aegilops* spp. can be partially fertile, with low male fertilities and some female fertility that allows for backcrosses with the parents to occur (Mujeeb-Kazi, 1995), raises the question of whether a wheat gene could be transferred when other wheat fields are grown near the hybrid zone. *Aegilops* x wheat hybrids showed some female fertility by backcrossing when placed inside a wheat plot. Seeds were found in *Ae. biuncialis* and *Ae. geniculata* x Deganit hybrid plants when they were placed inside 1 x 1 m wheat plots for backcrossing. Mean fertility rates were of 3.17% for *Ae. biuncialis* hybrids (0-9.26%) and 2.87%(0-8.33%) for *Ae. geniculata* hybrids, with great variability among plants (Loureiro et al., 2007). These backcrossing rates are in the range of that obtained by Snyder et al. (2000) for *Ae. cylindrica* in an experiment with one *Ae. cylindrica* x *T. aestivum* cv Madsen hybrid plant inside a 1 m^2 plot of wheat: they obtained average seed sets of 1.8% (1-2.5%) and 6% (3-9.2%) in each year. Morrison et al. (2002) found that a 44% of the 754 *Ae. cylindrica* x wheat hybrids produced BC_1 seeds at an average rate of 1%, but up to 8% can be achieved for some hybrid plants. Higher BC_1 seed set rates of near to 30% in some hybrid plants have been found for other wheat cultivars (Loureiro et al., 2009). Besides, BC_1 partial self-fertility can be restored to 37% in the second backcross generation using jointed goatgrass as the recurrent parent, indicating that only two backcrosses are needed to restore fertility (Wang et al., 2001).

Dose-response analysis was conducted on F_1 and BC_1 hybrids between *Ae. geniculata* (Loureiro et al., 2008) and *Ae. biuncialis* (Loureiro et al., 2009) and wheat. Herbicides (chlorotoluron and/or difenzoquat) were applied at 0, 0.5, 0.75, 1, 1.5 and 2X (X = recommended dose). The hybrids were extracted with their roots 15 days after treatment, washed with water and roots dried with paper to obtain the fresh weight. Three replicates and 3 seeds per replicate were used in each treatment. A log-logistic model (Seefeldt et al., 1995) was used to analyze the data to predict the trend of herbicide resistance. In this model, the equation

$$y = f(x) = C + (D - C) / (1 + (x/LD50)^{b})$$

was used to fit the data (LD_{50} = 50% inhibitory dose, b = slope of the curve at LD_{50}, C = lower limit and D = upper limit). Figure 5 shows the herbicide dose-response curves based on fresh weight 15 days after treatment of *Ae. geniculata*, F_1, BC_1 and wheat cultivars with the herbicides chlorotoluron and difenzoquat.

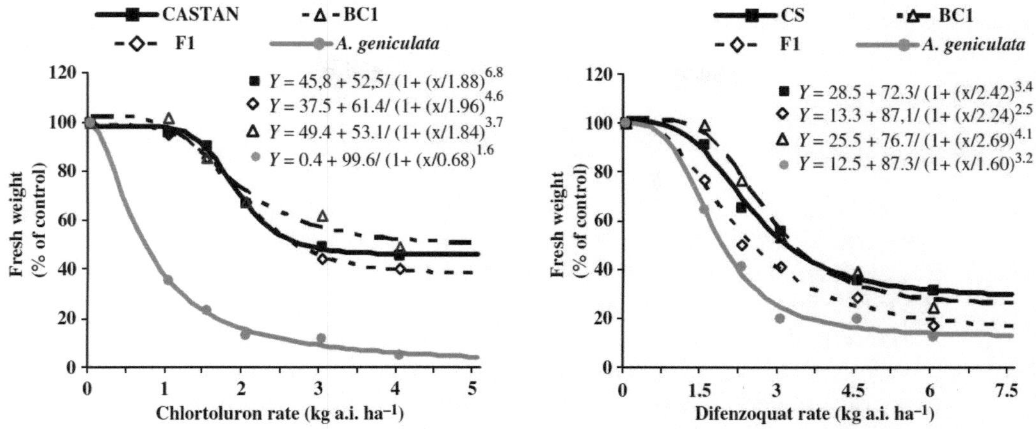

Fig. 5. Herbicide dose–response curves. *Ae. geniculata*, F_1, BC_1 and wheat cultivars with the herbicides (A) chlorotoluron and (B) difenzoquat.

As hybrids could maintain the herbicide resistance from wheat, as is shown by the LD_{50} values of the F_1s and BC_1s, the spread of these plants will be favoured by the use of the herbicide. At this point, herbicide resistance could be used as a good marker gene for hybrid detection and for the study of the herbicide resistance transference in the subsequent generations.

The hybridization ability, the partial fertility of *Aegilops*-wheat hybrids, the expression of herbicide tolerance from wheat in the cytoplasmic background of *Aegilops* and the successful backcross seed production indicate that hybrids could facilitate the transfer of herbicide resistance from cultivated wheat to *Aegilops* in the hypothesized case of backcrossing with *Aegilops* as male parent. Until now, no case of herbicide-resistance in *Ae. geniculata* or *Ae. biuncialis* harmful to farmers have been reported, which could be an indication of the real low level impact of hybridization. However, there is evidence of past gene-flow and natural, sporadic introgression from wheat into related *Aegilops* species (Weissman et al., 2005). This fact could give to the introgressed hybrids and successive generations a selective advantage and could increase the weediness of these species under an agronomic scenario of herbicide-resistant wheat, as is pointed out by Schoenenberger et al. (2006) for *Ae. cylindrica*. Broader research is needed on the fertility and fitness of the hybrids and their progenies when *Ae. geniculata* is the male parent in the backcrosses. This information could let us predict the relative advantage of hybridization on the adaptive ability of *Aegilops* spp. and hybrid derivatives and its impact on the environment and agricultural system.

6. Conclusions

Gene flow dynamics need to be considered in planning future field experiments with transgenic wheat. Agricultural reality shows that the degree of autogamy is high in wheat and that, generally, gene flow can be managed, provided that some precautionary measures are taken, such as keeping enough spatial isolation from other non GM wheat fields or from *Aegilops* wild relatives which wheat can hybridize. More research in this field is needed in order to establish coexistence measures to avoid unintended presence of GM in non-GM wheat, with cross-pollination being studied case by case and region by region. The fertility and fitness of the hybrids and their progenies must be also further evaluated in order to determine the potential introgression of the herbicide resistance genes into the wild species, a phenomenon that must be adequately assessed to avoid any potential risk derived of gene transfer.

7. References

Arias D.M. & Rieseberg L.H. (1994). Gene flow between cultivated and wild sunflower. *Theor Appl Genet* 89, 655-660.

Arriola P.E. & Ellstrand N.C. (1996). Crop-to-weed gene flow in the genus *Sorghum* (Poaceae): spontaneous interspecific hybridization between johnsongrass, *Sorghum halepense*, and crop sorghum, *S. bicolor*. *Am J Bot* 83, 1153-1160.

Bartsch D. & Pohl-Orf M. (1996). Ecological aspects of transgenic sugar beet: transfer and expression of herbicide resistance in hybrids with wild beets. *Euphytica* 91, 55-58.

Bitzer, M.J. & Patterson, F.L. (1967). Pollen dispersal and cross-pollination of soft red winter wheat (*T. aestivum* L.). *Crop Science* 7, 482-484.

Bozorgipour, R. & Snape, J.W. (1997). An assessment of somaclonal variation as a breeding tool for generating herbicide tolerant genotypes in wheat (*Triticum aestivum* L). *Euphytica* 94, 335–340.

Busch R., Behrens R., Ageez A. & Elakkad M. (1989). Inheritance of tolerance to, and agronomic effects of, difenzoquat herbicide in spring wheat. *Crop Sci.* 29, 47-50.

Darmency, H. (1994). The impact of hybrids between genetically modified crop plants and their related species—introgression and weediness. *Mol. Ecol.* 3, 37–40.

Doebley J. (1990). Molecular evidence for gene flow among *Zea* species. *BioSci* 40, 443-448.

D´Souza, V.L. (1970). Investigations concerning the suitability of wheat as pollen-donor for cross-pollination by wind as compared to rye, Triticale and Secalotricum. *Z Pflzücht.* 63, 246-269

de Vries, A.P. (1971). Flowering biology of wheat particularly in view of hybrid seed production- a review. *Euphytica* 20, 152-170.

de Vries, A.P. (1974). Some aspects of cross-pollination in wheat (*Triticum aestivum* L.). *Euphytica* 23, 601-622.

Farooq, S., Iqbal & N., Shah, T.M. (1989). Intergeneric hybridization for wheat improvement. I. Influence of maternal and paternal genotypes on hybrid production. *Cereal Res. Commun.* 17, 17–22.

Gandhi H.T., Mallory-Smith C.A., Watson C.J.W., Zemetra R.S. & Riera-Lizarazu O. (2006). Hybridization between wheat and jointed goat-grass (*Aegilops cylindrical* Host.) under field conditions. *Weed Sci.* 54, 1073-1079.

Griffin, W.B. (1987). Out-crossing in New Zealand wheats measured by occurrence of purple grain. *New Zealand J Agric Res* 30, 287-290.

Hanson, B.D.; Mallory-Smith, C.A.; Shafii, B.; Thill, D.C. & Zemetra, R.S. (2005). Pollen Mediated Gene Flow from Blue Aleurone Wheat to Other Wheat Cultivars. *Crop Science* 45, 1610-1617.

Hedge, S.G. & Waines, J.G. (2004). Hybridization and introgression between bread wheat and wild and weedy relatives in North America. *Crop Sci.* 44, 1145–1155.

Hucl, P. & Matus-Cádiz, M. (2001). Isolation distances for minimizing outcrossing in spring wheat. *Crop Science* 41, 1348-1351.

Hucl, P. (1996). Outcrossing rates for 10 Canadian spring wheat cultivars. *Can. J. Plant Sci.* 76, 423-427.

Isaaa (2010). http://www.isaaa.org

Jain, S.K. (1975). Population structure and the effects of breeding system. In: Frankel, O.H. & J.G. Hawkes (Eds.), Crop Genetic Resources for Today and Tomorrow, pp. 15-36. Cambridge

Jensen, N.F. (1968). Results of a survey on isolation requirement for wheat. *An.Wheat Newsl* 15, 26-28.

Johnson, V.A.; Schmidt, .W. & Mattern, P.J. (1967). Hybrid wheat in the United States. *Qual Plant Mater Veg* 14, 193-211.

Jørgensen R.B. & Andersen B. (1994). Spontaneous hybridisation between oilseed rape (*Brassica napus*) and weedy *B. campestris* (Brassicaceae): A risk of growing genetically modified oilseed rape. *Am J Bot* 81, 1620-1626.

Khan, M.N.; Heyne, E.G. & Arp, A.L. (1973). Pollen distribution and the seed set on *Triticum aestivum* L. *Crop Science* 13, 223-226.

Krugman, T., Levy, O., Snape, J.W., Rubin, B., Korol, A. & Nevo, E. (1997). Comparative RFLP mapping of the chlorotoluron resistance gene (Su1) in cultivated wheat (*Triticum aestivum*) and wild wheat (*Triticum dicoccoides*). *Theor. Appl. Genet.* 94, 46–51.

Loureiro, I., 2005. Estudio del riesgo potencial agrícola y medioambiental del cultivo de trigo tolerante a herbicidas. Tesis Doctoral Universidad Complutense.

Loureiro, I.; Escorial, M.C.; García-Baudín, J.M. & Chueca, M.C. (2005). Gene flow between wheat cultivars: *Triticum aestivum* and *Triticum turgidum*. Outcrossing in an individual recipient plant basis under field conditions. *GMCC-05*, 285-286.

Loureiro, I.; Escorial, M.C.; García-Baudín, J.M. & Chueca, M.C. (2006). Evidence of natural hybridization between *Aegilops geniculata* and wheat under field conditions in Central Spain. *Environ. Biosafety Res.* 5, 105–109.

Loureiro, I.; Escorial, M.C.; García-Baudín, J.M. & Chueca, M.C. (2007) Hybridization between wheat (*Triticum aestivum*) and the wild species *Aegilops geniculata* and *A. biuncialis* under experimental field conditions. *Agr. Ecosyst. Environ.* 120, 384–390.

Loureiro, I.; Escorial, M.C.; García-Baudín, J.M. & Chueca, M.C. (2008). Importance of the hybridization and the fertility of the hybrids between wheat and *Aegilops geniculata* in the herbicide resistance transference. *Weed Research* 48, 561-570.

Loureiro, I.; Escorial, M.C.; García-Baudín, J.M. & Chueca, M.C. (2009). Hybridization, fertility and herbicide resistance of hybrids between wheat and *Aegilops biuncialis*. *Agron. Sustain. Dev.* 29, 237–245.

Loureiro, I.; Escorial, M.C.; García-Baudín, J.M.; González-Andujar, J.L. & Chueca, M.C. (2007). Wheat pollen dispersal under semiarid field conditions. Potential outcrossing with *Triticum aestivum* and *Triticum turgidum*. *Euphytica* 156, 25–37.

Lu, A.Z.; Zhao, H.; Wang T.Y. & Wang, H.B. (2002). Study of possibility of target gene introgression from transgenic wheat into non-transgenic plants through pollens. *Acta Agric Bor Sin* 17, 1-6.

Matus-Cádiz, M.A.; Hucl, P.; Horak, M.J. & Blomquist, L.K. (2004). Gene flow in wheat at field scale. *Crop Science* 44, 718-727.

Morrison, L.A., Riera-Lizarazu, O., Cremieux, L. & Mallory-Smith, C.A. (2002). Jointed goatgrass (*Aegilops cylindrica* Host) x wheat (*Triticum aestivum* L.) hybrids: hybridization dynamics in Oregon wheat fields. *Crop Sci.* 42, 1863–1872.

Mujeeb-Kazi, A., (1995). Interspecific crosses: hybrid production and utilization. In: Mujeeb-Kazi, A., Hettel, G.P., (Eds.), Utilizing Wild Grass Biodiversity in Wheat Improvement: 15 Years of Wide Cross Research at CIMMYT. Me´xico D.F. CIMMYT Research Report No. 2, 14–21

Sanchez-Monge, E. (1955). Fitogenética. 521 pp. Ed. Omega

Schoenenberger N., Guadagnuolo R., Savovabianchi D., Küpfer P. & Felber F. (2006). Molecular analysis, cytogenetics and fertility of introgression lines from transgenic wheat to *Aegilops cylindrica* Host. *Genetics* 174, 2061-2070.

Seefeldt S.S., Jensen J.E. & Fuerst E.P. (1995). Log-logistic analysis of herbicide dose-response relationships. *Weed Technology* 9, 218–227.

Seefeldt, S.S., Zemetra, R., Young, F.L. & Jones, S.S. (1998). Production of herbicide resistant jointed goatgrass (*Aegilops cylindrica*) _ wheat (*Triticum aestivum*) hybrids in the field by natural hybridization. *Weed Sci.* 46, 632–634.

Sixto, H., Silvela, L., Escorial, C., García-Baudin, J.M. & Chueca, M.C. (1995). On the inheritance of tolerante to chlorotoluron application in wheat using a very efficient score test. *Weed Res.* 35, 7–13.

Snyder, J.R., Mallory Smith, C.A. & Balter, S. (2000). Seed production on *Triticum aestivum* by *Aegilops cylindrica* hybrids in the field. *Weed Sci.* 48, 588–593.

Stopkopf, N.C. & Rai, R.K. (1972). Cross-pollination in male-sterile wheat in Ontario. *Can. J. Plant Sci.* 52, 387-393.

van Acker, R.C.; Brûlé-Babel, A.L. & Friesen, L.F. (2003). An environmental safety assessment of Roundup ready wheat: risks for direct seeding systems in Western Canada. Report prepared for: The Canadian Wheat Board. 33pp.

van Slageren, M.W. (1994). Wild Wheats: a Monograph of *Aegilops* L. and *Ambylopyrum* (Jaub. & Spach.) Eig (Poaceae) Wageningen Agricultural University and ICARDA, Wageningen.

Waines, J.G. & Hegde, S.G. (2003). Intraspecific gene flow in bread wheat as affected by reproductive biology and pollination ecology of wheat flowers. *Crop Science* 43, 451-463.

Wang, Z., Zemetra, R.S., Hanson, J. & Mallory-Smith, C.A. (2001). The fertility of wheat _ jointed goatgrass hybrid and its backcross progenies. *Weed Sci.* 49, 340–345.

Weissman S., Feldman M. & Gressel J. (2005). Sequence evidence for sporadic intergeneric DNA introgression from wheat into a wild *Aegilops* species. *Molecular Biology and Evolution* 22, 2055-2062.

Zaharieva, M. & Monneveux, P. (2006). Spontaneous hybridization between bread wheat (*Triticum aestivum* L.) and its relatives in Europe. *Crop Sci.* 46, 512–527.

Zhao, H.; Lu, M.Y.; Wu, Z.M.; Wu, M.X.; Xie, X.L.; Ma, M.Q.; Sun, G.Z.; Cao, J.R.; Wang, T.Y. & Wang, H.B. (2000). Ecological safety assessment of herbicide resistant transgenic wheat. In: XIéme Colloque International sur la Biologie des Mauvaises Herbes, pp. 573-580.

Part 2

Analytical Techniques of Herbicide Detection

13

Overview of Analytical Techiques for Herbicides in Food

Hua Kuang, Libing Wang and Chuanlai Xu
Jiangnan University
China

1. Introduction

It is said that there are more than 5800 kinds of weeds, which significantly do harm to the agricultural production and weed control has always been an important issue in agrochemical practice. Chemical agents, that is herbicides, are used widely in the world in protecting crops from undue competition from weeds.

The main chemical classes of herbicides(Tadeo et al., 2000) include bipiridilium compounds, triazine derivatives containing three heterocyclic nitrogen atoms in the ring structure (atrazine, prometryn, propazin, etc.), chlorophenoxy acid derivatives (2, 4-D, 2, 4, 5-T), substituted chloro-acetanilides (alachlor, propachlor), derivatives of 2, 6-dinitroaniline (benfluralin, trifluralin), substituted phenylcarbamates (carbetamide, chlorbufam), urea derivatives (chlorbromuron, chlorotoluron), substituted sulphonylureas (amidosulfuron, trifusulfuron), etc. The intensive application of herbicides has resulted in the contamination of the atmosphere, soil and waste water, agricultural products (wheat, corn, fruits, vegetables, beans etc.) and, consequently, in the direct or indirect pollution of food and food products and biological system.

More studies have shown that herbicides or its metabolites can enter into the human body along food chain, which creates potential health risks to human. Growing concern has been taken for this issue and some herbicides have been banned to use (see table 1).

The development of a robust analytical method is a complex issue. All steps in the analytical process including sample preparation, extraction, cleanup and instrumental analysis are equally important. There are a vast series of techiniques to use in establishing analytical methods, however, some rules should be taken for the differences in polarity of herbicides and the type of sample matrix. The objective of this paper is to summarize the anlytical measures developed to detect the different classes of herbicides residues in various foods, and to review future trends.

2. Phenoxycarboxylic acid herbicides

The herbicidal effect of 2, 4-D was first discovered by Amchem company in 1942 (Kuang et al., 2006b), and more categories were developed by a lot of companies since 1945 based on the structure of 2, 4-D. The general formula of this class herbicides see fig 1 and the chemical structure of some most used phenoxycarboxylic acid herbicides were summarized in table 2. 2, 4-D is the world's largest broad-leaved weed herbicides. Phenoxycarboxylic acid

Class of Herbicides	EU	U. S. A	Japan	China
Phenols	Dinoterb 2-Methyl-4, 6-Dinitrophenol (DNOC) Pentachlorophenol	Dinoseb DNOC	Dinoterb Pentachlorophenol	Dinoterb DNOC Pentachlor ophenol
Ureas	Monolinuron Chloroxuron Difenoxuron Noruron Chlorbromuron Cycluron Dimefuron Momuron Neburon Tebuthiuron Thiazafluron Benzthiazuron Ethidimuron Metobromuron Metoxuron Fenuron	-	Chloroxuron Monolinuron Tebuthiuron Benzthiazuron	-
Amides	Metolachor Butachlor Monalide Diethatyl-Ethyl Mefenacet Tebutam Isocarbamide Diphenamide Chlorthiamid Pentanochlor Flamprop Flupoxam	Metolachor	Metolachor Butachlor Mefenacet Flamprop	
Triazine	Propazine Ametryn Aziprotryne Desmetryne Methoprothryne Trietazine Terbumeton Secbumeton Cyanazine Terbutryn Hexazinone Prometryn	Propazine Ametryn Cyanazine Hexazinone	Propazine Tebutryn	-

Class of Herbicides	EU	U. S. A	Japan	China
Dinitroaniline	Dinitramine Isopropalin Nitralin	Nitralin	-	-
Diphenyl Ethers	Fluoroglycofen Fluorodifen Acifluorfen Fomesafen Chlormethoxyfen	Acifluorfen	Fomesafen	Acifluorfen Fomesafen
Carbomates	Cycloate Vernolate Dimepiperate Dimexano Propham Butylate Chlorbufam Tiocarbazil Karbutilate Di-Allate Barban S-Ethyl-N, N-Dipropylthiocarbamate(EPTC) Orbencarb Pebulate	Propham Cycloate Butylate Pebulate	Cycloate Butylate Pebulate	-
Phenoxycarboxylic Acids	Fluazifop Quizalofop Fenoxaprop Haloxyfop 2, 4, 5-T Dichlorprop Fenoprop 2, 3, 6-Trichlorobenzoic Acid (2, 3, 6-TBA)	Fenoprop	2, 4, 5-T	2, 4, 5-T Fenoprop
Imidazolinones	Chloramben Imazamethabenz Imazapyr	-	Imazamethabenz Imazapyr	-
Cyclohexanediones	Sethoxydim Alloxydim	Sethoxydim	-	-
Others	Chlorfenprop-Methyl Allalacohol Benazolin Benzoylprop Bensulide Bromofenoxim Dalapon	Bensulide MSMA Norflurazon Benfuresate Bromacil	Bensulide Flamprop Pyrazoxyfen TCA Bromacil Naptalam	Bromacil

Class of Herbicides	EU	U. S. A	Japan	China
	Endothal			
	Flamprop			
	Fluridone			
	Flupoxam			
	Methazole			
	Sodium Hydrogen Methylarsenate (MSMA)			
	Norflurazon			
	Perfluidone			
	Pyrazoxyfen			
	Trichloroacetic Acid (TCA)			
	Tridiphane			
	Benfuresate			
	Bromacil			
	Naptalam			

Table 1. List of banned herbicides in various countries

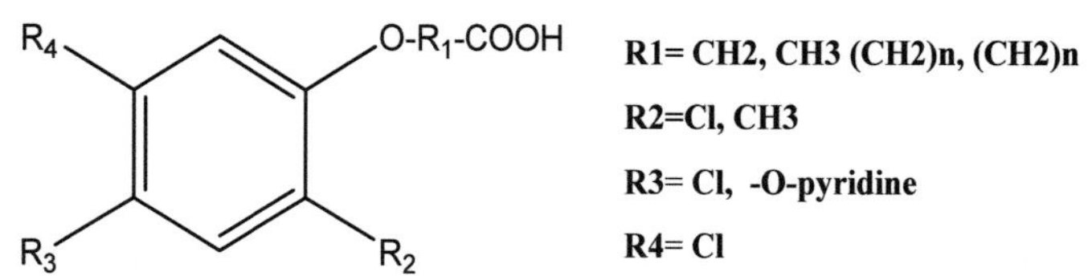

Fig. 1. Parent Chemical Structure of Phenoxycarboxylic Acid Herbicides

herbicides have been used intensively in the control of the growth of grass and the broad-leaf weeds in many crops such as paddyfield, wheat, soybean, etc.

Due to their solubility in water, these herbicides are easy to migrate in agricultural ecosystem causing the pollutions of soil, groundwaters, and air. Phenoxy acid herbicides are medium toxicity themselves, but their metabolic products (especially some halids)are harmful to the human and other creatures. Investigations indicate that they could induce the human parenchyma malignancy tumor and embryotoxicity in animals(Kuang et al., 2006a).

2.1 Sample extraction

Since phenoxy acid herbicides show high polarity and are easily dissolved in the water or aqueous-phase solution, phenoxyacids and benzonitriles are widely applied as salts or esters, but they are decomposed rapidly by hydrolysis, in the treated plants, to their respective phenols or acids. Residues of these acidic herbicides are best extracted from foods when a hydrolytic step is included to release the free acidic herbicide from the conjugated products formed with plant components(Rimmer et al., 1996). With this aim, acid or base hydrolysis has been used. For acid hydrolysis process, samples have to be acidified with acid solution transferring the analytical objective into the organic phase. It was reported(Baggiani et al., 2001) that the sample should be acidified with acidification water (pH < 2), then extracted with proper organic solvent-water mixture. Solvents, such as acetonitrile, toluene-ether, dichloromethane and etc., can be used to extract the phenoxy acid herbicides from matrix. For base hydrolysis, alkaline solution (0.1 M NaOH) was used mostly. Many extraction methods including ultrasonic extraction, shaking extraction, microwave-assisted solvent extraction(MASE) and supercritical fluid extraction(SFE) have been reported (table 3).

2.2 Cleanup

Many components in matrix can be co-extracted in sample extraction step and targeted compounds are generally present in very low concentration, they need to be separated from undesirable substances effectively. Some authors have previously summarized the primary cleanup process on these compounds(Cserháti et al., 2004). Liquid-liquid extraction (LLE) has frequently been used to remove co-extracts from sample constituents. The efficacy of the method is generally high but requires highly purified and expensive solvents. However, serious emulsifying phenomenon sometimes is present during the shaking process. Gel permeation chromatography (GPC) mainly used to remove lipids or colors from extracts based on the differences of molecule size between targeted compounds and interferences. Kuang et al., 2006 sucessefully purified 14 phenoxy acid herbicides (M. W. ranging from 180 to 327) from soybean extracts.

However, the most common approach to cleanup in herbicide analysis now is solid-phase extraction (SPE), sorbents such as aminopropyl(NH_2), reversed-phase (C_{18}), strong cation exchange(SCX) and normal-phase sorbents(florisil, alumina) are very useful for cleaning up complicated extracts (see table 3).

2.3 Detection

2.3.1 Gas chromatography (GC)

Phenoxy acid herbicides benefit poor volatility for its low pka (acid dissociation constant) values (see table 2) and derivatization process is needed when analysis by gas chromatography requires. The most frequently used derivatization reagent is diazomethane;however, due to its toxicity, carcinogenicity and explosiveness, other alternative esterification reagents such as sulphuric acid in 1-propanol or in methanol and boron trifluoride in methanol, n-butanol or 2-chloroethanol have been proposed.

Methylation and PFBBr (pentafluorobenzyl bromide) esterification are common approaches. Methylating agents such as boron trifluoride-methanol, chloroformate, trimethylsilyldiazomethane have been reported in detection of phenoxy acid herbicides (Table 4). Diazomethane was applied for methylation of 6 herbicides (Wei et al., 2005) and satisfying derivative effects obtained. Trimethylsilyldiazomethane, as a non-toxic non-mutagenicl alternative to diazomethane is widely used in methyl derivatization. The summary in table 4 showed that mainly mass spectrometry and electron capture detector (ECD) were used to detect phenoxy acid herbicides. Other detectors including hydrogen flame ionization detector (FID) and nitrogen-phosphorus detector(NPD) were also reported for analysis. Kuang(Kuang et al., 2006a) found that ECD response of methylated product of phenoxy acid herbicides, especially single-chlorine substituted molecules (MCPA, MCPP, MCPB etc.), was much lower than that of PFBBr ester. A comparison of the response factors between PFBBr ester and methyl ester of MCPA, 2, 4-D and 2, 4, 5-T had been made (Lee et al., 1991). The response factor of the chlorophenoxy herbicide of PFBBr ester was almost 600 times than that of methyl ester.

Name	Chemical Strucutre	CAS. No	pKa
Mecoprop (MCPP)	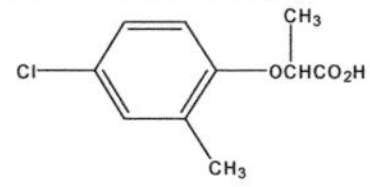	7085-19-0	3.78
2-Methyl(4-Chlorophenoxy) Acetic Acid (MCPA)	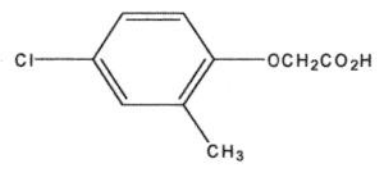	202-360-6	3.07
2-Methyl(4-Chlorophenoxy) Acbutyric Acid (MCPB)	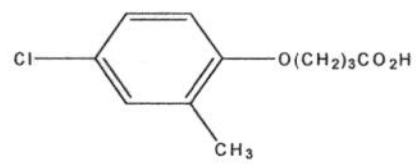	94-81-5	4.84
2, 4-Dichlorophenoxyacetic Acid(2, 4-D)	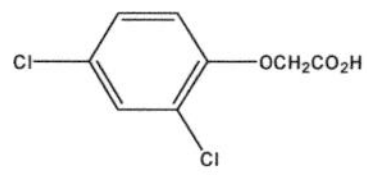	94-75-1	2.73
2, 4-Dichlorophenobutyric Acid	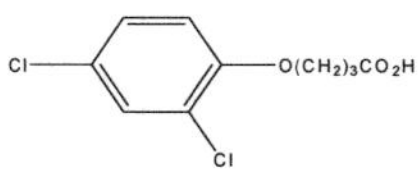	94-26-8	4.80
Dicamba	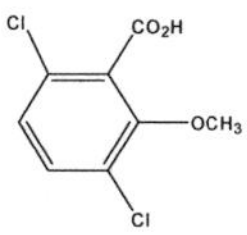	1918-00-9	1.97
Fluazifop	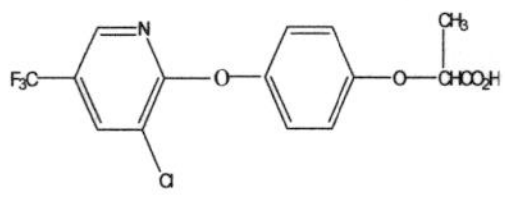	69335-91-7	3.20
4-Chlorophenoxyacetic Acid	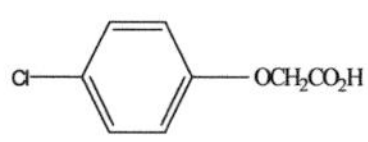	122-88-3	--
Dichlorprop	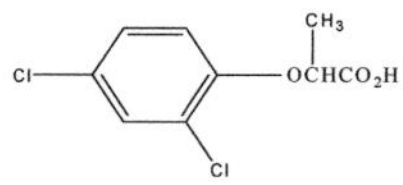	28631-35-8	3.00
2-(4-Chlorophenoxy) Propionic Acid	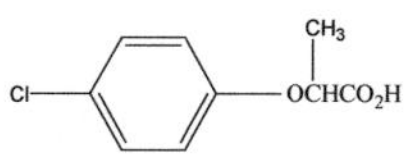	3307-39-9	--
3, 4-Dichlorophenoxyacetic Acid	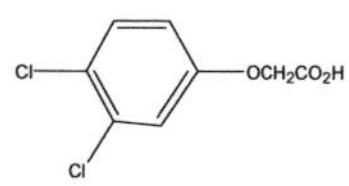	588-22-7	--
2, 4, 5-(Trichlorophenoxy) Propionic Acid (2, 4, 5-T)	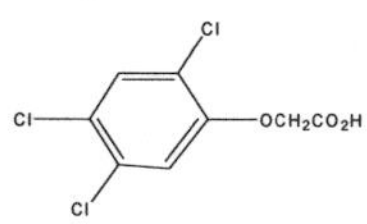	93-76-5	3.14
Fenoprop	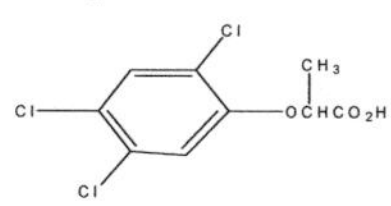	93-72-1	3.10
Phenoxy Butyric Acid	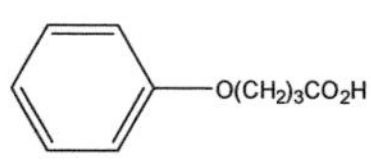	6303-58-8	--

Table 2. Information for 14 phenoxy acid herbicides

Matrix	Herbicide	Extraction	Clean-up	Ref
Oranges	2, 4-D	Methanol–homogeniser	-	(Williams et al., 1997)
Fruits, vegetables	2, 4-D	Diethyl ether–hexane (acidic pH), homogeniser	NH_2 cartridge	(TING & Kho, 1998)
Wheat	2, 4-D	Ethanol–water, homogeniser	LLE–Florisil column	(Cessna & Holm, 1993)
Onions	Fluazifop-butyl	CO_2 -SFE	-	(Wigfield & Lanouette, 1993)
Fruits, vegetables	2, 4-D	Methanol–water (basic pH), blender	C_{18} cartridge	(Richman et al., 1996)
Oranges, grapefruits	2, 4-D	Acetonitrile–water, homogeniser	LLE	(Rochette et al., 1993)
Citrus fruits	Dichlorprop	Methylene chloride–acetone, shaker	LC-SCX cartridge	(Peruzzi et al., 2000)
Barley, triticale	Mecoprop, 2, 4-D	0.1 *M* NaOH, blender Ethanol–water, homogeniser	LLE–Florisil column	(Cessna, 1992) (Sánchez-Brunete et al., 1994)
Wheat, barley	Phenoxyacids	Methanol, homogeniser	LLE–Florisil column	(Su, 1975)
Mushrooms	2, 4-D	Diethyl ether (acidic pH), homogeniser	Alumina column	(Siltanen, 1978)
Wheat	2, 4-D	0.1 *M* NaOH–diethyl ether–hexane (pH 1), blender	LLE–Florisil column	(Smith, 1984)
Potatoes, soybeans	Fluazipop-butyl	0.1 *M* NaOH, shaker	LLE–Florisil column	(Clegg, 1987)
Wheat	2, 4-D,	0.1 *M* NaOH, blender	LLE–Florisil column	(Cessna, 1980)
Soybean	Phenoxyacids	acetonitrile-50mM HCl (v/v 7:3)	LLE- anion exchange column GPC- anion exchange column	(Kuang et al., 2006a; Kuang et al., 2006b)

Table 3. Extraction and cleanup of phenoxy acid herbicides

Reagents	Matrix	Detection system	Ref
Diazomethane	Rice, Soil, water	GC-MS	(Wei et al., 2005); (Hodgeson et al., 1994)
CH_3I	Vegetables, water	GC-ECD	(Rompa, 2005)
Dimethyl sulfate	water	GC-MS	(Catalina et al., 2000)
Trimethylsulfonium hydroxide(TMSH)	water	GC-MS	(Neitzel et al., 1998)
Tetramethylammonium hydroxide (TMAH)	Standards	GC-MS	(Brondz & Olsen, 1992)
tetrabutyl ammonium salt，TBA	water	GC-MS	(Ding et al., 2000)
2-cyanoethylmethyldieth	Standards	GC-NPD	(Bertrand et al., 1987)
N, O-bis(trimethylsilyl) trifluoroacetamide, BSTFA	Standards	GC-MS	(Lou et al., 1999)
PFBBr	Water, Soil, rice, air	GC-MS, GC-ECD	(Cserháti & Forgács, 1998); (Tadeo et al., 2000)
Benzyl bromide	water	GC-MS, GC-FID	(Nilsson et al., 1998)
Chloromate	water	GC-MS, GC-ECD	(Butz & Stan, 1993)
Concentrated sulfuric acid	water	GC-MS	EPA Method 8151A
HCl- Acetic Anhydride	water	GC-MS	(Xing et al., 2002)
BF_3	Soil, Meat, Rice	GC-MS, GC-ECD	(Sánchez-Brunete et al., 1994)

Table 4. Derivatization method of phenoxy acid herbicides

The requirement of the maximum residue limits (MRLs) of phenoxy acid herbicides was critical, especially in Japan where 2, 4, 5-T can not be detected in foods. Most derivatization products can be separated on weakly polarity [stationary phase of column, (5%-Phenyl)-methylpolysiloxane and medium polarity [(14%-Cyanopropyl-phenyl)-methylpolysiloxane] capillary columns. Because of the similarity of these herbicides between their structures and polarities, slow temperature program-up was needed to acquire an effective separation. A typical programmed temperature is set as follows:

The oven initial temperature 60 °C holding 1 min and was programmed at 25 °C /min to 180 °C, (1min hold), then programmed at 2 °C /min to 205 °C, (3 min hold), finally programmed to 260 °C at 10 °C /min (5 min hold).

2.3.2 High performance liquid chromatogeaphy (HPLC)

Considersing weak volatility of phenoxy acid herbicides, liquid chromatographic separation seems more suitable than gas chromatography. Derivatization, not only is time consuming, but also affects the reproducibility and stability of the method.

Most phenoxy acid herbicides showed maximal UV absorption ranged from 200-220nm, where mighty interference existed and stable baseline often can't be gotten. Thus, some

analysts carried out derivatization process in analysis of these class compounds aimed to change their chromatographic behavior not to improve the detection sensitivity.
Phenoxy acid herbicides showed high polarity with pKa distributed in 2 to 5(Kuang et al., 2006a), the analysts need to adjust the pH of the mobile phase. Organic acids such as acetic acid, trifluoroacetic acid or inorganic acid can be used to adjust the acidity.
The great advantage of HPLC tandem mass spectrometry (HPLC-MS/MS) is its highly selectivity, which greatly reduce the false positive results in detection. Kim (Kim et al., 1991) applied HPLC-MS to detect 2, 4, 5-T, 2, 4-D and fenoprop residues in water, which was the first application of HPLC-MS techniques in phenoxy acid herbicide detection. Ultra Performance Liquid Chromatography (UPLC) employs 1.7 um particles, resulting in a very flat VanDeemter plot and a linear velocity faster than usual one with 5 um packings; consequently, improves resolution, speed and sensitivity for many HPLC methods. Chu, 2008(Chu et al., 2008) realized simultaneous determination of more than 100 herbicides in soybeans within 11 min by UPLC-MS/MS.

2.3.3 Other analytical methods

Capillary zone electrophoresis(CZE) and micellar electrokinetic chromatography(MEKC) (Farran & Ruiz, 2004) have been used by some researchers to separate phenoxy acid herbicides. Trace level analysis by electrophoresis meets some difficulties in detectors. UV-Vis(Nemoto & Lehotay, 1998) or fluorescence detector is common in the application. Besides, the process in separation with electrophoresis is greatly depending on the mobile phase (ionic strength, pH) and peak shif t sometimes is very serious, thus, quantitative analysis may be inaccurate.
Compared with instrumental separation methods, immunochemical determination technology exhibits remarkable specificity, sensitivity, rapidness and high throughput in detection. Moreover, immunochemical methods cost less and can be used in the field. I. A. Lyubavina, (Lyubavina et al., 2004) used monoclonal antibodies labeled with colloidal gold to detect 2, 4-D residues in aqueous samples.

3. Dinitroaniline herbicides

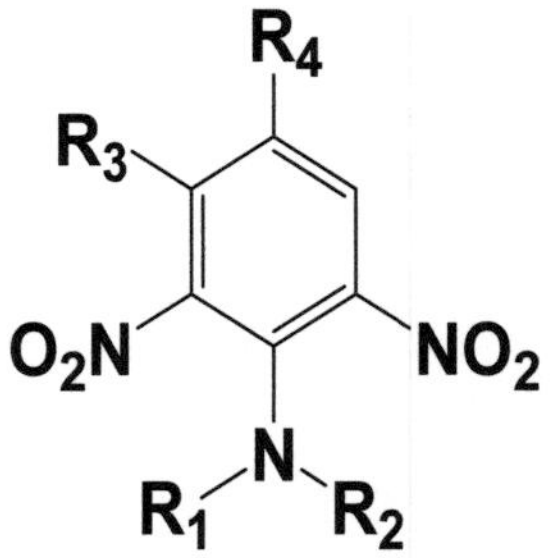

R_1=Akyl, Halogenated hydrocarbons,Naphthenic ;

R_2= Akyl, Halogenated hydrocarbons,Naphthenic , H;

R_3=NH_2, H,CH_3;

R_4=CF_3, CH_3, SO_2CH_3, SO_2NH_2, C_2-C_4Akyl;

Fig. 2. Chemical structure for dinitroaniline herbicides

Dinitroaniline herbicides are used to control some broad-leaved weeds and the major annual grasses(García-Valcárcel et al., 1996). There are two classes of dinitroaniline herbicides depending on different substituents at R^4 site (Fig 2). The R_4 is alkyl or halogenated hydrocarbon for class I dinitroaniline herbicides, that is methyl aniline herbicide. Trifluralin, pendimethalin and ethalfluralin are typical methyl aniline herbicides.

For class II, the R4 group contains sulfone structure and nitralin belongs to this class. Some toxicological experiments showed that dinitroaniline herbicides exhibits carcinogenicity and impaired the normal function of organs. MRLs of some dinitroaniline herbicides in agricutural products were listed in table 5.

3.1 Sample preparation

Because of the strong polarity of dinitroaniline herbicides, some slightly polarity organic solvents such as acetonitrile, methanol and acetic ether are most applied to extract these herbicides from various matrix by a single or mixed manner. Few reports were found using single non-polar solvents (e. g. n-hexane). For extraction procedure, MASE, SFE, sonication and pressurized liquid extraction (PLE) are reported (table 6). Some analysts applied solid phase microextraction (SPME), which is intensively used in headspace analysis, to analyze dinitroaniline herbicides, but recoveries were poor.

In nitrobenzene herbicide pre-treatment methods, SPE technique was used more often. Commonly used stationary phase was based on florisil and C_{18} sorbents depending on different nature of the targeted compounds and matrix. Florisil maily was used for removing lipophilic interfences in purification (Huo et al., 2006) procedure and usually florisil (25g, previously activated with 3% H_2O) can adsorbed 1g fat), so particularly suitable for oily substances Florisil. Some reports have showed that good purification effects (the average recovery rate was 74% or more) using florisil in cleanup step in food analysis. Another material - C_{18} sorbent is also widely used in purification step. Darcy D. Shackelford (Shackelford et al., 2000) successfully applied C_{18} sorbent to remove co-extracts in analysis (recovery> 80%)

3.2 Detection

In the residue analysis of dinitroaniline herbicides, chromatography detection was dominant, especially GC with high sensitivity and good separation effects based on the summary of recent 20 year literature. Detectors such as ECD, FID, NPD and MS were used widely (see table 7)

Herbicide	Agricutural product	USA	Japan	China	Canada	New Zealand	South Korea
Trifluralin	Grains, fruits, vegetables and vegetable oil	0.05	0.15	0.05-0.15	0.5	0.03	0.05
pendimethalin	Drinking water, fruits, nuts, vegetables	0.1	0.2	0.2	-	0.02	-
Benfluralin	Peanuts, lettuce	0.05	-	-	-	-	-
ethalfluralin	Soybean, peanuts, Sunflower seeds	0.05	-	-	-	-	-
Oryzalin	Apples, kiwi fruits, Pan pomegranate and drinking water	0.05	0.05-0.2	-	-	0.4	-

Table 5. MRLs of some dinitroaniline herbicides (mg/kg)

Matrix	Solvents for extraction	Cleanup	Recovery %
Carrots and fruit	Hexane + acetic ether (1:1)	SPE (Florisil)	—
Fruits, nuts, vegetables	Methanol, methanol-water, 2 - propionaldehyde and n-hexane	GPC& SPE（florisil）	72-126
Industrial wastewater and urban domestic water	Dichloromethane	—	73-99
Soil	Acetonitrile-water	SPE(Florisil)	90-120
Soil, plants and air	Methanol, acetic ether	SPE(Florisil)	75
Blood, urea and water	SPME	—	35-64
Peanuts	Methanol, Dichloromethane	SPE(Florisil)	75.6-80.4
Banana, cucumber, apple, lettuce and oranges	Acetonitrile	SPE(C_{18})	70-120
River water	-	SPE	>80
Canola seed, crude powder and Refined oil	Acetonitrile	SPE(C_{18})	89-96
Fruits and Vegetables	Acetonitrile	SPE	85-101
Soil	Acetone - water - acetic acid	—	96.6
Soil, water	Ether	SPE (C_{18})	89-104
water	-	SPE	50-77
Soil	Acetonitrile	—	—
Juice	Methanol	SPE (C18)	93.8～99.5
Buckwheat	n-hexane	SPE(Florisil)	>74

Table 6. Extraction and cleanup of initroaniline Herbicides

Targeted compounds	Analytical measure	Limit of Detection (LOD)	Ref
Benfluralin, Trifluralin	GC/FID	—	(Boyd-Boland & Pawliszyn, 1995)
Trifluralin, Benfluralin, ethalfluralin, isopropalin,	GC/ECD	0.01mg/kg	(West et al., 1988)
Benfluralin, ethalfluralin, isopropalin, profluralin, pendimethalin, fluchlorlin	GC/ECD	0.1 ng/mL (water, urea) 1 mg/mL(blood)	(Guan et al., 1998)
pendimethalin	GC/NPD	0.01 ppm (soil) 0.1 ppb(water)	(Sanchez-Brunete et al., 1994)
Trifluralin, ethalfluralin, profluralin,	GC/ECD	-	(Hsu et al., 1991)
Trifluralin	GC/ECD	2.5 pg/uL	(D'Amato, 1993)
pendimethalin	GC/ECD	0.022-0.045 mg /kg	(Engebretson et al., 2001)
pendimethalin	GC/NPD	0.1-4.4 µg/kg	(Fenoll José et al., 2007)
Trifluralin	GC/ECD	-	(Cessna & Kerr, 1993)
Trifluralin	Electrochemical analysis	2×10^{-9} mol/L	(Wen et al., 2008)
Benfluralin, pendimethalin, Trifluralin	GC/MS	0.05 -0.1mg /kg	(Tanabe et al., 1996)
Ethalfluralin, Trifluralin	GC/MS	0.1-4.6 ug/L	(Albero et al., 2005)
ethalfluralin, Benfluralin dinitramine	GC/MS GC/NPD	0.001-0.02 ug/g	(Sánchez-Brunete et al., 1998)
Trifluralin, ethalfluralin, pendimethalin, isopropalin	HPLC/UV	0.5µg/kg-0.02mg/kg	(Cabras et al., 1991)
Trifluralin, ethalfluralin, pendimethalin	HPLC/ UV	0.09-0.14ug/L	(Vitali et al., 1994)
nitralin	HPLC- UV	6.9 ng	(Ruiz de Erenchun et al., 1997)
Trifluralin	HPLC/ UV	1µg/kg	(Topuz et al., 2005)
Trifluralin	HPLC/ UV	0.025mg/kg	(Huang et al., 2004)
Trifluralin	ELISA	0.1-100ng/mL	(Gyöngyvér et al., 2000)
Trifluralin	Immunosensor	2×10^{-17}-3×10^{-5} ng/mL	(Szendr et al., 2003)

Table 7. Summary of analytical methods for dinitroaniline herbicides

4. Sulfonylurea herbicides

Sulfonylurea herbicides are one of the largest families of herbicides in the world. DuPont company first reported the herbicidal activity of sulfonylurea compounds and the first sulfonylurea herbicide- chlorsulfuron was marketed in 1976, which opened the era of super-efficient herbicide application(Mughari et al., 2007). Now the number of the patents related to sulfonylurea herbicides is more than 400. The information of some common sulfonylurea herbicides was shown in table 8.

These herbicides, which have low toxicity to mammals, are highly toxic to plants and, consequently, are used at low application rates (3-40 g ha^{-1}). The general structure of the sulfonylurea herbicides (R-SO2NH-CONH-R, fig) consists of two R groups attached to either side of the sulfonylurea linkage (fig 3). The R group attached to the sulfur atom of the sulfonyl moiety can be an aliphatic, aromatic, or heterocyclic group, whereas that attached to the terminal nitrogen atom of the urea moiety can be a substituted triazine or pyrimidine ring.

In recent years, sulfonylurea herbicides have become very popular worldwide because of their low application rates, low toxicity to mammals, and unprecedented herbicidal activity. These herbicides are non-volatile, and their water solubilities are pH dependent being greater in alkaline than in acidic solution

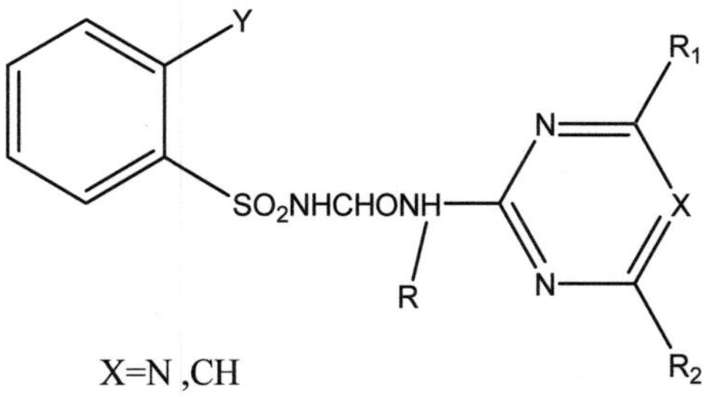

X=N ,CH

Y=Cl,F,Br,CH_3,$COOCH_3$,SO_2CH_3,SCH_3 ,$SO_2N(CH_3)2$,CF_3,CH_2Cl,OCH_3,OCF_3,NO_2

R=CH_3,Alkyl

R_1=CH_3,Cl

R_2=OCH,CH_3,Cl

Fig. 3. Parent chemical structure of Sulfonylurea herbicides

4.1 Sample preparation

As weak acids, sulfonylurea herbicides show a more rapid degradation in environment. Therefore, the concentration of this class herbicides usually found in environmental and food samples is about 100-1000-fold lower as compared to other herbicides. Generally, the trace analysis of complex environmental and food samples needs pretreatment steps in order to reduce matrix interferences and enrich trace level analytes.

Traditional liquid-liquid extraction (LLE) or more rapid and economic solid phase extraction (SPE) or dispersive solid phase extraction (DSPE) have been reported in sulfonylurea herbicide detection. Materials such as RP-C_{18}, ion exchangers, mixed mode phases, graphitized carbon, and polystyrene divinylbenzene supports have been shown to be valuable sorbents for sample enrichment of various sulfonylurea herbicides in different matrix. Acidified organic solvents such as acetonitrile, dichloromethane, ethyl acetate (pH=2) were often used to extract sulfonylurea herbicides from various matrix (table 9).

sulfonylureas herbicides	Structures	molecular formula	MW	pKa
oxasulfuron	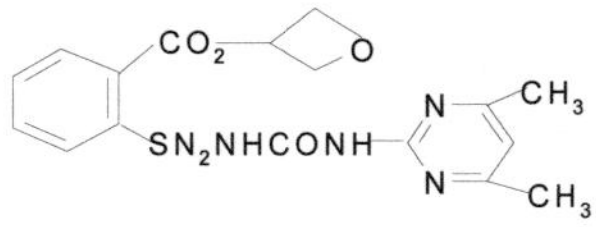	$C_{17}H_{18}N_4O_6S$	406.4	5.1
thifensulfuron-methyl	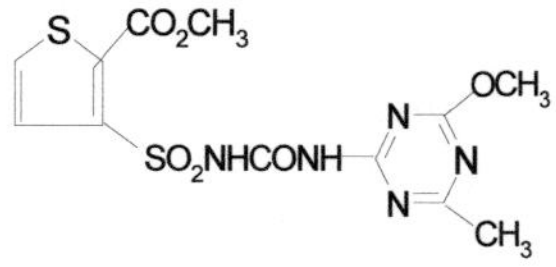	$C_{12}H_{13}N_5O_6S_2$	387.4	4.0
metsulfuron-methyl	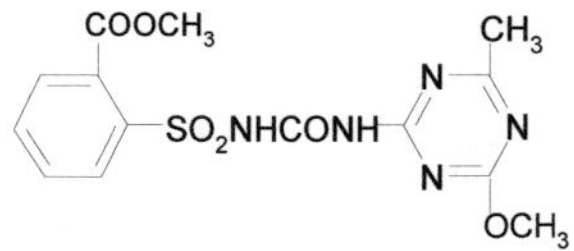	$C_{14}H_{15}N_5O_6S$	381.4	3.3
triasulfuron	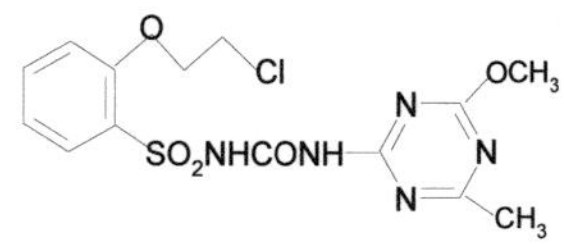	$C_{14}H_{16}ClN_5O_5S$	401.8	4.6
chlorsulfuron	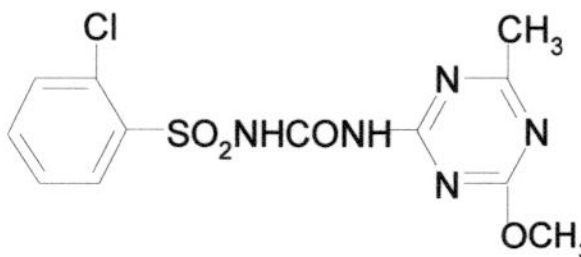	$C_{12}H_{12}ClN_5O_4S$	357.8	3.6
bensulfuron-methyl	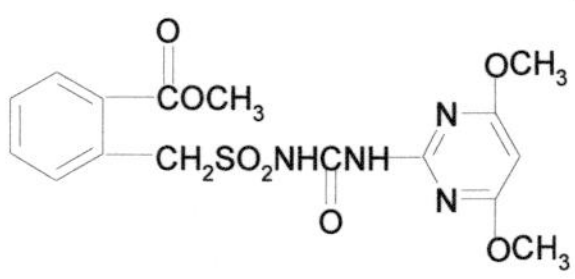	$C_{16}H_{18}N_4O_7S$	410.4	5.2
prosulfuron	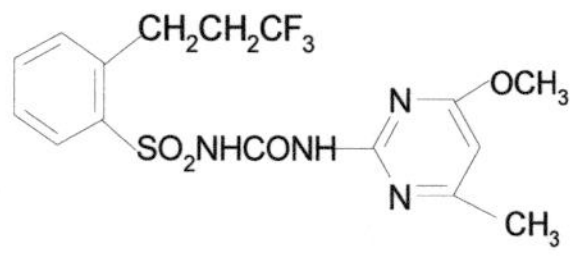	$C_{15}H_{16}F_3N_5O_4S$	419.4	3.8
pyrazosulfuron-methyl	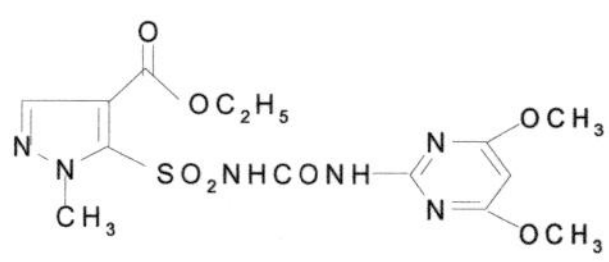	$C_{14}H_{18}N_6O_7S$	414.4	3.7
chlorimuron-ethyl	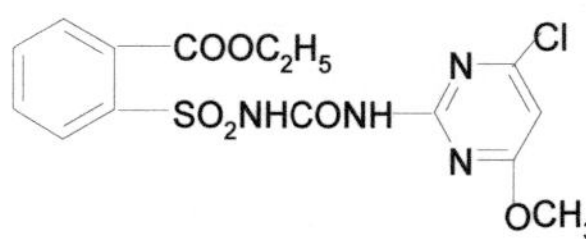	$C_{15}H_{15}N_4O_6S$	414.8	4.2
primisufuron-methyl	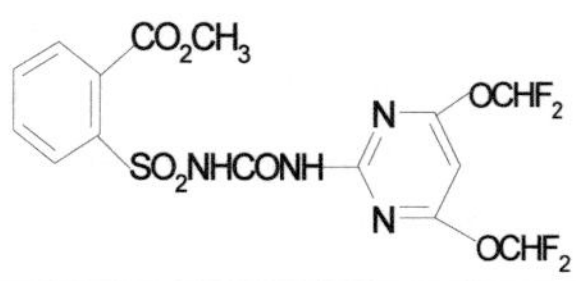	$C_{15}H_{12}F_4N_4O_7S$	468.3	5.1

Table 8. Infromation for some Sulfonylurea herbicides

Matrix	Herbicide	Extraction	Clean-up	Ref.
Carrots	Linuron	Hexane–diethyl ether, homogeniser	Florisil cartridge	(D'Amato, 1993)
Potatoes	Linuron	Acetone, homogeniser	LLE–Silica cartridge	(Miliadis & Vasilikiotis, 1990)
Cereals	Metsulfuron	Methanol, homogeniser	Liquid chromatography	(Zhou et al., 1994)
Rice	Bensulfuron	Methylene chloride, homogeniser	Silica cartridge	(Zhou et al., 1996)
Carrots	Linuron	Water (acidic pH), shaking	-	(Sojo et al., 1997)
Garlic	Linuron	Methanol, homogeniser	Alumina column	(Cessna, 1991a)
Asparagus	Linuron	Methanol, homogeniser	LLE–Florisil column	(Cessna, 1990)
Cereals	Chlortoluron	Ethanol–water, homogeniser	Silica column	(Pérez et al., 1993)
Potatoes	Isoproturon	Methanol, homogeniser	–	(Yaduraju, 1993)
Grains	Sulfonylureas	Acetonitrile, homogeniser	Cation–exchange cartridge	(Krynitsky & Swineford, 1995)
Potatoes	Linuron	Acetone, homogeniser	LLE–Florisil column	(Mattern, 1989)
Grains, cereals	Chlorsulfuron	Ethyl acetate, blender	LLE–GPC	(Slates, 1983)

Table 9. Extraction and clean-up for sulfonylurea herbicides

In order to determine the multiresidue of oxasulfuron, thifensulfuron-methyl, metsulfuron-methyl, triasulfuron, chlorsulfuron, bensulfuron-methyl, prosulfuron, pyrazosulfuron-methyl, chlorimuron-ethyl and primisufuron-methyl in soybeans, Qi tried various solvent system including acteone, acetonitrile, dichloromethane, ethyl acetate to optimize the extraction procedure. It showed that the serious emulsification occured when using dichloromethane and more interferences were extracted by acteone and ethyl acetate. Finally, they used acetonitrile to extract these compounds from soybean. For clean-up step, Qi tested the purification effects of SPE packed with different materials (C_{18} 500mg, Florisil 1000mg &3000mg, Al_2O_3-Neutral 500mg &1000mg) and satisfied results were obtained when using SPE columns packed with Florisil (3000mg).

4.2 Detection

Various methods for sulfonylurea herbicide determination have been published up to now. These compounds are not directly amenable to GC, because of their low volatility and thermal instability. Few is reported by GC analysis after derivatization.

Most of the applications known are based on HPLC using reversed phase columns followed either by ultraviolet (UV) or mass spectrometric (MS) detection. The typical conditions for HPLC separation were set as follows (table 10):

Column: C_{18} (250*4.6mm i.d., 5.0μm), temperature 45 °C; UV wavelength: 230nm
Mobile -phase: acetonitrile-water (pH=2.5, adjusted with 85% phosphoric acid); flow rate: 1.0mL/min
The gradient elution program of HPLC separate condition (table). Qi (Qi et al., 2004)applied this procedure to analyze the sulfonylurea herbicide residues in soybean samples.

Time (min)	Water acidfied with Phosphoric acid (pH=2.5)%	acetonitrile (%)
0.00	80	20
1.75	65	35
10.00	60	40
13.00	50	50
15.00	40	60
22.00	40	60
22.01	10	90
27.00	10	90

Table 10. Gradient elution program for HPLC

5. Triazine herbicides

Triazine herbicides are a class of herbicides used for protecting crops from weeds before emergence or during early stage after emergence. The history of their use can be traced back to1952 when J. R. Geogy synthesized and screened the first triazine derivatives. A great triazine herbicides are derived from *s*-triazine (fig 4) For R1 position, this is most often -Cl(the commercial names ending with ~azine), -SCH3(-tryn) and -OCH3(-ton). The substituents at R2 or R3 are usually amino groups. (See table 1)
Triazines and their degradation products are toxic and persistent in water, soil and organisms(Vitali et al., 1994). Moreover, atrazine is a member of the triazine family and has been classified as human carcinogen(Dean et al., 1996). From the view of their ecological and health hazards in use, some triazine herbicides have been banned in certain countries (e. g. atrazine banned to use in 1991, Germany). In the EU, the maximum allowed limit for each individual herbicide has been set at 0.1 ugL^{-1}, but the EPA of USA has set the maximum allowable level of atrazine at $3ug/L^{-1}$.

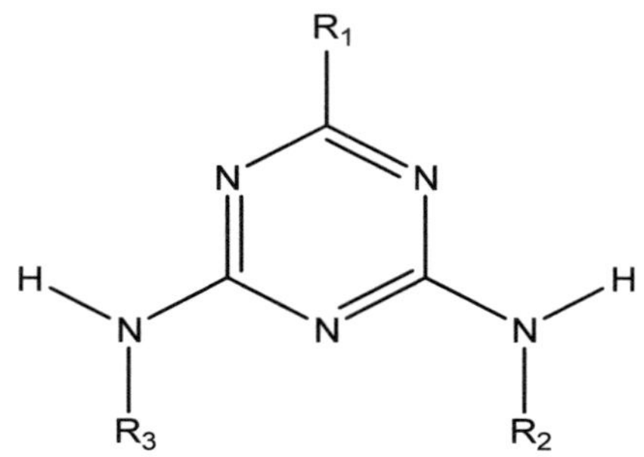

Fig. 4. Chemical structure for triazines

Compound	Substituents			Partition coefficient between octanol and water lg $P_{OC/W}$
	R_1	R_2	R_3	
Simazine	Cl	NHC_2H_5	NHC_2H_5	2.3
Atrazine	Cl	NHC_2H_5	$NHCH(CH_3)_2$	2.7
Propazine	Cl	$NHCH(CH_3)_2$	$NHCH(CH_3)_2$	2.91
Terbutylazine	Cl	NHC_2H_5	$NHC(CH_3)_3$	3.06
Trietazine	Cl	NHC_2H_5	$N(C_2H_5)_2$	3.07
Ipazine	Cl	$N(C_2H_5)_2$	$NHCH(CH_3)_2$	-
Deethylatrazine	Cl	NH_2	$NHCH(CH_3)_2$	1.6
Deisopropylatrazine	Cl	NHC_2H_5	NH_2	1.2
Deethyldeisopropylatrazine	Cl	NH_2	NH_2	0
Hydroxysimazine	OH	NHC_2H_5	NHC_2H_5	-
Hydroxyatrazine	OH	NHC_2H_5	$NHCH(CH_3)_2$	1.4
Hydroxypropazine	OH	$NHCH(CH_3)_2$	$NHCH(CH_3)_2$	-
Hydroxydeethylatrazine	OH	NH_2	$NHCH(CH_3)_2$	0.2
Hydroxydeisopropylatrazine	OH	NHC_2H_5	NH_2	-0.1
Simeton	OCH	NHC_2H_5	NHC_2H_5	-
Atrazon	OCH	NHC_2H_5	$NHCH(CH_3)_2$	2.69
Desmetryn	SCH_3	$NHCH_3$	$NHCH(CH_3)_2$	-
Simetryn	SCH_3	NHC_2H_5	NHC_2H_5	2.8
Ametryn	SCH_3	NHC_2H_5	$NHCH(CH_3)_2$	3.07
Prometryn	SCH_3	$NHCH(CH_3)_2$	$NHCH(CH_3)_2$	3.34
Terbutryn	SCH_3	NHC_2H_5	$NHC(CH_3)_3$	3.74

Table 11. Information for some triazine herbicides

5.1 Sample preparation

Numerous methods have also been published that examine a large variety of the triazines in many different matrices. Usually, the targeted compounds are extracted from foods by mechanical shaking or homogenisation with organic solvents, then clean-up of the extracts is carried out on SPE columns(Florisil, silica, alumina, cation-exchange cartridge). Triazine compounds are organic bases and very easy to be absorbed by cation exchange resin.

For the great differences in physical and chemical properties of different triazine herbicides, a wide array of solvents (acetone, ethanol, ether, chloroform, methanol, water et. c) have been used in analytical method development. (see table 12)

5.2 Detection

Different analytical methods, such as GC, HPLC and capillary electrophoresis, have been developed for the separation and quantification of triazine herbicides (table 13). Gas chromatography mainly with ECD, NPD and MS detection has been extensively employed for the measurement of triazine herbicide residues. DB-5 capillary column (5 % polydiphenyl- and 95 % polydimethylsiloxane; 30 m x 0.25 mm, film thickness 0.25 μm) or its analogue is suitable for triazine analysis.

Matrix	Herbicide	Extraction	Clean-up	Ref.
Vegetables, rye	Triazines	Dichloromethane maceration, shaker	Silica column	(Roseboom & Herbold, 1980)
Cereals, apples, celery	Triazines	Methanol, blender	LLE-Cation-exchange cartridge	(Pardue, 1995)
Vegetables	Triazines	Acetonitrile–water, homogeniser	Carbopack cartridge SCX column	(Battista et al., 1989)
Corn, vegetables, sugar beet	Simazine	Water, homogeniser Chloroform, shaker	Alumina column	(Pringle et al., 1978)
Cereals, vegetables	Metribuzine	Acetonitrile–water, reflux	LLE–Florisil column	(Thornton & Stanley, 1977)
Potatoes	Metribuzine	Water, steam distillation	LLE–Silica column	(Ohms, 1976)
Fruits, vegetables	Atrazine	Ethyl acetate, shaker	C18 column	(Wittmann & Hock, 1993)
Grape juice	Simazine	Diethyl ether (acidic pH), shaker	-	(Ortiz-Gomez et al., 1995)
Oil	Simazine	Acetonitrile, blender	-	(Montiel & Sánchez, 1996)
Olives	Simazine	Ethyl acetate, blender	-	(Cessna & Benoit, 1992)
Onions	Cyanazine	Ethanol–water, homogeniser	LLE–Florisil column	(Bailey et al., 1978)
Vegetables	Triazines	Acetone, blender	LLE–Florisil column	(Lawrence & Laver, 1974)
Cereals, fruits, vegetables	Triazines	Methanol, blender	Alumina column	(Mortimer et al., 1994)

Table 12. Extraction and clean-up for triazine herbicides

Tomkins and Ilgner (Tomkins & Ilgner, 2002) developed a GC-MS method for the detection of triazine herbicides (atrazine, cyanazine, simazine) and their decomposition products (deethylatrazine, deisopropylatrazine) in environmental waters. Balduini (Balduini et al., 2003) meaured the triazine herbicides in breast milk. Five triazines were adsorbed on a graphitized carbon black SPE cartridge, desorbed and analysed by GC/MS. Detection and quantification limits were 0.3 and 1 ppb from 1 mL of breast milk. Some triazine herbicides and their degradation products have been separated by reversed phase HPLC, and their atmospheric pressure chemical ionization (APCI) or electrospray mass spectra were measured. The APCI technique gives primarily [M+H]+ ions, but fragment ions are observed with electrospray and conditions that favor CID The LC/MS techniques are

Matrix	Herbicide	Analytical measure	LOD	Ref
Corn	Atrazine	GC-ECD	0.002 ppm	(Pylypiw et al., 1993)
Onion	Cyanazine	GC-NPD HP-1 Column	10 mg/kg	(Cessna, 1992)
Cereals, vegetables	Metribuzin	GC-ECD OV-225 Column	0.01 mg/g	(Ohms, 1976; Thornton & Stanley, 1977)
Vegetables, corn, sugar beet Oil, olives	Simazine	GC-NPD OV-101 Column HP-1 Column	mg/kg 0.01 ppm	(Pringle et al., 1978; Montiel & Sánchez, 1996)
Rye, vegetables Cereals, celery, apples	Triazines	GC-NPD Carbowax 20 M; OV-225 DB-17	0.01–0.02 mg/kg 0.02–1.0 ppm	(Roseboom & Herbold, 1980)
Breast milk	Triazines	GC-MS BPX-5 SGE	0.3-1 ppb	(Pardue, 1995)
Tap water, rice, maize and onion	Triazines	GC/MS CP-Sil 5 CB GC-FID CP-Sil 8 CB,	- 14-74 $ngmL^{-1}$	(Bailey et al., 1978)
Oranges, corn	Atrazine	HPLC Reversed-phase C18 Methanol–water UV 230 nm	0.015–0.300 ppm	(Wittmann & Hock, 1993)
Blueberries	Simazine	HPLC Reversed-phase C18 Acetonitrile–water UV	0.08–0.17 ppm	(Ely et al., 1993)
Grape juice	Simazine	HPLC Reversed-phase C18 Methanol–acetate buffer pH 5.0 UV 230 nm	20 mg/ L	(Ortiz-Gomez et al., 1995)
Vegetables	Triazines	HPLC Reversed-phase C18 Acetonitrile–phosphate buffer pH 6.7 UV 220 nm	10 ng/g	(Battista et al., 1989)
Oysters	Triazines	HPLC-MS/MS	-	(Wittmann & Hock, 1993)
Sediments and water	Triazines	HPLC-APCI-MS/MS	-	(Takats et al., 2001)
-	Triazines	ELISA	<1 ppb.	(Wittmann & Hock, 1993)
Surface water	Simazine	SPFIA	1.3±0.9 $ngmL^{-1}$	(Bruun et al., 2001)

Table 13. Extracton and cleanup for triazine herbicides

appropriate for triazine metabolites and their degradation products that are not amenable to GC/MS, but they may not provide advantages over GC/MS for most triazine herbicides and their dealkylated degradation products that are amenable to GC/MS. Hammock 's lab (Wortberg et al., 1995)developed immunoassy to detect four triazines in 1995 and the LOD of the ELISA was lower than 1ppb. Herranz (Herranz et al., 2008) developed solid-phase fluorescence immnunoassay (SPFIA) and applied it in simazine detection of surface water with higher sensitivity (LOD 1.3±0.9 ng/mL).

6. Amide herbicides

Amides, especially of chloroacetic acid and substituted anilines, have been and are popular herbicides since the first amide herbicide-allidochlor was found 60 years ago. Acetochlor, alachlor, butachlor, dimethenamide, metolachlor, and propachlor are amides of chloroacetic acid, and especially acetochlor, is used widely in the world for its high efficiency as the treatment agents before emergence. They are also in the list of chemical pollutants that need to be more heavily monitored due to their toxicity and accumulation in environment and their effects on the environment and human health. Acetochlor was listed as B-2 carcinogen by EPA (USA). Other acids used to form the amides include propanoic acid and several substituted benzoic acids(Nartova et al., 2008). An alkyl or alkyloxyalkyl group is usually substituted for the other hydrogen of the amide nitrogen. Some representative amide herbicides are shown in table 14.

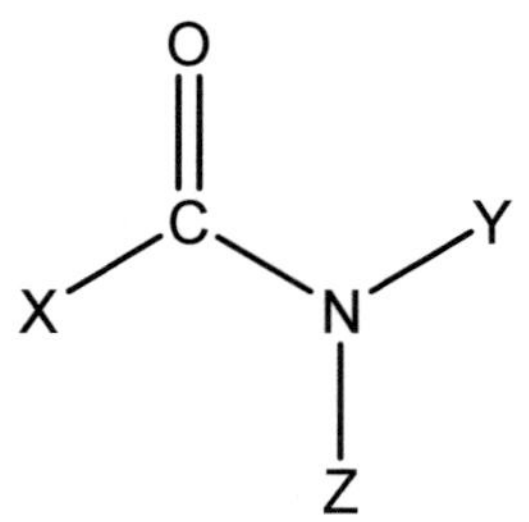

Fig. 5. Parent strucutre for amide herbicides

6.1 Sample pretreatment

For extraction of amide herbicides from food or agricutural products, solvents such as acetone, acetone-water, petroleum ether or acetonitrile were used widely (table 15). A typical sample treatment procedure was as follows: sample was extracted by acetone, then sulfate solution was added to the extracts, and finally LLE procedure was carried out with petroleum ether. But, the LLE isn't suitable for purification of some polar compounds (e. g. alachlor). For complex samples, further clean-up process is needed, usually based on SPE (florisil, alumina, silica or carbopack cartridge). C_{18} sorbents mainly used for the clean up of water samples before analysis solid-phase microextraction (SPME) considered as solventless analytical techniques, has been reported to detect the acetochlor, alachlor, and metolachlor residues in water samples.

6.2 Detection

GC was the most common method to detect the amide herbicides, usually equipped with selective detectors such as ECD, NPD or MS (Li et al., 2006). Acetochlor often can't be

Name	X	Y	Z
Acetochlor	$-CH_2Cl$	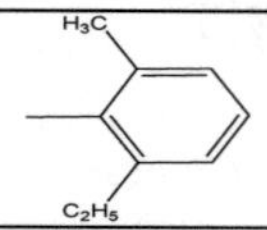	$-CH_2OCH_2CH_3$
Alachlor	$-CH_2Cl$	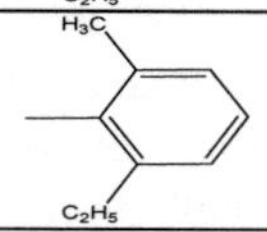	$-CH_2OCH_3$
Butachlor	$-CH_2Cl$	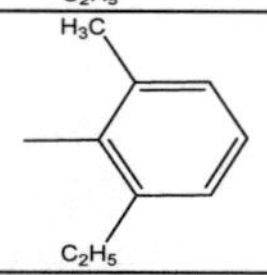	$-CH_2O(CH_2)_3CH_3$
Dimethachlor	$-CH_2Cl$	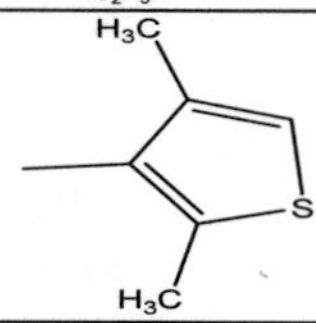	$-CH(CH_3)CH_2OCH_3$
Flufenacet	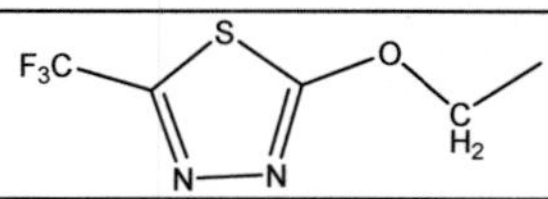		$- CH(CH_3)_2$
Isoxaben	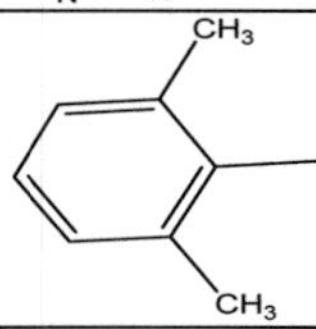	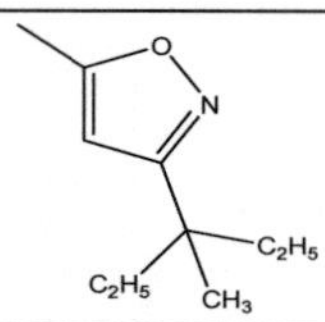	-H
Metolachlor	$-CH_2Cl$	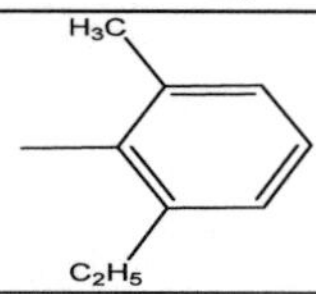	$-CH(CH_3)CH_2OCH_3$
Napropamide	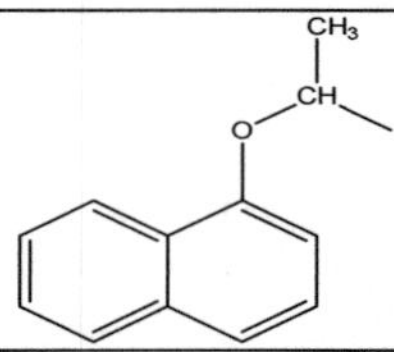	$-C_2H_5$	$-C_2H_5$
Pronamide	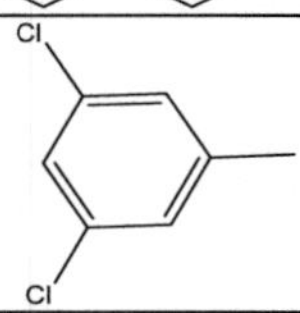	$-C(CH_3)_2C\equiv CH$	-H
Propachlor	$-CH_2Cl$	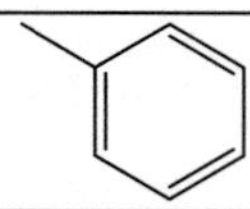	$- CH(CH_3)_2$
Propanil	$-C_2H_5$	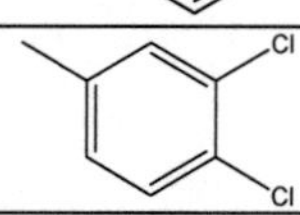	-H

Table 14. Information for some amide herbicides

separated with atrazine on the capillary column and thus, some analysts used NPD connected with ECD to realize the simutanious detection of the two compounds. What's more, the heated decomposition temperature of some amide herbicdes is low (metolachlor, 105 °C), which makes difficulties in detection of these compounds by GC.

The HPLC method, based on reversed phase C_{18} or C_8 column, came into being. The mobile phase often was methanol-water or acetonitrile-water (pH 3, adjusted with acetic acid). MS-MS techqiues further improved the analytical selectivity. Steeen, (Ling et al., 2006)used GC-MS/MS to detect the pesticide redidues in marine system with LOD ranging from 0.2 to 0.5 ng/L. Striley(Striley et al., 1999) developed ELISA to measure the putative major human metabolite of metolachlor, metolachlor mercapturate (MM) in human urea. Tessier, (Tessier & Marshall, 1998) developed immunoassay to detect alachlor in aqueous samples. Yakovleva, (Szendr et al., 2003) established ELISA and applied to analyze butachlor residues in mineral, ground and surface water. Other application of detection method was listed in table 16.

Matrix	Herbicide	Extraction	Clean-up	Ref
Tamotoes	metolachlor	Water(acidic pH), homogeniser	LLE	(Gaynor et al., 1993)
Carrots	metolachlor	Water(acidic pH), shaker	-	(Sojo et al., 1997)
Potatoes	metolachlor	Acetone-hexane, blender	LLE	(Singh, 1997)
Cereals	Chloroacetamides	Acetonitrile, homogeniser	LLE-florisil column	(Balinova, 1988)
Vegetables	metolachlor	Methanol, blender	LLE-silica cartridge	(Gaynor et al., 1992)
Tea leaves	Amide herbicides	ethyl acetate, shaker	An active carbon SPE column connected to a Florisil column	(Shen et al., 2007)
Soybean	Amide herbicides	Acetone, shaker	Florisil cartridge	(Li et al., 2006)
Onion	Amide herbicides	Acetonitrile microwave-assisted extraction(MAE)	Florisil cartridge	(Hans-Jürgen & Manfred, 1993)
Water	Amide herbicides	SPME	-	(Sauret-Szczepanski et al., 2006)
Water	Amide herbicides	water–acetonitrile MAE		(Fuentes et al., 2006)

Table 15. Extraction and clean-up for amide herbicides

Matrix	Herbicide	Analytical Method	LOD	Ref
Peanut, cereals	Alachlor	NPD, UC-W98	0.02–0.05 ug/g	(Conkin et al., 1978)
Corn	Alachlor, metolachlor	ECD	0.002 ppm	(Pylypiw et al., 1993)
Tomatoes	Metolachlor	Hydrolysis MS Supelcowax	10- 50 ppb	(Ely et al., 1993)
Potatoes	Metolachlor	ECD, OV-1	0.15 ng	(Gaynor et al., 1992)
Potatoes, tomatoes, maize	Chloroacetamides	ECD QF-11DC-200, Apiezon L	0.02–0.05 ng	(Singh, 1997)
Tea leaves	Amide herbicies	GC-NCI-MS GC-EI-MS	<2 ug/kg	(Balinova, 1988)
Soybean	Amide herbicies	HPLC-UV, C_{18} 210 nm	1-7.2 ppb	(Li et al., 2006)
Carrots	metolachlor	HPLC-UV 220nm	--	(Sojo et al., 1997)

Table 16. analytical methods for some amide hericides

7. Glyphosate

Glyphosate is the common name for *N*-(phosphonomethyl)-glycine, a total-kill herbicide (first found its herbicidal activity in 1971, introduced in 1974 by theMonsanto Company under the trade name "Roundup" and rapidly became one of leading herbicides in the world.), having the environmental advantages of low mammalian toxicity and rapid breakdown in the soil leaving no harmful residues. Having pKa values of 0.78, 2.29, 5.96 and 10.98, glyphosate is a very polar and amphoteric compound(fig 6).

Glyphosate is used to control grasses, herbaceous plants including deep rooted perennial weeds, brush, some broadleaf trees and shrubs, and some conifers(Tsui et al., 2005). Glyphosate does not control all broadleaf woody plants. Glyphosate applied to foliage is absorbed by leaves and rapidly moves through the plant. It acts by preventing the plant from producing an essential amino acid. This reduces the production of protein in the plant, and inhibits plant growth. Glyphosate is metabolized or broken down by some plants, while other plants do not break it down. Glyphosate dissolves easily in water. Aminomethylphosphonic acid (AMPA) is the main break-down product of glyphosate in plants (Zhao et al., 2009).

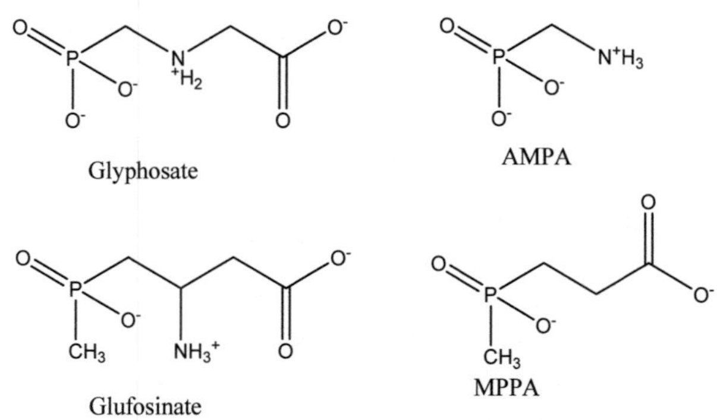

Fig. 6. Chemical structure for glyphosate, glufosinate and their main metabolites

Glufosinate [DL-homoalanine-4-yl(methyl)phosphinic acid], is another highly polar amino acid herbicides. A major breakdown product of glufosinate found in both plants and animals that have been exposed to glufosinate is 3-methylphosphinicopropioninc acid (MPPA) (Moye et al., 1983).
Based on the results of animal studies, glyphosate does not cause genetic damage or birth defects, and has little or no effect on fertility, reproduction, or development of offspring. There is not enough information available at this time to determine whether glyphosate causes cancer. There have been no reported cases of long term health effects in humans due to glyphosate exposure (Tsui et al., 2005). The Food and Agriculture Organization (FAO) of the United Nations has set a maximum residue limit (MRL) of glyphosate in wheat at 5 mg/kg. The world health organism evaluated glyphosate on its acceptable daily intake value and the data allocated for glyphosate was 0.3 mg/kg body mass.

7.1 Sample preparation

The main problem in glyphosate and its metabolite analysis is their recovery from biological or field samples. Glyphosate is a highly polar herbicide, very soluble in water and insoluble in most organic solvents, which does not allow extraction with organic solvents and makes the extraction difficult and the preconcentration step quite lengthy.
Due to the amphoteric character of glyphosate and AMPA, both anionic and cationic resins have been used for preconcentration and clean-up purposes. Another important aspect to be considered is the binding of glyphosate to organic matter. Some reports showed that humic substances adsorb glyphosate strongly because the hydrogen bonding interactions between the hydrogen acidic and the oxygen group of both substances. Glyphosate extraction is usually carried out with water or water with chloroform, sometimes at acidic pH (table 17). In this procedure, other water soluble components of foods, like amino acids, amino sugars, etc. are also extracted. These compounds interfere in the glyphosate determination making necessary the clean-up of extracts. More often used in this purification step is LLE or column chromatography on ion exchange columns.

7.2 Detection

Due to their very polar, and in most cases ionic character, Analytical methods for the analysis of glyphosate and its major metabolite, AMPA, include thin-layer chromatography, capillary electrophoresis (CE), gas (GC) and liquid chromatography (LC) after derivatisation.
The availability of derivatisation techniques compatible with an aqueous extract or sample and the chromatographic separation makes LC an attractive technique. However, for LC with conventional detection systems, such as UV-Vis or fluorescence detectors, glyphosate and AMPA need to be derivatised because of the lack of chromophore or fluorophore. Three different procedures are generally used for the determination of glyphosate with LC: (i) post-column ninhydrin derivatisation and UV detection; (ii) post-column fluorogenic labeling with o-phthalaldehyde and mercaptoethanol after oxidation of glyphosate to glycine; (iii) pre-column derivatization using 9-fluorenylmethyl chloroformate (FMOC-Cl) with fluorescence detection(FLD).
Post-column derivatisation was used in most of the previous studies for glyphosate analysis in water and has also been recommended by the US Environmental Protection Agency (EPA). Moyne and Boning were the first to use the FMOC-Cl reaction for derivatising glyphosate. A disadvantage of this reaction, however, is its reactivity with water, which

Matrix	Extraction	Clean-up	Ref.
Corn, fruits, soybeans	Water, blender	LLE–Cation-exchange column	(Alferness & Iwata, 1994)
Blueberries	Water, homogeniser	LLE–GPC–Cation-exchange column	(Guinivan, 1982)
Legumes, cereals	Water–chloroform, shaker	LLE–Cation-exchange column–Anion-exchange column	(Wigfield & Lanouette, 1991; Wigfield & Lanouette, 1991)
Fruits, field pea, barley and flax seed	0.1 *M* HCl–chloroform, blender	LLE–Ligand-exchange column–Anion-exchange column	(Cowell et al., 1986) (Cessna, 2002)
Berries	Water–chloroform, blender	LLE–Charcoal–Cation-exchange column	(Konar & Roy, 1990)
Kiwi fruit, asparagus	Water–chloroform, blender	LLE–Anion-exchange column–GPC	(Benfenati et al., 2006)
fruit juices	-	-	(Cláudia et al., 2007)
Fruits, vegetables	Water–chloroform, blender	LLE–Cation-exchange column	(Moye et al., 1983)
Cereals	Water, overnight standing extraction	100 mg C_{18} SPE	(Hogendoorn et al., 1999)
Rice, soybean sprouts	Water, acetone homogeniser	anion-exchange column Florisil Cartridge Cleanup.	(Tseng et al., 2004)

Table 17. Extraction and clean-up for glyphosate

leads to the formation of a FMOC-OH product (reaction of the acyl chloride with water) in the reaction mixture. To obtain quantitative yield in derivatisation, excess reagent has to be used. Different concentrations of FMOC-Cl have been reported in the literature for the derivatisation of glyphosate, still, there is little or no general agreement concerning the optimal molar ratio of glyphosate to FMOC-Cl to be used.

The common HPLC conditions for the separation of glyphosate and AMPA were using a single polymeric amino column and mobile phase at pH 10 which contained 55% (v/v) acetonitrile and 50mM phosphate buffer.

FMOC-OH by product make it difficult in separation by chromatography, which is represented by the large peak in front of the glyphosate chromatogram. The FMOC-OH product completely overlaps the glyphosate peak and creates difficulties in its detection (Tadeo et al., 2000). The removal of this FMOC product and separation of the glyphosate peak by column-switching technique using coupled C18 and amino columns was previously reported. However, these silica-based columns usually degrade under high alkaline conditions. Sancho (Sancho et al., 1994). reported that a gradual decrease in efficiency of the silica-based amino column after two months' use. Ion chromatography (IC) provides a useful tool in detecting ionic substance. Zhu used ion chromatography system equipped with anion exchange column and suppressed conductivity detector to determine the gylphosate in enviromental samples with LOD 0.042 ug/mL. Patsias (Patsias et al., 2001) developed an automated method based on the on-line coupling of anion-exchange solid-phase extraction (SPE) and cation-exchange liquid chromatography followed by post-column derivatization and fluorescence detection for the trace level determination of glyphosate and its primary conversion product aminomethyl phosphonic acid (AMPA) in water.

These ionic compounds were also determined in water by liquid chromatography with mass spectrometry (LC-MS) after derivatization with FMOC, achieving quite low detection limits. The coupling of ion chromatography (IC) with electrospray mass spectrometry (ES-MS) opens new ways for the determination of polar organic micropollutants in water samples. The technique of conductivity suppress ion has been found to reduce the background signal in the range of about two-orders of magnitude leading to a significant increase in sensitivity. In addition, the formation of salt adducts has been avoided. Bauer (Bauer et al., 1999) separated glyphosate and AMPA in water on an anion-exchange column without any erivatization and detected the signal by IC-ES-MS.

The GC method can be developed to analyze glyphosate through the preparation of N-heptafluorobutyrychloroethyl ester, N-trifluoroacetyltrifluoroalkyl ester, N-trifluoroacetylheptafluorobutyl ester and tert-butyldimethylsilyl derivatives. However, it is a time-consuming procdeure to prepare the derivatives under anhydrous conditions. The usual detector equipped with GC for glyphosate analysis can be flame photometric, msaa-selective detectors or the extreme sensitive electro-capture detector.

Capillary electrophoresis (Corbera et al., 2005), as an important separation technique due to its high resolving power and speed, was also reported for glyphosate anaysis. Some (Khrolenko & Wieczorek, 2005) used p-toluenesulfonyl chloride for derivatization prioe to CE separation, others (Cikalo et al., 1996) incorporates ribonucleotides into the background electrolyte to realize the indirect photometric detection. Chang (Chang & Liao, 2002) employed fluorescein as the buffer fluorophore and an argon-ion laser to induce the fluorescence background for detection of the glyphosate, AMPA, glufosinate and MPPA.

Mass spectrometry (MS) has the potential to be a rigorous direct detection method for these compounds, particularly in their ionic states. Utilising a simple microelectrospray interface, Goodwin (Goodwin et al., 2003) analyzed glyphosate, glufosinate and their metabolites on capillary electrophoresis–mass spectrometry (CE–MS) using a combination of electrical and pressure drive for interface. The observed concentration limit of detection for glyphosate in water is 1 mM and for a water–acetone extract of wheat is 2.5 mM, allowing the underivatised herbicide to be detected at 10% of the maximum residue limit in wheat.

8. Other herbicides

In addition to the above described types of herbicides, imidazolinone (imazethapyr, imazamox, imazapyr), imazapic, carbamate (isoprocarb, oxamyl, propoxur) and diphenyl ether herbicides (acifluorfen, chlornitrofen, aclonifen, bifenox and oxyfluorfen) are also popular in agricultural production.

8.1 Imidazolinone herbicides

This class of herbicide is used to protecing beans, peanuts, corn and other crops from weeds. These herbicides are used in a small amount for their long-acting effects and trace residues in soil may cause phytotoxicity on succeeding crop (Lewis et al., 2009). In 2005, canada set MRL for imazethapyr residue in soybean, 0.1mg/kg. USA regulated the MRL of imazethapyr residue in rice, 0.3 mg/kg (G/SPS/N/USA/1229). Japan set the MRLs ranging from 0.01 to 0.5 mg/kg of imazethapyr residue in foods depending on the food types.

There are carboxyl group and imino group in the chemical structure of imidazolinone herbicdie, which make imidazolinone herbicides show strong polarity, and thus the control of pH in sample extraction is critical.

Many analytical methods such as HPLC, GC-MS, LC/MS have been reported for imidazolinone herbicide detection. A typical HPLC method is as follows: the targeted molecules can be extracted from the matrix with mixed solution of ammonium bicarbonate(0.1 M, pH=5)- methanol(7:3, v/v). The extracts can be partitioned with dichloromethane and the organic layer was collected and condensed for further clean-up on cation exchange column. Separation of imidazolinone herbicides can be carried out by C_{18} column with acetonitrile-1% acetic acid as mobile phase. The detection UV length can be set 252-258 nm.

With the sensitivity and sepecificity of HPLC-MS (Chu et al., 2008), some analyzed the imidazolinone herbicides in various matrix. Under positive mode, [M+H]+ can be monitored for each compound(m/z262 for imazapyr, m/z275 for imazamethabenz acid, m/z306 for imazamox, m/z276 for imazapic, m/z290 for imazethapyr and m/z312 for imazaquin).

These compounds should be esterized before analysis by GC. Anisuzzaman (Anisuzzaman et al., 2000) detected the imidazolinone herbicides in soil, water and soybean by GC-NPD and GC-MS after synthesis of dimethyl derivatives.

8.2 Carbamate herbicides

Three classes of carbamate pesticides are known. The carbamate ester derivatives, used as insecticides (and nematocides), are generally stable and have a low vapour pressure and low water solubility. Carbamate fungicides contain a benzimidazole group. It is well known that carbamate pesticides are esters of carbamic acid, having the general structure $R_1NHC(O)OR_2$, in which R_1 and R_2 are aromatic and/or aliphatic moieties.

Carbamate herbicides (Vasilescu et al., 2005) are known to repress cell division as a consequence of their disturbing nucleic acid metabolism and protein synthesis. Clorpropham, sulfallate and phenmedipham are the representatives of this family herbicides. Some examples about extraction and clean-up of carbamates and thiocarbamates are shown in table.

The well-known thermal instability of carbamates has led to the use of HPLC, but its most usual detectors have a limited sensitivity. In the 1980s, some used post-column hydrolysis and derivatization with fluorescence detection to overcome these disadvantages. The carbamates were degraded into methylamine and then derivatized to a fluorescent isoindole product, which was widely used in carbamate residue analysis in fruits and vegetables. In addition, many references investigation showed that both ESI and/or APCI with HPLC/MS were used to analyze the carbamates and APCI can help to reduce matrix effects.

Although careful control of experimental conditions may allow direct determination of carbamates by GC, large number of experimental factors such as injector temperature, residence time in the injector, solvent nature and injection mode, are known to affect the results. Derivatization reactions are therefore required prior to GC analysis.

Matrix	Herbicide	Extraction	Clean-up	Ref.
Rice	Thiobencarb	Methanol or acetone, blender	-	(Au & Fung, 1988)
Potatoes	Chlorpropham	Tetrahydrofuran-water-acetonitrile -acetic acid, homogeniser	-	(Camire et al., 1995)
Fruits, vegetables	Chlorpropham	Methanol, blender	Alumina column	(Wilson et al., 1981)
Potatoes	Chlorpropham	Acetone, homogeniser	LLE	(Tsumura-Hasegawa et al., 1992)
Garlic	Triallate	Methanol, homogeniser	LLE-Florisil cartridges Alumina column	(Cessna, 1991b)
Potatoes	Chlorpropham	Dichloromethane (water), blender	-	(Mondy et al., 1992)
Lentils	Triallate	Acetonitrile, shaker	Alumina column	(Cessna, 1980)
Potatoes	Chlorpropham, propham	Dichloromethane (water), blender	Silica-TLC	(Corti et al., 1991)
Fruits, vegetables	Chlorpropham, propham, triallate	Ethyl acetate, homogeniser	LLE-Florisil column	(Blaicher et al., 1980)
Potatoes	Chlorpropham	Water suspension,	solid-phase microextraction	(Volante et al., 1998)
fruit and vegetables	carbamate herbicides	acetonitrile (MeCN) containing 1% acetic acid (HAc)	dispersive-SPE cleanup step (primary secondary amine+ C_{18})	(Martínez Vidal et al., 2006)

Table 18. Extraction and clean-up for some carbamate herbicides

The main derivatization reactions applied to the family of herbicides involve the N-protection for carbamates. Among them, silylation, acylation and alkylation, together with reactions of transformation into aniline have been used. An N-protection reaction for derivatization of compounds containing an NH-reactive group, based on the use of sodium hydride/dimethyl sulphoxide/methyl iodide (NaH/DMSO/CH_3I) has been frequently used.

8.3 Diphenyl ether herbicides

Among the herbicides being used, diphenylether compounds of herbicide are mainly introduced at pre- or post-emergence in controlling annual broad-leaved weeds and some types of grasses in numerous crops like rice, cereals, maize, etc(Murakami et al., 1988). This class of herbicides has proved to be an inhibitor of protoporphyrinogen oxidase, that leads to the accumulation of protoporphrin and therefore blocks the formation of chlorophyll. Molecules that inhibit protoporphyrinogen oxidase (Protox) have been among the most frequently patented class of herbicides over the past decade. Commercial Protox inhibitors can be classified in a major chemical group, the *p*-nitrodiphenyl-ethers, commercially known as the diphenyl-ethers (DPhE).

This class of herbicides is mainly composed of esters but few compounds are acids or have an acidic behavior, with pKa comprised between 2.7 and 3.8. There are two main metabolites that arise from the degradation of the DPhE herbicides studied, bifenox acid from the hydrolysis of bifenox and acifluorfen from the degradation of lactofen and fluoroglycofen. Bifenox and oxyfluorfen are reported to be carcinogenic or suspected to be carcinogenic compounds (Sabino et al., 2004).

The herbicides in this category have a 2-chlorodiphenyl ether nucleus in common, and most also have nitro and trifluoromethyl substituents. As this class of compounds is usually nonvolatile and thermally unstable, most of the direct methods have been performed by using LC. Acifluorfen and fomesafen can be separated on a C-18 column, with a slightly acidic mobile phase, followed by electrospray to give $[M-H]^-$ ions. Lactofen and oxyfluorfen were also separated on a C-18 column, but without acid in the mobile phase.

Oxyfluorfen is amenable to GC separation (Wong et al., 2003), and nitrofen, with a similar structure, should have favorable properties for GC. Shen (Shen et al., 2008) extracted DPhE from vegetables sample with acetonitrile, then the extract was cleaned up by Envi-Carb SPE column connected to Alumina Neutral SPE column, determined by gas chromatography-negative chemical ionization mass spectrometry. The lactofen esters may be amenable to GC separation, but the acid acifluorfen and the sulfonamide fomesafen require derivatization for GC.

DPhE showed good solubility in acetone and acetonitrile and both organic solvents can be miscible with water. However, acetone can extract more interferences from matrix, expecially from samples containing high fat and thus, acetonitrile is used a lot in DPhE extracting from agricutral products. Considering the polarity of DPhE, sorbents such as florisil or alumina are suitable for clean-up steps.

9. Conclusion

Food analysis entails important difficulties owing to the complexity of the sample matrix. Most methods for the analysis of pesticide residues described in the literature use a combination of some form of extraction with an organic solvent, with one or several clean-up and purification steps to remove coextractants before the sample is subjected to a further separation/detection technique.

One of the current trends of modern analytical chemistry is the miniaturization of the various tools daily used by a large number of researchers. Ultrafast separations, consumption of small amounts of both samples and reagents as well as a high sensitivity and automation are some of the most important goals desired to be achieved.

9.1 Sample treatment

Sample treatment has been recognized as the main bottleneck of the analytical process, especially when trace analysis is the purpose. For many years a large number of research laboratories and analytical instrument manufacturing companies have been investing their efforts in this field, which includes miniaturized extraction materials, sample pre-treatment procedures and separation techniques.

Solid-phase microextraction (SPME) is a relatively new technique introduced by Pawliszyn and coworkers in the early 1990s (Janusz, 1997). The feature of this technique is that it enables sample preparation and enrichment in one step. SPME is based on the partitioning of analytes between a coated fibre and a sample. The coated fibre consists of a small fused-silica rod coated with a thin layer of a sorbing material. Upon exposure to the vapour phase above a solution or upon direct immersion in the solution, a mass-transfer process begins, driven by the second law of thermodynamics, according to which the chemical potential of each compound should be equal throughout the system. If the analyte is in the gas phase and the extractant is liquid, dissolution of the gas in the liquid is the main process, and that is governed by Henry's law and Raoult's law. As solubility is the main concept, partition of the analyte between the gas and the liquid phases will take place and all variables affecting it will influence the extraction (Volante et al., 1998).
PLE (pressurized liquid extraction) is another extraction technique recently attracted considerable attention. PLE is a sample preparation technique that combines elevated temperature and pressure with liquid solvents to achieve fast and efficient extraction of the analytes from solid matrices (Marchese et al., 2009). In PLE, the variables that affect extraction efficiency are the nature of the solvent or mixture of solvents, the solvent volume/sample mass ratio, extraction pressure and temperature, the number of extraction cycles and the duration of each cycle. However, the temperature and type of solvent seem to be the two variables with the greatest bearing on the extraction process. The solvents commonly used in pesticide extraction from vegetables and fruit are acetone, n-hexane, ethyl acetate, dichloromethane and water, while those least used are acetonitrile, ethanol and 1-propanol (Nemoto & Lehotay, 1998).
Molecularly imprinted polymers (MIPs) with better specificities than those of traditional SPE adsorbents have recently been introduced as novel matrices for the extraction and clean-up of target compounds (Hu et al., 2010). To date, many papers describing the use of MIPs as SPE materials to clean-up and preconcentrate trace compounds from various matrices have been published (She et al., 2010; Baggiani et al., 2001; Sambe et al., 2007; Mhaka et al., 2009). She et al, 2010 prepared class-specific molecularly imprinted polymers for the selective extraction and determination of sulfonylurea herbicides in maize samples by high-performance liquid chromatography–tandem mass spectrometry.

9.2 Separation system

Among the separation techniques, capillary electromigration methods (which also include capillary electrochromatography, CEC), microchip and nano-LC/capillary LC have received especial attention. Besides their well known advantages over other separation tools, the role of these miniaturized techniques in food analysis is still probably in an early stage. In fact, applications in this field carried out by CEC, microchip, nano-LC and capillary LC are only a few when compared with other more established procedures such as conventional GC or HPLC (Myint et al., 2009).
In the last few years biosensors have shown great potential as analytical tools for the development of rather automatic, fast and direct analysis methods that in many cases avoid sample pretreatment or requireminimal sample preparation, allowing on-site field monitoring (Salmain et al., 2008). For example, an optical fiber based biosensor was developed for atrazine and endrine monitoring in water using *Scenedesmus subspicatus* cells, immobilized on filter paper and covered with a thin alginate layer hardened with calcium chloride(Védrine et al., 2003).

10. References

Albero, B.S., Nchez-brunete, C. & Tadeo, J.L. (2005). Multiresidue determination of pesticides in juice by solid-phase extraction and gas chromatography-mass spectrometry. *Talanta* 66(4), 917-924.

Alferness, P.L. & Iwata, Y. (1994). Determination of glyphosate and (aminomethyl)phosphonic acid in soil, plant and animal matrixes, and water by capillary gas chromatography with mass-selective detection. *Journal of Agricultural and Food Chemistry* 42(12), 2751-2759.

Anisuzzaman, A.K., Amin, M., Ogg, N., Hoq, F., Kanithi, M.R. & Jenkins, R.E. (2000). Synthesis of dimethyl derivatives of imidazolinone herbicides: their use in efficient gas chromatographic methods for the determination of these herbicides. *Journal of Agricultural and Food Chemistry* 48(12), 5893-5902.

Au, A.M. & Fung, W.D. (1988). Rapid technique for thiobencarb residue determination in rice samples. *Bulletin of Environmental Contamination and Toxicology* 40(5), 655-659.

Baggiani, C., Giovannoli, C., Anfossi, L. & Tozzi, C. (2001). Molecularly imprinted solid-phase extraction sorbent for the clean-up of chlorinated phenoxyacids from aqueous samples. *Journal of Chromatography A* 938(1-2), 35-44.

Bailey, R., Lebel, G. & Lawrence, J.F. (1978). Gas-liquid chromatography of triazine herbicides as heptafluorobutyryl derivatives and some applications to analysis in foods. *Journal of Chromatography A* 161, 251-257.

Balduini, L., Matoga, M., Cavalli, E., Seilles, E., Riethmuller, D., Thomassin, M. & Guillaume, Y.C. (2003). Triazinic herbicide determination by gas chromatography-mass spectrometry in breast milk. *Journal of Chromatography B* 794(2), 389-395.

Balinova, A. (1988). Gas chromatographic determination of chloroacetamide herbicides in plants and soil. *Journal of Chromatography A* 455, 391-395.

Battista, M., Di corcia, A. & Marchetti, M. (1989). Extraction and isolation of triazine herbicides from water and vegetables by a double trap tandem system. *Analytical Chemistry* 61(9), 935-939.

Bauer, K.H., Knepper, T.P., Maes, A., Schatz, V. & Voihsel, M. (1999). Analysis of polar organic micropollutants in water with ion chromatography-electrospray mass spectrometry. *Journal of Chromatography A* 837(1-2), 117-128.

Benfenati, E., Natangelo, M. & Tavazzi, S. (2006). *Gas Chromatography/Mass Spectrometry Methods in Pesticide Analysis*. John Wiley & Sons, Ltd.

Bertrand, M.J., Ahmed, A.H., Sarrasin, B. & Mallet, V.N. (1987). Gas chromatographic and mass spectrometric determination of chlorophenoxy acids and related herbicides as their (cyanoethyl)dimethylsilyl derivatives. *Analytical Chemistry* 59(9), 1302-1306.

Blaicher, G., Pfannhauser, W. & Woidich, H. (1980). Problems encountered with the routine application of HPLC to the analysis of carbamate pesticides. *Chromatographia* 13(7), 438-446.

Boyd-boland, A.A. & Pawliszyn, J.B. (1995). Solid-phase microextraction of nitrogen-containing herbicides. *Journal of Chromatography A* 704(1), 163-172.

Brondz, I. & Olsen, I. (1992). Intra-injector formation of methyl esters from phenoxy acid pesticides. *Journal of Chromatography A* 598(2), 309-312.

Bruun, L., Koch, C., Jakobsen, M.H., Pedersen, B., Christiansen, M. & Aamand, J. (2001). Characterization of monoclonal antibodies raised against different structures belonging to the s-triazine group of herbicides. *Analytica Chimica Acta* 436(1), 87-101.

Butz, S. & Stan, H.J. (1993). Determination of chlorophenoxy and other acidic herbicide residues in ground water by capillary gas chromatography of their alkyl esters formed by rapid derivatization using various chloroformates. *Journal of Chromatography A* 643(1-2), 227-238.

Cabras, P.M., Spanedda, M.L., & Tuberoso, C. (1991). High-performance liquid chromatographic determination of dinitroaniline herbicides in soil and water. *Journal of Chromatography A* 585(1), 164-167.

Camire, M.E., Bushway, R.J., Zhao, J., Perkins, B. & Paradis, L.R. (1995). Fate of thiabendazole and chlorpropham residues in extruded potato peels. *Journal of Agricultural and Food Chemistry* 43(2), 495-497.

Catalina, M.I., Dall ge, J., Vreuls, R.J.J. & Brinkman, U.A.T. (2000). Determination of chlorophenoxy acid herbicides in water by in situ esterification followed by in-vial liquid-liquid extraction combined with large-volume on-column injection and gas chromatography-mass spectrometry. *Journal of Chromatography A* 877(1-2), 153-166.

Cessna, A., Benoit, D.L. (1992). Weed control and herbicide residues in onion following use of chlorpropham and cyanazine. *Pesticide Science* 35(4), 355-362.

Cessna, A. (1991a). Residue analysis of garlic (Allium sarivum L.) cloves following a postemergence application of linuron. *Canadian J Plant Sci* 71, 951-955.

Cessna, A.J. (1991b). Residues of triallate in garlic (Allium sativum L.) cloves following preplant incorporation. *Canadian Journal of Plant Science* 71(4), 1257-1261.

Cessna, A.J. (1980). Simultaneous extraction and detection of residues of (2, 4-dichlorophenoxy)acetic acid and bromoxynil from wheat. *Journal of Agricultural and Food Chemistry* 28(6), 1229-1232.

Cessna, A.J. (1990). HPLC determination of linuron residues in asparagus following pre- and early postemergence applications. *Canadian Journal of Plant Science* 70, 591-597.

Cessna, A.J. (1992). Comparison of extraction/hydrolysis procedures for the determination of acidic herbicides in plants: residues of mecoprop in barley following postemergence application. *Journal of Agricultural and Food Chemistry* 40(7), 1154-1157.

Cessna, A.J., Darwent, A.L., Townley-smith, L., Harker, K.N. & Kirkland, K.J. (2002). Residues of glyphosate and its metabolite AMPA in field pea, barley and flax seed following preharvest applications. *Canadian Journal of Plant Science* 82, 485-489.

Cessna, A.J., Holm, F. A. (1993). Residues of 2, 4-d in wheat following application after heading. *Canadian Journal of Plant Science* 74(1), 199-203.

Cessna, A.J., Holt, N.W. & Drew, B.N. (1980). Tolerance and residue studies of triallate in lentils. *Canadian Journal of Plant Science* 60(4), 1283-1288.

Cessna, A.J. & Kerr, L.A. (1993). Use of an automated thermal desorption system for gas chromatographic analysis of the herbicides trifluralin and triallate in air samples. *Journal of Chromatography A* 642(1-2), 417-423.

Chang, S.Y. & Liao, C.H. (2002). Analysis of glyphosate, glufosinate and aminomethylphosphonic acid by capillary electrophoresis with indirect fluorescence detection. *Journal of Chromatography A* 959(1-2), 309-315.

Chu, X.G., Yong, W., Ling, Y., Qiu, W.L., Yao, H.Y. (2008). Herbicides in soybean by ultra performance liquid chromatography tandem mass spectrometry. *Chinese Journal of Analytical Chemistry* 36(3), 325-329.

Cikalo, M.G., Goodall, D.M. & Matthews, W. (1996). Analysis of glyphosate using capillary electrophoresis with indirect detection. *Journal of Chromatography A* 745(1-2), 189-200.

Claudia, F.B.C., Coutinho, L.F.M., Luiz, H.M., Suzana, L.N., Carlos, A.P.C., Fernando, M.L. (2007). Direct determination of glyphosate using hydrophilic interaction chromatography with coulometric detection at copper microelectrode. *Analytica Chimica Acta* 592(1), 30-35.

Clegg, B.S. (1987). Gas chromatographic analysis of fluazifop-butyl (Fusilade) in potatoes, soybeans, and soil. *Journal of Agricultural and Food Chemistry* 35(2), 269-273.

Conkin, R.A., In Zweig, G., Sherma, J. (EDS.), (1978). *Analytical Methods for Pesticides and Plant Growth Regulators.* New York, Academic Press.

Corbera, M., Hidalgo, M., Salvad, V. & Wieczorek, P.P. (2005). Determination of glyphosate and aminomethylphosphonic acid in natural water using the capillary electrophoresis combined with enrichment step. *Analytica Chimica Acta* 540(1), 3-7.

Corti, P., Dreassi, E., Politi, N. & Aprea, C. (1991). Comparison of an HPTLC and an HPLC procedure for the determination of chlorpropham, propham and thiabendazole residues in potatoes. *Food Additives and Contaminants* 8(5), 607 - 615.

Cowell, J.E., Kunstman, J.L., Nord, P.J., Steinmetz, J.R. & Wilson, G.R. (1986). Validation of an analytical residue method for analysis of glyphosphate and metabolite: an interlaboratory study. *Journal of Agricultural and Food Chemistry* 34(6), 955-960.

Cserhati, T. & Forgacs, E. (1998). Phenoxyacetic acids: separation and quantitative determination. *Journal of Chromatography B: Biomedical Sciences and Applications* 717(1-2), 157-178.

Cserhati, T., Forgacs, E., Deyl, Z., Miksik, I. & Eckhardt, A. (2004). Chromatographic determination of herbicide residues in various matrices. pp. 350-359. John Wiley & Sons, Ltd.

Damato, A., Semeraro, I., Bicchi, C. (1993). Simultaneous determination of linuron and trifluralin residues in carrots and their pulp by liquid chromatography and gas chromatography. *Journal of AOAC International* 76(3), 657-662.

Dean, J.R., Wade, G. & Barnabas, I.J. (1996). Determination of triazine herbicides in environmental samples. *Journal of Chromatography A* 733(1-2), 295-335.

Ding, W.H., Liu, C.H. & Yeh, S.P. (2000). Analysis of chlorophenoxy acid herbicides in water by large-volume on-line derivatization and gas chromatography-mass spectrometry. *Journal of Chromatography A* 896(1-2), 111-116.

Ely, C.B., Frans, R.E., Lavy, T.L., Talbert, R.E., Mattice, J.D. (1993). Determining diuron, simazine, and methiocarb residues in highbush blueberries (Vaccinium corymbosum). *HortScience* 28(1), 33-35.

Engebretson, J., Hall, G., Hengel, M. & Shibamoto, T. (2001). Analysis of pendimethalin residues in fruit, nuts, vegetables, grass, and mint by gas chromatography. *Journal of Agricultural and Food Chemistry* 49(5), 2198-2206.

Farran, A. & Ruiz, S. (2004). Application of solid-phase extraction and micellar electrokinetic capillary chromatography to the study of hydrolytic and photolytic degradation of phenoxy acid and phenylurea herbicides. *Journal of Chromatography A* 1024(1-2), 267-274.

Fuentes, E., Baez, M.E., Reyes, D. (2006). Microwave-assisted extraction through an aqueous medium and simultaneous cleanup by partition on hexane for determining pesticides in agricultural soils by gas chromatography: A critical study. *Analytica Chimica Acta* 578(2), 122-130.

Garcia-valcarcel, A.I., Sanchez-brunete, C., Martinez, L. & Tadeo, J.L. (1996). Determination of dinitroaniline herbicides in environmental samples by gas chromatography. *Journal of Chromatography A* 719(1), 113-119.

Gaynor, J.D., Hamill, A.S., Mactavish, D.C. (1993). Efficacy, fruit residues, and soil dissipation of the herbicide metolachlor in processing tomato. *J. Am. Soc. Hort. Sci.* 118, 68-72.

Gaynor, J.D., Mactavish, D.C. & Hamill, A.S. (1992). A GC/MSD method for the analysis of metolachlor in cabbage, broccoli, and tomato. *Communications in Soil Science and Plant Analysis* 23(13), 1549 - 1558.

Goodwin, L., Startin, J.R., Keely, B.J. & Goodall, D.M. (2003). Analysis of glyphosate and glufosinate by capillary electrophoresis-mass spectrometry utilising a sheathless microelectrospray interface. *Journal of Chromatography A* 1004(1-2), 107-119.

Guan, F., Watanabe, K., Ishii, A., Seno, H., Takeshi, K., Hattori, H. & Suzuki, O. (1998). Headspace solid-phase microextraction and gas chromatographic determination of dinitroaniline herbicides in human blood, urine and environmental water. *Journal of Chromatography B: Biomedical Sciences and Applications* 714(2), 205-213.

Guinivan, R.A., Thompson, N.P., Wheeler, W.B. (1982). Derivatization and Cleanup Improvements in Determination of Residues of Glyphosate and Aminomethylphosphonic Acid in Blueberries. *J. Assoc. Off. Anal. Chem.* 65, 35-39.

Gyongyver, H., Ivan, B., Andras, S. (2000). Development of an enzyme-linked immunosorbent assay (ELISA) for the herbicide trifluralin. *Analytica Chimica Acta* 421(2), 121-133.

Jose, F., Pilar, H., Martinez, C.M., Miguel, M. & Flores, P. (2007). Multiresidue method for analysis of pesticides in pepper and tomato by gas chromatography with nitrogen-phosphorus detection. *Food Chemistry* 105(2), 711-719.

Hans-jurgen, S., Manfred, L. (1993). Capillary gas chromatography—atomic emission detection: A useful instrumental method in pesticide residue analysis of plant foodstuffs. *Journal of High Resolution Chromatography* 16(9), 539-548.

Herranz, S., Ramon-azcon, J., Benito-pena, E., Marazuela, M., Marco, M. & Moreno-bondi, M. (2008). Preparation of antibodies and development of a sensitive immunoassay with fluorescence detection for triazine herbicides. *Analytical and Bioanalytical Chemistry* 391(5), 1801-1812.

Hodgeson, J., Collins, J. & Bashe, W. (1994). Determination of acid herbicides in aqueous samples by liquid--solid disk extraction and capillary gas chromatography. *Journal of Chromatography A* 659(2), 395-401.

Hogendoorn, E.A., Ossendrijver, F.M., Dijkman, E. & Baumann, R.A. (1999). Rapid determination of glyphosate in cereal samples by means of pre-column derivatisation with 9-fluorenylmethyl chloroformate and coupled-column liquid chromatography with fluorescence detection. *Journal of Chromatography A* 833(1), 67-73.

Hsu, R.C., Biggs, I. & Saini, N.K. (1991). Solid-phase extraction cleanup of halogenated organic pesticides. *Journal of Agricultural and Food Chemistry* 39(9), 1658-1666.

Hu, X., Dai, G., Huang, J., Ye, T., Fan, H., Youwen, T., Yu, Y. & Liang, Y. (2010). Molecularly imprinted polymer coated on stainless steel fiber for solid-phase microextraction of chloroacetanilide herbicides in soybean and corn. *Journal of Chromatography A* In Press, Accepted Manuscript.

Huang, L.Z., Dai, H., Li, Y.J., Gao, X.L. (2004). Determination of Trifluralin Residues in Garlic by HPLC. *Chinese Food Science* 25(7), 151-153.

Huo, J.L., LI, J., Ge, Y.Q., Chu, X.G. (2006). Determination of dinitroaniline multi-residues in soybean by gas chromatography-mass spectrometry. *Chinese Journal of Analytical Chemistry* 34, S63-S67.

Janusz., P. (1997). *Solid Phase Microextraction Theory and Practice*. NewYork, NyWiley-VCH. Inc.

Khrolenko, M.V. & Wieczorek, P.P. (2005). Determination of glyphosate and its metabolite aminomethylphosphonic acid in fruit juices using supported-liquid membrane preconcentration method with high-performance liquid chromatography and UV detection after derivatization with p-toluenesulphonyl chloride. *Journal of Chromatography A* 1093(1-2), 111-117.

Kim, I.S., Sasinos, F.I., Stephens, R.D., Wang, J. & Brown, M.A. (1991). Determination of chlorinated phenoxy acid and ester herbicides in soil and water by liquid chromatography particle beam mass spectrometry and ultraviolet absorption spectrophotometry. *Analytical Chemistry* 63(8), 819-823.

Konar, S.K. & Roy, D.N. (1990). Method for the determination of residues of the herbicide glyphosate and its principal metabolite, aminomethylphosphonic acid, in plant materials by nitrogen-selective gas chromatography. *Analytica Chimica Acta* 229, 277-280.

Krynitsky, A.J., Swineford, D.M. (1995). Determination of sulfonylurea herbicides in grains by capillary electrophoresis. *Journal of AOAC International* 78(4), 1091-1096.

Kuang, H., Chu, X.G., Hou, Y.X.,, Xu, C.L. (2006a). Simultaneous Determination of 13 Phenoxy Acid Herbicide Residues in Soybean by GC-ECD *Analytical Letters* 39, 2617-2627.

Kuang, H., Hou, Y.X., Chu, X.G., Xu, C.L. (2006b). Simultaneous Determination of Phenoxyacid Herbicides in Soybean by Gas Chromatography-Mass Spectrometry. *Chinese Journal of Analytical Chemistry* 34(12), 1733-1736.

Lawrence, J.F. & laver, G.W. (1974). Analysis of triazine herbicides by combined thin-layer chromatography and fluorimetry. *Journal of Chromatography A* 100(1), 175-179.

Lee, H.B., Te, P., John, M. (1991). Chemical derivation analysis of pesticides residues. Part . an improved method for the determination and confirmation of acidic herbicides in water. *J. Assoc Off. Anal. Chem* 74(3), 835-842.

Lewis, S.E., Brodie, J.E., Bainbridge, Z.T., Rohde, K.W., Davis, A.M., Masters, B.L., Maughan, M., Devlin, M.J., Mueller, J.F. & Schaffelke, B. (2009). Herbicides: a new threat to the great barrier reef. *Environmental Pollution* 157(8-9), 2470-2484.

Li, J.Z., Chu, X., Cai, H.X., An, J., Yang, Q. (2006). Simultaneous determination of residues of 12 amide herbicides in soybeans by high performance liquid chromatography. *Chinese Jour nal of Chromatography* 24(6), 585-588.

Li, J.Z., CuI, X., Cai, H.X., An, J., Yang, Q. (2006). Simultaneous determination of residues of 12 amide herbicides in soybeans by high performance liquid chromatography. *Chinese Jour nal of Chromatogr a phy* 24(6), 585-588.

Lou, X., Miller, D.J. & Hawthorne, S.B. (1999). Static subcritical water extraction combined with anion exchange disk sorption for determining chlorinated acid herbicides in soil. *Analytical Chemistry* 72(3), 481-488.

Lyubavina, I.A., Zinchenko, A.A., Salomatina, I.S., Zherdev, A.V. & Dzantiev, B.B. (2004). An immunochromatographic assay of 2, 4-dichlorophenoxyacetic acid and simazine using monoclonal antibodies labeled with colloidal gold. *Russian Journal of Bioorganic Chemistry* 30(2), 178-183.

Marchese, S., Perret, D., Bafile, E., Gentili, A., Caretti, F. & Berardino, M. (2009). Pressurized liquid extraction coupled with LC–ESI–MS–MS for the determination of herbicides chlormequat and mepiquat in flours. *Chromatographia* 70(5), 761-767.

Martinez Vidal, J.L., Frenich, A.G. & Lehotay, S.J. (2006). Quick, easy, cheap, effective, rugged, and safe approach for determining pesticide residues. In *Pesticide Protocols* (Ed J. M. Walker), pp. 239-261. Humana Press.

Mattern, G.C., Singer, G.M., Louis, J, Robson, M., Rosen, J.D. (1989). Determination of linuron in potatoes using capillary column gas chromatography/mass spectrometry. *Journal of the Association of Official Analytical Chemists* 72(6), 970-974.

Mhaka, B., Cukrowska, E., Tsesumbui, B., Ramstr, M.O., Haupt, K., Tutu, H. & Chimuka, L. (2009). Selective extraction of triazine herbicides from food samples based on a combination of a liquid membrane and molecularly imprinted polymers. *Journal of Chromatography A* 1216(40), 6796-6801.

Miliadis Ge, S.P., Vasilikiotis, G.S. (1990). Simplified cleanup and liquid chromatographic ultraviolet determination of linuron and three metabolites in potatoes. *J Assoc Off Anal Chem.* 73(3), 435-437.

Mondy, N.I., Sharada, D., Munshi, C.B. & Wurm, C.M. (1992). Effect of storage time, temperature, and cooking on isopropyl N-(3-chlorophenyl)carbamate levels in potatoes. *Journal of Agricultural and Food Chemistry* 40(2), 197-199.

Montiel, A.Y., Sánchez, I. (1996). Study of the presences of Simazine residues in fruit and oil obtanined from olives collected from the ground. . *Bol. San. Veg. Plagas,* 22(3), 571-575.

Mortimer, R.D., Black, D.B. & Dawson, B.A. (1994). Pesticide residue analysis in foods by NMR. 3. comparison of 19F NMR and GC-ECD for analyzing trifluralin residues in field-grown carrots. *Journal of Agricultural and Food Chemistry* 42(8), 1713-1716.

Moye, H.A., Miles, C.J. & Scherer, S.J. (1983). A simplified high-performance liquid chromatographic residue procedure for the determination of glyphosate herbicide and (aminomethyl)phosphonic acid in fruits and vegetables employing postcolumn fluorogenic labeling. *Journal of Agricultural and Food Chemistry* 31(1), 69-72.

Mughari, A.R., Vazquez, P.P. & Galera, M.M. (2007). Analysis of phenylurea and propanil herbicides by solid-phase microextraction and liquid chromatography combined with post-column photochemically induced fluorimetry derivatization and fluorescence detection. *Analytica Chimica Acta* 593(2), 157-163.

Murakami, Y., Matsumoto, H., Kuwabara, K., Nishimune, T. & Tanaka, R. (1988). Survey of diphenyl ether herbicides in dietary foods by the total diet study in Osaka, Japan. *Bulletin of Environmental Contamination and Toxicology* 40(1), 69-73.

Myint, K.T., Uehara, T., Aoshima, K. & Oda, Y. (2009). Polar Anionic Metabolome Analysis by Nano-LC/MS with a Metal Chelating Agent. *Analytical Chemistry* 81(18), 7766-7772.

Nartova, Y., Ermolaeva, T., Fleisher, M., Abuknesha, R. & Eremin, S. (2008). Determination of Alachlor in food products by a fluorescence polarization immunoassay. *Journal of Analytical Chemistry* 63(5), 499-505.

National foodsafety resource database. beijing www.fsr.org.cn

Neitzel, P.L., Walther, W. & Nestler, W. (1998). In-situ methylation of strongly polar organic acids in natural waters supported by ion-pairing agents for headspace GC-MSD analysis. *Fresenius' Journal of Analytical Chemistry* 361(3), 318-323.

Nemoto, S. & Lehotay, S.J. (1998). Analysis of multiple herbicides in soybeans using pressurized liquid extraction and capillary electrophoresis. *Journal of Agricultural and Food Chemistry* 46(6), 2190-2199.

Nilsson, T., Baglio, D., Galdo-migue, Z., Madsen, J. & Facchetti, S. (1998). Derivatisation/solid-phase microextraction followed by gas chromatography-mass spectrometry for the analysis of phenoxy acid herbicides in aqueous samples. *Journal of Chromatography A* 826(2), 211-216.

Ohms, R.E. (1976). Minutes Potato Pathology Section PAA Chicago, Ill. Feb. 18, 1976. *American Journal of Potato Research* 53(12), 463-470.

Ortiz-gomez, M.T., Perez-arribas, L.V., Leon-gonzalez, M.E. & Polo-diez, L.M. (1995). Liquid Chromatography Determination of Simazine and Antimycin A in Must. *Journal of Agricultural and Food Chemistry* 43(11), 2883-2886.

Pardue, J. (1995). Multiresidue method for the chromatographic determination of triazine herbicides and their metabolites in raw agricultural products. *J AOAC Int.* 78(3), 856-862.

Patsias, J., Papadopoulou, A. & Papadopoulou-mourkidou, E. (2001). Automated trace level determination of glyphosate and aminomethyl phosphonic acid in water by on-line anion-exchange solid-phase extraction followed by cation-exchange liquid chromatography and post-column derivatization. *Journal of Chromatography A* 932(1-2), 83-90.

Peruzzi, M., Bartolucci, G. & Cioni, F. (2000). Determination of phenoxyalkanoic acids and other herbicides at the ng/ml level in water by solid-phase extraction with poly(divinylbenzene-co-N-vinylpyrrolidone) sorbent and high-performance liquid chromatography-diode-array detection. *Journal of Chromatography A* 867(1-2), 169-175.

Prez, S., Matienzo, M. & Tadeo, J. (1993). GC-ITD analysis of chlortoluron and its metabolites in cereals. *Chromatographia* 36(1), 195-200.

Pringle, J.C., Anderson, L.W.J. & Raines, R.W. (1978). Residues in crops irrigated with water containing simazine. *Journal of Agricultural and Food Chemistry* 26(5), 1143-1147.

Pylypiw, H.M., Bugbee, G.J. & Frink, C.R. (1993). Uptake of pre-emergent herbicides by corn: Distribution in plants and soil. *Bulletin of Environmental Contamination and Toxicology* 50(3), 474-478.

Qi, Y., Li, S., Zhan, C.R., Peng, T. (2004). Simultaneous determination of sulfonylurea herbicides residues in soybeans by high performance liquid chromatography-mass spectrometry. *Chinese Journal of Analytical Chemistry* 32(11), 1436-1440.

Richman, S.J., Karthikeyan, S., Bennett, D.A., Chung, A.C. & Lee, S.M. (1996). Low-level immunoassay screen for 2, 4-dichlorophenoxyacetic acid in apples, grapes, potatoes, and oranges: circumventing matrix effects. *Journal of Agricultural and Food Chemistry* 44(9), 2924-2929.

Rimmer, D.A., Johnson, P.D. & Brown, R.H. (1996). Determination of phenoxy acid herbicides in vegetation, utilising high-resolution gel permeation chromatographic clean-up and methylation with trimethylsilydiazomethane prior to gas chromatographic analysis with mass-selective detection. *Journal of Chromatography A* 755(2), 245-250.

Rochette, E.A., Harsh, J.B. & Hillijr, H.H. (1993). Supercritical fluid extraction of 2, 4-D from soils using derivatization and ionic modifiers. *Talanta* 40(2), 147-155.

Rompa, M.K.E., Zygmunt, B. (2005). Screening of alkanocarboxylic and phenolicherbicides in water samples by meansof derivatization-based gas chromatography. *Polish Journal of Environmental Studies* 14(1), 81-86.

Roseboom, H. & Herbold, H.A. (1980). Determination of triazine herbicides in various crops by capillary gas chromatography with thermionic detection. *Journal of Chromatography A* 202(3), 431-438.

Ruiz De, E.N., Goicolea, M.A., Gomez debalugera, Z., Portela, M.J. & Barrio, R.J. (1997). Determination of herbicides by reductive amperometric detection in liquid chromatography. *Journal of Chromatography A* 763(1-2), 227-235.

Sabino, A., Bufo, M.D.A., Laura, S., Roberto, T. (2004). Determination of herbicide residues by laser microprobe mass analysis (Lamma). *International Journal of Environmental Analytical Chemistry* 84(1), 39 - 45.

Salmain, M., Fischer-durand, N. & Pradier, C.M. (2008). Infrared optical immunosensor: Application to the measurement of the herbicide atrazine. *Analytical Biochemistry* 373(1), 61-70.

Sambe, H., Hoshina, K. & Haginaka, J. (2007). Molecularly imprinted polymers for triazine herbicides prepared by multi-step swelling and polymerization method: Their application to the determination of methylthiotriazine herbicides in river water. *Journal of Chromatography A* 1152(1-2), 130-137.

Sanchez-brunete, C., Garcia-valcircel, A.I. & Tadeo, J.L. (1994). Determination of residues of phenoxy acid herbicides in soil and cereals by gas chromatography-ion trap detection. *Journal of Chromatography A* 675(1-2), 213-218.

Sanchez-brunete, C. Martinez, L. & Tadeo, J. L. (1994). Determination of corn herbicides by GC-MS and GC-NPD in environmental samples. *Journal of Agricultural and Food Chemistry* 42(10), 2210-2214.

Sanchez-brunete, C., Pirez, R.A., Miguel, E. & Tadeo, J.L. (1998). Multiresidue herbicide analysis in soil samples by means of extraction in small columns and gas chromatography with nitrogen-phosphorus and mass spectrometric detection. *Journal of Chromatography A* 823(1-2), 17-24.

Sancho, J.V. F. J., López, F., Hernández, F., Hogendoorn, E.A. & Van zoonen, P. (1994). Rapid determination of glufosinate in environmental water samples using 9-fluorenylmethoxycarbonyl precolumn derivatization, large-volume injection and coupled-column liquid chromatography. *Journal of Chromatography A* 678(1), 59-67.

Sauret-szczepanski, N., Mirabel, P. & Wortham, H. (2006). Development of an SPME-GC-MS/MS method for the determination of pesticides in rainwater: Laboratory and field experiments. *Environmental Pollution* 139(1), 133-142.

Shackelford, D.D., Ma, R., West, S.D., Turner, L.G. (2000). Determination of ethalfluralin in canola seed, meal, and refined oil by capillary gas chromatography with mass selective detection. *J Agric Food Chem.* 48(9), 4422-4427.

She, Y.X., Cao, W.Q., Shi, X.M., Lv, X.L., Liu, J.J., Wang, R.Y., Jin, F., Wang, J. & Xiao, H. (2010). Class-specific molecularly imprinted polymers for the selective extraction and determination of sulfonylurea herbicides in maize samples by high-performance liquid chromatography-tandem mass spectrometry. *Journal of Chromatography B* 878(23), 2047-2053.

Shen, W.J., Xu, J., Yang, W.Q., Shen, C.Y. (2007). Determination of acetanilide herbicide residues in tea by gas chromatograph- Mass Spectrometry with two different ionization techniques *Chinese Journal of Chromatography* 25(5), 753-757.

Shen, W.J., Xu, J.Z., Zhao, Z.Y., Ding, T., Jiang, Y., Chu, X.G., Shen, C.Y. (2008). Determination of 11 kinds of ether herbicide residues in vegetables by gas chromatography-negative chemical ionization mass spectrometry. *Chinese Journal of Analytical Chemistry* 36(5), 663-667.

Siltanen, H.R.C. (1978). Analysis of 2, 4-D and 2, 4, 5-T in lingonberries, wild mushrooms, birch and aspen foliage. *Bull Environ Contam Toxicol.* 19(2), 177-182.

Singh, S.B., Yaduraju, N.T., Kulshrestha, G. (1997). Residues of metolachlor herbicide in soil and potato tubers under Indian tropical conditions. *Bulletin of Environmental Contamination and Toxicology* 59(2), 216-221.

Slates, R.V. (1983). Determination of chlorsulfuron residues in grain, straw, and green plants of cereals by high-performance liquid chromatography. *Journal of Agricultural and Food Chemistry* 31(1), 113-117.

Smith, E. (1984). Gas chromatographic method for analysis of 2, 4-D in wheat: interlaboratory study. *J Assoc Off Anal Chem* 67(4), 794-798.

Sojo, L.E., Gamble, D.S. & Gutzman, D.W. (1997). Sorption and Bound Residue Formation of Linuron, Methylparathion, and Metolachlor by Carrot Tissues: Kinetics by On-Line HPLC Microextraction. *Journal of Agricultural and Food Chemistry* 45(9), 3634-3641.

Striley, C.A.F., Biagini, R.E., Mastin, J.P., Mackenzie, B.A. & Robertson, S.K. (1999). Development and validation of an ELISA for metolachlor mercapturate in urine. *Analytica Chimica Acta* 399(1-2), 109-114.

Su, K. (1975). Electron capture gas-liquid chromatographic method for the simultaneous analysis of 2, 4-D, dicamba, and mecoprop residues in soil, wheat, and barley. *J Assoc Off Anal Chem.* 58(5), 1027-1031.

Szendr, S.A.A., Trummer, N.A., Danyi, N., Varadi, M. & Szendro, I. (2003). Development of a non-labeled immunosensor for the herbicide trifluralin via optical waveguide lightmode spectroscopic detection. *Analytica Chimica Acta* 487(1), 31-42.

Tadeo, J.L., Sanchez-brunete, C., Perez, R.A., Fernandez, M.D. (2000). Analysis of herbicide residues in cereals, fruits and vegetables *Journal of Chromatography A* 882, 175-191.

Takats, Z., Vargha, M., Vekey, K. (2001). Investigation of atrazine metabolism in river sediment by high performance liquid chromatography/mass spectrometry. *Rapid Communications in Mass Spectrometry* 15, 1735-1742.

Tanabe, A., Mitobe, H., Kawata, K. & Sakai, M. (1996). Monitoring of herbicides in river water by gas chromatography-mass spectrometry and solid-phase extraction. *Journal of Chromatography A* 754(1-2), 159-168.

Tessier, D.M. & Marshall Clark, J. (1998). An enzyme immunoassay for mutagenic metabolites of the herbicide alachlor. *Analytica Chimica Acta* 376(1), 103-112.

Ting, K.C., Kho, P.K. (1998). *Determination of trace 2, 4-dichlorophenoxyacetic acid in fresh produce by gas chromatography with boron trichloride/2-chloroethanol derivatization.* Gaithersburg, MD, ETATS-UNIS: AOAC International.

Tomkins, B.A. & Ilgner, R.H. (2002). Determination of atrazine and four organophosphorus pesticides in ground water using solid phase microextraction (SPME) followed by gas chromatography with selected-ion monitoring. *Journal of Chromatography A* 972(2), 183-194.

Tong, L., Ma, X.D. & Li, C.J. (2006). Application of gas chromatography-tandem mass spectrometry (GC-MS-MS) with pulsed splitless injection for the determination of multiclass pesticides in vegetables. *Analytical Letters* 39(5), 985 - 996.

Topuz, S., Ozhan, G. & Alpertunga, B. (2005). Simultaneous determination of various pesticides in fruit juices by HPLC-DAD. *Food Control* 16(1), 87-92.

Thornton, J.S. & Stanley, C.W. (1977). Gas chromatographic determination of Sencor and metabolites in crops and soil. *Journal of Agricultural and Food Chemistry* 25(2), 380-386.

Tseng, S.H., LO, Y.W., Chang, P.C., Chou, S.S. & Chang, H.M. (2004). Simultaneous Quantification of Glyphosate, Glufosinate, and Their Major Metabolites in Rice and Soybean Sprouts by Gas Chromatography with Pulsed Flame Photometric Detector. *Journal of Agricultural and Food Chemistry* 52(13), 4057-4063.

Tsui, M. T. K. Wang, W. -X. & Chu, L. M. (2005). Influence of glyphosate and its formulation (Roundup? on the toxicity and bioavailability of metals to Ceriodaphnia dubia. *Environmental Pollution* 138(1), 59-68.

Tsumura-hasegawa, Y., Tonogai, Y., Nakamura, Y. & Ito, Y. (1992). Residue levels of dichlorvos, chlorpropham, and pyrethrins in postharvest-treated potatoes during storage or processing into starch. *Journal of Agricultural and Food Chemistry* 40(7), 1240-1244.

Vadrine, C., Leclerc, J.C., Durrieu, C. & Tran-minh, C. (2003). Optical whole-cell biosensor using Chlorella vulgaris designed for monitoring herbicides. *Biosensors and Bioelectronics* 18(4), 457-463.

Vasilescu, M.N., Medvedovici, A.V., Paul, W., Alan, T. & Colin, P. (2005). Herbicides. In *Encyclopedia of Analytical Science* pp. 243-260. Oxford: Elsevier.

Vitali, P., Venturini, E., Bonora, C., Calori, R. & Raffaelli, R. (1994). Determination of triazines and dinitroanilines in waters by high-performance liquid chromatography after solid-phase extraction. *Journal of Chromatography A* 660(1-2), 219-222.

Volante, M., Cattaneo, M., Bianchi, M. & Zoccola, G. (1998). Some applications of solid phase micro extraction (SPME) in the analysis of pesticide residues in food. *Journal of Environmental Science and Health, Part B: Pesticides, Food Contaminants, and Agricultural Wastes* 33(3), 279 - 292.

Wei, F., Li, D.Z., Yu, L. (2005). Simultaneous determination of phenoxy acid herbicides in rice. *Chinese Journal of Instrumental Analysis* 24, 304-308.

Wen, X., Fei, J., Chen, X., Yi, L., Ge, F. & Huang, M. (2008). Electrochemical analysis of trifluralin using a nanostructuring electrode with multi-walled carbon nanotubes. *Environmental Pollution* 156(3), 1015-1020.

West, S.D., Weston, J.H., Day, E.W.Jr. (1988). Gas chromatographic determination of residue levels of the herbicides trifluralin, benefin, ethalfluralin, and isopropalin in soil with confirmation by mass selective detection. *J Assoc Off Anal Chem.* 71(6), 1082-1085.

Wigfield, Y.Y. & Lanouette, M. (1991). A modified clean-up for the determination of glyphosate and its metabolite residues in lentils using high pressure liquid chromatography and post-column fluorogenic labelling. pp. 491-498. John Wiley & Sons, Ltd.

Wigfield, Y.Y. & Lanouette, M. (1993). Supercritical fluid extraction of the fortified residues of fluazifop-P-butyl (Fusilade II) and its major metabolite, fluazifop-P, in onions. *Journal of Agricultural and Food Chemistry* 41(1), 84-88.

Wigfielday, L.M. (1991). Residue analysis of glyphosate and its principal metabolite in certain cereals, oilseeds, and pulses by liquid chromatography and postcolumn fluorescence detection. *J Assoc Off Anal Chem.* 74(5), 842-847.

Williams, K.J., James, C.R., Thorpe, S.A. & Reynolds, S.L. (1997). Two analytical methods for the measurement of 2, 4-d in oranges: an ELISA screening procedure and a GC-MS confirmatory procedure. pp. 135-140. John Wiley & Sons, Ltd.

Wilson, A.M., Bushway, A.A. & Bushway, R.J. (1981). Residue analysis of isopropyl N-(3-chlorophenyl)carbamate in fruits and vegetables using high-performance liquid chromatography. *Journal of Agricultural and Food Chemistry* 29(4), 746-749.

Wittmann, C. & Hock, B. (1993). Analysis of atrazine residues in food by an enzyme immunoassay. *Journal of Agricultural and Food Chemistry* 41(9), 1421-1425.

Wong, J.W., Webster, M.G., Halverson, C.A., Hengel, M.J., Ngim, K.K. & Ebeler, S.E. (2003). Multiresidue pesticide analysis in wines by solid-phase extraction and capillary gas chromatography-mass spectrometric detection with selective ion monitoring. *Journal of Agricultural and Food Chemistry* 51(5), 1148-1161.

Wortberg, M., Kreissig, S.B., Jones, G., Rocke, D.M. & Hammock, B.D. (1995). An immunoarray for the simultaneous determination of multiple triazine herbicides. *Analytica Chimica Acta* 304(3), 339-352.

Xing, Q., Xi, R., Zhou, Z. (2002). *Basic organic chemistry*. Beijing: Higher Education Press.

Yaduraju, N.T.K.G., Sharma, R.P., Ahuja, K.N. (1993). Isoproturon for weed control in potato (Solanum tuberosum) and its residue in soil and tubers. *Indian Journal of Agricultural Sciences* 63(11), 731-773.

Zhao, B., Zhang, J., Gong, J., Zhang, H. & Zhang, C. (2009). Glyphosate mobility in soils by phosphate application: Laboratory column experiments. *Geoderma* 149(3-4), 290-297.

Zhou, M., Bramble, Q.F., Devine, J.T. & Norwood, I.G. (1996). *Determination of bensulfuron methyl in rice and crayfish by liquid chromatography with a column- and eluant-switching system and UV detection*. Gaithersburg, MD, ETATS-UNIS: AOAC International.

Zhou, M., Li, G.Y. & Whalen, A.S. (1994). Determination of metsulfuron methyl and its two metabolites in crops by liquid chromatography with ultraviolet detection. Gaithersburg, MD, ETATS-UNIS: AOAC International.

14

Enantioseparation and Enantioselective Analysis of Chiral Herbicides

Lixia Jin[1], Weiliang Gao[1], Ling Li[1], Jing Ye[2],
Chunmian Lin[1] and Weiping Liu[1,2]
[1]*Research Center of Environmental Science,*
College of Biological and Environmental Engineering,
Zhejiang University of Technology, Hangzhou 310032
[2]*Ministry of Education Key Laboratory of*
Environmental Remediation and Ecosystem Health,
College of Environmental and Resource Sciences,
Zhejiang University, Hangzhou 310029
China

1. Introduction

Many commercial agrochemicals in current use contain chiral structures and thus consist of enantiomers. Here chiral herbicide is one of the most important agrochemicals which are widely used. Enantiomers of a chiral compound have identical physical-chemical properties and appear as a single compound in standard analysis. However, the biological effects of enantiomers such as toxicity, mutagenicity, carcinogencity, and endocrine disruption activity, are generally different, due to the inherent enantioselectivity of biological interactions.

According to the chemical structure, the familiar chiral herbicides have been classified with amide herbicides, phenoxy herbicides, imidazolinone herbicides, organophosphorus herbicides and so on. The analysis and preparation of pure enantiomer herbicides have been summarized with HPLC, GC, CE and SFC methods. Finally, information concerning the stereoselective toxicity and degradation of chiral herbicides in environmental behavior has been offered.

2. Classification of chiral herbicides

2.1 Amide herbicides

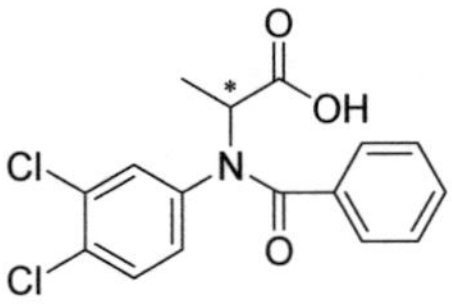

benzoylprop

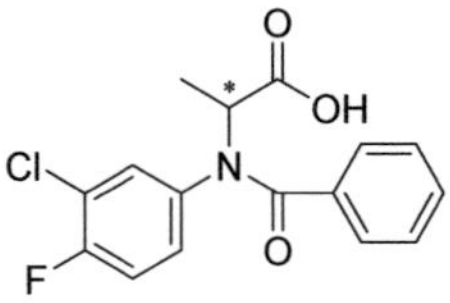

flamprop

arylalanine herbicides

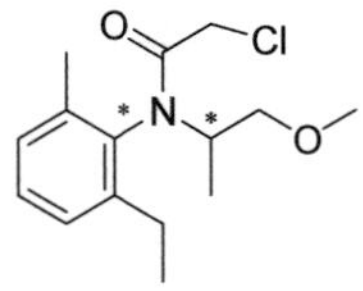

metolachlor

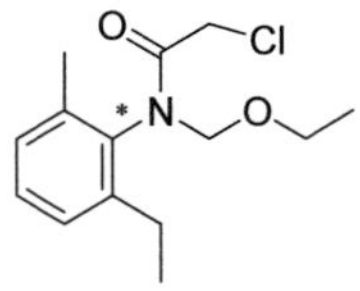

acetochlor

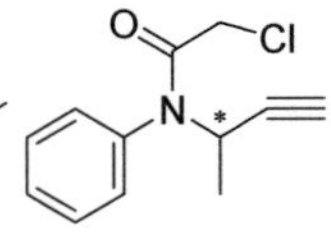

prynachlor

chloroacetanilide herbicides

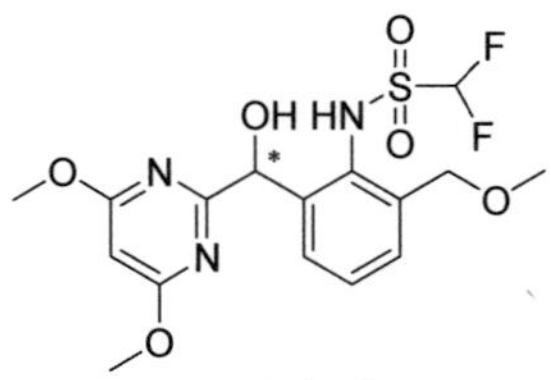

pyrimisulfan

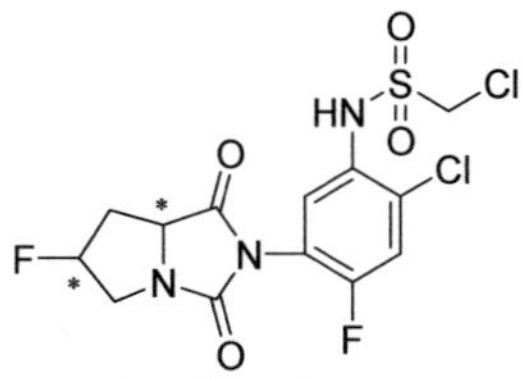

profluazol

sulfonanilide herbicides

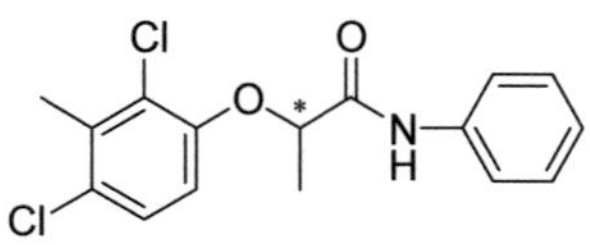

clomeprop

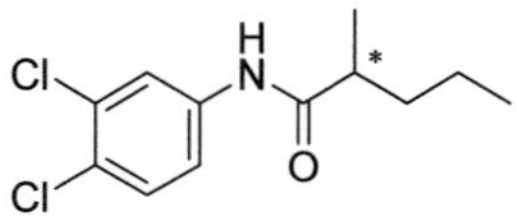

(RS)-3′,4′-dichloro-2-methylvaleranilide

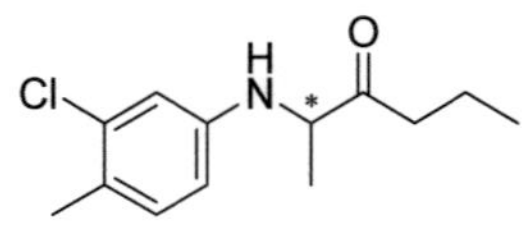

pentanochlor

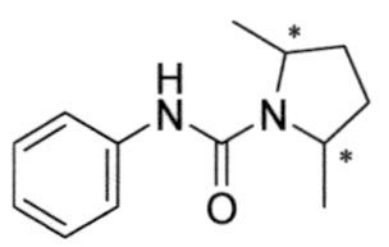

cisanilide

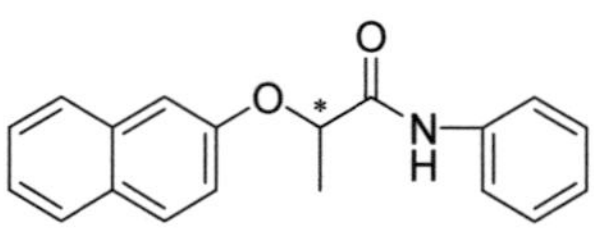

naproanilide

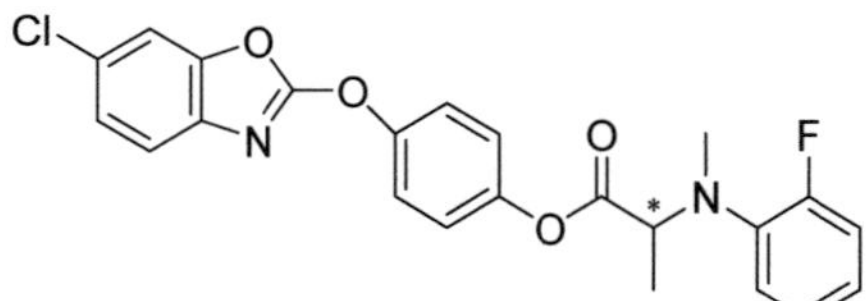

metamifop

anilide herbicides

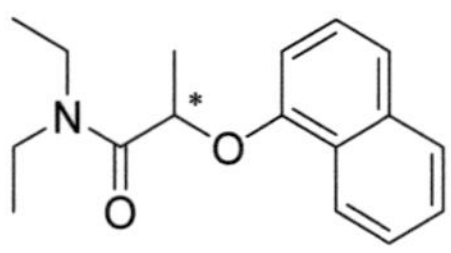

napropamide

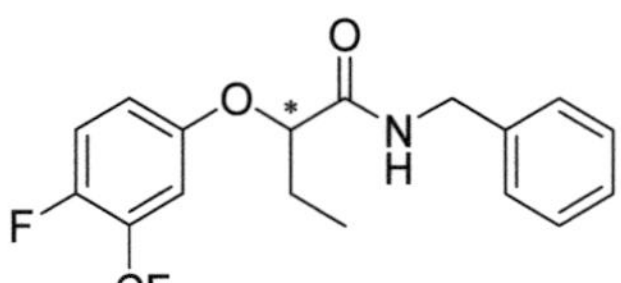

beflubutamid

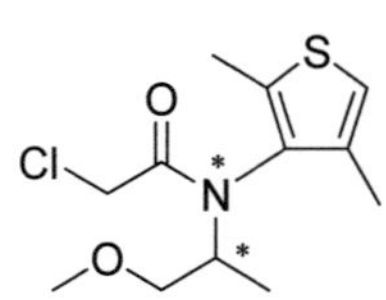

dimethenamid

2.2 Phenoxy herbicides

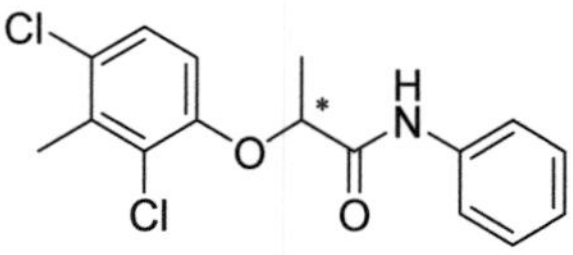
clomeprop

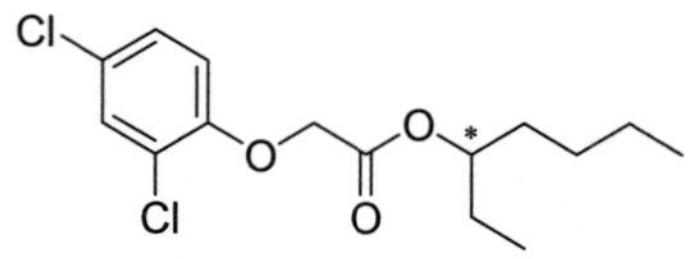
2,4-D-ethylhexyl

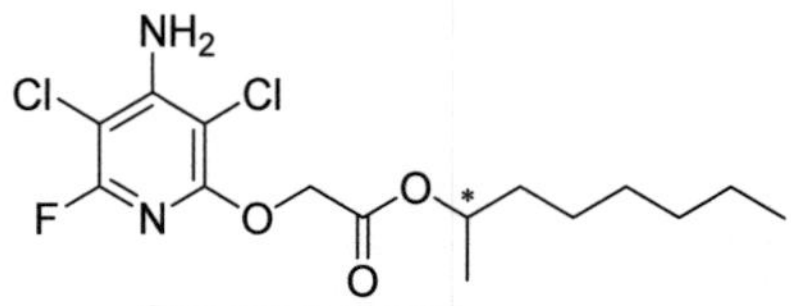
fluroxypyr-meptyl

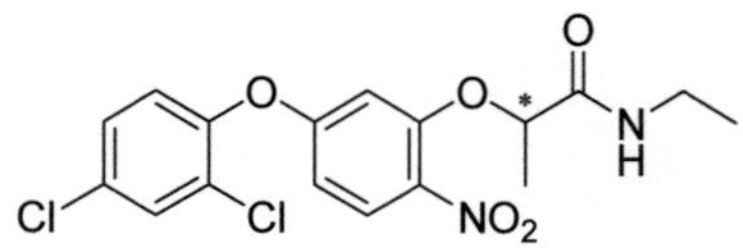
etnipromid

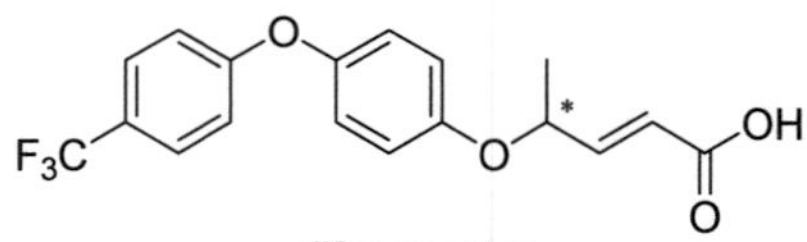
difenopenten

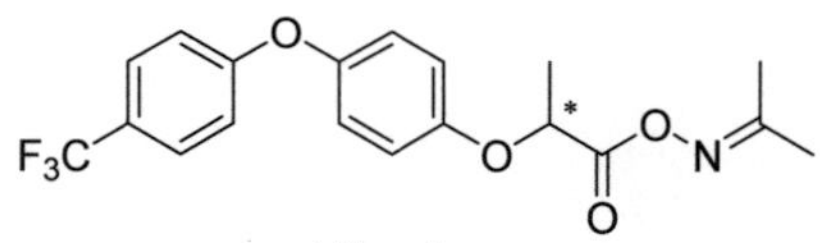
trifopsime

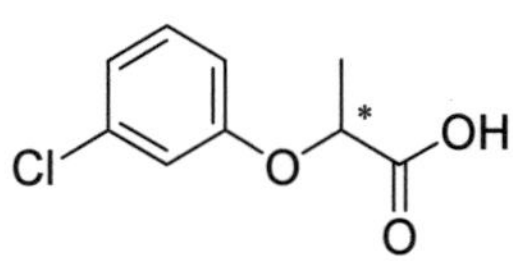
cloprop

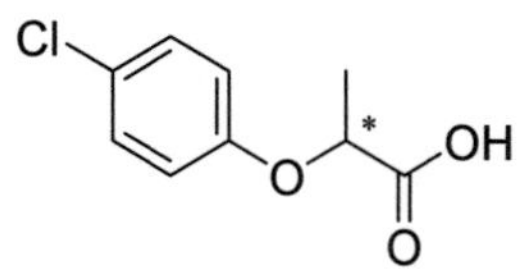
4-CPP

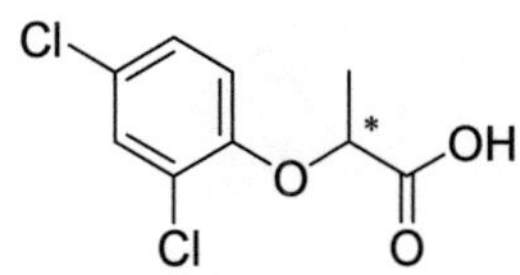
dichlorprop

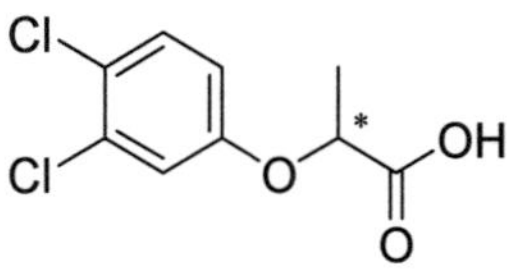
3,4-DP

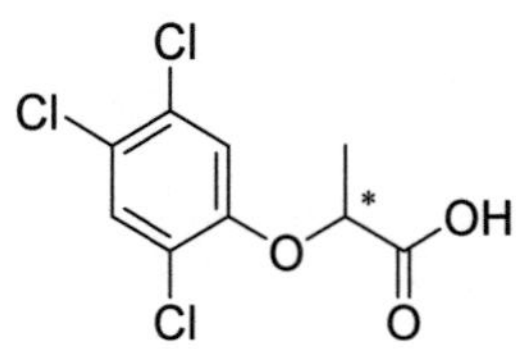
fenoprop

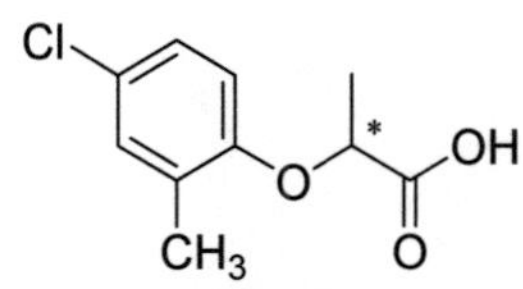
mecoprop

phenoxypropionic herbicides

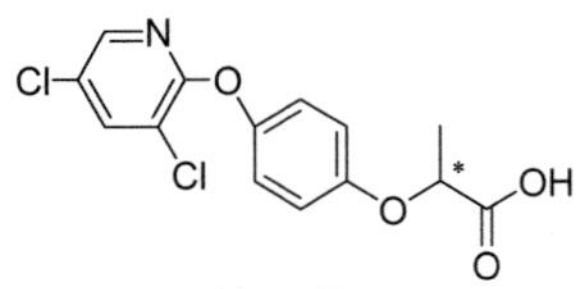
chlorazifop

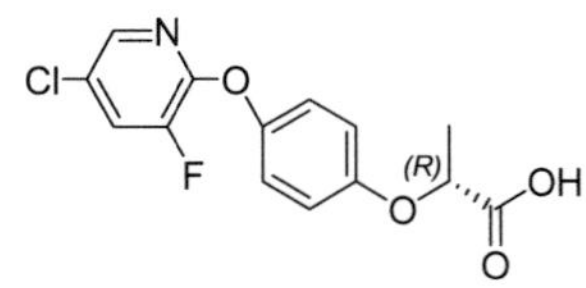
clodinafop

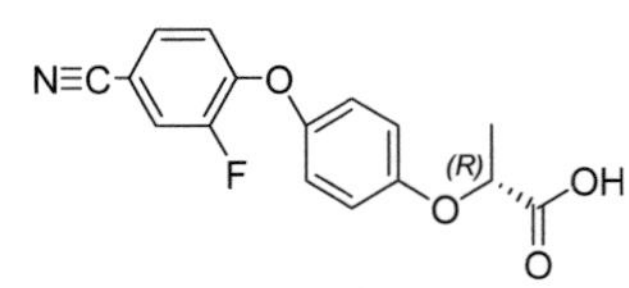
cyhalofop

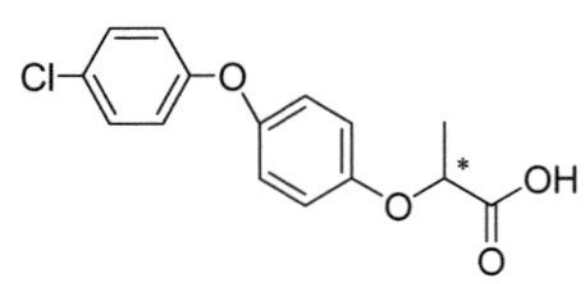
clofop

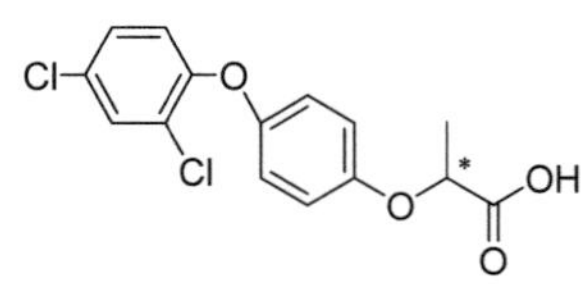
diclofop

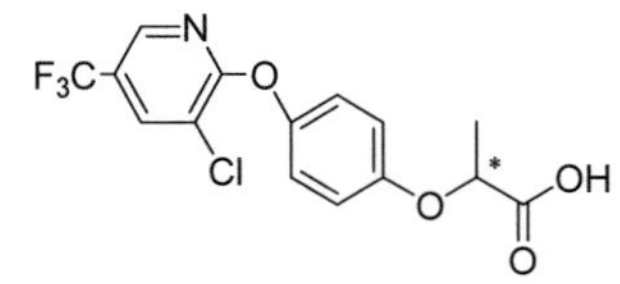
haloxyfop

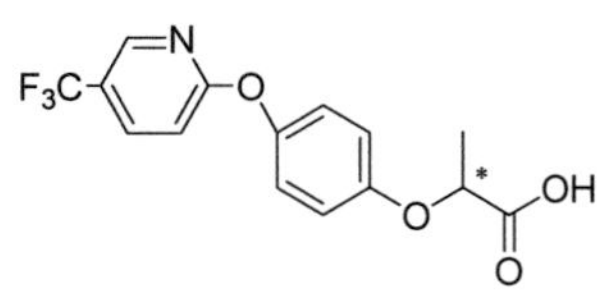
fluazifop

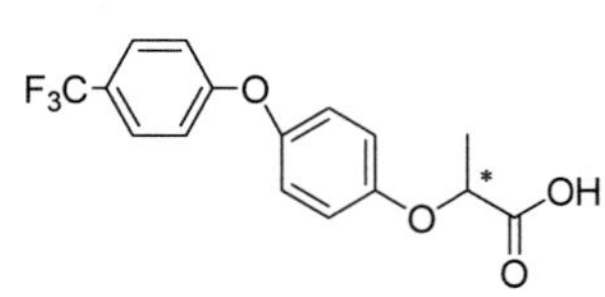
trifop

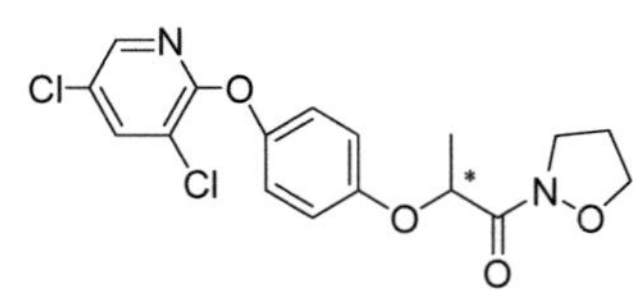
isoxapyrifop

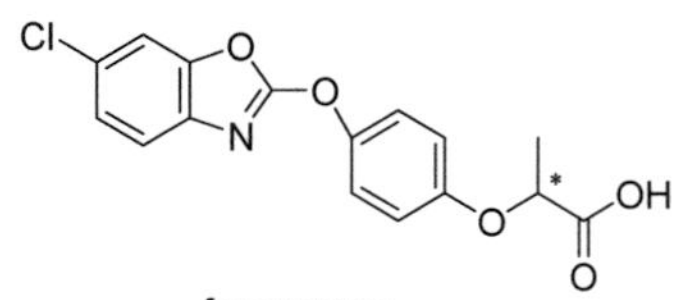
fenoxaprop

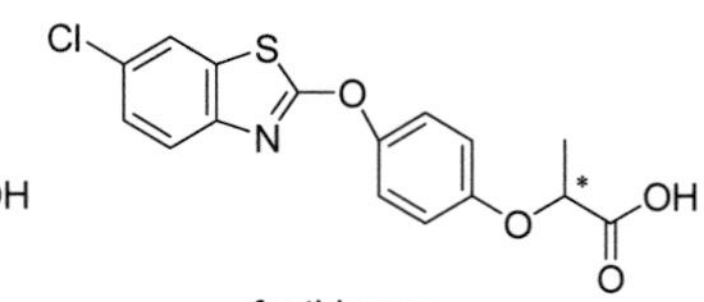
fenthiaprop

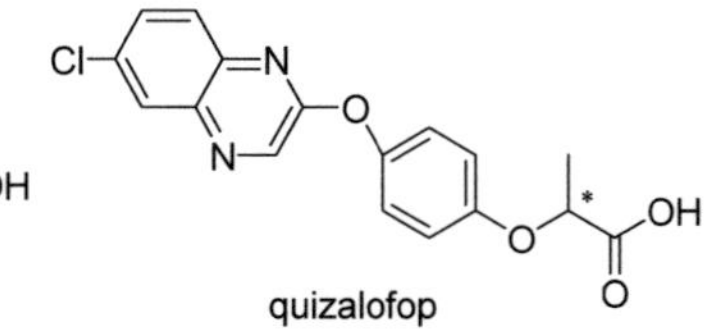
quizalofop

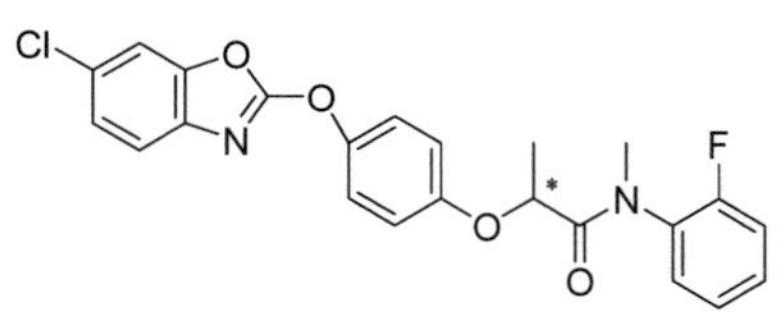
metamifop

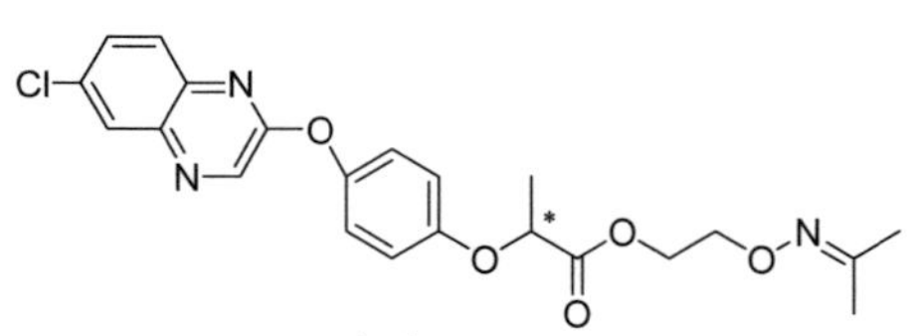
propaquizafop

aryloxyphenoxypropionic herbicides

2.3 Imidazolinone herbicides

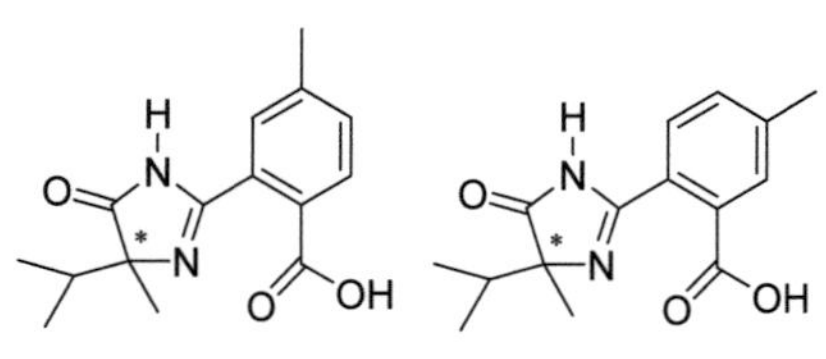
imazamethabenz

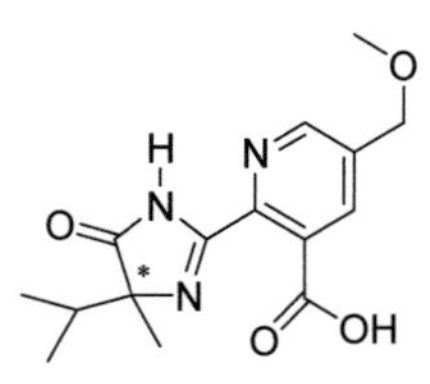
imazamox

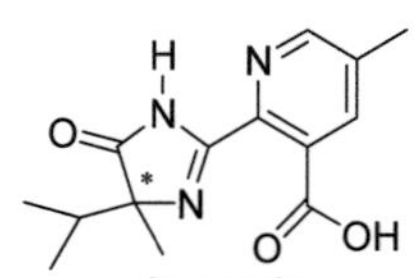
imazapic

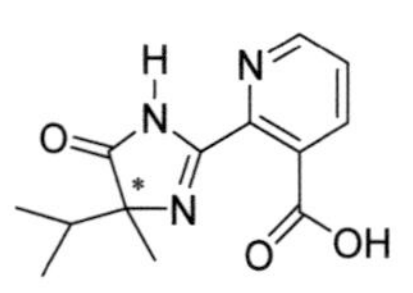
imazapyr

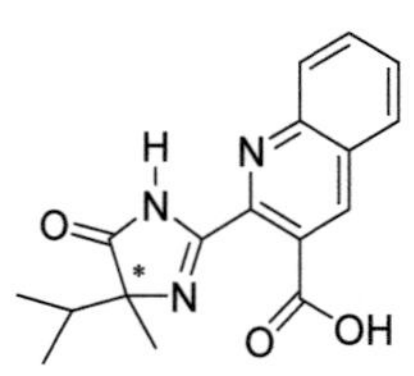
imazaquin

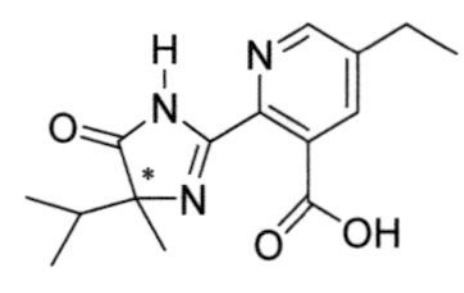
imazethapyr

2.4 Organophosphorus herbicides

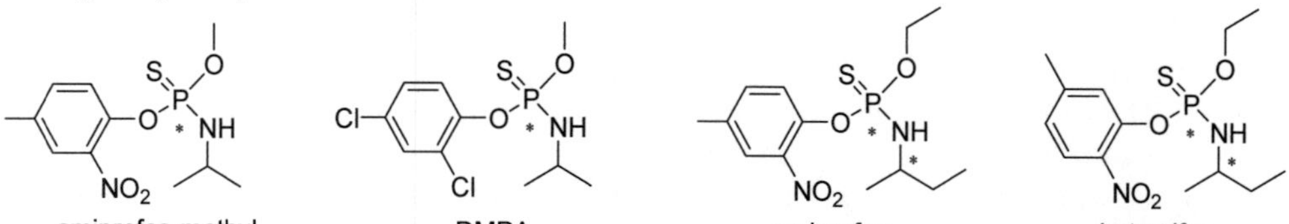

amiprofos-methyl DMPA amiprofos butamifos

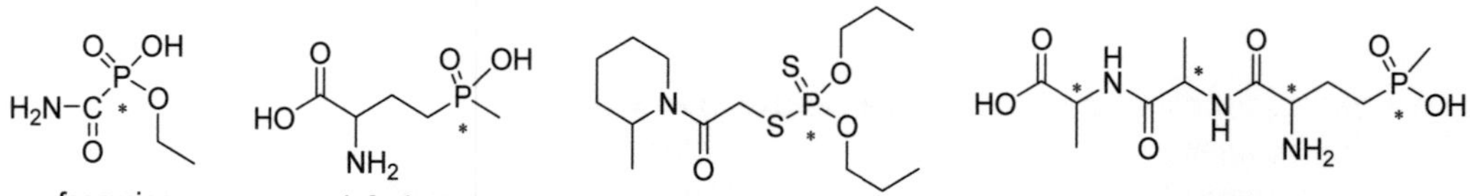

fosamine glufosinate piperophos bilanafos

2.5 Diphenyl ether herbicides

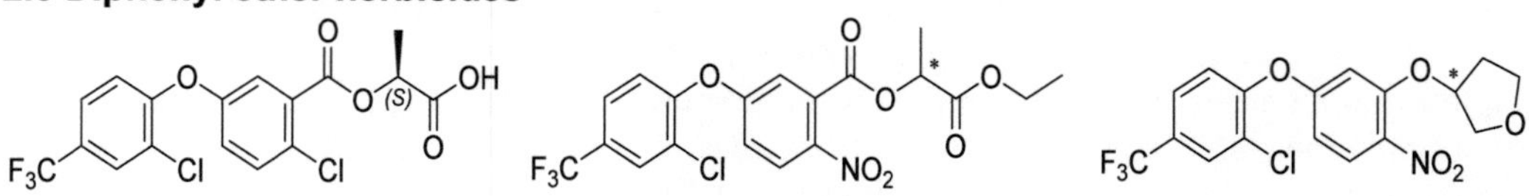

ethoxyfen lactofen furyloxyfen

2.6 Other herbicides

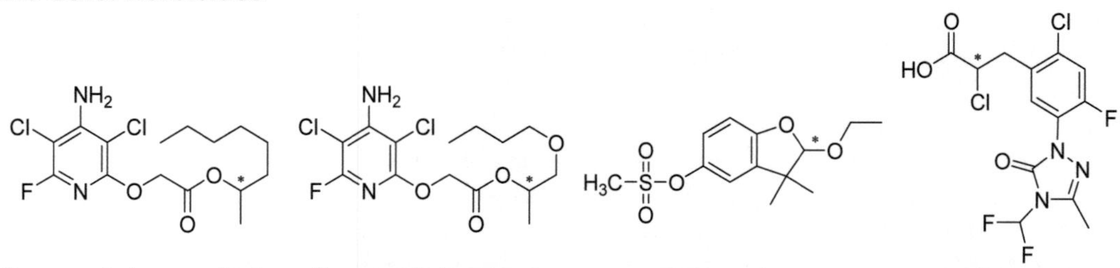

fluroxypyr butoxypropyl ester fluroxypyr-butoxypropyl ethofumesate carfentrazone

3. Chromatographic methods for chiral herbicides

3.1 Separation of chiral herbicides by HPLC

HPLC combined with kinds of CSPs is one of the most common and easily obtained approaches for enantiomer analysis and preparation. Today, CSPs have been developed at least seven classes, including Pirkle-type CSPs, polysaccharides CSPs, cyclodextrins CSPs, macrocyclic glycopeptide antibiotics CSPs, proteins CSPs, crown ethers CSPs and ligand exchange CSPs, etc. Profiting from the development of CSPs, chiral HPLC methods have held the balance both for determining optical purity of enantiomers and for preparing enantiopure standards.

A group of herbicides, diclofop-methyl, quizalofop-ethyl, lactofen, fluroxypyr-meptyl, acetochlor, ethofumesate, clethodim, napropamide, fenoxaprop-ethyl and carfentrazone-ethyl, were partial or near-baseline separated on self-prepared amylose tris-(S)-1-phenylethylcarbamate CSP by HPLC with η-hexane/isopropanol as mobile phase (Wang *et al.*, 2006).

Chiral pyrazole phenyl ethers (PPE) are highly active herbicides which were resolved by direct HPLC on commercially available CSPs derived from N-3,5-dinitrobenzoyl derivatives of α-amino acids or amines (Whelk-O 1). Chromatographic resolution obtained was suitable for determination of enantiomeric purities and, in some cases, for preparative resolution of the enantiomers with ee>99% (Hamper *et al.*, 1994). (+)- and (-)-enantiomers of thiobencarb sulfoxide were collected with purities more than 99.0% ee and 99.8% ee on a Chiralcel OB column at 25 °C, 1 mL/min 95/2.5/2.5 hexane/EtOH/MeOH as a mobile phase (Kodama *et al.*, 2002).

3.1.1 Enantioseparation of amide herbicides by HPLC

Amide herbicides are a group of important chiral herbicides, and metolachlor, which contains two chiral elements (an asymmetrically substituted carbon and a chiral axis), consists of four stereoisomers stable at ambient temperature with aSS-, aRS-, aSR-, and aRR-configurations. Two of the four metolachlor isomers were isolated from rac-metolachlor in enantio- (ee>98%) and diastereomerically pure forms by a combination of achiral Hypercarb PH and chiral chiralcel OD-H HPLC with 98/2 n-hexane/IPA. The enantiomer elution sequence is aS prior to aR (retention times, aSS<aRS and aSR<aRR) and 1′S prior to 1′R (retention times, aSS<aSR and aRS<aRR) (Muller *et al.*, 2001). Baseline separation of four metolachlor isomers by HPLC was achieved on Chiralcel OD-H using 91/9 Hex/diethyl ether as the mobile phase by Polcaro *et al.* (Polcaro *et al.*, 2004). Enantiomers and diastereomers of some acetamide pesticides, alachlor, acetochlor, metolachlor, and dimethenamid, were separated using achiral and chiral high-resolution GC/MS (HRGC/MS) and chiral HPLC. Chiral HPLC using modified cellulose and phenylglycine columns also showed some isomer resolution. A novel thermal equilibration procedure allowed distinction among axial-chiral and C-chiral enantiomers (Buser *et al.*, 1995). Additionally, acetochlor enantiomers were partial identified with a cellulose derivative fixed phase CDMPC by HPLC with n-hexane or petroleum ether with different percents alcohol (Peng *et al.*, 2005). And dimethenamid-P was completely resolved on a normal phase Chiralpak AD-H column (Saito *et al.*, 2008).

Another typical amide herbicide, napropamide, was separated both by normal phase HPLC and by reverse phase HPLC by Liu *et al.* (Chen *et al.*, 2006, Zhou *et al.*, 2006). In the former research, a method for the chiral separation and micro-determination of napropamide in water was established on a Chiralcel OJ-H column. The linearity of calibration curve for racemic mixture was 10-100 ng/mL and the correlation coefficient was 0.99 (Chen *et al.*, 2006). In the latter, the enantiomers were resolved using Chiralcel AD-RH and Chiralcel OD-RH with MeCN/H_2O as mobile phase. The stereoselectivity of Chiralcel AD-RH was better than Chiralcel OD-RH for napropamide (Zhou *et al.*, 2006). In a report by Zhou *et al.* (Tian *et al.*, 2010), napropamide was partially separated (R_s 1.05) under 40/60 MeCN/water reverse phase HPLC on amylose tris(3,5-dimethylphenylcarbamate) CSP (ADMPC).

Flamprop was resolved on 150×4.6 mm I.D. terguride-based CSP (selectivity factor α 1.09) by using 45% 0.02 M potassium acetate buffer (pH 3.5) and 55% MeCN as the elution solvent by HPLC (Padiglioni *et al.*, 1996).

3.1.2 Enantioseparation of phenoxy herbicides by HPLC

Phenoxy herbicides are a large group of chiral herbicides with widespread application in agriculture. The most representative herbicides are diclofop, mecoprop (MCPP), dichlorprop (DCPP) and their derivatives as classified with phenoxypropionic acids herbicides, which are widely applied to control broad-leaf weeds. In Padiglioni's study (Padiglioni *et al.*, 1996), MCPP, DCPP, diclofop, fenoxaprop, fenoprop, fluazifop, haloxyfop, quizalofop-ethyl ester and quizalofop were well resolved on 150×4.6 mm I.D. terguride-based CSP by using 0.02 M potassium acetate buffer (pH 3.5)-MeCN as the mobile phase by HPLC. Furthermore, a semipreparative-scale separation of fenoprop enantiomers was carried out on a 250×7.8 mm I.D. column, yielding approximately 1.0 mg of each enantiomer in a single chromatographic run, with a recovery of 88% and optical purity greater than 99%.

Several phenoxypropionic acid herbicides were separated on two CD-derivatived CSPs, Nucleodex α-PM and Nucleodex β-PM. Phenoxypropionic acids can be divided into three different groups. The first one has one or two small substituents such as methyl, chlorine or hydroxyl at the aromatic ring (e.g. MCPP, DCPP). The separation of MCPP and DCPP was possibly conducted using NUCLEODEX α-PM CSP, whereas the methyl ester of these compounds was resolved by both Nucleodex α-PM and Nucleodex β-PM. A further substitution (e.g. fenoprop R1, R2. R3=C1, R4=H) leads to the second group and results in the failure of the permethylated α-CD to achieve separation, but fenoprop can be sufficiently resolved by Nucleodex β-PM. The third group contains compounds like fenoxaprop or diclotop with large substituents at the aromatic ring. In this case only the methyl or ethyl esters can be separated by permethylated β-CD. No resolution can be obtained with Nucleodex α-PM (Riering *et al.*, 1996). Resolution of MCPP and DCPP and 2,4-D were proved to be obtained on Nucleodex-α-PM-CD CSP with 70% MeOH and 30% 50 mM NaH_2PO_4 as elutent by Kohler *et al.* (Zipper *et al.*, 1999) and Bjerg *et al.* (Rugge *et al.*, 2002).

MCPP and DCPP, and bromacil with a pyrimidinedione ring were better resolved on the native teicoplanin CSPs than the aglycone teicoplanin CSPs with 100% MeOH containing 0.1% TEA and 0.1% acetic acid (v/v) and 20/80 MeOH/water buffered at pH 4.1 by 1% TEAA for bromacil by HPLC (Berthod *et al.*, 2000a). Furthermore, MCPPM and DCPPM were better resolved on the native teicoplanin CSPs with 20% MeOH/80% aqueous buffer (pH 4.1 by TEAA, 1%). However, the resolution for bromacil with a pyrimidinedione ring was higher on teicoplanin structurally related A-40926 CSP than on teicoplanin CSP (R_s 2.8 vs. R_s 2.5) (Berthod *et al.*, 2000b). Rac-diclofop methyl and rac-diclofop acid were baseline separated on a chiralcel OJ-H column using chiral HPLC coupled with fluorescence detection with a mobile phase of Hex/IPA/HAc (90:10:0.2, v/v) at a flow rate of 0.5 mL/min under 20 °C (Lin *et al.*, 2006) while in a report by Zhou *et al.* (Gu *et al.*, 2010), they were completely resolved on CDMPC CSP with Hex/IPA(98:2) containing 0.1% TFA as mobile phase by HPLC-DAD.

A group of 2-aryloxypropionic acids (TR-1 to 13) and their esters (TR-19 to 20) were used to evaluated four new brush-type CSPs (CSP I-IV) comprising N-3,5,6-trichloro-2,4-dicyanophenyl-L-α-amino acids by HPLC. The best separation of these herbicides was obtained with CSP I, and the -(-)-S enantiomer were regularly eluted first. The mechanism of chiral recognition implies a synergistic interaction of carboxylic acid analyte with the chiral selector and achiral free γ-aminopropyl units on silica. (Vinkovic *et al.*, 2001) In a study by Badjah-Hadj-Ahmed (Tazerouti *et al.*, 2002), eleven 2-aryloxypropionic acids and esters herbicides were partially separated on the prepared phenylated β-CD CSP when using heptane and either IPA or chloroform as organic mobile phase modifier.

Enantioseparation of 2,4-DP and MCPP was obtained completely using enantioselective HPLC on a chirobiotic T column with 5:95 MeOH and 1% TEAA as mobile phase (Schneiderheinze *et al.*, 1999).

Fenoxaprop-ethyl could obtain baseline separation on ADMPC CSP by reversed phase HPLC with MeOH/water or MeCN/water at a flow rate of 0.5 mL/min, while the enantiomers of quizalofop-ethyl, fluroxypyr-meptyl and 2,4-D-ethylhexyl got partial separation (Tian *et al.*, 2010).

A group of chlorophenoxypropionic acid herbicides 2,2-CPPA, 2,3-CPPA and 2,4-CPPA were separated in capillary LC, while with 0.1 mM teicoplanin in the mobile phase was sufficient for the baseline enantioresolution of 2,2-CPPA and 2,4-CPPA (Kafkova *et al.*, 2005).

Eight commercially available herbicides, dimethenamid-P, dichlorprop-P, fluazifop-P butyl, mecoprop-P, quizalofop-P ethyl, were completely resolved by HPLC combined with a photodiode-array (PDA) detector and a circular dichroism (CD) detector on a normal phase Chiralpak AD-H column (Saito *et al.*, 2008). Optical purity measurement was developed. The enantiomeric excess (ee) of some herbicides investigated was approximately over 95%, while of quizalofop-P ethyl and fluazifop-P butyl was in the range 34.1-94.5%.

3.1.3 Enantioseparation of imidazolinone herbicides by HPLC

Imidazolinones are a class of chiral herbicides that are widely used. They inhibit branched-chain amino acid biosynthesis in plants by targeting acetolactate synthase (ALS). Five imidazolinone herbicides imazapyr, imazapic, imazethapyr, imazamox and imazaquin and their methyl derivatives were separated using reversed phase HPLC on Chiralcel OD-R and normal phase HPLC on Chiralcel OJ (Lao *et al.*, 2006a). Enantiomers of imazethapyr, imazaquin, and imazamox were separated on a Chiralcel OD-R column using 50 mM phosphate buffer-MeCN as mobile phase. Enantiomers of imazethapyr, imazaquin, and imazamox were separated on a Chiralcel OD-R column using 50 mM phosphate buffer-MeCN as mobile phase. Enantiomers of imazapyr, imazapic, imazethapyr, imazamox, imazaquin and their five methyl derivatives were resolved on a Chiralcel OJ column using Hex (0.1% TFAA)-alcohol as mobile phase. The described normal phase method was successfully applied for chiral analysis of two imidazolinone herbicides (imazapyr and imazaquin) in spiked soil samples. In a further report (Lao *et al.*, 2006b), temperature affects on enantioseparation of these five imidazolinone herbicides and conformation of CSP were conducted on Chiralcel OJ. The van't Hoff plots of retention factor (k'), distribution constant (K) and separation factor (α) for imazapyr, imazapic, imazethapyr, and imazamox were linear within 15-50 °C. Nonlinear van't Hoff plots of α were observed for imazaquin with mobile phase of Hex (0.1% TFA)-IPA at 70/30 or 60/40 (v/v). Chiralcel OJ column may yield satisfactory results at 15-50 °C but not at ≤ 15 °C.

Recently, Lin *et al.* (Lin *et al.*, 2007) also investigated the enantiomeric separation of imazethapyr, imazapyr, and imazaquin on Chiralpak AS, Chiralpak AD, Chiralcel OD, and Chiralcel OJ columns. Chiralcel OJ column showed the best chiral resolving capacity among the test columns. The optimal chromatographic conditions for complete separation of imidazolinone enantiomers were a mobile phase of Hex/EtOH/HAc (77/23/0.1, v/v/v), flow rate of 0.8 mL/min, and a column temperature in the range of 10–30 °C. It was showed that small amounts of enantiopure imidazolinones may be prepared by using the analytical chiral HPLC approach.

Enantiomers of imazethapyr were separated by HPLC on Chiralcel OJ with a Hex/EtOH/HAc solution (75/25/0.5 by volume) (Zhou *et al.*, 2009, Zhou *et al.*, 2010), and their absolute configurations were confirmed as S-(+)-IM and R-(-)-IM by the octant rule as shown in Fig. 3-1.

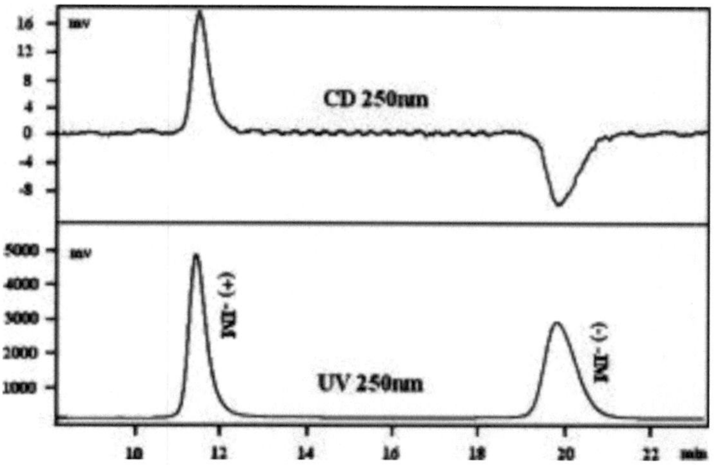

Fig. 3-1 HPLC chromatogram for the enantiomeric separation of imazethapyr on Chiralcel OJ. (Zhou *et al.*, 2009)

3.1.4 Enantioseparation of organophosphorus herbicides by HPLC

Five chiral O-aryl O-alkyl N-alkylphosphoramidothioates herbicides were nearly baseline separated on a pirkle-type column OA-4700 (Chirex(S)-LEU & (R)-NEA) by HPLC. The chromatographic elution order is S>R, and the S-enantiomer showed higher herbicidal activity than R-enantimer and/or racemates (Gao *et al.*, 2000).

In another report by our group (Li *et al.*, 2008), enantioselective separation and biological toxicity of a series of 1-(substituted phenoxyacetoxy)alkylphosphonates organophosphorous compounds (OPs compounds 1-5) were investigated on Chiralpak AD, Chiralpak AS, Chiralcel OD, and Chiralcel OJ. All the analytes investigated obtained baseline resolution (R_s>1.5) on Chiralpak AD, which showed best chiral separation capacity. The acute aquatic toxicity of enantiomers and racemate to *Daphnia magna* (*D. magna*) were assessed. The in vivo assays showed that compound 3 was about 2-148.5 times more toxic than the other four analogues to *D. magna*. The racemates of compounds 3 and 5 showed intermediate toxicity compare to their enantiomers, while those of compounds 1, 2, and 4 showed synergistic or antagonistic effect. These results suggest that the biological toxicity of chiral OPs to nontarget organisms is enantioselective and therefore should be evaluated with their pure enantiomers.

3.1.5 Enantioseparation of diphenyl ether herbicides by HPLC

Ethoxyfen-ethyl and lactofen were separated using HPLC on polysaccharide CSPs by Zhou *et al.* (Wang *et al.*, 2006, Tian *et al.*, 2010, Hou *et al.*, 2002, Diao *et al.*, 2009). Enantioseparation of a novel herbicide ethoxyfen-ethyl was conducted on self-prepared CDMPC, and with the content of IPA in hexane in mobile phase decreased to 1%, the resolution factors increased to 3.95 (Hou *et al.*, 2002). The two enantiomers of the herbicide lactofen in soils were baseline separated and semiprepared on CDMPC using a normal phase HPLC (n-Hex/IPA 95/5). However, the baselined separation was not obtained on a self-prepared tris-(S)-1-phenylethylcarbamate CSP (Wang *et al.*, 2006). And lactofen also could be completely resolved (R_s 2.07) by a reserved phase HPLC using 80/20 MeOH/H_2O as mobile phase on ADMPC (Tian *et al.*, 2010).

3.2 Separation of chiral herbicides by GC

GC is more suitable in analyzing because of its higher sensitivity, higher precision and less injection volume than HPLC system. Besides, contaminants and impurities usually can be separated from the analytes facilely by GC.

The most common chiral selectors used for GC are a group of CD and CD-derivatives. Enantiomers and diastereomers of some acetamide pesticides, alachlor, acetochlor, metolachlor, and dimethenamid, were separated using achiral and chiral high-resolution GC/MS (HRGC/MS) and chiral HPLC. Whereas alachlor is achiral, all other compounds are axial- and/or C-chiral and consist of two or four stereoisomers (enantiomers and diastereomers). Chiral HRGC using a β-CD derivative showed varied resolution of diastereomers and/or enantiomers; achiral HRGC showed no resolution of diastereomers. Resolution of C-chiral enantiomers was easier than resolution of axial-chiral enantiomers (atropisomers) (Buser *et al.*, 1995). And all four metolachlor isomers were identified by HRGC (Muller *et al.*, 2001).

Leachate samples from a waste disposal site in Switzerland and groundwater samples downstream of the landfill were analyzed for residues of MCPP, DCPP, and 2,4-D esterified with 2,3,4,5,6-Pentafluorobenzyl (PFB) by means of enantiomer-specific GC-MS (Zipper *et al.*, 1999, Zipper *et al.*, 1998). The PFB esters of MCPP and DCPP were nearly baseline separated (R_s=0.9) on a 15 m glass column (0.25 mm i.d.) with an OV1701 polysiloxane phase containing 35% heptakis(2,3-dimethyl-6-tert-butyldimethylsilyl)-β-CD (TBDM-β-CD) as the chiral selector.

A capillary column BGB-172 (20% tert-butyldimethylsilyl-β-CD dissolved in 15% diphenyl-polysiloxane and 85% dimethylpolysiloxane, GBG Analytik, Adliswil, Switzerland) was used for chiral GC separation of some herbicides by Liu *et al.* (Wen *et al.*, 2004, Ma *et al.*, 2006, Ma *et al.*, 2009). DCPP methylated by diazomethane in water was separated and determined with a recovery about 90% (Wen *et al.*, 2004). They also separated rac-metolachlor and S-metolachlor in soil. However, the baseline separation was not achieved because of the presence of two chiral elements (asymmetrically substituted carbon and chiral axis nitrogen) (Ma *et al.*, 2006). Furthermore, the enantiomeric separation of DCPPM was investigated by GC on BGB-172 and HPLC on Chiralcel OJ-H by this group. Baseline separation by both GC and HPLC was achieved (Ma *et al.*, 2009).

3.3 Separation of chiral herbicides by SFC

As complementary analytical techniques for HPLC, packed-column SFC with sub- and/or supercritical fluid contains kinds of organic polar solvents is becoming a very popular chromatographic technique, for both analysis and small-scale preparation of optically pure chemicals and enantiomers identification, especially as CSPs are becoming easily available and widely applied. Nearly all the conventional HPLC CSPs could be applied in SFC mode except the chiral crown ester CSPs and the protein-based CSPs. Sub- and supercritical carbon dioxide (CO_2) remains the most commonly used fluid for SFC. Mechanistically, SFC plays a unique role acting as a bridge between GC and LC. Owing to the good diffusibility and low viscosity of supercritical fluids, column equilibration is accomplished more rapidly and enables faster flow rates in SFC than in HPLC. Besides, the higher diffusivity between mobile phase and CSPs yields greater efficiency (smaller plate heights) in resolving a sample.

Generally, SFC shows notable advantages and superior developmental potential on enantiomer separation. The advantages contain environmental friendly with low organic solvent consumption of mobile phase, simple method development, high efficiency on enantioseparation, low column pressure drop besides ease of coupling with chiral columns or MS. However, the high investment of SFC apparatus restricts its widespread application in enantioseparation. To date, the research about chiral herbicides separation by SFC is very limited. One herbicide example that can be resolved by SFC is the diasteriomeric compound metolachlor. The ability to quickly detect and identify metolachlor and its isomeric ratios in low concentration samples is possible, via SFC (Cole *et al.*, 2007).

3.4 Separation of chiral herbicides by CE

CE is shown to be a simple, efficient, and inexpensive way with unique versatility to chiral separation because it can be applied to a wide variety of analytes flexibly with various modes. Hitherto, six separation modes of CE has been successfully used in chiral separation, including capillary zone electrophoresis (CZE), capillary electrochromatography (CEC), micellar electrokinetic chromatography (MECC or MEKC), capillary gel electrophoresis (CGE), capillary isoelectric focusing (CIEF), capillary isotachophoresis (CITP) (Li *et al.*, 2010), where CZE, CEC and MEKC are the most successful CE modes. For the enantioseparation of chiral herbicides by CE, CD and its derivatives are often added to the electrophoresis buffer as the chiral selectors.

Some chlorophenoxy acid herbicides and their enantiomers, 2,4-dichlorophenoxy-acetic acid (2,4-D), 2-(2,4-dichlorophenoxy)propionic acid (2,4-DP), 4-(2,4-dichlorophenoxy)butyric acid (2,4-DB), 4-chloro-2-methylphenoxyacetic acid (MCPA), were successfully swparated within 7 min by adding 4 mM α-CD and 1 mM β-CD in the buffer in CE (Hsieh *et al.*, 1996). Analyzing the herbicides by CE posed the advantages of a high resolution, high separation efficiency and good reproducibility.

A novel, selective precolumn derivatization reaction was introduced and evaluated in the fluorescence labeling of phenoxy acid herbicides including 2,4-D, (2,4,5-trichlorophenoxy)-acetic acid (2,4,5-T), 2-phenoxypropionic acid (2-PPA), MCPP, 2-(2-chlorophenoxy)propionic acid (2,2-CPPA), 2-(3-chlorophenoxy)propionic acid (2,3-CPPA), 2-(4-chlorophenoxy)propionic acid (2,4-CPPA), DCPP and silvex with 7-aminonaphthalene-1,3-disulfonic acid (ANDSA) by CE (Mechref *et al.*, 1996a). The ANDSA-phenoxy acid herbicide enantiomers exhibited higher chiral resolution than their underivatized counterparts in the presence of CD in the running electrolyte. The best enantioselectivity was achieved when 2,3,6-tri-O-methyl-β-CD (TM-β-CD) was used as the chiral selector. Mixed CDs based on β-CD and TM-β-CD proved to be the most effective as far as the enantiomeric resolution of the chiral analytes is concerned.

A novel chiral nonionic surfactant, namely octyl-*b*-*D*-maltopyranoside (OM), was evaluated in chiral CE of fluorescently labeled phenoxy acid herbicides (Mechref *et al.*, 1997a). The labeling of the analytes with 7-aminonaphthalene-1,3-disulfonic acid (ANDSA) permitted a concentration detection limit of 5×10^{-10} M using laser-induced fluorescence detection. The tagging of the phenoxy acid herbicides with ANDSA increased the hydrophobicity of the analytes, thus favoring an enhanced solubilization of the derivatized herbicides in the OM micellar phase. The net results of this effect were a much shorter analysis time and an improved enantiomeric resolution of the derivatives when compared to underivatized phenoxy acid herbicides. Baseline enantiomeric resolution of phenoxy acid herbicides including silvex, DCPP, MCPP, 2,4-CPPA, 2,3-CPPA, 2,2-CPPA and 2-PPA was attained without 30 min by CE using 200 mM sodium phosphate buffer, pH 6.5, containing 60 mM n-octyl-β-D-maltopyranoside (OM) (Mechref *et al.*, 1996b). Silvex, DCPP, MCPP, 2,4-CPPA, 2,3-CPPA, 2,2-CPPA and 2-PPA were baseline separated except silvex by performing the separation at 10 °C and using 250 mM sodium phosphate buffer, pH 6.5, containing 50 mM n-nonyl-β-glucopyranoside (NG) or 70 mM n-octyl-β-glucopyranoside (OG) in CE. (Mechref *et al.*, 1997b)

Vancomycin was used as chiral selector for the enantiomeric separation of several free acid herbicides including MCPP, fenoprop, DCPP, flamprop, haloxyfop, fluazifop, diclofop and fenoxaprop in CE (Desiderio *et al.*, 1997a). The increase of vancomycin concentration caused a general increase of migration time, resolution and selectivity. Baseline resolution was achieved when a 6 mM vancomycin was used. The CE separation of some herbicidal enantiomers was conducted applying 1-allylterguride as chiral selector (Ingelse *et al.*, 1997). Baseline separation was shown for the enantiomers of fluazifop, halossifop and fenoxaprop, whereas the optical isomers of flamprop could be partially resolved using 100 mM β-alanine-acetate, 50 mM TEA in 100% MeOH supported with 100 mM allyl-TER. Separation times are short compared to similar analyses, applying HPLC and a terguride CSP.

The enantiomers of a number of 2-aryloxypropionic acids and their ester and amide counterparts are readily separated on the commercially available β-GEM 1 and Whelk-O 1 CSPs. Of the analytes studied, the N,N-diethylamides typically show the greatest enantioselectivity. The enantiomers of several commercial herbicides from this family, including diclofop ethyl, devrinol, and MCPP were separated using the Whelk-O 1 CSP. β-Gem1 is a π-acceptor chiral stationary phase and is prepared by covalently bonding N-3,5-dinitrobenzoyl-3-amino-3-phenyl-2-(1,1-dimethylethyl)-propanoate, to 5 μm silica through an ester linkage. (Pirkle *et al.*, 1997)

Baseline enantiomeric separation of a mixture of six pairs of phenoxypropionic acid herbicides (PPAHs) including 2,3-CPPA, 2,2-CPPA, 2,4-CPPA, 2(2,4-DCPPA), 2(2,4,5-TCPPA) and 2-PPA was achieved in less than 30 min by CE with heptakis(6-methoxyethylamine-6-deoxy)-β-CD [β-CD-OMe (VII)] as chiral selector. The two most substituted herbicides [2(2,4-DCPPA) and 2(2,4,5-TCPPA)] were best resolved. One of the faster migrating antipodes of 2(2,4,5-TCPPA) co-eluted with one slower antipode of 2(2,4-DCPPA) while both baseline separation was obtained run separately (Fig. 3-2) (Haynes *et al.*, 1998).

DCPP and imazaquin was analyzed by CE as an anion (Jarman *et al.*, 2005). DCPP was separated using 25 mM sodium tetraborate (Na-TB), pH 8.5, with 25 mM trimethyl-β-CD as the chiral selector, while imazaquin was analyzed with 15 mM dimethyl-β-CD in 50 mM acetate, pH 4.5. Furthermore, sodium hydrogen phosphate (50 mM) at pH 10.1 containing 30 mM hydroxypropyl-β-CD (HP-β-CD) was found to be the suitable BGE to separate imazaquin enantiomers in field soils (Yi *et al.*, 2007). In another report (Han *et al.*, 2008), the two imazethapyr enantiomers were separated using 6% hydroxypropyl-β-CD as chiral selector in buffer at pH 11.0.

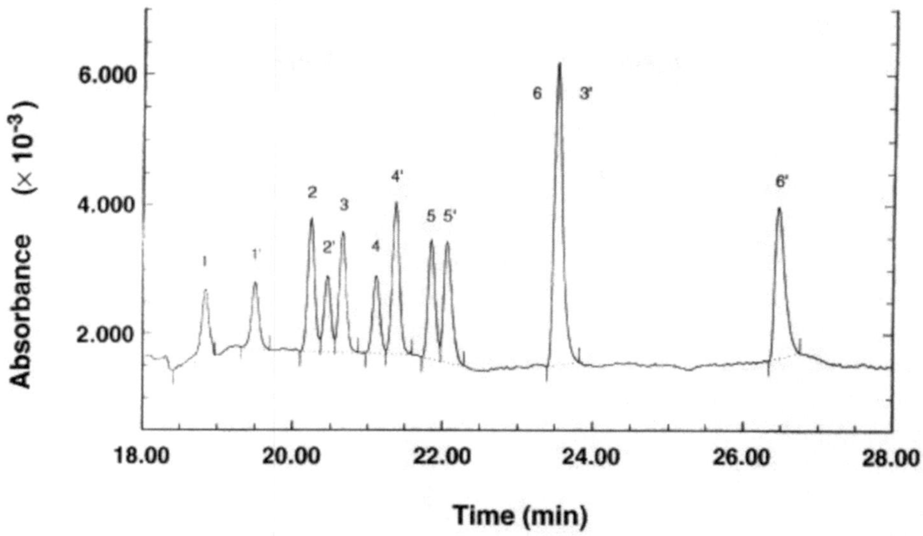

Fig. 3-2. Enantiomeric separation of a standard mixture of 12 (±) PPAH enantiomers. The BGE contains 50 mM NaH_2PO_4 adjusted to pH 6; 3 mM β-CD-OMe (VII); applied voltage was -15 kV, -25 μA; pressure injection 85 kPa ·s; sample concentration 0.1 mg/mL in methanol–water (1:1, v/v). 1,1′=2-PPA, 2,2′=2,4-CPPA, 3,3′=2(2,4-DCPPA), 4,4′=2,2-CPPA, 5,5′=2,3-CPPA, 6,6′=2(2,4,5-TCPPA). (Haynes *et al.*, 1998)

3.4.1 Separation of chiral herbicides by CZE

The separation mechanism for CZE is based on the differences about the charge/mass ratios. Uncoated fused-silica capillary is filled with some type of electrolyte solution (running buffer or BGE). An electric field is applied to the capillary, and then cations go to the cathode, whereas anions migrate to the anode (Pico *et al.*, 2003).

CD-CZE was applied successfully to the enantiomeric and isomeric separation of chiral herbicides.

Chiral separations of phenoxypropionic acid herbicides were achieved by adding a suitable CD-type chiral selector to the electrophoresis buffer (Nielen, 1993, Otsuka *et al.*, 1998, Zerbinati *et al.*, 2000). DCPP, fenoprop and MCPP, were baseline separated by the coupling of CE-MS with 20 mM TM-β-CD in 50 mM ammonium acetate (pH 4.6) (Otsuka*et al.*, 1998). Separation of the four enantiomers of MCPP and DCPP was conducted on an ethylcarbonate derivative of β-CD with three substituents per molecule, hydroxypropyl-β-CD and native α-CD as chiral selectors in CZE. Complete resolution of the four optical isomers was obtained with10 mM ethylcarbonate-β-CD in the running buffer of 45 mM NaH_2PO_4, pH 5.6 (Zerbinati *et al.*, 2000).

The separation and detection of 2,4-dichlorophenoxyacetic acid and three optically active phenoxy acid herbicides (DCPP, MCPP and fenoprop) was investigated in CZE (Garrison *et al.*, 1994). A 50 mM acetate buffer at pH 4.5 gave the best separation. Baseline separation of the two enantiomers of each three optically active herbicides, separately and in mixtures of the three, was accomplished by the addition of 25 mM tri-*O*-methyl-β-CD to the acetate separation buffer. Di-*O*-methyl-β-CD or α-CD separated enantiomers of DCPP and MCPP, but not those of fenoprop; β-CD provided very little separation and γ-CD gave no separation.

Several chiral herbicides, bromacil, chlorbufam, ethofumesate, imazapyr, flamprop-isopropyl, flamprop-free acid, fluazifop-free acid, haloxyfop-free acid, and napropamide, were separated in CZE (Desiderio *et al.*, 1997b). Different β-CD derivatives were investigated for chiral separations and among them the negatively charged sulfobutyl ether β-CD (SBE-β-CD) proved to be effective for the stereoselective resolutions of the investigated herbicides. Addition of SBE-β-CD (5-50 mg/mL) to the buffer at pH 9 resulted in a general increase of migration times as well as resolution. A CD concentration as low as 5 mg/mL was effective to completely resolve napropamide and ethofumesate enantiomers.
The enantiomeric and isomeric separation of imazaquin, diclofop and imazamethabenz was investigated in CD-CZE (DM-β-CD, TM-β-CD and HP-γ-CD) (Penmetsa *et al.*, 1997). The enantiomers of imazaquin and diclofop, and the isomers of imazamethabenz could be resolved with $R_s \geq 1.5$ (Fig. 3-3). By employing mixed CDs in the running buffer, the three herbicides were simultaneously separated in a single run (Fig. 3-4).
The separation of DCPP was reported in CZE with α-, β- and γ-CDs as well as their chemical derivatives C_6-capped-β-CD, ethylcarbonate-β-CD, ethylcarbonate-γ-CD, methyl-β-CD and hydroxypropyl-β-CD as chiral selectors. Several of the investigated CDs allowed DCPP enantiomer resolution. In particular, a newly synthesised ethylcarbonate derivative of β-CD showed the best enantiomer resolution properties among the tested compounds. (Zerbinati *et al.*, 1998)
Biological degradation of acetanilide herbicides in soil results in the formation of the ethanesulfonic acid (ESA) and oxanilic acid (OXA) derivatives. These molecules exist in two (alachlor), four (acetochlor), and eight (metolachlor) stereoisomeric forms. Using γ-CD as chiral selector in CZE, complete separation of all four isomers of enantiomerically enriched (5S)-metolachlor OXA was achieved. The enantiomers of acetochlor ESA, acetochlor OXA, and racemic metolachlor OXA were partially separated. (Aga *et al.*, 1999)
CZE was used for the chiral and mutual separation of four phenoxy acid herbicides, fenoprop, dicloprop, MCPP and 2,4-DB, using highly sulphated CD (HSCD) in the buffer. The CE runs were performed with reverse polarity (anode in the outlet vial) using the acidic ammonium formate buffer (20 mmol, pH 3.0). The chiral separation of dicloprop and MCPP were achieved with α-HSCD but it was not able to resolve fenoprop. With β-HSCD the required base line separation was achieved. The limit of detection (S/N= 3) achieved in present case is 0.15 ppm for fenoprop, 0.14 ppm for dicloprop and MCPP and 0.11 ppm for 2,4-DB. (Malik *et al.*, 2009)
Soil samples taken from a field plot at increasing time intervals after application of Foxtril, a commercial herbicide formulation, were solvent-extracted and analyzed for total DCPP by CZE, using an acetate buffer at pH 4.7. TM-β-CD, was then added to the buffer as chiral reagent to effect separation of the (+)- and (–)-enantiomers of DCPP. Baseline resolution allowed calculation of relative concentrations (enantiomer ratios) of the two isomers. The hydrolysis product [methyl 2-nitro-5-(2,4-dichlorophenoxy) benzoic acid] of bifenox methyl ester, another herbicide component of Foxtril, was detected in the soil samples taken at 17 and 31 d. The acetate separation buffer was 50 mM at pH 4.65 and was composed as follows: 0.05 M glacial acetic acid: 0.05 M sodium acetate, 1:1, v:v. The cyclodextrin-containing buffer for enantiomeric separation was prepared by dissolving TM-β-CD in the acetate separation buffer to a final concentration of 25 mM cyclodextrin. (Garrison *et al.*, 1996)

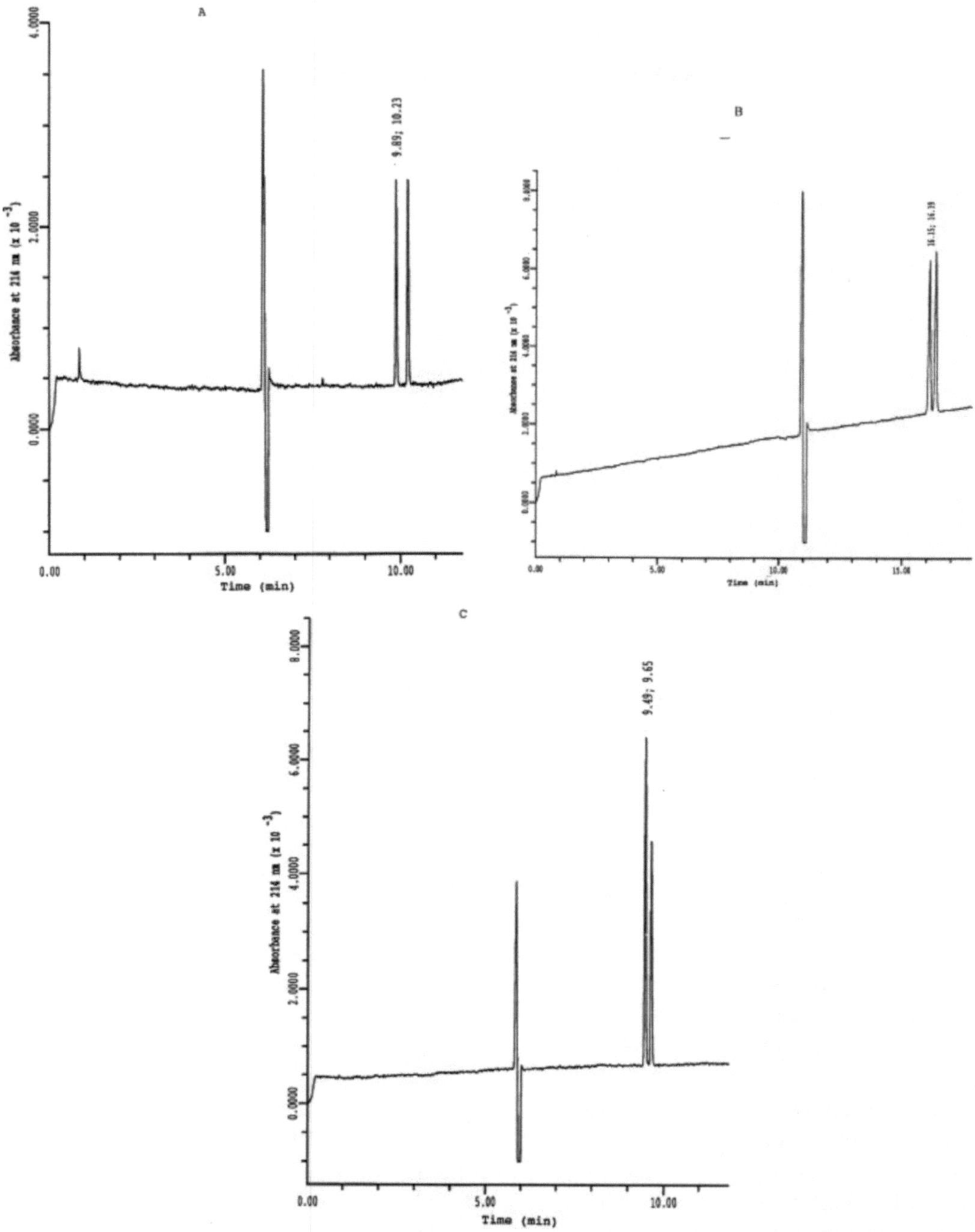

Fig. 3-3. Separation of (A) imazaquin enantiomers, (B) diclofop enantiomers and (C) imazamethabenz isomers (9.49 min, *para* and 9.65 min, *meta* isomers). Analysis conditions: 57 cm (50 cm to detector) × 50 µm I.D. capillary column; pressure injection (2 s=2.4 nl); 25 kV (35 µA); 214 nm UV absorbance. Buffer: (A) 50 mM sodium acetate + 10 mM DM-β-CD buffer, pH 4.6, (B) 50 mM sodium acetate + 10 mM TM-β-CD buffer, pH 3.6 and (C) 50 mM sodium acetate + 10 mM TM-β-CD buffer, pH 4.6.

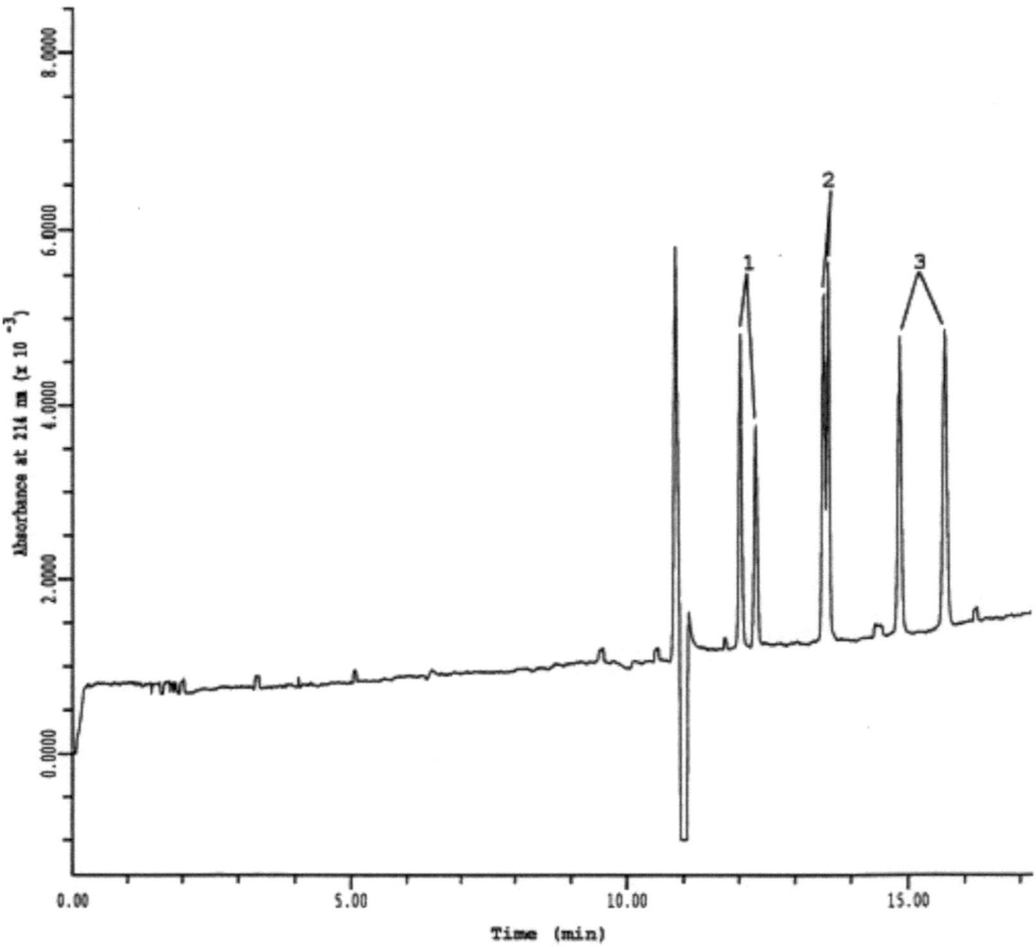

Fig. 3-4. Simultaneous separation of herbicides using mixed cyclodextrins. (1) Imazamethabenz isomers, (2) diclofop enantiomers and (3) imazaquin enantiomers. Analysis conditions: 57 cm (50 cm to detector) × 50 µm I.D. capillary column; pressure injection (2 s=2.4 nl); 50 mM sodium acetate + 10 mM DM-β-CD + 10 mM TM-β-CD buffer, pH 3.6; 25 kV (35 µA); 214 nm UV absorbance. (Penmetsa *et al.*, 1997)

3.4.2 Separation of chiral herbicides by CEC

CEC utilises a stationary phase rather than a micellar pseudo-stationary one. CEC is a hybrid technique that couples the selectivity of LC and the separation efficiency of CE. Both charged and uncharged compounds can be separated effectively using CEC.

A series of herbicide molecules (haloxyfop, fluazifop, fenoxaprop, and flamprop free acids, diclofop, MCPP, DCPP, fenoprop, 2-PPA) were separated using a CSP derived from an L-RNA aptamer by CEC after binding to biotin and grafting upon streptavidin-modified porous glass beads. (Andre *et al.*, 2006)

A porous monolithic chiral column was prepared by in situ copolymerization of glycidyl methacrylate, methyl methacrylate and ethylene glycol dimethacrylate in the presence of formamide and 1-propanol as the porogen solvents to analyze the DCPP enantiomers.

Subsequently, the epoxide groups at the surface of the monolith were reacted with (+)-1-(4-aminobutyl)-(5R,8S,10R)-terguride as the chiral selector. Optimum conditions for the herbicide resolution by CEC were found using mobile phases consisting of HAc/TEA mixtures in MeCN:MeOH (9:1 v/v). Under these conditions fully separation of DCPP enantiomers in the presence of clofibric acid (internal standard) was achieved in about 5 min. (Messina *et al.*, 2007)
A silica based monolithic capillary column derivatized with *O*-9-(tert-butylcarbamoyl) quinidine was prepared for CEC enantiomer separation of chiral 2-aryloxypropionic acid herbicides including *inter alia* DCPP, MCPP and fenoprop. Reasonable baseline separations of enantiomers were accomplished for all analytes after optimization of relevant mobile phase parameters in the anion-exchange CEC system, and the separations were comparable to such obtained on an optimized high density quinidine-carbamate modified organic polymer monolith column. (Buchinger *et al.*, 2009)

3.4.3 Separation of chiral herbicides by MEKC

MEKC separation mechanism is based on the differences between interactions of analytes with micelles present in the separation buffer, which can easily separate both charged and neutral solutes with either hydrophobic or hydrophilic properties.
Silvex was separated partially with 50.0 mM N,N-bis-(3-*D*-gluconamidopropyl)deoxycholamide as chiral selector, 400.0 mM borate treated fused-silica capillaries at pH 10.0, 15 °C, voltage 20.0 kv in MEKC. (Mechref *et al.*, 1996c)
Enantiomeric ratios of methyl esters of phenoxy acids herbicides and an acetamide herbicide metolachlor were being measured. Each of six CD, α-CD, β-CD, γ-CD, hydroxypropyl-β-CD, dimethyl-β-CD and trimethyl-β-CD, were then added to the borate-SDS buffer, with and without the organic modifier, to test for separation of the non-chiral compounds and the enantiomers of the chiral racemates by CD-MEKC. γ-CD with MeOH modifier allowed baseline separation of the three phenoxy acid methyl esters and of fenoprop methyl ester, but none of the CDs separated the enantiomers of MCPP and DCPPM. Finally, attempts were made to separate the four enantiomers of the herbicide metolachlor; three of the enantiomers were separated by γ-CD with methanol. (Schmitt *et al.*, 1997)
The enantiomeric resolution of chiral phenoxy acid herbicides was performed by MEKC using several neutral and charged CD as chiral pseudophase (CD-MEKC). Among the CDs tested, HP-β-CD was found to be the most appropriate for the enantioseparation of phenoxy acids. The use of a 50 mM electrolyte solution in ammonium formate at pH 5 containing 15 mM HP-β-CD and a temperature of 40 °C enabled the enantiomeric resolution of four of the six phenoxy acids investigated (2-PPA, 2,3-PPA, 2,4-CPPA, and 2-(2,4-DCPPA)) obtaining migration times ranging from 9 to 15 min. Mixtures of the two phenoxy acids not enantiomerically resolved (2-(4-chlorophenoxy)-2-methylpropionic acid and 2-(2,4,5-trichlorophenoxy)propionic acid) and up to three of the phenoxy acids enantiomerically resolved were separated in about 15 min. (Martin-Biosca *et al.*, 2001)
CD-MEKC was applied to the enantioseparation of thiobencarb sulfoxide, which is produced by S-oxygenation of thiobencarb, using γ-CD together with sodium dodecyl sulfate. The optimum running conditions were found to be 20 mM phosphate-5 mM borate buffer (pH 8.5) containing 60 mM hydroxypropyl-γ-CD and 100 mM sodium dodecyl sulfate with an effective voltage of +20 kV at 20 °C using direct detection at 220 nm with resolution (R_s) approximately 1.7. (Kodama *et al.*, 2002)

3.5 Separation of chiral herbicides by other chromatographic methods

A preparative enantiomer separation method of DCPP was developed utilizing a purposefully designed, highly enantioselective chiral stationary phase additive (CSPA) cinchona-derived chiral seleector derived from bis-1,4-(dihydroquinidinyl)phthalazine in centrifugal partition chromatography (CPC). A solvent system consisting of 10 mM CSPA in methyl tert-butyl ether and 100 mM sodium phosphate buffer (pH 8.0) was identified as a suitable stationary/mobile-phase combination. Complete enantiomer separations of up to 366 mg of racemic DCPP could be achieved, corresponding to a sample load being equivalent to the molar amount of CSPA employed. Comparison of the preparative performance characteristics of the CPC protocol with that of a HPLC separation using a silica-supported bis-1,4-(dihydroquinidinyl)phthalazine CSP revealed comparable loading capacities for both techniques but a significantly lower solvent consumption for CPC. Given that further progress in instrumental design and engineering of dedicated, highly enantioselective CSPAs can be achieved, CPC may offer a viable alternative to CSP-based HPLC for preparative-scale enantiomer separation. (Gavioli *et al.*, 2004)

4. Enantioselective herbicidal activity and toxicity of herbicide enantiomers

For the amide herbicides, the product enantiomerically enriched with the herbicidally active 1'S-metolachlor (aSS, aRS) has replaced the racimate worldwide after 2004 (Muller *et al.*, 2001). S-metolachlor was more toxic to C pyrenoidosa than rac-metolachlor, and the catalase activity of C pyrenoidosa treated by S-metolachlor was higher than that exposed to rac-metolachlor (Liu *et al.*, 2009). And enantioselective degradation and/or interconversion for metolachlor was determined, S-metolachlor degraded faster in soil than rac-metolachlor (Ma *et al.*, 2006, Kurt-Karakus *et al.*, 2010). After 42-day incubation, 73.4% of rac-metolachlor and 90.0% of S-metolachlor were degraded. However, due to the absence of biological processes the degradation process in sterilized soil showed no enantioselectivity. The results indicated that enantioselective degradations could greatly affect the environmental fate of metolachlor and should be considered when the environmental behavior of these compounds was assessed. Napropamide is a highly active preemergence herbicide whose R-enantiomer has high phytocidal activity to unifacial-leaf and broad-leaf weeds. It was found that R-napropamide was about eight fold more active than S-napropamide, and two more active than rac-napropamide (Chan *et al.*, 1975). The green *alga Scenedesmus* acutus growth was strongly inhibited and fatty acid was desaturated by S-alachlor and S-dimethenamid while the R isomer had no effect (Couderchet *et al.*, 1997). Furthermore, the comparable biological activities of dimethenamid and alachlor indicate that this target is common to both N-phenyl and N-thienyl chloroacetamide herbicides.

Enantioselective herbicidal activity and toxicity of the phenoxy herbicides has been reported profoundly and roundly. The in vivo inhibition of R-(+)- and S-(−)-diclofop-methyl affected on root growth was hardly enantioselective (Shimabukuro *et al.*, 1995), while in a report by Liu *et al.* (Ye *et al.*, 2009), the S-diclofop acid was more toxic to leaves and the R-diclofop acid was more toxic to roots of rice Xiushui 63 seedlings. Furthermore, absorption and translocation to the leaf axil of the two-leaf stage plants of diclofop-methyl enantiomers were similar in both susceptible and resistant biotypes, while the rate of metabolism was increased 1.5-fold in this resistant biotype compared to the susceptible (Maneechote *et al.*, 1997). More, the herbicidally inactive S-(−)-enantiorners of both diclofop-methyl and diclofop were similar to or higher than the corresponding R-(+)-forms in toxicity to algae, depending on specific species. Although no enantiomeric conversion occurred for diclofop-methyl and diclofop, the difference in the enantioselective degradation of these herbicides observed in algae cultures suggested that their application forms were an important factor determining their enantioselective environmental behavior. It was concluded that the enantioselective degradation of diclofop in algae cultures was governed primarily by the facilitated uptake by algae, whereas the enantioselective toxicity was primarily governed by the passive uptake (Cai *et al.*, 2008). And it was proved that the S-diclofop-methyl dissipated faster than R-diclofop-methyl while the generation and degradation rates of S-diclofop were higher than R-enantiomer in the plant by Zhou *et al.* (Gu *et al.*, 2010). However, in a former report of Zhou *et al.* (Diao *et al.*, 2010a), it was found that the degradation of diclofop-methyl in two soils was not enantioselective while the degradation of diclofop was enantioselective under both aerobic and anaerobic conditions, and the S-(–)-diclofop was preferentially degraded, resulting in relative enrichment of the R-(+)-form. To haloxyfop ethoxyethyl ester, the S-form was degraded faster than R-form (the enantiomeric fraction of R-form was about 72%) (Desiderio *et al.*, 1997a).

Phenoxypropionic acid (PPA) derivatives are widely used in agriculture as selective herbicides. R-enantiomer of PPAs is known for its herbicidal activity while S-isomer is inactive as herbicidal agent (Buser *et al.*, 1997a). A large number of papers have discussed the enantioselectivity of DCPP and MCPP (Zipper *et al.*, 1999, Rugge *et al.*, 2002, Schneiderheinze *et al.*, 1999, Zipper *et al.*, 1998, Ma *et al.*, 2009, Jarman *et al.*, 2005, Garrison *et al.*, 1996, Messina *et al.*, 2007, Kurt-Karakus *et al.*, 2010, Buser *et al.*, 1997b, Muller *et al.*, 1997, Harrison *et al.*, 2003, Williams *et al.*, 2003, Wen *et al.*, 2009, Wen *et al.*, 2010), thereinto Bidleman *et al.* (Kurt-Karakus *et al.*, 2010) reviewed the concentrations and stereoisomer ratios of DCPP, MCPP and metolachlor. Mostly, the S-enantiomer of these herbicides degraded faster than the R-enantiomer (Zipper *et al.*, 1999, Zipper *et al.*, 1998, Garrison *et al.*, 1996, Messina *et al.*, 2007, Buser *et al.*, 1997b, Muller *et al.*, 1997). Enantioselective microbial degradation increased the enantiomeric ratio of R- to S-MCPP during groundwater passage of the landfill leachate (Zipper *et al.*, 1998). The S-enantiomers of MCPP, DCPP and 2,4-D were preferentially degraded under aerobic conditions (Zipper *et al.*, 1999). The S-(–)-DCPP degraded significantly faster ($t_{1/2}$) = 4.4 d) than the R-(+)-isomer ($t_{1/2}$ = 8.7 d) in soil (Garrison *et al.*, 1996). No preferential degradation of the R- and S-enantiomers of MCPP and of DCPP took place in an aerobic field-injection experiment (Rugge *et al.*, 2002, Jarman *et al.*, 2005). However, in the nitrate-reducing microcosm S-MCPP did not degrade but R-MCPP degraded with zero order kinetics at 0.65 mg/(L·d) to produce a stoichiometric equivalent amount of 4-chloro-2-methylphenol while no biodegradation of MCPP was observed in the methanogenic, sulphate-reducing or iron-reducing microcosms. And in aerobic conditions S- and R-MCPP degraded with zero order kinetics at rates of 1.90 and 1.32 mg/(L·d), respectively (Harrison *et al.*, 2003, Williams *et al.*, 2003). Chitosan also changed the enantioselective bioavailability of DCPP (Wen *et al.*, 2010). The dissipation of S-enantiomer in Chlorella pyrenoidosa culture media without chitosan was faster than that of the herbicidally active R-enantiomer, whereas it was inversed to R-enantiomer being faster than S-enantiomer when chitosan was added into the media. In the absence of chitosan, the toxicity of R-enantiomer to Chlorella pyrenoidosa was more potent than that of the S-enantiomer. On the contrary, in the presence of chitosan, R-enantiomer was less toxic than S-enantiomer. R-DCPP interacted with penicillium expansum alkaline lipase the strongest, followed by Rac-DCPP, while S-DCPP had the weakest interaction (Wen *et al.*, 2009). R-DCPPM was preferentially degraded over the S-DCPPM in different pH solutions (Ma *et al.*, 2009).

Racemic mixtures of 2,4-DP and MCPP were applied to three species of turf grass, four species of broadleaf weeds, and soil. Both herbicides were degraded more quickly and completely by plants than by soil microbes. Preferential degradation of the S-(–)-enantiomer of each herbicide was observed in most species of broadleaf weeds and soil, while the degradation in all species of grass occurred without enantioselectivity. The biodegradation in all systems appeared to follow pseudo first-order kinetics will the fastest degradation occurring in broadleaf weeds, followed by the grasses. The slowest degradation was observed in soil. (Schneiderheinze *et al.*, 1999)

Enantioselective herbicidal activity and degradation of imidazolinone herbicides has been reported recently. Imazaquin exhibited nonselective enantiomer loss over its 3 months of incubation time, which could have been due to abiotic or nonselective microbial reactions (Jarman *et al.*, 2005). However, in another report (Yi *et al.*, 2007), the degradation rates of the two imazaquin enantiomers were slightly different, and the pH of the soil, combined with the moisture content in the soil, had a strong influence on the rate of degradation. And the first enantiomer imazethapyr-I, eluted by CE using 6% hydroxypropyl-β-CD as chiral selector in buffer at pH 11.0, degraded at a higher rate when compared with imazethapyr-II (Han *et al.*, 2008). The R-(+)-enantiomer of all three herbicides, imazapyr, imazethapyr and imazaquin, which has greater herbicidal activity (up to eight times), was found to degrade faster than the less active S-(–)-enantiomer (Ramezani *et al.*, 2010). Generally, the R former of imidazolinones was more active than S former. Imazethapyr inhibits elongation of primary roots and shoots, and reduces the number of adventitious roots and the density of root hairs. The maximal root relative inhibition rate reached 80.4%, 67.0%, and 73.5% for R-(–)-imazethapyr, S-(+)-imazethapyr and (+/–)-imazethapyr at the concentration of 0.5 mg/L, respectively, and the maximal shoot relative inhibition rate reached 77.7%, 26.9%, and 61.7%, respectively (Qian *et al.*, 2009). The inhibition abilities of (+/–)-imazethapyr to the root growth of maize seedlings was between S-(+)- and R-(–)-imazethapyr (Zhou *et al.*, 2009). Moreover, imazethapyr enantiomers enantioselectively suppressed the in vitro and in vivo acetolactate synthase (ALS) activity of maize leaves (Zhou *et al.*, 2010). The in vivo ALS activity study showed only a 2-fold difference between R-(–)-imazethapyr and S-(+)-imazethapyr, while the in vitro study showed that the difference in inhibition between the enantiomers fell sharply as concentration increased. At the lowest concentration of 40 μg/L, R-(–)-imazethapyr appeared 25 times more active than S-(+)-imazethapyr, but only 7 times at 200 μg/L. At the highest concentration of 25 mg/L, in vitro ALS activity was almost completely inhibited by S-(+)-, R-(–)- and (+/–)-imazethapyr, there was only 1.1 times differences between S-(+)- and R-(–)-imazethapyr.

Thiobencarb was treated with a rat liver microsomal fraction containing cofactors (known as S9mix) (Kodama *et al.*, 2002). The ratio between (+)- and (–)-thiobencarb sulfoxide was found to be 15:85. It was also found that the ratio between (+) and (–)-thiobencarb sulfoxide produced in soil spiked with thiobencarb was 3:7. These results indicated marked enantioselectivities for these metabolisms. The activities of thiobencarb, (+)- and (–)-thiobencarb sulfoxides on 5α-dihydrotestosterone- and 17 β-estradiol-induced transcriptions were also investigated. Thiobencarb and (+)-thiobencarb sulfoxide did not show any activities, (–)-thiobencarb sulfoxide showed significant anti-estrogenic and anti-androgenic activities, suggesting that thiobencarb sulfoxide can act as both an enantioselective anti-estrogen and an enantioselective anti-androgen.

Racemic and the enantiopure S-(+)- and R-(–)-lactofen were incubated under aerobic and anaerobic conditions. The data from sterilized controls indicated that the dissipation of lactofen was biological. The dissipation was shown to be enantioselective with S-(+)-enantiomer being degraded faster than the R-(–)-enantiomer, resulting in residues enriched with R-(–)-lactofen when the racemic compound was incubated. Lactofen was configurationally stable in soil, showing no interconversion of S-(+)- to R-(–)-enantiomer and vice versa (Diao *et al.*, 2009). The enantioselective degradation of lactofen enantiomers was proved in a report by Zhou *et al.* (Diao *et al.*, 2010b). In sediments, S-(+)-lactofen or S-(+)-desethyl lactofen was preferentially degraded, resulting in relative enrichment of the R-(–)-form. Lactofen and desethyl lactofen were both configurationally stable in sediment, showing no interconversion of S- to R-enantiomers or vice versa. Furthermore, the acute toxicities of lactofen and desethyl lactofen enantiomers to Daphnia magna were enantioselective. The calculated LC50 values of S-(+)-, rac-, and R-(–)-lactofen were 17.689, 4.308, and 0.378 μg/mL, respectively, and the calculated LC50 values of S-(+)-, rac-, and R-(–)-desethyl lactofen were 21.327, 13.684, and 2.568 μg/mL, respectively.

2-α-substituted benzylamino-4-substituted-amino-6-chloro-1,3,5-triazines are herbicidal compounds showing leaf-burning and/or growth inhibition with concomitant greening and stunting. The test compounds inhibited root growth due to interference with a system or systems other than photosynthesis. 4-(R)-sec-butylamino-2-(α,α-dimethylbenzyl)amino-6-chloro-1,3,5-triazine showed the highest inhibitory activity, and 4-methylamino-2-(R)-α-methylbenzylamino-6-chloro-1,3,5-triazine was second. The chiral requirement for a strong inhibition of root growth was the R-configuration, contrasting with the requirement for the S-configuration for an inhibition of photosystem II. (Omokawa *et al.*, 1992)

Differential chiral responses including enantioselectivity and cross intergenus response on root growth between Oryza and Echinochloa plants against optical active α-methylbenzyl p-tolylureas were indicated. Rice was more affected by the R-enantiomers and barnyard miller by the S-enantiomers (Omokawa *et al.*, 2001). Plants of the tribe Oryzeae respond enantioselectively and homogeneously to optically active 1-(C)-methylbenzyl-3-p-tolylurea (MBTU) in root growth inhibition. The root growth of the genus Oryza was inhibited more by R-MBTU than by S-MBTU (Omokawa *et al.*, 2004).

5. Conclusions

Over the last several decades, the enantioseparation of chiral herbicides has been widely studied and has made a great contribution for studying their stereoselectivity in biological target activity and non-target toxicity. The direct chromatographic separation approaches play a leading role in separation of chiral herbicides. HPLC combined with CSPs shows its superiority for the enantiomer analysis and enantiomer preparation of many common herbicides especially for the group of amide herbicides, phenoxy herbicides and imidazolinone herbicides. GC is powerful in the determination while CE with diversified modes is also useful for its maneuverability. The application of herbicides separation by SFC is relatively limited.

Many herbicides, related to amide herbicides, phenoxy herbicides and imidazolinone herbicides and so on, have shown the enantioselective herbicidal activity and phytotoxicity with their enantiomers. Many chiral herbicides have been commercialized with the pure enantiomer such as S-metolachlor, quizalofop-P-ethyl, haloxyfop-P-methyl, fluazifop-P-butyl, (R)-napropamide *etc.*. Additionally, more work should be conducted on researching enantioselectivity and environmental fate of herbicides.

6. References

Wang, P., Liu, D. H., Lei, X. Q., Jiang, S. R., Zhou, Z. Q., (2006), Enantiomeric separation of chiral pesticides by high-performance liquid chromatography on an amylose tris-(S)-1-phenylethylcarbamate chiral stationary phase, *Journal of Separation Science, 29*, pp. 265-271.

Hamper, B. C., Dukesherer, D. R., Moedritzer, K., (1994), Analytical and preparative separation of the enantiomers of pyrazole phenyl ether herbicides on three chiral stationary phases, *Journal of Chromatography A, 666*, pp. 479-484.

Kodama, S., Yamamato, A., Matsunaga, A., Okamura, K., Kizu, R., Hayakawa, K., (2002), Enantioselective analysis of thiobencarb sulfoxide produced by metabolism of thiobencarb by hydroxypropyl-γ-cyclodextrin modified micellar electrokinetic chromatography, *Journal of Separation Science, 25*, pp. 1055-1062.

Muller, M. D., Poiger, T., Buser, H. R., (2001), Isolation and identification of the metolachlor stereoisomers using high-performance liquid chromatography, polarimetric measurements, and enantioselective gas chromatography, *Journal of Agricultural and Food Chemistry, 49*, pp. 42-49.

Polcaro, C. M., Berti, A., Mannina, L., Marra, C., Sinibaldi, M., Viel, S., (2004), Chiral HPLC resolution of neutral pesticides, *Journal of Liquid Chromatography & Related Technologies, 27*, pp. 49-61.

Buser, H. R., Muller, M. D., (1995), Environmental Behavior Of Acetamide Pesticide Stereoisomers .1. Stereoselective And Enantioselective Determination Using Chiral High-Resolution Gas-Chromatography And Chiral High-Performance Liquid-Chromatography *Environmental Science & Technology, 29*, pp. 2023-2030.

Peng, W., Shu-Ren, J., Dong-Hui, L., Ping, W., (2005), Chiral Recognition of Acetochlor Enantiomers, *Chinese Journal of Pesticides, 44*, pp. 113-114.

Saito, K., Yato, M., Ito, T., Iwasaki, Y., Ito, R., Matsuki, Y., Nakazawa, H., (2008), Verification of the need for optical purity measurement of chiral pesticide standards as agricultural reference materials, *Accreditation and Quality Assurance, 13*, pp. 373-379.

Chen, S. W., Cai, X. Y., Xi, M., Zhang, A. P., Liu, W. P., (2006), Characterization of napropamide enantiomers by CD and determination of the enantiomeric ratios in water, *Spectroscopy and Spectral Analysis, 26*, pp. 1649-1652.

Zhou, Y., Zhang, S.-H., Liu, W.-P., Chen, Y.-H., Ye, L., (2006), Enantioresolution and study of thermodynamical property of napropamide by high performance liquid chromatography, *Chinese Journal of Pesticide Science, 8*, pp. 260-264.

Tian, Q., Lv, C. G., Ren, L. P., Zhou, Z. Q., (2010), Direct Enantiomeric Separation of Chiral Pesticides by LC on Amylose Tris(3,5-dimethylphenylcarbamate) Stationary Phase under Reversed Phase Conditions, *Chromatographia, 71*, pp. 855-865.

Padiglioni, P., Polcaro, C. M., Marchese, S., Sinibaldi, M., Flieger, M., (1996), Enantiomeric separations of halogen-substituted 2-aryloxypropionic acids by high-performance liquid chromatography on a terguride-based chiral stationary phase, *Journal of Chromatography A, 756*, pp. 119-127.

Riering, H., Sieber, M., (1996), Covalently bonded permethylated cyclodextrins, new selectors for enantiomeric separations by liquid chromatography, *Journal of Chromatography A, 728*, pp. 171-177.

Zipper, C., Bolliger, C., Fleischmann, T., Suter, M. J. F., Angst, W., Muller, M. D., Kohler, H. P. E., (1999), Fate of the herbicides mecoprop, dichlorprop, and 2,4-D in aerobic and anaerobic sewage sludge as determined by laboratory batch studies and enantiomer-specific analysis, *Biodegradation, 10,* pp. 271-278.

Rugge, K., Juhler, R. K., Broholm, M. M., Bjerg, P. L., (2002), Degradation of the (R)- and (S)-enantiomers of the herbicides MCPP and dichlorprop in a continuous field-injection experiment, *Water Research, 36,* pp. 4160-4164.

Berthod, A., Chen, X. H., Kullman, J. P., Armstrong, D. W., Gasparrini, F., D'Acquarica, I., Villani, C., Carotti, A., (2000a), Role of the carbohydrate moieties in chiral recognition on teicoplanin-based LC stationary phases, *Analytical Chemistry, 72,* pp. 1767-1780.

Berthod, A., Yu, T., Kullman, J. P., Armstrong, D. W., Gasparrini, F., D'Acquarica, I., Misiti, D., Carotti, A., (2000b), Evaluation of the macrocyclic glycopeptide A-40,926 as a high-performance liquid chromatographic chiral selector and comparison with teicoplanin chiral stationary phase, *Journal of Chromatography A, 897,* pp. 113-129.

Lin, K. D., Cai, X. Y., Chen, S. W., Liu, W. P., (2006), Simultaneous determination of enantiomers of rac-diclofop methyl and rac-diclofop acid in water by high performance liquid chromatography coupled with fluorescence detection, *Chinese Journal of Analytical Chemistry, 34,* pp. 613-616.

Gu, X., Lu, Y. L., Wang, P., Dang, Z. H., Zhou, Z. Q., (2010), Enantioselective degradation of diclofop-methyl in cole (Brassica chinensis L.), *Food Chemistry, 121,* pp. 264-267.

Vinkovic, V., Kontrec, D., Sunjic, V., Navarini, L., Zanetti, F., Azzolina, O., (2001), Mechanism of chiral recognition in the enantioseparation of 2-aryloxypropionic acids on new brush-type chiral stationary phases, *Chirality, 13,* pp. 581-587.

Tazerouti, F., Badjah-Hadj-Ahmed, A. Y., Meklati, B. Y., Franco, P., Minguillon, C., (2002), Enantiomeric separation of drugs and herbicides on a β-cyclodextrin-bonded stationary phase, *Chirality, 14,* pp. 59-66.

Schneiderheinze, J. M., Armstrong, D. W., Berthod, A., (1999), Plant and soil enantioselective biodegradation of racemic phenoxyalkanoic herbicides, *Chirality, 11,* pp. 330-337.

Kafkova, B., Bosakova, Z., Tesarova, E., Coufal, P., (2005), Chiral separation of beta-adrenergic antagonists, profen non-steroidal anti-inflammatory drugs and chlorophenoxypropionic acid herbicides using teicoplanin as the chiral selector in capillary liquid chromatography, *Journal of Chromatography A, 1088,* pp. 82-93.

Lao, W. J., Gan, J., (2006a), High-performance liquid chromatographic separation of imidazolinone herbicide enantiomers and their methyl derivatives on polysaccharide-coated chiral stationary phases, *Journal of Chromatography A, 1117,* pp. 184-193.

Lao, W., Gan, J., (2006b), Responses of enantioselective characteristics of imidazolinone herbicides and Chiralcel OJ column to temperature variations, *Journal of Chromatography A, 1131,* pp. 74-84.

Lin, K. D., Xu, C., Zhou, S. S., Liu, W. P., Gan, J., (2007), Enantiomeric separation of imidazolinone herbicides using chiral high-performance liquid chromatography, *Chirality, 19,* pp. 171-178.

Zhou, Q. Y., Xu, C., Zhang, Y. S., Liu, W. P., (2009), Enantioselectivity in the Phytotoxicity of Herbicide Imazethapyr, *Journal of Agricultural and Food Chemistry, 57,* pp. 1624-1631.

Zhou, Q. Y., Zhang, N., Zhang, C., Huang, L. D., Niu, Y. F., Zhang, Y. S., Liu, W. P., (2010), Molecular Mechanism of Enantioselective Inhibition of Acetolactate Synthase by

Imazethapyr Enantiomers, *Journal of Agricultural and Food Chemistry, 58,* pp. 4202-4206.

Gao, R. Y., Wang, H. F., Chen, Z. Y., Yang, H. Z., Wang, Q. S., (2000), Enantiomer separation and preparation of five organophosphorus compounds and their biological activities, *Chromatographia, 51,* pp. 322-324.

Li, L., Zhou, S. S., Zhao, M. R., Zhang, A. P., Peng, H., Tan, X. S., Lin, C. M., He, H. W., (2008), Separation and aquatic toxicity of enantiomers of 1-(substituted phenoxyacetoxy)alkylphosphonate herbicides, *Chirality, 20,* pp. 130-138.

Hou, S., Wang, M., Zhou, Z., Qiao, Z., Guo, H., Shi, X., (2002), Preparation of cellulose-based chiral stationary phase and enantioseparation of ethoxyfen-ethyl by high performance liquid chromatography, *Se Pu, 20,* pp. 537-539.

Diao, J. L., Lv, C. G., Wang, X. Q., Dang, Z. H., Zhu, W. T., Zhou, Z. Q., (2009), Influence of Soil Properties on the Enantioselective Dissipation of the Herbicide Lactofen in Soils, *Journal of Agricultural and Food Chemistry, 57,* pp. 5865-5871.

Zipper, C., Suter, M. J. F., Haderlein, S. B., Gruhl, M., Kohler, H. P. E., (1998), Changes in the enantiomeric ratio of (R)- to (S)-mecoprop indicate in situ biodegradation of this chiral herbicide in a polluted aquifer, *Environmental Science & Technology, 32,* pp. 2070-2076.

Wen, Y. Z., Cai, X. Y., Ma, Y., Liu, W. P., (2004), Enantioseparation and determination of dichlorprop in water by derivatization-chiral gas chromatography, *Chinese Journal of Analytical Chemistry, 32,* pp. 1492-1494.

Ma, Y., Liu, W. P., Wen, Y. Z., (2006), Enantioselective degradation of rac-metolachlor and S-metolachlor in soil, *Pedosphere, 16,* pp. 489-494.

Ma, Y., Xu, C., Wen, Y. Z., Liu, W. P., (2009), Enantioselective Separation and Degradation of the Herbicide Dichlorprop Methyl in Sediment, *Chirality, 21,* pp. 480-483.

Cole, J., Dolak, L. A., Lefler, J. L., (2007), The chiral resolution of an herbicidal product by SFC, *Chimica Oggi-Chemistry Today, 25,* pp. 32-36.

Li, L., Zhou, S. S., Jin, L. X., Zhang, C., Liu, W. P., (2010), Enantiomeric separation of organophosphorus pesticides by high-performance liquid chromatography, gas chromatography and capillary electrophoresis and their applications to environmental fate and toxicity assays, *Journal of Chromatography B-Analytical Technologies in the Biomedical and Life Sciences, 878,* pp. 1264-1276.

Hsieh, Y. Z., Huang, H. Y., (1996), Analysis of chlorophenoxy acid herbicides by cyclodextrin-modified capillary electrophoresis, *Journal of Chromatography A, 745,* pp. 217-223.

Mechref, Y., El., R. Z., (1996a), Capillary electrophoresis of herbicides. 1. Precolumn derivatization of chiral and achiral phenoxy acid herbicides with a fluorescent tag for electrophoretic separation in the presence of cyclodextrins and micellar phases, *Analytical Chemistry, 68,* pp. 1771-1777.

Mechref, Y., ElRassi, Z., (1997a), Capillary electrophoresis of herbicides .4. Evaluation of octylmaltopyranoside chiral surfactant in the enantiomeric separation of fluorescently labeled phenoxy acid herbicides and their laser-induced fluorescence detection, *Electrophoresis, 18,* pp. 220-226.

Mechref, Y., ElRassi, Z., (1996b), Capillary electrophoresis of herbicides .3. Evaluation of octylmaltopyranoside chiral surfactant in the enantiomeric separation of phenoxy acid herbicides, *Chirality, 8,* pp. 518-524.

Mechref, Y., ElRassi, Z., (1997b), Capillary electrophoresis of herbicides .2. Evaluation of alkylglucoside chiral surfactants in the enantiomeric separation of phenoxy acid herbicides, *Journal of Chromatography A, 757*, pp. 263-273.

Desiderio, C., Polcaro, C. M., Padiglioni, P., Fanali, S., (1997a), Enantiomeric separation of acidic herbicides by capillary electrophoresis using vancomycin as chiral selector, *Journal of Chromatography A, 781*, pp. 503-513.

Ingelse, B. A., Reijenga, J. C., Flieger, M., Everaerts, F. M., (1997), Capillary electrophoretic separation of herbicidal enantiomers applying ergot alkaloids, *Journal of Chromatography A, 791*, pp. 339-342.

Pirkle, W. H., Lee, W., Welch, C. J., (1997), Chromatographic separation of the enantiomers of 2-aryloxypropionic acids, esters and amides, *Enantiomer, 2*, pp. 423-431.

Haynes, J. L., Shamsi, S. A., O'Keefe, F., Darcey, R., Warner, I. M., (1998), Cationic beta-cyclodextrin derivative for chiral separations, *Journal of Chromatography A, 803*, pp. 261-271.

Jarman, J. L., Jones, W. J., Howell, L. A., Garrison, A. W., (2005), Application of capillary electrophoresis to study the enantioselective transformation of five chiral pesticides in aerobic soil slurries, *Journal of Agricultural and Food Chemistry, 53*, pp. 6175-6182.

Yi, F., Guo, B. Y., Peng, Z. L., Li, H. F., Marriott, P., Lin, J. M., (2007), Study of the enantioseparation of imazaquin and enantioselective degradation in field soils by CZE, *Electrophoresis, 28*, pp. 2710-2716.

Han, L., Guo, B. Y., Feng, J. H., Lu, X. M., Lin, J. M., (2008), Study on the Enantioselective Degradation of Imazethapyr in Soil by CE, *Chromatographia, 68*, pp. 1071-1073.

Pico, Y., Rodriguez, R., Manes, J., (2003), Capillary electrophoresis for the determination of pesticide residues, *Trac-Trends in Analytical Chemistry, 22*, pp. 133-151.

Nielen, M. W. F., (1993), (Enantio-)separation of phenoxy acid herbicides using capillary zone electrophoresis, *Journal of Chromatography, 637*, pp. 81-90.

Otsuka, K., Smith, C. J., Grainger, J., Barr, J. R., Patterson, D. G., Tanaka, N., Terabe, S., (1998), Stereoselective separation and detection of phenoxy acid herbicide enantiomers by cyclodextrin-modified capillary zone electrophoresis electrospray ionization mass spectrometry, *Journal of Chromatography A, 817*, pp. 75-81.

Zerbinati, O., Trotta, F., Giovannoli, C., (2000), Optimization of the cyclodextrin-assisted capillary electrophoresis separation of the enantiomers of phenoxyacid herbicides, *Journal of Chromatography A, 875*, pp. 423-430.

Garrison, A. W., Schmitt, P., Kettrup, A., (1994), Separation of phenoxy acid herbicides and their enantiomers by high-performance capillary electrophoresis, *Journal of Chromatography A, 688*, pp. 317-327.

Desiderio, C., Polcaro, C. M., Fanali, S., (1997b), Stereoselective analysis of herbicides by capillary electrophoresis using sulfobutyl ether beta-cyclodextrin as chiral selector, *Electrophoresis, 18*, pp. 227-234.

Penmetsa, K. V., Leidy, R. B., Shea, D., (1997), Enantiomeric and isomeric separation of herbicides using cyclodextrin-modified capillary zone electrophoresis, *Journal of Chromatography A, 790*, pp. 225-234.

Zerbinati, O., Trotta, F., Giovannoli, C., Baggiani, C., Giraudi, G., Vanni, A., (1998), New derivatives of cyclodextrins as chiral selectors for the capillary electrophoretic separation of dichlorprop enantiomers, *Journal of Chromatography A, 810*, pp. 193-200.

Aga, D. S., Heberle, S., Rentsch, D., Hany, R., Muller, S. R., (1999), Sulfonic and oxanilic acid metabolites of acetanilide herbicides: Separation of diastereomers and enantiomers by capillary zone electrophoresis and identification by H-1 NMR spectroscopy, *Environmental Science & Technology, 33,* pp. 3462-3468.

Malik, A. K., Aulakh, J. S., Fekete, A., Schmitt-Kopplin, P., (2009), Separation of the Phenoxy Acid Herbicides and Their Enantiomers by Capillary Zone Electrophoresis in Presence of Highly Sulphated Cyclodextrins, *Journal of the Chinese Chemical Society, 56,* pp. 1163-1167.

Garrison, A. W., Schmitt, P., Martens, D., Kettrup, A., (1996), Enantiomeric selectivity in the environmental degradation of dichlorprop as determined by high performance capillary electrophoresis, *Environmental Science & Technology, 30,* pp. 2449-2455.

Andre, C., Berthelot, A., Thomassin, M., Guillaume, Y. C., (2006), Enantioselective aptameric molecular recognition material: Design of a novel chiral stationary phase for enantioseparation of a series of chiral herbicides by capillary electrochromatography, *Electrophoresis, 27,* pp. 3254-3262.

Messina, A., Sinibaldi, M., (2007), Enantioseparations on chiral monolithic columns: A study of the stereoselective degradation of (R/S)-dichlorprop 2-(2,4-dichlorophenoxy)propionic acid in soil, *Electrophoresis, 28,* pp. 2613-2618.

Buchinger, S., Follrich, B., Lammerhofer, M., Lubda, D., Lindner, W., (2009), Chirally functionalized anion-exchange type silica monolith for enantiomer separation of 2-aryloxypropionic acid herbicides by non-aqueous capillary electrochromatography, *Electrophoresis, 30,* pp. 3804-3813.

Mechref, Y., ElRassi, Z., (1996c), Micellar electrokinetic capillary chromatography with in-situ charged micelles VI. Evaluation of novel chiral micelles consisting of steroidal glycoside surfactant borate complexes, *Journal of Chromatography A, 724,* pp. 285-296.

Schmitt, P., Garrison, A. W., Freitag, D., Kettrup, A., (1997), Application of cyclodextrin-modified micellar electrokinetic chromatography to the separations of selected neutral pesticides and their enantiomers, *Journal of Chromatography A, 792,* pp. 419-429.

Martin-Biosca, Y., Garcia-Ruiz, C., Marina, M. L., (2001), Enantiomeric separation of chiral phenoxy acid herbicides by electrokinetic chromatography. Application to the determination of analyte-selector apparent binding constants for enantiomers, *Electrophoresis, 22,* pp. 3216-3225.

Gavioli, E., Maier, N. M., Minguillon, C., Lindner, W., (2004), Preparative enantiomer separation of dichlorprop with a cinchona-derived chiral selector employing centrifugal partition chromatography and high-performance liquid chromatography: A comparative study, *Analytical Chemistry, 76,* pp. 5837-5848.

Liu, H. J., Xiong, M. Y., (2009), Comparative toxicity of racemic metolachlor and S-metolachlor to Chlorella pyrenoidosa, *Aquatic Toxicology, 93,* pp. 100-106.

Kurt-Karakus, P. B., Bidleman, T. F., Muir, D. C. G., Struger, J., Sverko, E., Cagampan, S. J., Small, J. M., Jantunen, L. M., (2010), Comparison of concentrations and stereoisomer ratios of mecoprop, dichlorprop and metolachlor in Ontario streams, 2006-2007 vs. 2003-2004, *Environmental Pollution, 158,* pp. 1842-1849.

Chan, J. H. H., Walker, F., Tseng, C. K., Baker, D. R., Arneklev, D. R., (1975), Synthesis and herbicidal activity of N,N-diethyl-2-(1-naphthyloxy)propionamide and its optical isomers, *Journal of Agricultural and Food Chemistry, 23,* pp. 1008-1010.

Couderchet, M., Bocion, P. F., Chollet, R., Seckinger, K., Boger, P., (1997), Biological activity of two stereoisomers of the N-thienyl chloroacetamide herbicide dimethenamid, *Pesticide Science, 50,* pp. 221-227.

Shimabukuro, R. H., Hoffer, B. L., (1995), Enantiomers of diclofop-methyl and their role in herbicide mechanism of action, *Pesticide Biochemistry and Physiology, 51,* pp. 68-82.

Ye, J., Zhang, Q. O., Zhang, A. P., Wen, Y. Z., Liu, W. P., (2009), Enantioselective Effects of Chiral Herbicide Diclofop Acid on Rice Xiushui 63 Seedlings, *Bulletin of Environmental Contamination and Toxicology, 83,* pp. 85-91.

Maneechote, C., Preston, C., Powles, S. B., (1997), A diclofop-methyl-resistant Avena sterilis biotype with a herbicide-resistant acetyl-coenzyme A carboxylase and enhanced metabolism of diclofop-methyl, *Pesticide Science, 49,* pp. 105-114.

Cai, X. Y., Liu, W. P., Sheng, G. Y., (2008), Enantioselective degradation and ecotoxicity of the chiral herbicide diclofop in three freshwater alga cultures, *Journal of Agricultural and Food Chemistry, 56,* pp. 2139-2146.

Diao, J. L., Xu, P., Wang, P., Lu, Y. L., Lu, D. H., Zhou, Z. Q., (2010a), Environmental Behavior of the Chiral Aryloxyphenoxypropionate Herbicide Diclofop-Methyl and Diclofop: Enantiomerization and Enantioselective Degradation in Soil, *Environmental Science & Technology, 44,* pp. 2042-2047.

Buser, H. R., Muller, M. D., (1997a), Enantioselective analyses of persistent and modern pesticides. A step toward sustainable agriculture, *Chimia, 51,* pp. 694-700.

Buser, H. R., Muller, M. D., (1997b), Conversion reactions of various phenoxyalkanoic acid herbicides in soil .2. Elucidation of the enantiomerization process of chiral phenoxy acids from incubation in a D2O/soil system, *Environmental Science & Technology, 31,* pp. 1960-1967.

Muller, M. D., Buser, H. R., (1997), Conversion reactions of various phenoxyalkanoic acid herbicides in soil .1. Enantiomerization and enantioselective degradation of the chiral 2-phenoxypropionic acid herbicides, *Environmental Science & Technology, 31,* pp. 1953-1959.

Harrison, I., Williams, G. M., Carlick, C. A., (2003), Enantio selective biodegradation of mecoprop in aerobic and anaerobic microcosms, *Chemosphere, 53,* pp. 539-549.

Williams, G. M., Harrison, I., Carlick, C. A., Crowley, O., (2003), Changes in enantiomeric fraction as evidence of natural attenuation of mecoprop in a limestone aquifer, *Journal of Contaminant Hydrology, 64,* pp. 253-267.

Wen, Y. Z., Yuan, Y. L., Shen, C. S., Liu, H. J., Liu, W. P., (2009), Spectroscopic Investigations of the Chiral Interactions Between Lipase and the Herbicide Dichlorprop, *Chirality, 21,* pp. 396-401.

Wen, Y. Z., Yuan, Y. L., Chen, H., Xu, D. M., Lin, K. D., Liu, W. P., (2010), Effect of Chitosan on the Enantioselective Bioavailability of the Herbicide Dichlorprop to Chlorella pyrenoidosa, *Environmental Science & Technology, 44,* pp. 4981-4987.

Ramezani, M. K., Oliver, D. P., Kookana, R. S., Lao, W., Gill, G., Preston, C., (2010), Faster degradation of herbicidally-active enantiomer of imidazolinones in soils, *Chemosphere, 79,* pp. 1040-1045.

Qian, H. F., Hu, H. J., Mao, Y. Y., Ma, J., Zhang, A. P., Liu, W. P., Fu, Z. W., (2009), Enantioselective phytotoxicity of the herbicide imazethapyr in rice, *Chemosphere, 76*, pp. 885-892.

Diao, J. L., Xu, P., Wang, P., Lu, D. H., Lu, Y. L., Zhou, Z. Q., (2010b), Enantioselective Degradation in Sediment and Aquatic Toxicity to Daphnia magna of the Herbicide Lactofen Enantiomers, *Journal of Agricultural and Food Chemistry, 58*, pp. 2439-2445.

Omokawa, H., Konnai, M., (1992), Inhibition of Echinochloa crus-galli var. frumentacea seedling root elongation by chiral 1, 3, 5-triazines in the dark, *Pesticide Science, 35*, pp. 83-86.

Omokawa, H., Ryoo, J. H., (2001), Enantioselective response of rice and barnyard millet on root growth inhibition by optically active alpha-methylbenzyl phenylureas, *Pesticide Biochemistry and Physiology, 70*, pp. 1-6.

Omokawa, H., Murata, H., Kobayashi, S., (2004), Chiral response of Oryzeae and Paniceae plants in alpha-methylbenzyl-3-p-tolyl urea agar medium, *Pest Management Science, 60*, pp. 59-64.

15

Residual Herbicide Dissipation in Vegetable Production

Timothy Grey and William Vencill
University of Georgia
United States

1. Introduction

The use of low density polyethylene mulch for fumigation, weed control, and soil cover has become the standard for production of many vegetables in the southeastern United States. Most low density polyethylene mulch laid for spring vegetable production is followed by a second crop in the autumn and potentially a third crop the following spring. These succeeding vegetable crops can be transplanted directly into the existing low density polyethylene mulch covered beds formed prior to spring fumigation. This allows for multiple crop production using the same beds. This is done in order to minimize expenses associated with low density polyethylene mulch and drip tape irrigation, by distributing costs over multiple crops (Fig. 1).

Fig. 1. Newly laid low density polyethylene mulch for spring planting of vegetables (left) and clean beds prior to autumn vegetable transplanting.

2. Important

Alternative methyl bromide fumigants have been investigated with varying levels of weed control success (Csinos et al., 2002; Webster et al., 2001; Gilreath et al., 2004). The major source of new herbicides for minor crops is the adaptation of herbicides registered for major crops. The process of registration of herbicides is expensive and time consuming for minor crops (Fennimore & Doohan, 2008). Yellow nutsedge (*Cyperus esculentus*) and purple nutsedge (*Cypresrotundus*) are the most common and troublesome vegetable weeds in vegetables throughout the southern United States (Webster, 2006; Webster & MacDonald, 2001). Even with polyethylene bed covers, nutsedge control in vegetable production is essential, as emerging shoots can grow though the cover (Adcock et al., 2008; Bullock, 1990; Locascio et al., 1994; Igbokwe, 1996; Stiles et al., 1999). Herbicides that could be soil incorporated into vegetable systems using low density polyethylene mulch must be effective on *Cyperus* species. Although extended residual control can be beneficial, it poses a threat to future crops (Johnson et al., 2010). If growers use the low density polyethylene mulch for multiple vegetable cropping systems, there is a potential for herbicide carryover to injury succeeding crops, which could increase, dependent on herbicide persistence (Johnson & Mullinix, 2005). In previous studies, it was determined that pesticide dissipation was affected by polyethylene mulch, which could influence weed control, crop injury, and pesticide persistence. The dissipation of linuron, pendimethalin, chlorobromuron, and flurochloridone was reduced when applied to soil under perforated polyethylene covers verses bare soil (Bond & Walker, 1989). Many vegetable producers in the southeastern United States often apply herbicides between crop plantings in order to destroy the previous crop and/or weed infestations (Gilreath et al., 2006). Herbicides that can be used between crops for low density polyethylene mulch vegetation control include glyphosate, paraquat, carfentrazone, and halosulfuron. Pesticides applied over the top of low density polyethylene mulch can leave residues on the mulch (Nerin et al., 1996). When herbicides are applied to the low density polyethylene mulch or row middles, and then when crops are transplanted soon afterwards, injury can occur (Culpepper et al., 2009; MacRae & Culpepper, 2007; Gilreath & Duranceau, 1986). The purpose of this chapter is to present current information about the chemical dissipation of herbicides used for vegetable production as alternatives to methyl bromide under different application scenarios.

3. Information

Herbicide dissipation is chemical and environmentally dependent. Soil incorporated herbicides are exposed to variable microbial, hydrolysis, soil pH, organic matter, and other factors that may limit their activity. However, herbicide adsorption to soil colloids with subsequent hysteresis may extend activity and thus potential for either weed control or carryover to subsequent crops. Post emergence applied herbicide dissipation can be influenced by chemical properties such as water solubility, photo degradation, volatility, and environmental aspects such as rainfall and irrigation volumes, plant interception and absorption. While herbicide dissipation differs with respect to application method, many of the same factors influence fate in the environment.

3.1 Soil applied herbicides

Halosulfuron (Grichar et al., 2003; Nelson & Renner, 2002; Vencill et al., 1995), sulfentrazone (Grichar et al., 2003; Wehtje et al., 1997), and metolachlor (Cornelius et al., 1985; Obrigawitch et al., 1980) provide soil residual activity on *Cyperus* species with control often extending for many weeks or months after applications. While these herbicides are viable alternatives to fumigation in vegetable production, they may cause injury to newly transplanted crops (Figure 2), and potential carryover issues to subsequent crops.

Dermiyati & Yamamoto (1997a) indicated that halosulfuron adsorption was highly correlated with soil organic carbon content and inversely related to soil pH. Degradation of halosulfuron increases with increasing temperature and lower soil pH but varied with soil moisture content and soil type (Dermiyati & Yamamoto, 1997b). Halosulfuron degradation

Fig. 2. Foreground: bed nontreated control transplanted cucumber; background bed herbicide injury on transplanted cucumber.

is primarily through chemical hydrolysis and microbial means. Carpenter et al. (1999) reported a positive relationship between organic matter and halosulfuron adsorption, that sorghum [*Sorghum bicolor* (L.) Moench] injury was less likely on soils with high organic matter content, and halosulfuron can exhibit soil hysteresis.

Metolachlor dissipation from soil has been extensively investigated (Bouchard et al., 1982; Braverman et al., 1986; Gaynor et al., 1993; Obrigawitch et al., 1981; Peter & Weber, 1985; Weber et al., 2003). Weber et al. (2003) reported that metolachlor sorption, mobility, and soil retention was related to organic matter, clay content, and surface area. As soil organic matter concentration increases, adsorption of metolachlor increases. Metolachlor mobility was inversely related to soil organic matter and clay content. Other studies came to the same conclusions and also indicated that metolachlor binding was by physical forces between metolachlor molecules and soil constituent surfaces (Weber et al., 2003). Half-life of metolachlor varies with soil temperature, moisture, and organic matter content (Parker et al., 2005; Vencill, 2002 a).

Previous research indicated that the adsorption and mobility of sulfentrazone is pH and soil type-dependent (Grey et al., 1997) and that it does exhibit hysteresis (Grey et al., 2000). Reddy & Locke (1998) confirmed the conclusion that sulfentrazone availability was both pH and soil series-dependent. They also concluded that sulfentrazone sorption was greater in no-till than in conventional tillage and attributed this to the higher organic matter content. Soil dissipation of sulfentrazone has varied with climatic factors. Ohmes et al. (2000) noted that sulfentrazone dissipation was slowed by dry soil conditions, leading to substantial residual activity in subsequent crops. There was little-to-no injury to rotational vegetable crops from sulfentrazone when it was applied in accordance to the product label (Garvey & Monks, 1998). For the vegetable crops they investigated, all exhibited little to no adverse effects as part of a rotation with sulfentrazone. The residual effects of sulfentrazone on rotational cotton have been established with rates exceeding 0.4 kg ha^{-1} applied the previous year (Main et al., 2004; Ohmes et al., 2000).

Persistence, dissipation, and degradation of halosulfuron (Kuwatsuka & Yamamoto 1997a, 1997b), metolachlor (Gaynor et al., 1993, Parker et al 2005; Weber et al., 2003), and sulfentrazone (Grey et al., 1997, 2000; Ohmes et al., 2000; Reddy & Locke, 1998) have been previously investigated in agronomic and/or vegetable soils. These investigations emphasized soil and/or organic cover scenarios in separate experiments. However, there is little information on these herbicides applied to bare-soil verses soil under polyethylene mulch situations (Figure 3).

Fig. 3. Bare soil and soil covered with low density polyethylene mulch prepared for vegetable planting.

3.2 Herbicides applied to low density polyethylene mulch

Low density polyethylene mulch has permeability to fumigants via diffusion through the matrix (Papiernik & Yates, 2001a; Papiernik & Yates, 2001b). This occurs as the fumigant dissolves into the surface of the low density polyethylene mulch facing the soil, then diffusion through the film, and eventual evaporation from the opposite surface (Rogers, 1985). However, little information exists about low density polyethylene mulch adsorptive properties with respect to pesticides.

Previous research noted that paraquat dissipation from low density polyethylene mulch, when post emergence surface applied, was achieved by photo degradation (Gilreath et al., 2006; Gilreath & Duranceau, 1986) or removal with an eluent such as water (Gilreath & Duranceau, 1986). The persistence, dissipation, and degradation of paraquat from low density polyethylene mulch were evaluated using bioassays (Gilreath & Duranceau, 1986) and colorimetric procedures (Gilreath et al., 2006). However, besides paraquat, few herbicides have been analytically quantified for dissipation from low density polyethylene mulch (Grey et al., 2009).

3.3 Research

These two factors, soil applied residual herbicides and herbicide residues remaining on low density polyethylene mulch, can potentially injure or kill vegetable crops in rotation. Understanding the impact of low density polyethylene mulch on residual herbicide soil dissipation when incorporated into vegetable production and the respective rotational issues will impact what herbicides, and crops, producers will apply and grow, respectively. Additionally the impact of dissipation of herbicides when surface applied to low density polyethylene mulch will impact which herbicides producers will utilize between vegetable crop plantings. The objectives were to review information which compares soil dissipation of residual herbicides in bare soil verses soil under low density polyethylene mulch, and herbicide dissipation from low density polyethylene mulch when topically applied, using field experiments and analytical chemical analysis.

4.1 Field studies

Field studies conducted to evaluate herbicide dissipation of herbicides had two distinct research objectives. However, all experiments were conducted similarly. Bed formation (20 cm raised bed), single drip irrigation tube, and lying of 32 *u*m-thick (1.25 mil) low density polyethylene mulchoccurred simultaneously. All studies were conducted on Tifton loamy sand (fine-loamy, kaolinitic, thermic PlinthicKandiudults) with 86 to 88% sand, 8% silt, 4 to 6% clay, 0.5 to 1.3% organic matter, and pH ranging from 6.3 to 6.9.

4.2 Soil dissipation research

The first experiments evaluated herbicide dissipation for bare soils verses soil under low density polyethylene mulch. For the soil dissipation experiments, herbicide treatments included halosulfuron, *S*-metolachlor, and sulfentrazone applied at recommended rates for weed control for the region.Surface soil was sampled with a plugger-type sampler with four samples were collected to a depth of 8 cm from each plot and combined into a single sample. Soil cores were collected at 1 hour, 1, 2, 14, 27, and 56 d after treatment for one experiment and 1 hour, 1, 2, 3, 7, 14, 21, 28, 44, and 66 days after treatment in the second experiment.All samples were immediately frozen upon collection and stored at -10 C prior to analysis. For soil herbicide analysis, soils were thawed, air dried on a lab bench for 8 hours, passed through a 3-mm sieve, and then stored at -10 C. Field plot replicate sample integrity was maintained throughout sample collection, preparation, and chemical analysis.

4.3 Low density polyethylene mulch research

For herbicide dissipation from the surface of low density polyethylene mulch experiments, dissipation was measured quantitatively using analytical techniques for two scenarios: under dry conditions (rain free and no irrigation) versus wash off using water as an eluent. Herbicide treatments included paraquat, glyphosate, carfentrazone, and halosulfuron. For the wash off studies, samples were collected at one hour after treatment, irrigated at three hours after treatment with one cm of water using an overhead irrigation system, then sampled again at five hours after treatment. This washing and sampling procedure was then repeated at 24, 48, 72 and 96 hours after treatment. Samples were collected from each plot. Samples were then carefully stored in brown glass jars. For all studies, care was taken to prevent contamination between samples and to collect a representative sample from each plot. All samples were immediately frozen upon collection and stored at -10 C prior to analysis.

4.4 Analytical herbicide quantification

Herbicide analytical methods differ by chemical due to variation in solubility, structure, volatility, etc. Therefore, various methods are used to quantify. One common method is the use of high pressure liquid chromatography in tandem with mass spectrometry (HPLC-MS). This procedure allows for accurate measures for many different compounds at extremely sensitive levels of detection. Proper sample handling includes solvent identification, solvent ratios, extraction methods, injection volumes, and equipment settings. Herbicides and methods of analysis discussed for the purpose of this chapter, using HPLC-MS, are listed in Table 1.

Herbicide	Method	Mobile phase	Column	Flow rate	Ion monitoring	Cone voltage	APCI /ESI
				ml/min		C	C
S-metolachlor	LC-MS SIR	ACN 40% MeOH with 0.2% acetic acid	Ymc ODS	0.2	ESI positive	20	380
Sulfentrazone	LC-MS SIR	ACN 25% MeOH with 0.2% acetic acid	Ymc ODS	0.2	ESI negative	51	370
Glyphosate	LC-MS SIR	ACN 50 mM ammonium acetate	Phenomene x C18	0.7	APCI positive	100	475
Paraquat	LC-MS SIR	60:40 ACN Buffer pH 4.5	Waters Si	0.4	ESI negative	41	380
Halosulfuron	LC-MS SIR	ACN 10% MeOH with 0.1 ml ammonium hydroxide	ymc ODS	0.2	ESI negative	32	370
Carfentrazone	LC-MS SIR	ACN 10 MeOH with 10 mM formic acid	ymc ODS	0.2	ESI positive	23	350
Flumioxazin	LC-MSMRM	ACN 1.25 mM TDFHA	Hypersil ODS	0.6	APCI positive	40	484

Table 1. Analytical methods for herbicide dissipation studies.

4.5 Herbicide dissipation kinetics

Herbicide dissipation data are often described by non-linear regression in addition to analysis of variance for the specific test. The intent is to determine if the responses can be described by using the exponential decay equation

$$y = B_o e^{-B_1(x)} \quad (1)$$

where y is herbicide concentration, B_0 is the initial concentration, B_1 is dissipation rate, and x is time in hours or days after treatment. After data is regressed against time, the output from the analysis includes first-order dissipation rate constant (k) (Ohmes et al., 2000). All data by herbicide for the exponential decay equations can be subjected to analysis of variance (ANOVA) using the general linear models procedures with mean separation using 95% asymptotic confidence intervals. Dissipation time (50%) is then determined using the equation

$$DT_{50} = \ln 0.50/k \quad (2)$$

(Dermiyati & Yamamoto, 1997b; Lui et al., 2002; Mueller et al., 1999). Data are then often presented with graphics software.

5.1 Herbicide dissipation research

For the following research information, all experiments were conducted at times when herbicide applications could potentially occur in the south-eastern vegetable production regions of the United States and are thus representative of producer practices.

5.2 Soil dissipation research

The exponential decay equation [1] effectively describes halosulfuron dissipation (Figure 3). First-order dissipation rate constants (k) for halosulfuron were less (i.e. slower dissipation) for soil under low density polyethylene mulch (0.07) than for bare soil (0.10). Halosulfuron dissipation for bare soil dropped to undetectable levels by 27 and 28 days after treatment in two studies, respectively. This trend was similar for soil under low density polyethylene mulch. From equation [2], the DT_{50} for bare-soil was 6 to 7 daysverses soil under low density polyethylene mulch which was 10 days. Although the first-order rate constants were not significantly different between bare soil and soil under LDPE mulch, the DT_{50} was 3 to 4 days longer for soil under low density polyethylene mulch. Dermiyati and Yamamoto (1997b) reported halosulfuron half-lives of 7 to 98 days depending on soil moisture and temperature regimes.

S-metolachlor dissipation was well described by the exponential decay equation [1] and for bare soil and soil under low density polyethylene mulch (Figure 3). First-order dissipation rate constants for *S*-metolachlor were less for soil under low density polyethylene mulch (0.2) than for bare soil (0.4). *S*-metolachlor dissipation was rapid for bare soil and soil under low density polyethylene mulch dropping to undetectable levels by 44 days after treatment. Rapid dissipation has been previously noted for metolachlor with sandy soil under moist soil conditions (Weber et al., 2003). In one experiment, *S*-metolachlor dissipation was biphasic, dropping to less than 400 *u*g/kg of soil at 7 days after treatment, yet was detectable at 44 days after treatment for both soil scenarios. While the $DT_{50}$was 2 and 5 days for bare soil and soil under low density polyethylene mulch, respectively, dissipation was slower in one year as compared to another. This could be attributed to an equilibrium that was reached with *S*-metolachlor where soil adsorption had occurred, and then desorption of the parent was observed over time (Patakioutas & Albanis 2002). Data indicated that low density polyethylene mulch decreased the rate of dissipation of *S*-metolachlor versus bare soil which could extend its herbicidal activity.

Sulfentrazone dissipation varied but was slower than halosulfuron and *S*-metolachlor (Figure3) and had the longer DT_{50}. Overall, the exponential decay equation [1] adequately described the sulfentrazone dissipation. Sulfentrazone dissipation first order rate constants were equal on average, with 0.055 for soil under low density polyethylene mulch and 0.050 for bare-soil. Half-lives were 16 days for bare-soil and 13 days for soil under low density polyethylene mulch. While counter intuitive, this could be due to increased temperature regimes that have been noted under polyethylene mulch (Peachey et al., 2001), that could have accelerated dissipation. Ohmes et al. (2000) previously noted that sulfentrazone dissipation followed first-order kinetics in Tennessee soils. They reported varying dissipation with DT_{50} ranging from 24 to 118 days. Variation in sulfentrazone DT_{50} has been noted from 2 (Collins et al., 1999) to 302 days (Vencill, 2002 b).

These studies indicate that halosulfuron-methyl and *S*-metolachlor dissipation was more rapid for bare-soil than soil under low density polyethylene mulch. However, sulfentrazone dissipation was variable. For bare-soil and soil covered with low density polyethylene mulch, dissipation of halosulfuron and *S*-metolachlor were biphasic (Figure 3).

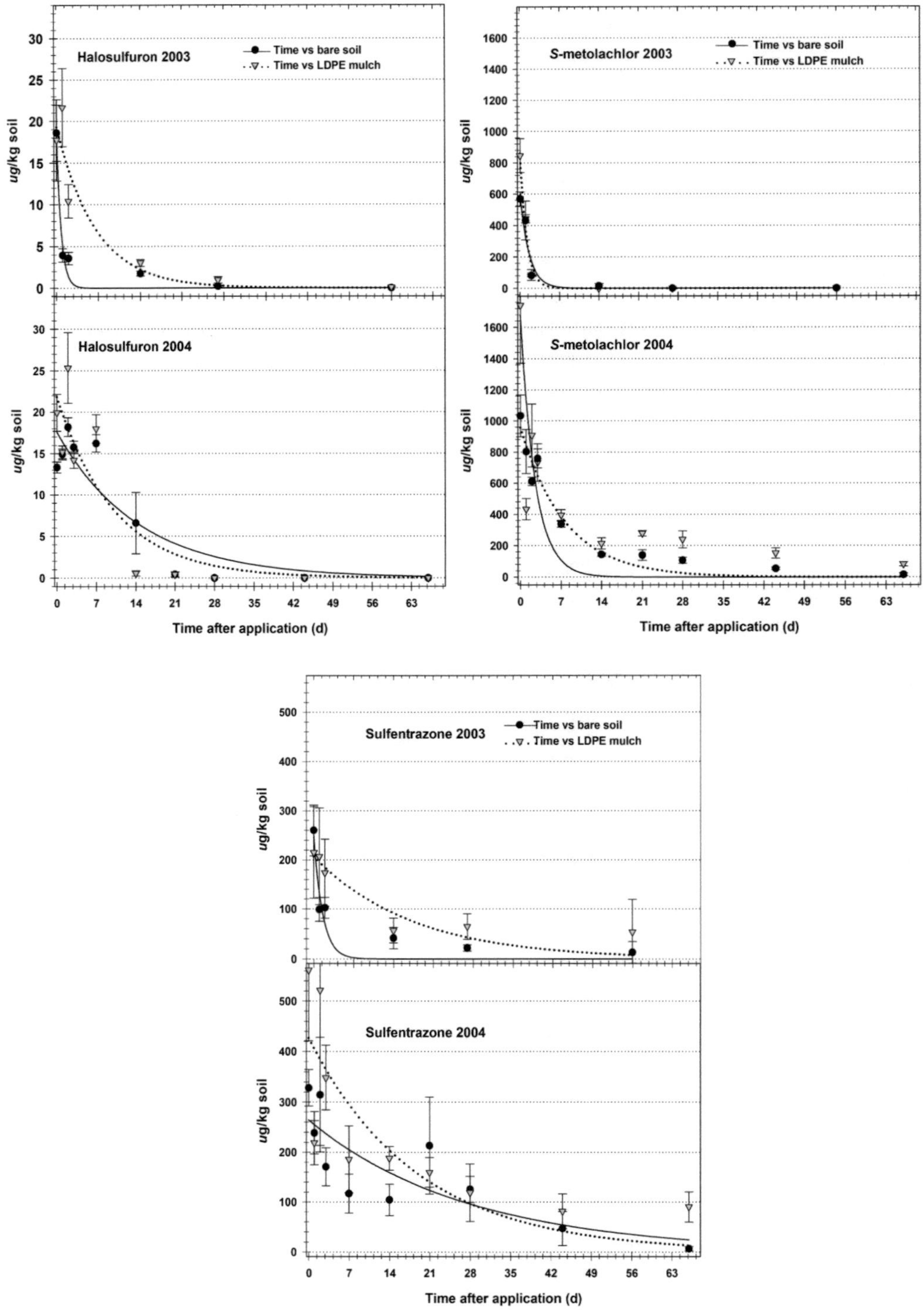

Fig. 3. Halosulfuron, *s*-metolachlor, and sulfentrazone dissipation for bare soil and soil covered with low density polyethylene mulch.

Sulfentrazone dissipation was slower (Figure 3) than halosulfuron or metolachlor. This indicates that sulfentrazone could provide residual *Cypress* species control when preemergence applied to vegetables but could also result in carryover problems to subsequent plantings.

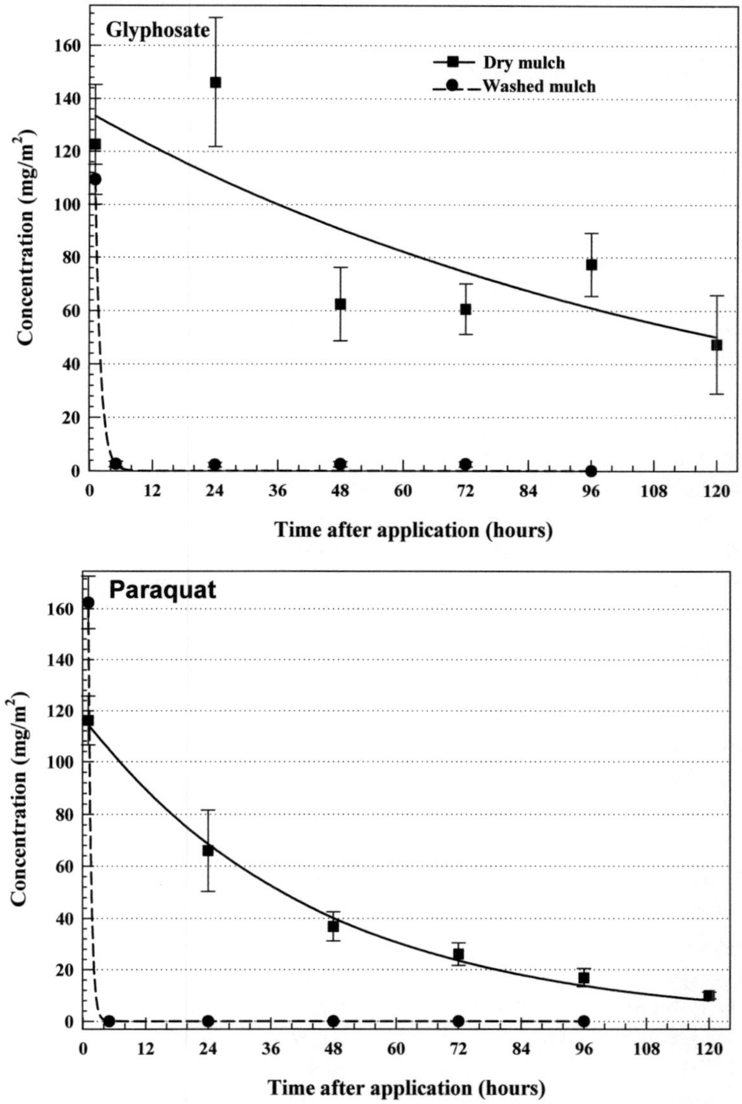

Fig. 4. Glyphosate and paraquat dissipation from low density polyethylene mulch for dry and wash off conditions over time.

6. 2 Low density polyethylene dissipation research

The exponential decay equation [1] effectively described dry and irrigated glyphosate dissipation (Figure 4). First-order dissipation rate constants (k) for glyphosate were (i.e. slower dissipation) less for the dry study at 0.008 than for the irrigated study at 0.933. For glyphosate, DT_{50} for the dry study was 84 hours, while it was 1 hour in the irrigated experiment. Glyphosate concentration dropped to less than 5 mg/m^2 levels by the 2nd irrigation event at 24 hours after treatment. Glyphosate dropped to undetectable levels by the 5th irrigation at 96 hours after treatment when greater than 4 cm of water had been applied. Glyphosate has negligible photo degradation losses, is tightly adsorbed to soil, and high water solubility (Senseman, 2007). Glyphosate adsorption to clay minerals is pH dependent and fluctuations can occur, depending upon the type of soil saturating cation (McConnell & Hossner, 1985). In contrast to the wash off experiments, glyphosate dissipation from low density polyethylene mulch for the dry study was linear and 50 mg/m^2 remaining 120 hours after treatment. For the dry study there was an 84 hour half-life for glyphosate, and it would require at least 28 d (eight half-lives) to reach less than 1 mg/m^2 on the mulch. Glyphosate can be persistent in low density polyethylene mulch as reported by Gilreath & Santos (2004). For their bioassay study with tomato (*Lycopersiconesculentum* L.) in a dry study, they indicated there was enough glyphosate remaining 16 days after application to reduce fresh plant weight by 73%.

First order rate dissipation constants for paraquat were significantly different for the irrigated (1.88) and dry studies (0.022). Paraquat is a cationic dichloride salt with a water solubility of 620,000 mg/L^1 (Senseman, 2007). After the 1st irrigation application of 1 cm of water, paraquat was undetectable on the mulch (Figure 4), which was further demonstrated with a 1 hour DT_{50}. Given the high water solubility of paraquat, rapid dissipation will occur with water. As previously noted, paraquat dissipation can also occur via photo degradation (Senseman, 2007). With each subsequent 24 hour sampling period, paraquat dissipation was reduced step-wise, falling to 10 mg/m^2 at 120 hours after treatment (Figure 2) with a DT_{50} of 32 hours for the dry study. Gilreath et al. (2006) reported similar findings for paraquat on low density polyethylene mulch using a colorimetric analysis procedure.

The exponential decay equation [1] described halosulfuron dissipation for the dry scenario with a first order rate constant of 0.038. In the irrigated study, however, the exponential decay equation did not accurately describe halosulfuron dissipation from low density polyethylene mulch, and actually under estimated the levels detected (Figure 5). Halosulfuron dissipation for the irrigated study appeared to be biphasic, with an initial rapid decline, and then little to no removal with each subsequent irrigation event. The first phase of halosulfuron dissipation is chemical hydrolysis and is abiotic in nature, whereas the second phase is microbial dependent. This would explain the biphasic nature for observed dissipation with irrigation. Halosulfuron dissipation in the irrigated study had first order rate constant of 0.24, which was significantly higher than that in the dry study. Halosulfuron is a weak acid with negligible photo-degradation losses (Senseman 2007). Halosulfuron has exhibited hysteresis in higher organic matter Japanese soils (Dermiyati & Yamamoto, 1997a). Given this previously noted hysteric soil affect, it is suspected halosulfuron is behaving similarly when applied to low density polyethylene LDPE mulch. Halosulfuron dissipation was linear and varied by less than 1.1 mg/m^2 from initial application with 3.5 mg/m^2 at 1 hour after treatment to 2.4 mg/m^2 120 hours after treatment for the dry study.

Dissipation of carfentrazone was well described by the exponential decay equation [1] with first order rate constants of 0.023 and 0.025 and DT_{50} values of 30 and 28 hours for the dry and wash off studies, respectively. Sampling of the dry and irrigation studies indicated nearly identical dissipation curves (Figure 5). Initial carfentrazone concentrations on the low density polyethylene mulch were 7.8 to 8.4 mg/m^2 at 1 hour after treatment. With each subsequent sampling at 24, 48, 72, and 96 hours after treatment, carfentrazone concentrations did not differ by more than 0.4 mg/m^2 for samples of the dry and wash off low density polyethylene mulch. Carfentrazone water solubility is 12,000 mg/L[1] and increases with temperature; it does not photo-degrade, is non-volatile, is not adsorbed to soil, but it is broken down via microbial break down (Senseman, 2007).

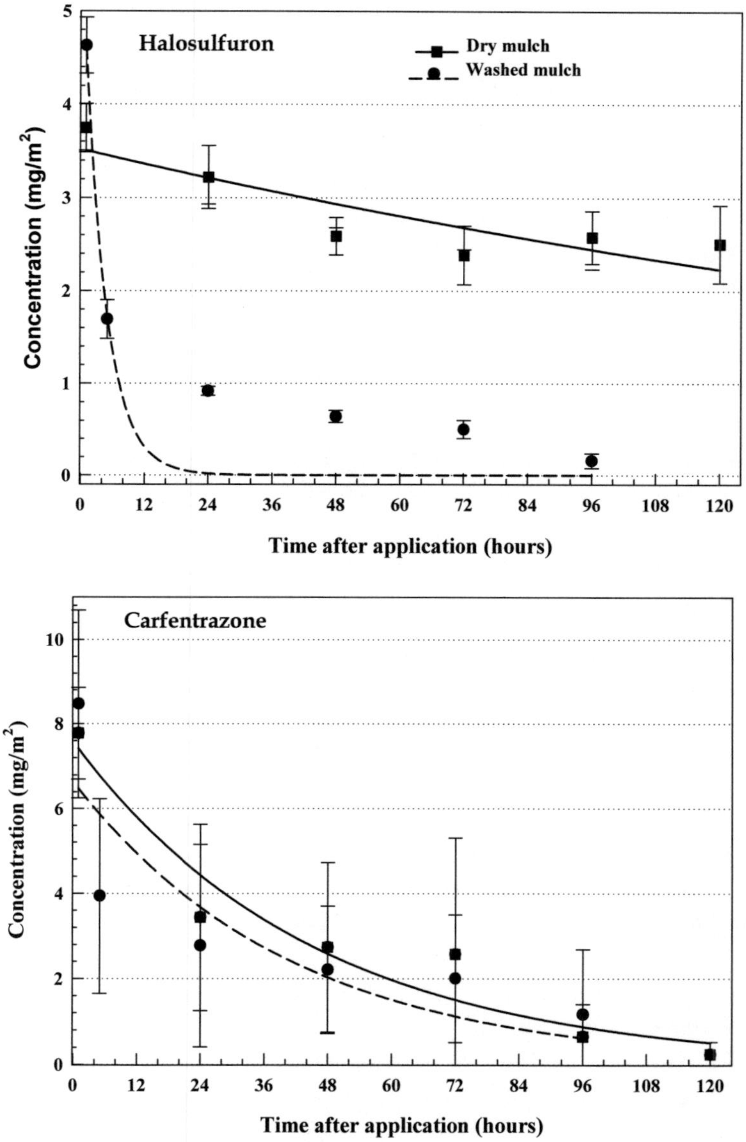

Fig. 5. Halosulfuron and carfentrazone dissipation from low density polyethylene mulch for dry and wash off conditions over time.

These studies indicate that glyphosate and paraquat dissipation was rapid from low density polyethylene mulch when irrigation water was used as a solvent. Halosulfuron and carfentrazone were detectable even after five wash off events, indicating some type of adsorption, or physical trapping within the matrix, maybe occurring with the low density polyethylene mulch, with subsequent release with each wash off event (Figures 4 and 5).

7. Conclusions

Herbicides will be an alternative to fumigants for weed control in low density polyethylene mulch vegetable production. However, given the persistent chemical nature of some herbicides, care must be taken in planning potential rotational crops to prevent any soil carryover issues that could injury or kill sensitive species. *S*-metolachlor soil dissipation data indicated it was less likely to persist than halosulfuron or sulfentrazone. These studies indicated that halosulfuron-methyl and *S*-metolachlor dissipation was more rapid for bare-soil than soil under low density polyethylene mulch.

Glyphosate and paraquat can be quickly dissipated by water when these herbicides are applied to the surface of low density polyethylene mulch. Carfentrazone and halosulfuron tended to adsorb to the mulch, increasing the potential for transplant injury. Glyphosate, paraquat, halosulfuron, and carfentrazone were all detectable at efficacious levels on the low density polyethylene mulch at 120 hours after treatment for the dry studies. These studies indicate that producers must be very conscious of the contact herbicide they apply between crops in low density polyethylene mulch production. They must also understand that using water as a dissipation mechanism may not totally remove the potential for herbicide injury to vegetable transplants, that failure to do so could result in significant plant injury and potential crop failure.

8. References

Adcock, C.W., Foshee III, W.G., Wehtje, G.R., & C.H. Gilliam. (2008). Herbicide combinations in tomato to prevent nutsedge (*Cyperus esulentus*) punctures in plstic mulch for multi-cropping systems. *Weed Technology*. Vol 22:136-141.

Bond, S. & Walker, A. (1989).Aspects of herbicide activity and persistence under low level polyethylene covers.*Annals Applied Biology*Vol. 114:133-140.

Bouchard, D.C., Lavy, T.L., & Marx, D.C. (1982). Fate of metribuzin, metolachor, and fluometuron in soil. *Weed Science* Vol. 30:629-632.

Braverman, M.P., Lavy, T.L., & Barnes, C.J. (1986). The degradation and bioactivity of metolachlor in the soil. *Weed Science* Vol. 34:479-484.

Bullock, F.D. (1990). Weed control. *Commercial pepper production*. Cooperative Extension Service, The University of Georgia College of Agriculture and Environmental Science, Athens, Ga. p 3-5.

Carpenter, A.C., Senseman, S.A., & Cralle, H.T. (1999). Adsorption-desorption of halosulfuron on selected Texas soils, *Proceedings of Southern Weed Science Society*, Greensboro, NC. p. 211

Csinos, A.S., Webster, T.M., Sumner, D.R., Johnson, A.W., Dowler, C.C., & Seebold, K.W. (2002). Application and crop safety parameters for soil fumigants. *Crop Protection*. Vol. 21:973-982.

Cornelius, A.J., Meggitt, W.F., & Penner, D. (1985). Activity of acetanilide herbicides on yellow nutsedge (*Cyperus esculentus*). *Weed Science*. Vol. 33:721-723.

Culpepper, A. S., Grey, T.L., & Webster, T.M. (2009). Vegetable response to herbicides applied to low density polyethylene mulch prior to transplant. *Weed Technology*. Vol. 23:444-449.

Dermiyati, S.K. & Yamamoto, I. (1997a). Degradation of the herbicide halosulfuron-methyl in two soils underdifferent environmental conditions. *Journal of Pesticide Science*. Vol. 22:282-287.

Dermiyati, S.K. & Yamamoto, I. (1997b). Relationships between soil properties and sorption behavior of the herbicide halosulfuron-methyl in selected Japanese soils. *Journal of Pesticide Science*. Vol. 22:288-292.

Fennimore, S.A. and Doohan, D.J. (2008). The challenges of specialty crop weed control, future directions. *Weed Technology*.Vol 22:364-372.

Garvey, P.V. & Monks, D.W. (1998). Response of vegetable crops grown in rotation to sulfentrazone treated soybeans, *Proceedings of Southern Weed Science Society*, Atlanta, GA.pp 91-92.

Gaynor, J.D., Hamill, A.S., & MacTavish, M.C. (1993). Efficacy, fruit residues, and soil dissipation of the herbicide metolachlor in processing tomato. *Journal of American Society Horticulture Science*. Vol. 118:68-72.

Gilreath, J. P. & Duranceau, S.J. (1986). Photodegradation of paraquat applied to polyethylene mulch film. *Horticulture Science*. Vol. 21:1145-1146.

Gilreath, J. P., Santos, B.M., & Duranceau, S.J. (2006). Seasonal variation of paraquat photodegradation rate on polyethylene mulch. *Weed Technology*. Vol. 20: 315-318.

Gilreath, J.P. & Santos, B.M. (2004). Bioassys on glyphosate residue toxicity and photodegradation on polyethylene mulch, *Soil and Crop Science Society of Florida Proceedings*, Tallahassee, FL.Vol. 63:21-24.

Gilreath, J.P., Noling, J.W., & Santos, B.M. (2004). Methyl bromide alternatives for bell pepper (*Capsicum annuum*) and cucumber (*Cucumissativus*) rotations. Crop Protection.Vol. 23:347-351.

Grey, T.L., Bridges, D.C., & NeSmith, D.S. (2002). Transplanted pepper (*Capsicum annum*) tolerance to selected herbicides and method of application. *Journal of Vegetable Crop Protection*.Vol. 8:27-39.

Grey, T.L., Culpepper, A.S., & Webster, T.M. (2007). Fall vegetable response to herbicidesspring applied under polyethylene mulch. *Weed Technology*. Vol. 21:496–500.

Grey, T. L., Vencill, W.K., Webster, T.M. & Culpepper, A.S. (2009). Herbicide dissipation from low density polyethylene mulch. *Weed Science*. Vol. 57:351-356.

Grey, T.L., Walker, R., Wehtje, G.R., Adams Jr., J., Kwon, O., Weete, J., Dayan, F., & Hancock, H.G. (2000). Behavior of sulfentrazone with ionic exchange resins, electrophoresis gels, & cation-saturated soils. *Weed Science*. Vol. 48:239-247.

Grey, T.L., Walker, R.H., Wehtje, G.R., & Hancock, H.G. (1997). Sulfentrazone adsorption andmobility as affected by soil and pH.*Weed Science*. Vol. 45: 733-738.

Grichar, W.J., Besler, B.A., & Brewer, K.D. (2003). Purple nutsedge control and potato (*Solanumtuberosum*) tolerance to sulfentrazone and halosulfuron. *Weed Technology*. 17:485–490.

Igbokwe, P.E., Tiwari, S.C., Collins, J.B., & Russell, L.C. (1990). Herbicides for weed control field-grown peppers.*Journal of Mississippi Academy of Science*.Vol. 35:27-29.

Johnson III, W.C., Grey, T.L., & Kissel, D.E. (2010). Interactive effects of soil pH, halosulfuron rate, and application method on carryover to turnip green and cabbage. *Weed Technology*. Vol24:160-164.

Johnson III, W.C. & Mullinix, Jr. B.G. (2005). Effect of herbicide application method on weed management and crop injury in transplanted cantaloupe production. *Weed Technology*. Vol. 19:108-112.

Kuwatsuka, S.S. & Yamamoto, I. (1997a). Degradation of the herbicide halosulfuron-methyl in two soils under different environmental conditions. *Journal of Pesticide Science*.Vol. 22:282-287.

Kuwatsuka, S.S. & Yamamoto, I.(1997b). Relationships between soil properties and sorption behavior of the herbicide halosulfuron-methyl in selected Japanese soils. *Journal of Pesticide Science*.Vol. 22:287.

Lacascio, S.J., Dickson, S.W., & Kuchare, T.A. (1994). Evaluation of multi-purpose soil fumigants for root-knot nematode and soilborne disease management on tomato in Florida. *Proceedings Annual International Research Conference on Methyl Bromide Alternatives and Emission Reductions*, Orlando, FL. p. 3

MacRae, A.W. & Culpepper, A.S. (2007). Weed management update for cantaloupe, cucumber, squash, and watermelon, *Proceedings of 2007 Southeast Regional Vegetable Conference*, Savannah, GA. p. 50.

McConnell, J.S. & Hossner, L.R. (1985). pH-dependent adsorption isotherms of glyphosate.*Journal of Agriculture Food Chemistry*.Vol33:1075-1076.

Main, C. L., Mueller, T.C., Hayes, R., Wilcut, J., Peeper, T., Talbert, R., & Witt, W.W.(2004). Sulfentrazone persistence in southern soils: bioavailable concentration and effect on a rotational cotton crop. *Weed Technology*. Vol. 18:346-352.

Mueller, T.C., Shaw, D.R. & Witt, W.W. (1999). Relative dissipation of acetochlor, alachlor, metolachlor and SAN 582 from three surface soils. *Weed Technology*. Vol. 13: 341-346.

Nelson, K.A. & Renner, K.A. (2002). Yellow nutsedge (*Cyperus esculentus*) control and tuberproduction with glyphosate and ALS-inhibiting herbicides. *Weed Technology*. Vol. 16:512-519.

Nerin, C., Tornes, A.R., Domeno, C. & Cacho, J. (1996). Absorption of pesticides on plastic films used as agricultural soil covers. *Journal of Agriculture and Food Chemistry*.Vol. 44:4009-4014.

Obrigawitch, T., Abernathy, J.R., & Gipson, J.R. (1980). Response of yellow (*Cyperus esculentus*) and purple (*Cyperus rotundus*) nutsedge to metolachlor. *Weed Science*. Vol. 28:708-715.

Obrigawitch, T., Hons, F.M., Abernathy, J.R., & Gipson, J.R. (1981). Adsorption, desorption, and mobility of metolachlor in soils. *Weed Science*. Vol. 29:332-336.

Ohmes, G.A., Mueller, T. & Hayes, R. (2000). Sulfentrazone dissipation in a Tennessee soil. *Weed Technology*. 14:100-105.

Papiernik, S.K. & Yates, S.R. (2001a). Transport of fumigant compounds through HDPE and virtually impermeable films, *Annual International Research Conference on Methyl*

Bromide Alternatives and Emissions Reductions Conference Proceedings: pp. 16-1 to 16-3 online at http://mbao.org/2001proc/mbrpro01.html.

Papiernik, S.K. and S.R. Yates. 2001b. An approach for estimating the permeability of agricultural films. *Environmental Science Technology*.Vol. 35:1240 -1246.

Parker, D.C., Simmons, F.W. & Wax, L.M. (2005). Fall and early preplant application timing effects on persistence and efficacy of acetamide herbicides. *Weed Technology*. Vol. 19:6-13.

Patakioutas, G. & Albanis, T.A. (2002). Adsorption-desorption studies of alachlor, metolachlor, EPTC, chlorothalonil and pirimiphos-methyl in contrasting soils. *Pest Management Science*.Vol. 58:352-362.

Peachey, R.E., Pinkerton, J.N., Ivors, K., Miller, M. & Moore, L. (2001). Effect of soil solarization, cover crops, and metham on field emergence and survival of buried annual bluegrass (*Poaannua*) seeds. *Weed Technology*. Vol. 15:81-88.

Peter, C.J. & Weber, J.B. (1985). Adsorption, mobility, and efficacy of alachlor, and metolachlor as influenced by soil properties. *Weed Science*. Vol. 33:874-881.

Reddy, K.N. & Locke, M.A. (1998). Sulfentrazone sorption, desorption, and mineralization insoils from two tillage systems. *Weed Science*. Vol. 46:494-500.

Rogers, C.E. 1985. Permeation of gases and vapors in polymers. In Comyn, J. (ed.), *Polymer Permeability*,Elevier: London, England.pp11-73.

Senseman, S.A. 2007. *Herbicide Handbook.9th edition*, Weed Science Society of America,Champaign, IL.

Stiles, C.L., Coffey, D.L., & Mueller, T.C. (1999). Yellow nutsedge control and tomato injury as affected by the interaction of vapam and the herbicides eptam, tillam, and matrix. *Proceedings of Southern Weed Science Society*. Greensboro, NC. pp82-83.

Vencill, W.K., Richburg III, J.S.,Wilcut, J.W., & Hawf, L.R. (1995). Effect of MON-12037 onpurple (*Cyperus rotundus*) and yellow (*Cyperus esculentus*) nutsedge. *Weed Technology*. Vol. 9:148–152.

Vencill, W. K. (2002a). *Herbicide Handbook.8th edition*, Weed Science Society of America, Champaign, IL.pp 299-300.

Vencill, W. K. (2002b). *Herbicide Handbook.8th edition*, Weed Science Society of America, Champaign, IL.pp. 405-406.

Weber, J.B., McKinnon, E.J. & Swain, L.R. (2003). Sorption and mobility of ^{14}C-labeled imazaquin and metolachlor in four soils as influenced by soil properties. *Journal of Agriculture Food Chemistry*. Vol. 51:5752-5759.

Webster, T.M. (2006). Weed survey—southern states: vegetable, fruit and nut crops subsection.*Proceedings of Southern Weed Science Society*.San Antonio, TX. pp 260-277.

Webster, T.M., Culpepper, A.S., & JohnsonIII, W.C. (2003). Response of squash and cucumbercultivars to halosulfuron. *Weed Technology*. Vol. 17:173-176.

Webster, T.M., Csinos, A.S., Johnson, A.W., Dowler, C.C., Sumner, D.R. & Fery, R.L.(2001). Methyl bromide alternatives in a bell pepper-squash rotation. *Crop Protection*. Vol. 20:605-614.

Webster, T.M. & MacDonald, G.E.(2001). A survey of weeds in various crops in Georgia. *Weed Technology*. Vol. 15:771-790.

Wehtje, G.R., Walker, R.H., Grey, T.L., &Hancock, H.G.(1997). Response of purple (*Cyperus rotundus*) and yellow (*Cyperus esculentus*) nutsedges to selectiveplacement of sulfentrazone. *Weed Science*. Vol. 45:382-387.

16

Solid-Phase Extraction for Enrichment and Separation of Herbicides

Pyrzynska Krystyna
University of Warsaw, Department of Chemistry
Poland

1. Introduction

Herbicides are widely used for of broad-leaved weeds and other vegetation. They are relatively inexpensive and very potent even at low concentrations. The majority of herbicides are directly applied to soil or sprayed over crop fields and as consequence of large production and high stability, they are released directly into environment. For that, herbicides can enter as contaminants into streams, rivers or lakes directly from drainage of agricultural areas. The need for herbicide monitoring in natural water is essential for achieving good water quality objectives, because in most countries groundwater is the major source of drinking water. Moreover, the monitoring of herbicides in soil and crops is important in order to control their impact on the environment.

Recent studies have focused more on herbicide degradation/transformation products (from hydrolysis, oxidation, biodegradation or photolysis) because they can be present at greater levels in the environment than the parent herbicide and can sometimes be as toxic or even more toxic. New compounds have also come on the market (such as glyphosate, organophosphorus herbicides) and studies are being conducted to understand their fate and transport in the environment (Richardson, 2009).

Thus, it is important to develop a reliable and sensitive method for the simultaneous determination of such compounds in different kinds of samples. High-performance liquid chromatography with mass spectrometry or array diode detection are good options for herbicides monitoring (Cheng et al., 2010; Maloschik et al., 2010). Tandem mass spectrometry is usually used to confirm identification of selected herbicides. While LC-MS/MS methods are now predominantly used for pesticides and their degradation products, GC-MS methods are still occasionally used. For example, GC-MS was used by Hildebrandt et al. (2007) to measure 30 priority pesticides and their transformation products in agricultural soils and an underlying aquifer in the Ebro River Basin in Spain. The sensitivity of detection, however, is still not high enough in many cases for direct determination of herbicides at the level required by different regulations. Therefore, a preconcentration procedure for the analytes and clean-up steps must be applied for complex samples.

Solid-phase extraction (SPE) is the most popular sample preparation technique of environmental, food and biological samples and it already replaced the classic liquid-liquid extraction as the reduction or complete elimination of solvent consumption in analytical procedures, which is very important according to the rules of green chemistry (Camel, 2003;

Pyrzynska, 2003; Fontales et al., 2007). The main goal for application of SPE is to achieve isolation, preconcentration and clean-up of the sample in a single step. This can be achieved by an appropriate selection of the type of sorbent or their combination. For this reason the properties of the analytes, nature of matrix, the required trace-level concentration and the type of chromatography involved later in the separation step should be taken into consideration. The strategy of sample pretreatment in SPE-HPLC system is also guided by the method of final detection after chromatographic separation. Application of a simple detection mode, *e.g.* diode array UV, requires more selective isolation and enrichment. When the more specific quantification is used, such as fluorescence, mass spectrometry or electrochemical methods, application of SPE sample pretreatment can improve the limit of detection.

The extraction process depends on the type of sorbent used and retention is due to reversible hydrophobic, polar and ionic interactions between the analyte and the sorptive material. Sorption can be non-specific, in that case weak dispersive interactions such as van der Waals forces will dominates. However, sorbents utilizing specific interactions resulting from analyte polarity, ionic nature or the presence of specific functional groups are preferred. The classical sorbents in SPE are silica-based (Spivakov et. al., 2006), carbonaceous materials (Kyriakopolous & Doulia, 2006; Pyrzynska, 2008) or polymeric, primarily styrene-divinylbenzene copolymers (Fontanals et al., 2004; Kyriakopolous & Doulia, 2006). The novel sorbents with improved selectivity towards the particular groups of compounds or even individual compounds includes immunosorbents (Haginaka, 2005) and molecularly imprinted polymers (MIP) (Dias et al., 2009; Lasákova & Jandera, 2009). Carbon nanotubes, a new form of carbon-based sorbents, are also promising materials in SPE of herbicides (Pyrzynska, 2008).

The objective of this chapter is to present the recent advances in the area of novel materials as solid phase extractors for herbicide analysis. The papers published over the last five years are discussed in more detail. The emphasis is also given to the application of several SPE systems for automated preparation of environmental, food and biological samples.

2. Classic sorbents

Silica chemically bonded with various groups has been the most common material for SPE. This sorbent can be classified as reversed-phase sorbent with octadecyl (C_{18}), octacyl (C_8), ethyl (C_2) and phenyl or as normal-phase sorbent with cyanopropyl, aminopropyl and diol functional groups. Their interaction mechanisms are mainly based on hydrophobic interaction (van der Waals forces), thus these SPE packing provide high recoveries for nonpolar analytes. Nevertheless, silica-based sorbents are unstable at extremes pH ($2 > pH > 8$), and they have relatively low capacity and low recovery for basic analytes. Several types of modifications were used to immobilize different compounds on the surface of classical silica-base sorbents to increase their selectivity (Parida et al., 2006; Kailasam et al., 2009). New materials based on poly(methyltetradecylsiloxane) and poly(methyloctylsiloxane) thermally immobilized onto the silica support have been tested for extraction of some herbicides (Vigna et al., 2006; Faria et al., 2007). Liu (2008) had used silica gel coated with gold nanoparticles self-assembled with alkanethiols for the extraction of steroidal compounds.

The bonded-silica sorbent may be packed in different formats: filled microcolumns, cartridges or discs. A variety of bonded-silica phases are commercially available in the cartridge format. Extraction could be also performed with membrane disks containing C_{18}-bonded silica (8 μm particles) on polytetrafluoroethylene or glass fiber supports (Spivakov et. al., 2006; Li et al., 2006). Disc provides shorter sample processing time on account of their larger cross-sectional area and decreased pressure drop, allowing higher sample flow rates. This is important for environmental samples, where larger sample volumes are usually employed to achieve adequate detection limits.

The polymeric sorbents based on styrene-divinylbenzene exhibit higher capacity and better chemical stability over the whole pH range in comparison with bonded silica. Due to the specific π-π interactions they are relatively selective for analytes with aromatic rings. The use of highly crosslinked polymeric sorbents with their specific surface up to 800 m^2/g or hypercrosslinked polymeric sorbents (over 1000 m^2/g) could improve the analytes retention as more π-π sites in the aromatic rings will then be accessible to interact with the analytes (Ahn et.al., 2006).

3. Hydrophilic and mixed-mode polymeric sorbents

The hydrophobic nature of classical sorbents leads to poor retention of polar compounds. To overcome these problems, the research in new SPE materials has been recently focused on the development of hydrophilic and mixed-mode polymeric materials. Such sorbents combine high specific surface area and polar interaction between sorbent and analyte due to introduction of the polar moiety to the polymer structure.

3.1 Hydrophilic polymeric sorbents

The hydrophilic polymeric sorbents are obtained by chemical modification of the existing hydrophobic materials or by copolymerisation of monomers that contain suitable functional groups. The polar substituents reduce the interfacial tension between the polymer surface and aqueous sample improving the wetting characteristics and increase contact between the analyte and polymeric sorbent. Strata-X (styrene skeleton modified with a pyrrolidone group) and Oasis HLB (macroporous poly(*N*-vinylpyrrolidone-divinylbenzene) copolymer) are the most common hydrophilic sorbent used in the herbicides extraction (Stoob et. al., 2005; D'Archivio et. al., 2007, Polati et al., 2006; Mazzella et al., 2008; Yu et al., 2009). Most of the studies investigate the performance of Oasis HLB in off-line SPE using different cartridge size available (from 30 to 500 mg). Other studies employ the direct coupling of on-line SPE to HPLC with column switching technique (Xu et al., 2007) or 96-well plate (Morihisa et al., 2008) to obtain high sample throughput. Abselut Nexus, the methacrylate and divinylbenzene copolymer has been recently applied in clean-up of complex samples, such as biological matrices with the subsequent extraction of analytes (Rodriguez-Gonzalo et. al., 2009).

Biesaga et al. (2005) compared the recoveries of chlorophenoxy acidic herbicides using various SPE cartridges (C_{18}, Strata-X, Oasis HLB, SAX and phenyl-silica). The better performance of Strata-X, Oasis and phenyl-silica sorbents in comparison with silica gel C_{18} can be attributed to their aromatic structure, which can interact with aromatic analytes *via* π-π interactions (Fig. 1). Additionally, Oasis HLB cartridges are water-wettable, and thus there is no need to ensure that it remains wet before loading the aqueous sample. The recovery of dicamba, the least hydrophobic compound evaluated, was much lower; only its sorption on Strata-X reached 74%.

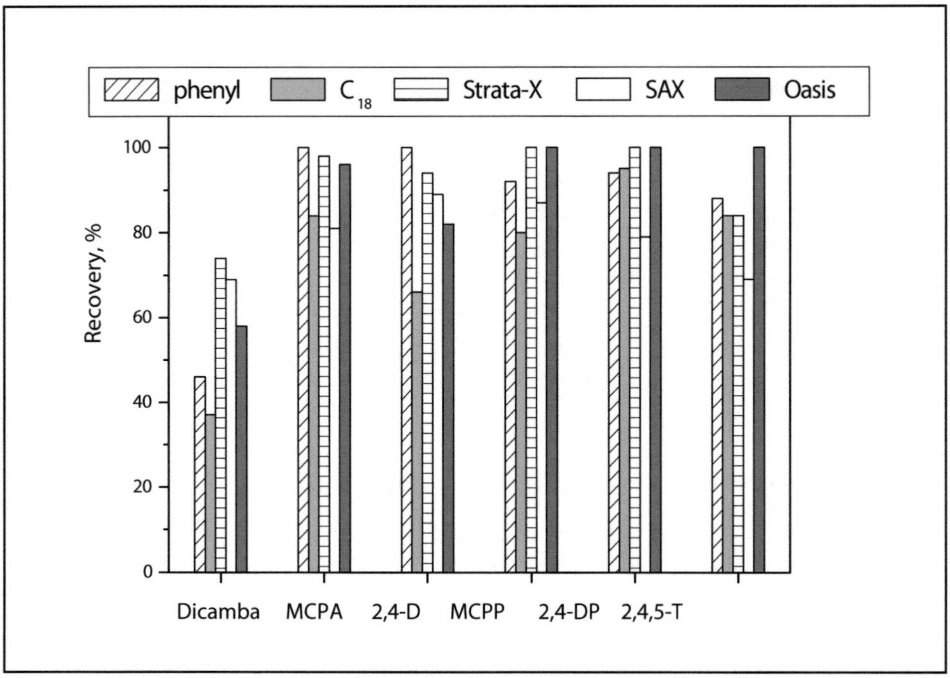

Fig. 1. Recoveries of chlorophenoxy acids extracted from 10 mL of deionized water spiked at the 5 µg/L level using various SPE cartridges. (Adopted from Biesaga et al., 2005).

3.2 Mixed-mode polymeric sorbents

Mixed-mode polymeric sorbents combine the polymeric skeleton with ion-exchange groups, thus these hybrid materials rely upon two types of interactions mechanism for their performance: reversed-phase and ion-exchange (Fontanals et al., 2010). Careful selection of the polymeric skeleton (which enhances the reversed-phase interactions) and the ionic groups (which tune the ion-exchange interactions) could give the combination of two highly desirable properties in solid phase extraction (i.e. retentivity and selectivity) in one single material. The benefit of the ion-exchange capacity is that either analytes, the matrix components or even the ionization state of the sorbent (in the case of weak-exchange resins) can be switched during the different steps in SPE procedure. It allows the interference elimination in the washing step and eluting the analytes more selectively, just by suitable pH combination in each step.

Mixed-mode sorbents are classified as cationic or anionic, and as strong or weak ion exchange, depending on the ionic group attached to the resin. Each of these groups is designed to extract selectively analytes with certain chemical properties (i.e. strong/weak acidic or basic). However, the selectivity of the extraction process depends on choosing not only a suitable sorbent but also a suitable SPE protocol (Fontanals et al., 2010).

Oasis MCX and Oasis MAX have the same as Oasis HLB skeleton (polyvinyl pyrrolidone-divinylbenzene) modified chemically with sulfonic acid and quaternary amine groups, respectively. These mixed-mode sorbents are mainly applied for extraction of analytes (charged or not) from complex biological and environmental matrices (Rosales-Conrado et al., 2005; Sorensen et al., 2008; Rodriguez-Gonzalo et al., 2009).

Lavén et al. (2009) proposed a novel solid phase extraction method whereby 15 basic, neutral and acidic compounds from wastewater were simultaneously extracted and subsequently separated into different fractions. This was achieved using mixed-mode cation- and anion-exchange SPE (Oasis MCX and Oasis MAX) in series. For less complex samples, e.g. the active-sludge-treatment effluent water, Oasis MCX used alone may be an alternative method. Although sewage treatment plant influent waters containing high loads of organic compounds, the clean-up step using only Oasis MCX was insufficient, leading to unreliable quantitation. Utilising the ability to separate compounds by mixed-mode SPE according to basic and acidic functionalities should be also very useful in the characterisation of unknown water contaminants.

4. Molecularly imprinted polymers

Molecularly imprinted polymers (MIPs) are highly crosslinked polymers with specific binding sites for a particular analyte. The print molecule - called the template - is chemically coupled with one or several functional monomers and then spatially fixed in a solid polymer by the polymerisation reaction. After template removal by extraction, polymers with imprints, which are complementary to the template in terms of size, shape and functionality are obtained. These polymers are able to rebind selectively the template molecule or its structural analogues. The right selection of functional monomers is important in molecular imprinting because the interactions with functional groups affect the affinity of MIPs (Lasákova & Jandera, 2009). Molecular modelling can be used to predict which functional monomers are capable to form effective polymers as some monomers have a natural affinity to some herbicides (Breton et al., 2007).

Two principally different approaches to molecular imprinting may be distinguished. In non-covalent (or self-assembly) approach the imprint molecule complexes the monomers by non-covalent or metal ion coordination interactions. The covalent imprinting employs reversible covalent bonds and usually involves a prior chemical synthesis step to link the monomers to the template. The first approach is more flexible in the range of templates that can be used but covalent imprinting yields better defined and more homogeneous binding sites. Moreover, the former is practically much easier, since complex formation occurs between template and monomers in a solution. Figure 2 shows this entire process schematically and more details on the preparation of imprints can be found elsewhere (Diaz-Garcia & Lamo, 2005; Qiao et al., 2006; Dias et al., 2009). It should be stressed that some monomers have natural affinity to some herbicides (Breton et al., 2007). The retention on blanks seems to be a good reflection of the relative affinity of monomers to the herbicides, and this interaction must be naturally strong enough to allow the binding enhancement by a MIP. Proper selection of reagents, reaction medium and conditions should take into consideration the complexity of selective sites formation in the polymer structure to obtain a material capable of not only highly selective recognition of target analytes but also having good kinetic parameters (Kloskowski et al., 2009). Kopohpaei et al. (2008) proposed a chemometric approach for the optimization of the main factors affecting the material structure and the molecular recognition properties of the MIPs.

Tamayo et al. (2005) found that the use of 2-(trifluoromethyl) acrylic acid as functional monomer leads to the synthesis of polymers with higher capacities and affinity constants for phenylurea herbicides in comparison with metacrylic acid when isoproturon was used as template. Thus, the simultaneous extraction of several herbicides was possible since each compound was able to interact with specific binding sites in the presence of related compounds. However, both linuron and metabromuron were clearly displaced by the other analytes in the competition experiments and were able to interact only with a very small number of binding sites.

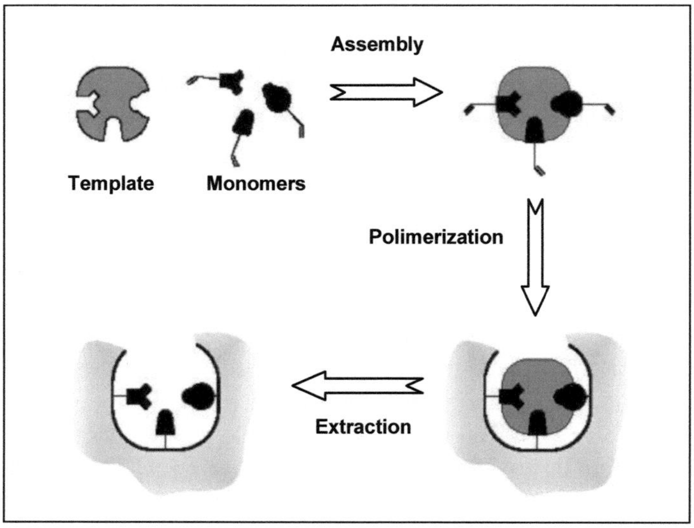

Fig. 2. Schematic representation of molecular imprinting principle.

Most of the reported studies concern the development of MIPs for one target analyte only, but basically similar compounds that are present in samples, can also be recognized and extracted (Chapuis et al., 2003). Herrero-Hernandez et al. (2007) demonstrated the applicability of an imprinted polymer obtained using bisphenol-A as template for the determination of several xenobiotic compounds in honey samples. It was found that MIP was able to extract selectivity phenols and several phenoxyacids, while no-specific recognition of other compounds such as atrazine, chlotoluron, carbaryl and diuron herbicides was also observed.

MIPs can be obtained in the format of particles, coatings, monolayers of selective compounds bound to the surface of support, monolithic packings or fibers (Oxelbark et al. 2007). A fast and straightforward method for preparation and binding study of solid phase microextraction (SPME) fiber on the basis of atrazine- and ametryn-imprinted polymers has been proposed (Djozan & Ebrahimi, 2008; Djozan et al., 2009). The fabricated fibers were thermally and chemically stable and flexible enough to be placed in home-made SPME syringe and to be inserted directly into GC injection port.

Porous self-supported MIP membranes with developed inner surface have been proposed for atrazine enrichment (Sergeyeva et al.; 2007). It was shown that the MIP particles demonstrated significantly less pronounced imprinting effect and lower adsorption capabilities as compared to the MIP membranes of the same composition. MIPs could be also incorporated into the acceptor phase of a microporous membrane liquid-liquid extraction system for preconcentration and clean-ups step before chromatographic analysis (Mhaka et al., 2009; Hu et al., 2009).

Recent applications of MIP-SPE technique for herbicide analysis are presented in Table 1.

Template	Monomer/cross linker/solvent	Analytes	Sample	References
2,4,5-trichloro phenoxyacetic acid	4-vinyl pyridine/ EGDMA/ methanol -water (3+1, v/v)	Chlorinated phenoxyacids	River water	Baggiani et al., 2004
Metsulfuron methyl	4- or 2-vinyl pyridine/EGDMA/ acetonitrile	Sufonylurea herbicides	Tap water	Bastide et al., 2005
Linuron or isoproturon	MAA or TFMAA/ EDMA/toluene	Phenylurea herbicides	Corn sample extracts	Tamayo et al., 2005
Propazine	MAA/EGDMA/ CH_2Cl_2	Triazines	Soil, vegetable extracts	Cacho et al., 2006
Cyanazine	MAA/EGDMA/ toluene	Cyanazine, atrazine	Waters	Breton et al., 2006
Atrazine or ametryn	MAA or TFMAA or 4-vinyl pyridine /EGDMA/toluene	Chlorotriazine and methyl thiotriazine herbicides	River water	Sambe et al., 2007
Atrazine	MAA/EGDMA/ toluene	Atrazine	Ground waters	Prasad et al., 2007
Phenoxyacetic acid	4-vinyl pyridine /methanol+water (1+1, v/v)	Phenoxyacetic herbicides	Waters	Zhang et al., 2007
Atrazine	MAA/ TEDMA/ DMF	Triazine herbicides	Waters	Sergeyeva et al., 2007
Ametryn	MAA/EGDMA/ acetonitrile	Ametryn	Standards	Koohpaei et al., 2008
Atrazine	MAA/ EGDMA/ acetonitrile	Triazine herbicides	Waters, rice, onion	Djozan et al., 2008
Bisphenol-A	4-vinyl pyridine /EGDMA/toluene	Phenoxyacetic herbicides	Honey	Herrero-Hernández et al., 2009
Ametryn	MAA/ EGDMA/ acetonitrile	Triazine herbicides	Drinking waters	Koohpaei et al., 2009
Atrazine	MAA/ EGDMA/ acetonitrile	Triazine herbicides	Food samples	Mhaka et al., 2009

MAA - methacrylic acid; EGDMA – ethylene glycol dimethacrylate; TFMAA – 2-(trifluoromethyl) acrylic acid; DVB – divinylbenzene; CH_2Cl_2 – dichlooromethane; TEDMA – tri(ethylene glycol) dimethacrylate; DMF – dimethylformamide

Table 1. Recent applications of MIP-SPE technique for herbicide analysis

The analytical procedure based on molecularly imprinted SPE was developed for the determination of several triazine herbicides in soil and vegetable samples (Cacho et al., 2006). These samples has proven to be difficult to clean with a non-covalent imprinted polymer, making necessary the inclusion of an additional clean-up step to remove polar matrix components that prevented the final accurate quantification of target analytes. Figure 3 shows the chromatograms obtained with and without SPE procedure of soil (Fig.

3A) and potato (Fig. 3B) sample extracts spiked with 50 ng/g and 20 ng/g of triazine herbicides, respectively. As can be observed, the direct determination of triazines without clean-up was not possible due to interferences appearing in the chromatograms whereas it could be easily determined after cleaning sample extract using MIPs. The detection limits for the analysis ranged from 0.4 to 2.4 ng/g depending upon the herbicide, low enough to allow the environmental monitoring of triazines at concentration level below the established maximum residue limits by current legislation.

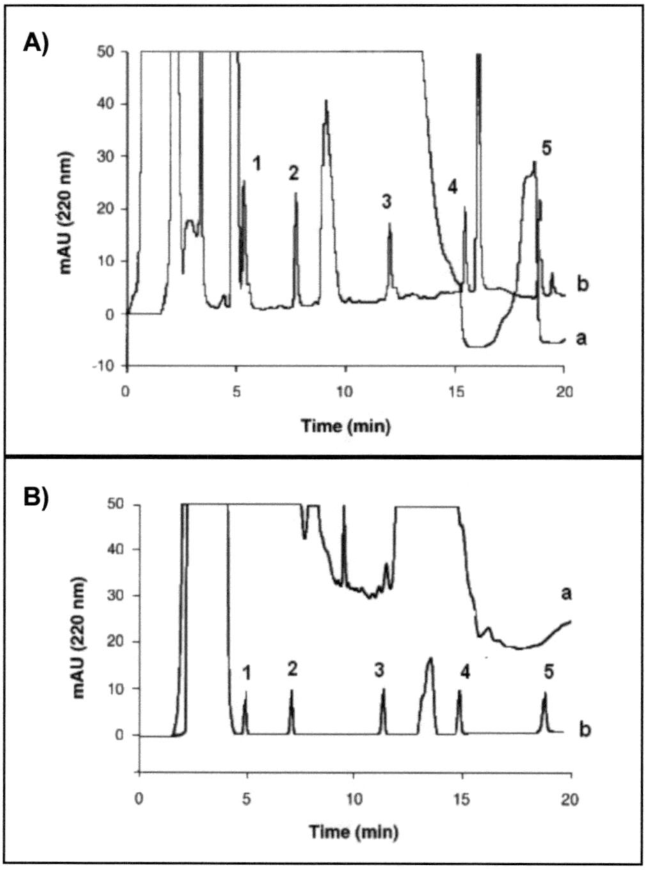

Fig. 3. Chromatograms obtained without (a) and with (b) SPE-MIP of soil (A) and potato (B) samples extracts spiked with triazine herbicides (50 and 20 ng/g, respectively). Peaks: 1-desisopropylatrazine; 2-desethylatrazine; 3-simazine; 4-atrazine; 5-propazine. Adopted from Cacho et al. (2006).

5. Carbon nanotubes

Carbon nanotubes (CNTs) represent the novel carbon-based nanomaterials with unique properties such as high surface areas, large aspect ratios, remarkably high mechanical strength as well as electrical and thermal conductivities. They can be described as a graphite sheet rolled up into a nanoscale-tube. Two structural forms of CNTs exist: single-walled (SWCNTs) and multi-walled (MWCNTs) nanotubes. CNT lengths can be as short as a few hundred nanometers or as long as several microns. SWCNT have diameters between 1 and 10 nm and are normally capped at the ends. In contrast, MWCNT diameters are much larger (ranging from 5 nm to a few hundred nanometers) because their structure consists of many concentric cylinders held together by van der Waals forces (Wepasnik et al., 2010).

The characteristic structures and electronic properties of carbon nanotubes allow them to interact strongly with organic molecules, *via* non-covalent forces, such as hydrogen bonding, π- π stacking, electrostatic forces, van der Waals forces and hydrophobic interactions. These interactions as well as hollow and layered nanosized structures make them a good candidate for application as a sorbent. The surface, made up of carbon atoms hexagonal arrays in graphene sheets, interacts particularly strongly with the benzene rings of aromatic compounds.

Oxidation of CNTs with nitric acid is an effective method to remove the amorphous carbon, carbon black and carbon particles introduces by their preparation process (Yang et al., 2006). It is known that oxidation of carbon surface can offer not only more hydrophilic surface structure, but also a larger number of oxygen-containing functional groups, which increase the ion-exchange capability of carbon material. Gas phase oxidation of activated carbon increases mainly the concentration of hydroxyl and carbonyl surface groups, while oxidation in the liquid phase increases particularly the content of carboxylic acids (Dastgheib & Rockstraw, 2002). The amount of carboxyl and lactone groups on the CNTs treated with nitric acid was higher in comparison to the process conducted using H_2O_2 and $KMnO_4$ (An & Zeng, 2003). Datsyuk et al. (2008) found that the nitric acid (65%) treated carbon nanotubes under reflux conditions for 48h suffered very high degree of degradation such as nanotube shortening and additional effect generation in the graphitic network. Functional groups can change the wettability of CNTs surfaces and consequently make them more hydrophilic and suitable for sorption of relatively low molecular weight and polar compounds. On the other hand, functional groups may increase diffusional resistance and reduce the accessibility and affinity of CNTs surfaces for organic compounds (Cho et al., 2008).

Recent applications of carbon nanotubes for removal and enrichment of herbicides in different types of samples are presented in Table 2. Earlier reports were discussed in the review papers (Pang & Xing, 2008; Pyrzynska, 2008).

The comparison of carbon nanotubes, activated carbon and C_{18} silica in terms of analytical performance, application to environmental water, cartridge re-use, adsorption capacity and cost of adsorbent has been made for propoxur, antrazine and methidation herbicides (El-Skeikh et al., 2008). The adsorption capacity of CNTs was almost three times higher than that of activated carbon and C_{18}, while activated carbon was superior over the other sorbents due to its low cost.

A comparative study suggested that carbon nanotubes had a higher extraction efficiency than Oasis HLB for the extraction of methamidophos and acephate, particularly for seawater samples (Li et al., 2009). Figure 4 presents the chromatograms of six organophosphorus pesticides in the spiked seawater sample extracted using CNTs and Oasis HLB sorbent. For other tested polar organophosphorus pesticides (dichlorvos, omethoate, monocrotophos and dimethoate) improvement was not significant, thus CNTs could supplement Oasis HLB for these compounds extraction.

Zhou et al. (2007) compared the trapping efficiency of CNTs and C_{18} packed cartridge using sulfonylurea herbicides as the model compounds. When the matrices of the samples were very simple, such as tap water and reservoir water, the enrichment performance between

Analytes	Sample	Eluent	Recovery %	Reference
Sulfonylurea herbicides	Waters	Acetonitrile	80 - 105	Zhou et al., 2007
Atrazine and its metabolites	Water, soil	Ethyl acetate	72-109	Min et al., 2008
Organophosphorous herbicides	Fruit juices	Dichloromethane	73 –103	Ravelo-Perez et al., 2008
Various herbicides	Natural waters	Acetonitrile	81 – 108	El-Sheikh et al., 2008
Pirimicarb, pyrifenox, penconazol, cyprodynil, carbendazim,	Mineral water	Dichloromethane with formic acid (5% v/v)	53 – 94	Awensio-Ramos et al., 2008
Chloroacetanilide herbicides	Tap, river water	Ethyl acetate	77 –104	Dong et al., 2009
Triazine herbicides	Water	Acetonitrile/ methanol (50%, v/v)	84-104	Al-Degs et al., 2009
Sulfonylurea herbicides	Environmental waters	Acetonitrile + 1% acetic acid	79 - 102	Niu et al., 2009
Organophosphorus herbicides	Seawater	Acetone or methanol	79 - 102	Li et al., 2009

Table 2. Recent applications of carbon nanotubes for removal and enrichment of herbicides

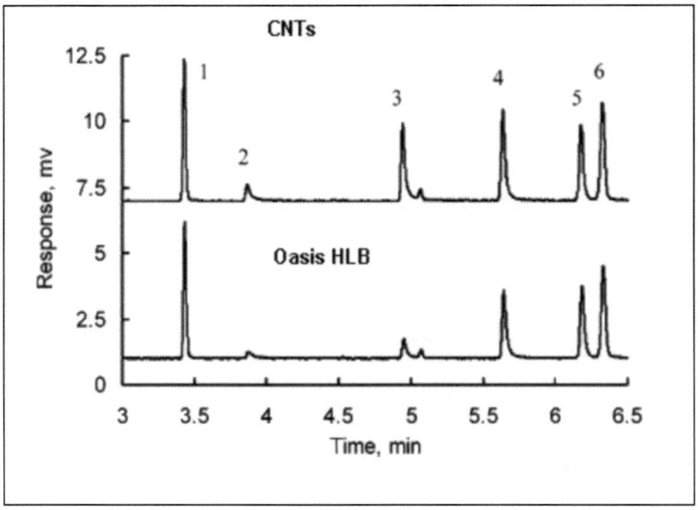

Fig. 2. Chromatograms of organophosphorus pesticides (1.0 µg/L) in the spiked seawater extracted with CNTs and Oasis HLB. Peaks identification: 1-dichlorvos, 2-methamidophos, 3-acephate, 4-omethoate, 5-monocrotophos, 6-dimethoate. Adapted from Li et al. (2009).

these two adsorbents had no significant difference. However, carbon nanotubes become much more suitable to extract herbicides from complex matrices (seawater and well-water). Carbon nanotubes could be also used in a format of disc. Incorporating sorbents of small particle size, the disc format possesses a larger surface area than the cartridge, resulting in good mass transfer and fast flow rates (Niu et al., 2009). To enhance the sorption capacity of the disks, double or even triple disks were used together. A comparison study showed that the double-disk system (comprising two stacked disks with 60 mg of CNTs) exhibited extraction capabilities that were comparable to those of a commercial C_{18} disk with 500 mg sorbent for nonpolar or moderately polar compounds. The triple layered CNTs disk system showed good extraction efficiency when the sample volume was up to 3 000 mL (Niu et al., 2008).

Carbon nanotubes with high porosity and large adsorption area seems to be a good candidate for solid phase microextraction coating. Rastkari et al. (2009) proposed a novel coating by attaching CNTs onto a stainless steel wire through organic binder. The results showed that the CNTs fiber exhibited higher sensitivity and longer life span (over 150 times) than the commercial carboxen/polydimethylsiloxane coating.

6. On-line preconcentration

Solid phase extraction could be performed on-line by direct connection to the chromatographic system, therefore fully automated technique could be utilised. Hyphenated on-line SPE-HPLC systems are designed to improve not only sensitivity and selectivity of determination but also reduced sample manipulation and time, better intra- and inter-day reproducibility, higher sample throughputs as well better precision due to lower human participation, but typically requires the use of program controlled switch valves and column reconfiguration (Segura et al., 2007; Viglino et al., 2008). The extraction sorbents include mainly disposable cartridges, restricted access media,, large-size particle and monolithic materials (Xu et al., 2007).

The valve setup for on-line SPE is presented in Fig. 3. The column-switching valve is used to direct the flow from the extraction column either to waste or to the HPLC analytical column. At the beginning of each run, the SPE column is conditioned. In the load position, sample is directly loaded in the loop and then preconcentrated, while matrix components are removed during the washing step. The valve is then switched, so that appropriate solution can elute the analytes from the extraction column onto the analytical column, when they are separated pior detection. After elution, the valve is switched back to its original position to wash and re-equilibrate the extraction column.

To improve the detection limit of column-switching system, the analytes should be preconcentrated from larger sample volume. Nevertheless, this would only be achieved if the analytes do not break through the SPE column. Garcia-Ac et al. (2009) estimated the breakthrough volumes of three herbicides (atrazine, cyanazine, simazine) and two of their transformation products (deethylatrazine and deisopropylatrazine) for several on-line SPE columns made of different sorbent materials. It was found that Strata-X was the best candidate for the preconcentration of large volume samples and all studied polymeric phases showed higher breaktrough volume than silica-based phases. The preconentration of 10 mL sample lowered the limit of detection by a factor of 5 for atrazine, deethylatrazine and simazine, while for deisopropylatrazine the improvement factor was > 10.

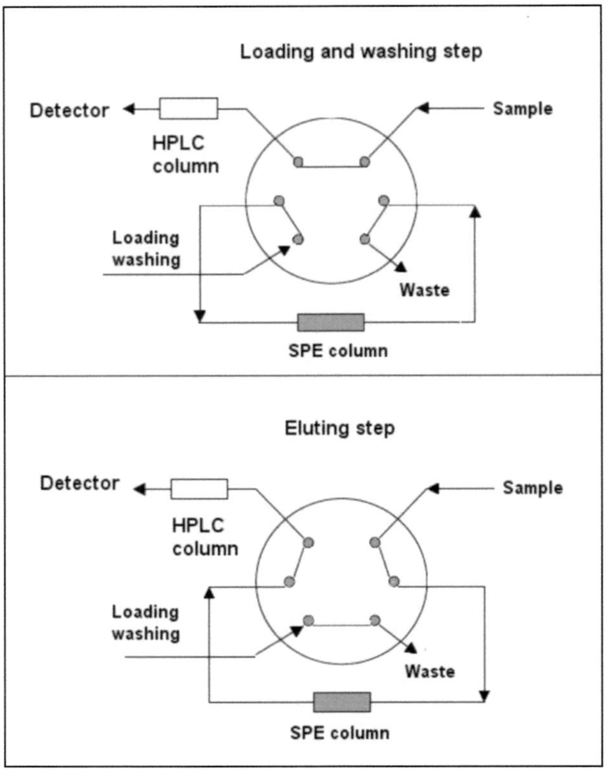

Fig. 3. Schematic diagram of valves configuration for on-line SPE-HPLC system.

The extraction column is treated as a permanent component of the flow network, being used repeatedly for the sample-loading and elution sequences, and being replaced or repacked only after long-term operation. The repeated use of sorbents may progressively affect their retention capabilities due to contamination or deactivation. Also, if the retained species are not totally eluted from the sorbent medium, this leads to carry-over effects between consecutive runs. An alternative to overcome these drawbacks relies on a surface-renewal scheme, the so-called SI-bead injection, where the contents of the SPE column are withdrawn on-line and replaced for each analytical run (Miró & Hansen, 2006). This approach was used for determination of chlorotriazine herbicides and primary monodealkylated metabolites in untreated complex environmental samples (e.g, ground waters from domestic rural wells and soil extracts). An automatic tandem-column multimodal-bead injection approach combining two types of sorbent beads (water-compatible MIP and reversed-phase mixed-mode Oasis HLB) was developed prior to on-line LC separation (Boonjob et al., 2010). The limit of detection for analysis of spiked water at the 0.5 μg/L level was in the range of 0.02 - 0.04 μg/L and overall procedure reproducibility within 1.4 – 5.5% RSD.

7. Quality control

Together with the fast development of analytical methodologies, the great importance is now attached to the quality of the measurement data. Many important decisions are based on measurements, thus good-quality analytical results are essential. The key property of reliable results is their metrological traceability to stated references with a well established evaluation of the measurement uncertainty (Quevauville, 2004). In practice, method validation is done by evaluating series of method-performance characteristics, such as linearity, operating range, recovery, limit of detection and quantification, precision, selectivity and calibration. The relevant information in the fields of analytical method validation and quality assurances have been published (Taverniers et al., 2004; Gonzalez et al., 2004).

Matrix Reference Materials (MRM) are essential tools for the analytical protocols validation. The feasibility study of a MRM for the analysis of triazines and phenylurea herbicides in water was carried out (Deplagne et al., 2006). Different types of candidates MRM were prepared: solutions of pesticides diluted in acetonitrile and stored in sealed vials or stored at the dry state after the solvent evaporation to dryness, pesticides stored on two different types of polymeric sorbents (Oasis HLB and ENVI-Chrom P) after the percolation of drinking or river water spiked with herbicides. The stability of compounds stored at various temperatures was studied over a period of approximately one year. During the storage, some samples of each different MRM candidate were monthly analyzed by HPLC. Regarding the choice of materials for storage, it was found that a careful control of the temperature of evaporation to dryness is not necessary and similar results were obtained for recovery of herbicides for both used sorbents All herbicides, except simazine, stored as a dry residue at room temperature exhibited a decrease in concentration of more than 20%. The stability seemed to be better when vials were stored at 0.5 °C and at -18 °C neither degradation nor loss of herbicides was observed. This study showed satisfactory long term stability (more than one year) at low temperature for herbicides stored in acetonitrile in vials and for herbicides concentrated on SPE cartridge obtained after passing through a water sample containing these analytes.

To evaluate behavior of these materials containing herbicides, a collaborative study including 15 laboratories has been organized (Mrabet et al., 2006). Observed reproducibility on candidate materials (after the removal of extreme results) was 16.1% for the vials with pesticides in acetonitrile (at around 0.125 mg/L) directly analyzed, 29.2% for a water sample spiked with the pesticides (at around 0.5 μg/L) analyzed after preconcentration on the cartridge and 26.7% for the cartridges previously percolated with the water containing the pesticides (250 mL at around 0.5 μg/L for each pesticide) analyzed after elution.

8. Conclusion

Several hundred herbicides of different chemical structure are used world-wide in agriculture. Due to their persistence, polar nature and water solubility, they are dispersed in the environment and their residues and transformation products are present in several environmental matrices. With increasing public concerns for agrochemicals and their potential movement in the ecosystem, many countries have severely restricted the maximum acceptable concentration of herbicides in drinking water and in vegetable foods. Therefore, the availability of sensitive, selective, precise and rapid analysis methods is essential. Herbicide residue analysis generally requires several steps such as extraction from the sample of interest, removal of interfering co-extractives, analytes enrichment and quantification of their content.

Solid-phase extraction is the top sample-extraction technique for liquid samples, since it can efficiently extract different types of analytes from their matrices and enrich them. Among other advantages, SPE is versatile because a variety of sorbents is available, and the extraction can be tuned depending on how these sorbents interact with the analytes. In recent years, research into new kind of sorbents has focused on improving their capacity and selectivity. Mixed-mode polymeric sorbents, molecularly imprinted polymers and carbon nanotubes are among the new kind of sorbents, which could be useful in enrichment and clean-up purposes in herbicide analysis. MIPs are more selective than mixed-mode sorbents; however, mixed-mode sorbents have greater capacity than MIPs. Carbon nanotubes have a strong adsorption affinity for a wide variety of organic compounds, including pesticides, and are also characterized by their high sorption surface. The use of carbon-encapsulated magnetic nanoparticles avoids the time-consuming column passing and filtration operation and shows great analytical potential in preconcentration of large volumes of real water samples (Zhao et al., 2008). Application of carbon nanostructures have been facilitated by the improvement in their production as the cost has been a main factor in limiting commercialization. However, it is widely believed that if production volumes increase, cost would decrease markedly, thereby significantly increasing the utilization of the excellent properties of nanostructured carbon. Recently, new solvent-free process for producing CNTs from used polymers *via* thermal dissociation in the closed reactor under the inert or air atmosphere has been proposed (Pol & Thiyagarajan, 2010).

9. References

Ahn, J.H.; Jang, J.E.; Oh, C.G.; Ihm, S.K.; Cortez, J. & Sherrington D.C. (2006). Rapid generation and control of microporosity bimodal pore size distribution and surface area in Davankov-type hyper-cross-linked resins. *Macromolecules* Vol. 39, No. 2, 627-632

Al-Degs, Y.D.; Al-Ghouti, M.A. & El-Sheikh, A.H. (2009). Simultaneous determination of pesticides at trace levels in water using multiwalled carbon nanotubes as solid-phase extractant and multivariate calibration. *Journal of Hazardous Materials*, Vol. 169, No. 1-3, 128-135.

An, X. & Zeng, H. (2003). Functionalization of carbon nanobeads and their use as metal ion adsorbents. *Carbon*, Vol. 41, No. 14, 2889-2896

Asensio-Ramos, M.; Hernandez-Borges, J.; Ravelo-Perez, L.M. & Rodriguez-Delgado, M.A. (2008). Simultaneous determination of seven pesticides in waters using multi-walled carbon nanotube SPE and NACE. *Electrophoresis*, Vol. 29, No. 21, 4412-4421

Baggiani, C.; Anfossi, L.; Giovannoli, C. & Tozzi, C. (2004) Binding properties of 2,4,5,-trichlorophenoxyacetic acid-imprinted polymers prepared with different molar ratios between template and monomer. *Talanta*, Vo. 62, No. 5, 1029-1034

Bastide, J.; Cambon, J.-P.; Breton, F.; Piletsky, S.A. & Rouillon, R. (2005). The use of molecularly imprinted polymers for extraction of sulfonylurea herbicides. *Analytica Chimica Acta*, Vol. 542, No. 1, 97-103

Boonjob, W.; Yu, Y.; Miró, M.; Segundo, M.A.; Wang, J. & Cerda, V. (2010). Online hyphenation of multimodal microsolid phase extraction involving renewable molecularly imprinted and reversed-phase sorbents to liquid chromatography for automatic multiresidue assays. *Analytical Chemistry*, Vo. 82, No. 7, 3052-3060

Breton, F.; Euzet, P.; Piletsky, S.A.; Giardi, M.T. & Rouillon, R. (2006). Integration of photosynthetic biosensor with molecularly imprinted polymer-based solid phase extractrion cartridge. *Analytica Chimica Acta*, Vol. 569, No. 1, 50-57

Breton, F., Rouillon, R., Piletska, E.V., Karim, K., Guerreiro, A., Chianella, I. & Piletsky, S.A. (2007). Virtual imprinting as a tool to design efficient MIPs for photosynthesis-inhibiting herbicides. *Biosensors and Bioelectronics* Vol. 21, No. 10, 1948-1954

Cacho, C.; Turiel, E.; Martin-Esteban, A.; Ayala, D. & Pėrez-Conde, C. (2006). Semi-covalent imprinted polymer using propazine methacrylate as template molecule for the clean-up of triazines in soil and vegetable samples. *Journal of Chromatography* A, Vol. 1114, No. 2, 255-262

Camel, C. (2003). Solid phase extraction of elements. *Spectrochimica Acta* Part B, Vol. 58, No. 7, 1177-1233

Chapuis, F.; Pichon, V.; Lanza, F.; Sellergren, S.; Hennion, M.-C. (2003). Optimization of the class-selective extraction of triazines from aqueous samples using a molecularly imprinted polymer by a comprehensive approach of the retention mechanism. *Journal of Chromatography* A, Vo. 999, No. 1-2, 23-33

Cheng, X.L.; Shi, H.L.; Adams, C.D.; Timmons, T. & Ma, Y.F. (2010). Simultaneous screening of herbicide degradation byproducts in water treatment plants using high performance liquid chromatography-tandem mass spectrometry. *Journal of Agricultural and Food Chemistry*, Vol. 58, No. 8, 4588-4593

Cho, H.H.; Smith, B.A.; Wnuk, J.D.; Fairbrother, D.H. & Ball, W.P. (2008). Influence of surface oxides on the adsorption of naphthalene onto multiwalled carbon nanotubes. *Environment Science & Technology*, Vol. 422, No. 8, 2899-2905

D'Archivio, A.A.; Fanelli, M.; Mazzeo, P. & Ruggieri, F. (2007). Comparison of different sorbents for multiresidue solid-phase extraction of 16 pesticides from groundwater coupled with high-performance liquid chromatography. *Talanta* Vol. 71, No. 1, 25-30

Dastgheib, S.A. & Rockstraw, D.A. (2002). Model for the adsorption of single metal ion solutes in aqueous solution onto activated carbon produced from pecan shells. *Carbon*, Vol. 40, No. 11, 1843-1851

Datsyuk, V.; Kalyva, M.; Papagelis, K.; Parthenios, J.; Tasis, D.; Siokou, A.; Kallitsis, I. & Galiiotis, C. (2008). Chemical oxidation of multiwalled carbon nanotubes. *Carbon*, Vol. 46, No. 6, 833-840

Deplagne, J.; Vial, J.; Pichon, V.; Lalere, B.; Hervouet, G. & Hennion, M.-C. (2006). Feasibility study of reference material for water chemistry: Long term stability of triazine and phenylurea residues stored in vials or on polymeric sorbents. *Journal of Chromatography* A, Vol. 1123, No. 1, 31-37

Dias, A.C.B.; Fingueiredo, E.C.; Grassi, V.; Zagatto, E.A.G. & Arruda, M.A.Z. (2009). Molecularly imprinted polymer as a solid phase extractor in flow analysis. *Talanta* Vol. 76, No. 5, 988-996

Diaz-Garcia, M.E. & Lamo, R.B. (2005). Molecular imprinting in sol-gel materials: recent developments and applications. *Microchimica Acta*, Vol. 149, No. 1-2, 19-36

Djozan, D. & Ebrahimi, B. (2008). Preparation of new solid phase micro exctraction fiber on the basis of atrazine-molecular imprinted polymer: application for GC and GC-MS screening of triazine herbicides in water, rice and onion. *Analytica Chimica Acta*, Vol. 616, No. 1, 152-159.

Dong, M.; Ma, Y.; Zhao, E.; Qian, C.; Han, L. & Jiang, S. (2009). *Microchimica Acta*, Vol. 165, No. 1-2, 123-128

El-Mrabet, K,; Poitevin, M.; Vial, J.; Pichon, V.; Amarouche, S.; Hervouet, G. & Lalere, B. (2006). An interlaboratory study to evaluate potential matrix reference materials for herbicides in water. *Journal of Chromatography* A, Vol. 1134, No. 1-2, 151-161

El-Skeikh, A.H.; Sweileh, J.A.; Al-Degs, Y.S.; Insisi, A.A. & Al-Rabady, N. (2008). Critical evaluation and comparison of enrichment efficiency of multi-walled carbon nanotubes, C_{18} silica and activated carbon towards some pesticides from environmental waters. *Talanta*, Vol. 74, No. 5, 1675-1680

Faria, A.M.; Maldaner, L.; Santana, C.C.; Jardim, I.C. & Collins, C.H. (2007). Poly(methyltetradecylsiloxane) immobilized onto silica for extraction of multiclass pesticides from surface waters. *Analytica Chimica Acta* Vol. 582, No. 1, 34-40

Fontanals, N.; Marcé, R.M. & Borrull, F. (2007). New materials in sorptive extraction techniques for polar compounds. *Journal of Chromatography* A Vol. 1152, No. 1-2, 14-31

Fontanals, N.; Cormack, P.A.G.; Rosa M.; Marcé, R.M. & Borrull, F. (2010). Mixed-mode ion-exchange polymeric sorbents: dual-phase materials that improve selectivity and capacity. *Trends in Analytical Chemistry*, doi: 10.1016/j.trac.2010.03.015

González, F.J.E.; Torres, M.E.H.; Frenich, A.G.; Vidal, J.L.M. & Campaña, A.M.G. (2004). Internal quality-control and laboratory-management tools for enhancing the stability of results in pesticide multi-residue analytical methods. *Trends in Analytical Chemistry*, Vo. 23, No 5, 361-369

Haginaka, J. (2005). Selectivity of affinity media in solid phase extraction of analytes. *Trends in Analytical Chemistry*, Vo. 24, No. 5, 407-415

Herrero-Hernádez, E.; Carabias-Martinez, R. & Rodriguez-Gonzalo, E. (2009). Use of a bisphenol-A imprinted polymer as a selective sorbent for the determinaton of phenols and phenoxyacids in honey by liquid chromatography with diode array and tandem mass spectrometric detection. *Analytica Chimica Acta*, Vol. 650, No. 2, 195-201

Hildebrandt, A.; Lacorte, S.; Barceló, D. (2007). Assessment of priority pesticides, degradation products, and pesticide adjuvants in groundwaters and top soils from agricultural areas of the Ebro river basin. *Analytical and Bioanalytical Chemistry*, Vol. 387, No. 4, 1459–1468

Kailasam, K.; Fels, A. & Müller, K. (2009). Octadecyl grafted MCM-41 silica spheres using trifunctionalsilane precursors - preparation and characterization. *Microporous and mesoporous materials*, Vol. 117, No. 1-2, 136-147

Kloskowski, A.; Pilarczyk, M.; Przyjazny, A. & Namieśnik, J. (2009). Progress in development of molecularly imprinted polymers as sorbents for sample preparation. *Critical Reviews in Analytical Chemistry*, Vol. 39, No. 1, 43-58

Koohpaei, A.R.; Shahtaheri, S.J.; Ganjali, M.R.; Forushani, A.R. & Golbabaei, F. (2008). Application of multivariate analysis to the screening of molecularly imprinted polymers (MIPs) for ametryn. *Talanta*, Vol. 75, No. 4, 978-986

Koohpaei, A.R.; Shahtaheri, S.J.; Ganjali, M.R.; Forushani, A.R. & Golbabaei, F. (2009). Optimization of solid-phase extraction using develped modern sorbent for trace determination of ametryn in environmental matrices. *Journal of Hazardous Materials*, Vol. 170, No. 2-3, 1247-1255

Kyriakopoulos, G. & Doulia, D. (2006). Adsorption of pesticides on carbonaceous and polymeric materials from aqueous solutions. *Separation and Purification Reviews*, Vol. 35, No. 3, 97-191

Lasákova, M. & Jandera, P. (2009). Molecularly imprinted polymers and their application in solid phase extraction. *Journal of Separation Science*, Vol. 32, No. 5-6, 799-812

Lavén, M.; Alsberg, T.; Yu, Y.; Adolfsson-Erici, M. & Sun, H.W. (2009). Serial mixed-mode cation- and anion-exchange solid-phase extraction for separation of basic, neutral and acidic pharmaceuticals in wastewater and analysis by high-performance liquid chromatography-quadrupole time-of-flight mass spectrometry. *Journal of Chromatography* A, Vol. 1216, No. 1, 49-62

Li, Y.; George, J.E.; MsCarty, C.L. & Wendelken, S.C. (2006). Compliance analysis of phenylurea and related compounds in drinking water by liquid chromatography/ electrospray ionization/mass spectrometry coupled with solid-phase extraction. *Journal of Chromatography* A, Vol. 1134, No. 1, 170-176

Li, Q.; Wang, X. & Yuan, D. (2009). Solid-phase extraction of polar organophosphorous pesticides from aqueous samples with oxidized carbon nanotubes. *Journal of Environmental Monitoring*, Vol. 11, No. 2, 439-444

Liu, F.K. (2008). Preconcentration and separation of neutral steroid analytes using a combination of sweeping micellar electrokinetic chromatography and a Au nanoparticle-coated solid phase exctraction sorbent. *Journal of Chromatography* A, Vol. 1215, No. 1-2, 194-202

Li, Q.; Wang, X. & Yuan, D. (2009). Solid-phase extraction of polar organophosphorous pesticides from aqueous samples with oxidized carbon nanotubes. *Journal of Environmental Monitoring*, Vol. 11, No. 2, 439-444

Maloschik, E.; Mortl, M. & Szekacs, A. (2010). Novel derivatisation technique for the determination of chlorophenoxy acid type herbicides by gas chromatography-mass spectrometry. *Analytical and Bioanalytical Chemistry*, Vol. 397, No. 2, 537-548

Mazzella, N.; Debenest, T. & Delmas, F. (2008) Comparison between the polar organic chemical integrative sampler and the solid-phase extraction for estimating herbicide time-weighted average concentrations during a microcosm experiment. *Chemosphere*, Vol. 73, No. 4, 545-550

Mhaka, B.; Cukrowska, E.; Tse Sum Bui, B.; Ramström, O.; Haupt, K.; Tutu, H. & Chimuka, L. (2009). Selective extraction of triazine herbicides from food samples based on a combination of a liquid membrane and molecularly imprinted polymers. *Journal of Chromatography* A, Vol. 1216, No. 41, 6796-6801

Min, G.; Wang, S.; Zhu, H.; G. Fang, G. & Zhang, Y. (2008). Multi-walled carbon nanotubes as solid-phase extraction adsorbents for determination of atrazine and its principal metabolites in water and soil samples by gas chromatography-mass spectrometry. *Science of the Total Environment*, Vol. 398, No. 1-3, 79-85

Miró, M. & Hansen, E.H. (2006). Solid reactors in sequential injection analysis: recent trends in the environmental field. *Trends in Analytical Chemistry*, Vo. 25, No. 3, 267-281

Morihisa, H.; Fukata, F.; Muro, H.; Nishimura, K.I. & Makini, T. (2008). Determination of indapamide in human serum using 96-well solid phase extraction and high-performance liquid chromatography–tandem mass spectrometry. *Journal of Chromatography* B, Vol. 870, No. 1, 126-130

Niu, H.Y.; Cai, Y.Q.; Shi, Y.L.; Wei, F.S.; Liu, J.M. & Jiang, G.B. (2008). A new solid-phase extraction disk based on a sheet of single-walled carbon nanotubes. *Analytical and Bioanalytical Chemistry*, Vol. 392, No. 5, 927-935

Niu, H.; Shi, Y.; Cai, Y.; Wei, F. & Jiang, G. (2009). Solid phase extraction of sulfonylurea herbicides from water samples with single-walled carbon nanotube disk. *Microchimica Acta*, Vol. 164, No. 3-4, 431-438

Oxelbark, J.; Legido-Quigley, C.; Aureliano, C.S.A.; Titirici, M.M.; Schillinger, E.; Sellergren, B.; Courtois, J.; Irgum, K;, Dambies, L.; Cormack, P.A.G.; Sherrington, D.C. & De Lorenzi, E. (2007). Chromatographic comparison of bupivacaine imprinted-polymers prepared by crushed monolith, microsphere, silica-based composite and capillary monolith formats. *Journal of Chromatography* A, Vol. 1160, No. 1-2, 215-226.

Quevauviller, P. (2004) Traceability of environmental chemical measurements. *Trends in Analytical Chemistry*,, Vol. 23, No. 3, 171-177

Pan, B. & Xing, B. (2008). Adsorption mechanisms of organic chemicals on carbon nanotubes. *Environmental Science & Technology*, Vol. 42, No. 24, 9065-9113

Parida, S.K.; Dash, S.; Patel, S. & Mishra, B.K. (2006). Adsorption of organic molecules on silica surface. *Advances in Colloid and Interface Science*, Vol. 121, No. 1-3, 77-101

Pol, V.G. & P. Thiyagarajan, P. (2010). Remediating plastic waste into carbon nanotubes. *Journal of Environmental Monitoring*, Vol. 12, No. 2, 455-459

Polati, S.; Bottaro, M.; Frascarolo, P.; Gosetti, F.; Gianotti, V. & Gennaro, M.C. (2006). HPLC-UV and HPLC-MS^n multiresidue determination of amidosulfuron, azimsulfuron, nicosulfuron, rimsulfuron, thifensulfuron methyl, trobenuron methyl and azoxystrobin in surface waters. *Analytica Chimica Acta*, Vol. 579, No. 2, 146-151

Prasad, K.; Prathish, K.P.; Gladis, J.M.; Naidu, G.R.K. & Rao, T.P. (2007). Molecularly imprinted polymer (biomimetic) based potentiometric sensor for atrazine. *Sensors and Actuators*, Vol. 123, No. 1, 65-70

Pyrzynska, K. (2003). Novel selective sorbents for solid-phase extraction. *Chemia Analityczna - Chemical Analysis (Warsaw)* Vol.48, No. 5, 781-795

Pyrzynska, K. (2008). Carbon nanotubes as a new solid-phase extraction material for removal and enrichment of organic pollutants in water. *Separation and Purification Reviews*, Vol. 37, No. 4, 375-392

Qiao, F.; Sun, H.; Yan, H. & Row, K.H. (2006). Molecularly imprinted polymers for solid phase extraction. *Chromatographia*, Vol. 64, No. 2, 625-634

Rastkari, N.; Ahmadkhaniha, R. & M. Yunesian, M. (20009). Single-walled carbon nanotubes as an effective adsorbent in solid-phase microextraction of low level methyl *tert*-butyl ether, ethyl *tert*-butyl ether and methyl *tert*-amyl ether from human urine. *Journal of Chromatography* B, Vol. 877, No. 14-15, 1568-1574

Ravelo-Perez, L.M.; Hernandez-Borges, J. & Rodriguez-Delgado, M.A. (2008). Multi-walled carbon nanotubes as efficient solid-phase extraction materials of organophosphorus pesticides from apple, grape, orange and pineapple fruit juices. *Journal of Chromatography* A, Vol. 1211, No. 1-2, 33-42

Richardson, S.D. (2009) Water analysis: emerging contaminants and current issues. Analytical Chemistry, Vo. 81, No. 12, 4645-4677

Rodriguez-Gonzalo, E.; Carabias-Martinez, R.; Cruz, E.M.; Dominguez-Alvarez, J. & Hernandez-Mendez, J. (2009). Ultrasonic solvent extraction and nonaqueous CE for the determination of herbicide residues in potatoes. *Journal of Separation Sciences*, Vol. 32, No. 4, 575-584

Rosalez-Conrado, N.; León-Gonzalez, M.E.; Pérez-Arribas, L.V. & Polo-Diez, L.M. (2005). Capillary liquid chromatography of chlorophenoxy acid herbicides and their esters in apple juice samples after preconcentration on a cation exchanger based on polydivinylbenzene-*N*-vinylpyrrolidone. *Journal of Chromatography* A, Vol. 1076, No. 1-2, 202-206

Sambe, H.; Hoshina, K. & Haginaka, J. (2007). Molecularly imprinted polymers for triazine herbicides prepared by multi-step swelling and polymerization method. Their application to the determination of methylthiotriazine herbicides in river water. *Journal of Chromatography* A, Vol. 1152, No. 1-2, 130-137

Sergeyeva, L.V.; Brovko, O.O.; Piletska, E.V.; Piletsky, S.A.; Goncharova, L.A.; Karabanova, L.V.; Sergeyeva, L.M. & El`skaya, A.V. (2007). Porous molecularly imprinted membranes and polymeric particles. *Analytica Chimica Acta*, Vol. 582, No. 2, 311-31

Segura, P.A.; Gagnon, C. & Sauvé, S. (2007). A fully automated on-line preconcentration and liquid chromatography-tandem mass spectrometry method for the analysis of anti-infectives in wastewaters. *Analytica Chimica Acta*, Vol. 604, No. 2, 147-157

Sorensen, L.M.; Nielsen, K.F.; Jacobsen, T.; Koch, A.G.; Nielsen, P.V. & Frisvad, J.C. (2008). Determination of mycophenolic acid in meat products using mixed mode reversed phase-anion exchange clean-up and liquid chromatography-high-resolution mass spectrometry. *Journal of Chromatography* A, Vol. 1205, No. 1-2, 103-108

Spivakov, B.Y.; Malofeeva, G.L. & Petruskhin, O.M. (2006). Solid-phase extraction on alkyl-bonded silica gels in inorganic analysis. *Analytical Sciences*, Vol. 22, No. 4, 503-519

Stoob, K.; Singer, H.P.; Goetz, C.W.; Ruff, M. & Mueller, S.R. (2005). Fully automated online solid phase extraction coupled directly to liquid chromatography–tandem mass spectrometry: Quantification of sulfonamide antibiotics, neutral and acidic pesticides at low concentrations in surface waters. *Journal of Chromatography* A, Vol. 1097, No. 1-2, 138-147

Tamayo, F.G., Casillas, J.L. & Martin-Esteban, A. (2005). Evaluation of new selective molecularly imprinted polymers prepared by precipitation polymerisation for the extraction of phenylurea herbicides. *Journal of Chromatography* A, Vol. 1069, No. 2, 173-181

Taverniers , I.; De Loose, M. & Van Bockstaele, E. (2004). Trends in quality in the analytical laboratory. II. Analytical method validation and quality assurance. *Trends in Analytical Chemistry*, Vol. 23, No. 8, 535-552

Viglino, L.; Aboulfadl, K.; Mahvelat, A.D.; Prevost, M. & Sauvé, S. (2008). On-line solid phase extraction and liquid chromatography/tandem mass spectrometry to quantify pharmaceuticals, pesticides and some metabolites in wastewaters, drinking and surface waters. *Journal of Environmental Monitoring*, Vol. 10, No. 4, 482-489

Vigna, C.R.M.; Morais, L.S.; Collins, C.H. & Jardim, I.C. (2006). Poly(methyloctylsiloxane) immobilized on silica as a sorbent for solid-phase extraction of some pesticides. *Journal of Chromatog*raphy A, Vol. 1114, No. 2, 211-215

Wepasnick, K.A.; Smith, B.A.; Bitter, J.L. & Fairbrother, D.H. (2010). Chemical and structural characterization of carbon nanotube surfaces. *Analytical and Bioanalytical Chemistry*, Vol. 396, No. 3, 1003-1014

Xu, R.N.; Fan, L.; Rieser, M.J. & El-Shourbag, T.A. (2007). Recent advances in high-throughput quantitative bioanalysis by LC-MS/MS. *Journal of Pharmaceutical and Biomedical Analysis*, Vol. 44, No. 2, 342-355

Yang, K.; Zhu, L. & Xing, B. (2006). Adsorption of polycyclic aromatic hydrocarbons by carbon nanomaterials. *Environmental Science & Technology*, Vol. 40, No. 6, 1855-1861

Yu, Z.; Liu, B.; Jiang, Z.H. & Zhang, G. (2009). Simultaneous determination of herbicide mefenacet and ist metabolities residues in river water by solid phase extraction and rapid resolution liquid chromatography-mass spectrometry with pre-column derivatization. *Journal of Chromatography* A, Vol. 1216, No. 15, 3090-3097

Zhang, H.; Song, T.; Zhang, W.; Hua, W. & Pan, C. (2007). Retention behavior of phenoxyacetic herbicides on a molecularly imprinted polymer with phenoxyacetic acid as a dummy template molecule. *Bioorganic & Medicinal Chemistry*, Vol. 15, No. 18, 6089-6095

Zhao, X.; Y. Shi, Cai, Y. & Mou, S. (2008). Cetyltrimethylammonium bromide-coated magnetic nanoparticles for the preconcentration of phenolic compounds from environmental water samples. *Environmental & Science Technology*, Vo. 42, No. 4, 1201-1206

Zhou, Q.; Xiao, J. & Wang, W. (2007). Comparison of multiwalled carbon nanotubes and a conventional absorbent on the enrichment of sulfonylurea herbicides in water samples. *Analytical Sciences*, Vol. 23, No. 2, 189-192

Chemometric Strategies for the Extraction and Analysis Optimization of Herbicide Residues in Soil Samples

Cristina Díez[1], Enrique Barrado[1] and José Antonio Rodríguez[2]
[1]University of Valladolid
[2]Hidalgo State Autonomous University Pachuca-Hidalgo
[1]Spain
[2]Mexico

1. Introduction

1.1 Herbicide benefits and concerns

The two cereal crops which are grown most abundantly in the EU are wheat and barley (Document No. SANCO/D3/SI2.396179, 2005). Herbicides play a very important role in effectively controlling annual grasses and broad-leaved weeds affecting these crops. Their use cannot be neglected due to the enormous benefits in agricultural outputs. Among them, acidic herbicides are widely used for control of broad-leaved weeds and other vegetation because they are relatively inexpensive and very potent even at low concentrations (Wells & Yu, 2000).

However, due to the herbicide widespread and possible toxicity in the environment, it is important to monitor their residues. Under realistic field situations there is a potential exposure to agricultural soils by these products indirectly through spray drift and run-off from crop vegetation surfaces, and directly through soil treatment practices (Document No. D/00/SuM/5277, 2000). They have harmful effects on the microflora of the soil when they are not degraded quickly enough (Santos-Delgado et al., 2000).

1.2 Herbicide characteristics

There are many compounds registered as herbicides intended for their use in cereal crops, which can be classified into several chemical classes in accordance with their chemical structures.

Substituted ureas are one of the oldest herbicide groups used in agriculture, being two of the most important, phenylureas, employed since early fifties, and sulphonylureas, developed more recently with a high herbicidal activity resulting in low applications doses (Tadeo et al., 2000). Among the basic herbicides, triazines are the most important selective herbicides (Pinto & Jardim, 2000). Triazinic herbicides have been widely used in the last years for crop protection in agriculture and weed removal in different lands. Among neutral herbicides, dinitroaniline herbicides are usually soil applied in a wide variety of agronomic crops and particularly, in winter and spring cereals. Thiocarbamates have been used as herbicides in maize and wheat for several decades (Tadeo et al., 2000).

Acidic herbicides consist of several families of compounds that are related by similarities in biological activity and chemical properties which influence the way they are extracted and analyzed. These families of compounds are derivatives of acidic functional groups including benzoic acid (dicamba), acetic acid [2,4-dichlorophenoxyacetic acid (2,4-D) and 4-chloro-2-methylphenoxyacetic acid (MCPA)], propanoic acid [dichlorprop, diclofop, fenoxaprop p and 2-(4-chloro-2-methylphenoxy)propanoic acid (MCPP)], picolinic acid (clopyralid) and pyridinecarboxylic acid (fluroxypyr) among others (Wells & Yu, 2000). Acidic herbicides can be applied in the form of free acids, salts or esters (Analytical Methods for Pesticide Residues in Foodstuffs, Part I, 1996). Several studies have shown that in the environment, acidic herbicides formulated as esters undergo fast hydrolysis, on the order of 24–48 h, depending on pH and other conditions and in the presence of vegetable tissues and soil bacteria yielding the corresponding free acids (Budde, 2004; Marchese, 2001; Tadeo et al., 2000). Therefore, they are generally present as the corresponding acids and most frequently exist in ionized form at most environmental pH values (Wells & Yu, 2000).

1.3 Multiresidue determination and analysis

Since spray history or environmental background of most soil samples is unknown, method development efforts have concentrated on multiresidue methods (Regulation EC No 1107/2009, 2009; Document No. SANCO/825/00, 2004). They require universality of the isolation and clean-up procedure and, as far as possible, unification of the conditions of the chromatographic separation (Tekel & Kovacicová, 1993).

Acetone, acetonitrile and ethyl acetate, sometimes at acidic pH, are the most usual organic solvents employed in the extraction of a large number of herbicide residues belonging to different groups (Kremer et al., 2004; Jiménez et al., 2000; Mastovska & Lehotay, 2004; Papadopoulou-Mourkidou et al., 1997; Sánchez-Brunete & Tadeo, 1996). The neutral (dinitroanilines, phenylureas, thiocarbamates) and basic (triazines) multiresidue herbicide extractions from soils are usually carried out with organic solvents (Tadeo et al., 1996). The addition of water has been reported, in some cases, to increase desorption of herbicides from the matrix because it is both a solvent for the analyte and a solute that can compete for adsorption sites (Tadeo et al., 1996).

Acidic herbicides (phenoxyacids, benzoic acids, sulfonylureas) are reported in most cases to be extracted at low pH conditions that suppress the ionization of acids and make them neutral and more apt to be extracted with an organic solvent (Macutkiewicz et al., 2003; Marchese, 2001; Nolte & Kruger, 1999; Sánchez-Brunete & Tadeo, 1996; Wells & Yu, 2000). They are usually extracted from soils with solid-liquid extraction with organic solvent–water mixtures at an acid pH with a solvent of medium polarity or with an alkaline solution with sodium hydroxide 0.5 M (EPA Method 8151A, 1996). Afterwards, the extract is acidified and partitioned into an organic solvent immiscible with water or concentrated with solid-phase extraction (SPE) (Crespín et al., 2001; Menasseri & Koskinen, 2004; Patsias, 2002). Many procedures have been shown to effectively extract acid herbicides with organic solvents of medium polarity in mixtures with water and acetic acid (Ahmed & Bertrand, 1989; Crescenzi et al., 1999; Menasseri & Koskinen, 2004; Smith & Milward, 1981; Smith, 1995). In the same way, ammonium hydroxide has been reported to enhance basic herbicide recoveries (Smith & Milward, 1981).

No pH adjustment for non-ionic and basic analyte water removal by liquid partition has been reported (Papadopoulou-Mourkidou et al., 1997), meanwhile methods without previous acidification (Ahmed & Bertrand, 1989) and with previous pH adjustment to pH 2 (Crespín et al., 2001; EPA Method 8151A, 1996; Sánchez-Brunete & Tadeo, 1996) have been found for acid analytes.

Herbicides formulated as esters have been reported to rapidly hydrolyze in contact with soil to their corresponding acids and phenols (Budde, 2004). The analyst, therefore, must either evaluate the herbicides in both the ester and hydrolyzed acid forms, or convert all components present to their free acids before analysis. For example, some analytical methods specify a strong base hydrolysis of any residual esters before conversion of the acids to methyl esters for GC (Wells & Yu, 2000). However, this approach is unsuitable for multi-class herbicide residue analysis because other analytes will be destroyed under such strong conditions. To avoid this loss, a unique multiresidue extraction and a simultaneous analysis for both esters and their corresponding free acids is intended.

The use of GC-MS is a very versatile and sensitive method for residue analysis due to the high sensitivity obtained and MS is a very valuable detection technique, because it provides information on the compound molecular structure and it is also highly sensitive and selective when used in the single ion monitoring (SIM) mode (Tadeo et al., 2000).

The acidic compounds, because of their polar nature, suffer from peak asymmetry and tailing in the GC stationary phases. Masking of these acidic hydrogens by derivatization to their corresponding esters is essential in order to yield products with enhanced volatility that can undergo analysis by GC (Catalina et al., 2000). The typical reactions of derivatization of phenoxyacids are trans-esterification, esterification, silylation, alkylation and extractive and pyrolytic alkylation (Rompa et al., 2004). The formation of methyl esters/ethers is particularly preferred, because they can be easily prepared and have reasonably short GC retention times (Macutkiewicz et al., 2003). The most useful reagent of pyrolytic alkylating reagents is TMSH ($(CH_3)_3SOH$ (Yamauchi et al, 1979) which provides an efficient methylation by pyrolysis of the previously formed salts of nucleophiles, e.g. carboxylic acids and phenols to their corresponding methyl esters and methyl ethers (Butte, 1983). It can be used in two ways, i.e. to methylate free acids by pyrolysis of the salt in the heated injection port of a GC, or to effect base-catalyzed trans-esterification of other esters to their methyl esters (Butte, 1983; Christie, 1993). As a result, methyl esters are the final product of both reactions. This reaction is very elegant and convenient, because it is just necessary to add the reagent to the sample solution with little or no work-up and reacts very rapidly (Butte, 1983; Halket & Zaikin, 2004). In addition, removal of excess reagent is not required, as in other derivatization reactions, because the only by-products of this reaction are dimethylsulphide (b.p. 37ºC), and methanol that elute with the solvent peak and do not disturb the chromatographic separation of analytes.

2. Introduction. Method development. Chemometric strategies

Since the extraction process for a number of analytes occurs, more or less, in a single run within multiresidue methods, the efficiency of the recovery of each individual component differs from each other, due to their different chemical structures. A detailed optimization of these multiresidue procedures would, therefore, help to adjust the applied conditions in a way to obtain the maximum recovery percentage for most of the constituents of the sample. These multiresidue methods, however, are in principle rather costly for implementation on a large scale, so they require the use of chemometric strategies applied to method development in order to ensure an efficient recovery.

A frame of integration between analytical procedures and chemometric methods has made the extraction of relevant underlying analytical information possible, largely applied in the environmental science where data interpretation is of great interest (Einax et al., 1997). Chemometrics is a chemical discipline that uses mathematics, statistics and formal logic to design or select optimal experimental procedures, to provide maximum relevant chemical information by analyzing chemical data; and to obtain knowledge about chemical systems (Massart et al., 1998). Some of these chemometric strategies are detailed in this chapter.

2.1 Pattern recognition: multivariate analysis

Pattern recognition is the scientific discipline whose goal is the classification of objects into a number of categories or classes. It "reveals" the organization of patterns into "sensible" clusters (groups), which will allow to discover similarities and differences among patterns and to derive useful conclusions about them.

Classification is synonymous with pattern recognition, and scientists have turned to it and PCA and cluster analysis to analyze the large data sets typically generated in environmental studies that employ computerized instrumentation. The set of measurements that describe each sample in the data set is called a pattern. The determination of the property of interest by assigning a sample to its respective category is called recognition, hence the term pattern recognition. Clustering and classification are the major subdivisions of pattern recognition techniques.

In a typical pattern recognition study, samples are classified according to a specific property using measurements that are indirectly related to that property. An empirical relationship or classification rule is developed from a set of samples for which the property of interest and the measurements are known. The classification rule is then used to predict this property in samples that are not part of the original training set (Lavine, 2000; McLachlan, 1992).

2.1.1 Cluster analysis

Cluster analysis (Kaufman & Rousseeuw, 1990; Massart, 1983) is the name given to a set of techniques whose basic objective is to discover sample groupings within data. For cluster analysis, each sample is treated as a point in an n-dimensional measurement space. The coordinate axes of this space are defined by the measurements used to characterize the samples. Cluster analysis assesses the similarity between samples by measuring the distances between the points in the measurement space. A basic assumption is that the distance between pairs of points in this measurement space is inversely related to the degree of similarity between the corresponding samples. Points representing samples from one class will cluster in a limited region of the measurement space distant from the points corresponding to the other class. Samples that are similar will lie close to one another, whereas dissimilar samples are distant from each other (Lavine, 2000). Samples within the same group are more similar to each other than samples in different groups.

Clustering methods are divided into three categories, hierarchical, object-functional, and graph theoretical. The hierarchical methods are the most popular. The results of a hierarchical clustering study are usually displayed as a dendogram, which is a treeshaped map of the intersample distances in the data set. The dendogram shows the merging of samples into clusters at various stages of the analysis and the similarities at which the clusters merge, with the clustering displayed hierarchically (Lavine, 2000).

Clustering has a lot of applications:

1. Data reduction: Many times, the amount of the available data, N, is very large and, as a consequence, its processing becomes very demanding. Cluster analysis can be used in order to group the data into a number of sensible clusters, m (<<N) and to process each cluster as a single entity.
2. Prediction based on groups: the resulting clusters are characterized based on the characteristics of the patterns by which they are formed. In the sequel, if an unknown pattern is given, it can be determined the cluster to which it is more likely to belong and it can be characterized based on the characterization of the respective cluster.
3. Hypothesis generation: cluster analysis is applied to a data set in order to infer some hypotheses concerning the nature of the data. Thus, clustering is used as a vehicle to suggest hypotheses. These hypotheses must then be verified using other data sets.
4. Hypothesis testing: In this context, cluster analysis is used for the verification of the validity of a specific hypothesis.

2.1.2 Principal component analysis

PCA (Brown, 1995; Joliffe, 1986, Wold et al., 1987) aims to reduce the dimensionality of a data set, while simultaneously retaining the information present in the data. It allows the transformation and visualization of complex data sets into a new perspective in which the more relevant information is made more obvious. PCA extracts maximal information from large data matrices containing numerous columns and rows because it calculates the correlations between the columns of the data matrix and classifies the variables according to the coefficients of correlations (Cserháti; 2010; Kaliszan, 1997; Mardia et al., 1979; Vandeginste et al., 1998).

The original measurement variables are transformed into new conceptually meaningful variables called principal components which account for most of the variation providing reduction of the dimensionality of the dataset. By plotting the data in a coordinate system defined by the two or three largest principal components, it is possible to identify key relationships in the data, that is, find similarities and differences among objects in a data set. The first component is the linear combination of variables that contribute most to the total variance. The second principal component is orthogonal to the first and accounts for most of the residual variance. Each principal component describes a different source of information because each defines a different direction of scatter or variance in the data (the scatter of the data points in the measurement space is a direct measure of the data's variance). Hence, the orthogonality constraint imposed by the mathematics of PCA ensures that each variance-based axis will be independent (Lavine, 2000).

One measure of the amount of information conveyed by each principal component is the variance of the data explained by the principal component. The variance explained by each principal component is expressed in terms of its eigenvalue. For this reason, principal components are usually arranged in order of decreasing eigenvalues or waning information content. The most informative principal component is the first and the least informative is the last. By examining the eigen vector for those variables that load heavily to the component axis, it is possible to give the principal axis a physical interpretation. The closer the values are to 1 or -1, the more they contribute to that component, i.e. the axis aligned to the variable is also closely aligned to the component axis. If the value is closer to 0, the axis for the variable is at a right angle to the component axis and does not influence it greatly.

Due to its versatility and its easy-to-use multivariate mathematical-statistical procedure, PCA is frequently used in many fields of up-to date research, such as environmental protection studies (Cserháti; 2010; Hildebrandt et al., 2008).

2.2 Optimization experimental designs. Orthogonal Arrays

The optimization of any process can be tried either by the trial and error method, the one-at-a-time design or achieved by experimental design methods. The one-at-a-time design is a classical **Univariate method** which consists of investigating the response for each factor while all other factors are held at a constant level. Therefore, the variation of response can be attributed to the variation of the factor. They are time-consuming methods which do not take interactive effects between factors into account because the real optimum cannot be achieved. In this case, the use of factorial designs, which are based in blocking, is very useful because the response is measured for all possible combination of the chosen factor levels.

Blocking is one of the fundamental principles of good experimental design because it reduces the variability from the most important sources and hence increases the precision of experimental measurements. Essentially, experimental units are grouped into homogeneous clusters in an attempt to improve the comparison of treatments by randomly allocating the treatments within each cluster or "block" (Hanrahan et al., 2008).

Screening techniques such as **Factorial Designs** allow the analyst to select which factors are significant and at what levels. Such techniques are vital in determining initial factor significance for subsequent optimization. The most general (two-level design) is a full factorial design and described as 2k designs, where the base 2 stands for the number of factor levels and k is the number of factors each with a high and a low value (Bruns et al., 2006; Otto, 1999). One obvious disadvantage of factorial designs is the large number of experiments required when several variables are examined. However, this number can be considerably reduced by the use of **Fractional Factorial Designs**, such as **Orthogonal Array** designs **(OA)** (Lan et al., 1994; Lan et al., 1995), orthogonal meaning balanced (Wan et al., 1994). The theory and methodology of OA, as a chemometric method for the optimization of the analytical procedure, have been described in detail elsewhere (Lan et al., 1994; Lan et al., 1995). They imply the use of a strategically designed experiment which deliberately introduces changes in order to identify factors affecting the procedure, and estimate the factor levels yielding the optimum response with minimal experimental investment (Oles, 1993; Wan et al., 1994). They assign factors to a series of experiment combinations whose results can then be analyzed by using a common mathematical procedure. The main effects of the factors and preselected interactions are independently extracted.

Although the optimization by factorial designs is regarded as a simultaneous method, the optimum is actually located step by step as in sequential approaches. Therefore, previous knowledge of the variables, past experience and intuition are very helpful in arranging the variables and levels of the experiment because OA only cover a predefined region (Wan et al., 1994).

Taguchi Parameter Design, which uses OA, introduces, in addition, the concept of the signal-to-noise ratio to evaluate the variation of the response around the mean value due to experimental noise, which makes the optimum response robust against uncontrollable external variability, named noise factors (Barrado et al., 1998; Bendell et al., 1989; Ross, 1988; Taguchi, 1991). It allows separating the effect of each factor on the output variable in terms of mean response (regular analysis) and signal-to-noise ratio analysis. It has the following aims: to identify factors affecting the procedure, to estimate the factor values leading an optimum response and to decrease the process variability without controlling or eliminating causes of variation, which yields a process robust against noise factors.

The steps for implementing the experimental design are the following:

1. To select the output variable to be optimised,

2. To identify factors and their interactions affecting the output variable and to choose the levels to be tested,
3. To select the adequate orthogonal array,
4. To assign factors and interactions to the columns of the array,
5. To perform the experiments,
6. To carry out an statistical analysis of the data and determine the optimum factor levels, and
7. To conduct a confirmatory experiment.

Different OA have been applied in analytical method development allowing the identification of the principal and interaction effects of the extraction conditions on the recovery of pollutants (Mostert et al., 2010), and more specifically to pesticides from various environmental samples, such as vegetables (Pena et al., 2006; Quan et al., 2004; Wan et al., 2010), soils (Delgado-Moreno et al., 2009; Fuentes et al., 2007; Sun et al., 2003) or water samples (Bagheri et al., 2000; Chee et al., 1995; Lin & Fuh, 2010; Pasti el al, 2007; Wells et al., 1994; Wan et al., 1994). OA have also been applied to the optimization of derivatization procedures to analyse pesticides by GC (Stalikas & Pilidis; 2000).

3. Experimental procedures

3.1 Principle of the experimental method

The application of some chemometric strategies in order to develop a multiresidue extraction and analysis method for nearly 40 herbicides, belonging to very different chemical families, in agricultural soils of barley crops is shown.

The influence of some variables in recovery was studied by a set of previous experiments analyzed by PCA and Clustering techniques. Then, the most important factors affecting the multiresidue herbicide extraction were optimized by an OA.

The acidic and phenol herbicide methylation by TMSH in order to analyse their methyl esters/ethers by GC, was also optimised by an OA.

3.2 Reagents, equipment and analysis

Reagents

- The herbicides studied in this work are summarised in Table 1 together with some important physicochemical properties (The FOOTPRINT Pesticide Properties DataBase, 2006). All herbicide standards were obtained from Dr. Ehrenstorfer (Augsburg, Germany). Individual stock standard solutions (1000 mg/l) were prepared in acetone and stored in the dark at -20ºC. They were kept for 1 hour at ambient temperature previously to their use. Working standard mixtures in acetone, containing 10 mg/l of each pesticide were prepared by dilution.
- Calibration standards were prepared by dilution in acetone acidified with 1% acetic acid. The internal standard was prepared by dissolving Alachlor (a sunflower herbicide) in acetone to make stock solutions of 1000 mg/l and diluted in acetone acidified with 1% acetic acid to 1 mg/l before the addition of 20 µl to samples.
- Organic solvents intended for extraction, were at least HPLC grade and were provided by Labscan (Dublin, Ireland) together with the glacial acetic acid and the ammonium hydroxide (28% in water).
- Trimethylsulfonium hydroxide (TMSH) purum 0.25 M in methanol, was purchased from Fluka (Buchs, Switzerland) and stored at 4ºC.

- Bulk quantities of Na_2SO_4, obtained from Merck (Darmstadt, Germany), were heated to 500°C for more than 5 hours to remove phthalates and any residual water prior to its use in the laboratory.
- The same soil was used for all the tests: 46% sand, 37% silt, 17% clay; 0.69% organic matter, 8.5 pH (H2O) and 9.2 meq/100 g ion exchange capacity. Soil samples were allowed to dry at room temperature in the dark, sieved and frozen at -20ºC till extraction.

Equipment and analysis

- An Agilent Technologies 6890N Network GC System Chromatograph (Waldbronn, Germany) equipped with an Agilent Technologies 7683 Series Splitless Injector and an Agilent Technologies 5973 Quadrupole Mass Selective Detector operated in the SIM mode was used. Injector temperature was set at 250ºC and the transfer line temperature at 280ºC. Splitless injection volume was 1 µl.
- A J & W Scientific, DB-17, (30 m × 0.25 mm I.D.), 0.25 µm film thickness column, was employed with helium (99.999% purity) as carrier gas at a constant flow of 1 ml/min.
- The oven temperature for neutral and basic analytes, was maintained at 60ºC for 1 min and then programmed at 6ºC/min to 165ºC, then at 12ºC/min to 215ºC, then at 2ºC/min to 230ºC and finally at 8ºC/min to 280ºC, held for 10 min.
- The oven temperature for acids analysed as their methyl esters/ethers, was maintained at 60ºC for 1 min and then programmed at 22ºC/min to 290ºC, held for 4.55 min.
- Acidic herbicides were compared with procedural standards, i.e. mixtures of acid standards of known concentration derivatized in the same way as samples.

3.3 Working procedure

3.3.1 PCA and cluster analysis. Previous experiments for soil extraction OA

Table 2 shows the previous tests designed to characterize the influence of the variables that would be further optimized with the OA after their analysis by PCA and Cluster techniques. These experiments were designed taking into account the Kovacs series of extraction solvents (Kovacs, 1996), together with the use of water and different modifiers (acetic acid and ammonium hydroxide) in order to increase recoveries of ionic herbicides as already detailed in section 2.3. Acetone was chosen as the unique organic solvent in these previous experiments because it has been widely used in herbicide extraction (Sánchez-Brunete & Tadeo, 1996), and its medium polarity and water miscibility provided a general overview. All quantities were made equivalent in order to compare recoveries. The same sample:solvent ratio was used in all these previous tests (1:3.2). A fixed water volume of 7.5 ml, enough to adequately wet 15 g of the spiked soil, was added in all the experiments where water addition was tested.

After shaking 15 g of blank soil samples, spiked at 0.05 mg/l, with the corresponding extraction mixture for 1 hour, and centrifugation at 2500 rpm for 5 min, an extract volume equivalent to 8 g of soil was recovered and concentrated until near dryness in a turbo vap at 35°C. Then, the concentrated extract was filled up with acetone:1% acetic acid until an equivalent concentration of 8 g/ml, filtered through a 0.45 µm PTFE filter and added the internal standard previously to the GC–MS analysis. In case water was present in the extraction mixture, the supernatant was previously partitioned after the centrifugation with 30 ml of dichloromethane and enough Na_2SO_4 to bind the water. No pH adjustment and pH

No.	Compound	MF	Structural group	MW	log Kw	pKa
1	Dicamba	$C_8H_6Cl_2O_3$	Benzoic acid	221.0	0.55	1.87
2	2,4-D	$C_8H_6Cl_2O_3$	Aryloxyalkanoic acid	225.7	-0.83	2.87
3	MCPP	$C_{10}H_{11}ClO_3$	Aryloxyalkanoic acid	214.6	0.64	3.11
4	Dichlorprop p	$C_9H_8Cl_2O_3$	Aryloxyalkanoic acid	235.1	-0.56	3.67
5	MCPA	$C_9H_9ClO_3$	Aryloxyalkanoic acid	200.6	2.80	3.73
6	Amidosulfuron	$C_9H_{15}N_5O_7S_2$	Sulfonylurea	369.4	1.63	3.58
7	Tribenuron methyl	$C_{15}H_{17}N_5O_6S$	Sulfonylurea	395.4	0.78	4.70
8	Fenoxaprop p	$C_{16}H_{12}ClNO_5$	Aryloxyphenoxypropionic acid	333.8	1.83	4.60
9	Diclofop	$C_{15}H_{12}Cl_2O_4$	Aryloxyphenoxypropionic acid	326.2		3.60
10	Flamprop	$C_{16}H_{13}ClFNO_3$	Aryaminopropionic acid	221.0	2.90	3.70
11	Bromoxynil	$C_7H_3Br_2NO$	Hydroxybenzonitrile	276.9	1.04	3.86
12	Ioxynil	$C_7H_3I_2NO$	Hydroxybenzonitrile	370.9	2.20	4.10
13	Cyanazine	$C_9H_{13}ClN_6$	Chlorotriazine	240.7	2.10	0.63
14	Terbuthylazine	$C_9H_{16}ClN_5$	Chlorotriazine	229.7	3.21	2.00
15	Terbutryn	$C_{10}H_{19}N_5S$	methylthiotriazine	241.4	3.65	4.30
16	Metribuzin	$C_8H_{14}N_4OS$	Triazinone	214.3	1.65	
17	Carfentrazone ethyl	$C_{13}H_{10}Cl_2F_3N_3$	Triaolinone	412.2	3.36	
18	Metoxuron	$C_{10}H_{13}ClN_2O_2$	Phenylurea	228.7	1.60	
19	Isoproturon	$C_{12}H_{18}N_2O$	Phenylurea	206.3	2.50	
20	Chlortoluron	$C_{10}H_{13}ClN_2O$	Phenylurea	212.7	2.50	
21	Methabenzthiazuron	$C_{10}H_{11}N_3OS$	Urea	221.3	2.64	
22	Linuron	$C_9H_{10}Cl_2N_2O_2$	Phenylurea	249.1	3.00	
23	Tralkoxydim	$C_{20}H_{27}NO_3$	Cyclohexadione oxime	329.4	2.10	
24	Flamprop isopropyl	$C_{19}H_{19}ClFNO_3$	Aryaminopropionate	363.8	3.69	
25	Mefenpyr diethyl	$C_{16}H_{18}Cl_2N_2O_4$	Herbicide safener	373.2	3.83	
26	MCPA tioethyl	$C_{11}H_{13}ClO_2S$	Phenoxyacid	244.7	4.05	
27	Bifenox	$C_{14}H_9Cl_2NO_5$	Diphenyl ether	342.1	4.48	
28	Fenoxaprop p ethyl	$C_{18}H_{16}ClNO_5$	Aryloxyphenopropionate	361.8	4.58	
29	Diclofop methyl	$C_{16}H_{14}Cl_2O_4$	Aryloxyphenopropionate	341.2	4.60	
30	Prosulfocarb	$C_{14}H_{21}NOS$	Thiocarbamate	251.4	4.65	
31	Triallate	$C_{10}H_{16}Cl_3NOS$	Thiocarbamate	304.7	4.66	
32	Diflufenican	$C_{19}H_{11}F_5N_2O_2$	Carboxamide	394.3	4.90	
33	Pendimethalin	$C_{13}H_{19}N_3O_4$	Dinitroaniline	281.3	5.18	
34	Trifluralin	$C_{13}H_{16}F_3N_3O_4$	Dinitroaniline	335.3	5.27	
35	MCPP isoctylic	$C_{18}H_{27}ClO_3$	Phenoxypropionate	327.6		
36	Bromoxynil octanoate	$C_{15}H_{17}Br_2NO_2$	Hydroxybenzonitrile	403.0	5.40	
37	Ioxynil octanoate	$C_{15}H_{17}I_2NO_3$	Hydroxybenzonitrile	497.1	6.12	

Table 1. Chemical characteristics of the herbicides of study: Molecular Formula (MF), Structural group, Molecular Weight (MW), Octanol-water coefficient (log Kw), and Acid dissociation constant (pKa 25ºC).

Previous Tests	% HAc or NH_3	Solvent (ml)	Water (ml)	% Rec. Total	%RSD (n=5)	% Rec. Acids	%RSD (n=5)	% Rec. Basic-Neutrals	%RSD (n=5)
Ac		47.50							
NaOH 0.5N DCM pH 2			47.50	97.55a	0.49	95.09a	0.90	98.72a	0.64
Ac:H_2O DCM		40.0	7.50	81.12d	0.86	49.62c	3.69	96.24ab	1.41
Ac:H_2O DCM pH2		40.0	7.50	86.55c	1.16	97.34a	1.88	81.37d	0.90
Ac:HAc	1.0	47.5		87.37c	2.51	60.81b	4.65	100.12a	1.90
Ac:H_2O:HAc DCM	1.0	40.0	7.50	92.70b	0.77	92.42a	0.86	92.84b	0.75
Ac:H_2O:HAc DCM pH2	1.0	40.0	7.50	89.72bc	0.61	97.08a	0.57	86.19c	1.21
Ac:NH_3	0.1	47.5		69.77e	0.60	20.79d	8.33	93.29b	1.04
Ac:H_2O:NH_3 DCM	0.1	40.0	7.50	78.58d	1.09	47.19c	2.13	93.64b	0.85

Table 2. Previous tests recoveries (% Rec.) and relative standard deviations (RSD) at 50 µg kg^{-1} spiking level soil samples. Means followed by different letters in the same column are significantly different at $p < 0.01$ level according to Tukey honest test for equal number of replicates. Acetone (Ac), Acetic Acid (HAc), DCM (dichloromethane).

adjustment to pH 2 were developed previously to the dichloromethane partition to study whether polar herbicides were lost in the aqueous phase with the different solvent combinations tested.

3.3.2 Optimization of operational variables. Herbicide soil extraction OA

The average recoveries were used as the output variable to optimize. Different OA were developed for acidic analytes and for basic and neutral herbicides due to dual methyl ester formation from the TMSH derivatization of acids and their ester forms prior to their GC analysis, in order to know which form the methyl ester came from.

After carefully studying the results obtained from the previous experiments analyzed by PCA, the following variable values were selected for the multiresidue extraction OA: solvent type and ratio, pH (percentage of acetic acid) and shaking time as showed in Table 3. Acetone, ethyl acetate and acetonitrile were selected as organic solvents because they are among the most used extractants in neutral and basic multiresidue herbicide procedures in soils. Ammonium hydroxide was not suitable for acids and showed no effect in basic recoveries, therefore it was not further used. The same water volume used in the previous experiments was taken for the OA due to its utility in mixtures with acetic acid and acetone. Due to different solvent volumes, water percentages changed from 14.3% to 33.3%, acetic acid percentages changed from 0.3% to 1.7%, and organic solvent percentages changed from 65.3% to 85.3%, covering the values found in the references (Ahmed & Bertrand, 1989; Crescenzi et al., 1999; Smith & Milward, 1981; Sutherland et al., 2003; Thorstensen & Christiansen, 2001).

Three levels for each control factor instead of two were chosen to detect any quadratic or non-linear relation between the factors and the output variable, and to obtain information over wider ranges of the variables. Four control factors at three levels contain eight degrees of freedom, and can be fitted to the $L_9(3^4)$ OA. The nine different trials resulting from this design were duplicated to calculate the residual error, and randomized to minimize the effects of uncontrolled factors that may introduce a bias on the measurements (Table 6).

Notation	Factor	Level 1	Level 2	Level 3
S	solvent + water	acetone	ethyl acetate	acetonitrile
A	% acetic acid	0.5	1	2
V	volume (ml)	15 (1:1.5)	30 (1:2.5)	45 (1:3.5)
T	shaking time (seg.)	15	30	60

Table 3. Factors and levels for the herbicide soil extraction $L_9(3^4)$ OA optimization.

A 15 g amount of blank soil spiked at 0.05 mg/l was added 7.5 ml water, shacked the corresponding time with the appropriate solvent mixture, centrifuged and partitioned with dichloromethane. A fixed extract volume equivalent to 8 g of soil was evaporated to dryness in every experiment, dissolved in 1 ml of acetone:1% acetic acid and split in two aliquots. One of them was directly analyzed by GC-MS and the second one was derivatized before the acidic analyte analysis with an optimized procedure described afterwards, which consists on adding 100 µl of TMSH derivatization reagent to 500 µl of final extract directly in the vial. The effect of the presence of substance/s in the matrix in the chromatographic determination, was corrected with the use of calibration lines prepared in 900 µl of blank soil extracts obtained in the same way as samples in each trial, i.e. matrix-matched standard calibration (Analytical Methods for Pesticide Residues in Foodstuffs, 1996).

3.3.3 Optimization of operational variables. Acidic herbicide analysis OA

Acidic herbicides were divided in two groups, those only present in their acidic form and those also esterified. These esters were called "original" to differentiate them from the methyl esters produced after derivatization. Due to dual methyl ester formation, different OA were developed for the acidic herbicides (named "Acid matrix") and for the original esters (named "Ester matrix") in order to know which form the methyl ester came from and the way factors affected both esterification and trans-esterification reactions.
The total peak area value, defined as the total sum of peak areas, was used as variable to optimize because the formation of peaks as high as possible was the goal, therefore no calibration was necessary. Two output variables were chosen to be optimized due to dual methyl ester formation and the separately OA for acidic and original ester herbicides. TMEPA (total methyl ester peak area) was calculated in both matrices to study methyl ester formation meanwhile, TOEPA (total original ester peak area) was only evaluated in the "Ester matrix" to know the amount of remaining non-trans-esterified original esters.

Notation	Factor	Level 1	Level 2	Level 3
S	solvent	acetone	ethyl acetate	acetonitrile
T	time of incubation (min)	5	30	45
C	temperature of incubation (ºC)	20	40	70
P	pH	–	1 % acetic acid	1 % phosphoric acid

Table 4. Factors and levels for the acidic herbicide analysis $L_9(3^4)$ OA optimization.

Organic solvents alone, slightly and strongly acidified (added 1% acetic acid and 1% phosphoric acid, respectively) were selected as reaction media because they are usually employed for acidic herbicide extraction as already detailed in section 2.3. Subsequently, final extract derivatization reactions were affected by pH values, which have been reported to play an important role in the process (Catalina et al., 2000).

The direct injection of analytes and TMSH mixtures into the hot injection port of the GC has been reported (Zapf & Stan, 1999). For some weak acids deprotonation and thermally decomposition of the resulting salts after derivatization have been reported to occur simultaneously in a heated GC injector (Rompa et al., 2004), meanwhile other authors recommend pre-heating in an oven in a closed sample vial previously to injection (Halket & Zaikin, 2004). In order to evaluate the usefulness of pre-heating, standard mixtures were incubated for 5-30-45 min at three different temperatures: 40 °C (recommended maximum heating temperature recommended in the TMSH label), 70 °C, both maintained in an oven, and 20 °C, kept constant in an incubation chamber to simulate the absence of pre-heating.
Consequently, the following variables were selected: temperature and time of incubation, solvent and pH (composition of reaction mixture) (Table 4).
All experiments were carried out with standards diluted in the tested solvent at a concentration of 250 μg/l in order to avoid the possibility of finding matrix derivatized interferences.
Previously, the optimum quantity of TMSH was studied and 100 μl of a solution of TMSH 0.25 M in methanol added to 500 μl standard solutions were shown enough to provide a high excess of derivatizing reagent and to ensure the complete derivatization of all compounds present in the sample.
Four control factors at three levels contain eight degrees of freedom, and can be fitted to the $L_9(3^4)$ OA. The nine different trials resulting from this design were randomized and duplicated in order to calculate the residual error, so a total number of 18 standard solutions were derivatized and analysed by GC-MS to determine the corresponding total peak area values as described above (Table 8).

3.4 Results and discussions

3.4.1 PCA and cluster analysis results. Previous experiments for soil extraction OA

PCA was applied to the average herbicide recovery values of 5 replicates obtained from the previous experiments (Table 2) in order to provide a global overview and clarify the relationships among the several variables related to the extraction procedure and their effects on extractability. Both average recoveries for basic and neutral herbicides with acetone extraction and for acidic herbicides with alkaline extraction were taken together as the specific method results.
Statistical analyses were performed using the Minitab v.13.0 program package, with the Ward linkage method, and using none rotation option.
From the PCA it was found that 93.80% of the variation of the dataset could be explained using four factors. From the loading on the four factors of the PCA (Table 5) some conclusions can be drawn. The factor pattern of component 1 showed contributions from a set of procedures intended for neutral and basic herbicides while those more specific for acid herbicides formed the component 2. Component 3 and 4 consisted on both the specific methods and the acetone-water-acetic acid combination for all the herbicides of study.
The PCA showed groupings of the herbicides based on their chemical nature (Fig. 1). Acidics are grouped separately from the basics and neutrals, which did not show a different trend between them implying that both types of analytes could be extracted with the same procedures. However, acids are very different in nature and needed specific extraction methods. Loadings for both methods with pH adjustment before the partitioning step, lay near the acidic analytes (dotted lines) while loadings for acetone in combination with water and ammonium hydroxide (striped lines) are orientated to the basic and neutral grouping.

	Components			
Variables	1	2	3	4
Specific		-37.89	-82.52	-35.91
Ac:H_2O DCM	-92.51			
Ac:H_2O DCM pH2		-90.25		
Ac:HAc	-92.58			
Ac:H_2O:HAc DCM		-50.72	51.77	-64.08
Ac:H_2O:HAc DCM pH2		-90.82		
Ac:NH_3	-94.85			
Ac:H_2O:NH_3 DCM	-88.78			
%Variance	45.80	26.20	12.40	9.40

Table 5. Loading of variables on the four first components resulting from the PCA of extraction procedures with different solvent combinations with water and modifiers. Component loading less than |0.35| are omitted.

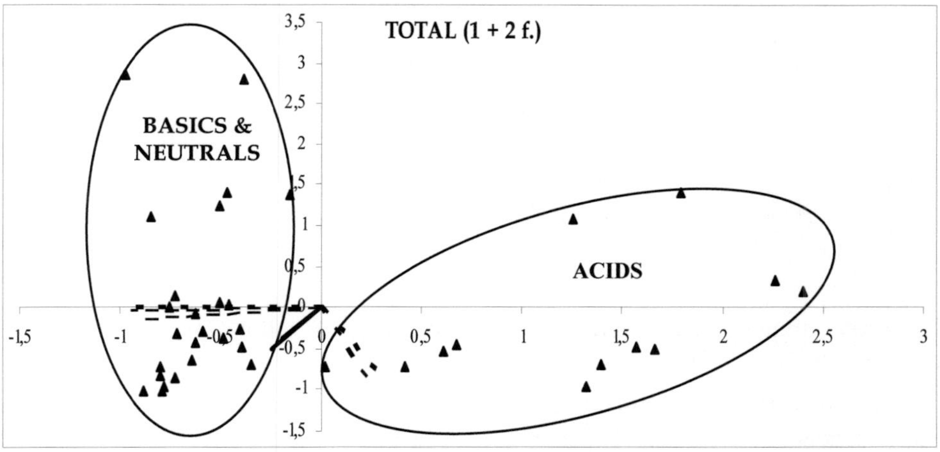

Fig. 1. Score plot for first two factors for all the herbicides. Loadings for the 8 different methods tested have been represented as lines.

Loadings for the specific methods and the acetone-water-acetic acid combination (black lines) are directed in the same way, their direction of maximum dispersion laying between the acid and basic and neutral groupings, what it could indicate the suitability of this combination of solvents and modifiers as multiresidue methods.

This result was also found in the cluster analysis (Fig. 2), where procedures were grouped in similarity in this way: neutral and basic herbicides, liquid-liquid partitioning with dichloromethane at pH 2 for acids and, specific methods and acetone-water-acetic acid for all of the studied herbicides.

The acetone-water-acetic acid combination was significantly the more efficient in extracting the whole range of different herbicides apart from the specific methods. The best acidic average recoveries were found for those combinations using water-acetic acid and those using a partitioning step with prior pH adjustment to pH 2. However, these both last methods were exactly the less effective in extracting basic and neutral analytes, although they were significantly recovered by the rest of the tested extraction methods. Basic recovery

showed no enhancement with the use of ammonium hydroxide as expected (Smith & Milward, 1981).
Basic herbicides behaved in the same way as neutral analytes; therefore their recoveries were averaged together. The significance of differences among the procedure recoveries were examined by applying analysis of variance (ANOVA). Values represent means for the average recovery replicates for all the spiked blank soil samples extracted with the different procedures tested. Means followed by different letters in the same column are significantly different at $p < 0.01$ level according to Tukey honest test for equal number of replicates (Table 2).

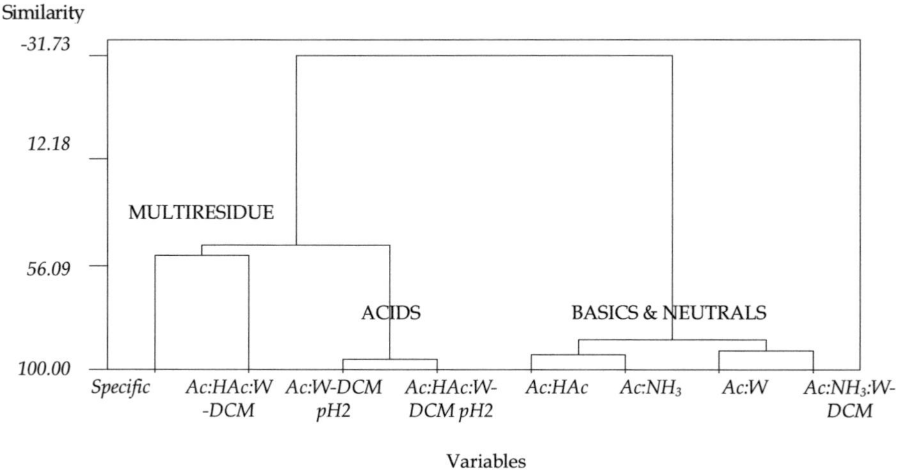

Fig. 2. Cluster variables for the eight extraction solvent combinations tested in the previous experiments.

The addition of water alone did not significantly recover more residues than the organic solvent as previously reported (Tadeo et al., 1996). However, the addition of acetic acid to the water and acetone combination enhanced significantly the acidic recovery with no detrimental in the basic and neutral extraction, and no pH adjustment prior to the dichloromethane partition was needed.
After carefully studying the results obtained from the previous experiments by PCA and Cluster analysis, the following variables were selected for the subsequent OA design: solvent type and ratio, pH (percentage of acetic acid) and shaking time.

3.4.2 Herbicide soil extraction OA results

Table 6 shows the average recovery data obtained by duplicate for each of the 9 experiments. For the regular analysis, an ANOVA table with pooled errors was calculated from these experimental data in order to identify individual sources of variation and to calculate the contribution of each factor to the response variation (Table 7).
ANOVAs of the recovery data obtained for both matrices revealed that factor S, the type of solvent, contributed by the highest percentage to the variability of the recoveries (49.1% for acids and 67.2% for basics and neutrals). Maximum recovery for all the analytes was obtained for level S1, acetone (Fig. 3). In contrast to acetone and acetonitrile, ethyl acetate was practically immiscible with water which could be easily removed by using only

anhydrous Na_2SO_4 as a drying agent. However, the dichloromethane partition was also carried out in order to develop all experiments in the same way and to benefit from the polar interference removal provided by the partitioning. In addition to the variability in recoveries due to the immiscibility of ethyl acetate with water and acetic acid, pesticides with a thioether group (ureas) have been reported to degrade in the ethyl acetate (Mastovska & Lehotay, 2004), what explains the lower recoveries observed when using this solvent. Acidic herbicides were very influenced by the acetic acid percentage (46.1%), meanwhile the contribution for basic and neutral was low (4.1%). Maximum recovery of the acid herbicides was obtained for level A3 (2% acetic acid), meanwhile level A1 (0.5% acetic acid) provided the maximum recovery for basics and neutrals (Fig. 3).

	Control Factors and Levels				Acids		Basics & Neutrals	
Trial	*S*	*A*	*V*	*T*	*1*	*2*	*1*	*2*
1	1	1	1	1	80.13	77.64	100.54	100.84
2	1	2	2	2	93.34	92.12	93.65	94.40
3	1	3	3	3	94.18	94.30	83.12	83.34
4	2	1	2	3	69.67	69.51	76.73	76.59
5	2	2	3	1	74.86	74.35	67.68	68.83
6	2	3	1	2	81.77	82.54	78.15	79.28
7	3	1	3	2	79.73	80.23	84.77	85.91
8	3	2	1	3	88.13	88.21	93.67	93.84
9	3	3	2	1	93.83	92.97	86.46	87.08

Table 6. Experimental average recoveries obtained for each duplicated trial in the herbicide soil extraction $L_9(3^4)$ OA optimization.

Variation source		S. Solvent + Water	A. % Acetic acid	V. Volume	T. Shaking time	Residual	Total
Degrees of freedom		2	2	2			8
Basics & Neutrals	Sum of squares	1055.34	65.30	441.03	3.99		1567.96
	Variance ratio (F)[a]	923.58	57.15	385.96			
	Pool	No	No	No	Yes	Yes	
	Pooled sum of squares	1054.20	64.16	439.89		9.71	
	Contribution (%)[b]	67.23	4.09	28.05		0.62	100.00
Acids	Sum of squares	313.14	293.88	10.01	10.88		1260.59
	Variance ratio (F)[a]	87.42	82.05				
	Pool	No	No	Yes	Yes	Yes	
	Pooled sum of squares	619.11	580.59			60.89	1260.59
	Contribution (%)[b]	49.11	46.06			4.83	100.00

[a]Critical variance ratio for a 95% confidence level is 19.00.
[b]Contribution is defined as 100 x (pooled sum of squares/total sum of squares).

Table 7. Pooled ANOVA for the regular analysis of the mean average recoveries obtained for acidic, basic and neutral herbicide soil extraction $L_9(3^4)$ OA optimization.

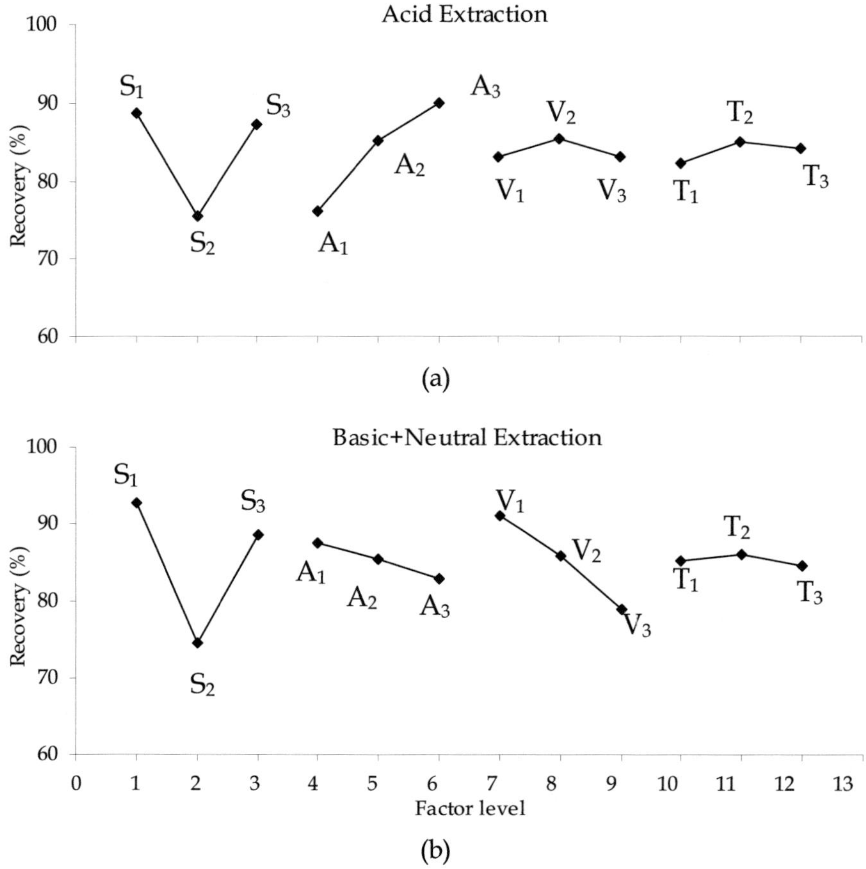

Fig. 3. Effect of the interaction of control factors on the mean response obtained for acidic (a), and basic and neutral (b) herbicide soil extraction $L_9(3^4)$ OA optimization.

Solvent volume was significant for basics and neutrals (28.1%) due to their volatility; however, this factor was pooled for acids. The maximum recovery was obtained for level V1, 15 ml, for basics and neutrals. An equivalent volume of 8 g of soil was taken to near dryness in all the tests, therefore more extract volume was concentrated when using 30 and 45 ml of solvent, leading to a higher loss of more volatile analytes, even after adding 20% of ethylene glycol in acetone as a holder solution when evaporating. However, solvent volume had not a statistically significant effect (at 95% confidence level) on the acid recoveries. Acids are not lost during evaporation in the same way as basics and neutrals, because they are taken to near dryness in their non-volatile acidic form and converted to the more volatile methyl esters/ethers just before analysis. As a compromise between both types of analytes, level V2, 30 ml, was selected. The main advantage of acetone over ethyl acetate and acetonitrile was its greater volatility, having the smallest boling point (56.2ºC for acetone, 77.1ºC for ethyl acetate and 81.6ºC for acetonitrile) and therefore, minimizing volatile losses due to evaporation.

The time of extraction was negligible indicating that there were no significant differences (at 95% confidence level) among the levels tested, its contribution being pooled for all the analytes. Level T2, 30 min, was chosen, because it gave a slightly higher response than 15 min.

The contribution of the residual error to the recovery variability (4.8% for acids and 0.6% for basics and neutrals) indicates the experimental design took into account all the variables affecting the response, the levels tested were fit for the purpose and the variance of the experimental data was explained by the effect of factors and interactions
Fig. 3 shows the effects of control factor levels on the output variable. Factor T variation (shaking time) had a slight influence on recoveries, and a change in their level produced very small variation in the multiresidue herbicide extraction. However, the significant influence of the solvent type (S) for all analytes, the acetic acid percentage (A) for acids and the solvent volume (V) for basics and neutrals can be observed by the statistically different recoveries obtained when changing these variables.

3.4.3 Acidic herbicide analysis OA results

Table 8 shows the output variables, TMEPA and TOEPA, obtained by duplicate for each of the 9 experiments. For the regular analysis, an ANOVA table with pooled errors was calculated from these experimental data in order to identify individual sources of variation and to calculate the contribution of each factor to the response variation (Table 9).
ANOVAs of the TMEPA and TOEPA for both matrices revealed that factor P (pH) contributed by the highest percentage to the variability of the signal (93.78 % for methyl ester formation, 78.56 % for methyl ester conversion and 97.04 % for original ester permanence).
Although very small, contribution made by the other variables for methyl ester trans-esterification was the only one that could not be neglected. In both the cases of methyl ester formation and permanence of original esters, the rest of factors were negligible indicating that there were no significant differences (at 95% confidence level) among the levels tested.
The pH of the solution (P) during both esterification and trans-esterification processes has been shown to play an important role. The presence of the anionic form of the acids was essential for the formation of the trimethylsulfonium salts as well as for the previous saponification in trans-esterification. Both esterification and trans-esterification reactions were enhanced in a strong basic environment provided by the addition of TMSH that yielded a solution pH value of 9. However, the presence of 1 % acids neutralized this strong

Control Factors and Levels					*TMEPA* (x 10^5)				*TOEPA* (x 10^5)	
					Acids		*Esters*		*Esters*	
Trial	*S*	*T*	*C*	*P*	*1*	*2*	*1*	*2*	*1*	*2*
1	1	1	1	1	69.3	76.0	95.6	140.1	63.7	67.5
2	1	2	2	2	106.8	115.5	37.1	32.6	419.7	412.6
3	1	3	3	3	13.8	12.4	9.0	7.4	503.4	450.2
4	2	1	2	3	11.4	12.3	9.3	7.4	480.8	448.1
5	2	2	3	1	89.3	83.3	101.0	92.2	79.4	70.9
6	2	3	1	2	91.2	117.6	52.2	45.6	491.1	494.3
7	3	1	3	2	108.9	105.4	42.7	48.9	543.5	480.9
8	3	2	1	3	14.3	13.0	5.9	6.0	510.8	543.6
9	3	3	2	1	110.6	104.8	208.1	207.0	66.5	68.7

Table 8. Experimental average recoveries obtained for each duplicated trial in the acidic herbicide analysis $L_9(3^4)$ OA optimization.

	Variation source	S. Solvent	T. Time of incubation	C. Tª of incubation	P. pH	Residual	Total
	Degrees of freedom	2	2	2	2		8
TMEPA (Acid Matrix)	Sum of squares (x 10^4)	3.81	3.77	5.41	301.80		319.30
	Variance ratio (F)[a]				129.20		
	Pool	Yes	Yes	Yes	No	Yes	
	Pooled sum of squares				294.40	19.90	319.30
	Contribution (%)[b]				93.78	6.22	100.00
TMEPA (Ester Matrix)	Sum of squares (x 10^4)	47.94	56.20	36.40	566.51		718.05
	Variance ratio (F)[a]	19.61	22.99	14.89	231.75		
	Pool	No	No	No	No	Yes	
	Pooled sum of squares	45.50	53.75	33.95	564.07	20.78	718.05
	Contribution (%)[b]	6.34	7.49	4.73	78.56	2.89	100.00
TOEPA (Ester Matrix)	Sum of squares (x 10^4)	69.00	1.58	70.98	6778.50		6960.01
	Variance ratio (F)[a]				280.03		
	Pool	Yes	Yes	Yes	No	Yes	
	Pooled sum of squares				6754.25	205.75	
	Contribution (%)[b]				97.04	2.96	100.00

Table 9. Pooled ANOVA for the regular analysis of total methyl ester peak area (*TMEPA*) in the Acid Matrix and *TMEPA* and total original ester peak area (*TOEPA*) in the Ester Matrix obtained for acidic herbicide analysis $L_9(3^4)$ OA optimization.

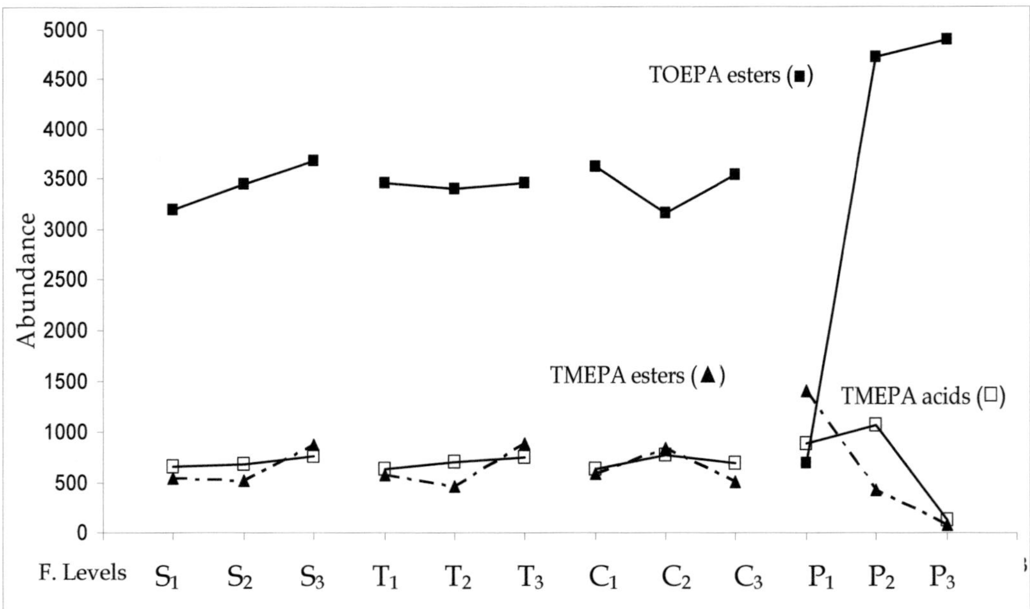

Fig. 4. Comparison of the effect of the interaction of control factors (f. levels) on the mean response for Original remaining Esters (■) and Methyl Esters in the Acid matrix (□) and in the Ester matrix (▲) obtained for acidic herbicide analysis $L_9(3^4)$ OA optimization.

basic TMSH media, and as a result, anionic forms of acids were not promoted and methylation yields decreased. A solution containing 1% of acetic acid had a pH value of 6 after adding TMSH meanwhile the strongest phosphoric acid decreased TMSH solution pH value till 2. Data in Table 8 clearly showed the effect of pH. All experiments developed at the same pH conditions had near TMEPA and TOEPA values regardless to the solvent, incubation time and temperature used.

Maximum methylation of acidic herbicides was obtained for P2, (pH) 1 % acetic acid (pH value of 6). The other three factors had not a statistically significant effect (at 95% confidence level) on the signal ratio; however, level S3 (solvent), acetonitrile; T3 (incubation time), 45 min; and C2 (incubation temperature), 40ºC, gave a slightly higher ratio. A slightly acidic environment gave the highest methyl ester formation but results were very close to those obtained in a basic medium. The very low methyl ester peak areas obtained with 1% of phosphoric acid, suggest that TMSH reaction was more influenced by very acidic pH values and the reaction worked properly from a neutral to a basic pH.

The contribution of the residual error to the TMEPA and TOEPA variability (6.22 %, 2.89 % and 2.96 % respectively) indicates the goodness of the experimental design used.

Fig. 4 shows the effects of control factor levels on the output variable. It can be observed that control factors different than pH (P) had a slight influence on the TMEPA and TOEPA value, and a change in their level produced very small variation in the conversion or permanence efficiency.

Fig. 4 also shows the effect of control factors on trans-esterification. TMEPA esters and TOEPA esters representations were obviously found to be opposite, the highest the methyl ester conversion, the smallest the permanence of remaining original esters. Both esterification and trans-esterification methyl ester formation were affected in the same way by pH being very diminished at strongly acidic pH values, although it seemed that trans-esterification needed a stronger basic media and did not work properly at a pH value of 6 (1% acetic acid) as esterification.

5. Conclusions

Herbicides play a very important role in agriculture but the toxicity and widespread of their residues pose a potential risk for the environment. In addition, their determination in soils is of primary importance because their dispersion in the environment depends on their behaviour in soils. The integration between analytical procedures and chemometric strategies has proved very valuable in the always difficult herbicide multiresidue extraction and analysis optimization development. The optimized methods have been applied to environment soils where herbicide residue data interpretation is of great interest.

The statistical analysis of the OA data revealed that all the factors were significant being the most important, the type and ratio of solvent for basic and neutral herbicides and the acetic acid percentage for acid herbicides. The final optimized method consisted of shaking previously wet soil samples for 30 min with 30 ml of acetone acidified with 1% acetic acid.

As a result, any organic solvent acidified with 1 % acetic acid was suitable for methylation with TMSH and as, pre-heating was shown not to improve derivatization yield, it was just necessary to add the derivatizing reagent to the sample vial and methylation was completely carried on in the injector port of the GC system.

6. Acknowledgement

This study was founded by JCyL project (VA 023A10-2).

7. References

Ahmed, A.W. & Bertrand, M.V.N. (1989). Determination of acidic herbicides and related compounds in water and soil by capillary gas chromatography using a nitrogen-phosphorus detector. *J. Assoc. Off. Anal. Chem.*, 72, 365-367

Analytical Methods for Pesticide Residue in Foodstuffs. General Inspectorate for Health Protection, Ministry of Health, Welfare and Sport. (1996). 6th Ed. The Hague

Bagheri, H.; Saraji, M.; Chitsazan, M.; Mousavi, S.R. & Naderi, M. (2000). Mixed-level orthogonal array design for the optimization of solid-phase extraction of some pesticides from surface water. *J. Chromatogr. A*, 888, 197–208

Barrado, E.; Prieto, F.; Vega, M. & Fernández-Polanco, F. (1998). Optimization of the operational variables of a medium-scale reactor for metal-containing wastewater purification by ferrite formation. *Water Res.*, 32 (10), 3055-3061

Bendell, A.; Disney, J. & Pridmore, W.A. (1989). Taguchi Methods: Applications, In: *World Industry*. IFS Publications, UK

Brown, S.D. (1995). Chemical Systems Under Indirect Observation: Latent Properties and Chemometrics. *Appl. Spectrosc.*, 49(12), 14A–31A

Bruns, R.E.; Scarminio, I.S. & de Barros-Neto, B. (2006). Statistical design—chemometrics. Elsevier, Amsterdam

Budde, W.L. (2004). Analytical mass spectrometry of herbicides. *Mass Spectr. Rev.* 23, 1-24

Butte, W. (1983). Rapid method for the determination of fatty acid profiles from fats and oils using trimethylsulphonium hydroxide for transesterification. *J. Chromatogr.*, 261, 142-145

Catalina, M.I.; Dalluge, J.; Vreuls, R.J.J. & Brinkman, U.A.T. (2000). Determination of chlorophenoxy acid herbicides in water by in situ esterification followed by in-vial liquid-liquid extraction combined with large-volume on-column injection and gas chromatography-mass spectrometry. *J. Chromatogr. A*, 877, 153-166

Christie, W.W. (1993). Advances in Lipid Methodology. W.W. Christie (Ed.), Oily Press, Dundee

Chee, K.K.; Wong, M.K. & Lee, H.K. (1995). Optimization by orthogonal array design of solid-phase extraction of organochlorine pesticides from water. *Chromatogr.*, 41, (3-4) 191-196

Crescenzi, C.; D'Ascenzo, G.; Di Corcia, A.; Nazzari, M.; Marchese, S. & Samperi. R. (1999). Multiresidue herbicide analysis in soil: Subcritical water extraction with an on-line sorbent trap. *Anal. Chem.*, 71, 2157-2163

Crespín, M.A.; Gallego, M.; Valcárcel, M. & González, J.L. (2001). Study of the degradation of the herbicides 2,4-D and MCPA at different depths in contaminated agricultural soil. *Environ. Sci. Technol.* 35, 4265-4270

Cserháti, T. (2010). Data evaluation in chromatography by principal component analysis. Biomed. *Chromatogr.*, 24, 20–28

Delgado-Moreno, L.; Peña, A. & Mingorance, M.D. (2009). Design of experiments in environmental chemistry studies: Example of the extraction of triazines from soil after olive cake amendment. *J. Hazard. Mater.*, 162, 1121–1128

Regulation (EC) No 1107/2009 of the European Parliament and of the Council of 21 October 2009 concerning the placing of plant protection products on the market and repealing Council Directives 79/117/EEC and 91/414/EEC

Document No. SANCO/D3/SI2.396179, Final Report. (2005). European Commission, Brussels

Document No. D/00/SuM/5277 (2000). European Crop Protection Association, Brussels

Document No. SANCO/825/00. rev. 7. (2004). European Commission, Brussels

Einax, J.W.; Zwanzinger, H.W. & Geiss, S. (1997). Chemometrics in Environmental Chemistr,. VCH, Weinham, Germany

EPA Method 8151A. Chlorinated Herbicides by GC using methylation or pentafluorobenzylation derivatization (1996)

FOOTPRINT Pesticide Properties DataBase. (2006). Database collated by the University of Hertfordshire as part of the EU-funded FOOTPRINT project (FP6-SSP-022704)

Fuentes, E.; Báez, M.E. & Labra, R. (2007). Parameters affecting microwave-assisted extraction of organophosphorus pesticides from agricultural soil. *J. Chromatogr. A*, 1169, 40-46

Halket, J.M. & Zaikin, V.G. (2004). Derivatization in mass spectrometry-3. Alkylation (arylation). *Eur. J. Mass Spectrom.*, 10, 1-19

Hanrahan, G.; Montes, R. & Gomez, F.A. (2008). Chemometric experimental design based optimization techniques in capillary electrophoresis: a critical review of modern applications. *Anal. Bioanal. Chem.*, 390, 169–179

Hildebrandt, A.; Guillamon, M.; Lacorte, S.; Tauler, R. & Barcelo, D. (2008). Impact of pesticides used in agriculture and vineyards to surface and groundwater quality (North Spain). *Water Res.*, 42, 3315–3326

Jiménez, J.J.; Bernal, J.L.; del Nozal, M.J. & Martin, M.T. (2000). Use of a particle beam interface combined with mass spectrometry/negative chemical ionization to determine polar herbicide residues in soil by liquid chromatography. *J. AOAC Int.*, 83, 756-761

Joliffe, I.T. (1986). Principal Component Analysis, Springer, New York

Kaliszan, R. (1997) Structure and Retention in Chromatography: a Chemometric Approach. Harwood Academic, Melbourne

Kaufman, L. & Rousseeuw, P.J. (1990) Finding Groups in Data, John Wiley & Sons, New York

Kovacs, M. H. Jr. (1986). Regulatory aspects of bound residues (chemistry). *Res. Rev.*, 97, 1-17

Kremer, E.; Rompa, M. & Zygmunt, B. (2004). Extraction of acidic herbicides from soil by means of accelerated solvent extraction. *Chromatogr.*, 60, 169-174

Lan, W.G.; Wong, M.K.; Chen, N. & Sin, Y.M. (1994). Orthogonal array design as a chemometric method for the optimization of analytical procedures. 1. 2-level design and its application in microwave dissolution of biological samples. *Analyst*, 119, 1659-1667

Lan, W.G.; Chee, K.K.; Wong, M.K.; Lee, H.K. & Sin, Y.M. (1995). Orthogonal array design as a chemometric method for the optimization of analytical procedures. 4. Mixed-level design and its application to the high performance liquid chromatographic determination of polycyclic aromatic-hydrocarbons. *Analyst*, 120, 281-287

Lavine, B.K. (2000). Clustering and Classification of Analytical Data, In; Encyclopedia of Analytical Chemistry: Instrumentation and Applications, John Wiley & Sons Ltd., Chichester, 9689-9710

Lin, S.L. & Fuh, M.R. (2010). Orthogonal array optimization of ultrasound-assisted emulsification-microextraction for the determination of chlorinated phenoxyacetic acids in river water. *J. Chromatogr. A*, 1217, 21, 3467-3472

McLachlan, G.L. (1992). Discriminant Analysis and Statistical Pattern Recognition, John Wiley & Sons, New York

Macutkiewicz, E.; Rompa, M. & Zygmunt B. (2003). Sample preparation and chromatographic analysis of acidic herbicides in soils and sediments. *Crit. Rev. Anal. Chem.*, 33, 1-17

Marchese, S. (2001) Development of a method based on accelerated solvent extraction and liquid chromatography/mass spectrometry for determination of arylphenoxypropionic herbicides in soil. *Rapid Commun. Mass Spectrom.*, 15, 393-400

Mardia, K.V.; Kent, J.T. & Bibby, J.M. (1979). Multivariate Analysis. Academic Press: London

Massart, D.L. & Kaufman, L. (1983). The Interpretation of Analytical Chemical Data by the Use of Cluster Analysis, John Wiley & Sons, New York

Massart, D.L.; Vandeginste, B.G.M.; Buydens, L.M.C.; de Jong, S.; Lewi, P.J. & Smeyers-Verbeke, J. (1998). Handbook of Chemometrics and Qualimetrics, Elsevier Science, Amsterdam

Mastovska, K. & Lehotay, S. (2004). Evaluation of common organic solvents for gas chromatographic analysis and stability of multiclass pesticide residues. *J. Chromatogr. A*, 1040, 259-272

Menasseri, S.; Koskinen, W. & Yen, P.Y. (2004). Sorption of aged dicamba residues in soil. *Pest. Manag. Sci.*, 60, 297-304

Mostert, M.M.R.; Ayoko, G.A. & Kokot, S. (2010). Application of chemometrics to analysis of soil pollutants. *Trends Anal. Chem.*, 29 (5), 430-445

Nolte, J. & Kruger, R. (1999). Direct methylation at the surface of Carbopack B - Part II: Determination of phenoxyalkanoic acid herbicides. *Fresen. J. Anal. Chem.*, 365, 610-614

Oles, P.J. (1993). Fractional factorial design approach for optimizing analytical methods. *J. AOAC Int.*, 76, 615-620

Otto, M. (1999). Chemometrics: statistics and computer applications in analytical chemistry. Wiley-VCH, Chichester

Papadopoulou-Mourkidou, E.; Patsias, J. & Kotopoulou, A. (1997). Determination of pesticides in soils by gas chromatography-ion trap mass spectrometry. *J. AOAC Int.*, 80, 447-454

Pasti, L.; Nava, E.; Morelli, M.; Bignami, S. & Dondi, F. (2007). GC/MS analysis of pesticides in the Ferrara area (Italy) surface water: a chemometric study. *Annali di Chimica*, 97

Patsias, J.; Papadakis, E.N. & Papadopoulou-Mourkidou, E. (2002). Analysis of phenoxyalkanoic acid herbicides and their phenolic conversion products in soil by microwave assisted solvent extraction and subsequent analysis of extracts by on-line solid-phase extraction-liquid chromatography. *J Chromatogr. A*, 959, 153-161

Pena, A.; Ruano, F. & Mingorance, M.D. (2006). Ultrasound-assisted extraction of pesticides from olive branches: a multifactorial approach to method development. *Anal. Bioanal. Chem.*, 385, 918-925

Pinto, G.M.F. & Jardim, I.C.S.F. (2000). Use of solid-phase extraction and high-performance liquid chromatography for the determination of triazine residues in water: validation of the method. *J. Chromatogr. A*, 869, 463-469

Quan, L.; Li, S.F.; Tian, S.J.; Xu, H.; Lin, A.Q. & Gu, L. (2004). Determination of organochlorine pesticides residue in ginseng root by orthogonal array design soxhlet extraction and gas chromatography. *Chromatogr.*, 59 (1-2), 89-93

Rompa, M.; Kremer, E. & Zygmunt, B. (2004). Derivatization of acidic herbicides with selected tetraalkyl ammonium and trimethyl sulfonium hydroxides for their GC analysis. *Anal. Lett.*, 37, 3299-3312

Ross, P.J. (1988). Taguchi Techniques for Quality Engineering, McGraw-Hill Book Company, New York

Sánchez-Brunete, C. & Tadeo, J.L. (1996). Multiresidue analysis of herbicides in soil by GC-MS. *Quim. Anal.*, 15, 53-57

Santos-Delgado, M.J.; Crespo-Corral, E. & Polo-Díez, L.M. (2000). Determination of herbicides in soil samples by gas chromatography: optimization by the simplex method. *Talanta*, 53, 367-377

Smith, A.E. & Milward, L.J. (1981). Comparison of Solvent Systems for the Extraction of Diclofop Acid, Picloram, Simazine, and Triallate from Weathered Field Soils. *J. Agric. Food Chem.*, 31, 633-637

Smith, A.E. (1995). A review of analytical methods for sulfonylurea herbicides in soil. *Int. J. Environ. Anal. Chem.*, 59, 97-106

Stalikas, C.D. & Pilidis, G.A. (2000). Development of a method for the simultaneous determination of phosphoric and amino acid group containing pesticides by gas chromatography with mass-selective detection - Optimization of the derivatization procedure using an experimental design approach. *J. Chromatogr. A*, 872 (1-2) 215-225

Sun, L. & Lee, H.K. (2003). Optimization of microwave-assisted extraction and supercritical fluid extraction of carbamate pesticides in soil by experimental design methodology. *J. Chromatogr. A*, 1014 (1-2) 165-177

Sutherland, D.J.; Stearman, G.K. & Wells, M.J.M. (2003). Development of an analytical scheme for simazine and 2,4-D in soil and water runoff from ornamental plant nursery plots. *J Agr. Food Chem.*, 51, 14-20

Tadeo, J.L.; Sánchez-Brunete, C.; García-Valcárcel, A.I.; Martínez, L. & Pérez, R.A. (1996). Determination of cereal herbicide residues in environmental samples by gas chromatography. *J Chromatogr. A*, 754, 347-365

Tadeo, J.L.; Sánchez-Brunete, C.; Pérez, R.A. & Fernández, M.D. (2000). Analysis of herbicide residues in cereals, fruits and vegetables. *J. Chromatogr. A*, 882, 175-191

Taguchi, G. (1991). System of Experimental Design. Vol I-II; A.S.I.I., New York

Tekel, J. & Kovacicová, J. (1993). Chromatographic methods in the determination of herbicide residues in crops, food and environmental samples. *J. Chromatogr. A*, 643, 291-303

Thorstensen, C.W. & Christiansen, A. (2001). Determination of bentazone, dichlorprop, and MCPA in different soils by sodium hydroxide extraction in combination with solid-phase preconcentration. *J. Agric. Food Chem.*, 49, 4199-4202

Vandeginste, B.G.M.; Massart, D.L.; Buyden, L.M.C.; De Jong, S.; Lewi, P.J. & Smeyers-Verbeke, J. (1998). Handbook of Chemometrics and Qualimetric. Data Handling in Science and Technology, Vol. 20. Elsevier, Amsterdam

Wan, H.B.; Lan, W.G.; Wong, M.K.; Mok, C.Y. & Poh, Y.H. (1994). Orthogonal array designs for the optimization of SPE. *J. Chromatogr. A*, 677, 255-263

Wan, Y.Q.; Mao, X.J. & Yan, A.P. (2010). Simultaneous determination of organophosphorus and organonitrogen pesticides residues in Angelica sinensis. *J. Environ. Sci. Health, Part B: Pestic., Food Contam., Agric. Wastes*, 45 (4) 315-324

Wells, M.J.M. & Yu, L.Z. (2000). Solid-phase extraction of acidic herbicides. *J. Chromatogr. A*, 885, 237-250

Wold, S; Esbensen, K & Geladi, P. (1987). Principal component analysis. *Chemom. Intell. Lab. Syst.*, 2, 37-52

Yamauchi, K.; Tanabe, T & Kinoshita, M. (1979). Trimethylsulfonium hydroxide: a new methylating agent. *J. Org. Chem.*, 44, 638-639

Zapf, A. & Stan, H.J. (1999). GC analysis of organic acids and phenols using on-line methylation with trimethylsulfonium hydroxide and PTV solvent split large volume injection. *J. High Resolut. Chromatogr.*, 22, 83-89

18

Membrane Treatment of Potable Water for Pesticides Removal

Anastasios Karabelas and Konstantinos Plakas
Laboratory of Natural Resources and Renewable Energies Utilization, Chemical Process Engineering Research Institute, Centre for Research and Technology – Hellas, Thessaloniki Greece

1. Introduction

Over the last 50 years, plant protection products (PPPs), which are commonly referred to as "pesticides" (a term used henceforth in this chapter), are indispensable agents for the sustainable production of high quality food and fibres. The significant role of pesticides in controlling weeds (herbicides), insects (insecticides) and plant diseases that interfere with the growth, harvest, and marketability of crops has rendered the pesticide industry a significant economic player in the world market. At the same time, the widespread use of pesticides for agricultural and non-agricultural purposes has resulted in the presence of their residues in various environmental compartments. Traces of these products are frequently detected in surface water and in some cases in groundwater, which is the major source of drinking water around the world (Novotny, 1999; Martins et al., 1999; Loos et al., 2009). The frequent detection of many types of pesticide residues (including herbicides) in natural waters is of great concern to the public, to authorities and to all those involved in potable water production, wastewater treatment, and water reuse applications, due to potentially adverse health effects associated with these compounds even at very small concentrations (pg/L to ng/L). Specifically, potential health risks identified in toxicological and epidemiological studies include cancer, genetic malformations, neuro-developmental disorders and damage of the immune system (Skinner et al., 1997; Sanborn et al., 2004; McKinlay et al., 2008).

Regarding the potential for exposure of humans to pesticides residues, a strict regulatory framework is in force today. To ensure a high level of protection of both human and animal health and of the environment, the European Union (EU) developed and implemented a Thematic Strategy for Pesticides lately. The strategy is comprised of four elements:

- the Regulation (EC) 1107/2009, concerning the placing of plant protection products on the market (repealing Council Directives 79/117/EEC and 91/414/EEC),
- the Directive 2009/128/EC, establishing a framework for Community action to achieve the sustainable use of pesticides,
- the Regulation (EC) 1185/2009, concerning statistics on pesticides, and
- the Directive 2009/127/EC, regarding the equipment for pesticide application.

Moreover, EU implemented the Regulation (EC) No 396/2005 on maximum residue levels of pesticides in or on food and feed of plant and animal origin, in order to control the end of the life cycle of such products. Regarding the quality of water intended for human consumption, the Drinking Water Directive (98/83/EC) sets a limit of 0.1 μg/L for a single active ingredient of pesticides, and 0.5 μg/L for the sum of all individual active ingredients detected and quantified through monitoring, regardless of hazard or risk. In contrast, the residue limits and guideline levels set by the World Health Organisation (WHO) or the U.S. Environmental Protection Agency (USEPA) depend on the toxicity of the active substances and are determined using a risk-based assessment. The broad spectrum of legislation makes clear that pesticides are amongst the most thoroughly controlled substances in use today.

In parallel with appropriate regulatory controls and best pesticide-use practices, there is an urgent need for determination and removal of pesticides from potable water sources. These are in themselves difficult tasks, which are further complicated by the fact that a very large number of these synthetic chemical compounds are spread in the environment for crop protection. Conventional methods for potable water treatment, still widely employed, comprising particle coagulation–flocculation, sedimentation and dual media filtration, are ineffective for removing pesticide residues. The addition of more advanced final treatment steps (usually involving oxidation by H_2O_2 or O_3, and granular activated carbon - GAC - filtration) is generally considered to be effective, although significant problems still arise, mainly related to saturation of activated carbon, and to toxic chemical by-products, which may develop in the GAC filters under some conditions.

In view of the problems inherent in presently used processes, for removing various pesticides as well as the multitude of other synthetic organic micropollutants frequently encountered in drinking water sources (e.g. persistent organic pollutants-POPs, pharmaceutically active compounds-PhACs, endocrine disrupters-EDCs, etc), significant research effort has been invested to develop effective treatment methods, based on pressure-driven membrane processes. The growing interest in such processes is justified on account of the high and stable water quality they can achieve, although their cost effectiveness needs improvement. Therefore, influenced also by social and legislative pressure for more stringent potable water quality regulations, membrane processes, such as nanofiltration or low pressure reverse osmosis, are under development for broad applicability. To underpin these efforts, special attention is required for clarifying the attributes and limitations of membrane processes for pesticides removal as well as for prioritizing related R&D.

In view of the above considerations, the scope of this chapter is to review our current understanding and knowledge, gained from laboratory research, pilot and industrial-scale activity, regarding pesticides removal by membrane based processes. A fairly thorough discussion of pesticides retention by membranes will be provided, highlighting the prevailing mechanisms and the main factors involved. Particular attention will be paid to the role played by the dissolved organic matter (DOM), commonly present in the raw feed-water. The relevant physico-chemical properties of typical herbicides, of DOM, and of the active membrane surface will be assessed in an effort to clarify the significant membrane – organic species interactions. For a better understanding of the terminology used for membranes and membrane processes, some fundamental relations describing the function of a membrane and the basic principles of membrane processes will be briefly reviewed. Finally, future R&D needs for trace organic contaminants removal from potable water will be discussed, both at the scientific and the technological level.

2. Membrane technology – A short review of potable water treatment

2.1 Membrane processes in water treatment

Since the early 1990's membrane filtration has gained momentum and is now considered mainstream technology for removing a broad spectrum of contaminants from water and effluents. Advances in materials science and membrane manufacturing technology have shaped this trend, together with the increased regulatory pressures as well as an increased demand for drinking water originating from water sources of inferior quality (surface water, other). Moreover, membrane technologies have emerged as a very attractive option, in the production of clean and safe drinking water, due to their significant advantages over the conventional water treatment methods. Specifically:

- membrane treatment takes place at ambient temperature without phase change; this explains, for example, the success of reverse osmosis for water desalination;
- membrane separations occur without accumulation of substances inside the membranes; thus, membranes are well adapted to be ran continuously without a regeneration cycle as, for example, in ion-exchange resin operations;
- membrane separations do not involve addition of chemical additives; this affords advantages regarding the quality of treated water and leads to reduced environmental load;
- most membrane systems are compact (with reduced plant footprint), modular in nature, allowing retrofitting of existing processes;
- membrane processes are often technically simpler and more energy efficient than conventional separation techniques and are equally well suited for large-scale continuous operations as for batch-wise treatment of very small quantities,
- advances in polymer chemistry have led to the development of low pressure membranes, less prone to fouling, which are associated with reduced energy requirements, reduced chemical cleaning frequency, longer membrane life, and thereof, reduced operating costs.

A disadvantage of membrane processes is the usually required costly feed-water pre-treatment to avoid membrane fouling caused by various species. Furthermore, membranes are structurally not very robust and can be damaged by deviations from their normal operating conditions. However, significant progress has been made in recent years, especially in seawater reverse osmosis desalination, in developing membranes which have not only significantly better overall performance but also exhibit better chemical and thermal stability and are less sensitive to operating upsets.

The technically and commercially established membrane processes, for water treatment, are reverse osmosis (RO), nanofiltration (NF), ultrafiltration (UF) and microfiltration (MF). Although there is no sharp distinction, these processes are defined mainly according to the pore size of the respective membranes, and to a lesser extent by the level of driving force for permeation, i.e. the pressure difference across the membrane (Table 1). With decreasing porosity (i.e. from MF to UF and NF to RO) the hydrodynamic resistance of the respective membranes increases and consequently higher pressures are applied to obtain required water fluxes. MF and UF systems generally operate at a pressure of ~25 to ~150 psi, while some operate under vacuum at less than 12 psi. These systems can be operated in dead-end or cross-flow mode. The dead-end mode resembles conventional sand filter operation, where the feed solution flows perpendicular to the membrane surface. Unlike crossflow filtration, there is normally no reject stream, only a feed and a permeate stream, as shown in

Fig. 1. The crossflow system, which has gained wider acceptance in recent years, operates in a continuous manner where the feed solution flows tangentially across the membrane surface, thus generating a continuous exiting stream (defined as "retentate" or "concentrate") capable of partly sweeping the rejected substances, away from the membrane surface (Fig. 1). NF and RO operate almost exclusively in the crossflow mode and the operating pressure depends on the type of membrane used and the required water quality characteristics. Typical operating pressure for a NF system ranges from 100 to 200 psi, while for RO the pressure may vary between 100 and 400 psi, depending on ionic strength. For seawater desalination, RO plants operate at even higher pressures, between 800 to 1000psi.

Membrane process	Typical pore size (nm)	Pressure (bar)	Permeability ($Lm^{-2}h^{-1}bar^{-1}$)
Microfiltration (MF)	50-1000	0.1-2.0	> 50
Ultrafiltration (UF)	10-50	1.0-5.0	10 – 50
Nanofiltration (NF)	< 2	5.0-20	1.4 – 12
Reverse Osmosis (RO)	< 1	10-100	0.05 – 1.4

Table 1. Comparison of pressure-driven membrane processes (Mulder, 1998; Singh, 2006)

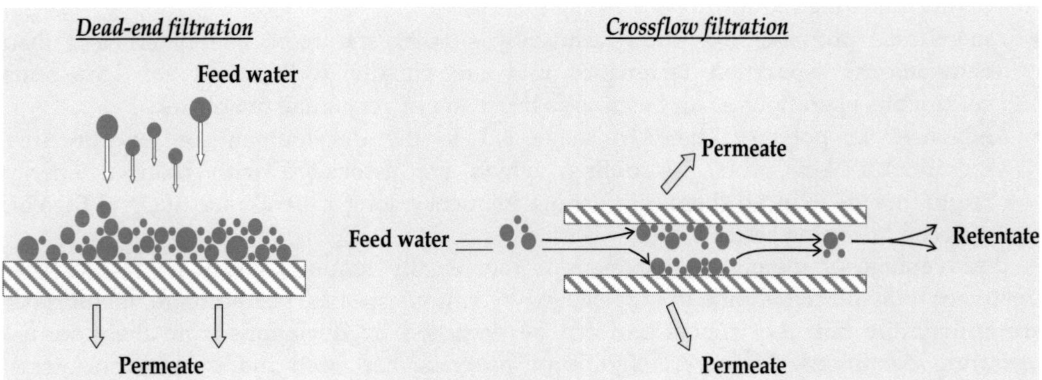

Fig. 1. Dead-end versus crossflow filtration

The porous MF and UF membranes are characterized by the molecular weight cut-off (MWCO), which is expressed in Dalton indicating the molecular weight of a hypothetical non-charged solute that is 90% rejected (Mulder, 1996). NF can be characterized either by MWCO or ionic retention of salts such as NaCl or $CaCl_2$; RO membranes being dense are characterized by salt retention, although some researchers have modeled molecular retention to determine a MWCO (Kimura et al., 2004). The percentage retention (R%) of species in solution is defined as:

$$R(\%) = \left(1 - \frac{C_p}{C_f}\right) x100 \quad (1)$$

where C_p and C_f are the permeate and feed concentration, respectively. Other common performance parameters are the permeate recovery and flux, given as follows:

$$\text{Recovery} = \frac{Q_p}{Q_f} \qquad (2)$$

$$J_w = L_p\,(\Delta P - \Delta \pi) \qquad (3)$$

Recovery is defined as the ratio of permeate production rate Q_p over the feed flow rate Q_f. J_w is the permeate water flux, L_p the membrane permeability, ΔP the applied transmembrane pressure and $\Delta \pi$ the osmotic pressure difference between feed and permeate.
From Table 1 it is evident that the selection of a particular membrane type mainly depends on the contaminant size to be removed. MF is usually applied to separation from an aqueous solution of particles of diameter greater than 100nm (usually 0.05-1μm), while UF to separation of macromolecules (of size down to 30nm), with molecular weights varying from about 10^4 to more than 10^6. Examples of species that can be removed with MF and UF processes include assorted colloids (frequently referred to as "turbidity"), iron and manganese precipitates, coagulated organic matter, and pathogens such as *Giardia* and *Cryptosporidium* cysts. UF membranes are also capable of removing viruses. RO membranes are used to remove from the feed stream even smaller species, of diameter as small as 0.1nm, such as hydrated ions and low molecular weight solutes. On the other hand, NF, also called "loose RO", lies between RO and UF in terms of selectivity of the membrane as it is designed for removal of multivalent ions (typically calcium and magnesium) in water softening operations and for organic species control. The feed water to NF plants can be any non-brackish, ground or surface water. For treatment of brackish water, nanofiltration is usually not the most suitable process, since Cl^- and Na^+ are among the ions with the lowest retention rates. A simplified decision tree for selecting the suitable membrane process for treatment of potable water is shown in Fig. 2.

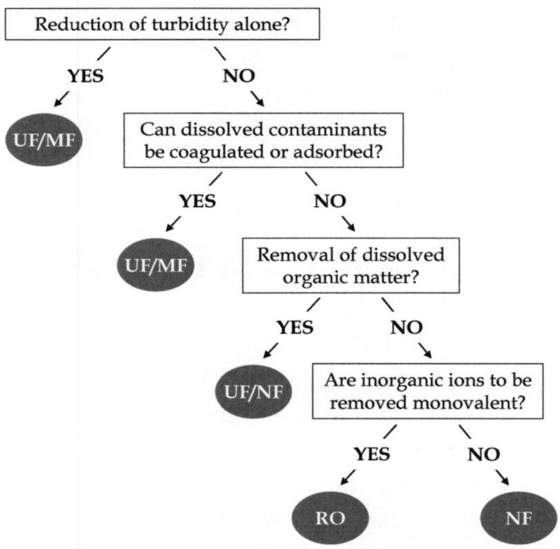

Fig. 2. Simplified decision tree for selecting a membrane process for treatment of potable water.

Taking into consideration that the majority of the compounds categorized as pesticides have molecular weights (MW) greater than 200 Da and a size in the range of ions (close to 1 nm), reverse osmosis and nanofiltration are promising options for their removal from contaminated water sources. However, RO is generally more expensive, regarding both investment and operating costs, due to the required greater pressures (lower permeability membrane). For these reasons scientists and all those involved in potable water production have turned their attention to the application of NF and ultra low-pressure RO membranes (ULPRO). Related R&D has resulted in the development of an advanced type of NF/ULPRO membranes, the so called thin film composite membranes (TFC or TFM) which have been successfully applied for the removal of pesticides in past 10-20 years (Hofman et al., 1997; Wittmann et al., 1998; Bonné et al., 2000; Cyna et al., 2002).

TFC are multi-layer membranes comprising a very thin and dense active layer (of cross-linked aromatic polyamide) which is formed in situ on a porous support layer, usually made of polysulfone (Fig.3). Their broad applicability is attributed to their unique characteristics such as the high salt retention capacity, the good chemical stability and mechanical integrity as well as to the fact that they can achieve high specific water fluxes at lower operating pressures (AWWA, 1996; Filteau & Moss, 1997). A list of the TFC membranes studied for the removal of pesticides from potable water is given in the Appendix, together with their retention performance and their characteristic surface properties (MWCO).

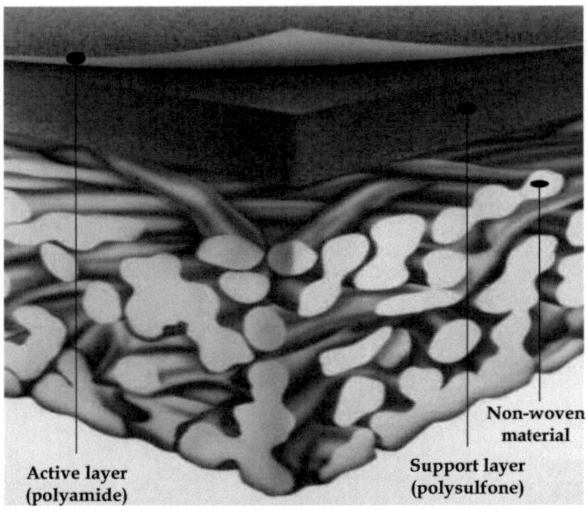

Fig. 3. Schematic representation of a thin film composite (TFC) membrane (Dow, 2010)

2.2 Examples of water treatment plants using NF/ULPRO membranes

A list of significant water treatment plants using nanofiltration or ultra-low pressure RO membranes is shown in Table 2. An outstanding example of nanofiltration for the removal of pesticides and other organic residues, for the production of drinking water, is the Méry-sur-Oise plant in the northern part of Paris, in France. The Méry-sur-Oise plant has been successfully producing water from the river Oise, using NF technology, since 1999. Its performance indicators are very satisfactory, especially with regard to the two main objectives; i.e., elimination of organic matter and of pesticides, which renders nanofiltration a very successful technology (Ventresque et al., 2000).

The design of a membrane water treatment plant may vary depending on the feed water conditions, the required final water quality, the water recovery ratio, the membrane module configuration (spiral wound, hollow fiber, tubular) and the material of membrane active surface layer (asymmetric cellulosic or non-cellulosic membranes, thin film ether, or amidic composite membranes). In general, a conventional NF/RO treatment system includes

Location	Capacity (m^3/d)	Application	Reference
Boca Raton, Florida, US	152,000	Groundwater softening	Suratt et al., 2000
Méry-sur-Oise, Paris, France	140,000	Pesticide removal for drinking water supply	Cyna et al., 2002
Heemskerk, Holland	~57,000	Surface water treatment for drinking water supply	Kamp et al., 2000
Bajo Almanzora, Andalusia, Spain	30,000	Groundwater softening	Redondo & Lanari, 1997
Debden Road, Saffron Walden, England	3,000	Pesticide removal for drinking water supply	Wittmann et al., 1998

Table 2. Case studies of water treatment plants using NF/ULPRO membranes

pre-treatment, membrane filtration and post-treatment, as schematically shown in Fig. 4. Pretreatment of the feed is required to protect the membranes and to improve their performance, while post-treatment includes several unit operations common to drinking water treatment such as aeration, disinfection, and corrosion control. The pre-treatment should be carefully designed, mainly to cope with the fouling propensity of the feed water and aims to (Redondo & Lomax, 2001):

- reduce suspended solids and minimise the effect of colloids
- reduce the microbiological fouling potential of the feed water
- condition the feed by adding chemicals (antiscalant, pH adjustment)
- remove oxidising compounds in the feed if required (to protect the membranes)

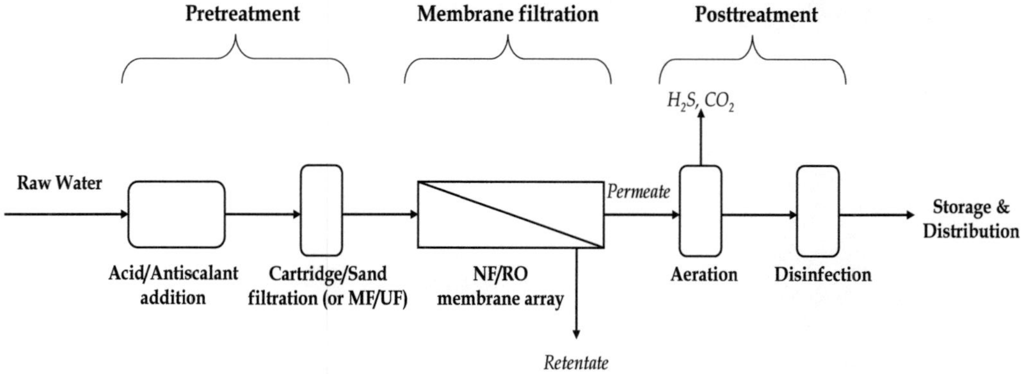

Fig. 4. A typical NF/RO membrane water treatment process.

In the case of the Méry-sur-Oise plant, the full scale facility consists of the following treatment steps (Ventresque et al., 2000):

- ACTIFLO® clarifiers (coagulation using polyaluminium chloride and an anionic polyelectrolyte at pH 6.9, flocculation)
- Ozonation
- Dual-media filtration (two-layer sand and anthracite bed, preceded by a second injection of coagulant)
- Cartridge filtration (6 μm micro-filters, back-washable and chemically cleanable)

- Nanofiltration
- CO_2 stripping (degassing towers)
- UV disinfection

Pretreatment plays a critical role in the performance, life expectancy and the overall operating costs of NF/RO systems. R&D in this direction includes studies on new technologies and/or new design concepts on feed pretreatment, membrane washing and chemical cleaning (to restore membrane fluxes) and extensive studies on membrane performance improvement, focused on development of low fouling membranes. More information on these matters can be found in various publications, in scientific articles as well as in technical reports issued by several membrane manufacturers (Tanninen et al., 2005; Al-Amoudi & Lovitt, 2007; Dow, 2010). In the following, for the sake of completion and to facilitate the discussion in sub-section 3.5, a brief introduction to fouling is presented and of the related phenomena occurring at the membrane surface.

2.3 Membrane fouling

Membrane performance can be negatively affected by a number of species whose concentration and/or presence in the feed water must be controlled. As indicated in Fig. 5, these species are divided in two categories: substances capable of damaging the membranes and species with potential for membrane fouling or scaling. The discussion is concentrated on fouling, which is the major problem faced in any membrane separation. Membrane fouling, if not controlled, is detrimental to the overall process efficiency because of the increased energy requirements, reduced plant productivity and increased cost of chemicals due to cleaning as well as the shorter lifetime of the membranes, which also lead to an increase of the total production cost. Moreover, membrane fouling may alter the surface characteristics of NF/RO membranes, which in turn could potentially influence the removal of undesirable dissolved species, including pesticides.

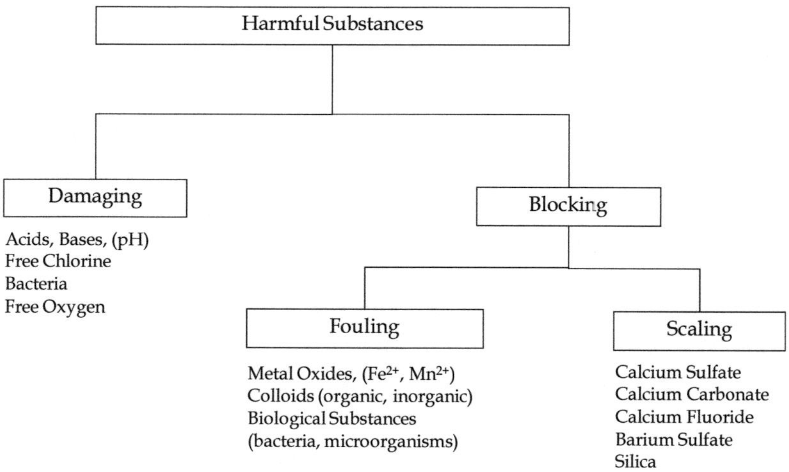

Fig. 5. Substances potentially harmful to membranes (Rautenbach & Albrecht, 1989)

The main fouling categories are organic, inorganic, particulate and biological fouling. Metal complexes and silica are also important. In operating plants all types of fouling may occur (Yiantsios et al., 2005), depending on the feed water composition. Research on understanding fouling and applying appropriate control strategies are important endeavours aiming at improvement of NF/RO membrane processes. Among the different kinds of fouling, emphasis is given here to fouling by organic matter, naturally occurring in source waters in concentrations ranging from 2 to 40mgC/L, which are roughly 10,000 times greater than pesticide concentrations encountered in surface waters.

Extensive research on fouling of NF membranes by natural organic matter (NOM) has shown that it can be influenced by membrane characteristics, including surface structure as well as surface physico-chemical properties, composition of feed solution including ionic strength, pH and concentration of divalent ions, NOM properties, including molecular weight and polarity, as well as hydrodynamic and operating conditions including permeate flux, pressure, concentration polarization, and the mass transfer properties of the fluid boundary layer (Al-Amoudi, 2010). The effect of the aforementioned factors on NOM fouling is summarized in Table 3. The significant role of feed-water chemical composition (ionic strength, pH, divalent cations) on NOM fouling, as well as the fouling mechanisms involved in the case of humic substances (Hong & Elimelech, 1997) are illustrated in Fig. 6.

	Value	NOM fouling rate	Cause
Ionic strength concentration	Increased	Increased	Electrostatic repulsion
pH	High pH Low pH	Increased Increased	Hydrophobic forces Electrostatic repulsion
Divalent cations	Presence	Increased	Electrostatic repulsion and bridging between NOM and membrane surface
NOM fraction	Hydrophobic Hydrophilic	Increased Decreased	Hydrophobicity
Molecule or membrane charge	High charge	Increase	Electrostatic repulsion
Concentration polarization	High	Increased	
Surface morphology	Higher	Increased	"Valley" blocking
Permeate flux (high recovery)	Higher	Increased	Hydrophobicity
Pressure	Higher	Increased	Compaction

Table 3. Factors affecting natural organic matter fouling of NF membrane (Al-Amoudi, 2010)

The term concentration polarization (CP) mentioned earlier describes the process of accumulation of retained solutes in the membrane boundary layer where their concentration will gradually increase. Such a concentration build-up will generate a diffusive flow back to the bulk of the feed, but after a certain period of time steady-state conditions will be established. The consequences of CP can be summarised as follows (Mulder, 1996):

- Flux will be reduced.
- Retention of low molecular weight solutes, such as salts, can be reduced.

- Retention can be higher: this is especially true in the case of mixtures of macromolecular solutes where CP can have a strong influence on the selectivity. The higher molecular weight solutes that are retained completely form a kind of second or dynamic membrane. This may result in a higher retention of the lower molecular weight solutes.

Concentration polarization is considered to be reversible and can be controlled in a membrane module by means of velocity adjustment, pulsation, ultrasound, or an electric field. Most membrane suppliers recommend a minimum feed flow rate (i.e. minimum superficial velocity at the retentate side) and a maximum allowable water recovery rate to minimize the effects of CP. Membrane fouling, on the other hand, is more complicated in that it is considered as a group of physical, chemical, and biological effects, which lead to irreversible loss of membrane permeability (Sablani et al., 2001).

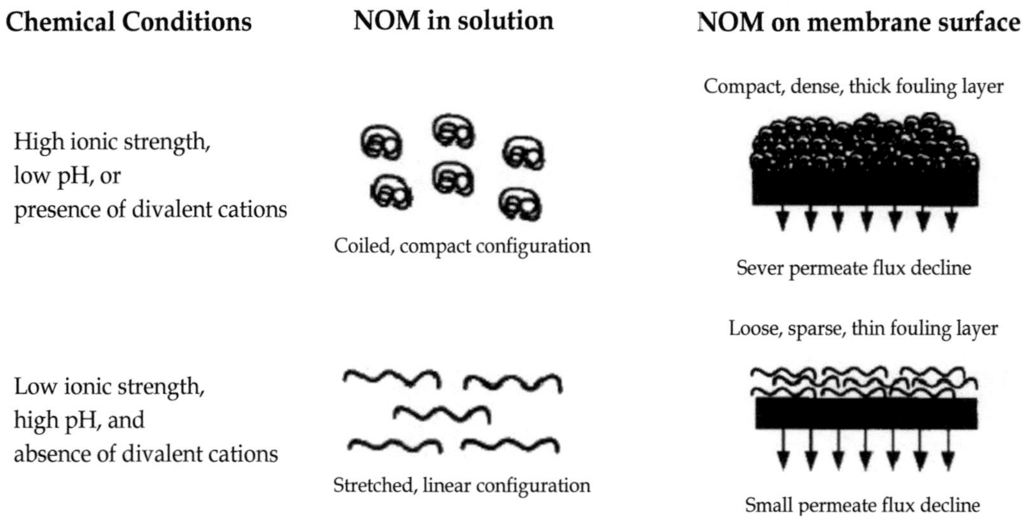

Fig. 6. Schematic description of the effect of solution chemistry on the conformation of NOM macromolecules in the solution and on the membrane surface and the resulting effect on membrane permeate flux. The NOM fouling described in the diagram is applicable for permeation rates above the critical flux. The difference, between the two chemical conditions shown, becomes less clear at very high permeate flux. At low permeate flux (below the critical flux), no significant fouling is observed for both conditions (adapted from Hong & Elimelech, 1997)

2.4 Retention mechanisms in NF/RO processes

There is a great deal of published work on the basic retention mechanisms and the various applications of NF/RO processes (Mulder, 1996; Scott, 1998; Nghiem & Schäfer, 2005). In general, the separation process involves several mechanisms such as size exclusion or charge repulsion. Moreover, a sorption-diffusion mechanism can also contribute to the separation process, attributed to hydrophobic interactions or hydrogen bonding between the contaminants and the membrane surfaces (solute-membrane affinity) (Nghiem & Schäfer, 2005). Depending on the physicochemical characteristics of the contaminant and the membrane, separation can be achieved by one or several mechanisms. The word 'physicochemical' implies that separation can be attributed either to physical selectivity (charge repulsion, size exclusion or steric hindrance) or to chemical selectivity (solvation energy, hydrophobic interaction or hydrogen bonding). Consequently, the separation process can be strongly influenced by the physicochemical interaction between the solute and the membrane polymer and/or with water (Nghiem & Schäfer, 2005). In the case of trace organic contaminants, like pesticides, such interactions are complicated and their transport across the membrane is still a topic of extensive research.

For non-charged solutes, the distribution at the boundary layer/membrane interface is considered to be determined by a steric exclusion mechanism. Steric exclusion is not typical for nanofiltration but applies to ultrafiltration and microfiltration, where solutes larger than the pore size of the membranes are retained. This is comparable to a sieving phenomenon except that in membrane filtration, neither pores nor solutes have a uniform size. For instance, dissolved organic species may change their configuration due to changes in solution chemistry or interactions with other molecules or surfaces. For example, the combined nanofiltration of triazine herbicides and naturally occurring humic substances facilitates the formation of complexes with triazines resulting in an increased steric congestion or reduction of the diffusivity of the NOM–triazine pseudo-complex (Plakas & Karabelas, 2009).

For charged solutes, an additional mechanism can be recognised, the Donnan exclusion, which has a pronounced effect on the separation by NF. Due to the slightly charged membrane surface, solutes with an opposite charge compared to the membrane (counter-ions) are attracted, while solutes with a similar charge (co-ions) are repelled. At the membrane surface, a distribution of co- and counter-ions will occur, thereby influencing separation. The relative importance of Donnan exclusion in solute retention by NF membranes is still debated in the scientific community since steric hindrance appears to be capable of significantly influencing such retention. For instance, Van der Bruggen et al., (1999) suggest that the charge effect can be important when the molecules are much smaller than the pores; when the molecules have approximately the same size as the pores, charge effects can exert only a minor influence, as the molecules are mainly retained by a sieving effect.

In the case of polar organic species, separation by NF/RO membranes is even more complicated as the process is not only affected by charge repulsion and size exclusion but it is also influenced by polar interactions between solutes and the membrane polymeric suface. Research in this direction has led to the conclusion that retention may be negatively affected by the polarity of a molecule (Van der Bruggen et al., 1999; Agenson et al., 2003; Kimura et al., 2003a). A possible explanation for this behaviour is related to electrostatic interactions; specifically, the dipole can be directed towards the charged membrane in such a way that the side of the dipole with the opposite charge is closer to the membrane (Van der Bruggen et al., 1999). The dipole is thus directed towards the pore and enters more easily into the membrane structure; moreover, once the molecule is in an open (straight-through) pore, it will follow the permeate. The polarity effect is expected to be the same for positively and negatively charged membranes, since the only change occurring is the direction of the dipole (Van der Bruggen et al., 1999).

Adsorption of organic species to membrane materials is an important aspect of trace organic matter removal using NF/RO. Organic contaminants, which can adsorb onto the membrane, are usually hydrophobic (high $\log K_{ow}$) or present high hydrogen bonding capacity. In addition, experimental results have shown that the adsorption of hydrophobic compounds is significant for neutral compounds and for ionizable compounds when electrostatically neutral (Kimura et al., 2003b). Also, operating conditions such as the permeate flux can have a significant effect on the degree of compound adsorption (Kimura et al., 2003b). Although adsorption contributes to an initial retention, an increased surface concentration as a result of adsorption, favouring species diffusion through the membrane, can reduce process effectiveness to some extent (Nghiem & Schäfer, 2005). Moreover, adsorption, resulting in the accumulation of organic molecules on the membrane surfaces, can cause several problems leading to overall performance deterioration.

3. Factors affecting the removal of pesticides by NF/RO treatment

3.1 Introduction

The idea of applying membrane processes for the removal of pesticide residues from potable water is not new. It originates back in the late '60s when Hindin et al. (1969) studied the removal of a few chlorinated pesticides, including DDT, TDIE, BHC, and lindane, by reverse osmosis using an asymmetric cellulose acetate (CA) membrane. The initial results of their study have shown that RO filtration, employing a CA membrane, is a promising treatment process for producing water low in organic substances, including pesticides. The excellent performance of RO membranes in removing a variety of pesticides, including chlorinated hydrocarbons, organophosphorous, and miscellaneous pesticides, was also shown in an early study by Chian et al. (1975) in which a number of non-cellulosic membranes, such as aromatic polyamide and cross-linked polyethylenimine membranes exhibited far better performance in pesticides removal and resistance to pH than conventional CA membranes. Because of this advances in membrane technology, RO has been gradually finding applications in the treatment of a variety of domestic, industrial, and hospital wastewaters.

In the past three decades, the need for a complete assessment of the RO, and of the later developed NF process, regarding removal of pesticide residues from various aquatic matrices, led to an extensive research effort in many laboratories (Berg et al., 1997; Devitt et al., 1998a; Van der Bruggen et al., 1998, 2001; Kiso et al., 2000, 2001a, 2002; Košutić et al., 2002, 2005; Zhang et al., 2004; Causserand et al., 2005; Bhattacharya et al., 2006; Plakas et al., 2006; Sarkar et al., 2007; Plakas & Karabelas, 2008, 2009; Ahmad et al., 2008a, 2008b; Comerton et al., 2008; Caus et al., 2009; Benítez et al., 2009; Pang et al., 2010; Wang et al., 2010), pilot (Baier et al., 1987; Duranceau et al., 1992; Agbekodo et al., 1996; Berg et al., 1997; Hofman et al., 1997; Wittmann et al., 1998; Bonné et al., 2000; Boussahel et al., 2000, 2002; Chen et al., 2004) as well as to industrial scale experiments (Agbekodo et al., 1996; Wittmann et al., 1998; Ventresque et al., 2000; Cyna et al., 2002). A fairly large number of commercially available NF/RO membranes have been tested for the removal of an even larger number of herbicides, insecticides, fungicides and miscellaneous pesticides from various water matrices. The results of the respective literature review are summarized in the Appendix, in which the NF/RO membranes employed are listed together with their pesticide rejection performance.

A critical review of the rejection mechanisms and of the main parameters involved in pesticide removal by NF/RO processes is made in the following. Specifically, the findings of a comprehensive literature review are reported together with the results obtained from the experimental work performed by the authors.

3.2 The role of membrane characteristics

The success of pesticides removal from potable water by membrane processes is strongly related to the type of membrane selected. Important aspects to consider when choosing an appropriate membrane are MWCO, porosity, degree of ionic species rejection, surface charge and membrane type (polymer composition). The significance of each parameter on pesticides removal is directly related to the solute properties (molecular weight, molecular size, acid disassociation constant-pKa, and hydrophobicity/hydrophilicity-logK_{ow}) which determine the strength of the pesticide-membranes physicochemical interactions.

Membrane molecular weight cut-off

Based on the molecular weight of the majority of the pesticide residues detected in potable water sources (usually greater than 200Da), membranes with a MWCO varying from 200 to 400Da are promising options for the successful removal of such solutes from water. These are reverse osmosis and tight nanofiltration membranes which are characterized by pore sizes close to those of pesticides (<1nm). It is evident that the larger the pesticide molecule the greater the sieving effect, resulting in greater retention. On the other hand, the retention of small pesticide molecules by wider pore membranes can be influenced not only by the sieving parameters (pesticide and membrane pore size) but also by the physicochemical interactions taking place between the pesticides and the membrane surfaces. For example, in pilot studies (Boussahel et al., 2000; 2002), among the two membranes tested, Desal DK membranes achieved the best retention results for all pesticides and water matrices tested due to their lower MWCO value (150-300Da) compared to NF200 (300Da) membranes. The low MWCO of Desal DK membranes provided an explanation for the similar percentage removal for all pesticides (except from the polar diuron), something that was not observed in the case of NF200 membranes, for which the retention capacity was found to be dependent both on the size and the polarity of the pesticide molecules (Boussahel et al., 2000). In a recent work (Zhang et al., 2004), the retention of two triazine herbicides (atrazine and simazine) by four nanofiltration membranes was also related to their MWCO. Specifically, the smaller MWCO of UTC-20 (180Da) and UTC-60 (150Da) membranes resulted in significantly greater removal than that achieved by DESAL 51 HL (150-300Da) and DESAL 5 DL (150-300Da) membranes (Table 5).

Some deviations from the aforementioned trends have been also reported. For instance, in a study by Van der Bruggen et al. (1998), the MWCO of the employed NF membranes was poorly correlated with the removal of two classes of herbicides; i.e. triazines (atrazine, simazine) and phenyl-ureas (isoproturon, diuron). Specifically, the NF70 membrane, with a MWCO 200Da, presented greater retention capability than the seemingly somewhat tighter UTC-20 membrane (MWCO 180Da). On the other hand, a NTR-7450 membrane exhibited the worst performance (<20% retention) due to the larger pore sizes, indicated by its high MWCO (600-800Da) (Van der Bruggen et al., 1998). Similar observations were also made in another study (Mohammad & Ali, 2002), where the rejection of uncharged solutes and salts did not conform to the expected trend of reduced rejection with increasing MWCO of the NF membranes used.

Membrane porosity

The above results support the commonly held belief that the characterization of NF and ULPRO membranes by a nominal MWCO value may be convenient in practice, but it is questionable on physical grounds since the molecular weight of a model compound, used to determine MWCO, cannot be representative of all molecular species (i.e. the pollutants to be separated) of the same molecular weight but differing in conformation and in other physical properties, which affect molecule-membrane interaction and permeation; thus, MWCO provides only a rough estimate of the membrane capability to retain dissolved uncharged compounds. However, other quantities such as the nominal pore size of a membrane, which refers to the smallest pore size in the membrane matrix, and the porosity, expressed as pore density, pore size distribution (PSD), or effective number of pores (*N*) in the membrane top layer (skin) have been regarded as representative parameters for predicting the rejection of different organic compounds or particles (Van der Bruggen et al., 1999; Lee et al., 2002; Košutić et al., 2002, 2005, 2006). For instance, the rejection of uncharged pesticide molecules was positively correlated with membrane porosity parameters (PSD and *N*) (Košutić et al., 2002, 2005). The apparent sensitivity of rejection, to accurate characterization of the membrane porosity, is in itself an indication of the dominant role played by the sieving mechanism; this is also consistent with findings that the membrane pore size is a crucial parameter for pesticide removal by a specific membrane (Van der Bruggen et al., 1998). It should be pointed out that, although in these studies the physicochemical effects on the rejection of pesticides may be of lesser importance, they cannot be neglected as they can contribute to final rejection achieved for specific membrane-pesticides systems. This issue is subsequently discussed.

Degree of membrane desalination

The separation capability of tight NF and RO membranes is commonly characterized by their salt rejection performance, rather than by MWCO which is often not reported by the manufacturers. The desalination degree of a membrane is usually reported as the stabilized salt rejection of a 2000 mg/L sodium chloride or magnesium sulfate solution, and/or a 500 mg/L calcium chloride solution. The desalination degree can be a useful parameter in roughly estimating the rejection of pesticides, because the MWCO of a membrane is often unknown and manufacturer-specific, whereas PSD and porosity determination require the performance of specific filtration experiments or the application of special analytical techniques (atomic force microscopy, bubble point, gas adsorption/desorption, thermoporometry, etc). The usefulness of salt rejection has been demonstrated in studies (Kiso et al., 2000, 2001a) where the rejection of aromatic and non-phenylic pesticides was positively correlated with the desalination degree of commercial NF membranes; indeed, rejection was greatest in the case of the highest desalting membranes. Specifically, the order of rejection followed that of the nominal salt rejection capability of the membranes; i.e., NTR-729HF > NTR-7250 > NTR-7450 > NTR-7410, with 92%, 60%, 51% and 15% NaCl rejection, respectively. It is interesting to notice that only the highest desalting membrane was found to reject effectively almost all pesticides. However, rejection was again found to be strongly influenced by the pesticide properties (hydrophobicity, charge), regardless of the membrane salt rejection performance. In general, the reliability of the membrane desalination degree as an accurate indicator for assessing the removal of hydrophobic organic micro-pollutants is doubtful.

Membrane material

Membrane material is also identified as an important factor of the system pesticide-water-membrane that affects the membrane rejection performance through physicochemical interactions in that system. For example, a number of studies confirm that composite polyamide (PA) membranes exhibit far better rejection performance for several mixtures of micropollutants, including pesticides, compared to the cellulose acetate (CA) membranes (Chian et al., 1975; Hofman et al., 1997; Causserand et al., 2005). This behavior has been attributed to the higher polarity of CA membranes which is responsible for the poor rejection of the highly polar pesticides (Chian et al., 1975). On the contrary, the relatively nonpolar aromatic PA membranes exhibit better rejection performance as well as high water fluxes attributed to the very small thickness characterizing their effective active layer (skin), which varies between 10nm and 500nm for various TFC NF and ULPRO membranes. It has been also reported (Kiso et al., 2000, 2001a) that membranes made of sulfonated polyethersulfone display lower rejection of pesticides compared to poly(vinyl alcohol)/ polyamide ones, even though their desalination capabilities are similar.

Membrane charge

The majority of the commercial TFC membranes is characterized by a negative charge which tends to minimize the adsorption of negatively charged foulants present in membrane feed waters and to enhance the rejection of dissolved salts (Xu & Lebrun, 1999; Deshmukh & Childress, 2001). The electrostatic repulsion of negatively charged pesticides (pH>pKa) at the membrane surface is expected to enhance the overall rejection performance. This is in agreement with results obtained by Berg et al. (1997) where the rejection of the negatively charged mecoprop (at neutral pH) was greater than the one measured for non-charged herbicides of the same size. Specifically, rejection experiments with mecoprop in dissociated and undissociated form were conducted with five different NF membranes; in this study, it was estimated that less than 10% of mecoprop was dissociated at pH 3. Mecoprop, in the dissociated form, was rejected more than in the undissociated form, by all five NF membranes at levels between 10% and 90%. The rejection of the undissociated form of mecoprop was comparable to the uncharged diuron which is of similar size, providing additional evidence that rejection of undissociated organic molecules is due to steric effects.

3.3 Effect of pesticides properties on retention

According to the preceding discussion, the selection of an appropriate membrane is primarily made on the basis of key pesticide parameters, like the molecular weight, the molecular dimensions (length and width), the polarity (dipole moment), the hydrophobicity /hydrophilicity ($logK_{ow}$), and the acid dissociation constant (pKa). Several research groups have systematically studied the role of one or more of the aforementioned pesticide parameters on membrane rejection, and their results are summarised here.

Pesticide molecular weight and size

Researchers agree that size exclusion is the most important mechanism of pesticide retention. Various size parameters used in the literature to correlate pesticide rejection include the molecular weight (MW), the Stokes diameter (d_s), the diameter derived from the molar volume (d_m), the molecular length and molecular width (calculations based on molecular STERIMOL parameters), and the diameter which is calculated from the molecular structure by using special computer software (HyperChem, ChemOffice) (Van der Bruggen et al., 1998, 1999; Kiso et al., 2001a; Agenson et al., 2003; Chen et al., 2004). Typical values of size parameters for selected pesticides are listed in Table 6, where it is clearly shown that the dimensions of a pesticide are not directly related with its MW. Small MW pesticides can be characterized by a larger molecular length and/or width compared to other pesticides of larger MW. This is attributed (Chen et al., 2004) to the structure and the small range of molecular weights of the specific pesticides (198-286Da).

Pesticide	Molecular weight (gr/mol)	Molecular length (Å)	Molecular width (Å)
Atrazine	215	10.36	8.02
Bentazone	240	9.31	5.42
Cyanazine	240	10.38	8.33
Diuron	233	9.19	4.87
Mecoprop	214	9.43	4.88
Metribuzin	214	10.43	4.43
Pirimicarb	238	10.30	7.93
Simazine	201	10.34	7.49

Table 4. Size of selected pesticides; calculations using the HyperChem software (Chen et al., 2004)

Since MW is the most easily accessible parameter (though only indicative of molecular size), in the majority of studies attempts are made to relate the retention of uncharged pesticides to this quantity. It has been reported (Chen et al., 2004) that a positive correlation exists between the rejection of eleven pesticides with their molecular weights, from which a MWCO of 200Da was determined for the membrane tested (Dow Filmtec NF70). In pilot studies (Boussahel et al., 2002), the higher rejection of atrazine and cyanazine was attributed to their molecular weight, which is larger than the one characterizing the other three herbicides tested (DEA, simazine and isoproturon). Significant efforts were also made (Van der Bruggen et al., 1999) to correlate the rejection of miscellaneous organic molecules with their molecular weight values as well as with other size parameters with physical meaning (d_s, d_m, molecular diameter calculated with the HyperChem software). Interestingly, it was found that the correlation of retention was only slightly improved by employing size parameters, as compared to correlation with MW; this implies that MW is a useful indicator for correlating retention (Van der Bruggen et al., 1999). Nevertheless, MW cannot be recommended for modeling efforts, since it is not representative of the geometry of the molecules that affects their rejection or transfer through the membrane.

Molecular length and molecular width are also reported in the literature to be realistic measures of molecular size and good parameters for predicting the rejection of different groups of organic compounds by NF/RO membranes. For example, the rejection of aromatic pesticides was found (Chen et al., 2004) to be best correlated with their molecular length rather than their molecular width (theoretical calculations by HyperChem based on their structures and orientation). The molecular length in this case represented the cross-sectional diameter due to structural orientation. On the other hand, the molecular width (MWd) was suggested (Kiso et al., 2001b) as a useful descriptor of the steric hindrance effect on the rejection of alcohols and carbohydrates. In addition to MWd, Kiso et al. (2001b) developed another molecular size parameter which correlated the rejection of alcohols and carbohydrates better than the MWd or the Stokes diameter; specifically, they calculated a mean molecular size (MMS) by taking half of the length of the edge of the cube encompassing the molecule (Kiso et al., 2001b). Better correlations with MMS where observed for high MWCO membranes (>500Da), while for low MWCO membranes (<250Da) MWd was found to be a better descriptor than MMS (which is the case for most pesticides) (Kiso et al., 2001b).

Regarding the aromatic (phenylic) and the non-phenylic pesticides, it was found (Kiso et al., 2000, 2001a) that rejection cannot be correlated solely with a molecular size parameter. This is attributed to the sorption capacity of these molecules on the membrane polymer which together with the molecule planarity (size) explain the solute permeability through the nanofiltration membranes. In an effort to combine steric hindrance effects with adsorption, Kiso et al. (2001a) developed an alternative molecular width parameter (P-MWd) which was used in the statistical processing of their experimental results. A regression analysis showed that the permeability of an aromatic compound through a membrane can be reduced due to both its sorption capacity and its molecular width. Similar observations were also made for alkyl phthalates and mono-substituted benzenes (Kiso et al., 2001b) with the rejection being strongly affected by their hydrophobic properties. These results indicate the significance of the solute-membrane affinity on rejection, and that solute transport predictions should not be based only on steric exclusion effects (Verliefde et al., 2009a).

Pesticide hydrophobicity/hydrophilicity

The significance of adsorption on the rejection of pesticides during membrane applications has been first reported by Chian and his coworkers (Chian et al., 1975). They claimed that the interaction between the hydrocarbon (nonpolar) segments of pesticide molecule and membranes is due to hydrophobic bonding. Since then, many researchers have reported significant adsorption of pesticides and of other organic micropollutants onto the membrane polymer (Kiso et al., 2000, 2001a; Nghiem & Schäfer, 2002; Agenson et al., 2003; Kimura et al., 2003a, 2003b; Comerton et al., 2007; Plakas & Karabelas, 2008). A literature review shows that except from the hydrophobic interactions, adsorption may also take place through hydrogen bonding between the organic molecules and the hydrophilic groups of the membrane material (Nghiem et al., 2002). Hydrogen bonding and hydrophobic interactions can apparently act either independently or together. In the latter case, it is often difficult to distinguish the two effects. Regarding pesticides, the literature review suggests that the hydrophobic interactions are mostly responsible for pesticide adsorption onto membrane surfaces, which is considered to be the first step of the rejection mechanism. This observation led researchers to the conclusion that the rejection of hydrophobic compounds should be experimentally evaluated after the tested membrane is saturated with the target compounds; otherwise, the rejection is likely to be overestimated, with adsorption misinterpreted as some kind of high initial rejection (Kimura et al., 2003b).

A measure of solute hydrophobicity/hydrophilicity is the octanol/water partition coefficient ($logK_{ow}$ or logP), while the hydrophobic nature of a membrane is characterized by its contact angle value (Mulder, 1998). $LogK_{ow}$ values of trace organic molecules vary between -3 and 7, with the higher values characterizing hydrophobic compounds (usually for $logK_{ow}>2$). Kiso et al. (2000, 2001a, 2002) systematically investigated the relationship between $logK_{ow}$ versus retention and adsorption of a number of aromatic and non-phenylic pesticides, using flat sheet and hollow fiber nanofiltration membranes. While no significant correlation was identified between retention and $logK_{ow}$, there was a rather good correlation between the adsorption and the characteristic $logK_{ow}$ values of the pesticides tested (Kiso et al., 2000, 2001a, 2002). Moreover, it was found that the presence of a phenyl group in a molecule increases its adsorption capacity (aromatic pesticides), while alkyl groups can have negative effects on the interaction between a phenyl group and the membrane (Kiso et al., 2001a). In a recent study (Comerton et al., 2007), static adsorption experiments with 22 endocrine disrupting species and pharmaceutically active compounds (including the pesticides alachlor, atraton, metolachlor, DEET), and UF, NF and RO membranes, showed that adsorption was strongly correlated with compound $logK_{ow}$ and membrane pure water permeability, and moderately correlated with compound solubility in water. Kimura et al.

(2003b) reported also the negative effect of solute charge on adsorption, since adsorption was found to be greater for electrostatically neutral hydrophobic compounds.
Finally, in a systematic study on the effect of coexisting herbicides on rejection (Plakas & Karabelas, 2008), a competition was identified for adsorption sites on the membrane surfaces between the different solutes present in the feed-waters. This phenomenon resulted in different rejection values, since herbicides were better rejected in single solute solutions than in mixed solute systems. This effect was particularly pronounced in the case of tight membranes (NF90, XLE), since the more porous membrane (NF270) showed an increased retention of the herbicides atrazine and isoproturon when treated together with prometryn or in triple-solute solutions (Table 5). A pore restriction effect, due to the larger prometryn molecule, could be responsible for this trend, which seems to positively influence the retention of the smaller molecules (Plakas & Karabelas, 2008).

Membrane	Herbicide	Single solute system	Double solute system			Triple solute system
			A	I	P	
NF270	Atrazine	78.9 (18.8)	-	73.2 (20.2)	86.1 (16.5)	81.2 (17.1)
	Isoproturon	73.1 (25.0)	63.8 (26.0)	-	85.0 (15.1)	82.4 (17.0)
	Prometryn	90.8 (23.7)	87.7 (27.5)	82.7 (33.6)	-	83.1 (32.5)
NF90	Atrazine	99.3 (21.1)	-	93.1 (19.2)	86.2 (30.5)	87.5 (26.8)
	Isoproturon	95.1 (25.6)	93.1 (23.1)	-	91.8 (25.3)	92.1 (23.2)
	Prometryn	99.8 (28.3)	96.6 (26.2)	96.8 (29.0)	-	96.3 (27.3)
XLE	Atrazine	97.6 (24.8)	-	88.2 (27.0)	94.9 (23.0)	90.1 (22.5)
	Isoproturon	96.6 (5.1)	83.2 (11.3)	-	84.1 (8.2)	87.0 (9.0)
	Prometryn	98.1 (31.2)	95.5 (29.5)	94.0 (32.4)	-	94.9 (31.3)

Table 5. Herbicide retention results (%) and percentage adsorption data (values in the brackets) in the case of single and multi-solute nanofiltration experiments; A, I and P designate solutions with Atrazine, Isoproturon and Prometryn, respectively (Plakas & Karabelas, 2008).

Pesticide polarity

One of the most important physicochemical criteria governing nanofiltration and reverse osmosis separation of trace organic compounds in aqueous solution is the "Polar Effect" of the solute molecule (Matsuura & Sourirajan, 1973). As outlined in paragraph 2.4, the passage of polar organic molecules to the permeate side is facilitated by the polar interactions with the membrane charge, which leads to a reduced solute rejection. Van der Bruggen et al. (1998) have successfully combined size exclusion and polarity effects to explain the retention of four pesticides. Specifically, the retention of the two phenyl-urea derivatives, diuron and isoproturon, was lower than the one measured for the two triazine compounds, atrazine and simazine (Van der Bruggen et al., 1998). Diuron and isoproturon are not smaller than the two triazines, but they have a higher dipole moment (a measure of polarity) which favors the sorption, and consequently the diffusion of these molecules into the membrane polymer. The effect of the dipole moment was also confirmed by comparing the retentions of the two polar herbicides with those measured for a series of non-polar carbohydrates. The filtration results showed that a greater dipole moment leads to a lower retention (Van der Bruggen et al., 1998). In general, it has been concluded that solute polarity is important for membranes with an average pore size that is larger than the size of compounds to be retained (Van der Bruggen et al., 1999, 2001; Košutić et al., 2002).

3.4 Effect of the feed water composition

Membrane filtration experiments with real or simulated raw waters (i.e. solutions containing salts, organic matter and pesticides) have shown that pesticide rejection can vary greatly, depending on the feed water composition. Specifically, pH, ionic strength, and the presence of organic matter are identified as having an influence on pesticide rejection. The respective literature results are discussed next.

Influence of water pH

The role of pH on pesticide rejection is related mainly to the changes taking place in the membrane surface structure and charge. It has been determined that pH has an effect upon the charge of a membrane due to the dissociation of functional groups. Zeta potential for most membranes has been observed in many studies to become increasingly more negative as the pH is increased and functional groups deprotonate (Childress & Elimelech, 1996; Deshmukh & Childress, 2001; Afonso et al., 2001). Moreover, pore enlargement or shrinkage can occur depending upon the electrostatic interactions between the dissociated functional groups of the membrane material (Freger et al., 2000). In a study performed by Berg et al. (1997) the rejection of uncharged organic compounds (atrazine, terbuthylazine) at pH 3 and 7 was relatively constant. However, higher pH values resulted in reduced rejection rates together with an increased permeate flux. This was attributed to the pore enlargement at higher pH values.

Experiments with the uncharged simazine molecule showed that rejection attained the highest value at pH 8, and consistently lower values at pH 4 and 11 (Zhang et al., 2004). These results were attributed to ion adsorption on the membrane surface; specifically, at higher pH, OH^- ions adsorption increased, resulting in an increase of the membrane charge. Polar components such as pesticides exhibit a reduced rejection with increasing membrane charge, because such molecules tend to preferentially orient themselves so that the dipole with a charge opposite to that of the membrane charge is the closest to the membrane surface. Consequently, this preferential orientation results in an increased attraction, an increased permeation and thus a lower rejection. At lower pH, the same effect might occur with H^+ ions (Zhang et al., 2004).

Finally, it was recently reported (Ahmad et al., 2008b) that increasing the solution pH led to enhanced atrazine and dimethoate rejection, but degraded the permeate flux performance for NF200, NF270 and DK membranes. However, the NF90 membrane exhibited relatively consistent performance in both rejection and permeate flux, regardless of the solution pH (Ahmad et al., 2008b).

Influence of solute concentration

Filtration experiments with atrazine and prometryn in different concentrations (10–700 μg/L) showed small variations in rejection by NF/ULPRO membranes (Plakas et al., 2006; Plakas & Karabelas, 2008). Specifically, the differences in retention values varied between 7 and 13%. This is in agreement with observations made by other researchers (Agbekodo et al., 1996; Van der Bruggen et al., 1998; Zhang et al., 2004; Ahmad et al., 2008a), in that herbicide concentration does not significantly affect their retention. The fact that the filtration of fluids with smaller feed concentrations led to a slight reduction of triazine retention (especially in the case of a ULPRO membrane) could be attributed to the amount of triazines adsorbed on the selected membranes; more specifically, the smaller triazine concentration may be associated with a slightly smaller adsorption, in comparison to the results obtained with greater feed concentrations, something that was more pronounced in the case of the less tight NF membrane (Plakas et al., 2008).

Influence of the ionic environment

A number of studies have shown that the retention of pesticides can be moderately influenced by the presence of dissolved salts in the feed solution due to the interactions taking place between the ions and the membrane surfaces. Specifically, it has been suggested (Yoon et al., 1998) that, at high ionic concentrations, there may be a reduction in the electrostatic forces inside the membrane (i.e. reduced repulsion) which may cause a reduction of the actual size of the pores, leading to a reduced membrane permeability; consequently, a better rejection of pesticides accompanied by a reduced water flux could be observed. Based on these considerations, an explanation can be also provided for the higher rejection of pesticides by nanofiltration membranes with ground water (Van der Bruggen et al., 1998), tap and/or river water (Zhang et al., 2004). It should be noted, however, that the presence of natural organic matter in the natural water samples employed may have also positively affected the rejection of pesticides (Zhang et al., 2004).

In an earlier study (Boussahel et al., 2002), the presence of divalent cations (calcium) in the feed solution appeared to exercise little influence on pesticide rejection, whereas rejection was found to be related to the membrane type. Specifically, an improvement in pesticide rejection by approx. 5% (in the presence of $CaCl_2$) and 10% (in the presence of $CaSO_4$) was reported for a NF200 membrane, while for the Desal DK membrane very little change was noted, i.e. a slight drop in the percent removal (5%) for DEA and simazine with $CaCl_2$ (Boussahel et al., 2002). These results are in agreement with those from a recent study (Plakas & Karabelas, 2008), where a moderate influence of calcium ions on herbicide retention was obtained; this influence, was either positive or negative depending on the membrane type. For example, the effect of calcium ions on pesticide removal by relatively dense and neutral NF/ULPRO membranes was found to be negative. This was not observed in the case of dense and negatively charged membranes which were not significantly influenced by the presence of calcium. On the other hand, the retention of pesticides by relatively porous NF membranes was found to increase with the presence of calcium ions, possibly due to the mechanism of pore blockage described earlier (Plakas & Karabelas, 2008).

In the case of elevated ionic strength, due to the presence of sodium chloride in the feed solution, rejection was reduced for all herbicides and membranes tested (Plakas & Karabelas, 2008). This was explained by the reduction of the hydrodynamic radius of herbicides in the presence of NaCl, especially of the hydrophobic triazines, with a likely contribution of concentration polarization on the membrane surface. Regarding the effect of herbicides on salt rejection, there was an increase observed in sodium chloride rejection only for the wide-pore NF membranes, something that was not observed in the case of calcium ion retention which remained constant. However, the calcium retention was reduced somewhat, by approximately 7% and 13% for the tight NF90 and XLE membranes, respectively. Furthermore, the presence of calcium ions had no influence on herbicide adsorption on all membranes tested, as also observed by previous researchers (Boussahel et al., 2002).

Pesticide retention in the presence of organic matter

A number of studies performed with either NF/RO membranes (Agbekodo et al., 1996; Berg et al., 1997; Devitt et al., 1998a; Boussahel et al., 2000; Zhang et al., 2004) or dialysis membranes (Devitt & Wiesner, 1998b; Dalton et al., 2005) have shown that the retention of pesticides is significantly influenced by the presence of natural organic matter (NOM) in water. This fact is of considerable importance since a large percentage of pesticide residues is present in surface and ground waters together with organic matter; i.e. humic and fulvic acids, polysaccharides, etc. (Kulikova & Perminova, 2002). In general, humic substances (HS) are a ubiquitous component of natural water systems that may function as an auxiliary phase to alter the speciation and transport behaviour of other xenobiotic compounds present in water (Wersaw, 1991). Thus, organic micropollutants, like pesticides, may exist either as free dissolved species or as a complex with HS.

A literature review on the effect of NOM on pesticide retention by membranes, suggests that there is a dependence on the type of NOM present in the water. NOM is composed of an extremely diverse group of compounds, including humic acids, carbohydrates, alcohols, amino acids, carboxylic acids, lignins, and pigments, whose origin greatly influences its character and behaviour. The majority of the published works agree on the fact that the retention of pesticides in membrane-based systems tends to increase in the presence of NOM (Agbekodo et al., 1996; Devitt et al., 1998a, 1998b; Zhang et al., 2004; Dalton et al., 2005), which is generally attributed to a variety of factors; e.g., the size, shape, and surface chemistry of compounds involved. On the other hand, the use by various researchers of NOM of different origin, and the inadequate information regarding their physicochemical properties (elemental analysis, functional groups), hinder the systematic comparison of experimental results as well as the correlation of the pesticide/NOM membrane retention with the characteristic properties of the organic matter naturally occurring in water.

To identify the variability introduced by the different properties of humic substances on pesticide rejection, Plakas & Karabelas (2009) performed systematic studies with well-characterized HS in order to improve the understanding of mechanisms of NOM–pesticide retention by membranes. Specifically, they used four different types of HS; i.e. three of them were typical water-born HS (humic acid, fulvic acid, and a mixture of NOM) whereas the fourth one was a HS surrogate (tannic acid). The results of this study show that the combined nanofiltration of triazines (atrazine, prometryn) and naturally occurring humic substances facilitates the formation of complexes with triazines which in turn enhance their removal by nanofiltration (Fig. 7). This complexation appeared to be related not to the characteristic acidity (phenolic, carboxylic) of the HS used, but rather to their molecular conformation (Plakas & Karabelas, 2009). More specifically, a preferential binding was observed between triazines and low molecular weight fractions of humic compounds (especially of fulvic acid and tannic acid), which resulted in higher retention values for the two triazines. Under all conditions, tannic acid exhibited the greatest effect on triazine retention, among the four standard HS compounds used, leading to an almost complete removal of the two triazines (95-100%) for all three membranes tested (Fig 7).

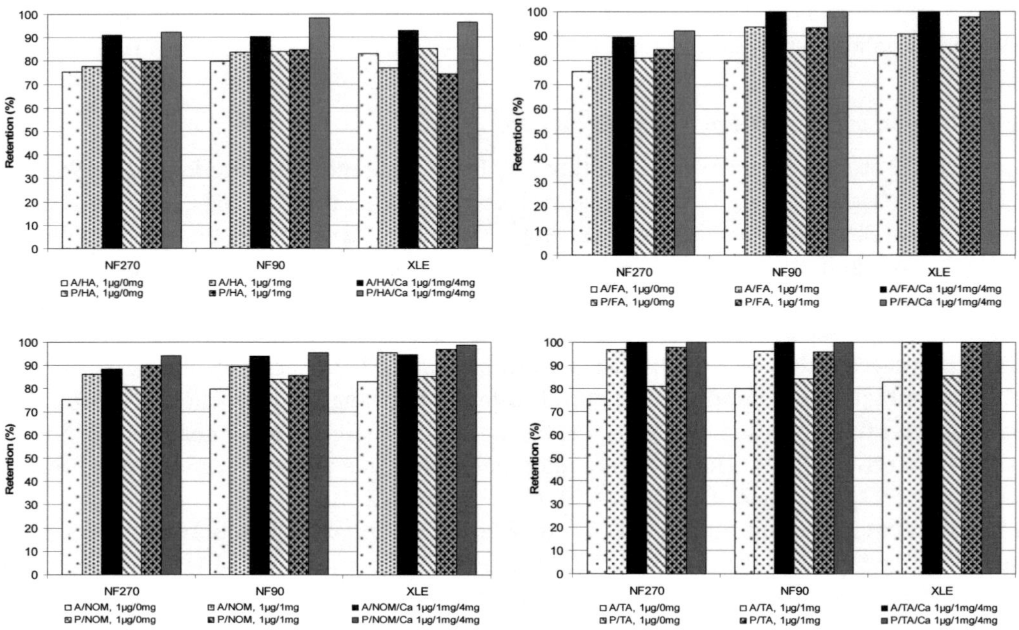

Fig. 7. Retention of atrazine (A) and prometryn (P) by three NF/ULPRO membranes in the absence or presence of humic substances (HA, FA, NOM, TA) and/or calcium ions (Plakas & Karabelas, 2009)

Moreover, triazine retention was found to increase with increasing HS concentration, to a degree depending on the type of HS; additionally, removal of triazines was improved in the presence of calcium which displayed a tendency to enhance the interaction between HS and triazines (Plakas & Karabelas, 2009). In parallel, it is noted that a number of studies with dialysis membranes (Devitt et al., 1998a, 1998b, Dalton et al., 2005) have reported reduced values of atrazine retention when divalent calcium is present together with naturally occurring organic matter, including the NOM surrogate, tannic acid. According to Devitt et al. (1998a, 1998b), this trend is due to the reduced association of atrazine and NOM, as a result of the occupation of interaction sites by calcium and/or the reduced access of atrazine to NOM sites due to changes in molecular conformation. However, gel permeation chromatography experiments (Plakas & Karabelas, 2009) have shown that this is not the case, since the presence of calcium had the tendency to increase the interaction of humic substances with triazine compounds. These conflicting results could be attributed to the different types of membranes and filtration techniques used. In particular, the use of cellulose ester membranes, as well as the experimentation on batch dialysis systems by Devitt et al. (1998a, 1998b), where concentration and osmotic pressure difference serve as the driving force for solute transport (absence of hydrodynamic forces), may justify the seemingly different calcium effect on triazine retention.

3.5 Effect of membrane fouling

The significant number of parameters affecting pesticide retention is indicative of the complicated interactions taking place, which can be further influenced by the changes occurring in membrane surface properties as a result of fouling. This is especially true in the case of the organic micropollutants (EDCs, PhACs, pesticides, etc), since their retention is determined by electrostatic, steric and hydrophobic/hydrophilic solute-membrane interactions, which can be modified due to foulants depositing on the membrane surface. The effect of fouling on organic micropollutant retention has been the subject of rather extensive research in the past decade (Ng & Elimelech, 2004; Xu et al., 2006; Plakas et al., 2006; Steinle-Darling et al., 2007; Agenson & Urase, 2007; Nghiem & Hawkes, 2007; Bellona et al., 2010; Nghiem & Coleman, 2008; Verliefde et al., 2009; Yangali-Quintanilla et al., 2009). Systematic investigations on the influence of colloidal and/or organic fouling on various trace organic species suggests that solute retention can be distinguished in two different cases, depending on the relative solute selectivities of the fouling layer and the membrane. First, if the membrane rejects solutes better than the deposited layer, hindered back diffusion of solutes (by the fouling layer) would cause solute accumulation near the membrane surface. This cake-enhanced concentration polarization results in greater concentration gradient across the membrane and, hence, a decrease in solute retention. Second, if solutes are rejected better by the deposited layer than the membrane, the fouling layer controls solute retention which tends to improve.

The literature review suggests that membrane fouling may significantly affect the retention of low MW organic compounds depending on the concentration and characteristics of the foulants, the membrane properties, and the chemical composition of feed water. Regarding pesticides, it has been shown (Plakas et al., 2006) that the differences in retention between fouled and virgin membranes are related to the diffusion capacity of herbicides across the membranes. When a rather loose humic layer is formed on the membrane surfaces, especially when membranes are fouled by humic substances alone, in the absence of calcium ions, herbicides retention can be reduced due to their increased diffusion through the membrane polymeric matrix, which is further facilitated by the cake-enhanced concentration polarization effect. In the case of rather dense fouling layers formed through HS-Ca complexation, herbicide retention may improve; indeed, these layers can serve as additional barriers which enhance the sieving effect, resulting in higher retention values (Plakas et al., 2006).

3.6 Influence of the operating parameters

Rejection of pesticides is also found to be influenced by operating parameters, such as the water flux and the feed-stream velocity in the cross-flow mode of filtration. In a study conducted by Chen et al. (2004) rejection of pesticides was shown to be dependent on operating flux and recovery. In particular, the highest percent rejection occurred at high flux and low recovery, whereas the lowest percent rejection took place at low flux and high recovery, which is in accord with the solution-diffusion theory (Chen et al., 2004). This finding is in agreement with the work performed by Ahmad et al. (2008a), where the retention of both dimethoate and atrazine was found to be better when the pressure was increased from 6 to 12×10^5 Pa (increased water flux).

It is interesting to note that in an early study (Chian et al., 1975), the effect of pressure on pesticide separation was negligible in the case of a high-desalting membrane. However, it was anticipated that rejection of the more polar molecules would increase somewhat with increasing pressure, especially for membranes exhibiting inferior rejection performance (Chian et al., 1975). Finally, in a pilot study (Duranceau et al., 1992), no effect on pesticide mass transfer was observed for varied feed-stream velocity, which was estimated to vary between 0.07 and 0.16m/s. This is in agreement with the crossflow experiments performed by the authors (paper in preparation) where the cross-flow velocity had a minimum effect on atrazine and prometryn rejection by a relatively porous NF membrane. It will be added that ongoing work in the authors Laboratory, shows that an increase in applied pressure results in a more pronounced increase in herbicides retention.

Finally, a cascade of NF stages was recently proposed (Caus et al., 2009) to attain high removal of organic pollutants, combined with low salt rejection; to achieve the latter, loose commercial nanofiltration membranes were selected (Desal51HL, N30F and NF270). Through modelling, it was shown that the separation could be significantly improved by a design involving cascade of NF membrane stages. Moreover, researchers have suggested the use of a Desal51HL membrane for an almost complete pesticide rejection combined with moderate salt passage (Caus et al., 2009).

3.7 Summary

By reviewing the literature, one is led to the conclusion that pesticides removal by nanofiltration and low-pressure reverse osmosis membranes is a complicated process in which several membrane and solute parameters, including feed water composition and process conditions play a role. In general, there is ample evidence that size exclusion (sieving) by the membrane pores is one of the main mechanisms determining the retention of pesticides; the pesticides molecular mass, in comparison to the MWCO of the membrane used, appears to be a very rough, albeit frequently convenient, criterion for assessing the effectiveness of the separation process. For the relatively small size uncharged pesticides, molecular mass in combination with the hydrophobic character of the molecules (commonly characterized by $\log K_{ow}$) seem to determine the retention. For instance, hydrophobic pesticides (with a large value of $\log K_{ow}$) are not well retained by nanofiltration membranes; this is attributed to the increased adsorption on the membrane surfaces that promotes their subsequent diffusion to the permeate side. For charged pesticides, both size exclusion and electrostatic interactions appear to control the degree of separation. In the case of polar pesticides, rejection may be reduced due to polar interactions with the charged membranes; this is especially true for membranes with an average pore size larger than the compounds to be retained. In general, pesticides characterized by increased affinity for the membrane tend to be rejected to a lesser extent than those of a similar size but with reduced tendency for adsorption on the membrane.

The aforementioned results can form the basis for recommending general rules for selecting membrane type for efficient separation of pesticides, taking also the composition of feed-water into account. In principle, a nonpolar membrane surface would be preferable for improved, overall, pesticides rejection. However, it should be recognized that the presently widely employed polyamide NF/RO membranes are characterized by surface hydrophilicity (desirable as it resists organic fouling) and by rather small negative charge. Regarding porosity, dense membranes are definitely preferable, for effective removal of even small pesticides molecules. However, membranes characterized by reduced porosity and polarity are associated with reduced flux, thus requiring increased operating pressure (and energy expenditure) to achieve a given clean water production rate.

Another aspect to be considered in purification of water from organic micro-pollutants, like pesticides, is membrane fouling. Systematic studies on the effect of organic fouling on pesticide rejection have shown that fouling alters the membrane surface properties and, as a consequence, rejection of pesticides can drastically change in comparison with virgin membranes. Therefore, it is of paramount importance in membrane applications to identify the type of foulants with potential to deposit on the membrane surface, in order to predict the influence of these deposits on membrane surface properties and thus on rejection. In this direction, an adequate characterisation of the membrane surface as well as of the composition of the feed water is necessary.

4. Current trends and R&D needs for removal of trace organic contaminants from potable water

Regarding the design and operation of modern water treatment processes, to remove toxic pollutants including pesticides, there are two major issues with very significant technological, economic and (above all) environmental and human health impact, that have to be successfully addressed by the scientific community : (a) Production of safe potable water. This target entails the design of effective, environmentally friendly and economically attractive processes capable of meeting the stringent drinking water standards, even in cases of feedwater with variable load of pollutants (including pesticides) of uncertain type and concentration. (b) Elimination or disposal of liquid and solid wastes from the water treatment process, after appropriate treatment to render them safe for humans and the environment; this problem is especially acute due to the high concentration of pollutants retained in the wastes. It is evident that development of *integrated processes*, successfully coping with the above problems should be pursued, and that R&D activities should support these efforts.

Considering the first issue, as discussed in this chapter, NF has emerged as a reliable operation that provides the basis for developing effective potable water treatment processes. However, in general NF may not be possible (and perhaps should not be assigned) to handle alone the water purification task. Indeed, NF has to be combined with other complementary operations, in the context of an effective integrated design. The main considerations and current trends regarding the design of such integrated processes, taking advantage of the NF attributes, should be stressed:

- NF alone can achieve three technical objectives, on the basis of its characteristics; (i) partial hardness removal (i.e. water conditioning) by reducing the concentration of Ca and Mg salts, (ii) practically total removal of NOM and of assorted colloidal species, with the unavoidable penalty of membrane fouling, (iii) removal of pesticides and of other toxic compounds, to a rather high degree depending on many factors.
- The currently favored approach of coping with pesticides and the multitude of toxic substances, at very small concentration, is to incorporate in an integrated process sequential operations (akin to successive "lines of defense"), ensuring adequate final removal of all these pollutants. The key role of NF in this scheme is to perform as best as possible, and at least to remove most of the toxic pollutants, so that a final purification can be achieved in one or two subsequent steps; e.g. by employing granular activated carbon. This approach affords significant advantages over the currently employed conventional treatment processes, which tend to rely mostly on activated carbon.

In view of the above considerations, it appears that priority should be mainly given to the following R&D areas:

- To maximize the rejection of pesticides (and of other micro-pollutants) by the NF membranes. Particular attention deserve the improved understanding of the physico-chemical interactions between pesticides (and other such species) and various types of NF and LPRO membranes, as well as the clarification of the interaction between common organic matter (humic and fulvic acids, polysaccharides, etc) and the micro-pollutants. As the latter cannot be avoided, it may have to be facilitated (possibly by adjusting conditions) to maximize pesticides removal.
- In connection with the above areas, further investigation of the role of membrane fouling layers on the adsorption and/or rejection of pesticides.
- Development of processes for pesticides degradation that may be combined with, and complement, NF for optimum overall performance. Typical cases currently studied include Advanced Oxidation Processes (AOP); photo-catalytic and electro-Fenton processes, belonging in this category, need further study as they may offer significant advantages in conjunction with NF.
- Design of novel integrated process schemes, including NF; e.g. a combination of NF and AOP with final activated carbon treatment, could be pursued for developing optimum solutions. Structural (flow-sheet) and parameter optimization of these processes is necessary. One of the design objectives of the integrated processes should be the minimization of liquid and solid wastes, thus reducing the load of the following waste treatment stage. It should be pointed out that, due to social and legislative pressure, major stake-holders in the water treatment sector are very concerned about this waste treatment problem, and are taking steps to address it at the R&D and demonstration levels [e.g. Bozkaya-Schrotter et al., 2009].

5. References

Afonso, MD., Hagmeyer, G. & Gimbel, R. (2001). Streaming potential measurements to assess the variation of nanofiltration membranes surface charge with the concentration of salt solutions. *Separation & Purification Technology*, Vol. 22-23, pp. 529–541, ISSN 1383-5866.

Agbekodo, KM.; Legube, B. & Dard, S. (1996). Atrazine and simazine removal mechanisms by nanofiltration: influence of natural organic matter concentration. *Water Research*, Vol. 30, pp. 2535-2542, ISSN 0043-1354.

Agenson, KO.; Oh, JI. & Urase, T. (2003). Retention of a wide variety of organic compounds by different nanofiltration/reverse osmosis membranes: controlling parameters of process. *Journal of Membrane Science*, Vol. 225, pp. 91-103, ISSN 0376-7388.

Agenson, KO. & Urase, T. (2007). Change in membrane performance due to organic fouling in nanofiltration (NF)/reverse osmosis (RO) applications. *Separation & Purification Technology*, Vol. 55, pp. 147-156, ISSN 1383-5866.

Ahmad, AL.; Tan, LS. & Shukor, SRA. (2008a). Dimethoate and atrazine retention from aqueous solution by nanofiltration membranes. *Journal of Hazardous Materials*, Vol. 151, pp. 71–77, ISSN 0304-3894.

Ahmad, AL.; Tan, LS. & Shukor, SRA. (2008b). The role of pH in nanofiltration of atrazine and dimethoate from aqueous solution. *Journal of Hazardous Materials*, Vol. 154, No. 1-3, pp. 633-638, ISSN 0304-3894.

Al-Amoudi, A. & Lovitt, RW. (2007). Fouling strategies and the cleaning system of NF membranes and factors affecting cleaning efficiency. *Journal of Membrane Science*, Vol. 303, pp. 4–28, ISSN 0376-7388.

Al-Amoudi, AS. (2010). Factors affecting natural organic matter (NOM) and scaling fouling in NF membranes: A review. *Desalination*, Vol. 259, pp. 1–10, ISSN 0011-9164.

Baier, JH.; Lykins Jr, BW.; Fronk, CA. & Kramer SJ. (1987). Using reverse osmosis to remove agricultural chemicals from ground water. *Journal of American Water Works Association*, Vol. 79, pp. 55-60, ISSN 0003-150X.

Becker, FF.; Janowsky, U.; Overath, H. & Stetter, D. (1989). Die Wirksamkeit von Umkehrosmosemembranen bei der Entfernung von Pestiziden. *GWF Wasser Abwasser*, Vol. 130, No. 9, pp. 425-431, ISSN 0016-3651.

Benítez, FJ.; Acero, JL.; Real, FJ. & García, C. (2009). Nanofiltration processes applied to the removal of phenyl-ureas in natural waters. *Journal of Hazardous Materials*, Vol. 165, pp. 714–723, ISSN 0304-3894.

Bellona, C.; Drewes, JE.; Xu, P. & Amy, G. (2004). Factors affecting the rejection of organic solutes during NF/RO treatment - a literature review. *Water Research*, Vol. 38, pp. 2795-2809, ISSN 0043-1354.

Bellona, C.; Marts, M. & Drewes, JE. (2008). The effect of organic membrane fouling on the properties and retention characteristics of nanofiltration membranes. *Separation & Purification Technology*, doi: 10.1016/j.seppur.2010.05.006.

Bellona, C.; Marts, M. & Drewes, JE. (2010). The effect of organic membrane fouling on the properties and retention characteristics of nanofiltration membranes. *Separation & Purification Technology*, Vol. 74, pp. 44-54, ISSN 1383-5866.

Berg, P.; Hagmeyer, G. & Gimbel, R. (1997). Removal of pesticides and other micropollutants by nanofiltration. *Desalination*, Vol. 113, pp. 205-208, ISSN 0011-9164.

Bhattacharya, A.; Ray, P.; Brahmbhatt, H.; Vyas, KN.; Joshi, SV.; Devmurari, CV. & Trivedi, JJ. (2006). Pesticides removal performance by low-pressure reverse osmosis membranes. *Journal of Applied Polymer Science*, Vol. 102, pp. 3575–3579, ISSN 0021-8995.

Bonné, PAC.; Beerendonk, EF.; van der Hoek, JP. & Hofman, JAMH. (2000). Retention of herbicides and pesticides in relation to aging of RO membranes. *Desalination*, Vol. 132, pp. 189-193, ISSN 0011-9164.

Boussahel, R.; Bouland, S.; Moussaoui, KM. & Montiel, A. (2000). Removal of pesticide residues in water using the nanofiltration process. *Desalination*, Vol. 132, pp. 205-209, ISSN 0011-9164.

Boussahel, R.; Montiel, A. & Baudu, M. (2002). Effects of organic and inorganic matter on pesticide rejection by nanofiltration. *Desalination*, Vol. 145, pp. 109-114, ISSN 0011-9164.

Bozkaya-Schrotter, B.; Daines, C.; Brunel, A.; Schrotter, JC. & Breant, B. (2009). Treatment of nanofiltration membrane concentrates : Organic micro-pollutant and NOM removal. *Desalination and Water Treatment*, Vol. 9, pp. 36-42, ISSN 1944-3994.

Caus, A.; Vanderhaegen, S.; Braeken, L. & Van der Bruggen, B. (2009). Integrated nanofiltration cascades with low salt rejection for complete removal of pesticides in drinking water production. *Desalination*, Vol. 241, pp. 111-117, ISSN 0011-9164.

Causserand, C.; Aimar, P.; Cravendi, JP. & Singlande, E. (2005). Dichloroaniline retention by nanofiltration membranes. *Water Research*, Vol. 39, pp. 1594-1600, ISSN 0043-1354.

Chen, SS.; Taylor, JS; Mulford, LA. & Norris, CD. (2004). Influences of molecular weight, molecular size, flux, and recovery for aromatic pesticide removal by nanofiltration membranes. *Desalination*, Vol. 160, pp. 103-111, ISSN 0011-9164.

Chian, ESK.; Bruce, WN. & Fang, HHP. (1975). Removal of pesticides by reverse osmosis. *Environmental Science & Technology*, Vol. 9, No. 1, pp. 52-59, ISSN 0013-936X.

Childress, AE. & Elimelech, M. (1996). Effect of solution chemistry on the surface charge of polymeric reverse osmosis and nanofiltration membranes. *Journal of Membrane Science*, Vol. 119, pp. 253-268, ISSN 0376-7388.

Comerton, AM.; Andrews, RC.; Bagley, DM. & Yang P. (2007). Membrane adsorption of endocrine disrupting compounds and pharmaceutically active compounds. *Journal of Membrane Science*, Vol. 303, pp. 267–277, ISSN 0376-7388.

Comerton, AM.; Andrews, RC.; Bagley, DM. & Hao, C. (2008). The rejection of endocrine disrupting and pharmaceutically active compounds by NF and RO membranes as a function of compound and water matrix properties. *Journal of Membrane Science*, Vol. 313, pp. 323–335, ISSN 0376-7388.

Cyna, B.; Chagneau, G.; Bablon, G. & Tanghe, N. (2002). Two years of nanofiltration at the Méry-sur-Oise plant, France. *Desalination*, Vol. 147, pp. 69-75, ISSN 0011-9164.

Dalton, SK.; Brant, JA. & Wiesner, MR. (2005). Chemical interactions between dissolved organic matter and low-molecular weight organic compounds: impacts on membrane separation. *Journal of Membrane Science*, Vol. 266, pp. 30–39, ISSN 0376-7388.

Deshmukh, SS. & Childress, AE. (2001). Zeta potential of commercial RO membranes: influence of source water type and chemistry. *Desalination*, Vol. 140, pp. 87–95, ISSN 0011-9164.

Devitt, EC.; Ducellier, F.; Côté, P. & Wiesner, MR. (1998a). Effects of natural organic matter and the raw water matrix on the rejection of atrazine by pressure-driven membranes. *Water Research*, Vol. 32, pp. 2563-2568, ISSN 0043-1354.

Devitt, EC. & Wiesner, MR. (1998b). Dialysis investigations of atrazine–organic matter interactions and the role of a divalent metal. *Environmental Science & Technology*, Vol. 32, pp. 232–237, ISSN 0013-936X.

Dow Water & Process Solutions. (2010). Library of materials on water chemistry and pretreatment requirements and methods. Available from: http://www.dow.com/liquidseps/service/lm_pretreat.htm, Accessed 2010 August 10.

Duranceau, SJ.; Taylor, JS. & Mulford, LA. (1992). SOC removal in a membrane softening process. *Journal of American Water Works Association*, Vol. 84, pp. 68-78, ISSN 0003-150X.

Filteau, G. & Moss, P. (1997). Ultra-low pressure RO membranes: an analysis of performance and cost. *Desalination*, Vol. 113, pp. 147-152, ISSN 0011-9164.

Freger, V.; Arnot, TC. & Howell, JA. (2000). Separation of concentrated organic/inorganic salt mixtures by nanofiltration. *Journal of Membrane Science*, Vol. 178, pp. 185–193, ISSN 0376-7388.

Hindin, H.; Bennett, PJ. & Narayanan, SS. (1969). Organic compounds removed by reverse osmosis. *Water & Sewage Works*, Vol. 116, pp. 466-470, ISSN 0043-1125.

Hofman, JAMH., Beerendonk, EF.; Folmer, HC. & Kruithof, JC. (1997). Removal of pesticides and other micropollutants with cellulose-acetate, polyamide and ultra-low pressure reverse osmosis membranes. *Desalination*, Vol. 113, pp. 209-214, ISSN 0011-9164.

Hong, SK. & Elimelech, M. (1997). Chemical and physical aspects of natural organic matter (NOM) fouling of nanofiltration membranes. *Journal of Membrane Science*, Vol. 132, No. 2, pp. 159–181, ISSN 0376-7388.

Kamp, PC.; Kruithof, JC. & Folmer, HC. (2000). UF/RO treatment plant Heemskerk: from challenge to full scale application. *Desalination*, Vol. 131, pp. 27-35, ISSN 0011-9164.

Kimura K.; Amy, G.; Drewes, JE.; Heberer, T.; Kim, TU. &. Watanabe, Y. (2003a). Rejection of organic micropollutants (disinfection by-products, endocrine disrupting compounds, and pharmaceutically active compounds) by NF/RO membranes. *Journal of Membrane Science*, Vol. 227, pp. 113-121, ISSN 0376-7388.

Kimura, K.; Amy, G.; Drewes, J. & Watanabe, Y. (2003b). Adsorption of hydrophobic compounds onto NF/RO membranes: an artifact leading to overestimation of rejection. *Journal of Membrane Science*, Vol. 221, pp. 89-101, ISSN 0376-7388.

Kimura, K.; Toshima, S.; Amy, G. & Watanabe, Y. (2004). Rejection of neutral endocrine disrupting compounds (EDCs) and pharmaceutical active compounds (PhACs) by RO membranes. *Journal of Membrane Science*, Vol. 245, pp. 71–78, ISSN 0376-7388.

Kiso, Y.; Nishimura, Y.; Kitao, T. & Nishimura, K. (2000). Rejection properties of non-phenylic pesticides with nanofiltration membranes. *Journal of Membrane Science*, Vol. 171, pp. 229-237, ISSN 0376-7388.

Kiso, Y.; Sugiura, Y.; Kitao, T. & Nishimura, K. (2001a). Effects of hydrophobicity and molecular size on rejection of aromatic pesticides with nanofiltration membranes. *Journal of Membrane Science*, Vol. 192, pp. 1-10, ISSN 0376-7388.

Kiso, Y.; Kon, T.; Kitao, T. & Nishimura, K. (2001b). Rejection properties of alkyl phthalates with nanofiltration membranes. *Journal of Membrane Science*, Vol. 182, pp. 205–214, ISSN 0376-7388.

Kiso, Y.; Mizuno, A.; Othman, RAA.; Jung, YJ.; Kumano, A. & Ariji, A. (2002). Rejection properties of pesticides with a hollow fiber NF membrane (HNF-1). *Desalination*, Vol. 143, pp. 147-157, ISSN 0011-9164.

Košutić, K. & Kunst, B. (2002). Removal of organics from aqueous solutions by commercial RO and NF membranes of characterized porosities. *Desalination*, Vol. 142, pp. 47-56, ISSN 0011-9164.

Košutić, K.; Furač, L.; Sipos, L. & Kunst, B. (2005). Removal of arsenic and pesticides from drinking water by nanofiltration membranes. *Separation & Purification Technology*, Vol. 42, pp. 137-144, ISSN 1383-5866.

Košutić, K.; Dolar, D. & Kunst, B. (2006). On experimental parameters characterizing the reverse osmosis and nanofiltration membranes' active layer. *Journal of Membrane Science*, Vol. 282, pp. 109–114, ISSN 0376-7388.

Kulikova, NA. & Perminova, IV. (2002). Binding of atrazine to humic substances from soil, peat, and coal related to their structure. *Environmental Science & Technology*, Vol. 36, pp. 3720–3724, ISSN 0013-936X.

Lee, S.; Park, G.; Amy, G.; Hong, SK.; Moon, SH.; Lee, DH. & Cho, J. (2002). Determination of membrane pore size distribution using the fractional rejection of nonionic and charged macromolecules. *Journal of Membrane Science*, Vol. 201, pp. 191–201, ISSN 0376-7388.

Loos, R.; Gawlik, BM.; Locoro, G.; Rimaviciute, E.; Contini, S. & Bidoglio, G. (2009). EU-wide survey of polar organic persistent pollutants in European river waters. *Environmental Pollution*, Vol. 157, pp. 561–568.

Martin, JD.; Gilliom, RJ. & Schertz, TL. (1999). *Summary and Evaluation of Pesticides in Field Blanks Collected for the National Water-Quality Assessment Program 1992–1995*, National Water-Quality Assessment Program, Open-File Report 98-412, Indianapolis, IN.

Matsuura, T. & Sourirajan, S. (1973). Reverse osmosis separation of hydrocarbons in aqueous solutions using porous cellulose acetate membranes. *Journal of Applied Polymer Science*, Vol. 17, No. 12, pp. 3683-3708, ISSN 0021-8995.

McKinlay, R.; Plant, JA.; Bell, JNB. & Voulvoulis, N. (2008). Endocrine disrupting pesticides: Implications for risk assessment. *Environment International*, Vol. 34, No. 2, pp. 168–183, ISSN 0160-4120.

Mohammad, AW. & Ali, N. (2002). Understanding the steric and charge contributions in NF membranes using increasing MWCO polyamide membranes. *Desalination*, Vol. 147, pp. 205-212, ISSN 0011-9164.

Mulder, M. (1998). *Basic Principles of Membrane Technology*, 2nd edition, Kluwer Academic Publishers, ISBN 0-7923-4247-X, Dordrecth, The Netherlands.

Ng, HY. & Elimelech, M. (2004). Influence of colloidal fouling on retention of trace organic contaminants by reverse osmosis. *Journal of Membrane Science*, Vol. 244, pp. 215-226, ISSN 0376-7388.

Nghiem, LD. & Hawkes, S. (2007). Effects of membrane fouling on the nanofiltration of pharmaceutically active compounds (PhACs): Mechanisms and role of membrane pore size. *Separation & Purification Technology*, Vol. 57, pp. 176-184, ISSN 1383-5866.

Nghiem, DL. & Shäfer, AI. (2002). Adsorption and transport of trace contaminant estrone in NF/RO membranes. *Environmental Engineering Science*, Vol. 19, No. 6, pp. 441-451, ISSN 1092-8758.

Nghiem, LD. & Shäfer, AI. (2005). Trace contaminant removal with nanofiltration, In: *Nanofiltration, Principles and Applications*, Schäfer, AI.; Fane, AG. & Waite, TD. (Eds), pp. 479-520, Elsevier Advanced Technology, ISBN 1-85617-405-0, Oxford, UK.

Nghiem, LD. & Coleman, PJ. (2008). NF/RO filtration of the hydrophobic ionogenic compound triclosan: Transport mechanisms and the influence of membrane fouling. *Separation & Purification Technology*, Vol. 62, pp. 711-718, ISSN 1383-5866.

Novotny, V. (1999). Diffuse pollution from agriculture—a worldwide outlook. *Water Science & Technology*, Vol. 39, No. 3, pp. 1-13, ISSN 0273-1223.

Pang, W.; Gao, N. & Xia, S. (2010). Removal of DDT in drinking water using nanofiltration process. *Desalination*, Vol. 250, pp. 553–556, ISSN 0011-9164.

Plakas, KV.; Karabelas, AJ.; Wintgens, T. & Melin, T. (2006). A study of selected herbicides retention by nanofiltration membranes—The role of organic fouling. *Journal of Membrane Science*, Vol. 284, pp. 291–300, ISSN 0376-7388.

Plakas, KV. & Karabelas, AJ. (2008). Membrane retention of herbicides from single and multi-solute media: The effect of ionic environment. *Journal of Membrane Science*, Vol. 320, pp. 325–334, ISSN 0376-7388.

Plakas, KV. & Karabelas, AJ. (2009). Triazine retention by nanofiltration in the presence of organic matter: The role of humic substance characteristics. *Journal of Membrane Science*, Vol. 336, pp. 86–100, ISSN 0376-7388.

Rautenbach, R. & Albrecht, R. (1989). *Membrane Processes*, John Wiley & Sons, ISBN 0-471-91110-0, New York.

Redondo, JA. & Lanari, F. (1997). Membrane selection and design considerations for meeting European potable water requirements based on different feedwater conditions. *Desalination*, Vol. 113, pp. 309-323, ISSN 0011-9164.

Redondo, JA. & Lomax, I. (2001). Y2K generation FILMTEC RO membranes combined with new pretreatment techniques to treat raw water with high fouling potential: summary of experience, *Desalination*, Vol. 136, pp. 287-306, ISSN 0011-9164.

Sablani, SS.; Goosen, MFA.; Al-Belushi, R. & Wilf, M. (2001). Concentration polarization in ultrafiltration and reverse osmosis: a critical review. *Desalination*, Vol. 141, pp. 269-289, ISSN 0011-9164.

Sanborn, M.; Cole, D.; Kerr, K.; Vakil, C.; Sanin, KH. & Bassil, K. (2004). *Systematic review of pesticide human health effects*, The Ontario College of Family Physicians, Toronto, Ontario, Canada. Available from: http://www.ocfp.on.ca/local/files/Communications/Current%20Issues/Pesticides/Final%20Paper%2023APR2004.pdf Accessed 2010 August 10.

Sarkar, B.; Venkateshwarlu, N.; Nageswara Rao, R.; Bhattacharjee, C. & Kale, V. (2007). Potable water production from pesticide contaminated surface water - A membrane based approach. *Desalination*, Vol. 204, pp. 368-373, ISSN 0011-9164.

Scott, K. (1998). *Handbook of industrial membranes*, 2nd edition, Elsevier Advanced Technology, ISBN 1-85617-233-3, Oxford, UK.

Singh, R. (2006). *Hybrid Membrane Systems for Water Purification: Technology, Systems Design and Operations*, Elsevier Science & Technology Books, ISBN 1856174425, Oxford, UK.

Skinner, JA.; Lewis, KA.; Bardon, KS.; Tucker, P.; Catt, JA. & Chambers, BJ. (1997). An overview of the environmental impact of agriculture in the UK. *Journal of Environmental Management,* Vol. 50, No. 2, pp.111–128, ISSN 0301-4797.

Steinle-Darling, E.; Zedda, M.; Plumlee, MH.; Ridgway, HF. & Reinhard, M. (2007). Evaluating the impacts of membrane type, coating, fouling, chemical properties and water chemistry on reverse osmosis retention of seven nitrosoalklyamines, including NDMA. *Water Research*, Vol. 41, pp. 3959-3967, ISSN 0043-1354.

Suratt, WB.; Andrews, DR.; Pujals, VJ. & Richards, SA. (2000). Design considerations for major membrane treatment facility for groundwater. *Desalination*, Vol. 131, pp. 37-46, ISSN 0011-9164.

Tanninen, J.; Kamppinen, L. & Nyström, M. (2005). Pretreatment and hybrid processes, In: *Nanofiltration, Principles and Applications*, Schäfer, AI.; Fane, AG. & Waite, TD. (Eds), pp. 481, Elsevier Advanced Technology, ISBN 1-85617-405-0, Oxford, UK.

Taylor, JS. & Jacobs, EP. (1996). Reverse Osmosis and Nanofiltration, In: *Water Treatment Membrane Processes,* American Water Works Association (AWWA) Research Foundation; Lyonnaise des Eaux.; Water Research Commission of South Africa (Eds), pp. 9.1-9.70, McGraw Hill, ISBN 0-07-001559-7, New York.

Van der Bruggen, B.; Schaep, J.; Maes, W.; Wilms, D. & Vandecasteele, C. (1998). Nanofiltration as a treatment method for the removal of pesticides from ground water. *Desalination*, Vol. 117, pp. 139-147, ISSN 0011-9164.

Van der Bruggen, B.; Schaep, J.; Wilms, D. & Vandecasteele, C. (1999). Influence of molecular size, polarity and charge on the retention of organic molecules by nanofiltration. *Journal of Membrane Science*, Vol. 156, pp. 29-41, ISSN 0376-7388.

Van der Bruggen, B.; Everaert, K.; Wilms, D. & Vandecasteele, C. (2001). Application of nanofiltration for removal of pesticides, nitrate and hardness from ground water: rejection properties and economic evaluation. *Journal of Membrane Science*, Vol. 193, pp. 239-248, ISSN 0376-7388.

Ventresque, C.; Gisclon, V.; Bablon, G. & Chagneau, G. (2000). An outstanding feat of modern technology: the Mery-sur-Oise nanofiltration treatment plant (340,000 m^3/d). *Desalination*, Vol. 131, pp. 1-16, ISSN 0011-9164.

Verliefde, ARD.; Cornelissen, ER.; Heijman, SGJ.; Hoek, EMV.; Amy, GL.; Van der Bruggen, B. & van Dijk, JC. (2009a). Influence of solute−membrane affinity on rejection of uncharged organic solutes by nanofiltration membranes. *Environmental Science & Technology*, Vol. 43, No. 7, pp 2400–2406, ISSN 0013-936X.

Verliefde, ARD.; Cornelissen, ER.; Heijman, SGJ.; Petrinic, I.; Luxbacher, T.; Amy, GL.; Van der Bruggen, B. & van Dijk, JC. (2009b). Influence of membrane fouling by (pretreated) surface water on retention of pharmaceutically active compounds (PhACs) by nanofiltration membranes. *Journal of Membrane Science*, Vol. 330, pp. 90-103, ISSN 0376-7388.

Wang, Y.; Shu, L.; Jegatheesan, V. & Gao, B. (2010). Removal and adsorption of diuron through nanofiltration membrane: The effects of ionic environment and operating pressures. *Separation and Purification Technology*, Vol. 74, pp. 236–241, ISSN 1383-5866.

Wershaw, RL. (1991). The importance of humic substance–mineral particle complexes in the modeling of contaminant transport in sediment–water systems, In: *Organic Substances and Sediments in Water: Humics and Soils*, Baker, RA. (Ed.), pp. 23–34, Lewis Publishers, ISBN 0-87371-342-7, Chelsea, Michigan, USA.

Wittmann, E.; Côté, P.; Medici, C.; Leech, J. & Turner, AG. (1998). Treatment of hard borehole water containing low levels of pesticide by nanofiltration. *Desalination*, Vol. 119, pp. 347-352, ISSN 0011-9164.

Xu, Y. & Lebrun, RE. (1999). Investigation of the solute separation by charged nanofiltration membrane: effect of pH, ionic strength and solute type. *Journal of Membrane Science*, Vol. 158, pp. 93–104, ISSN 0376-7388.

Xu, P.; Drewes, JE.; Kim, TU.; Bellona, C. & Amy, G. (2006). Effect of membrane fouling on transport of organic contaminants in NF/RO membrane applications. *Journal of Membrane Science*, Vol. 279, pp. 165-175, ISSN 0376-7388.

Yangali-Quintanilla, V.; Sadmani, A.; McConville, M.; Kennedy, M. & Amy, G. (2009). Retention of pharmaceutically active compounds and endocrine disrupting compounds by clean and fouled nanofiltration membranes. *Water Research*, Vol. 43, pp. 2349-2362, ISSN 0043-1354.

Yiantsios, SG.; Sioutopoulos, DC. & Karabelas, AJ. (2005). Colloidal fouling of RO membranes - An overview of key issues and efforts to develop improved prediction techniques. *Desalination* , Vol. 183, pp 257-272, ISSN 0011-9164.

Yoon, SH.; Lee, CH.; Kimand, KJ. & Fane, AG. (1998). Effect of calcium ion on the fouling of nanofilter by humic acid in drinking water production. *Water Research*, Volume 32, No. 7, pp. 2180-2186, ISSN 0043-1354.

Zhang, Y.; Van der Bruggen, B.; Chen, GX.; Braeken, L. & Vandecasteele, C. (2004). Removal of pesticides by nanofiltration: effect of the water matrix. *Separation & Purification Technology*, Vol. 38, pp. 163-172, ISSN 1383-5866.

Appendix

Membrane	Specifications	Remarks	Pesticides	Retention (%)	Reference
YC 05	Amicon MWCO 500Da	Lab scale (dead-end)	Atrazine	~10	Devitt et al., 1998a
HR95PP	Dow Filmtec	Lab scale (crossflow)	Atrazine	99.0	Košutić et al., 2002
			MCPA	93.6	
			Propham	96.8	
			Triazimefon	82.9	
NFc	Dow Filmtec	Lab scale (crossflow)	Atrazine	80-85	Košutić et al., 2005
			Diazinon	86-94	
			Dichlorvos	56-62	
			Triadimefon	63-67	
NF45	Dow Filmtec MWCO 300Da	Lab scale (dead-end)	Atrazine	~31	Devitt et al., 1998a
		Lab scale (crossflow)	Atrazine	91.6-91.8	Van der Bruggen et al., 1998
			Diuron	59.4	
			Isoproturon	81.0	
			Simazine	84.8-85.9	
		Lab scale (crossflow)	Atrazine	87.0	Van der Bruggen et al., 2001
			Diuron	51.0	
			Isoproturon	75.0	
			Simazine	64.5	
NF70	Dow Filmtec MWCO 200-300Da	Pilot and industrial scale	Atrazine	50-90	Agbekodo et al., 1996
			Simazine	50-100 (Dissolved organic carbon present: 0.4-3.6 mg/L)	
		Lab scale (dead-end)	Atrazine	~65	Devitt et al., 1998a
		Lab scale (crossflow)	Atrazine	89.9-92.0	Van der Bruggen et al., 1998
			Diuron	85.9	
			Isoproturon	90.3	
			Simazine	88.5-89.2	
		Lab scale (crossflow)	Atrazine	93.5	Van der Bruggen et al., 2001
			Diuron	92.0	
			Isoproturon	90.0	
			Simazine	90.1	

Table A. Rejection characteristics of pesticides by commercially available NF/RO membranes (alphabetical listing of membrane manufacturers).

Membrane	Specifications	Remarks	Pesticides	Retention (%)	Reference
NF70	Dow Filmtec MWCO 200-300Da	Pilot scale (retention for two different water recoveries: 50 and 15%)	Atrazine Bentazone Cyanazine Diuron DNOC Mecoprop Metamitron Metribuzin Pirimicarb Simazine Vinclozolin	86.1/93.5 100/100 92.2/93.6 50.1/71.4 60.8/87.2 93.0/100 -/53.4 87.5/93.7 100/100 71.6/86.4 100/100	Chen et al., 2004
NF90	Dow Filmtec MWCO 200Da	Lab scale (dead-end)	Atrazine Prometryn Isoproturon	86.2-99.3 96.3-99.8 91.8-95.1 in single or multi-solute solutions	Plakas et al., 2008
		Lab scale (dead-end)	Atrazine Dimethoate	>95 ~90	Ahmad et al., 2008a
NF200	Dow Filmtec MWCO 300Da	Lab scale (dead-end)	Atrazine	~39	Devitt et al., 1998a
		Industrial scale	Atrazine Chlorotoluron Simazine	<<0.1μg/L permeate concentration	Wittmann et al., 1998
		Pilot scale	Atrazine Cyanazine DEA Diuron Isoproturon Simazine	~82 ~81 ~70 ~45 ~75 ~70	Boussahel et al., 2000, 2002
		Lab scale (dead-end)	Atrazine Prometryn Isoproturon	83.3 97.0 82.0	Plakas et al., 2006
		Lab scale (dead-end)	Atrazine Dimethoate	75-78 ~55	Ahmad et al., 2008
NF270	Dow Filmtec MWCO 200-400Da	Lab scale (crossflow)	Atrazine Diazinon Dichlorvos Triadimefon	81-85 90-93 ~40 >99.0	Košutić et al., 2005
		Lab scale (dead-end)	Atrazine Prometryn Isoproturon	73.2-86.1 82.7-90.8 63.8-85.0 in single or multi-solute solutions	Plakas et al., 2008

Table A. Continued

Membrane	Specifications	Remarks	Pesticides	Retention (%)	Reference
NF270	Dow Filmtec MWCO 200-400Da	Lab scale (dead-end)	Atrazine Dimethoate	65-70 25-35	Ahmad et al., 2008
		Lab scale (crossflow)	Alachlor Atraton DEET Metolachlor	13.4±11.0 11.6±1.8 11.5±2.2 21.7±11.3	Comerton et al., 2008
TFC-8821ULP	Fluid Systems Co.	Lab scale (crossflow)	Atrazine MCPA Propham Triazimefon	89.6 89.4 89.8 78.5	Košutić et al., 2002
BQ-01	GE Water Technol. (Osmonics)	Lab and pilot scale	Atrazine Diuron Melazachlorine Simazine Terbutylazine	~50 ~68 ~35 ~20 ~45	Berg et al., 1997
CK	GE Water Technol. MWCO 200Da	Lab scale (crossflow)	Dichloroaniline	<25	Causserand et al., 2005
Desal 5 DK	GE Water Technol. MWCO 150-300Da	Lab and pilot scale	Atrazine Diuron Melazachlorine Simazine Terbutylazine	~47 <10 ~73 ~35 ~53	Berg et al., 1997
		Pilot scale	Atrazine Cyanazine DEA Diuron Isoproturon Simazine	>95 >95 >95 ~75 ~95 ~95	Boussahel et al., 2000, 2002
		Lab scale (dead-end)	Dichloroaniline	60-95	Causserand et al., 2005
		Lab scale (dead-end)	Atrazine Dimethoate	75-82 62-75	Ahmad et al., 2008
Desal 5 DL	GE Water Technol. MWCO 150-300Da	Lab scale (crossflow)	Atrazine Simazine	~58 ~45	Zhang et al., 2004
Desal 51HL	GE Water Technol. MWCO 150-300Da	Lab scale (crossflow)	Atrazine Simazine	~71 ~70	Zhang et al., 2004

Table A. Continued

Membrane	Specifications	Remarks	Pesticides	Retention (%)	Reference
NF-CA 50	Hoechst	Lab and pilot scale	Atrazine	<10	Berg et al., 1997
			Diuron	<10	
			Melazachlorine	~20	
			Simazine	<10	
			Terbutylazine	~15	
CPA2	Hydranautics	Lab scale (crossflow)	Atrazine	95.9	Košutić et al., 2005
			Dichlorvos	94.7	
			Triadimefon	78.3	
		Lab scale (crossflow)	Atrazine	88.9	Košutić et al., 2002
			MCPA	82.3	
			Propham	80.7	
PVD1	Hydranautics	Lab and pilot scale	Atrazine	~89	Berg et al., 1997
			Diuron	~83	
			Melazachlorine	>95	
			Simazine	>90	
			Terbutylazine	>95	
NTR-7250	Nitto Denko	Lab and pilot scale	Atrazine	>95	Berg et al., 1997
			Diuron	~67	
			Melazachlorine	>95	
			Simazine	>90	
			Terbutylazine	>95	
		Lab scale (dead-end)	Anilazine	72.8	Kiso et al., 2000
			Atrazine	68.4	
			Chlorpyrifos	>99.95	
			Diazinon	95.1	
			Dichlorvos	46.2	
			Imidacloprid	54.6	
			Isoprothiolane	93.7	
			Malathion	88.1	
			Molinate	60.7	
			Pyridine	5.52	
			Simazine	59.8	
			Simetryn	57.6	
			Thiram	56.4	
			2,3,5-Trichloropyridine	88.9	

Table A. Continued

Membrane	Specifications	Remarks	Pesticides	Retention (%)	Reference
		Lab scale (dead-end)	Carbaryl (NAC)	40.3	Kiso et al., 2001a
			Chloroneb	53.3	
			Chlorothalonil (TPN)	70.5	
			Esprocarb	99.6	
			Fenobucarb (BPMC)	79.4	
			Isoxathion	99.8	
			Mefenacet	94.9	
			Methyldymron	95.9	
			Propiconazole	97.6	
			Propyzamide	81.8	
			Tricyclazole	26.5	
NTR-7450	Nitto Denko MWCO 600-800Da	Lab scale (crossflow)	Atrazine	19.2-19.8	Van der Bruggen et al., 1998
			Diuron	2.8	
			Isoproturon	15.5	
			Simazine	14.6-15.5	
		Lab scale (dead-end)	Anilazine	29.3	Kiso et al., 2000
			Atrazine	14.9	
			Chlorpyrifos	99.32	
			Diazinon	44.8	
			Dichlorvos (DDVP)	13.0	
			Imidacloprid	3.70	
			Isoprothiolane	36.3	
			Malathion	42.0	
			Molinate	20.4	
			Simazine	9.15	
			Simetryn	6.95	
			Thiram	18.7	
			2,3,5-Trichloropyridine	96.5	
		Lab scale (dead-end)	Carbaryl (NAC)	23.2	Kiso et al., 2001a
			Chloroneb	98.6	
			Chlorothalonil (TPN)	69.7	
			Esprocarb	98.7	
			Fenobucarb (BPMC)	14.6	
			Isoxathion	99.6	
			Mefenacet	90.0	
			Methyldymron	32.9	
			Propiconazole	72.4	
			Propyzamide	16.9	
			Tricyclazole	1.7	

Table A. Continued

Membrane	Specifications	Remarks	Pesticides	Retention (%)	Reference
		Lab and pilot scale	Atrazine	~53	Berg et al., 1997
			Diuron	~25	
			Melazachlorine	~73	
			Simazine	~45	
			Terbutylazine	~58	
CE 100	Spectrum MWCO 100Da	Lab scale (dead-end)	Atrazine	~48	Devitt et al., 1998a
CE 500	Spectrum MWCO 500Da	Lab scale (dead-end)	Atrazine	~13	Devitt et al., 1998a
NTC-60	Toray	Lab and pilot scale	Atrazine	~90	Berg et al., 1997
			Diuron	~58	
			Melazachlorine	~90	
			Simazine	~85	
			Terbutylazine	~93	
NTR-729 HF	Nitto Denko	Lab scale (dead-end)	Carbaryl (NAC)	92.4	Kiso et al., 2001a
			Chloroneb	93.9	
			Chlorothalonil(TPN)	96.1	
			Esprocarb	99.94	
			Fenobucarb (BPMC)	94.8	
			Isoxathion	99.84	
			Mefenacet	99.1	
NTR-729 HF	Nitto Denko	Lab scale (dead-end)	Methyldymron	98.4	Kiso et al., 2001a
			Propiconazole	96.9	
			Propyzamide	98.6	
			Tricyclazole	79.6	
NTR-729 HF	Nitto Denko	Lab scale (dead-end)	Carbaryl (NAC)	92.4	Kiso et al., 2001a
			Chloroneb	93.9	
			Chlorothalonil (TPN)	96.1	
			Esprocarb	99.94	
			Fenobucarb (BPMC)	94.8	
			Isoxathion	99.84	
			Mefenacet	99.1	
			Methyldymron	98.4	
			Propiconazole	96.9	
			Propyzamide	98.6	
			Tricyclazole	79.6	

Table A. Continued

Membrane	Specifications	Remarks	Pesticides	Retention (%)	Reference
		Lab scale (dead-end)	Anilazine	99.3	Kiso et al., 2000
			Atrazine	97.5	
			Chlorpyrifos	>99.95	
			Diazinon	99.52	
			Dichlorvos (DDVP)	86.7	
			Imidacloprid	97.6	
			Isoprothiolane	99.76	
			Malathion	99.64	
			Molinate	98.5	
			Pyridine	18.5	
			Simazine	96.7	
			Simetryn	98.6	
			Thiram	97.7	
			2,3,5-Trichloropyridine	96.8	
NTR-7410	Nitto Denko	Lab scale (dead-end)	Anilazine	21.8	Kiso et al., 2000
			Atrazine	10.9	
			Chlorpyrifos	99.51	
			Diazinon	44.6	
			Dichlorvos (DDVP)	4.28	
			Imidacloprid	2.92	
			Isoprothiolane	28.1	
			Malathion	41.4	
			Molinate	20.0	
			Simazine	6.40	
			Simetryn	6.69	
			Thiram	8.42	
			2,3,5-Trichloropyridine	95.6	
		Lab scale (dead-end)	Carbaryl (NAC)	24.7	Kiso et al., 2001a
			Chloroneb	98.6	
			Chlorothalonil (TPN)	61.6	
			Esprocarb	94.6	
			Fenobucarb (BPMC)	17.8	
NTR-7410	Nitto Denko	Lab scale (dead-end)	Isoxathion	99.5	Kiso et al., 2001a
			Mefenacet	72.5	
			Methyldymron	22.6	
			Propiconazole	77.0	
			Propyzamide	22.4	
			Tricyclazole	1.8	

Table A. Continued

Membrane	Specifications	Remarks	Pesticides	Retention (%)	Reference
UTC-20	Toray MWCO 180Da	Lab scale (crossflow)	Atrazine	74.3-80.4	Van der Bruggen et al., 1998
			Diuron	39.7	
			Isoproturon	72.3	
			Simazine	67.2-89.2	
		Lab scale (crossflow)	Atrazine	84.2	Berg et al., 1997
			Diuron	50.0	
			Isoproturon	73.0	
			Simazine	71.4	
		Lab scale (crossflow)	Atrazine	~95	Zhang et al., 2004
			Simazine	~80	
UTC-60	Toray MWCO 150Da	Lab scale (crossflow)	Atrazine	83.2	Van der Bruggen et al., 2001
			Diuron	49.0	
			Isoproturon	79.0	
			Simazine	71.4	
		Lab scale (crossflow)	Atrazine	~85	Zhang et al., 2004
			Simazine	~75	
TS80	TriSep Co. MWCO <200Da	Lab scale (crossflow)	Atrazine	81.2	Košutić et al., 2002
			MCPA	91.2	
			Propham	84.3	
			Triazimefon	58.1	
		Lab scale (crossflow)	Alachlor	41.8±2.8	Comerton et al., 2008
			Atraton	21.7±9.4	
			DEET	18.1±6.2	
			Metolachlor	50.5±7.9	
X20	TriSep Co. MWCO <200Da	Lab scale (crossflow)	Alachlor	97.3±1.4	Comerton et al., 2008
			Atraton	96.9±2.7	
			DEET	96.1±0.9	
			Metolachlor	97.2±0.6	
HNF-1	Hollow fiber composite membrane	Lab scale (crossflow)	Alachlor	88.7	Kiso et al., 2002
			Aldicarb	43.2	
			Atrazine	61.4	
			Methoxychlor	99.2	
			Metolachlor	93.9	
			Pirimicarb	89.9	
			Simazine	42.2	
			Thiobencarb	88.7	

Table A. Continued

19

Electrochemical Oxidation of Herbicides

Sidney Aquino Neto and Adalgisa Rodrigues De Andrade
Departamento de Química, Faculdade de Filosofia Ciências e Letras de Ribeirão Preto, University of São Paulo
Brazil

1. Introduction

Remediation of water polluted by toxic organic compounds such as herbicides, dyes, pesticides, pharmaceuticals, detergents, among many other highly toxic compounds, even at very low concentration, has been the subject of many investigations (Brillas et al., 2007; Weiss et al., 2007). These organic pollutants are responsible for a lot of environmental damage, especially when they accumulate in the environment, both in landfields and water (Relyea, 2005). Since water contamination affects not only aquatic species, but also humans and animals consuming it, one of the major concerns of environmentalists is the contamination of groundwater, which receives much of the slurry containing organic compounds. This serious situation has engaged environmental councils worldwide in the supervision of both water consumption and infection. In Brazil, the Environmental Sanitation Technology Company (CETESB) began to register the contaminated areas in the state of São Paulo in 2002. The number of contaminated areas has increased continuously over the last 8 years, reaching a total of 2500 contaminated areas in 2009. A considerable portion of the contamination process can be attributed to the activity of oil refineries, as well as chemical, textile and pharmaceutical industries, not to mention the large contribution from agriculture. The worldwide growth of agricultural production has led to increased demand for agrochemicals. The ever-growing need for food and fibers requires an agricultural system with high productivity per cultivated area, consequently giving rise to a situation in which the consumption of agrochemicals is escalating. In recent years, the intensive use of herbicides has significantly increased environmental concern, mainly because of the adverse effects of these pollutants on soil and aquatic microorganisms.

Generally, water contamination by herbicides can occur through:

- transport from aerial or ground spraying;
- leaching through the soil and water erosion;
- disposal of commercial packaging;
- cleaning of spray-contaminated tanks.

A problem of major concern is the resistance of these organic compounds to the available wastewater treatment techniques, which culminates in lower efficiency of pollutant removal from water streams. In the current scenery in which water resources are continuously diminishing while both population and consumption are increasing, many research teams worldwide have focused on alternatives and new technologies for the treatment of persistent toxic organic compounds in the environment (Brillas et al., 2007; Comninellis, 1994). Some approaches for the treatment of these substances are available, and the main goal is to obtain a viable process that will allow for complete removal of contaminants or at least lead to the formation of biodegradable compounds. Several processes aim at the degradation of organic pollutants in the environment or at least at their oxidation to less toxic compounds. The choice of methodology involves factors related mainly to cost and efficiency, so each method offers advantages as well as limitations. The methods currently available for the treatment of effluent can be basically separated into two broad classes. The first involves the classic physical-chemical treatments such as sedimentation, filtration, centrifugation, flotation, and adsorption onto activated carbon. Although these methods display highly efficient rates for the removal of contaminants, they consist of phase-transfer process, i. e., further disposal is required after treatment, which is a major drawback In the second class are the oxidative methods, i. e., methods in which there is not only pollutant phase-transfer, but also its oxidation to inert compounds. Nowadays, the majority of industries generating large amounts of effluent opt for remediation using biological treatment, since it is relatively inexpensive. However, its degradation kinetics is very slow; limiting its action to compounds with low toxicity and effluents with low concentration of contaminants (Freire et al., 2000). Because conventional chemical and biological methods are no longer efficient due to the resistance gained by many compounds to wastewater treatment, some approaches for the treatment of toxic organic materials that will remove or convert these pollutants to biodegradable compounds have been developed. Electrochemical (Comninellis, 1994), electro-Fenton (Sires et al., 2007), ozonation (Canizares et al., 2007), Fenton, and photo-Fenton (Brillas et al., 2007) processes have been frequently proposed for the treatment of organic pollutants. Regardless of the selected methodology, the generation of high oxidative species such as hydroxyl radical (E^o = 2.80 V vs. SCE) must take place, in order to ensure elimination of the toxic organic compound. Table 1 shows the generation of hydroxyl radical in the most common advanced oxidation technologies investigated to date (Martínez-Huillé & Brillas, 2009).

Tecnique	•OH production
Electrochemical	$MO_x + H_2O \rightarrow MO_x(\bullet OH) + H^+ + e^-$
Fenton reaction, electro-fenton, and photoelectro-fenton	$H_2O_2 + Fe(II) \rightarrow \bullet OH + Fe(III)$
Photocatalysis	$TiO_2 + h\upsilon \rightarrow TiO_2(h + e)$ $TiO_2(h) + OH^- \rightarrow \bullet OH$
UV-peroxide	$H_2O_2 + h\upsilon \rightarrow 2\bullet OH$
Sonolysis	$H_2O +))) \rightarrow \bullet OH + \bullet H$
Radiolysis	$H_2O + \gamma \rightarrow e_{aq,} \bullet OH, \bullet H$

Table 1. Mechanism of generation of hydroxyl radical in the most common advanced oxidation technologies

2. Electrochemical treatment of organic pollutants

2.1 Electrochemical process for organic mineralization

The concern about the increasing environmental contamination, which consequently affects water quality, forces the development of many alternatives for the treatment of organic pollutants. In this context, electrochemical processes have always been claimed to be an attractive approach due to their versatility, easier operation, effectiveness, and lower cost (Gandini et al., 2000). The efficiency of the electrochemical process depends on many factors; however, the main focus is placed on electrode activity and lifetime and, consequently, on the electrode material. Besides catalytic activity, the choice of electrode material will also consider characteristics such as mechanical strength, physical and chemical stability under drastic operational conditions (high current density and high potential), and cost. Metallic oxide electrodes containing RuO_2 have been widely employed in environmental electrochemistry because of their mechanical resistance, inexpensiveness, and successful scale-up in the electrochemical industry. Apart from leading to chloro-alkali production, dimensionally stable anode (DSA®) electrodes are also a good alternative to the oxidation of various organic compounds (Trasatti, 2000). Another material good candidate material for the electrochemical oxidation of organic compounds is the boron-doped diamond (BDD), because of its generally large potential window and feasibility of the produced hydroxyl radicals (Panizza et al., 2008b). Although the BDD electrode is an extremely efficient material for organic mineralization, its use in large-scale operational conditions is still limited due to the high cost of BDD production.

2.2 DSA® electrodes

The metallic oxide electrodes introduced by Beer in 1966 consist of an inert metal support coated with noble metal oxides such as RuO_2 and IrO_2. In addition to the noble metal oxides, oxide electrodes also contain the so-called modulators, which are oxides such as SnO_2, TiO_2, Ta_2O_5, and PbO_2, whose function is to enhance the electrochemical characteristics related to electrode lifetime, mechanical stability, and catalytic activity, not to mention cost reduction. Electrode preparation is usually performed by thermal decomposition of the precursor's salts, which are then deposited onto an inert support material (titanium is the most commonly employed material for this purpose, due to its relatively low cost). There are several studies focusing on the use of oxide electrodes in the electrochemical degradation of toxic organic compounds. These materials have been shown to display excellent catalytic activity, resistance to corrosion, dimensional stability, high electrochemically active area, low maintenance cost, and low power consumption (Trasatti, 2000). Moreover, the use of DSA® with photoactive surface enables accomplishment of heterogeneous photocatalysis (Pelegrini et al., 1999). The mechanism of organic compound oxidation by electrochemical processes, as described by Comninellis (1994), can occur directly at anodes through generation of physically adsorbed hydroxyl radicals (Eq. 1). These processes may ultimately result in fully oxidized reaction products such as CO_2 (Kapałka et al., 2008). The •OH radical undergoes a fast reaction to form higher oxide (Eq. 3) on DSA-type anodes. Although this mechanism has been proposed long ago, only recently has experimental evidence from Electrochemical Differential Mass Spectroscopy been able to confirm participation of the higher oxide species (Fierro et al., 2007)

$$MO_x + H_2O \rightarrow MO_x(\bullet OH) + H^+ + e^- \quad (1)$$

$$R + MO_x(\bullet OH) \rightarrow CO_2 + \text{inorganic ions} + MO_x + H^+ + e^- \quad (2)$$

$$MO_x(\bullet OH) \rightarrow MO_{x+1} + H^+ e^- \quad (3)$$

It is well known that the higher oxide species (MO_{X+1}) are much milder oxidants than the weakly bound radical formed in reaction 1. However, many modifications to material design, such as preparation methodology and changes in the modulator oxide have been introduced, in order to enhance the catalytic activity of the electrode material. The oxygen evolution reaction (OER) is an undesirable side reaction responsible for the lower current efficiency of organic compound oxidation during the electrochemical process. The mechanism proposed for this reaction involves the discharge of water molecules at the metal oxide surface (Eq. 1). Depending on the characteristic of the anode material, oxygen evolution proceeds via two different pathways: oxidation of weakly adsorbed hydroxyl radicals (Eq. 4) or formation of the higher oxide followed by oxygen evolution (Eq. 5 and 6).

$$MO_x(\bullet OH) \rightarrow MO_x + \tfrac{1}{2}O_2 + H^+ + e^- \quad (4)$$

$$MO_x(\bullet OH) \rightarrow MO_x^{+1} + H^+ + e^- \quad (5)$$

$$MO_{x+1} \rightarrow \tfrac{1}{2}O_2 + MO_x \quad (6)$$

In order to increase the oxidation rate of DSA-kind materials, electrochemical remediation frequently calls for more powerful oxidizing conditions as well as electrogeneration of large amounts of hydroxyl radical or other oxidizing species such as Cl_2, ClO^-, or O_3. These requirements can be met by changing the electrode material (SnO_2, PbO_2, BDD) or the supporting electrolyte (SE).

2.3 Electrolysis in chloride media and formation of organochlorinated compounds

Knowing that electrochemical remediation seeks more powerful oxidizing conditions, the electrolysis in chloride medium is a good alternative when more efficienty organic compound oxidation is desirable. NaCl is one of the most attractive media in the field of indirect oxidation owing to its straightforward impact on electrochemical technology. Oxide electrodes such as DSA® anodes are very active for Cl_2 evolution, so many studies have reported on quite advantageous features concerning the use of this medium in the oxidation of organic pollutants. Considering the standpoint of thermodynamics, electrolysis in chloride medium should favor OER in detriment of chlorine evolution reaction (ClER), since the reversible thermodynamic potential for oxygen evolution (1.23 V) is below the potential for ClER (1.36 V). However, the kinetics of ClER is favoured for DSA-type materials and, in practice, it occurs at lower overpotential. As described in the literature, the mechanism of ClER on metal oxide electrodes proceeds as follows (Trasatti, 1987):

$$2Cl^- \leftrightarrows Cl_{2(el)} + 2e^- \quad (7)$$

$$Cl_{2(el)} \leftrightarrows Cl_{2(sol)} \quad (8)$$

Adsorbed chlorine, $Cl_{2(el)}$, will form free species in solution, $Cl_{2(sol)}$, which will further react to form reactive species such as hypochlorous acid (HClO) and hypochlorite (ClO^-), which are responsible for faster organic compound degradation:

$$Cl_{2(sol)} + H_2O \rightarrow HClO + Cl^- + H^+ \quad (9)$$

$$HClO \leftrightarrows H^+ + ClO^- \qquad (10)$$

Nevertheless, despite its several advantages, besides promoting faster oxidation of the organic compound (Eq. 11), electrolysis in chloride medium also enables formation of organochlorine compounds (RCl), as seen in Eq. 12:

$$ClO^-_{(sol)} + R \rightarrow CO_2 + \text{inorganic ions} + H^+ + e^- \qquad (11)$$

$$ClO^-_{(sol)} + RCl \rightarrow R \qquad (12)$$

The main concern of different groups is that the presence of a C—Cl bond affects the chemical properties of the organic compound and therefore its toxicological behavior. In general, the introduction of chlorine in the organic molecule increases the chemical and biological reactivity of the chlorinated compound, thereby significantly enhancing the toxicity of these substances. Due to their high reactivity, these compounds usually have large lipophilicity, favoring interactions with enzymes and promoting biotransformations, for example. Furthermore, organochlorine compounds display several genotoxic effects depending on the chemical structure of the generated compound. Some literature studies have also shown that these compounds are responsible for different mutagenic and carcinogenic effects (Henschler, 1994). Nevertheless, the formation of organochloride species in solution during electrolysis in chloride medium has not received much attention; in fact, only a few studies have investigated the influence of experimental parameters on the formation of RCl compounds (Comninellis & Nerini, 1995). In this context, the possible formation of RCl during electrolysis makes the determination and evaluation of these compounds extremely important, mainly when electrolysis in chlorine medium is proposed as an alternative route for wastewater treatment. On the basis of the large number of literature papers focusing on the electrochemical degradation of herbicides, we are going to present the results on the degradation profile of different herbicides. Based on the results obtained by our research group, we will also focus on data regarding the electrochemical degradation of the glyphosate herbicide (GH) (Aquino Neto & De Andrade, 2009a; Aquino Neto & De Andrade, 2009b). Regarding the electrochemical oxidation behavior of GH, we are willing to discuss the performance of the RuO_2-based anode. Moreover, the effect of different parameters such as SE, pH, and current density on electrode activity will be presented. It is known that metallic oxide electrodes display good performance for the anodic mineralization of organic pollutants in chloride medium; however, the formation of organochloride compounds during electrolysis in the latter medium has not received much attention, since just a few reports have investigated the influence of this parameter. For this reason, a full discussion regarding the formation of these toxic species will be presented here. Besides the standard sample degradation behavior, the oxidation of commercial GH formulation will also be discussed. Finally, the very recent literature focusing on the degradation of different herbicides will be presented.

3. Glyphosate contamination

In recent years, the intensive use of herbicides has increased environmental concern, partly because of the adverse effects of these chemicals on soil and aquatic microorganisms (Farah et al., 2004). One of the most commonly employed agrochemicals is glyphosate (Fig. 1). N-(phosphonomethyl) glycine is a highly effective broad-spectrum, post-emergence, non-selective herbicide widely used in agriculture worldwide (De Amarante et al., 2002). GH is highly soluble in water (12 g L^{-1} at 25 °C), and it is currently utilized in more than 30 types of crops, to control a wide variety of annual weeds, mainly in the case of sugarcane and soybean plantations.

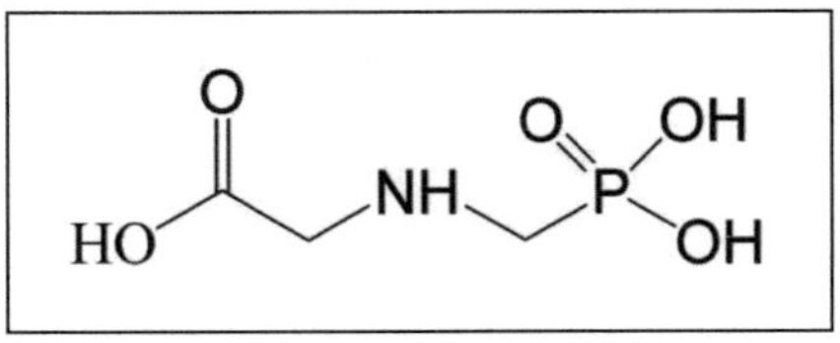

Fig. 1. Chemical structure of GH.

GH is commercialized as an isopropylamine salt. Formulations found in the market generally contain surfactants, to prevent formation of drops and to reach larger sprayed areas. Some of these components are serious irritants and toxic to fish; moreover, the commercial formulations display more toxicity and are more persistent than the active ingredient. The most common commercial formulations contain surfactants based on ethylamine alone, which is significantly more dangerous than GH itself and very toxic for invertebrates and fish (De Amarante et al., 2002). The half-life of the commercial formulations is relatively long, about 7-70 days (Giesy et al., 2000). According to resolution 375 of the National Brazilian Environmental Council (CONAMA), the maximum value allowed for GH in sweet water is 0.280 mg L^{-1}. The US Environmental Protection Agency also classifies GH as "extremely persistent". The main reason for the widespread use of this chemical worldwide is its relatively low toxicity to humans and animals. However, despite low toxicity, its quick biodegradation to the main metabolite aminomethylphosphonic acid (AMPA) is a matter of concern because this compound is considered more toxic and persistent than the original herbicide (Williams et al., 2000). Increased use of GH is expected due to the development of transgenic plants tolerant to this compound (Owen & Zelaya, 2005). Even though commercial GH formulations are considered to have low toxicity, there are evidences of noxious effects on the environment after its prolonged use, mainly because of the resistance gained by the annual weeds. In this context, the exposure of non-target aquatic organisms to this herbicide is the concern of many ecotoxicologists. Several in vivo and in vitro studies on animals have revealed the mutagenic and carcinogenic effect of GH (Lin & Garry, 2000) as well as its impact on the environment and aquatic life (Tsui & Chu, 2003). Some studies conducted with GH commercial formulations have demonstrated the potential toxicity of these formulations to the environment. Electron microscopy studies on fish of the Cyprinus carpio species have shown that this herbicide causes disruption of the inner mitochondrial membrane (Tsui & Chu, 2003). Another investigation has pointed out that some formulations are largely responsible for the toxicity in the energy levels of mitochondrial oxidative phosphorylation in rat livers (Peixoto, 2005). Due to the great concern about GH contamination, many studies have focused on degradation of this compound. Shifu and Yunzhang (2007) have reported the photocatalytic degradation of GH using TiO_2 as photocatalyst and a mercury lamp of 375 W, with the concentration of herbicide being maintained at 42 mg L^{-1}. The results showed that 92% GH were mineralized after 3.5 h of illumination. Chen et al. (2007) have investigated the photodegradation of GH in a system using ferrioxalate as Fe^{2+} source, a metal halide lamp of 250 W, and a constant concentration of GH of 5 mg L^{-1}. The efficiency of GH mineralization reached values around

60%. Barret & Mcbride (2005) have evaluated the oxidative degradation of GH on manganese oxide. There was no significant herbicide degradation (10 mg L^{-1}), and much of the herbicide simply adsorbed onto the manganese oxide. Huston & Pignatello (1999) have investigated the degradation of several active ingredients of pesticides as well as several commercial herbicides formulations via photo-Fenton reaction. The experiments were carried out in a solution containing 10^{-5} mol L^{-1} Fe (III) and 10^{-2} mol L^{-1} H_2O_2, and the initial GH concentration was 33 mg L^{-1}. A reactor with 16 black light lamps of 14 W was employed, and the results showed that 35% GH were mineralized after 2h of irradiation. Castro et al. (2007) have examined the biodegradation of GH (500 mg L^{-1}) using the Fusari fungus. Efficiency of removal of the active principle reached values around 40%. Speth (1993) investigated different processes for GH removal from drinking water, such as adsorption onto activated carbon at various pH values, as well as the efficiency of different oxidants for GH degradation. The initial amount of treated herbicide was 1.75 mg L^{-1}. Adsorption of the herbicide on activated carbon proved to be quite high. Results from treatment by coagulation, sedimentation, and sand filtration revealed a removal efficiency of 7% only. Degradation with chlorine led to higher efficiency, with almost complete removal of the contaminant. Munner and Boxall (2008) have investigated the photocatalytic degradation of GH in TiO_2 suspension with 12 black lamps of 15 W, using initial concentration of GH equal to 170 mg L^{-1}. Good degradation rates were achieved, and 90% GH removal was observed. Bazot and Lebeau (2008) have evaluated the efficiency of GH oxidation using the bacterium Pseudomonas 4ASW. The initial concentration of GH was 80 mg L^{-1}, and 80% of active ingredient removal was obtained after 80 h of treatment. In this chapter, results from the electrochemical degradation of GH using oxide electrodes obtained by our research group will be presented.

4. Experimental procedures

4.1 Preparation of DSA® electrodes

The choice of a methodology for the preparation of DSA® electrodes is very important, since the final properties and characteristics of the electrode are highly dependent on the method of preparation. The preparation techniques aim at films with adequate mechanical stability as well as oxide mixtures with high catalytic activity. The methodology usually employed in the preparation of oxide electrodes with greater stability is the thermal decomposition of suitable precursors. This procedure consists of successive steps in which thin layers of the precursor solutions are applied onto the inert support, followed by calcination at elevated temperatures (T > 400 °C) and appropriate O_2 flow (Trasatti & Lodi, 1981). Several methods for the production of oxide electrodes are found in the literature: the thermal decomposition of chlorides (traditional methodology; Beer, 1966), spray-pyrolysis method (De Battisti et al., 1997), the sol-gel method (Diaz-Flores et al., 2003), and the thermal decomposition of polymeric precursors (also known as Pechini method, Pechini, 1967). The traditional method is still the most frequently employed. It consists of the thermal decomposition of inorganic precursors, usually employed in the form of chlorides. The traditional method has been adopted since the 60 s, and has been used in the preparation of several films in various areas. In this methodology, the deposition of thin layers of the precursor salt solutions (usually chlorides of the desired metals in a solution of HCl and water 1:1) occurs via brushing or dipping into the precusor solution (dip-coating method). After this step, the supports are calcined at high temperatures (usually above 400 °C), to obtain the respective metal oxides. The great advantage of this method is the easy preparation, prompt availability, and relatively lower cost of the employed salts. The main feature of the films obtained by this kind of preparation is the "mud-cracked" morphology (Trasatti & Lodi, 1981). Recent work employing the traditional method using isopropanol as solvent instead of an acid solution has also furnished very promising results, since the films displayed excellent mechanical stability and catalytic activity (Coteiro & De Andrade, 2007). Another more recent method of preparation is the thermal decomposition of polymeric precursors. In this method, the formed polymeric resin "captures" the metal atoms, which provides better control of both stoichiometry and particle size (Pechini, 1967). Because the polymer is formed before the calcination process, metallic atoms such as tin are trapped in this matrix, thereby avoiding evaporation and consequent losses, and enabling production of films with better handled composition. Film deposition on the inert support material occurs in the same way as in the case of the traditional methodology, with the difference that a polyester is initially formed. Preparation of oxide electrodes by this methodology has been shown to provide uniform films with more homogeneous surfaces. Moreover, the decomposition of polymeric precursors culminates in largely reduced particle size, thus furnishing materials with high surface area. These features make this a promising method for the preparation of oxide films, with potential application in different scientific areas (Santos et al., 2005).

4.2 Electrolytic system and electrodes

The results from GH degradation presented in this chapter were obtained using the experimental conditions described below. The electrochemical measurements were conducted in an open system, using a three-compartment electrolytic cell consisting of a main body (50 mL solution) and two smaller compartments containing the counterelectrodes, which were isolated from the main body by coarse glass frits. The electrolyses experiments were accomplished in the galvanostatic mode, under magnetic stirring. Electrochemical experiments (cyclic voltammetry and galvanostatic electrolyses) were performed using a potentiostat/galvanostat Autolab, mode SPGSTAT30. All experiments were carried out at 25 ± 1 °C. The working anodes were 2 cm^2 large and were prepared by thermal decomposition. The precursor mixtures were applied on both sides of the pre-treated Ti support by brushing, as described in previous works (Coteiro & De Andrade, 2007; Aquino Neto & De Andrade, 2009a). For the electrochemical oxidation of GH, the composition $Ti/Ru_{0.30}Ti_{0.70}O_2$ was employed. Details about the preparation, methodologies, and the physical and electrochemical characterization of the anode are given elsewhere (Aquino Neto & De Andrade, 2009a). Two spiraled platinized platinum wires (15 cm), placed parallel to each other, were used as counterelectrodes. All potentials are referred to the saturated calomel electrode (SCE). H_2SO_4 and NaOH were employed to adjust the pH values of the solutions. In all experiments, the ionic strength was kept constant (μ = 1.5) by adjusting the Na_2SO_4 and NaCl concentrations. Solutions were prepared with high-purity water from a Millipore Milli-Q system, and pH measurements were carried out with a pH electrode coupled to a Qualxtron model 8010 pH meter.

4.3 Quantification of glyphosate

The chemical structure of GH does not display a chromophore group, so spectroscopic determinations have to be performed only after its derivatization reaction. Hereafter, two different derivatization reactions for GH determination were used. The first derivatization consisted of the reaction of ninhydrin in the presence of the Na_2MoO_4 catalyst at 100 °C

(Bhaskara & Nagaraja, 2006), which produces the Ruhemann´s purple product with a maximum absorption at 570 nm (Fig. 2A). GH degradation was also followed by nitrosation reaction in acidic media (Food and agriculture organization of the United Nations, 2001), which produces a UV spectrophotometrically active compound at 243 nm (Fig. 2B).

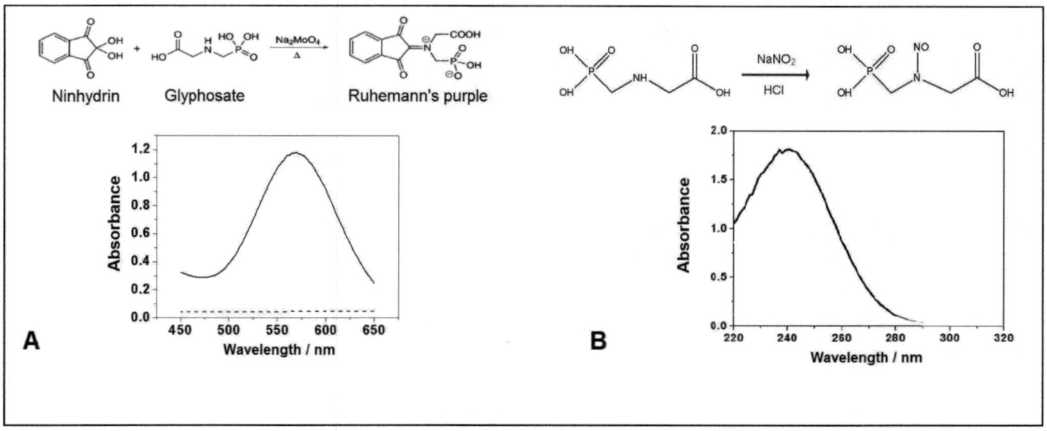

Fig. 2. Derivatization reaction of GH with ninhydrin produces the Ruhemann´s purple product with a maximum absorption at 570 nm (A). Nitrosation reaction of GH in acidic media produces a UV spectrophotometrically active compound at 243 nm (B)

As depicted in Fig. 3, both spectrophotometric methods (ninhydrin and nitrosation) give exactly the same degradation rate, indicating the good accuracy of the proposed methods. Therefore, from now on all the results presented here will be related to the determination using the nitrosation reaction, which is simple and employs readily available materials.

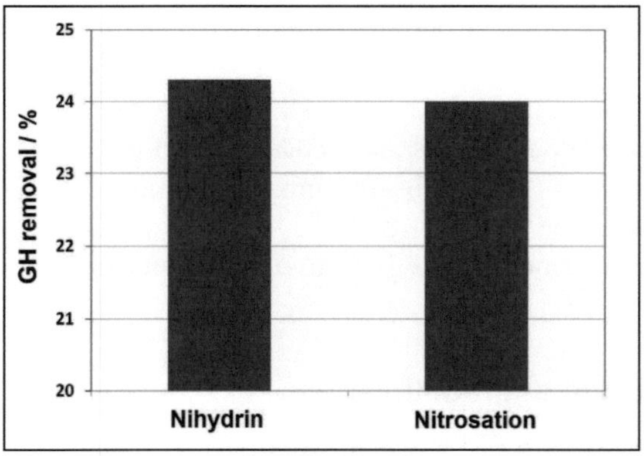

Fig. 3. GH removal as a function of both derivatizations reactions employed after 4 electrolysis at 50 mA cm^{-2}, $\mu = 1.5$ (Na_2SO_4 + NaCl, pH = 3)

As shown in Eq. 13, GH total oxidation produces PO_4^{-3} ions as one of the final degradation products. For this reason, the PO_4^{-3} release rate is a good indication of complete GH degradation. Its determination can be easily performed colorimetrically by the molybdenum blue method, according to the standard method (American Public Health Association, 1998).

$$C_3H_8O_5NP + 8H_2O \rightarrow 3CO_2 + PO_4^{-3} + NO^{-3} + 24H^+ + 20e^- \quad (13)$$

For analysis of GH total combustion, both chemical oxygen demand (COD) and total organic carbon (TOC) were also performed after electrolysis.

4.4 Organochlorine analysis

The amount of adsorbable organic halide (AOX) encountered in water samples is generally very low; for this reason, the analytical procedures involve a pre-analytical step for sample concentration. To this end, the constituents are loaded onto activated charcoal before the assay or, in the case of organic material extraction from mud, organic solvents are employed. To convert the AOX compounds into an analyzable form, the dissolved organic material is initially concentrated via adsorption onto activated charcoal. Then, nitrate ions are added to the mixture, to eliminate the inorganic chloride also adsorbed onto the activated charcoal by competition. Next, the "loaded" activated charcoal containing the sample is burned in a furnace at ~ 950 ° C; in this step, hydrogen halides, carbon dioxide, and water are formed. After drying of the generated gases, the halides are determined by microcoulometry, which involves a process that occurs in acetic acid medium according to the following equation:

$$Ag^+ + X^- \rightarrow AgX \quad (14)$$

The silver ions used for the precipitation of the halides are electrolytically generated at the silver anode. After quantitative conversion of the halide, concentration of the silver ions in the electrolyte increases indicating the the end point of the titration, as determined by a pair of polarized indicators electrodes. The amount of halide present in the sample is then measured using Faraday's law. AOX analysis during the electrochemical oxidation of GH was performed by means of the multi X 2000 analyzer (Analytik Jena, Germany) and the shaking method. This equipment allows for the direct determination of organochlorinated compounds adsorbed previously onto activated carbon from an aqueous solution containing AOX over 10 µg L^{-1}. However, the sample should meet the following criteria prior to the analysis:

- TOC should be less than 10 mg L^{-1};
- The amount of inorganic chloride should be less than 1 g L^{-1};
- Oxidant species wastes (ClO^{3-}, Cl_2 etc.) must be removed by addition of the proper amount of sodium sulfite.

All experiments were performed at 25 ± 1 ° C, and the results are presented as the average of triplicate measurements.

5. Results of herbicides degradation

5.1 Electrochemical oxidation of the GH herbicide

We began our investigation by carrying out the electrochemical oxidation of an standard GH sample. Once the most efficienty electrolysis conditions had been established, the electrooxidation of a commercial formulation (Roundup®) was also investigated. Cyclic voltammograms in both the absence and presence of GH were conducted. Electrochemical characterization showed that GH is not electroactive in the potential window 0.2 -1.2 V vs. SCE, so its oxidation hindered by OER. This is a very common characteristic of DSA-type anodes once they are very active for OER and this reaction occurs simultaneously with the oxidation of organic compounds in aqueous medium. The competition between the oxidation of the organic compound and OER is responsible for a significant reduction in the efficiency of the electrochemical process (Aquino Neto & De Andrade, 2009a).

Establishment of the best degradation conditions

Aiming at finding the best conditions for the electrolysis, the preliminary stage of the investigation consisted of selecting the most suitable experimental setup, so that the highest rate possible of electrochemical degradation would occur. In this step, the evaluation of pH, current density, and supporting electrolyte should be performed. There are many ways to measure the real efficiency of a treatment technology. In general, both energy consumption and organic combustion are evaluated. An easy way for judging the performance of DSA® anodes in electrochemical degradation studies is to determinate the current that is effectively used for oxidation of the organic compound. The instantaneous current efficiency (ICE) is obtained considering that during electrochemical incineration two parallel reactions (organic compound oxidation and OER) takes place. So, ICE is defined as the current fraction used for the organic oxidation (Comninellis & Pulgarin, 1991; Pacheco et al., 2007) and was calculated considering the values of chemical oxygen demand (COD) of the wastewater before and after the electrolysis, using the relation

$$ECI = \frac{FV}{8I}\frac{\left[(DQO)_t - (DQO)_{t+\Delta t}\right]}{\Delta t}, \tag{15}$$

where F is the Faraday constant (C mol^{-1}), V is the volume of the electrolyte (m^3), I is the applied current (A), and $(COD)_t$ and $(COD)_{t+\Delta t}$ are the chemical oxygen demand (g O_2 m^{-3}) at times t and t + Δt (s), respectively.

After 4h of electrolysis at a constant current density of 50 mA cm^{-2} in Na_2SO_4 medium, 24% of the starting material (1000 mg L^{-1}) was oxidized, and the mineralization rate reached c.a. 16%. When one compares this value with the rate reported for the degradation of phenol (Comninellis & Pulgarin, 1991), which is a compound generally referred as a model for organic degradation, we can confirm the recalcitrant behavior of herbicides in aqueous solution. Due to the low degradation rate of GH, the ICE in these conditions was very low, less than 5 %, indicating that OER is an important side reaction in the electrochemical process. The difference between the data obtained from spectrophotometric methods (24%) and TOC removal (16 %) has been explained by us previously (Aquino Neto & De Andrade, 2009a) and is related to the formation of recalcitrant intermediate products such as AMPA (metabolite aminomethylphosphonic acid) and sarcosine (n-methylglycine).

To understande the degrability of GH as a function of time, long-term electrolyses (12 h in Na_2SO_4 medium, pH 3, at 50 mA cm^{-2}) were performed. The results of GH degradation as a function o time showed that after 12 h of electrolysis only 43 % GH had been oxidized. In order to improve the degradation rate, pH and concentration effects must be investigated. The best results for the electrochemical oxidation of GH were found in acidic medium (Aquino Neto & De Andrade, 2009a). The low oxidation rates obtained in Na_2SO_4 medium can be explained by the general mechanism of organic compound oxidation (Comninellis, 1994). Briefly, the oxidation power of the anode is directly related to the overpotential for oxygen evolution. For DSA-like anodes, the •OH radicals strongly bind to the surface, eventually leading to the indirect oxidation of organics via formation/decomposition of an oxide of higher valence (De Oliveira et al., 2008). In order to increase the oxidation rate of DSA-kind materials, different approaches have been proposed in the literature, such as the use of PbO_2 (Cestarolli & De Andrade, 2003; Aquino et al., 2010; Panizza et al., 2008a) and BDD (Panizza et al., 2008b), and changes in the supporting electrolyte (Aquino Neto & De Andrade, 2009b).

Electrolysis in chloride medium

The electrolyses in chloride medium were performed as a function of chloride concentration. The assays were carried by varying the amount of chloride from 200 to 3500 mg L^{-1}. An increase in the initial concentration of chloride ion leads to a significant enhancement in the rate of the oxidation reaction. In the case of GH oxidation, there is an increase of 42% PO_4^{-3} release and 53% GH removal even at a very low NaCl concentration (220 mg L^{-1}). When a high concentration of chloride ions is employed (1000 mg L^{-1}), over 80% PO_4^{-3} release is obtained (Aquino Neto & De Andrade, 2009b). It is noteworthy that as the medium becomes more active toward organic compound oxidation, as in the case of chloride medium, there is no significant influence of the anode composition or current density on the oxidation rate. Therefore, one can improve the electrolysis by changing the supporting electrolyte, which culminates in less drastic conditions. This procedure offers two main advantages, namely a decrease in total energy consumption and maximized oxidation rate and larger electrode lifetime, which both contribute to diminishing the cost of the electrolytic system. Figure 4 shows the electrochemical oxidation profile of standard GH and of a commercial formulation of this herbicide as a function of time.

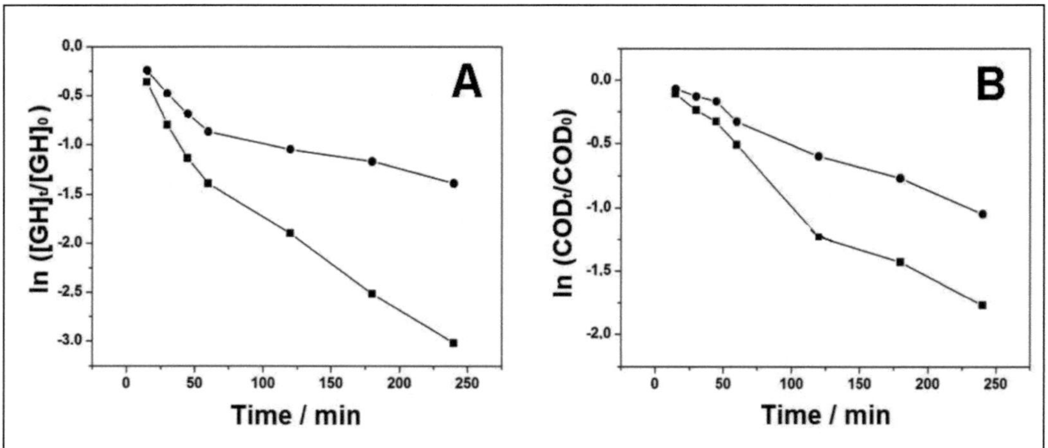

Fig. 4. Linearized removal (A) and COD removal (B) as a function of electrolysis time. Electrode composition $Ti/Ru_{0.30}Ti_{0.70}O_2$, i = 30 mA cm^{-2}, $[Cl^-]$ = 2662 mg L^{-1}, μ = 1.5 (Na_2SO_4 + NaCl, pH 3). ■ = standard GH sample; • = commercial GH formulation

The kinetic data reveal a complex oxidation profile. This behavior can be explained considering the competition between the oxidation of the starting material and of the byproducts formed within the first minutes of electrolysis. However, a linear decay is obtained within the first 60 min of electrolysis, as depicted in Fig. 4A and Fig. 4B. A pseudo first-order kinetic behavior is achieved, so assuming that this is the case in the first 60 min (Pelegrino et al., 2002), the oxidation rate can be written as:

$$dC_{(t)}/dt = -kC_{(t)} \qquad (16)$$

$$C_{(t)} = C_{(0)}\, e^{-kt}\,, \tag{17}$$

where $C_{(0)}$ corresponds to the initial GH concentration and k is the constant of velocity of GH oxidation. The k value is obtained by the following equation:

$$k_{HG} = k\, V/A, \tag{18}$$

where V is the solution volume (m^3) and A is the electrode area (m^2). In the same way, the kinetic constant for COD removal was obtained. Table 2 summarizes the kinetic constants for standard GH and its commercial formulation:

Herbicide	k_{GH} / 10^{-3} m s^{-1}	k_{COD} / 10^{-3} m s^{-1}
GH commercial formulation	3.48 ± 0.10	2.40 ± 0.12
GH standard sample	5.75 ± 0.11	1.37 ± 0.15

Table 2. Kinetic constants of standard GH and its commercial formulation for the first 60 min of electrolysis. Composition $Ti/Ir_{0,30}Sn_{0.70}O_2$, i = 30 mA cm^{-2}, [Cl^-] = 2662 mg L^{-1}, μ = 1.5 (Na_2SO_4 + NaCl, pH 3)

Data from Table 2 demonstrate the higher oxidation rate of the active ingredient in relation to its commercial formulation (the kinetic constant is 1.6 times larger). Considering the COD decay, the commercial formulation displays the largest kinetic constant, because of the higher organic load of commercial formulations, which, apart from the active ingredient, present "inert compounds" such as carriers, wetting agents, antifreezes, and other compounds employed to facilitate handling and application. Most of the components of commercial formulations are surfactants that increase the spreading and the penetration power. Taking into account the kinetic data it can be inferred that these compounds are much less recalcitrant than the active principle. Also, these data provide some interesting information concerning GH oxidation compared to the electrochemical oxidation of other organic compounds. The k_{COD} values of GH presented here are far superior to the ones found for the electrochemical oxidation of phenols (Coteiro & De Andrade, 2007) particularly, 4-chlorophenol (Alves et al., 2004). It is clear that some experimental conditions must also be considered, in order to evaluate the real efficiency of the electrode material. However, the data presented here demonstrate that the electrochemical process is really satisfactory for the treatment of organic pollutants in chloride medium. Finally, the anodic mineralization of organic pollutants in chloride medium perhaps may open the possibility of using DSA-type materials under mild oxidation conditions for the treatment of organic waste in water.

5.2 Study of AOX formation

As reported before in an early investigation (Aquino Neto & De Andrade, 2009b), the electrode material has a great effect on the amount of organic chloride species formed along of the electrolysis. The choice of material depends on the structure and complexity of the sample to be electrolyzed. For this reason, we cannot make a straightforward generalization for the best electrode material. However, our investigation has pointed $Ti/(RuO_2)_{0.70}(Ta_2O_5)_{0.30}$ or $Ti/Ru_{0.30}Sn_{0.70}O_2$ as compositions displaying longer lifetime and environmental friendly electrode composition (Aquino Neto & De Andrade, 2009b). The results also showed that an increase in the applied current leads to an increase in the amount of ClO^- (Eq. 10), so the yields of AOX species are enhanced. The AOX formation behavior is exponential in the case of commercial formulations. For the standard sample, the formation of organochloride compounds during electrolysis increases almost linearly with the applied current, and the AOX concentration remains below the allowed values almost in the entire investigated current window. Although some papers have shown that once the R—Cl compound is formed, it is quickly consumed before the end of the electrolysis, giving a volcano type curve (Comninellis & Nerini, 1995; Rajkumar et al., 2005), the results of AOX formation as a function of time reported by our research group reveal a different behavior for AOX formation. Even in the case of long-term electrolysis (14 h), there still is a mild increase in the production of organochloride compounds. These results may be attributed to the lower quantity of AOX compounds formed along the electrolysis, compared with results found elsewhere (Comninellis & Nerini, 1995). These data may be understood by considering the competition between the degradation of the generated AOX species and the organic compounds present in the solution. Although the formation of AOX is faster in the beginning of the electrolysis, the quantity formed within the first hours of treatment is very low compared with the remaining organic load. For this reason, an ascending behavior of AOX formation is observed for the entire investigated time window (Aquino Neto & De Andrade, 2009b). Finally, the results of AOX formation as a function of many experimental parameters show that the formation of organochloride compounds is straightly related with factors like chloride concentration, anode material, and current density, among others. For this reason, it is very important that these parameters are carefully evaluated when an alternative treatment that allows the formation of this species in solution is proposed.

6. Electrochemical oxidation of different formulations

A reduced number of papers deal with the oxidation of herbicides using electrochemical methods; indeed, the majority of the works focuses on photochemical processes. However, a literature search reveals that there is increased interest in the former subject. Table 3 lists the latest papers dealing with the electrochemical degradation/treatment of herbicides. Studies of Diuron photocatalytic degradation show that the herbicide mineralization rate reaches almost 97% after 8 hours of irradiation with light of 280nm. Studies combining photo-Fenton treatment afford 82.5% COD removal from wastewater generated by the sugarcane industry (Katsumata et al., 2009). Recently, Oturan et al. (2010) have used the electro-Fenton process to oxidize a group of phenylurea herbicides. The results showed that the degradation rate increases with the number of chlorine groups, being Diuron the most reactive herbicide. The authors observed that even with pronounced COD reduction, the treatment in mild conditions also produces several aromatic byproducts (Oturan et al., 2010). An interesting approach involving combination of biological and electrochemical oxidation processes has been proposed by Liu et al. (2010). They showed that the open ring byproducts formed during the electrochemical process can maintain the activity of microorganisms on a biofilter, consequently enhancing the activity of the process. A comparison between different electrochemical methods has been reported by Yatmaz & Uzman (2009), who used organophosphorus pesticides as a model compound. The authors claimed that the degradation of the pesticides proceeds with the following decreasing selectivity: indirect electrooxidation processes using Ti electrodes > electrocoagulation using Fe electrodes > electro-Fenton process using Fe electrodes. The herbicide Alachlor is also frequently investigated as a model of chloroacetamide compounds. A variety of effective techniques for the treatment of effluents containing alachlor are available. Herbicides such as 2,4-D have been efficiently degraded by several advanced oxidation processes in which the oxidizing hydroxyl radical (•OH) is produced by chemical, photochemical, and photocatalytic systems, such as H_2O_2/Fe^{+2}. Alternative procedures such as anodic oxidation and electrochemical methods with indirect electro-oxidation by generation of H_2O_2 are also employed for 2,4-DP removal from water (Brillas et al., 2007). Badellino et al. (2007) have studied the degradation of 2,4-DP in an electrochemical flow reactor with generation of H_2O_2 and Fenton's reagent, and obtained open ring acids as the main final products.

Herbicide	Methodology	Reference
Desethyl atrazine and desethyl terbutylazine	Electrochemical reduction	Colombini et al., 1998
Ethylene thiourea	Electrochemical and Fenton treatment	Saltmiras & Lemley, 2000
2,4-D (2,4-dichlorophenoxyacetic acid)	Electro-Fenton	Wang & Lemley, 2001
S-Triazine	Electrochemical reduction	Galvin et al., 2001
Bendiocarb, pirimiphos-methyl, coumatetralyl and chlorophacinon	Photoelectro-Fenton	Aaron & Oturan, 2001
Phenoxyacetic acid (4-chlorophenoxyacetic acid (4-CPA), 4-chloro-2-methylphenoxyacetic acid (MCPA)	Electron-Fenton e BDD	Brillas et al., 2004
Thiocarbamate herbicides	Electroflotation, Electrochemical and photocatalytic	Mogyorody, 2006
2,4-DP (2-(2,4-dichlorophenoxy)-propionic acid)	Electrochemical reactor for H_2O_2 and Fe production	Badellino et al., 2007
2,4-DP (2-(2,4-dichlorophenoxy)-propionic acid)	Electro-Fenton	Brillas et al., 2007
Atrazine	Electrochemical oxidation $Ti/\beta\text{-}PbO_2$	Vera et al., 2009
Organophosphorus pesticides	Electrochemical, electron-Fenton and electrocoagulation	Yatmaz & Uzman, 2009
Diuron and fenuron	Electro-Fenton	Oturan et al., 2010

Table 3. Latest papers dealing with the electrochemical degradation/treatment of herbicides

7. Conclusion and perspectives on the electrooxidation of toxic organic compounds

The electrochemical technology is potentially useful for the treatment of organic pollutants. Some important features such as easy automation, rare need for addition of reagents, robustness, versatility, and operation at mild temperatures make this technique quite promising for wastewater decontamination. Although several studies have focused on the use of phenolic intermediates, it can also be noted that the number of papers proposing electrochemical treatment of herbicide is still modest. In this scenery, electrochemical treatment using Fenton's reaction stands out. This is a straightforward result of the great interest shown by the use of Fenton's reagent in the photochemical investigation of herbicides degradation. As stated recently by Anglada et al. (2009), an efficient treatment of contaminated effluents is rarely performed by a single process; indeed, usually two or three associated processes must be involved, so that a reduction in the energy consumption and low level of organic material can be achieved. The main conclusion is that the oxidation of herbicides through electrochemical technology represents a viable technique for reducing the toxicity of wastewater. The use of this type of treatment may provide a breakthrough in the handling of toxic waste in the coming years. This is the reason why great efforts have been made in order to couple electrochemical treatments with established methodologies such as the biological ones.

8. References

Aaron, J.J. & Oturan, M.A. (2001). New photochemical and electrochemical methods for the degradation of pesticides in aqueous media: environmental applications. *Turkish Journal of Chemistry*, Vol. 25, No. 4, 509-520, ISSN 1303-6130

Alves, P.D.P.; Spagnol, M.; Tremiliosi-Filho, G. & De Andrade A.R. (2004). Investigation of the influence of the anode composition of DSA-type electrodes on the electrocatalytic oxidation of phenol in neutral medium. *Journal of the Brazilian Chemical Society*, Vol. 15, No. 5, 626-634, ISSN 1678-4790

American Public Health Association. American Water Works Association. Water Pollution Control Federation. (1998). *Standard Methods for the Examination of Water and Wastewater*, 20th ed., American Public Health Association, ISBN 0875532357, Washington, DC.

Anglada, A.; Urtiaga, A. & Ortiz, I. (2009). Contributions of electrochemical oxidation to waste-water treatment: fundamentals and review of applications. *Journal of Chemical Technology & Biotechnology*, Vol. 84, No. 12, 1747-1755, ISSN 0268-2575

Aquino, J. M.; Rocha-Filho, R. C.; Bocchi, N. & Biaggio, S.R. (2010). Electrochemical degradation of the reactive red 141 dye on a β-PbO2 anode assessed by the response surface methodology. *Journal of the Brazilian Chemical Society*, Vol. 21, No. 2, 324-330, ISSN 1678-4790

Aquino Neto, S. & De Andrade, A.R. (2009a). Electrooxidation of glyphosate herbicide at different DSA® compositions: pH, concentration and supporting electrolyte effect. *Electrochimica Acta*, Vol. 54, No. 7, 2039-2045, ISSN 0013-4686.

Aquino Neto, S. & De Andrade, A.R. (2009b). Electrochemical degradation of glyphosate formulations at DSA anodes in chloride medium: an AOX formation study. *Journal of Applied Electrochemistry*, Vol. 39, No. 10, 1863-1870, ISSN 1572-8838

Badellino, C.; Rodrigues, C.A. & Bertazolli, R. (2007). Oxidation of herbicides by in situ synthesized hydrogen peroxide and fenton's reagent in an electrochemical flow reactor: study of the degradation of 2,4-dichlorophenoxyacetic acid. *Journal of Applied Electrochemistry*, Vol. 37, No. 4, 451-459, ISSN 1572-8838

Barret, K.A. & Mcbride, M.B. (2005). Oxidative degradation of glyphosate and aminomethylphosphonate by Manganese oxide. *Environmental Science & Technology*, Vol. 39, No. 23, 9223-9228, ISSN 1520-5851

Bazot, S. & Lebeau, T. (2008). Simultaneous mineralization of glyphosate and diuron by a consortium of three bacteria as free and/or immobilized-cells formulations. *Applied Microbiology and Biotechnology*, Vol. 77, No. 6, 1351-1358, ISSN 1432-0614

Beer, H.B. (1966), in *U.S. Patent*. 3: New York. p. 199.

Bhaskara, B.L. & Nagaraja, P. (2006). Direct sensitive spectrophotometric determination of glyphosate by using ninhydrin as a chromogenic reagent in formulations and environmental water samples. *Helvetica Chimica Acta*, Vol. 89, No. 11, 2686-2693, ISSN 1522-2675

Brillas, E.; Boye, B.; Sires, I.; Garrido, J.A.; Rodríguez, R.M. & Arias, C. (2004). Electrochemical destruction of chlorophenoxy herbicides by anodic oxidation and electro-Fenton using a boron-doped diamond electrode. *Electrochimica Acta*, Vol. 49, No. 25, 4487-4496, ISSN 0013-4686

Brillas, E.; Banos, M.A.; Skoumal, M.; Cabot, P.L.; Garrido, J.A. & Rodríguez R.M. (2007). Degradation of the herbicide 2,4-DP by anodic oxidation, electro-Fenton and photoelectro-Fenton using platinum and boron-doped diamond anodes. *Chemosphere*, Vol. 68, No. 2, 199-209, ISSN 0045-6535

Canizares, P.; Paz, R.; Sáez, C. & Rodrigo, M.A. (2007). Electrochemical oxidation of wastewaters polluted with aromatics and heterocyclic compounds: a comparison with other AOPs. *Journal of The Electrochemical Society*, Vol. 154, No. 11, E165-E171, ISSN 0013-4651

Castro, J.V.; Peralba, C.R.M. & Ayub, A.Z.M. (2007). Biodegradation of the herbicide glyphosate by filamentous fungi in platform shaker and batch bioreactor. *Journal of Environmental Science and Health, Part B*, Vol. 42, No. 8, 883-886, ISSN 1532-4109

Cestarolli, D.T. & De Andrade, A.R. (2003). Electrochemical and morphological properties of $Ti/Ru_{0.3}Pb_{(0.7-x)}Ti_xO_2$-coated electrodes. *Electrochimica Acta*, Vol. 48, No. 28, 4137-4142, ISSN 0013-4686

Chen, Y.; Wu, F.; Lin, Y.; Deng, N.; Bazhin, N. & Glebov, E. (2007). Photodegradation of glyphosate in the ferrioxalate system. *Journal of Hazardous Materials*, Vol. 148, No. 1-2, 360-365, ISSN 0304-3894

Colombini, M.P.; Fuoco, R.; Giannarelli, S.; Pospisil, L. & Trskova, R. (1998). Protonation and Degradation Reactions ofs-Triazine Herbicides. *Microchemical Journal*, Vol. 59, No. 2, 239-245, ISSN 0026-265X

Comninellis, C. & Pulgarin, C. (1991). Anodic oxidation of phenol for waste water treatment. *Journal of Applied Electrochemistry*, Vol. 21, No. 8, 703-708, ISSN 1572-8838

Comninellis C. (1994). Electrocatalysis in the electrochemical conversion/combustion of organic pollutants for waste water treatment. *Electrochimica Acta*, Vol. 39, No. 11-12, 1857-1862, ISSN 0013-4686

Comninellis, C. & Nerini, A. (1995). Anodic-oxidation of phenol in the presence of nacl for waste-water treatment. *Journal of Applied Electrochemistry*, Vol. 25, No. 1, 23-28, ISSN 1572-8838

Coteiro, R.D. & De Andrade A.R. (2007). Electrochemical oxidation of 4-chlorophenol and its by-products using $Ti/Ru_{0.3}M_{0.7}O_2$ (M = Ti or Sn) anodes: preparation route versus degradation efficiency. *Journal of Applied Electrochemistry*, Vol. 37, No. 6, 691-698, ISSN 1572-8838

De Amarante, O.P.; dos Santos, T.C.R.; Brito, N.M. & Ribeiro, M.L. (2002). Glyphosate: properties , toxicity, use and legislation. *Química Nova*, Vol. 25, No. 4, 589-593, ISSN 0100-4042

De Battisti, A.; Lodi, G.; Nanni, L.; Battaglin, G. & Benedetti A. (1997). Preparation and characterization of oxide film electrodes. *Canadian Journal of Chemistry/Revue Canadiene de Chimie*, Vol. 75, 1759-1765, ISSN 0008-4042

De Oliveira R.T.S.; Salazar-Banda, G.R; Machado, S.A.S. & Avaca, L.A. (2008). Electroanalytical determination of N-Nitrosamines in aqueous solution using a boron-doped diamond electrode. *Electroanalysis*, Vol. 20, No. 4, 396-401, ISSN 1521-4109

Diaz-Flores, L.L.; Ramirez-Bon, R.; Mendoza-Galvan, A.; Prokhorov, E. & Gonzalez-Hernandez J. (2003). Impedance spectroscopy studies on SnO_2 films prepared by the sol-gel process. *Journal of Physics and Chemistry of Solids*, Vol. 64, No. 7, 1037-1042, ISSN 0022-3697

Farah, M.A.; Ateeq, B.; Ali, M.N.; Sabir, R. & Ahmad, W. (2004). Studies on lethal concentrations and toxicity stress of some xenobiotics on aquatic organisms. *Chemosphere*, Vol. 55, No. 2, 257-265, ISSN 0045-6535

Fierro, S.; Nagel, T.; Baltruschat, H. & Comninellis, C. (2007). Investigation of the oxygen evolution reaction on Ti/IrO_2 electrodes using isotope labelling and on-line mass spectrometry. *Electrochemistry Communications*, Vol. 9, No. 8, 1969-1974, ISSN 1388-2481

Food and Agriculture Organization of the United Nations. (2001). *Glyphosate: N-(phosphonomethyl)glycine*. Specifications and evaluations for plant protection products.

Freire, R.S.; Pelegrini, R.; Kubota, L.; Duran, N. & Peraltazamora, P. (2000). Novas tendências para o tratamento de resíduos industriais contendo espécies organocloradas. *Química Nova*, Vol. 23, No. 4, 504-511, ISSN 0100-4042

Galvin, R.M.; Mellado, J.M.R.; Higuera, M.J. & Montoya, M.R. (2001). Reductive detoxification of the s-Triazine herbicides simazine and propazine at laboratory scale. *Bulletin of Electrochemistry*, Vol. 17, No. 2, 49-55, ISSN 0256-1654

Gandini, D.; Mahé, E.; Michaud, P.A.; Haenni, W.; Perret, A. & Comninellis, C. (2000). Oxidation of carboxylic acids at boron-doped diamond electrodes for wastewater treatment. *Journal of Applied Electrochemistry*, Vol. 30, No. 12, 1345-1350, ISSN 1572-8838

Giesy, J.P.; Dobson, S. & Solomon K.R. (2000). Ecological Risk Assessment for Roundup Herbicide. *Reviews of Environmental Contamination & Toxicology*, Vol. 167, 35-120, ISSN 0179-5953

Henschler, D. (1994). Toxicity of chlorinated organic compounds: effects of the introduction of chlorine in organic molecules. *Angewandte Chemie International Edition in English*, Vol. 33, No. 19, 1920-1935, ISSN 1521-3773

Huston, P.L. & Pignatello, J.L. (1999). Degradation of selected pesticide active ingredients and commercial formulations in water by the photo-assisted Fenton reaction. *Water Research*, Vol. 33, No. 5, 1238-1246, ISSN 0043-1354

Kapałka, A.; Fóti, G. & Comninellis C. (2008). Kinetic modeling of the electrochemical mineralization of organic pollutants for wastewater treatment. *Journal of Applied Electrochemistry*, Vol. 38, No. 12, 7-16, ISSN 1572-8838

Katsumata, H.; Sada, M.; Nakaoka, Y.; Kaneco, S.; Suzuki, T. & Ohta, K. (2009). Photocatalytic degradation of diuron in aqueous solution by platinized TiO_2. *Journal of Hazardous Materials*, Vol. 171, No. 1-3, 1081-1087, ISSN 0304-3894

Lin, V. & Garry, V. (2000). In vitro studies of cellular and molecular developmental toxicity of adjuvants, herbicides, and fungicides commonly used in Red River Valley, Minnesota. *Journal of Toxicology and Environmental Health, Part A*, Vol. 60, 423-439, ISSN 1087-2620

Liu, L.; Zhao, G.; Pang, Y.; Lei, Y.; Gao, J. & Liu, M. (2010). Integrated biological and electrochemical oxidation treatment for high toxicity pesticide pollutant. *Industrial & Engineering Chemistry Research*, Vol. 49, No. 12, 5496-5503, ISSN 1520-5045

Martínez-Huitle, C.A. & Brillas, E. (2009). Decontamination of wastewaters containing synthetic organic dyes by electrochemical methods: A general review. *Applied Catalysis B: Environmental*, Vol. 87, No. 3-4, 105-145, ISSN 0926-3373

Mogyorody, F. (2006). Reaction pathways in the electrochemical degradation of thiocarbamate herbicides in NaCl solution. *Journal of Applied Elctrochemistry*, Vol. 36, No. 6, 635-642, ISSN 1572-8838

Munner, M. & Boxall, C. (2008). Photocatalyzed degradation of a pesticide derivative glyphosate in aqueous suspensions of titanium dioxide. *International Journal of Photoenergy*, Vol. 2008, 1-7, ISSN 1110-662X

Oturan, M.A.; Edelahi, M.C.; Oturan, N.; kacemi, K. & Aaron, J. (2010). Kinetics of oxidative degradation/mineralization pathways of the phenylurea herbicides diuron, monuron and fenuron in water during application of the electro-Fenton process. *Applied Catalysis B: Environmental*, Vol. 97, No. 1-2, 82-89, ISSN 0926-3373

Owen, M.D.K. & Zelaya, I.A (2005). Herbicide-resistant crops and weed resistance to herbicides. *Pest Management Science*, Vol. 61, 301-311, ISSN 1526-4998

Pacheco, M.J.; Morão, A.; Lopes, A.; Ciríaco, L. & Gonçalves, I. (2007). Degradation of phenols using boron-doped diamond electrodes: A method for quantifying the extent of combustion. *Electrochimica Acta*, Vol. 53, No. 2, 629-636, ISSN 0013-4686

Panizza, M.; Sires, I. & Cerisola, G. (2008a). Anodic oxidation of mecoprop herbicide at lead dioxide. *Journal of Applied Electrochemistry*, Vol. 38, No. 7, 923-929, ISSN 1572-8838

Panizza, M.; Kapalka, A. & Comninellis, C. (2008b). Oxidation of organic pollutants on BDD anodes using modulated current electrolysis. *Electrochimica Acta*, Vol. 53, No. 5, 2289-2295, ISSN 0013-4686

Pechini, M.P. (1967). *Method of preparing lead and alkaline earth Titanates and Niobates and coated method using the same to form a capacitor*. US Patent 3, 330697, 7 nov.

Peixoto, F. (2005). Comparative effects of the Roundup and glyphosate on mitochondrial oxidative phosphorylation. *Chemosphere*, Vol. 61, No. 8, 1115-1122, ISSN 0045-6535

Pelegrini, R.; Peralta-zamora, P.; De Andrade A.R.; Reyes, J. & Duran, N. (1999). Electrochemically assisted photocatalytic degradation of reactive dyes. *Applied Catalysis B: Environmental*, Vol. 22, No. 2, 83-90, ISSN 0926-3373

Pelegrino, R.L.; Di Iglia, R.A.; Sanches, C.G.; Avaca, L.A. & Bertazzoli, R. (2002). Comparative study of commercial oxide electrodes performance in electrochemical degradation of organics in aqueous solution. *Journal of the Brazilian Chemical Society*, Vol. 13, No. 1, 60-65, ISSN 1678-4790

Rajkumar, D.; Kim, J.G. & Palanivelu, K. (2005). Indirect Electrochemical Oxidation of Phenol in the Presence of Chloride for Wastewater Treatment. *Chemical Engineering & Technology*, Vol. 28, No. 1, 98-105, ISSN 1521-4125

Relyea, R.A. (2005). The Impact of Insecticides and Herbicides on the Biodiversity and Productivity of Aquatic Communities. *Ecological Applications*, Vol. 15, No. 2, 618-627, ISBN 1051-0761

Saltmiras, D.A. & Lemley, A.T. (2000). Degradation of Ethylene Thiourea (ETU) with Three Fenton Treatment Processes. *Journal of Agricultural and Food Chemistry*, Vol. 48, No. 12, 6149-6157, ISSN 0021-8561

Santos, A.L.; Profeti, D. & Olivi, P. (2005). Electrooxidation of methanol on Pt microparticles dispersed on SnO_2 thin films. *Electrochimica Acta*, Vol. 50, No. 13, 2615-2621, ISSN 0013-4686

Sires, I.; Garrido, J.A; Rodriguez, R.M.; Brillas, E.; Oturan, N. & Oturan, M.A. (2007). Catalytic behavior of the Fe^{3+}/Fe^{2+} system in the electro-Fenton degradation of the antimicrobial chlorophene. *Applied Catalysis B: Environmental*, Vol. 72, No. 3-4, 382-394, ISSN 0926-3373

Shifu, C. & Yunzhang, L. (2007). Study on the photocatalytic degradation of glyphosate by TiO_2 photocatalyst. *Chemosphere*, Vol. 67, No. 5, 1010-1017, ISSN 0045-6535

Speth, T.F. (1993). Glyphosate removal from drinking water. *Journal of Environmental Engineering*, Vol. 119, No. 6, 1139-1157, 0301-4797

Trasatti, S. & Lodi, G. (1981). *Studies in Physical and Theoretical Chemistry*, In: *Electrodes of conductive metallic oxides, Part B.*, Trasatti, S., (Ed.), pp. 521, Elsevier, Amsterdam, The Netherlands.

Trasatti, S. (1987). Progress in the understanding of the mechanism of chlorine evolution at oxide electrodes. *Electrochimica Acta*, Vol. 32, No. 3, 369-382, ISSN 0013-4686

Trasatti, S. (2000). Electrocatalysis: understanding the success of DSA®. *Electrochimica Acta*, Vol. 45, No. 15-16, 2377-2385, ISSN 0013-4686

Tsui, M.T.K. & Chu, L.M. (2003). Aquatic toxicity of glyphosate–based formulations: comparison between different organisms and the effects of environmental factors. *Chemosphere*, Vol. 52, No. 7, 1189-1197, ISSN 0045-6535

Vera, Y.M.; De Carvalho, R.J.; Torem, M.L. & Calfa, B.A. (2009). Atrazine degradation by in situ electrochemically generated ozone. *Chemical Engineering Journal*, Vol. 155, No. 3, 691-697, ISSN 1385-8947

Wang, Q.Q. & Lemley, A.T. (2001). Kinetic Model and Optimization of 2,4-D Degradation by Anodic Fenton Treatment. *Environmental Science & Technology*, Vol. 35, No. 22, 4509-4514, 1452-3981, ISSN 1520-5851

Weiss, E.; Groenen-Serrano, K. & Savall, A. (2007). Electrochemical mineralization of sodium dodecylbenzenesulfonate at boron doped diamond anodes. *Journal of Applied Electrochemistry*, Vol. 37, No. 11, 1337-1344, ISSN 1572-8838

Williams, G.M.; Kroes, R. & Munro, I.C. (2000). Safety evaluation and risk assessment of the herbicide Roundup and its active ingredient, glyphosate, for humans. *Regulatory Toxicology and Pharmacology*, Vol. 31, No. 2, 117-165, ISSN 0273-2300

Yatmaz, H.C. & Uzman, Y. (2009). Degradation of Pesticide Monochrotophos from Aqueous Solutions by Electrochemical Methods. *International Journal of Electrochemical Science*, Vol. 4, No. 5, 614-626, ISSN 1452-3981

Part 3

Herbicide Toxicity and Further Applications

20

The Bioassay Technique in the Study of the Herbicide Effects

Pilar Sandín-España, Iñigo Loureiro, Concepción Escorial, Cristina Chueca and Inés Santín-Montanyá
Instituto Nacional de Investigación y Tecnología Agraria y Alimentaria (INIA)
Spain

1. Introduction

Strategies for weed control are based primarily on chemical control, since the last decades the use of synthetic chemical products has been dramatically increased. The use of plant protection products is a source of concern for the society of developed countries, which has a growing interest in the environment, nature conservation and public health in general. This situation has led to deep changes in the objectives of the research on agriculture. The development and implementation of sustainable agriculture conduct to a rational use of plant protection products. The regulatory organisms (national and international) and the chemical industry of pesticides have taken steps to reduce the environmental impact of such organic compounds. In this context, there is now a great concern about the chemical nature of the products used in agriculture and its impact on adjacent ecosystems and the toxicity of these substances in ground and surface water.

The widespread use of herbicides create also concern about the possibilities of the risk of phytotoxicity on other species which are not direct object of the treatment. On the one hand, the risk involved in rotational crops due to of the accumulation in the field of herbicides that have a high persistence and are applied repeatedly each year, and on the other hand, the crops or plants adjacent to the treated crop may be affected by herbicide drift during the application of the product (Pestemer & Zwerger, 1999).

On the basis of these considerations, the risk assessment of the use of plant protection products on non-target plants should focus taking into account the agronomic use of the product. In this context, the bioassay technique is a useful tool that complements the analytical methods and provides information regarding herbicide bioavailability for the plant and its possible phytotoxicity (Kotoula-Syka et al., 1993; Stork & Hannah, 1996). Therefore, in the case of herbicides, we can define two groups according to good agricultural practices: the vegetation adjacent of agriculture areas and successive crops in the rotation.

2. The role and application of bioassay techniques on the impact assessment of herbicides

Bioassays or biological tests applied to the study of herbicides are based on the response of different species, chosen as controls, to the application of the herbicide under study

(Horowitz, 1976). They represent a valuable and necessary tool that provides an overview of soil-plant-herbicide relationships (Rahman et al., 1993; Hernández-Sevillano et al., 1999). Although there are chemical methods of analysis accurate and simple to use, bioassays have certain advantages in the study of herbicides:

- Phytotoxicity bioassays detect both the active substance and the possible degradation products of the herbicide.
- The biological assays provide practical information, being based on observation of the response of the plant to the herbicide (Blacklow & Pheloung, 1991).
- The methodology and materials necessary to carry out bioassays are generally simple and inexpensive.

The sensitivity, low cost and reproducibility of bioassays fulfil the criteria for a good technique (Günther et al., 1993). However, the bioassays by themselves cannot provide complete information on the environmental performance of these substances. Although reveal potential problems that residues of these products may present, as the effect on non-target species or in successive crops do not provide information needed to relate these effects with the chemical nature of the residue. Therefore, it is also necessary to study the nature of this residue by conventional analytical methods to identify potential causes of environmental problems and possible solutions. In this sense, it is very important to know whether any phytotoxic effects were detected in the bioassay due to active substance applied or to some of its metabolites or degradation products. In this case, it is necessary to know the route and rate of degradation not only for the active substance but also for the products of degradation (Parrish et al., 1995).

There is a need for evaluating and assessing the risk of the use of Plant Protection Products on non-target plants. The requirements for non target plants testing of pesticides vary among international agencies and their member countries. The risk assessment of the use of plant protection products on non-target terrestrial plants has been included until now as a generic assessment for registration of the plant protection product in the European Union (EU). However, generally state that there is a need to report all potentially adverse effects and undertake additional studies where there are indications of such effects.

The European Commission recommends the use of bioassays as an acceptable method to detect low levels of herbicide residues in soil. Such recommendation has been published in the European Commission Guidance Document Residue Analytical Methods (Anonymous, 2000) has accepted bioassays as suitable screening test that can be useful to exclude the occurrence of low levels of residues of phytotoxic compounds. Bioassays have become a necessary tool to detect herbicide soil residues and the results of these bioassays are now used to guarantee non-injury to the succeeding crop in crop rotation (Pestemer et al., 1980). Additionally, available pesticide phytotoxicity data on crop and non-crop species included in dossiers submitted to EU Member States for evaluation of active substances that could be commercialised in Europe according to the requirements of the Directive 91/414/EEC (Anonymous, 1991), concerning the placing of plant protection products within the European Union provides a harmonized procedure for the approval of these products. This directive requires that plant protection products marketed and used in the EU meet in their normal use, the following requirements: a) they do not produce harmful effects on human or animal health nor an unacceptable effect on the environment, and b) waste resulting from their application does not have harmful effects on human or animal health or on groundwater or any unacceptable influence on the environment, and that can be measured by methods in general. It is not acceptable the use of products with toxic effects on its target species, whether caused by themselves or their products applied waste. This directive has been extended by a set of guidelines detailing the information required in each of the sections, to be submitted by notifiers to justify the inclusion of active substances in Annex I to that directive. Therefore, the results of herbicide bioassays are essential input for many herbicide optimization programs, and for many studies of plant biomechanisms.

To assess the acute risk for terrestrial plants, the EU Guidance Document (SANCO, 2002) suggested starting with a first tier. If negative effects on terrestrial plants occur in the screening test or if results indicate a hazard potential for further risk assessment, then specific information on the toxicity of the substance to terrestrial plants should be requested. It is recommended to conduct dose-response test on 6-10 plant species representing as many taxonomic groups as possible. Germany Federal Environmental Agency propose phytotoxicity test with at least six plant species, three monocotyledonae, three dicotyledonae, including one leguminose and *Avena* and *Brassica,* (Füll et al., 2000). Other researchers (Gong et al., 1999) suggested four species of higher plants used for testing, two monocotyledons and two dicotyledons. We propose that selection of plant species would be specific for each situation, and it does depend of each herbicide. All selected species should present a useful tool in laboratory and are representative of important families in the agrarian ecosystem.

2.1 The use of Bioassays to detect herbicide phytotoxicity

The power of bioassays to detect bioavailable residues has led to success (Streibig et al., 1993). Bioassay methods have been developed to determine the residue level of many herbicides in soil and water. There are different types of bioassays, depending on the species, the type of herbicide used, its mode of action, substrate and other environmental conditions, as well as the measured parameter. Essentially, the biological test requires the choice of an indicator organism or specie that in the study of herbicides´ effects is very often a terrestrial plant. After the herbicide treatment one or more biological parameters of the plants that were affected will be assessed. At this point, visual assessment is recommended, but more rigorous measures are needed such as germination percentage, size or weight of the plants, or changes in physiological activities like photosynthesis or respiration (Horowitz, 1976).

The relationship between herbicide dose and plant response is of fundamental importance in understanding herbicide efficacy and mode of action. A thorough understanding of this relationship is essential for the design and interpretation of research in the field, greenhouse, or laboratory. The results of the measurements should be statistically analyzed to determine whether the observed effects are due to herbicide treatment or there is a response to increasing doses of the herbicide. The results of bioassays show the potential risk to sensitive crops after treatment, and provide information about the phytotoxicity of herbicide residue in the soil at sowing time. The classical bioassay, often used to quantify the amount of herbicide in soil, employs a single "standard" dose-response curve. This standard curve show the plant response to different herbicide concentrations and report information of different concepts related to herbicide efficacy, such as selectivity, tolerance and resistance.

A typical dose-response curve is sigmoid in shape. One example of such a curve is the log-logistic curve (Seefeldt et al., 1995). The mathematical expression relating the response y to de dose x is:

$$\text{LOG-LOGISTIC: } Y=C+ ((D-C)/ (1+\exp (b.\ln(X)-\ln(EC_{50}+1)))))$$

(C=lower limit, D= upper limit, b= slope, and EC_{50}= dose giving 50% response)

The log-logistic is the most common model used in bioassays to describe dose-response relations. Other relevant sigmoid curves might be the Gompertz (Streibig et al., 1993), is used sometimes, for instance, in cases where a log-logistic model did not fit well to the data.

$$\text{GOMPERTZ: } \quad Y= \exp[\ln(A). \exp(-rX)]$$

(A is the upper asymptote and r the slope of the linearized function)

The methodology of the risk assessment consist in comparing the toxicity with the predicted exposure applying a safety factor to this ratio in order to cover the uncertainty of the extrapolation from laboratory data to the field. The toxicity factor used in the risk assessment is usually the EC_{50} (concentration required to give 50% reduction of the plant growth with respect to the control) and NOEC (No Observable Effect Concentration) of representative species assayed in laboratory.

In order to mitigate the adverse effects of the use of plant protection products have been developed in recent decades molecules effective at low doses in order to fulfill environmental requirements, water and soil pollution set by international legislation.

There are scientific references concerning the effect of low-dose herbicides belonging to the family of sulfonylureas, in susceptible crops in the rotation (Blacklow & Pheloung, 1991; Alonso-Prados et al., 2001). Sulfonylurea herbicides inhibit plant metabolism by inhibiting acetolactate synthase, a crucial enzyme for the biosynthesis of branched-chain amino acids. These herbicides have a high specific activity in cereal crops and can be used effectively to control a wide range of grass and broad-leaved weeds at low doses (between 4 and 20 g a.i./ha). Previous studies have shown that sugar beet, sorghum, barley, pea or oilseed rape were injured in field assays when they were grown after wheat treated with sulfonylurea herbicides during the preceding spring or even autumn (Günther et al., 1993; Szmigielska et al., 1998; Shinn et al., 1999). It has been described a 20% barley growth inhibition caused by 0.0015 mg/L sulfosulfuron in growth chamber bioassay and also some injury on this crop in field assay (Parrish et al., 1995); according to other authors, crops like lucerne, oilseed rape, flax and sugar beet respond to sulfonylurea herbicides residues similarly in the field and growth chamber experiments with soils (Moyer, 1995). Also, Hernández-Sevillano (2001) found that quantities between 0.008 and 0.003 mg/L of sulfosulfuron and triasulfuron reduced sunflower root length by 50% in soil bioassays carried out in growth chamber.

Several bioassay methods for sulfonylurea herbicides have been reported using lentil (*Lens culinaris* Med.), lettuce (*Lactuca sativa* L.), sunflower (*Helianthus annuus* L.), corn (*Zea mays* L.), pea (*Pisum sativum* L.) and lupin (*Lupinus angustifolius* L.). In some studies, plant height or dry or fresh weight has been found to be a sensitive response parameter to sulfonylurea exposure (Blacklow & Pheloung, 1991; Günther et al., 1993; Junnila et al. 1994; Stork and Hannah, 1996; Vicari et al., 1994; Walker & Welch, 1989). Root growth effects have also been assessed after sulfonylurea treatments due to improved precision and sensitive (Blacklow & Pheloung, 1991). Root responses to sulfonylurea exposure have been measured by root dry weight, but previous studies in our laboratory shown that the most sensitive biological parameter used in bioassay with sulfosulfuron was root length (Hernández-Sevillano et al., 2001); therefore we have used the inhibition of root growth as susceptible parameter that indicate injuries in plants. Landi & Catizone (1989) found that response of corn cultivars to soil-applied chlorsulfuron in the field correlated better with root length than with root dry weight. Sunflower root dry weight was used by Kotoula-Syka et al. (1993) to study the persistence and phytotoxicity of several sulfonylureas in three different soils.

Also, some authors have found that plant response to the total herbicide residue in soil is site-specific; Stadler & Pestemer (1980) found that herbicide damage in crops is related to water-extractable (plant-available) residues. Thus, relationships between crop response and herbicide dose could be determined in a soil-free system bioassay to eliminate the confounding effects of soil adsorption and degradation of the herbicide (Ferris & Haigh, 1992; Jettner et al., 1999). Hydroponics' conditions promote herbicide activity because allow the maximum herbicide bioavailability to the plant, all the roots were confined within the solution with the plant at maximum water uptake. Plants are growing in most uniform conditions in the soil-free system bioassays, and the variability of the results due to environmental conditions was reduced with this method.

On the basis of these considerations, it has been studied the response of seven species (flax, corn, onion, vetch, lepidium, tomato and barley) to different doses of sulfosulfuron in hydroponic culture (Santín-Montanyá et al., 2006), in order to use this system as a rapid bioassay to detect phytotoxic levels of herbicide in crops and non-target plants and determine its effect in the most susceptible specie and also the most sensitive biological parameter for each species. We generated the dose-response curves of root growth 7 days after treatment for sulfosulfuron with the susceptible species, in order to estimate the EC_{50}. The most susceptible biological parameter was root growth for all species studied; this parameter permit us knowing the effect of sulfosulfuron on plants and it was used for obtain the dosis-response curves for each specie studied that have been treated with sulfosulfuron. The results showed that all species were susceptibles to sulfosulfuron, therefore the injuries caused in shoot fresh weight and shoot dry weight were growing with the doses of herbicide for all species. Additionally, root system control and less injured plants were increasingly deformed (main tap root twisted and lack of secondary roots); we could see how root growth was increasingly affected with increasing doses for all species, causing between 60 % and 98 % of root growth reduction with doses of 5.10^{-4} and 0.1 ppm a.i. of sulfosulfuron respectively applied on flax. These experiments with known concentrations of sulfosulfuron on the eight bioassays species showed that flax was the most susceptible specie to this herbicide.

Log-logistic and Gompertz model were tested for all species and the root length estimated by non-linear regression in the fitted model (Table 1). The Gompertz was considering the better model for the response in flax, corn, tomato and onion to sulfosulfuron; and Log-logistic regression model describe the data for lepidium, vetch and barley. The $EC_{10,30,50}$ were calculated in flax, maize, onion, vetch and *Lepidium sativum* according the estimated equations of each bioassay varied from 0.000053 mg/L to 0.0017 mg/L. Therefore, we could see that flax, corn, onion, vetch and lepidium root growth proved sensitive enough to detect very low phytotoxic level of sulfosulfuron; while tomato and barley were the less susceptible species.

Ciclohexanodione herbicides are also a family used at low-dose rate as they are biologically active at very low concentration (0.2-0.5 kg a.i./ha). These herbicides inhibit the activity of acetyl CoA carboxilase, a crucial enzyme in fatty acid synthesis. Furthermore, their polar character makes them easily to leach and potentially contaminate groundwater. However,

Specie	Upper asymptote (cm)	Lower asymptote (cm)	Slope [cm/(mg.m.L-1)]	EC_{10} (mg/L)	EC_{30} (mg/L)	EC_{50} (mg/L)	R^2 (%)
Flax [a]	21.69	-	629.52	0.000053	0.00019	0.0004	97
Maize [a]	10.83	-	526.40	0.000085	0.0003	0.00065	92
Onion [a]	9.48	-	271.46	0.00017	0.00063	0.0013	83
Tomato [a]	8.37	-	33.05	0.0015	0.0055	0.011	98
Vetch [a]	11.83	-	171.34	0.00025	0.009	0.0019	98
Lepidium sativum [a]	4.55	-	151.90	0.00047	0.0017	0.004	93
Barley [b]	5.86	0.92	0.99	0.042	0.16	0.39	93

[a & b] Regression Equations by Gompertz and Seefeldt models respectively

Table 1. Parameters of regression equations that describe the relationship between sulfosulfuron and root growth of different species and plant response for 10%, 30% and 50% inhibition of root growth (EC_{10}, EC_{30}, EC_{50})

due to high phytotoxicity, small amounts of residual herbicide in soil may affect sensitive succeeding crops. In this context, there is some information about the mobility, degradation and persistence in soil and water. These studies were performed with a variety of analytical techniques like gas chromatography, liquid chromatography, mass spectroscopy, photodegradation studies, studies with ^{14}C, immunoassays, etc. However, most studies have been made in water and soil, occasionally there is some bioassays in microalgae (Santín-Montanyá et al., 2007). The last results obtained confirm that could be a susceptible specie capable to detect the presence of some herbicides (Fig. 1 & Table 2).

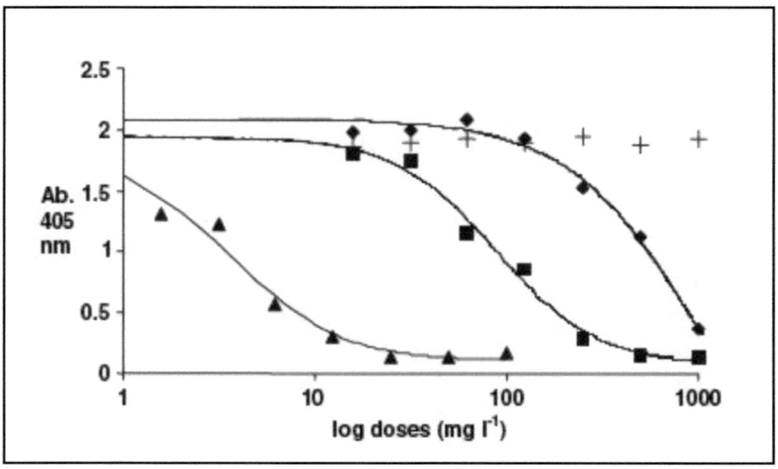

Fig. 1. Dose-response relationships of microalgae *Dunaliella primolecta* growth in the presence of different concentrations of alloxydim (♦), sethoxidim (■), metamitron (▲) and clopyralid (+)

Herbicides	**D** (cm)	**C** (cm)	b [cm/(mg.m.L^{-1})]	**EC_{10}** (mg/L)	**EC_{50}** (mg/L)	**R^2** (%)
Alloxydim [a]	2.09	-1.28	1.29	177.20	973.20	87.1
Sethoxydim [a]	1.95	0.072	1.66	23.32	87.63	87.6
Metamitron [a]	2.12	0.103	1.67	0.76	2.87	89.6
Clopyralid	Not adjusted to regression equation. No inhibitory effect					

[a] Regression equation by Seefeldt model

Table 2. Parameters of regression equations that describe the relationships between increasing rates of herbicides and growth of *Dunaliella primolecta*

Previous bioassays have been developed to detect phytotoxic residues of herbicide sethoxidim (Hsiao & Smith, 1983). In our group, initial attempts to obtain a practical hydroponic bioassay that allowed us to quantify tepraloxydim were frustrated due to the lack of repeatability and random results. Therefore, an investigation was carried out to determine the fate of tepraloxydim under bioassay conditions in order to clarify the reason for poor bioassay repeatability. The presence of residual chlorine in water was identified as a key factor on the repeatability of the bioassay. Finally, an extensive research was conducted to develop and optimize a bioassay based on the high sensitivity of wheat (*Triticum aestivum* L) to tepraloxydim in hydroponic culture using chlorine free mineral water (Sandín-España et al., 2003). Afterwards, similar studies were carried out with tralkoxydim (Fig. 2).
It has been demonstrated that water chlorination with disinfection purposes degrades completely any possible residue of herbicide clethodim (Sandín-España et al., 2005a). This degradation is very rapid, giving rise to different degradation products.

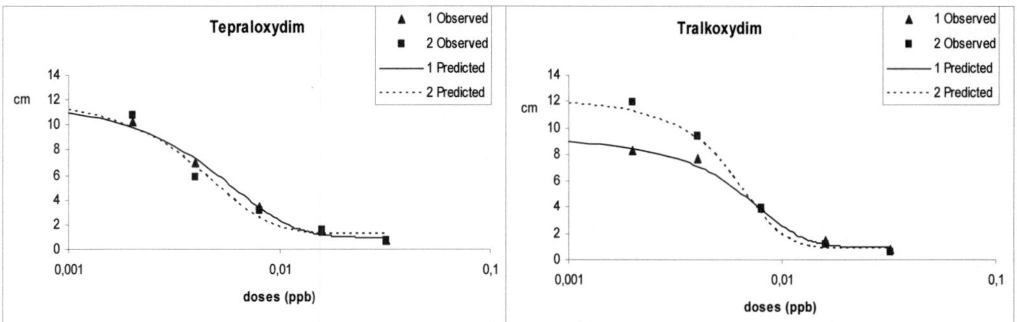

	Bioassay	EC_{50} (µg/L)	R^2 (%)
Tepraloxydim	B1	4.6	99.8
	B2	3.8	97.2
Tralkoxydim	B1	6.8	98.4
	B2	7.0	99.6

Fig. 2. Dose-response curves and EC_{50} to ciclohexanodione herbicides in hydroponic culture of wheat

The foregoing results suggest that the use of low dose herbicides can produce damage on succeeding crops, neighbouring crops and on non target plants. Overall, there is no one species or endpoint that is consistently the most sensitive for all species or all chemicals in all soils, and differences in bioavailability among compounds may confound comparison of test results (Clark et al., 2004). Therefore, bioassays can provide additional information, with acceptable reproducibility (Nyffeler et al., 1982; Streibig et al., 1995) on herbicide uptake and translocation (Horowitz 1976; Best et al., 1975). Besides, bioassays are employed in studies of persistence and mobility of herbicide soil residues (Ragab, 1974).

3. Bioassays in selectivity and resistance to herbicides

The basis for much of the work done in crop-weed management is weed control. In areas of well-developed agriculture weed control is mainly based on chemical control by herbicides. The extensive and redundant use of herbicides could present problems both in agricultural systems and in the surrounding environment. To detect any possible effect of herbicides in the plant and to test herbicides efficacy, response assays and tests must be carried out at various levels in the laboratory, greenhouse and field. Field studies are the best way of studying herbicide effect but accurate and efficient greenhouse and laboratory tests could be of the up most importance. The quicker and more simple the testing is, the more effective it will be. Because most laboratory research work utilizes large numbers of plants, a simple and rapid method is desirable. This is the case in determining crop selectivity, herbicide resistance in weeds or selecting individuals resistant to a particular herbicide, as a part of an improvement, mutagenic and/or gene flow process.

3.1 Crop selectivity

Conventional breeding programs frequently don't consider the herbicide response of cultivars during the selection process, it is why some cultivars show problems when treated with herbicides in culture in field. This is particularly true for crop response to new herbicides or new use of an herbicide. Cultivars show wide differences in response to herbicides and in many cases the concentration of herbicide needed to control weeds, or a particular weed, is deleterious if not lethal to the crop. For example, the control of *Bromus diandrus* in cereals is of concern. Bromes are vigorous competitors in winter cereals in many parts of the world (Blacksshaw, 1993) and cultural methods are the basis for their control because the herbicides used for weed control in cereals are not effective in controlling brome grass. The development of a sulfonilurea herbicide allowed a good control of Bromus spp in wheat. Hydroponic *in vitro* herbicide treatments were carried out. In those assays, germinated seeds were disposed on a grid in a black beaker filled with nutrient solution at the grid level (Fig. 3). When plants were 10 days old they were placed during 24 hours in another vessel filled with herbicide solution.

Six days after herbicide treatment plants were weighted. The results obtained from hydroponic herbicide treatment of wheat, barley and *B.diandrus* besides glasshouse spread of plants allowed to confirm the varietal selectivity of *Triticum aestivum* L.and T*riticum turgidum* L. cultivars and the susceptibility of barley cultivars to the herbicide doses that controls *B.diandrus* (Villarroya et al., 1997).

There are herbicides as glyphosate that when applied on the plant leaf, damaged the plant but the effect is relatively slow; several days will elapse before symptoms of damage appear (Duke, 1988). For cereals four to six weeks will be necessary; during this time the seedling

Fig. 3. Wheat plants in an *in vitro* herbicide response test

will need both soil or hydroponic support and space in a room or greenhouse. Quick methods to detect glyphosate's effect have been developed, (Harring et al., 1998; Madsen et al., 1995) including methods based on root absorption of glyphosate (Duke, 1978). Root application of the herbicide presents fewer problems than foliar application, especially with regard to interaction with ambient conditions (Hull et al., 1975). Trial assays in sorghum (Hensley et al., 1978), peas (Yenne et al., 1988) and corn (Racchi et al., 1995) have been carried out using root absorption. A method to evaluate the response of wheat and barley to glyphosate by measuring coleoptile length allows for the rapid detection of the more sensitive cereal lines and the selection of the more tolerant ones. Two barley cultivars (*Hordeum vulgare* L), "Jeff' and 'Amaji Nijo' (AN) as well as two wheat cultivars (*T.aestivum*), 'Chinese Spring' (CS) and 'Pavon' were used. Seeds were germinated in glyphosate solution in Petri dishes. After 24 hours, the dishes were opened and placed on a tray lined with water-moistened filter paper and covered with a transparent plastic film to maintain humidity. The tray containing the dishes was kept in a culture chamber under controlled conditions. The length of the coleoptile was measured four days after treatment (Fig. 4). The barley cultivars tolerated a higher dose of glyphosate than the wheat cultivars allows this method to evince differences in the responses of the cultivars as is shown by the log-logistic regression model applied. This method correlated with plant responses has provided an accurate model for describing the data with a good estimation of dose response (Fig. 5), equations for each cultivar by both methods.

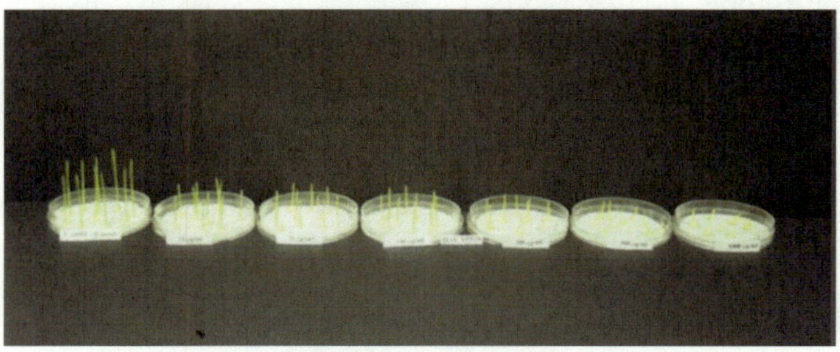

Fig. 4. Effect of herbicide dosage on wheat coleoptile length

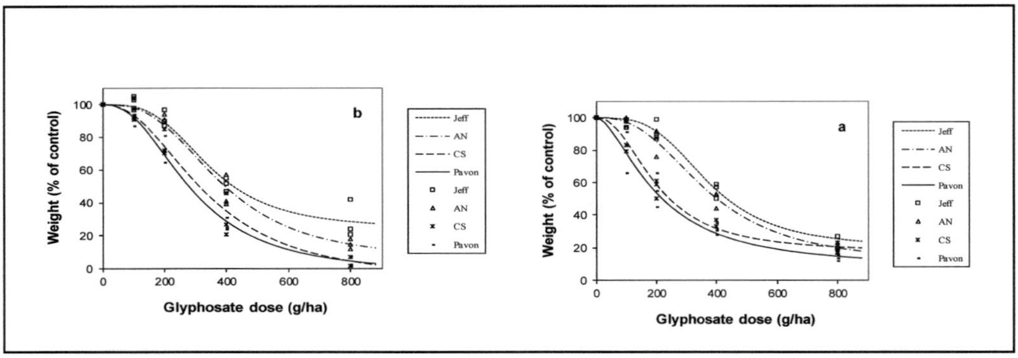

Fig. 5. Response in fresh weight of wheat (CS and Pavon) and barley (AN and Jeff) to glyphosate

The seed assay proved an accurate and rapid method to evaluate glyphosate efficacy. The seed assay can be completed in four to five days while the plant assay requires up to 30 to 45 days. The possible resistant plants detected by this method can be grown out after treatment in a greenhouse or in the field, where their resistance will be confirmed. This method is highly useful to detect tolerance to other herbicides as dalapon (Loureiro et al., 2001) and to detect populations of resistant weeds (Barroso et al., 2010) in the field as well as to initially select the lines obtained after mutagenic treatments or *in vitro* regeneration (Escorial et al., 2001).

Crop selectivity is related with the genetic control of herbicide response. Knowledge of the sources and genetic control of tolerance to herbicides should always be taken into account in the development of new improved crop varieties and in implementing a weed management system. Although genetic control of tolerance to herbicides was not largely been investigated in wheat, authors have already reported cytoplasmic, poligenic nuclear control as well as monogenic nuclear control of the response of different crops to the herbicides.

The bread wheat cultivars 'Castan' and 'Recital' are tolerant and susceptible respectively to chlorotoluron herbicide (Sixto & Garcia-Baudin, 1988). However, while the distribution of responses among wheat cultivars to chlorotoluron reported so far are discrete, some papers report only two classes, tolerant and susceptible (Tottman et al., 1975). A single seedling, non-destructive, easy to handle, cheap, fast and efficient assay was developed to score wheat responses to herbicides and to investigate the genetic control of the differences in response to chlortoluron between the cultivars 'Castan' and 'Recital' Its efficiency makes possible the detection not only of differences due to major genes but also to minor or modifier genes (Sixto et al., 1995). The results not only confirm the presence of a major tolerant allele controlling the differences in response between the two cultivars, but also, show the contributions of modifier genes present in 'Castan', 'Recital' and other related cultivars. This assay is applied nondestructively to single individuals plantlets that are scored in vitro in a herbicide solution of clortoluron (Fig. 6) and, if selected, can be transplanted, grown to maturity and cross-fertilized if desired. The test may save up to two generations in genetic schemes where scoring is done in large samples grown to maturity. This test was also used to conclude that in the inheritance of durum wheat (*T.turgidum* var *durum*) to metribuzin (Villarroya et al 2000) the tolerance is dominant and relatively few genes (around four) are involved in tolerance for this character. Heretability of this trait was very high with value of 0.60 in narrow sense and of 0.86 in broad sense. The results of this

work can help in the selection techniques employed to obtain durum wheat with increased tolerance to metribuzin, that could increase the margin of safety in Brome control in wheat.
If selectivity is not present for a herbicide-crop couple in vitro selection could be used to detect herbicide tolerance over mutations produced by somaclonal variation. *In vitro* culture has been used to select herbicide-tolerant plants of dicotyledonous (Aviv & Galum, 1977; Wersuhn et al., 1987) and of monocotyledoneous species between them *T.aestivum* L. tolerant to chlortoluron and to difenzoquat (Bozorgipour & Snape, 1991), and *H.vulgare* L. tolerant to chlorsulfuron (Baillie et al., 1993) and glyphosate (Escorial et al., 1996). In the last case, *in vitro* culture of barley calluses from immature embryos of barley (*H.vulgare* L. 'Jeff') were cultured for some months on medium with glyphosate. Plants were regenerated and the progeny of each regenerated plant was analyzed for response to glyphosate. An herbicide test was adapted to detect plants tolerant to glyphosate. Plants between 3 and 6 cm tall were treated with one drop of 1µl of glyphosate solution applied on the base of the third leaf. The length of the third leaf was measured at the time of herbicide treatment and seven and fourteen days after the treatment. Some progenies showed increased tolerance to glyphosate and show that glyphosate tolerance in barley can be increased by *in vitro* culture selection.

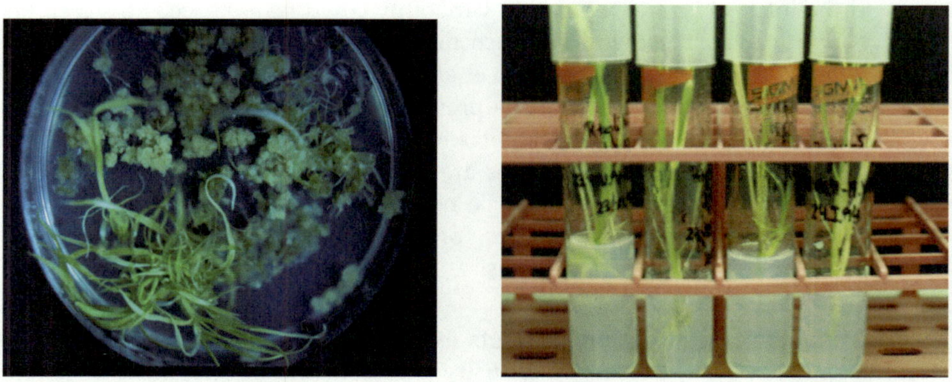

Fig. 6. Barley regenerants obtained after *in vitro* culture in medium with glyphosate herbicide

3.2 Herbicide resistance in weeds

The widespread use of herbicides for weed control over the past decades has exposed huge weed populations to strong selection pressures that lead to the appearance and proliferation of weeds resistant to different chemical classes of herbicides. The adoption of genetically modified crops will promote in the future a greater use of monoculture systems and generate a higher risk of possible appearance of resistance through the selection pressure produced by the continued use of a single herbicide (Powles, 2008). Thus, studies of weed resistance are important to stop or mitigate it. Herbicide resistance could be a field concern if it is spread in a field or in an area, or in a previous stage (not apparent in the field) in which an increase of proportion of resistant plants and/or a decrease of response in a given population.
To detect herbicide resistance, several authors have adjusted short, quick and cheap bioassays to evaluate herbicide effect, which have allowed to detect responses of biotypes to

diclofop-methyl, trifluralin, acetyl-coenzyme A carboxylase inhibitors, dalapon or glyphosate herbicides (Beckie et al., 2001; Barroso et al., 2010).
The above mentioned Petri dish bioassay cereal method adapted for weeds (*B.diandrus* and *L.rigidum*) is a proper method to detect weed resistant populations as well as to establish a baseline sensitivity, although caution is needed with the results obtained by this method if the resistance mechanisms are unknown. Baseline sensitivity gives information about the level of resistance to a particular plant protection product in a weed population and allow comparisons among different populations and between the same populations at different times, allowing the evaluation in sensitivity changes both between populations and along a period of time. The method was validated positively for dalapon and *B. diandrus and L. rigidum* (Barroso et al., 2010). This method once validated is much more practical than others methods used for herbicide resistance evaluation (Carrera de la et al., 1999; Carrera de la et al., 2000) as were methods based on mortality of the plant, number of leaves developed by the plant in a period of time and/or the length of the third or fourth leaf.
Before the molecular identification of resistant weeds (Sherwood & Jaseniuk, 2009) was largely used, this Petri dish bioassay could be of interest in the study of the structure of the populations in terms of evaluating the response of a population as the integrated response of each of the individuals that belongs to the population. By this way resistant plants can be detected and the inter and intra-population variability could be assessed. It is well known that the evolution of resistance will be much more rapid if a population carries resistant alleles before selection is imposed (Loureiro et al., 2010). The variation of the response of a population to herbicides is the result of the previous management of the fields and is the starting point for future weed-management strategies. It is likely that the frequency of resistant plants increase further if measures are not taken and control rely on herbicides with the same mode of action. It will thus be necessary to diversify the managements by rotating herbicides with different modes of action, by alternating crops and by implementing diversified cropping programs.

3.3 The importance of degradation products in the study of herbicides

The current tendency in agrochemical industry points to the development of herbicides more selective, less environmentally persistent, with less toxicity and bioaccumulation. In this regard, families of herbicides like sulfonilurea and cyclohexanedione oxime that belong to the third generation of pesticides, have appeared in the period 1970-1980 to fulfil these environmental requirements.
Although little attention has been paid to degradation products and metabolites in the past, by-products need to be considered to gain complete understanding of the environmental impact of these xenobiotics, otherwise herbicides' fate could be substantially underestimated. Determination of degradation products of organic compounds, such herbicides, is nowadays one of the major challenges in analytical chemistry of environmental pollutants.
In many cases, parent compound and transformation by-products possess different physico-chemical properties. The higher polarity and hence solubility in water of some degradation products increase the risk to contaminate the aquatic media. Data available show that concentration of degradation products presents in water is sometimes higher than those of the parent compound. Besides, degradation products are often more toxic and/or persistent in environmental matrices than their parent (Barceló & Hennion, 1997).

However, determination of transformation products is sometimes difficult to carry out. In many cases they have never been identified nor characterized before and the availability of analytical standards is scarce.
Therefore, to predict their fate in the natural environment and to assess their risk, it is necessary to improve our knowledge on the reactions under environmental conditions.

3.2.1 Processes of herbicide degradation

Degradation of herbicides can begin as soon as they are synthesized. Formulation processes, transport and/or storage can initiate degradation of the active substance. As well, once the herbicide is prepared in the tank mix, further transformation can take place because of the reactions with other substances present in the water or due to interactions with other herbicides.
Once applied to the field, most of the herbicides applied do not immediately enter the plant, but remains in soil, water, air and surface of the plant leaves where are subject to different agents capable of transforming by abiotic and/or biotic processes into one or more transformation products.
Most pesticides applied to the environment are ultimately degraded into universally present materials such as carbon dioxide, ammonia, water, mineral salts and humic substances. Different chemicals, however, are formed before the herbicides are completely degraded. If the products are results of biological degradation, they are referred to as metabolites.
Agents responsible for the transformation of herbicides in the field can be physical, chemical and biological. The influence of each agent in the herbicide depends on the physical properties and chemical structure of the herbicide molecule.
The two main physical agents involved in the degradation process are light and temperature. Solar radiation is responsible for the photolysis and thermal degradation of the herbicides in the surfaces of soil, plant and water. It is known that photodegradation is one of the main abiotic processes that take place for many herbicides (Dimou et al., 2004; Scrano et al., 1999; Saha & Kulshrestha 2002; Ibáñez et al. 2004). For this to occur in water, the emission spectrum of the sun needs to fit the adsorption spectrum of the pollutant. Cyclohexanedione oxime herbicides photodegrade rapidly when they are exposed to simulated or natural solar irradiation in different types of water showing a dependence both on the irradiation energy and on the composition of the water sample (Sevilla-Morán et al., 2010a).
The effect of temperature on degradation has been studied in tropical ecosystems (Sahid & Teoh, 1994). High temperatures encountered in the tropics will lead to enhance degradation of herbicide.
Chemical degradation can take place when the herbicide gets in contact with water that possesses substances that promote its degradation. It is known that the presence of substances employed for the disinfection of water such as hypochlorite and chloramines degrade herbicide to compounds more or less toxic that the active substance. Rapid degradation of herbicide tepraloxydim was observed in the presence of chlorine. In the same way, clethodim was degraded completely in a few minutes when is exposed to chlorinated water, giving rise to the formation of various oxidation by-products (Sandín-España et al., 2005a).
It is worth noting that natural substances present in aquatic systems (dissolved organic matter (DOM), nitrate and metal ions, ...) may influence the photochemical behaviour of

organic compounds (Mazellier et al., 1997; Quivet et al., 2006; Sevilla-Morán et al., 2008). Diverse studies are available from literature where humic acids act enhancing (Santoro et al., 2000; Vialaton & Richard, 2002) or inhibiting (Dimou et al., 2005; Dimou et al., 2004) the degradation of herbicides. For instance the irradiation of sethoxydim (Sevilla-Morán et al., 2010a), alloxydim (Sevilla-Morán et al., 2008) and clethodim (Sevilla-Morán et al., 2010b) solutions containing humic acids slowed down the rate of the photodegradation, suggesting a strong "filter effect", while the presence of nitrate ions had no effect on the degradation.

In general, by increasing the organic-matter content ant the temperature, the degradation of herbicides in soils is enhanced. When the organic-matter content increases, the biomass of the active microbial population also increases and so does the degradation.

The role of organic matter in soils is very important. It has been shown that the most persistent complexes result from the direct covalent binding of pesticides to soil humic matter or clay. The pesticides most likely to bind covalently to the soil have chemical functionalities similar to the components of humus. The humic material is derived from the remains of decomposing plants, animals and microorganisms, and is composed primarily of humic and fulvic acids.

In order to investigate the soil degradation of pesticides, laboratory incubation studies with ^{14}C-labelled pesticides are required. These allow one to assess the likely rate of degradation of parent pesticides in soil, and provide information on the structure and likely degradability of metabolites.

In the same way, a variance of pH can accelerate the degradation of herbicides. The soil pH, for example, is an important parameter affecting the persistence of chemically unstable herbicides. The mobilities of acidic herbicides are related to pH, with higher mobility in soils with higher pH (Brown, 1990; Scrano et al., 1999; Boschin et al., 2007). Microorganisms are the most important group of biological agents present in the soil that degrade herbicides.

3.2.2 Biological activity of degradation products

All these degradations imply different reactions before the active substances are completely degraded or mineralized and one or two transformations are sometimes sufficient to alter the biological activity of the parent compound. For some herbicides a change that takes place in its molecular structure can change the physicochemical properties and also the toxicity to different species. Herbicides alachlor and metolachlor showed that the toxicity to the bacteria *V. Fisheri* was enhanced upon degradation (Osano et al., 2002). On the contrary, other studies showed that herbicides and its degradation products cannot be considered a risk for the environment. This is the case of some sulfonilureas herbicides, where neither the active substance nor the metabolites are toxic to *D. Magna* and *V. Fisheri*. (Martins et al., 2001; Vulliet et al., 2004).

Major degradation products of some herbicides also have herbicidal activity against target and/or non-target weeds. However, few studies have documented the level of herbicidal activity. Some pesticide degradation products are of significance in crop protection by being effective against the target weeds. It has been demonstrated that the formation of the sulfoxide by-product of thiocarbamate herbicides like butylate (Fig. 7), increased the herbicidal activity (Tuxhorn et al., 1986). On the contrary, some can be responsible for inadequate weed control by inducing rapid degradation of their parent compounds.

Evidence shows, however, that for some pesticides, the herbicidal activity attributed to parent compound is partly due to the products formed (Tuxhorn et al., 1986; Bresnahan et al., 2004). In some cases, herbicides are formed as degradation products of other herbicides for instance, chlorthiamid, a benzonitrile herbicide (Fig. 7), is the parent compound and the precursor of dichlobenil that is a degradation product formed in soil and also an herbicide.
Oxidation reactions occur frequently in the soil and are extremely important transformation pathway. S-containing herbicides are often rapidly oxidized to sulfoxide and afterwards more slowly to sulfones. Sulfoxidation can occur in soil and water mediated chemical or biologically (López et al., 1994; Hsieh et al., 1998; Ankumah et al., 1995).
This oxidation is so rapid and complete that sulfoxides are often the compounds found in soil shortly after application of the parent sulfide compound. Furthermore, in some cases, sulfoxides and sulfones are suspected to have the herbicidal activity (Campbell & Penner, 1985).
The herbicidal activity of carbamothiate herbicides sulfoxides has been previously reported. (Tuxhorn et al., 1986). In soils treated with butylate (Fig. 7), herbicide residues of the parent

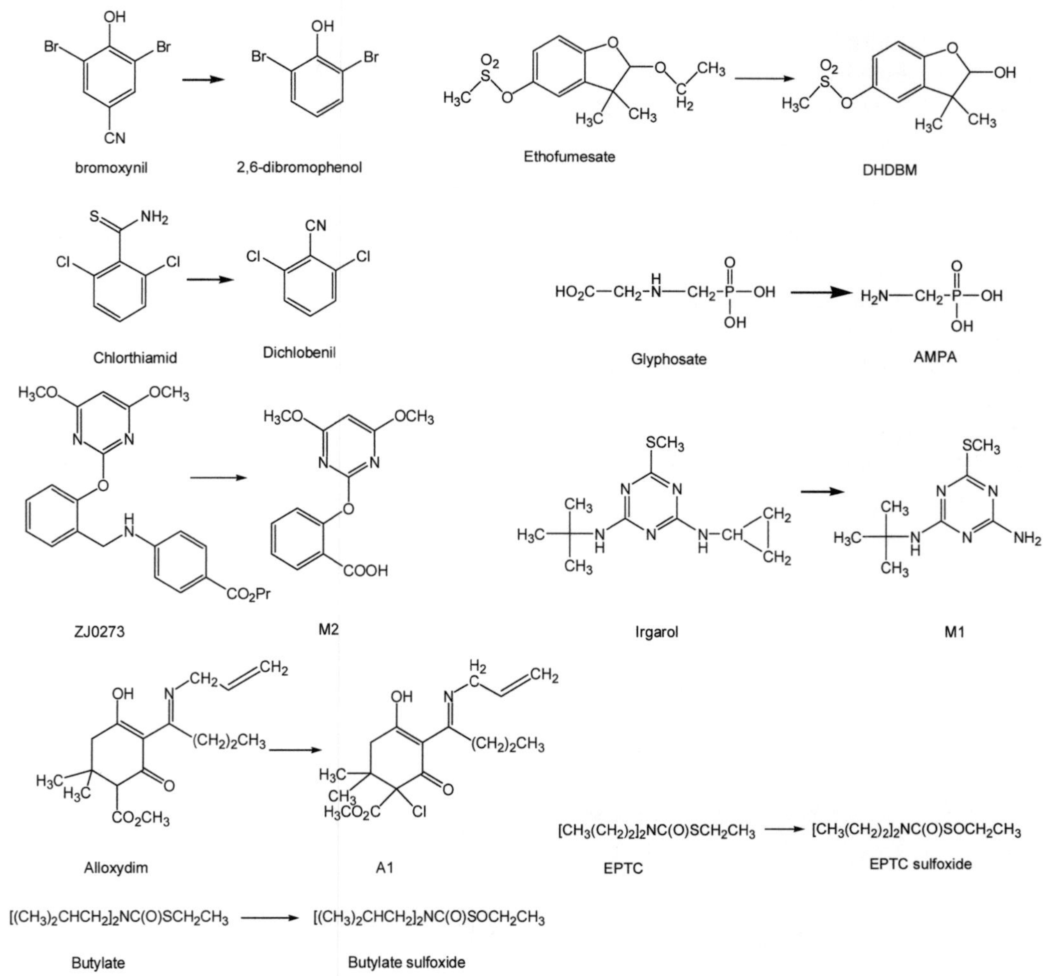

Fig. 7. Chemical structures of herbicides and degradation products discussed

compound were not detected in significant amounts within a few weeks after application. However, good control of weeds was observed in these fields. The good performance of this herbicide despite its lack of persistence was probably due to the by-products formed. Other degradation products effective on controlling target weeds are ETCP sulfoxide, which is the oxidation degradation product of the herbicide thiocarbamate EPTC (Fig. 7) (Somasundaram & Coats, 1991).

Relatively little is known of the potential phytotoxicity of degradation products and little literature exists on this topic.

Just as herbicides can be selective between plant species, metabolites can differ in their phytotoxicity pattern. Metabolites can have different mechanism of action and selectivities than the parent compound. For instance, bromoxynil (Fig. 7) is biological degraded in soil into 2,6-dibromophenol that is a potent growth regulates (Frear, 1976).

Kawahigashi et al., (2002) showed that the phytotoxicity of the de-ethylated metabolite of ethofumesate, DHDBM (Fig. 7), to rice plants was at least four times greater than that of the parent compound. Reddy et al., (2004) suggested that soybean injury to glyphosate-resistant soybean from glyphosate is due to its degradation product formed in plants, aminomethylphosphonic acid (AMPA). The degradation product of Irgarol 1051, M1 (Fig. 7) in the root elongation inhibition bioassay, showed a phytotoxicity at least 10 times greater than that of Irgarol and six other triazine herbicides (Okamura et al., 2000).

In many cases, degradation products are not phytotoxic as in the case of herbicide metsulfuron-methyl where the phytotoxicity of metsulfuron-methyl bound residues was mainly caused by the parent compound that became available during plant growth and no other metabolites detected (Ye et al., 2003).

As it has been explained before in this chapter, bioassays are important tools to screen herbicide residues and can be useful to exclude the occurrence of low levels of phytotoxic residues in soil (Hsiao & Smith, 1983; Sandín-España et al., 2003). In this sense, we have studied the phytotoxicity of alloxydim and its main metabolite with hydroponic bioassays on wheat (Sandín-España et al., 2005b).

Degradation product of alloxydim (Fig. 7) was the main product obtained in its degradation with chlorine, one of the most common disinfectant agents employed in water treatment.

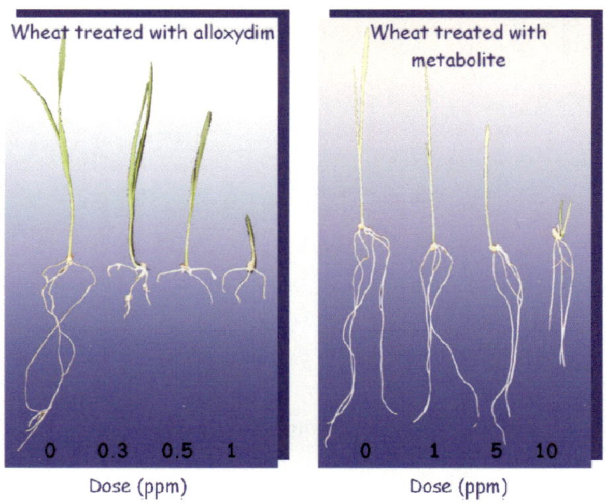

Fig. 8. Response of wheat plant to different doses of alloxydim and its metabolite

Results showed that after seven days of treatment the most sensitive biological parameter for alloxydim was root length, causing in the root growth of plants a 40% of significative reduction at the dose of 0.3 mg.l^{-1} and 94% of reduction at the highest dose. However, the effect of metabolite on root growth only occurred at the highest metabolite dose (10 mg.l^{-1}), causing a 32% of reduction in root growth. Root system control presented normal growth (main tap root plus secondary roots), while those from injured plants were increasingly deformed (main tap root twisted and lack of secondary roots). Root growth was increasingly affected with doses from 0.1 mg. l^{-1} to the highest dose (Fig. 8).

It is also important to highlight that a part of the degradation products formed in the soil from the herbicides remains as bound residues (Bresnahan et al., 2004; Albers et al., 2008; Rice et al., 2002). This non-extractable residue retained by organic matter in soil is bioavailable to plants. Therefore this portion of residue of degradation products and/or metabolites is underestimated if bound residues may be released from soil and absorbed by plants.

A study on phytotoxicity of soil bound residues of herbicide ZJ0273, a novel acetolactate synthase potential inhibitor, to rice and corn, revealed that one of his main metabolite (M2) (Fig. 7) played a dominant role in the inhibition effect on the growth of rice seedlings. In the extractable residues released from bound residues, the most biologically active M2 accounted for the largest fraction in all soils. Therefore, it was concluded that the main cause of phytotoxicity from exposure to soil bound residues of ZJ0273 is related to the release of ZJ0273 and its degradation products and the subsequent inhibition on ALS by M2 (Han et al., 2009).

In recent years it has been revealed the lack of data on the phytotoxic effects of herbicide residues. In this sense, it is necessary to study and develop simple methods for evaluating the environmental impact of these products based on hard scientific data. Besides, though most degradation products of herbicides are converted into less toxic or nontoxic compounds, some degradation products, because of their characteristics, may be biologically and/or environmentally active. Thus, major degradation products should be also considered in evaluating the potential bioactivity and environmental contamination of the parent compound. From these studies should be able to derive recommendations for agricultural practices for the use of these products to be environmentally friendly in general and in particular the agricultural environment capable of guaranteeing the future productivity of farms in the context of sustainable agriculture.

4. Acknowledges

This research was supported in part by the CICYT projects RTA 2008-00021-00-00 and RTA 2008-00027-00-00.

5. References

Albers, CN; Banta, GT; Hansen, P.E. & Jacobsen, O.S. (2008). Effect of Different Humic Substances on the Fate of Diuron and Its Main Metabolite 3,4-Dichloroaniline in Soil. *Environmental Science & Technology*, 42, 23, 8687-8691.

Alonso-Prados, J. L.; Hernández -Sevillano, E.; Llanos, S.; Villarroya, M. & García-Baudín, J. M. (2001). Effects of sulfosulfuron soil residues on barley (Hordeum vulgare), sunflower (Helianthus annuus) and common vetch (vicia sativa). *Crop Protection,* 21, 1061-1066.

Ankumah, R.; Dick, W. & Mcclung, G. (1995). Metabolism of carbamothioate herbicide, eptc, by rhodococcus strain je1 isolated from soil. *Soil science society of american journal,* 59, 4, 1071-1077.

Anonymous (1991). Council Directive 91/414/EEC concerning the commercialisation of plant protection products. Official Journal of the European Communities, L 230, 19 August 1991, pp.32.

Anonymous (2000). Guidance Document on Residue Analytical Methods. European Commission Document, Directorate General Health and Consumer Protection. SANCO/825/00 rev 6

Aviv, D. & Galum, E. (1977). Isolation of tobacco protoplasts in the presence of isopropyl-N-phenylcarbamate and their culture and regeneration into plants. *Z. Pflanzenphysiol.* 83, 267-273.

Baillie, A.M.R.; Rossnagel, B.G. & Kartha, K.K. (1993). *In vitro* selection for improved chlorsulfuron tolerance in barley (*Hordeum vulgare* L.). *Euphytica.* 67, 151-154.

Barceló, D. & Hennion, M.C. (1997). Pesticides and their degrdadation products: Characteristics, usage and environmental behaviour, In: *Trace determination of pesticides and their degradation products in water*. Barceló, D. & Hennion, M.C.,(Ed), 1-89, Elsevier, 0-444-81842-1.

Barroso, J.; Loureiro, I.; Escorial M.C. & Chueca, M.C. (2010). The response of *Bromus diandrus* and *Lolium rigidum* to dalapon and glyphosate I: baseline sensitivity *Weed Research* in press DOI: 10.1111/j.1365-3180.2010.00782.x

Beckie, H.J.; Hall, L.M. & Warwick, S.I. (2001). Impact of herbicide-resistant crops as weeds in Canada. *The BCPC Conference – Weeds,* 135-142.

Best, J.A.; Weber, J.B. & Monaco, T.J. (1975). Influence of soil pH on s-triazine availability to plants. *Weed Science,* 23, 378-382.

Blacklow, W. M. & Pheloung, P. C. (1991). Sulfonylurea herbicides applied to acidic sandy soils: a bioassay for residues and factors affecting recoveries. *Australian Journal of Agricultural Research,* 42, 1205-1216.

Blackshaw, R.E. (1993). Downy brome (*Bromus tectorum*) interference in winter rye (Secale cereale) *Weed Science,* 41, 557-562.

Boschin, G.; DÁgostina, A.; Antonioni, C.; Locati, A. & Arnoldi, A. (2007). Hydrolytic degradation of azimsulfuron, a sulfonylurea herbicide. *Chemosphere,* 68, 1312-1317.

Bozorgipour, R. & Snape, J.W. (1991) "In vitro" selection of herbicide tolerant variants of wheat.in : Caseley, J.C., Cussans, G.W. & Atkin, R.K. (Eds.)" *Herbicide Resistance in weeds and crops", XI Long Ashton Int. Symp. Sep.1989.* Butterworth. (pp. 422-423).

Bresnahan, G.A.; Koskinen, W.C.; Dexter, A.G. & Cox L. (2004). Sorption–desorption of 'aged' isoxaflutole and diketonitrile degradate in soil. *Weed Research,* 44, 397-403.

Brown, H. M. (1990). Mode of action, crop selectivity and soil relations of the sulfonylurea herbicides. *Pestic. Sci.* 29, 263-281.

Campbell J.R. & Penner D. (1985). Abiotic transformation of sethoxydim. *Weed Sci.* 33, 435-439

Carrera de la, N., Villarroya, M., Chueca, M.C. Garcia-Baudin, J.M. (1999). Evaluation of *Lolium rigidum* biotypes resistance to chlorsulfuron: useful parameters. *Proceedings of BCPC – Weeds* 153-158.

Carrera de la, N., Villarroya, M., Chueca, M.C. Garcia-Baudin, J.M. (2000). Relation entre des parametres phenologiques de plantules et la reponse de trois biotypes de *Lolium rigidum* Gaud à l'herbicide diclofop-méthyl *XIème Colloque International sur la Biologie des Mauvaises Herbes* . Dijon. 531-538.

Clark, J.; Ortego, L.S. & Fairbrother, A. (2004). Sources of variability in plant toxicity testing. *Chemosphere,* 57, 11, 1599-1612

Dimou, A.D.; Sakkas, V.A. & Albanis, T.A. (2004). Photodegradation of trifluralin in natural waters and soils: degradation kinetics and influence of organic matter. *Intern. J. Environ. Anal. Chem.* 84, 173-182.

Dimou, A.D.; Sakkas, V.A. & Albanis, T.A. (2005). Metolachlor photodegradation study in aqueous media under natural and simulated solar irradiation *J. Agric. Food Chem.* 53, 694-701.

Duke, S.O. (1988). Glyphosate. In: *Herbicides. Chemistry, Degradation, and Mode of Action.Vol.3.* Kearney P.C.; Kaufman D.D. Marcel Dekker, Inc. New York and Basel.pp.1-70.

Escorial, M.C.; Sixto, H.; Garcia-Baudin, J.M. & Chueca, M.C. (2001) A rapid method to determine cereal response to glyphosate. *Weed Technology* 15, 697-702.

Escorial, M.C.; Sixto,. H.: Garcia-Baudin, J.M. & Chueca, M.C. (1996). In vitro culture selecction increases glyphosate tolerance in barley *Plant Cell Tissue Organ Culture* 46, 179-186.

Ferrish, I. G. & Haigh, B. M. (1992). Prediction of herbicide persistence and phytotoxicity of residues. In: *Proceedings First International Weed Control Congress,* Melbourne 1, 193-207.

Frear, D. S. In *Herbicides – Chemistry, Degradation, and Mode of Action;* Kearney, P. C.; Kaufman, D. D., Eds; Marcel Dekker, Inc.: New York, NY, 1976, Vol. 2. pp. 541-607.

Füll, C.; Jung, S. & Schulte, C. (2000). Plant Protection products: assessing the risk for terrestrial plants. *Chemosphere,* 41, 625-629.

Gong, P.; Wilke, B. M. & Fleischmann, S. (1999). Soil-Based Phytotoxicity of 2,4,6-Trinitrotoluene (TNT) to Terrestrial Higher Plants. *Archives of Environmental Contamination and Toxicology,* 36, 152-157.

Günther, P.; Pestemer, W.; Rahman, A. & Nordmeyer, H. (1993). A bioassay technique to study the leaching behaviour of sulfonylurea herbicides in different soils. *Weed Research,* 33, 177-185.

Han, A.; Yue, L.; Li, Z.; Wang, H.; Wang, Y.; Ye, Q.; Lu, L. & Gan, J. (2009). Plant availability and phytotoxicity of soil bound residues of herbicide ZJ0273, a novel acetolactate synthase potential inhibitor. *Chemosphere,* 77, 955-961.

Harring, T.; Streibig, J.C. & Husted S. (1998). Accumulation of shikimic acid: A technique for screening glyphosate efficacy. *J. Agric. Food Chem.*m 46, 4406-4412.

Hensley D.L., Beuerman, D.S.N. & Carpenter, P.L. (1978). The inactivation of glyphosate by various soils and metal salts. *Weed Res.* 18:287-291.

Hernández- Sevillano, E.; Villarroya, M.; Alonso-Prados, J. L. & García Baudín, J. M. (2001). Bioassay to detect sulfosulfuron and triasulfuron residues in soil. *Weed Research*, 15, 447-452.

Hernández-Sevillano, E.; Villarroya, M.; Chueca, M. C.; Alonso-Prados, J. L. & García Baudín, J. M. (1999). A rapid, sensitive bioassay method for sulfonylurea herbicides. *Brighton Crop Protection Conference. Weeds*, 2, 711-716.

Horowitz, M. (1976). Application of bioassay techniques to herbicide investigations. *Weed Research*, 16, 209-215.

Hsiao, AI. & Smith, AE. (1983). A root bioassay procedure for the determination of chlorsulfuron, diclofop acid and sethoxydim residues in soils. *Weed Research*, 23, 231-236.

Hsieh, Y.N.; Liu, L.F. & Wang, Y.S. (1998). Uptake, translocation and metabolism of the herbicide molinate in tobacco and rice. *Pesticide Science*, 53, 2, 149-154.

Hull, H.M., Morton H.L. & Whanie, J.R. (1975). Environmental influences on cuticle development and resultant foliar penetration. *Bot. Rev.* 41, 421-452.

Ibáñez, M.; Sancho J.V.; Pozo, O. & Hernández, F. (2004). Use of quadrupole time-of-flight mass spectrometry in environmental analysis: elucidation of transformation products of triazine herbicides in water after UV exposure. *Anal. Chem.* 76, 5, 1328-1335.

Jettner, R. J.; Walker, S. R.; Churchett, J. D.; Blamey, F. P. C.; Adkins, S. W. & Bell, K. (1999). Plant sensitivity to atrazine and chlorsulfuron residues in a soil-free system. *Weed Research*, 39, 287-295.

Junnila, S.; Heinonen-Tanski, H.; Erviö, L. R. & Laitinen, P. (1994). Phytotoxic persistence and microbiological effects of chlorsulfuron and metsulfuron in Finnish soils. *Weed Research*, 34, 431-423.

Kawahigashi, H; Hirose, S; Hayashi, E; Ohkawa, H & Ohkawa, Y. (2002). Phytotoxicity and metabolism of ethofumesate in transgenic rice plants expressing the human CYP2B6 gene. *Pesticide biochemistry and physiology*, 74, 3, 139-147.

Kotoula-Syka, E.; Eleftherohorinos, I. G.; Gagianas, A. & Sticas, A. G. (1993). Phytotoxicity and persistence of chlorsulfuron, metsulfuron-methyl, triasulfuron and tribenuron-methyl in three soils. *Weed Research*, 33, 355-367.

Landi, P.A. & Catizone, P. (1989). Response of maize inbred lines and hybrids to chlorsulfuron. *Weed Research*, 29, 265-271.

Lopez, A.; Mascolo, G.; Tiravanti, G.; Santori, M. & Passino, R. (1994). Oxidation of sulfur-containing s-triazines during groundwater hypochlorination. *Water science and technology*, 30, 7, 53-59.

Loureiro I.; Rodriguez, E.; Gonzalez Andujar, J.L.; Escorial, M.C. & Chueca, M.C. (2010). Distribution and frequency of resistance to four herbicide modes of action in *Lolium rigidum* Gaud. Accessions randomly collected in winter cereal fields in Spain. *Crop Protection* doi:10.1016/j.cropro.2010.07.005.

Loureiro, I.; Avendaño, N.; Escorial, M.C.; Sixto.H.; Garcia-Baudin, J.M. & Chueca, M.C. (2001). Método rápido de evaluación del efecto del herbicida dalapon en cereales. *Investigación Agraria* 16, 463-469.

Madsen K.H., Heithol,. J.J., Duke, S.O., Smeda, R.J. & Streibig, J.C. (1995). Photosynthetic parameters in glyphosate-treated sugarbeet (*Beta vulgaris* L.). *Weed Res.* 35:81-88.

Martins, J.; Chevre, N.; Sapck, L.; Tarradellas, J.; & Mermoud, A. (2001). Degradation in soil and water and ecotoxicity of rimsulfuron and its metabolites. *Chemosphere,* 45, 4-5, 515-522.

Mazellier, P.; Jirkovsky, J. & Bolte, M. (1997). Degradation of diuron photoinduced by iron (III) in aqueous solution. *Pestic. Sci.*, 49, 3, 259-267.

Moyer, J. R. (1995). Sulfonylurea Herbicide Effects on Following Crops. *Weed Technology,* 9,373-379.

Nyffeler, A.; Gerber, H.R.; Hurle, W.; Pestemer, W. & Schmidt, R.R. (1982). Colaborative studies of dose-response curves obtained with different bioassays methods for soil-applied herbicides. *Weed Research,* 22, 213-222.

Okamura, H.; Aoyama, I; Liu, D.; Maguire, R.J.; Pacepavicius, G.J. & Lau, Y.L. (2000). Fate and ecotoxicity of the new antifouling compound Irgarol 1051 in the aquatic environment. *WaterResearch,* 34, 14, 3523-3530.

Osano, O.; Admiraal, W.; Klamer, D.; Pastor, D. & Bleeker, E.A.J. (2002). Comparative toxic and genotoxic effects of chloroacetanilides, formamidines and their degradation products on Vibrio fischeri and Chironomus riparius. *Environ. Pollut.* 119, 195-202.

Parrish, S. K.; Euler, J. P.; Gnogna, R,; Spirlet, M.; Walker, A., McVicar, F. M. & Cullinngton, J. E. (1995). Field, Glasshouses and laboratory investigations into the rate of degradation of MON-37500 in European Soils. *British Crop Protection Conference. Weeds,* 687-672.

Pestemer W. & Zwerger, P. (1999). The application of a standardized bioassay to estimate the phytotoxic effects of frequently used herbicides on non-target plants. *Human and Environmental Exposure to Xenobiotics,* 763-770.

Pestemer, W.; Stalder, L. & Eckert, B. (1980). Availability to plants of herbicide residues in soil. Part II: Data for use in vegetable crop rotations. *Weed Research,* 20, 349-353.

Powless, S.B. (2008). Evolved glyphosate-resistant weeds around the world: lessons to be learnt . *Pest Management Sciences* 64, 360-365.*Proceedings of BCPC – Weeds 2001,* 769-774.

Quivet, E.; Faure, R.; Georges, J.; Païssé, J. O.; Herbreteau, B. & Lantéri, P. (2006). Photochemical degradation of imazamox in aqueous solution: Influence of metal ions and anionic species on the ultraviolet photolysis. *J. Agric. Food Chem.*, 54, 10, 3641-3645.

Racchi, M.L., Rebecchi, M., Todesco, G. Nielsen, E. & Forlani G. (1995). Glyphosate tolerance in maize (Zea mays L.)2. Selection and characterization of a tolerant somaclone. *Euphytica,* 82, 165-173.

Ragab, M.T.H. (1974). Simazine persistence in soil and effects of its residue on crops. *Canadian Journal of Plant Science,* 54, 713-716.

Rahman, A.; James, T.K. & Günter, P. (1993). Bioassays of soil applied herbicides. *Proceedings Internationsl Symposium Indian Society of Weed Science,* 1, 95-106.

Reddy, K.; Rimando, A. & Duke, S. (2004). Aminomethylphosphonic acid, a metabolite of glyphosate, causes injury in glyphosate-treated, glyphosate-resistant soybean. *J. Agric. Food Chem.*, 52, 5139-5143.

Rice, P.J.; Anderson, T.A. & Coats, J.R. (2002). Degradation and persistence of metolachlor in soil: Effects of concentration, soil moisture, soil depth, and sterilization. *Environmental toxicology and chemistry*, 21, 12, 2640-2648.

Saha, S. & Kulshrestha, G. (2002). Degradation of sulfosulfuron, a sulfonylurea herbicide, as influenced by abiotic factors. *J. Agric. Food Chem.* 50, 16, 4572-4575.

Sahid, I.B., & Teoh, S-S., 1994. Persistence of terbuthylazine in soils. *Bull. Environ. Contam. Toxicol.* 52, 226-230.

SANCO. (2002). Guidance Document on Terrestrial Ecotoxicology (2002). *European Commission Health & Consumer Protection Directorate-General,* SANCO/10329/2002, pp. 37.

Sandín-España, P.; Llanos, S.; Magrans, J.O.; Alonso-Prados, J.L. & García-Baudín, J.M. (2003). Optimization of hydroponic bioassay for herbicide tepraloxydim by using water free from chlorine. *Weed Research*, 43, 451-457.

Sandín-España, P.; Magrans, J.O. & García-Baudín J.M. (2005a). Study of clethodim degradation and by-product formation in chlorinated water by HPLC. *Chromatographia,* 62, 3-4, 133-137.

Sandín-España, P.; Santín-Montanyá, I.; Magrans, J.O.; Alonso-Prados, J.L. & García-Baudín, J.M. (2005b). Degradation of alloxydim in chlorinated water. *Agronomy for Sustanaible Development*, 25, 331-334.

Santín-Montanyá, I.; Alonso-Prados J.L.; Villarroya M. & García-Baudín J.M. (2006). Bioassay for determining sensitivity to sulfosulfuron on seven plant species. *Journal of Environmental Science and Health. Part B.*, 41, 781-793.

Santín-Montanyá, I.; Sandín-España, P.; García-Baudín, J.M. & Coll-Morales, J. (2007). Optimal growth of Dunaliella primolecta in axenic conditions to assay herbicides. *Chemosphere*, 66, 1315-1322.

Santoro, A. Scopa, S.A. Bufo, M. Mansour & H. Mountacer. (2000). Photodegradation of the triazole fungicide hexaconazole. *Bull. Environ. Contam. Toxicol.* 64, 475-480.

Scrano, L.; Bufo, S.A.; Perucci, P.; Meallier, P. & Mansour, M. (1999). Photolysis and hydrolysis of rimsulfuron. *Pestic. Sci.* 55, 9, 955-961.

Seefeldt, S. S.; Jensen, J. E. & Fuerst, E. P. (1995). Log-logistic analysis of herbicide dose-response relationships. *Weed Technology*, 19, 218-227.

Sevilla-Morán, B.; Mateo-Miranda M. M.; Alonso-Prados, J.L.: García-Baudín, J.M. & Sandín-España P. (2010a). Sunlight transformation of sethoxydim-lithium in natural waters and effect of humic acids. *Intern. J. Environ. Anal. Chem* 90, 3-6, 487-496.

Sevilla-Morán, B.; Alonso-Prados, J.L.; García-Baudín J.M. & Sandín-España P. (2010b). Indirect Photodegradation of Clethodim in Aqueous Media. By-products identification by quadrupole time-of-flight mass spectrometry. *Journal of Agricultural and food chemistry*. 58, 3068-3076.

Sevilla-Morán, B.; Sandín-España, P.; Vicente-Arana, M. J.; Alonso-Prados, J. L. & García-Baudín, J. M. (2008). Study of alloxydim photodegradation in the presence of natural substances: Elucidation of transformation products. *J. Photochem. Photobiol. A*, 198, 2-3, 162-168.

Sherwood, A.M. & Jaseniuk, M. (2009) Molecular identification of weedy glyphosate-resistant Lolium (Poaceae) in California. *Weed Research*, 49, 354-364.

Shinn, S. L.; Thill, D. C. & Price, W. J. (1999). Volunteer Barley (*Hordeum vulgare*) Control in Winter Wheat *(Triticum aestivum)* with MON 37500. *Weed Technology*, 13, 88-93.

Sixto H. & Garcia-Baudin J.M. (1988) Diferentes respuestas a los herbicidas clortolurón e isoproturón de tres cultivares de trigo blando. *Investigación Agraria: Producción y Protección Vegetal* 3, 243-253.

Sixto, H., Silvela, L., Escorial, M.C., Garcia-Baudin, J.M. & Chueca, M.C. (1995) On the inheritance of tolerance to chlorotoluron application in wheat using a very efficient score test. *Weed Research*, 35, 7-13.

Somasundaram, L. & Coats, J.R. (1991). Pesticide transformation products in the environment, In: *Pesticide transformation products. Fate and significance in the environment.* L. Somasundaram and J. R. Coats. Washington D.C. (USA), 2-9, 0-8412-1994-X.

Stadler, L. & Pestemer, W. (1980). Availability to plants of herbicide residues in soil. Part I: a rapid method for estimating potentially available residues of herbicides. *Weed Research*, 20, 341-347.

Stork, P. & Hannah, M.C. (1996). A bioassay method for formulation testing and residue strudies of sulfonylurea and sulfonalide herbicides. *Weed Research*, 36, 271-278.

Streibig, J. C.; Rudemo, M. & Jensen, J. E. (1993). Dose-response curves and statistical models. In J. C. Streibig and P. Kudsk, eds. *Herbicide Bioassays*. Boca Raton, FL: CRC Press. pp. 29-55.

Streibig, J. C.; Walker, A.; Blair, A. M.; Anderson-Taylor, G. & Eagle, D. J. (1995). Variability of bioassays with metsulfuron-methyl in soil. *Weed Research*, 35, 215-224.

Szmigielska, A. M.; Schoenau, J. J. & Greer Ken (1998). Comparison of chemical extraction and bioassay for measurement in soil. *Weed Science*, 46, 487-493.

Tottman, D.R., Holroy, D.J., Lupton, F.G.H., Oliver, R.H., Barnes, T.R. & Tysoe, R.H (1975) The tolerance to isoproturon by varieties of winter wheat. *Proceeding of the European Weed Research Society Symposium Status and Control of Grass Weeds* 360-368.

Tuxhorn, G.L.; Roeth, F.W.; Martin, A.R. & Wilson, R.G. (1986). Butylate persistence and activity in soils previously treated with thiocarbamates. *Weed Sci.* 34, 961-965.

Vialaton, D. & Richard, C. (2002). Phototransformation of aromatic pollutants in solar light: photolysis versus photosensitized reactions under natural water conditions. *Aquat. Sci.* 64, 207-215.

Vicari, A.; Catizone, P. & Zimdahl, R. L. (1994). Persistence and movility of chlorsulfuron and metsulfuron under different soil and climatic conditions. *Weed Research*, 34, 147-155.

Villarroya, M., Escorial, M.C., Garcia-Baudin, J.M. & Chueca, M.C., (2000) Inheritance of tolerance to metribuzin in durum wheat". *Weed Research*, 40, 293-300.

Villarroya, M., Escorial, M.C., Sixto, H., Chueca, M.C., Garcia-Baudin, J.M. (1997) Glasshouse and laboratory response of some species of cereals and Bromus diandrus to the new herbicide MON 37500". *Brighton Crop Protection Conference Weeds* 3, 1037-1042. ISBN 1-90139-48-7

Vulliet, E.; Emmelin, C.; Chovelon, J.; Chouteau, C. & Clement, B. (2004). Assessment of the toxicity of triasulfuron and its photoproducts using aquatic organisms. *Environ. Toxicol. Chem.* 23, 12, 2837-2843.

Walker, A. & Welch, S. J. (1989). The relative movement and persistence in soil of chlorsulfuron, metsulfuron-methyl and triasulfuron. *Weed Research*, 29, 375-383

Wersuhn, G., Kirsch, K. & Gienapp, R. (1987) Herbicide tolerant regenerants of potato. *Theor. Appl. Genet.* 74, 480-482.

Ye, Q.; Sun, J. & Wu, J. (2003). Causes of phytotoxicity of metsulfuron-methyl bound residues in soil. *Environmental Pollution*, 126, 417-423.

Yenne, S.P., Thill, D.C., Letourneau, D.C. , Auld, D.L. and Haderlie,. L.C. (1988) Techniques for selection of glyphosate-tolerant field pea (Pisum sativum) cultivars. *Weed Tech.* 2, 286-290.

21

Plasmodesmata: Symplastic Transport of Herbicides within the Plant

Germani Concenco[1] and Leandro Galon[2]
[1]D.Sc in Crop Production, International Researcher at Valmont Industries, Omaha, NE,
[2]D.Sc. in Crop Production, Professor at Federal University of Pampa - UNIPAMPA
[1]USA
[2]Brazil

1. Introduction

When studying herbicide absorption, translocation, metabolism, and mode of action, transport pathways are usually referred to as apoplast (dead cells) and symplast (living cells) as simple synonyms of xylem and phloem. However, the behavior of an herbicide within a plant greatly depends upon several factors and its movement accomplished by different routes and processes.

If an herbicide takes too long to be absorbed after application, it will be more available for processes that would greatly reduce its absorption - rain, hot sun, and wind, among others. After the herbicide is absorbed, it needs to be quickly translocated from the point it was absorbed to the site of action. If it is not, chemical processes will take care of transforming the herbicide into non-toxic or less-toxic metabolites.

For a quick and efficient translocation, several pathways act together in a relatively dependent manner - everything is connected at different degrees of the plant's metabolic rate by the time reactions occur. For example, when a plant is under water stress, it may react differently to the same dose of herbicide usually applied to that species. In addition, phloem will only translocate an herbicide quickly if this compound is efficiently loaded into the phloem. From the leaf surface to the site of action, herbicide movement involves passage through the apoplast and symplast by several pathways, one of which is via plasmodesmata.

In the classical concept of Munch (1930), plasmodesmata are considered to form simple cytoplasmic bridges between neighboring plant cells in order to create the symplasm. This concept has dominated, if not monopolized, the thinking of plant biologists and, in particular, plant physiologists over the last few decades. Recent advances in ultra-structural, physiological, and molecular studies on plasmodesmata indicate that this simple view is in need of revision (Lucas, 1993). Plasmodesmata are plasma channels connecting neighboring cells that allow the exchange of informational, functional, and structural molecules and xenobiotics among cells of the same "group" (domain) , both apoplastically and symplastically. Cells of the same domain behave as functional units, and substances are able to move between them at rates above the observed for trans-membrane movement. Plasmodesmata participate symplastically in long-distance movement, both by association with phloem and interchange between neighboring domains. When the plant is under stress, and xylem and phloem flux is slower, plasmodesmata could be more participative in

long-distance translocation of systemic herbicide molecules. Plasmodesmata play a crucial role in transporting materials and signaling molecules intercellularly in higher plants. In the last decade, it has been discovered that plasmodesmal function is much more complex than previously thought, and more molecular and *in vivo* studies are necessary to discern the absolute structure and function of these interesting cytoplasmic channels (Hannahs, 1997).
The importance and role plasmodesmata play on intercellular transport of molecules - including molecules of herbicides - are explored in this chapter.

2. Herbicide absorption and translocation

The biological activity of herbicide within a plant is a function of absorption, translocation, metabolism, and susceptibility of the plant to the herbicide and/or to its metabolites. Because of that, the simple act of an herbicide reaching the leaf surface - or roots, in the case of a soil-applied herbicide - does not guarantee its effective action (Silva et al., 2007). The herbicide needs to be absorbed and translocated to reach the organelle where it will express its herbicidal activity. An active ingredient may affect several metabolic processes within the plant; however, the first biophysical lesion it causes will usually characterize its mechanism of action (Ferreira et al., 2008). The place where an herbicide effectively inhibits a biological process receives the name "site of action" (Hager & Sprague, 2002).
The main route of herbicide penetration in plants depends on a series of factors related to the plant, environment, and characteristics of the herbicide formulation and chemistry (Silva et al., 2007). After the herbicide passes through the first barrier - usually the cuticle - it should be moved to the site of action. Young plants incapable of regenerating from buried organs (tubercules, bulbs, rhizomes) after an herbicide application may easily be killed by a contact herbicide, once adequate coverage of the plant is reached during the application. For plants capable of regenerating from reserve organs, however, a given amount of herbicide must be able to move from the point where it was absorbed to the buried organs to ensure that the plant will be killed as a whole (Silva et al., 2007). In this case, the long-distance transport is even more important for efficient herbicide activity.
In a simplified way, the movement of an herbicide within the plant can be accomplished by two main routes: apoplast and symplast. **Apoplast** is a group of dead cells - including cell walls, intercellular spaces, and xylem - which form a *continuum* where water and solutes have the ability to move (Jachetta, 1986). **Symplast** is defined as the total mass of living cells in a plant, which forms a long and complex net along the plant both through phloem and through direct connections between neighboring cells that are usually in the same organ - plasmodesmata (Hay, 1976). These structures are also responsible for connecting neighboring cells in the phloem to form the *continuum* vase along the plant.
The main representatives of the apoplast and symplast are respectively xylem and phloem, and transportation through these routes is not completely independent - xylem-to-phloem transfer cells usually occur in specific parts of the plant (Figure 1). Since the translocation by xylem is unidirectional (from roots to leaves), it may be considered secondarily important for translocation of leaf-applied herbicides to fast-growing organs with low rates of respiration, such as buds, flowers, or fruits (Neumann, 1988). This task is fulfilled by the phloem. Some herbicides may present completely distinct behavior in relation to translocation as a function of the way they are being translocated. For example, atrazine behaves as a contact herbicide when applied to leaves (not translocated through phloem), but assumes a systemic behavior (meaning it moves within xylem) when applied to roots (Silva et al., 2007).

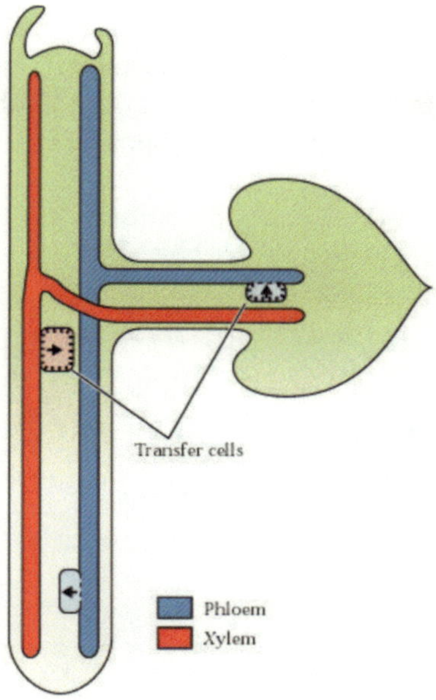

Fig. 1. Important sites of xylem-to-phloem solute transfer occur at leaf traces and minor veins of leaves. Source: Buchanan (2005).

The phloem is a network comprised of living cells, which goes from the tip of the root to the end of the leaves; within this network, translocation of photosynthates and many compounds occurs via sieve plates on both ends of the cells. Translocation through phloem is fundamental in the distribution of either natural or synthetic chemical compounds from mature leaves to growth regions in roots and stem (Vidal, 2002). There are mathematical models that allow efficient calculation of translocation rates of xenobiotics via phloem, according to membrane permeability, size of the phloem loading area, and other parameters (Tyree et al., 1979).

After the herbicidal molecule is translocated via phloem and entered into an adjacent cell, neighboring cells also need to be achieved to allow for proper action of the herbicide. This movement, which is usually a short distance, can be done through four primary ways: (1) apoplastic distribution or mass flow; (2) passive diffusion in favor of an electro-chemical gradient; (3) active translocation involving protein carriers at the expense of ATPs; and (4) movement and broadcast via plasmodesmata (Figure 2).

Both apoplastic and passive diffusion, in favor of an electro-chemical gradient, allow for relatively slow rates of movement for both molecules bigger than simple ions, whose size typically reaches only a few dozen Daltons (Da), and small molecules with an electric charge. These diffusions include passage by the plasma membrane of the cell of origin, cell wall (a tangle of cellulosic fibers stabilized normally by hemicellulose and pectin), middle lamella, and the cell wall and plasma membrane of the destination cell (Buchanan et al., 2005). Maximum rates of translocation through the membrane are approximately $1.0x10^{-8}$ cm s^{-1} for ions such as K^+ and Na^+ (Taiz & Zeiger, 2004). The actual rate of movement observed for glyphosate through the membranes is $1.7x10^{-8}$ cm s^{-1}, or 0.0006 mm h^{-1}, which is very similar to the observed rates for K^+ and Na^+ (Gougler & Geiger, 1981). Each membrane has a characteristic composition of proteins and lipids, making translocation also dependent on tissue or organs (Alberts et al., 1999). Active translocation can contribute to the movement of glyphosate to the interior of the cell. Inside this category of xenobiotics, this herbicide possesses the rare abilityof transposing the plasma membrane via a protein carrier. Due to phosphate carriers contained in the plasma membrane, a link is created between glyphosate and the carrier, translocating it to the cytoplasm (Denis & Delrot, 1993).

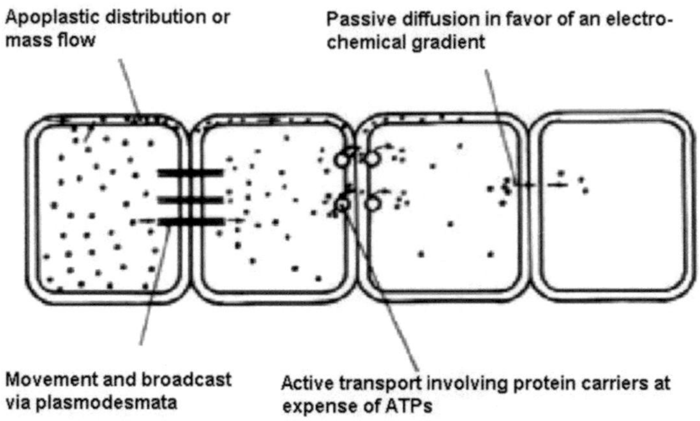

Fig. 2. Possible ways of herbicide translocation between plant cells. Among the depicted processes, only the active translocation at the expense of ATP demands energy. Source: Neumann, 1988.

3. Plasmodesmata

In a simple way, plasmodesmata are plasma membrane channels that pass through the cell wall, which not only allow for communication between plant cells, but also facilitate direct intercellular translocation of ions, photosynthates, growth regulators, and macromolecules of xenobiotics with similar characteristics (Robards, 1976). They provide a direct cytoplasmic connection between neighboring cells through cell walls. The properties of these communication channels are a factor in the establishment of the so-called "symplastic domains" - a group of cells that communicate and act as a physiological development unit with the ability of translocating macroproteins and RNA (Figure 3). Cells of the same domain are able to freely exchange information with each other while the communication is restricted between domains, occurring by translocation through the cell wall (Oparka & Roberts, 2001).

A typical plant cell may have between 10^3 and 10^5 plasmodesmata connecting it with adjacent cells, equaling between 1 and 10 per μm^2. Plasmodesmata are approximately 40-60nm in diameter at the mid-point and are constructed of three main layers: the plasma membrane, the cytoplasmic sleeve, and the desmotubule (central rod). They can transverse cell walls that are up to 90nm thick (Robards, 1976). There are three classical schematic models which try to clarify the plasmodesma structure (Figure 4).

The symplastic transport of substances through plasmodesmata can occur in two ways: via cytoplasmic connection or via endoplasmic reticulum. In Figure 3, two distinct regions can be seen at the canal of plasmodesma: (1) a cytoplasmic sleeve, which connects the cytoplasm of neighboring cells, and (2) a central rod, which connects the endoplasmic reticulum of neighboring cells. Concerning the herbicide transport, the cytoplasmic connection is the important route for symplastic transport of herbicides from cell to cell, via plasmodesmata.

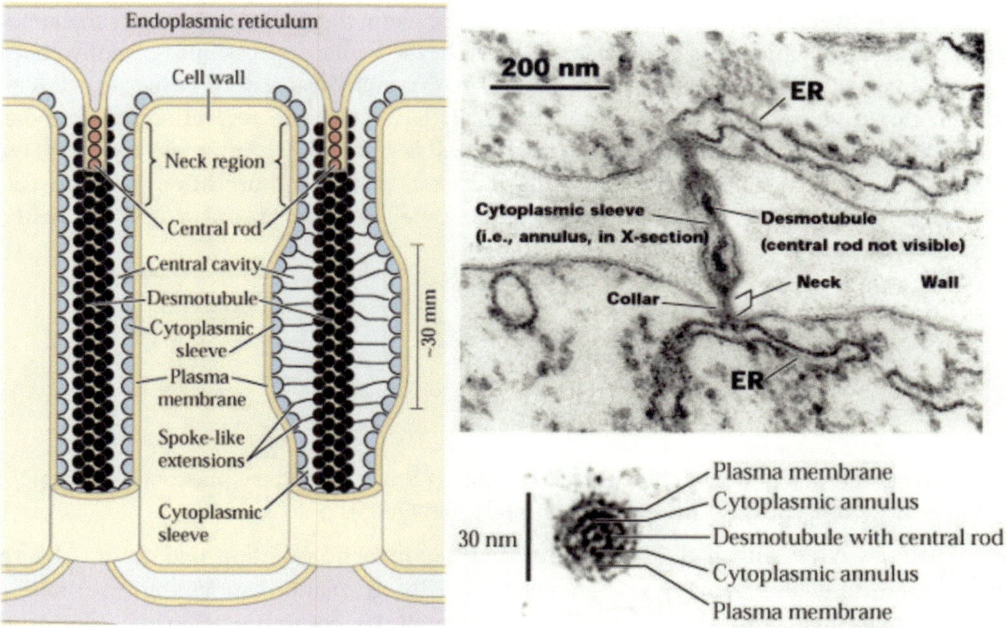

Fig. 3. Schematics depicting how plasmodesmata connect cytoplasms of neighboring cells. A plasmodesmata's pore diameter averages 50nm and allows diffusion of water and small molecules among cells. In order to allow translocation of molecules that are larger than the exclusion limit, the diameter of the pore can be modified by rearranging proteins connected to the inner surface of the pore. Affinity between some compounds and proteins in the inner surface of the canal make possible the flux of molecules larger than the Size Exclusion Limit (SEL). Source: Buchanan (2005).

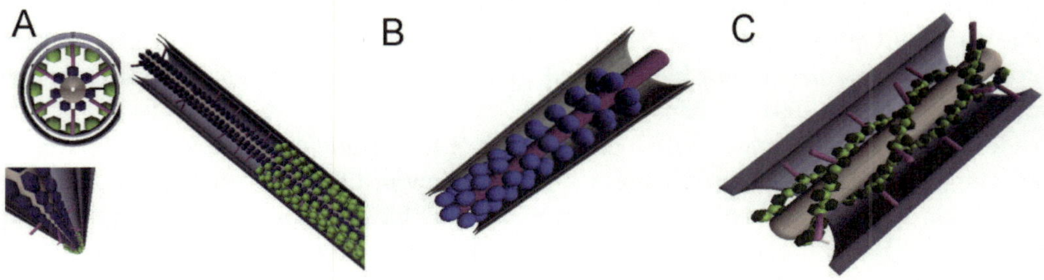

Fig. 4. Computer-generated models of the plasmodesmata structure, used for describing the three-dimensional characteristics of the plasmodesmal canal. (A) Ding model; (B) Overall model; (C) Radford model. All models depict the central rod (endoplasmic reticulum) and the space between the central rod and the cell wall (filled with cytoplasm). Source: Hannah (1997).

Plasmodesmata are not randomly scattered in a cell wall, but are rather grouped in specific points called "punctuations" or "pits." Plasmodesmata are formed when portions of the endoplasmic reticulum are trapped across the middle lamella as new cell wall is laid down between two newly divided plant cells; these eventually become the cytoplasmic connections between cells (primary plasmodesmata). Here, the wall is not further thickened, and depressions, or thin areas ("pits"), are formed in the walls; depressions normally couple between adjacent cells. Alternatively, plasmodesmata can be inserted into existing cell walls between non-dividing cells (Lucas et al., 1993). It is usually the formation of secondary plasmodesmata, which appear after a secondary wall is created. In grafting for example, two mature cells are side-by-side and are obliged to establish communication between each other (Figure 5). In this situation, the secondary wall is thinned and a new pit field is formed (Buchanan, 2005).

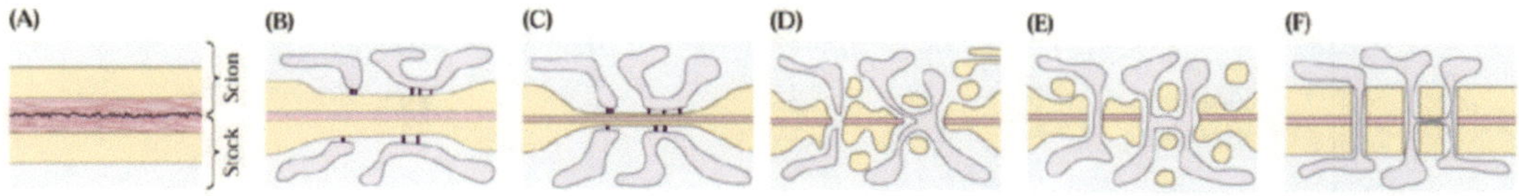

Fig. 5. Formation of a new punctuation field - "pit" (C) and secondary plasmodesmata (D, E, F) in mature cells. Source: Adapted from Buchanan (2005).

Recent studies have launched new visions about symplastic isolation and traffic of large molecules during the growth and development of a plant, confirming the role of plasmodesmata in controlling and mediating intercellular communication (Kragler et al., 1998; Tan et al., 2005). The greatness of the flow via plasmodesmata is usually measured by either a comparison between independent fluxes, which are estimated by concentration gradients, and diffusion coefficients, or by the hydrodynamic radius of the molecule, which is the amount of water a molecule carries around itself (Hatch & Slack, 1970; Terry & Robards, 1987). The size exclusion limit (SEL) of plasmodesmata corresponds to the maximum size of a "non-favored" molecule that is capable of crossing the plasmodesma and is both linked directly to the diameter of the canal and the affinity between the molecule and proteins embedded in the interior of the pore (Figure 3). The most accepted theory among researchers is only molecules smaller than 1 kDa (kiloDalton) move freely among cells of the same domain (Oparka & Roberts, 2001). However, molecules greater than 1 kDa can pass through plasmodesmata if they have some degree of affinity with proteins embedded in the canal (Taiz & Zeiger, 2004; Buchanan, 2005).

In addition, the SEL decreases with increasing age of the organ; for example, newer parts of the plant have the ability to carry larger molecules (Crawford & Zambrysky, 2001). This may help explain why plants become less susceptible to herbicides at more advanced stages of development. Formation of a less permeable, thicker secondary wall, among other factors, also limits the translocation of herbicides in older plants. These factors contribute to larger herbicide doses that are required to control older plants, until a certain point of development (Chamel, 1988). At maturity, plasmodesmata present very low conductance and contribute to a small extent for systemic distribution of large molecules. In addition, the conductance depends not only on the diameter of the canal, but also on the affinity between the molecules being conducted and the proteins embedded in the interface of the canal.

In addition to the presence of symplastic transport through plasmodesmata, apoplastic transport is also present, as shown in Figure 6. However, at the time of this publication, no studies were found to explore this pathway of transport via plasmodesmata in relation to apoplastic herbicide translocation.

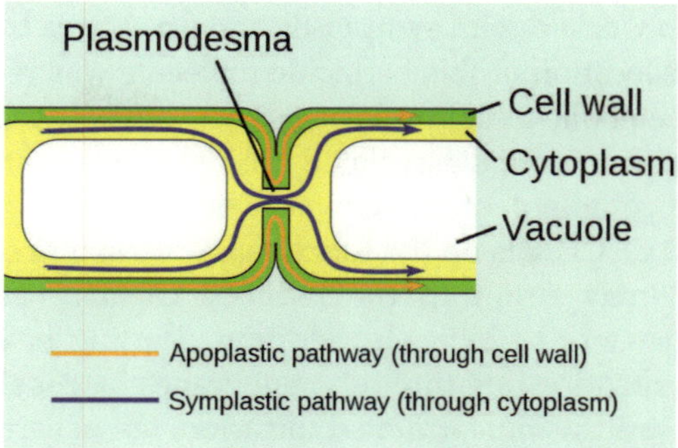

Fig. 6. Schematics showing symplastic and apoplastic movement pathways via cell-to-cell communication (plasmodesmata). Source: Wikipedia, licensed under GPL terms (2010).

4. Herbicide translocation through plasmodesmata

All herbicides applied to the leaves of plants with C_4 metabolism must penetrate the vascular bundle sheath cells in order to achieve xylem and/or phloem (Vidal, 2002). Once these cells are highly lignified (suberin may also occur in some monocots), the movement of the herbicide molecules from cells in the mesophyl to the cells of the vascular bundle sheath occurs exclusively by plasmodesmata present this interface (Osmond & Smith, 1976). This is the only way the herbicide can reach the phloem, which is located internally compared to the sheath cells (Taiz & Zeiger, 2004). The movement of larger molecules, such as herbicides, would likely be limited through membranes; even molecules with only four carbon (ie: malate or aspartate), which are responsible for translocating CO_2 fixed in the mesophyll to the sheath cells in the vascular bundle, are dependent on translocation via plasmodesmata (Figure 7). The translocation of malate from mesophyll to cells at the vascular bundle sheath is between 100 and 1,000 times greater than the maximum allowed translocation via biological membranes (Buchanan et al., 2005).

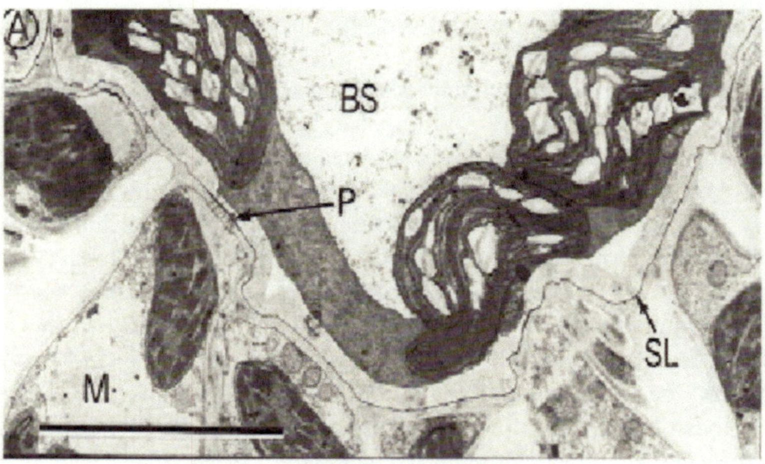

Fig. 7. Vascular bundle sheath (BS) and mesophyll (M) cells of sugarcane. It is possible to observe the suberized layer (SL) in the primary wall of the bundle sheet cells and punctuated spot with a plasmodesmata(P) crossing it. Source: Osmond & Smith (1976).

Sterling et al. (1990), who worked with symplastic and apoplastic translocation of bentazon, determined that the translocation of this herbicide molecule was reduced by an application of exogenous CCCP, a metabolic inhibitor (Wagatsuma, 1983; Zhang & Taylor, 1991; Rincon & Gonzales, 1992). These researchers associated the reduction in translocation of bentazon with the increase in the gradient of protons between neighboring cells. Presently, it is common knowledge that CCCP acts on the functional control of a cell as a whole; increased concentration of CCCP may result in the reduced capacity of translocation through plasmodesmata (Buchanan et al., 2005). In addition, Sterling et al. (1990) discussed the possibilities of bentazon broadcasting through membranes by considering possible ways of movement, such as simple diffusion, facilitated diffusion, use of carriers, and competition by a carrier between different subtracts. However, these types of translocation usually occur when cells of the same domain show similar concentrations of the herbicide between them and higher concentrations than the external environment. In this scenario, the herbicide is translocated through membranes, reaching cells from other symplastic domains to all others in the same domain by direct cytoplasmic connections. It would be incorrect to assume that a relatively large organic molecule, such as an herbicide, would prefer to cross two cell walls and two plasma membranes instead of being carried by a continuous cytoplasmic tube between cells in the same domain.

Another study describes the application of 10 droplets of chlorsulfuron on the third definitive leaf of a seedling of field pennycress *(Thlaspi arvense)* with five definitive leaves (Bestman et al., 1990). Chlorsulfuron is an herbicide that belongs to the group of sulfonylurea, inhibitor of the enzyme acetolactate synthase (ALS), involved in the synthesis of branched-chain amino acid valine, leucine, and isoleucine (Leite et al., 1998). The rate of efflux of photosynthates from leaves that did not receive a direct application of the herbicide in treated plants was reduced only 24 hours after the application of chlorsulfuron. Movement of this herbicide was slow, indicating either that xylem and phloem were probably not the preferential translocation pathways, or that the herbicide took too long to reach xylem and phloem. Translocation probably occurred between cells due to the fact that cells from different sheets are relatively distant, which means they would not belong to the same symplastic domain. In this case, translocation across membranes associated with the phloem may have been significant.

ALS-inhibiting herbicides can use characteristics of dissociation to act more efficiently in cells belonging to the same domain. As the pH in the exterior of the cell is approximately 5.5, ALS-inhibiting herbicides are in a non-dissociated form and are thus able to penetrate the cell more easily. Once at the cytoplasm, where the medium is more alkaline (pH approximately 7.5), these herbicides disassociate and turn into the most active, and less capable, translocating form of the molecule. Because the dissociated form is less capable of spreading to the exterior of the cell, the molecules of these herbicides get "stuck" in the cytoplasm; this behavior receives the name "ionic trap" (Vidal, 2002). In this situation, the herbicide moves freely among cells of the same domain because, in essence, "a single cytoplasm occurs" between cells of the same domain (Crawford & Zambryski, 2001). Penetration of these herbicides in a single cell enables their distribution to all other cells belonging to the same symplastic domain (Jachetta et al., 1986). In addition, plasmodesmata may have significant participation in translocation of molecules that have pKa (dissociation constant) below the pH of the xylem, which is approximately 5.5; it may also assist symplastic movement of these molecules by other routes besides phloem (Vidal, 2002).

The movement through plasmodesmata is relevant also in translocation of macromolecules that carry information (RNA) to neighboring cells. In this way, the behavior of cells in a given domain is not isolated, which can thus act as a functional unit (Jorgensen & Lucas, 2006; Lucas et al., 2009). Although studies related to plasmodesmata and herbicides are limited, information of movement of other macromolecules similar in size to herbicidal molecules can be adapted. In a trial that grafted tomato plants, the data determined that a macromolecule which carries information responsible for a leaf's deformation, known as "Mouse's Ear", encoded at the roots of the rootstock, reached the meristem of the graft, and caused the deformation (Kim et al., 2001). Since this compound was not translocated via the transpiratory pathway, the most probable route of translocation identified by the researchers was through cytoplasmic connections of cells both within the same domain and between interactions of different domains (Figure 8). The data highlighted the participation of plasmodesmata in the long-distance translocation of the molecule.
Being known the capacity of molecules to move cell-by-cell, the acropetal long-distance translocation of herbicides can occur by apoplast (xylem) or via symplast (plasmodesmata), provided that the characteristics of the molecule in relation to polarity, electric charge and dimensions, allow this translocation. Even the translocation in the phloem is accomplished through plasmodesmata present in the interface between phloem cells. If the pesticide was applied to the surface of the plant and was correctly absorbed, and this plant later is submitted to moderate water stress that causes closure of stomata and a consequent reduction of translocation by xylem, herbicide can still be translocated at some rate via plasmodesmata (Alberts et al., 1999). In most plants, translocation of glyphosate is typically fast as it is essential for herbicidal activity. After penetration in leaves, glyphosate may be translocated both by phloem sieve tubes, which also involves plasmodesmata, and cell-to-cell in the same symplastic domain via plasmodesmata, reaching all cells of the domain quickly (Franz et al., 1997; Jachetta et al., 1986). In fact, both translocations are complementary and non-competitive due to the fact that plasmodesmata act on loading and unloading the phloem (Sowinski et al., 2003). Glyphosate is one of the few studies where herbicide translocation via plasmodesmata is considered.

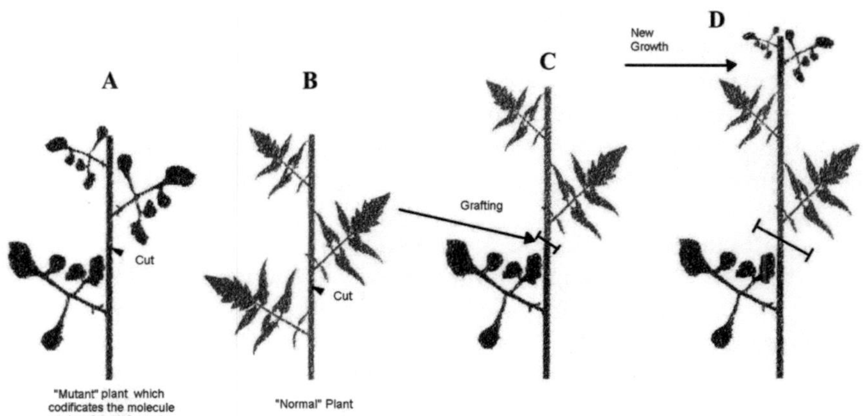

Fig. 8. Study showing the movement of informational substances codified at the roots of the rootstock and carried to the meristem of the graft, where the mutation known as "Mouse's Ear" was present on the newly developed leaves. Source: Adapted from Kim et al. (2001).

There are also translocation differences depending on the pH of the solution in which the herbicide is diluted. In one study involving absorption and translocation of sulfentrazone and glyphosate by plant roots, research determined that with the solution's decrease in pH, absorption of sulfentrazone increased, along with its solubility (Ferrell et al., 2003). In turn, glyphosate was not as dependent on pH. While the pKa of sulfentrazone was 6.5, glyphosate had a sequence of pKa (0.8, 2.3, 6.0, and 11.0), which shows different configurations as a function of the pH (Grey et al., 2000; Sprankle et al., 1975, Coutinho & Mazo, 2005). It is believed that at physiological pH, glyphosate is considered a zwitterion, behaving as a divalent anion with the possibility of being strongly complex with some divalent metal cations (Devine et al., 1993). This molecule has the ability of changing poles when it reaches the cytoplasm, acquiring a net negative charge promoted by deprotonation due to physiological pH, and contributing to its retention in the symplast (Wauchope, 1976). Even with this behavior, absorption and translocation of glyphosate was less affected by the pH than sulfentrazone. The role of plasmodesmata in glyphosate translocation is known, indicating that they act not only in conjunction with vascular system, but also in a semi-autonomous way (Jachetta et al., 1986). These characteristics are particularly important for the translocation of soil-applied herbicides where xylem has an important - but not exclusive - role in the translocation of these macromolecules.

5. Herbicide translocation and plasmodesmata in older plants

Older plants have greater dry mass, leaf area, and, consequently, a greater transpiratory rate. For example, consider a barnyard grass (*Echinochloa crusgalli*) seedling with 3 - 4 leaves, when herbicide is typically applied, and another plant at the stage of 2 - 3 tillers; both plants may present high metabolic rates because they are in full growth, but the numerical volume of acropetal water flow is surely higher in the plant at the stage of 2 - 3 tillers under the same environmental conditions (Taiz & Zeiger, 2004). When using a soil-applied herbicide, which is absorbed by the roots of both plants, it would be efficiently translocated via xylem due to the high transpiratory flow in both plants. However, the barnyard grass plant at the stage of 2 - 3 tillers is less susceptible to the herbicide than the plant with 2 - 4 leaves. Besides the causes already discussed by Vidal et al. (2002), in mature cells in a plant in active growth, the SEL of plasmodesmata is smaller and herbicide movement is more dependent upon the transpiratory flux. Symplastic translocation of molecules via plasmodesmata in these cells is severely reduced; in comparison to when the cell is developing and has the ability to exchange essential informational molecules, ions, and regulators, the SEL may be up to 50 times lower when the cell is mature and does not require a large influx of molecules (Oparka & Roberts, 2001).

As previously discussed, the SEL of a plasmodesma allows relatively free passage of molecules as large as 1 kDa through young organs. Considering the SEL can be reduced to approximately 50 times lower, a mature plant plasmodesmatal SEL may only, in general terms, allow passage of molecules around 20 - 50 Daltons (Da). Most herbicide molecules are bigger than 100 Da and smaller than 500 Da (Table 1). This information strongly suggests that plasmodesmatal reduction in SEL is one of the great responsibilities due to lower susceptibility of older plants to herbicides. As previously discussed, the SEL can be of smaller importance if the herbicidal molecule presents some degree of affinity with carrier proteins embedded in the inner surface of the plasmodesmata. This affinity may allow distinct rates of movement through plasmodesmata for herbicidal molecules of similar size.

Movement deficiency, which is caused by reduced absorption and/or translocation, of a given herbicide within the plant may be the reason for herbicide tolerance and/or selectivity in many crops and weed species (Hess, 1985; Ladlie, 1991). In addition, different rates of herbicide translocation, acropetaly or basipetaly, may result from alterations made at the genetic level and confer resistance to a plant due to an active ingredient usually lethal to that species. Reduced translocation of herbicides as a mechanism of resistance is extensively researched, and was identified for example, in Italian ryegrass *(Lolium multiflorum)* and Wimmera ryegrass *(Lolium rigidum)* (Ferreira et al, 2006; Lorraine-Colwill et al, 2002). Other plants, however, do not show differences in relation to the absorption and translocation of herbicides between resistant and susceptible biotypes (Carey et al., 1995; Dias et al., 2003). In cases where reductions on herbicide translocation occur, it is essential to investigate whether there is a reduction of plasmodesmata SEL or reduction in the association with its function of phloem loading/unloading. In addition, for herbicides carried via xylem and/or phloem, the role of plasmodesmata in acropetal translocation is essential in the same way it is for other organic molecules (Taiz & Zeiger, 2004).

The size of herbicidal molecules usually is not a limiting factor for translocation via plasmodesmata in younger plants because it generally lies between 150 Da and 450 Da (Table 1), and molecules up to 1 kDa usually have relatively free passage through plasmodesmata in younger plants (Taiz & Zeiger, 2004). However, size can be a limiting factor in older plants, as previously discussed. Herbicidal molecules are typically smaller than many proteins or enzymes translocated via plasmodesmata, and proteins that have a role in translocating other substances through the plasmodesmata, such as protein MP30, connected to the v-RNA (viral RNA), which makes traffic possible (Kragler et al., 2003). Besides size, other characteristics of molecules, such as electrical charges, can be important and can allow the passage of certain molecules in lieu of others (Devine & Hall, 1990).

6. Plasmodesmata and herbicide translocation under water stress

When plants are subjected to stress, metabolic reactions tend to decrease proportionally. Because herbicides are less translocated and, as a consequence, become more available to reactions of metabolization, conjugation, or trapping, many herbicides have their action strongly reduced if plants are under stress before or after application (Cataneo et al., 2003). In auxin-like herbicides, the herbicidal activity is typically resumed when metabolism is increased after water stress. Auxin-like substances previously either applied (synthetic) or produced (natural) are able to reach the site of action after stress is removed and the plant reaches the usual state of turgescence (Drake & Carr, 1978). Although plasmodesmata are not the only route of translocation these substances take, they can play an important role in the translocation of auxin-like herbicides, and possibly other classes of herbicides, under moderate water stress.

7. Conclusions and new insights

In order to determine if a given compound, or chemical group, has the ability to manipulate the size exclusion limit of the canal, mainly when it is reduced as the plant ages, more studies are needed to clarify the existence of affinity between certain herbicidal molecules and proteins embedded in the inner surface of plasmodesmata. Proved existence of such an affinity may favor molecular translocation, regardless of its size. Studies also need to be

Chemical Structure	Common Name And Formula	Size (Da)[a]	Chemical Structure	Common Name And Formula	Size (Da)[a]
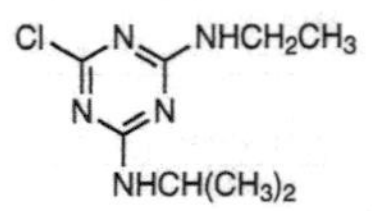	Atrazine $C_8H_{14}ClN_5$	215.7	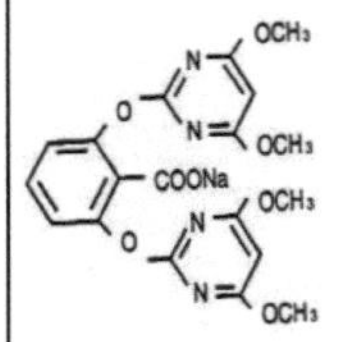	Bispyribac-sodium $C_{19}H_{18}N_4O_8$	430.2
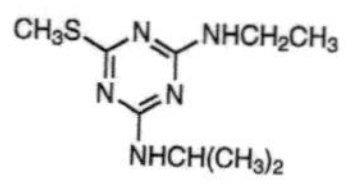	Ametrine $C_9H_{17}N_5S$	227.3	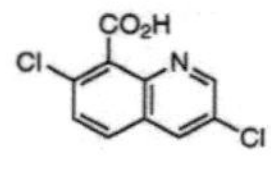	Quinclorac $C_{10}H_5Cl_2NO_2$	242.0
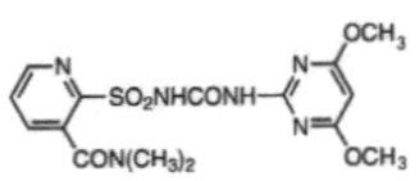	Nicosulfuron $C_{15}H_{18}N_6O_6S.H_2O$	428.4	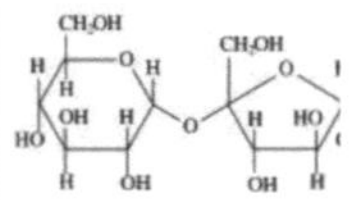	Sucrose $C_{12}H_{22}O_{11}$	342.3
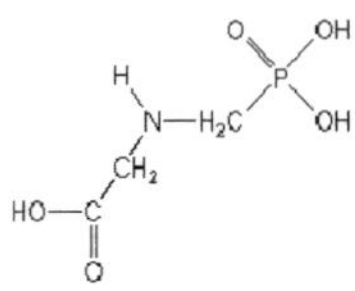	Glyphosate (acid eqv.) $C_3H_8NO_5P$	169.1	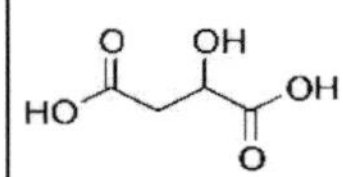	Malate $C_4H_6O_5$	134.1
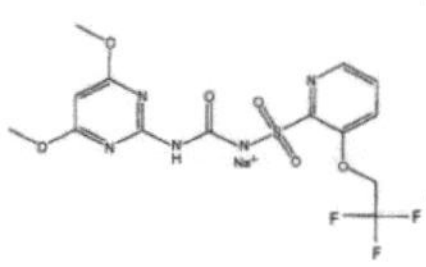	Trifloxysulfuron - sodium $C_{14}H_{13}F_3N_5O_6SNa$	459.3	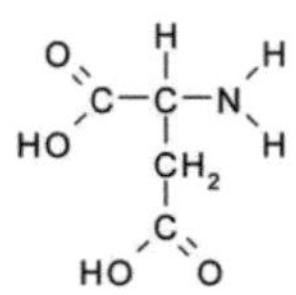	Aspartate $C_4H_7NO_4$	133.0
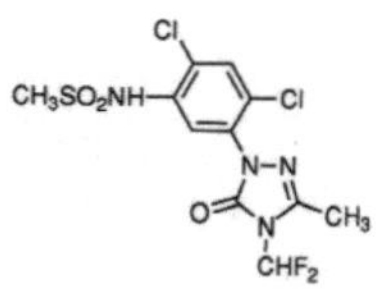	Sulfentrazone $C_{11}H_{10}Cl_2F_2N_4O_3S$	387.1	---	PEP carboxilase[b]	2.7×10^5
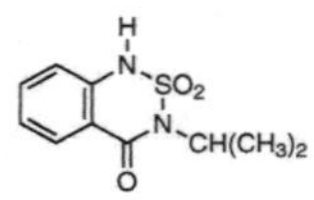	Bentazon $C_{10}H_{12}N_2O_3S$	240.3	---	v-RNA MP30[c]	3×10^4
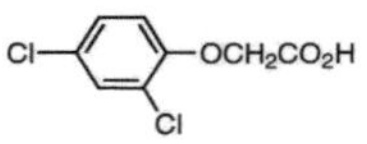	2,4-D (eq. ácido) $C_8H_6Cl_2O_3$	221.0	---	K	39.1
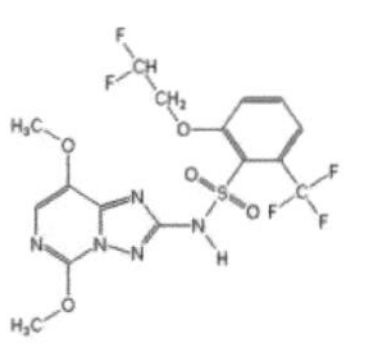	Penoxsulam $C_{16}H_{14}F_5N_5O_5S$	483.2	---	Na	23.0

[a] Original data calculated from the chemical formulas; [b] Source: Patel et al. (2004); [c] Source: Wolf et al. (1989); Kragler et al. (2003).

Table 1. Dimensions of some herbicidal molecules, compounds, proteins, and ions. Federal University of Viçosa, Brazil, 2010 devoted to determine if the degree of similarity between a given herbicidal molecule and a natural plant compound (such as for auxin-like herbicides) results in higher translocation rates through symplast. More research should also be given to the participation of plasmodesmata in the movement of systemic herbicides within the plant. Studies with radioactively-marked products, and the intensification of research on herbicide physiology, will help explain many aspects not fully understood involving herbicidal translocation via xylem and phloem, and their association with the apoplast and symplastic domains.

8. References

ALBERTS, B. et al. Fundamentos da biologia celular. Porto Alegre: Artes Médicas Sul, 1999. 757 p.

BESTMAN, H. D.; DEVINE, M. D.; BORN, W. H. V. Herbicide chlorsulfuron decreases assimilate translocation out of treated leaves of field pennycress (*Thlaspi arvense* L.) seedlings. Plant Physiol., v. 93, p. 1441-1448, 1990.

BUCHANAN, B. B.; GRUISSEM, W.; JONES, R. L. Biochemistry & molecular biology of plants. Rockville: American Society of Plant Physiologists, 2005. 1367 p.

CAREY, V. F. et al. Resistance mechanism of propanil-resistant barnyardgrass. I. Absorption, translocation, and site of action studies. Pestic. Biochem. Physiol., v. 52, p. 182-189, 1995.

CATANEO, A. C. et al. Glutathione S-transferase activity on the degradation of the herbicide glyphosate in maize (*Zea mays*) plants. Planta Daninha, v. 21, p. 307-312, 2003.

CHAMEL, A. Foliar uptake of chemicals studied with whole plants and isolated cuticles. In: NEUMANN, P. M. (Ed.). Plant growth and leaf-applied chemicals. Boca Raton: CRC Press, 1988. p. 27-50.

COUTINHO, C. F. B.; MAZO, L. H. Metallic complexes with glyphosate: a review. Quím. Nova, v. 28, p. 1038-1045, 2005.

CRAWFORD, K. M.; ZAMBRYSKI, P. C. Non-targeted and targeted protein movement through plasmodesmata in leaves in different developmental and physiological status. Plant Physiol., v. 125, p. 1802-1812, 2001.

DENIS, M. H.; DELROT, S. Carrier-mediated uptake of glyphosate in broad bean (*Vicia faba*) via a phosphate translocationer. Plant Physiol., v. 87, p. 569-575, 1993.

DEVINE, M. D.; HALL, L. M. Implications of sucrose translocation mechanisms for the translocation of herbicides. Weed Sci., v. 38, p. 299-304, 1990.

DEVINE, M. D.; DUKE, S. O.; FEDTKE, C. Physiology of herbicide action. Englewood Cliffs: Prentice Hall, 1993. 441 p.

DIAS, N. M. P. et al. absorption and translocation of herbicide diuron by species susceptible and tolerant grass mattress (*Digitaria* spp.). Planta Daninha, v. 21, p. 293-300, 2003.

DRAKE, G.; CARR, D. J. Plasmodesmata, tropism, and auxin translocation. J. Cat. Bot., v. 29, p. 1309-1318, 1978.

FERREIRA, F. A.; SILVA, A. A.; GALON, L.; CONCENÇO, G.; FERREIRA, E. A. Mecanismos de ação de herbicidas. In: ZAMBOLIN, L.; PINANÇO, M. C.; SILVA, A. A.; FERREIRA, L. R.; FERREIRA, F. A.; JESUS Jr., W. C. J. (Eds.). Produtos fitossanitários. Viçosa: UFV, 2008. p.349-383.

FERREIRA, E. A. et al. Translocation of glyphosate in ryegrass biotypes (*Lolium multiflorum*). Planta Daninha, v. 24, p. 365-370, 2006.

FERRELL, J. A.; WITT, W. W.; VENCILL, W. K. Sulfentrazone absorption by plant roots increases the soil or solution pH decreases. Weed Sci., v. 51, p. 826-830, 2003.

FRANZ, J. E.; MAO, M. K.; SIKORSKI, J. A. Uptake, translocation and metabolism of glyphosate in plants. In: Glyphosate: a unique global herbicide. Washington: American Chemical Society, 1997. p. 143-181.

GOUGLER, J. A.; GEIGER, D. R. Uptake and distribution of N-phosphonomethylglycine in sugar beet plants. Plant Physiol., v. 68, p. 668-672, 1981.

GREY, T. L. et al. Behavior of sulfentrazone in ionic exchange resins, electrophoresis gels, and cation-saturated soils. Weed Sci., v. 48, p. 239-247, 2000.

HAGER, A. G.; SPRAGUE, C. L. A contemporary, color-enhanced herbicide site of action bulletin. Journal of Extension, v.40, n.2, ON-LINE, ISSN 1077-5315.

HANNAHS, G. Plasmodesmata. Gambier: Kenyon College, 1997. On-line at http://biology.kenyon.edu/edwards/project/greg/pd.htm

HATCH, M. D.; SLACK, C. R. The C4-carboxylic acid pathway of photosynthesis. In: REINHOLD, L.; LIWSCHITZ, Y. (Eds.). Progress in phytochemistry. London: Wiley-Interscience, 1970. p. 35-106.

HAY, J. R. Herbicide translocation in plants. In: AUDUS, L. J. (Ed.). Herbicides physiology, biochemistry, ecology. V.1, 1976. p.365-393.

HESS, F. D. Herbicide absorption and translocation and their relationship to plant tolerance and susceptibility. In: DUKE, S. O. (Ed.). Weed physiology. Boca Raton: CRC Press, 1985. p. 192-214.

JACHETTA, J. J.; APPLEBY, A. P.; BOERSMA, L. Apoplastic and symplastic pathways of atrazine and glyphosate translocation in shoots of seedling sunflower. Plant Physiol., v. 82, p. 1000-1007, 1986.

JORGENSEN, R. A.; LUCAS, W. J. Movement of macromolecules in plant cells through plasmodesmata. Sci. STKE, 2006. <http://www.ncbi.nlm.nih.gov/entrez/> query.fcgi?cmd=Retrieve&db=PubMed&list_uids /=16493020&dopt=Abstract.

KIM, M. et al. Developmental changes due to long-distance movement of a homeobox fusion transcript in tomato. Science, v. 293, p. 287-289, 2001.

KRAGLER, F.; LUCAS, W. J.; MONZER, J. Plasmodesmata: dynamics, domains and patterning. Ann. Bot., v. 81, p. 1-10, 1998.

KRAGLER, F. et al. MPB2C, a microtubule-associated plant protein binds to and interferes with cell-to-cell translocation of Tobacco Mosaic Virus movement protein1. Plant Physiol., v. 132, p. 1870-1883, 2003.

LADLIE, J.S. Guide to herbicide injury symptoms in soybean with "look-alike" symptoms. Hollamdale: Agri-Growth Research, 1991. 86 p.

LEITE, C. R. F.; ALMEIDA, J. C. V.; PRETE, C. E. C. Aspectos fisiológicos, bioquímicos e agronômicos dos herbicidas inibidores da ALS (AHAS). Londrina: Grafmark, 1998. 68 p.

LORRAINE-COLWILL, D. F. et al. Investigations into the mechanism of glyphosate resistance in *Lolium rigidum*. Pestic. Biochem. Physiol., v. 74, p. 62-72, 2002.

LUCAS, W. J.; HAM, B. K.; KIM, J. Y. Plasmodesmata - bridging the gap between neighboring plant cells. Trends in Cell Biology, v.19, n.10, p.495-503, 2009. ISSN 0962-8924.

LUCAS, W. J.; DING, B.; SCHOOT, C. V. D. Plasmodesmata and the supracellular nature of plants. New Phytologist, v.125, p.435-476, 1993.

MÜNCH, E. Die Stoffbewegunen in der Pflanze. Jena: Verlag von Gustav Fischer, 1930. 234p.

NEUMANN, P. M. Agrochemicals: plant physiological and agricultural perspectives. In: NEUMANN, P. M. (Ed.). Plant growth and leaf-applied chemicals. Boca Raton: CRC Press, 1988. p. 1-13.

OPARKA, K. J.; ROBERTS, A. Plasmodesmata. A not so open-and-shut case. Plant Physiol., v. 125, p. 123-126, 2001.

OSMOND, C. B.; SMITH, F. A. Symplatic translocation of metabolites during C_4-photosynthesis. In: GUNNING, B. E. S.; ROBARDS, A. W. (Eds.). Intercellular communication in plants: studies on plasmodesmata. New York: Springer-Werlag, 1976. p. 229-241.

PATEL, H. M.; KRASZEWSKI, J. L.; MUKHOPADHYAY, B. The phosphoenolpyruvate carboxylase from *Methanothermobacter thermautotrophicus* has a novel structure. J. Bacteriol., v. 186, p. 5129-5137, 2004.

RINCON, M.; GONZALES, R.A. Aluminum partitioning in intact roots of aluminum-tolerant and aluminum-sensitive wheat (*Triticum aestivum* L.) cultivars. Plant Physiol., v. 99, p. 1021-1028, 1992.

ROBARDS, A. W. Plasmodesmata in higher plants. In: GUNNING, B. E. S.; ROBARDS, A. W. (Eds.). Intercellular communication in plants: studies on plasmodesmata. New York: Springer-Werlag, 1976. p. 15-57.

SILVA, J. F.; SILVA, J. F.; FERREIRA, L. R.; FERREIRA, F. A. Herbicidas: absorção, translocação, metabolismo, formulação e misturas. In: SILVA, A. A.; SILVA, J. F. (Eds.). Tópicos em manejo de plantas daninhas. Viçosa: UFV, 2007. p.149-188.

SOWINSKI, P.; RUDZISKA-LANGWALD, A.; KOBUS, P. Changes in plasmodesmata frequency in vascular bundles of maize seedling leaf induced by growth at sub-optimal temperatures in relation to photosynthesis and assimilate export. Environ. Exp. Bot., v. 50, p. 183-196, 2003.

SPRANKLE, P.; MEGGITT, W. F.; PENNER, D. Adsorption, mobility, and microbial degradation of glyphosate in the soil. Weed Sci., v. 23, p. 229-234, 1975.

STERLING, T.M.; BALKE, N. E.; SILVERMAN, D. S. Uptake and accumulation of the herbicide bentazon by cultured plant cells. Plant Physiol., v. 92, p. 1121-1127, 1990.

TAIZ, L.; ZEIGER, E. Fisiologia vegetal. Porto Alegre: ARTMED, 2004. 750 p.

TAN, B.L. et al. Localization and movement of mineral oil in plants by fluorescence and confocal microscopy. J. Exp. Bot., v. 56, p. 2755-2763, 2005.

TERRY, B. R.; ROBARDS, A. W. Hydrodynamic radius alone governs the mobility of molecules through plasmodesmata. Planta, v. 171, p. 145-157, 1987.

TYREE, M. T.; PETERSON, C. A.; EDGINGTON, L. V. A simple theory regarding ambimobility of xenobiotics with special reference to nematicide oxamyl. Plant Physiol., v. 63, p. 367-374, 1979.

VIDAL, R. A. Ação dos herbicidas. Porto Alegre: Universidade Federal do Rio Grande do Sul, 2002. 89 p.

WAGATSUMA, T. Effect of non-metabolic conditions on the uptake of aluminum by plant roots. Soil Sci. Plant Nutr., v. 29, p. 323-333, 1983.

WAUCHOPE, D. Acid dissociation constants of arsenic acid, methylarsonic acid (MAA), dimethylarsinic acid (cacodylic acid), and n-(phosphonomethyl)glycine (glyphosate). J. Agric. Food Chem., v. 24, p. 717-721, 1976.

ZHANG, G. C.; TAYLOR, G. J. Effects of biological inhibitors on kinetics of aluminum uptake by excised roots and purified cell-wall material of aluminum-tolerant and aluminum-sensitive cultivars of Triticum aestivum L. J. Plant Physiol., v. 138, p. 533-539, 1991.

WOLF, S. et al. Movement protein of tobacco mosaic virus modifies plasmodesmatal size exclusion limit. Science, v. 246, p. 377-349, 1989.

22

7-Keto-8-Aminopelargonic Acid Synthase as a Potential Herbicide Target

In-Taek Hwang[1], Dong-Hee Lee[2] and No-Joong Park[1]
[1]*Korea Research Institute of Chemical Technology*
[2]*Genomine Inc.*
Korea

1. Introduction

Agrochemicals are compounds that selectively kill or arrest the growth of pests and weeds. They have played a significant role in agricultural production that provided for about 600 million people during the past 50 years. And also, the increasing world population seems to be a major driving force for the need to enhance the output of food production per area (Joseph, 2004). The agrochemical industry has been very successful in developing new herbicides. New chemicals with improved properties, especially providing significantly reduced application rates, and often new modes-of-action have been discovered, developed and launched for diverse crops. This success has positively influenced agriculture as a whole. However, in these days the introduction of new herbicides with either a new mode of action or novel chemical classes has lingered. After launch of sulcotrione, a HPPD herbicide at 1991, any herbicide with new mode-of-action has not been commercialized in Europe, while there were 10 new modes-of-action commercialized between 1970 and 1985 and five new ones between 1986 and 1991 (Schulte, 2004; Rüegg, 2007).
Are there still opportunities for new herbicides, and what are the main search targets? Is there still an incentive to invest into herbicide research? Many factors adding complexity are agronomic, structural and technological changes, including the introduction of herbicide-tolerant crops, and the high costs of development for new active ingredients, mainly due to increasing regulatory requirements. In the light of increasing weed resistance to widely used herbicides, securing diversity in agronomy as well as weed management is a key to efficient crop production in future. Further problems to be addressed are the expectations regarding weed shifts and/or the occurrence of (new) weed problems, due to the introduction of new weed species by global travel or international transport of goods. Will certain plants profit from climatic changes like global warming? The increase in the global population has already led to an intensification of crop production and this must continue in order to secure world food supply. In order to secure crop yields, chemical solutions for weed management will continue to be the preferable choice for the predictable future because apparent alternatives are not in sight. In order to support this objective, new herbicides, preferably with new modes-of-action, need to be discovered and developed.
Until recently, the first step in agrochemical discovery was to take a collection of chemicals, apply each to a small population of representative pests, and assess their efficacy by visual inspection. This approach, sometimes impolitely called 'spray and pray', has its strengths

(Wolfgang et al. 2004, Manjula et al. 2010). It takes advantage of the obvious fact that it is easier to wreck a system than to fix it. The testing of chemicals for efficacy on whole plants is direct and integrates several important attributes that are needed for a crop protecting, including its uptake, transport, metabolism and ability to inhibit an important target protein. This approach does not require any detailed knowledge about the biochemical or cellular target, which was important in an era when our understanding of biology was poor. Following the initial identification of a lead chemical, intensive research and testing followed to optimize its structure to understand its action, and provide data on its environmental compatibility. However, the traditional approach depended on serendipitously discovering a chemical structure that could enter the pest, be transported within it, inhibit a key target, get away from detoxification, and also be modified to allow it to fulfill increasing-regulatory criteria with respect to environmental compatibility. Testing chemicals on whole organisms is logistically demanding, especially if the organisms are relatively large, relatively large amounts of chemicals required, so the ability of chemists to synthesize sufficient quantities of new structures becomes another serious limitation on the number of chemicals that can be tested (Wolfgang et al. 2004).

Currently, studies of environmental compatibility have become an increasingly large part of the entire research effort because the market was progressively occupied by effective agrochemicals and because hurdles have been enlarged with respect to environmental compatibility. Typically, 10-12 years are necessary to develop a lead chemical structure into a market product. Since the discovery of the auxinic herbicides in the late 1940s, empirical screening has led to the commercialization of around 270 active ingredients, representing 17 modes of action (Ott et al. 2003). Of these herbicides, approximately 50% act on one of only three targets: acetolactate synthase, photosystem II, or protoporphyrinogen oxidase. In addition, 10 herbicides account for 45% of the total market value. Thus, the major herbicides on the market act on only a handful of targets, whereas it is quite evident that there are many more ways to kill a plant.

In the past 10 years, strategies for the first steps of herbicide discovery have switched from the testing of chemicals for efficacy on whole plants towards a target-orientated approach using *in-vitro* assays against molecular targets, it is obviously essential to choose appropriate targets. Therefore, target-directed high throughput screening (HTS) systems are implemented as additional tools in addition to greenhouse screening. This requires the identification of proteins whose inhibition will lead to the death or a severe growth arrest of the objective organism. Many different approaches have been developed to identify *bona fide* targets for *in-vitro* screening (Wolfgang et al. 2004, Manjula et al. 2010). Developments in functional genomics could aid the development of assay systems for the evaluation of chemicals for their suitability as lead structures in herbicide discovery (Ott et al. 2003).

2. How can we select *bona fide* target?

Most of the herbicides attack to the unique biochemistry of plants causing severe disruption of the plant metabolism. These are usually inhibitors of specific enzymes binding either at the active site of the enzyme or at some domain apart from the active site (Berg et al., 1999; Dayan et al. 2009). Among the strategies to identify suitable targets, one strategy was to assume that if an enzyme in a pathway or process is a target, then others in the same pathway or process might be too. The problems and limitations of this 'copy cat' approach have been nicely reviewed (Abell, 1996). Another strategy is to use literature survey to identify 'key' or 'limiting' protein in the essential process that would catalyze irreversible reactions and is highly regulated. Third approach, a revolutionary tool in herbicide discovery, is to provide genetic evidence that the gene encodes the essential target protein (Wolfgang et al., 2004). An antisense technology was used to demonstrate that dehydroquinate dehydrase/ shikimate dehydrogenase constitutes an herbicide target (Freund et al., 2002). Genetic pre-validation of targets in a systematic manner started in the early 1990s, soon after routine methods for plant transformation were established. In its first phase, this approach was focused on specific pathways that were thought to be essential for the plant. The relevance of selected candidates was tested by partial inhibition of their activity using co-suppression or antisense strategies, which resulted in the variable inhibition of expression at the protein level. Typically, about 10% of the plants show a significant decrease in protein expression, with the extent of the decrease varying from 30~90% depending on the transgenic line. The inhibition of expression at the protein level can be quantified using measurements of enzyme activity in standardized conditions, and compared with the inhibition of growth and other phenotypical or biochemical changes in the plants. About 20% of the enzymes in these central pathways qualified as potential herbicide targets. Crucially, they would not have been reliably predicted by the traditional criteria for identifying 'key' regulated enzymes. Many highly regulated enzymes that catalyze irreversible reactions could be strongly inhibited without a significant impact on growth, whereas some of the experimentally validated targets were transporters or enzymes that catalyze readily reversible reactions. The accumulation of large amounts of sequence information from the late 1990s onwards, first as a result of expressed sequence tag (EST) sequencing and later from full-genome sequencing, made it possible to use unbiased and genome-wide strategies to identify targets. Nevertheless, the function of a large proportion of genes is either only vaguely annotated (around 50%) or completely unknown (more than 30%).

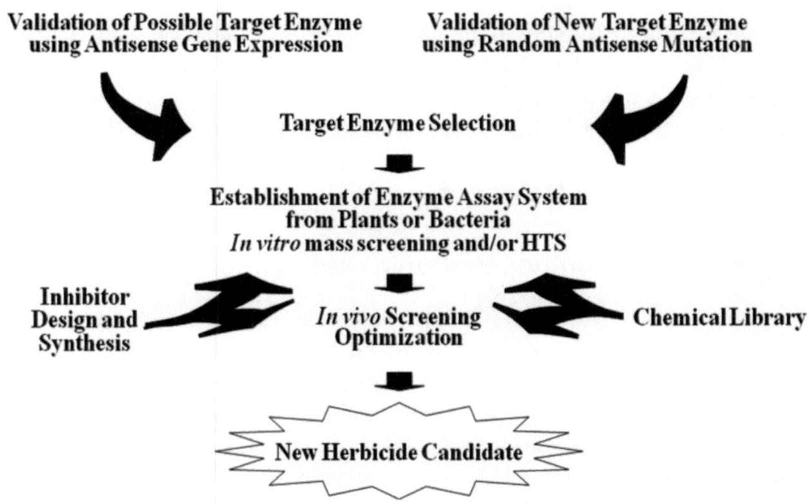

Fig. 1. Process of target search with antisense technology

Genetic approaches include studies of conditionally lethal bacterial and plant mutants and use of antisense technology (Fig. 1). In the absence of chemical leads with known sites of action, targets for validation may be selected by the following criteria: the target is essential to plants and, preferably, inhibition leads to multiple deleterious effects; the target is not present in mammals; the target has low intracellular concentration, i.e., has potential for low use rates; and the proposed inhibitors of the target are synthetically accessible. Potent inhibition of the selected target may still not produce an effective herbicide. Studies of the uptake, translocation and metabolism of the inhibitor are needed to determine if the cause of poor *in vivo* performance is due to these factors or to an intrinsically poor target. Without full appreciation of each of these aspects of herbicide design, the chances for success with the target-site directed approach are reduced. Promising target enzymes were established as 4-hydroxyphenylpyruvate dioxygenase, adenylosuccinate synthetase, AMP deaminase, anthranilate synthase, ascorbate peroxidase, asparagine synthetase, auxin transport, cytosolic glutamine syhthetase, dihydro dipicolinate synthase, dihydrodipicolinate reductase, carboxypeptidase A, chloroplast NADH dehydrogenase, cinnamyl-alcohol dehydrogenase, geranylgeranyl diphosphate synthase, glutamate dehydrogenase, glutamate synthetase, glutamate-1-semialdehyde aminotransferase, glutamate synthase, histidine biosynthesis, imidazoleglycerol phosphate dehydratase, isopropylmalate dehydrogenase, isopropylmalate isomerase, pheophorbidase, farnesyl transferase, *p*- hydroxyphenyl pyruvate dioxygenase, plasma membrane H+-ATPase, pyruvate orthophosphate dikinase, threonine dehydratase etc. (Kishore and Shah, 1988; Schloss and Aulabaugh, 1990; Abell et al., 1993; Rendina and Abell, 1994; Pillmoor et al., 1995; Abell, 1996; Kleier and Hsu, 1996; Subramanian et al., 1997; Bartley et al., 1999; Coulter, 1999; Cromartie et al., 1999; Ficarelli et al., 1999; Grossmann and Schiffer, 1999; Saari, 1999; Hwang et al., 2001).

This chapter, which focuses mainly on antisense technology, assesses progress being made and points to areas of research and new technologies regarding validation of the target KAPAS that have the potential to further increase the effectiveness of KAPAS inhibitor research. Successful design of novel herbicides based on the specific inhibition of selected enzyme targets requires careful consideration of the choice of the target, mechanism of the enzyme, design of potent inhibitors, delivery of the inhibitor to the target and metabolic fate of the inhibitor. Validated targets, those that produce phytotoxic effects upon partial inhibition, can be identified by genetic methods or by obtaining chemical leads. The aim of our investigation is to confirm that a particular enzyme chosen is indeed essential for a plant growth, and to validate the successful inhibition of the enzyme can lead to an herbicidal effect. Herein, we describe the genetic validation of KAPAS as a potential herbicide target enzyme, and chemical validation of TPTA as a lead compound for the potential KAPAS inhibiting herbicide derivatives *in vitro* and *in vivo*.

3. Discovery of 7-keto-8-aminopelargonic acid synthase

In a pioneering pilot study (Jun et al., 2002), *Arabidopsis* antisense lines were created using randomly selected cDNAs. These lines were then scored for mutant phenotypes and analyzed genetically to exclude mutants that were clearly not caused by antisense inhibition of gene expression. At present, about 10,000 genes have been put through the entire process, including confirmation by independent retransformation, and 46 potential herbicide targets have been identified. These are genes whose partial inhibition leads to chlorosis, necrosis, and concomitant growth defects. They contain both known herbicide targets (e.g. glutamine synthetase) and genes for which antisense has already been reported to mimic herbicidal phenotypes (e.g. Rubisco and foredooming: NADP oxidoreductase) (Stitt et al., 1999; Palatnik et al., 2003).

Among them, we have already described expressing antisense RNA of cloned plant genes encoding for a potential herbicide target enzyme, 7-keto-8-aminopelargonic acid synthase

(EC 2.3.1.47, KAPAS, also known as 8-amino-7-oxononanoate synthase) in stably transformed transgenic test plants (Hwang et al., 2003; 2010). Individual biotin auxotrophs for KAPA synthase, transformed with antisense *A. thaliana* KAPAS (*AtKAPAS*) construct, exhibited considerable phenotypic alterations such as growth inhibition, severe growth retardation, yellow-green cotyledons and leaves as well as lethal phenotype (Fig. 1). We performed the database screening of *Arabidopsis* genome sequence with *bioF* sequence of *E. coli, B. subtilis,* and *B. sphaericus.* Through the analysis of cDNA isolated by PCR amplification, *AtKAPAS* gene (TAIR accession number 3443298) contained an open reading frame (ORF) of 1,410 base pairs encoding a putative protein of 469 amino acids with a predicted molecular mass of 51.3 kDa. *AtKAPAS* contains the domain of predicted aminotransferase class I and II in the C-terminal region, as well as the domain of putative plasma membrane spanning region.

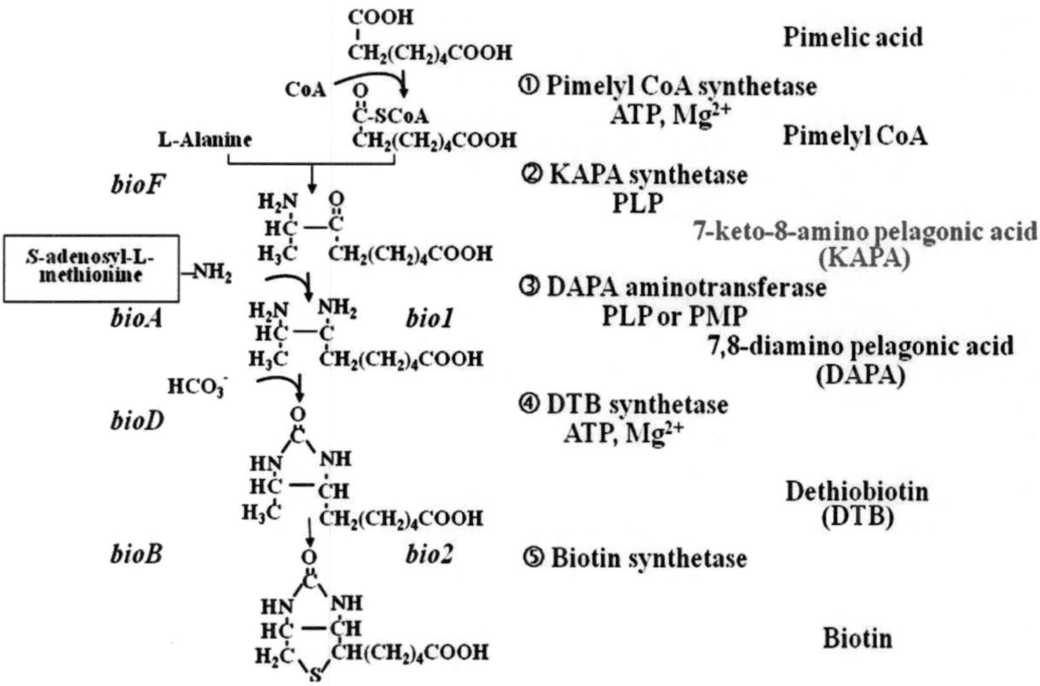

Fig. 2. Biosynthetic Pathway of Biotin in Microorganisms

Biotin is an essential vitamin and acts as a cofactor for a number of enzymes involved in facilitation of CO_2 transfer during carboxylation, decarboxylation, and transcarboxylation reactions that are related to fatty acid and carbohydrate metabolism (Dakshinamurti and Bhagavan, 1985; Alban et al., 2000). Bacteria, plants, and some fungi make their own biotin directly from endogenous biochemical intermediates, whereas other organisms such as most fungi and animals must obtain it from their surrounding environments. Therefore, the studies on the inhibition of the enzymes involved in the biotin pathway will potentially offer an attractive target for herbicide development. Since Eisenberg and Star (1968), Eisenberg and Stoner (1971) and Pai (1975) have first investigated on the biosynthetic pathway of biotin in *Escherichia coli* and *Bacillus subtilis* through biochemical studies including the analysis of auxotrophic mutants.

A number of researchers have widely investigated the biosynthetic pathway of biotin through combined biochemical and genetic studies in the bacteria (Ploux and Marquet, 1992; Alexeev et al., 1998; Huang et al., 1995). Also, there have been numerous studies on the related gene for biotin synthesis in a variety of other microorganisms (Zhang et al., 1994; Fleischmann et al., 1995; Bult et al., 1996). The ordinary pathway for biotin biosynthesis in microorganisms and plants is shown in Fig. 1 (Patton et al., 1998). Most steps in the pathway have been clearly investigated in the microorganism, particularly *E. coli* and *Bacillus* spp. *E. coli* and *B. subtilis* have only one *bio* cluster consisted of five genes: *bioABFCD* and six genes: *bioWAFDBI*, respectively (Bachman, 1990; Bower et al., 1996), whereas *B. sphearicus* has two separate clusters consisted of seven genes: *bioXWF* and *bioDAYB* (Gloeckler et al., 1990). In *E. coli*, *BirA* protein is known to act as a negative regulator for the expression of the biotin operon by interaction in a region between *bioA* and *bioB* (Barker and Campbell, 1981). The decarboxylative condensation of L-alanine and pimeloyl CoA into 7-keto-8-aminopelargonic acid (KAPAS, also known as AON, 8-amino-7-oxononanoate), which is catalyzed by KAPA synthase (EC 2.3.1.47), ~~is~~ the first committed step in the pathway of biotin biosynthesis (Fig. 2), and that was first identified in *E. coli* (Eisenberg and Star, 1968). KAPA synthase, the product of the *bioF* gene from *E. coli*, is a homodimeric and pyridoxal 5′-phosphate (PLP)-dependent enzyme. The molecular mass of the enzyme subunit is about 42 kDa. The enzyme is structurally related to dialkylglycine decarboxylase, a type II aminotransferase when compared to other PLP-dependent enzymes in the amino acid sequence and tertiary structure (Toney et al., 1993). Recent studies by spectroscopic, kinetic, and crystallographic techniques have shown the KAPA synthase from *E. coli* was structurally the apo- and holoform (Alexeev et al., 1998), and the enzyme generates external aldimine complex (Webster et al., 2000). The biotin biosynthesis is understood in detail in microorganisms, but it is relatively poorly understood in plants. Particularly, there is little study on the KAPA synthase of plants. Moreover, of particular interest is the evidence that plants synthesize biotin using the same route as that in *E. coli* (Baldet et al., 1993). Biotin synthesis and utilization in plants have been mainly investigated through analysis of biotinylated proteins (Nikolau et al., 2003; Tissot et al., 1997), and isolation and characterization of auxotrophic mutants (Meinke, 1994). Detailed mutational analysis such as that of auxotrophic mutants has led to an inclusive understanding of biotin synthesis and regulation. The *bio1* auxotroph of *Arabidopsis*, first identified among the collection of recessive embryo-defective mutants, has been shown to be defective in the early step of biotin synthesis, the conversion of KAPA to 7,8-diaminopelargonic acid (DAPA) (Meinke, 1985). The *bio2* mutants have shown to be embryo-defective in the final step of biotin synthesis, the conversion of dethiobiotin to biotin (Patton et al., 1998). These results suggest that the antisense disruption of *AtKAPAS* gene cause lethality in the early stage of plant development. 7-keto-8-aminopelargonate synthase is a pyridoxal 5′-phosphate-dependent enzyme which catalyzes the decarboxylative condensation of L-alanine with pimeloyl-CoA in a stereospecific manner to form KAPA, coenzyme A, and carbon dioxide in the first committed step of biotin biosynthesis. Perhaps the most important role of biotin is in the carboxylation of acetyl-CoA to give malonyl-CoA, which is the first step in fatty acid biosynthesis. Since fatty-acid synthesis is essential for the growth and development of most organisms, biotin is thus an essential nutrient for plants and animals. Plants, microorganisms, and some fungi biosynthesize their own biotin, while animals necessarily require trace amounts of the vitamin in their diet. Therefore, inhibition of the enzymes involving in the biotin biosynthesis pathway can cause irreparable damage to plants, and for this reason, such enzymes can be useful targets for the rational design of inhibitors in the hopes of finding new herbicides (Webster et al., 2000; Nudelman et al., 2004). The aim of our investigation is to confirm that a particular enzyme chosen is indeed essential for a plant growth, and to validate the successful inhibition of the enzyme can lead to an herbicidal effect. Herein, we describe the genetic validation of KAPAS as a potential herbicide target enzyme, and chemical validation of TPTA as a lead compound for the potential KAPAS inhibiting herbicide derivatives *in vitro* and *in vivo*. We have described the effects of expressing anti-sense RNA of cloned plant genes encoding for potential herbicide target enzyme 7-keto-8-aminopelargonic acid synthase (EC 2.3.1.47, KAPAS, also known as 8-amino-7-oxononanoate synthase) in stably transformed transgenic test plants. Individual biotin auxotrophs for KAPA synthase, transformed with anti-sense *Arabidopsis thaliana* KAPAS (*AtKAPAS*) construct, exhibited considerable phenotypic alterations such as growth inhibition, severe growth retardation, yellow-green cotyledons and leaves as well as lethal phenotype (Fig. 3).

These results suggest that the anti-sense disruption of *AtKAPAS* gene causes lethality in the early stage of plant development. 7-Keto-8-aminopelargonate synthase is a pyridoxal 5'-phosphate dependent enzyme which catalyzes the decarboxylative condensation of L-alanine with pimeloyl-CoA in a stereospecific manner to form KAPA, coenzyme A, and carbon dioxide in the first committed step of biotin biosynthesis. Perhaps the most important role of biotin is in the carboxylation of acetyl-CoA to give malonyl-CoA, which is the first step in fatty-acid biosynthesis. Since fatty-acid synthesis is essential for the growth and development of most organisms, biotin is thus an essential nutrient for plants and animals. Plants, micro-organisms, and some fungi biosynthesize their own biotin, while animals necessarily require trace amounts of the vitamin in their diet. Therefore, inhibition of the enzymes involved in the biotin biosynthesis pathway can cause irreparable damage to plants, and for this reason, such enzymes can be useful targets for the rational design of inhibitors in the hopes of finding new herbicides (Webster et al., 2000; Nudelman et al., 2004). The aim of our investigation is to confirm that a particular enzyme chosen is indeed essential for a plant growth, and to validate the successful inhibition of the enzyme can lead

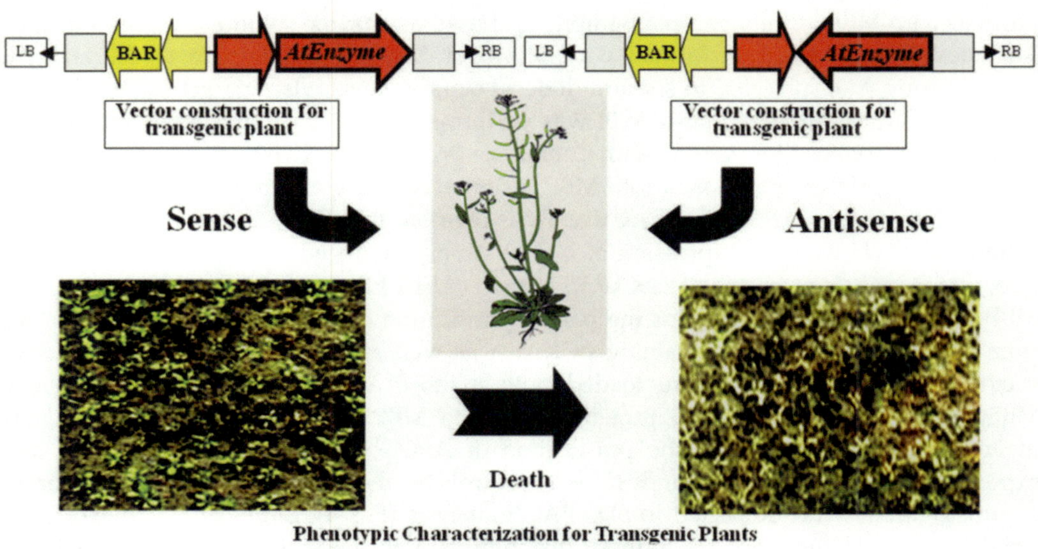

Fig. 3. Sense and Anti-sense Expression of Target Gene in Arabidopsis

to an herbicidal effect. Herein, we describe the genetic validation of KAPAS as a potential herbicide target enzyme, and chemical validation of TPTA as a lead compound for the potential KAPAS inhibiting herbicide derivatives *in vitro* and *in vivo*.

4. Genetic and chemical validation

4.1 *AtKAPAS* from transgenic *E. coli*

Total RNA isolated from leaf tissues of *A. thaliana* was used for preparation of poly(A)+mRNA. Double-stranded cDNA was constructed from 5 μg of poly(A)$^+$mRNA with the Time Saver cDNA synthesis kit (Pharmacia, Piscataway, NJ, USA), using Oligo(dT)18 as a primer. By performing PCR (polymerase chain reaction) with the two primers, the full-length AtKAPAS cDNA was amplified and isolated from *A. thaliana* cDNA library prepared. The primers encompassing the full-length cDNA of *AtKAPAS*, KAPAFB (5'-CAAAAAGAATTCGACGACGACGACAAGATGGCGGATCATTCGTGG GATAAA-3') and KAPARH (5'-GTGCACCTCGAGTTATAATTTGGGAAATAGAAAGGA-3'), were synthesized to include *Eco*RI and *Xho*I restriction site, respectively. Primers of KAPAFB and KAPARH were used in a PCR reaction to amplify the *AtKAPAS*-encoding region. The resulting PCR fragment was digested with *Eco*RI and *XhoI*, and cloned into MBP (maltose binding protein) fusion vector (Bioprogen Co., Ltd., Korea) to generate construct pEMBPek-KAPAS (Fig. 2). *E. coli* BL21-Gold(DE) (Stratagene, USA) was transformed with expression vector pEMBPek-KAPAS and than cultured in LB (Luria–Bertani broth, USB, USA) medium containing 100 μg·mL^{-1} of ampicillin at 37°C (150 rpm) until the value of $_{OD}$600 reached 0.6. In order to induce the expression of the target protein in *E. coli* cells, isopropyl-D-thiogalactoside was added to the suspension at a final concentration of 1 mM, and further cultured for 3 h. The culture cells were washed with 50 mM Tris–HCl buffer, pH 8.0, containing 1 mM EDTA, after centrifugation at 9000g for 10 min. The cell pellets were resuspended and pooled in 50 mL of buffer solution (50 mM Tris–HCl, pH 8.0, 200 mM NaCl). The sample was sonicated for 30 s and cooled on ice for 3–5 min, and the procedure was repeated three times. After centrifugation at 1000g for 30 min, the supernatant was purified with MBP affinity chromatography and used as enzyme solution. Eluting fractions separated from *E. coli* transformed with pEMBPek-KAPAS recombinant vector and the control group was subjected to sodium dodecyl sulfate-polyacrylamide gel electrophoresis (SDS–PAGE), respectively. SDS–PAGE was performed on a 12% running gel and protein bands were visualized by staining with Coomassie Brilliant Blue G250.

The *AtKAPAS* cDNA was cloned into MBP fusion vector to generate the *E. coli* expression construct *pEMBPek-KAPAS*. SDS-PAGE analysis revealed that *E. coli* transformed with MBP fusion vector showed the expression of a very strongly induced fusion protein of ca. 98.2 kDa, which may be consisted of *AtKAPAS* protein of 51.3 kDa, and maltose binding peptide MBP affinity tag of 46.9 kDa. For the partial purification of *AtKAPAS* protein, the lysates from IPTG-induced *E. coli* containing pCKAPA as well as from *E. coli* harboring control vector MBP fusion vector were loaded onto maltose affinity column (1.1cm x 30cm, Millipore, USA). The *AtKAPAS* protein binding to MBP resin was eluted with 10 mM maltose solution. To confirm the purification of *AtKAPAS* protein, elutes with *E. coli*-expressed *AtKAPAS* protein and *E. coli* control in the various fractions of affinity chromatography were subjected to SDS-PAGE analysis (Fig. 4). Elutes of *E. coli*-expressed *AtKAPAS* protein contained the induced fusion protein of ca. 98.2 kDa while those of *E. coli* control didn't contain *AtKAPAS* protein.

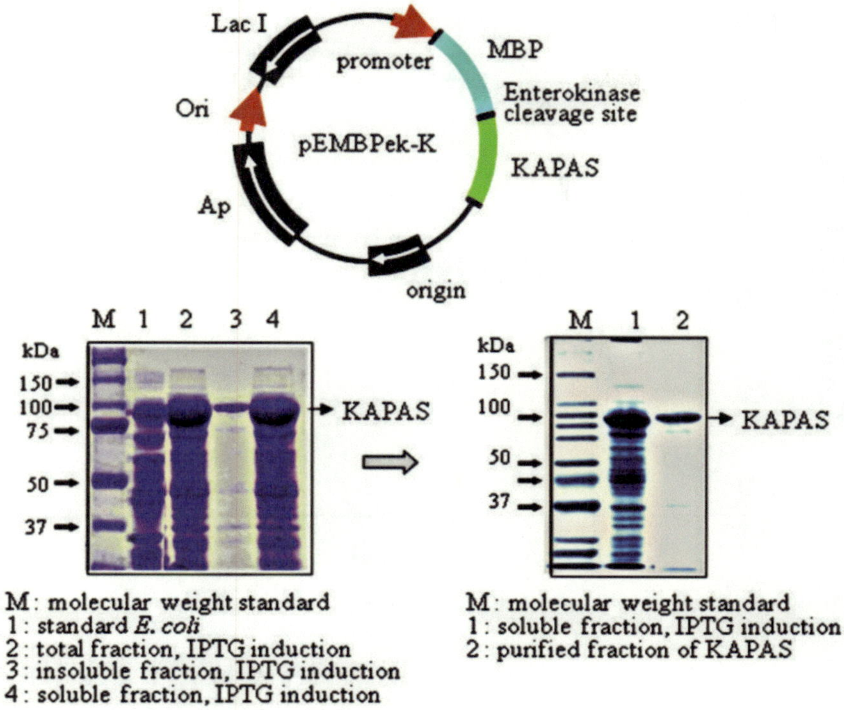

Fig. 4. KAPAS over expression and purification from transgenic *E. coli*.

4.2 *AtKAPAS* inhibition *in vitro* treated with TPTA

For substrate synthesis and enzyme assay *in vitro*, substrate pimeloyl CoA was synthesized according to the method of Ploux and Marquet (1992). TPTA was purchased from Sigma (USA) and used as a KAPAS-inhibitor. KAPAS activity was determined according to the method of Webster et al. (2000) using a linked assay by monitoring the increase in absorption of NADH at 340 nm using a microplate spectrophotometer (Benchmark Plus, Bio-Rad, USA), thermostatically controlled at 37°C. The procedure was the same apart from the reaction volume of 250 µL instead of 1 mL. L-Alanine and pimeloyl-CoA were added to give the desired final concentrations. Prior to analysis, enzyme samples were dialyzed for 2 h at 4°C against 20 mM potassium phosphate (pH 7.5) containing 100 µM pyridoxal 5'-phosphate (PLP). The KAPAS concentration in all analysis was 10 µM in 20 mM potassium phosphate (pH 7.5) and the concentrations of TPTA were 3.125, 6.25, 12.5, 25, 50, and 100 µM. Reference cuvettes contained all other compounds except inhibitor.

Enzyme activity was assayed with the partially purified *AtKAPAS* protein extracted from transgenic *E. coli*. *AtKAPAS* protein was expressed in *E. coli* at a very high level, and a significant portion of these proteins was soluble, and their affinity-purified preparations contained a single major polypeptide. The dose-dependent *in vitro* inhibition of KAPAS activity by TPTA was noticeably examined and the IC_{50} was calculated as 19.85 µM (Fig. 5).

4.3 Herbicidal activity of TPTA under greenhouse condition

Seeds of *A. thaliana* were sown in plastic pots (24 cm^2 surface area) filled with artificial nursery soil (Boo-Nong Soil, Seoul, Korea), and the plants were grown to the required

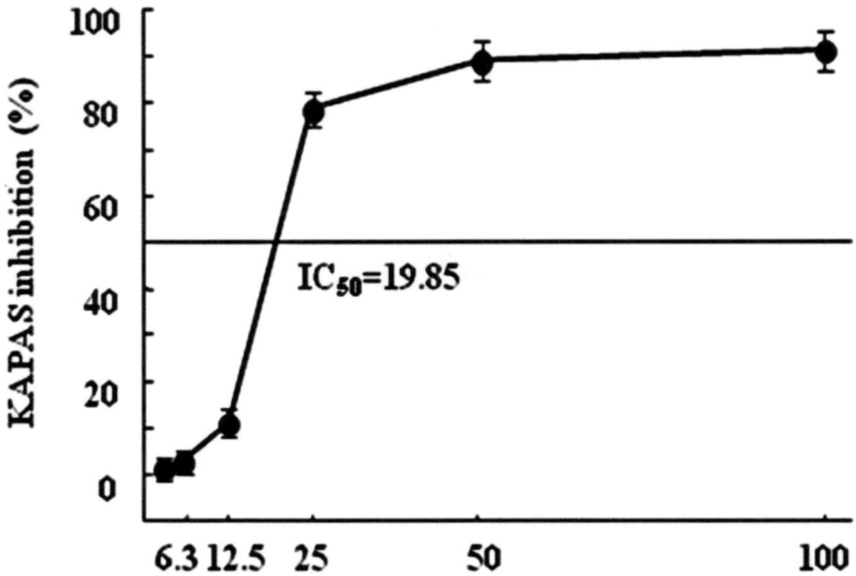

Fig. 5. KAPAS inhibition treated with triphenyltin acetate *in vitro*. Data was expressed as a mean ± S.D.

growth stage for application in a greenhouse maintained at 30~35°C during the day and 20~25°C at night. Application was conducted at 40 days after seeding for foliar application of 16, 32, 62.5, 125, 250, and 500 g·ha^{-1} with laboratory spray gun (spray volume of 1000 L·ha^{-1}). The TPTA was used as a solution in acetone/water (60:40 by volume) containing 1.0 g·L^{-1} of Tween-20. The plants were photographed at 1 week after application. The herbicidal spectrum of TPTA was investigated to 10 weed species, *Sorghum bicolor, Echinochloa crus-galli, Agropyron smithii, Digitaria sanguinalis, Panicum dichotomiflorum, Solanum nigrum, Aeschynomene indica, Abutilon avicennae, Xanthium strumarium, Calystegia japonica* with foliar application. Foliar application of 0.25, 0.5, 1, 2, and 4 kg·ha^{-1} with laboratory spray gun (spray volume of 1000 L·ha^{-1}) was conducted at 2 weeks after sowing each seeds in plastic pot (350 cm2 surface area) filled with upland soil. Visual injury was determined at 2 weeks after application with a scale of 0 (no injury) to 100 (complete death).

The foliar-treatment of 16, 32, 62.5, 125, 250, and 500 g·ha^{-1} TPTA to the 40-day old *A. thaliana* plants has caused herbicidal effects of 8.3, 20, 47, 90, 97, and 100%, respectively. The herbicidal activity was increased as time passed after application. The application rate of more than 125 g·ha^{-1} was shown almost complete death at 1 week after application (Fig. 6). The main symptoms were desiccation and burning effect. Symptoms begun to appear within several hours after application, and the applied region of the leaf was desiccated at 1 day after treatment of more than 250 g·ha^{-1}.

Foliar application of TPTA to 10 weed species was showed good herbicidal activity. The most sensitive species was *Xanthium strumarium* which was completely dead at 250 g·ha^{-1} of TPTA foliar application. *Abutilon avicennae, Calystegia japonica,* and *Aeschynomene indica* were also controlled by 500 g·ha^{-1} of TPTA foliar application (Table 1). However, grass weed such as *Sorghum bicolor, Echinochloa crus-galli, Agropyron smithii, Digitaria sanguinalis,* and *Panicum dichotomiflorum* was tolerant to TPTA foliar application comparing to the broad-leaf weeds.

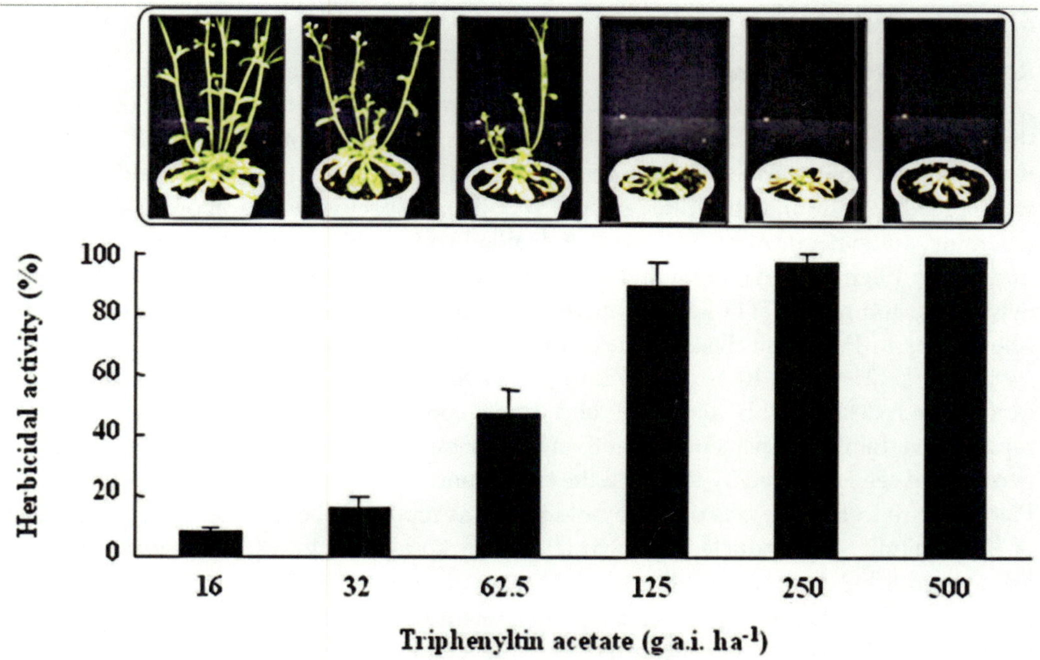

Fig. 6. Herbicidal activity of triphenyltin acetate foliar application at 40 days after seeding under greenhouse condition on the *Arabidopsis thaliana*.

Rate (kg/ha)	Control value (%)									
	SORBI	ECHCG	AGRSM	DIGSA	PANDI	SOLNI	AESIN	ABUTH	XANSI	CAGEH
0.25	30	40	0	30	50	30	70 C	40 C	100	30
0.5	40	40	0	50	60	80	95	100	100	100
1	40	40	0	50	60	80	95	100	100	100
2	40 C	60	60	50	80	100	100	100	100	100
4	70 CI	70 BC	50 BC	60 N	100	100	100	100	100	100

SORBI, ***Sorghum bicolor***; ECHCG, ***Echinochloa crus-galli***; AGRSM, ***Agropyron smithii***; DIGSA, ***Digitaria sanguinalis***; PANDI, ***Panicum dichotomiflorum***; SOLNI, ***Solanum nigrum***; AESIN, ***Aeschynomene indica***; ABUTH, ***Abutilon avicennae***; XANSI, ***Xanthium strumarium***; CAGEH, ***Calystegia japonica***. **Pre, pre-emergence application; Post, post-emergence application. Description of footnotes: B, stunting; C, desiccation; I, chlorosis or abnormal color of plant; N, bleaching (lack of pigmentation). Visual injury was determined at 2 weeks after application with a scale of 0 (no injury) to 100 (complete death).**

Table 1. Herbicidal activity of triphenyltin acetate on the several weed species under a greenhouse condition

5. Reversal study

Reversal effect was estimated via chlorophyll contents to foliar application and via % germination. Germination test: Seeds of *A. thaliana* were germinated in 55 mm plastic Petri-dish lined with one-layer filter paper (Advantec No. 2). About 1 mL of each TPTA solution dissolved in absolute acetone with various concentrations of 0, 0.063, 0.0125, and 0.025 mM was spread evenly onto the filter paper (Ø 5 cm), respectively and allowed to dry in a laboratory fume hood. After that, 1 mL of distilled water with or without supplement of 0.5 mM biotin (Sigma, USA), dethiobiotin (Sigma, USA), 7,8-diaminopelargonic acid (DAPA, Synthesis), and KAPA (TRC, Inc., Canada) was added, and 30-seeds were placed onto the filter paper in Petri-dish. Each Petri-dish was sealed with laboratory film and held in an incubator at 25°C, 14/10 h (Light/Dark). The assays were conducted in a completely randomized design with a control and three concentrations of chemicals with three replications. Inhibition percentages at 8 days after treatment were calculated the number of germinated seeds divided by total and the significance level was 0.05 for all analysis.

Plant growth test: *A. thaliana* of 40-day-old plants as reported above were used. Supplement of 1 mM biotin was conducted by foliar laboratory spray gun with spray volume of 5000 L ha^{-1} at each 1 or 2 days before 100 $g{\cdot}ha^{-1}$ TPTA application. At 5 days after TPTA application, plant leaves were harvested and chlorophyll content was determined following the method reported by Hiscox and Israelstam (1979). One gram of leaf tissue was placed in a vial containing 7 mL of dimethyl sulfoxide (DMSO, Sigma-Aldrich, USA) and chlorophyll was extracted into the fluid without grinding at room temperature for 24 h in darkness. The extract liquid was transferred to a graduated tube and made up to a total volume of 10 mL with DMSO, and 1.5 mL of the 10 mL aliquots was transferred to microcentrifuge tubes. After centrifuged at 5000g for 10 min, total chlorophyll amount in extracts was determined by the absorbance measurement at 645 nm and 663 nm for each sample using a Microplate Spectrophotometer (Benchmark Plus, Biorad, USA) against DMSO blank. Chlorophyll content was calculated following the equation used by Arnon (1949).

The germination of *A. thaliana* seeds was almost completely inhibited by 0.05 mM TPTA. Also, more than 0.125 mM of TPTA treatment completely inhibited the germination and significantly reduced the plant growth of early stage plants after seed germination. However, the inhibited germination by 0.05 mM TPTA was recovered to 85~92% with the supplement of 0.5 mM biotin, dethiobiotin, and DAPA, except KAPA, one of the biotin biosynthesis intermediates (Fig. 7). Additional supplement of 0.5 mM SAM with 0.5 mM KAPA increased up to 91% of the germination previously inhibited by 0.05 mM TPTA. At 5 days after TPA application, plant leaves were harvested and chlorophyll content was determined following the method reported by Hiscox and Israelstam (1979).

The chlorophyll content in *A. thaliana* plant treated TPTA without biotin pretreatment was 10.7 $mg{\cdot}L^{-1}$. The chlorophyll content of the untreated control *A. thaliana* plant was 20.5 $mg{\cdot}L^{-1}$, however, the amount of chlorophyll extracted from the *A. thaliana* plant treated with TPTA at 1 and 2 days after biotin pretreatment was 19.5 and 19.8 $mg{\cdot}L^{-1}$, respectively. The chlorophyll loss of *A. thaliana* plant treated TPTA was reversed by biotin pretreatment at 1 and 2 days before TPTA application. Consequently, biotin pretreatment reversed the growth inhibition of *A. thaliana* plant treated TPTA at the same extent to the untreated control plants (Fig. 8).

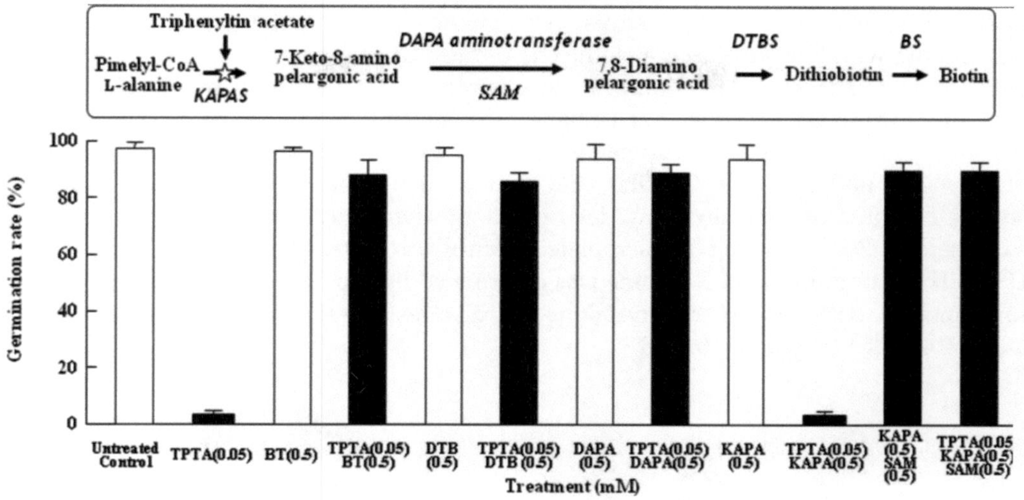

Fig. 7. **Reversal of *A. thaliana* seed germination with biotin biosynthesis intermediates supplement.** KAPAS, 7-keto-8-aminopelargonic acid synthase; DAPA, 7,8-diaminopelargonic acid synthase; DTBS, dithiobiotin synthase; BS, biotin synthase; TPTA, Triphenyltin acetate; BT, Biotin; DTB, dithiobiotin; DAPA, 7,8-diaminopelargonic acid; KAPA, 7-keto-8-aminopelargonic acid; SAM, S-adenosyl-L-methionine.

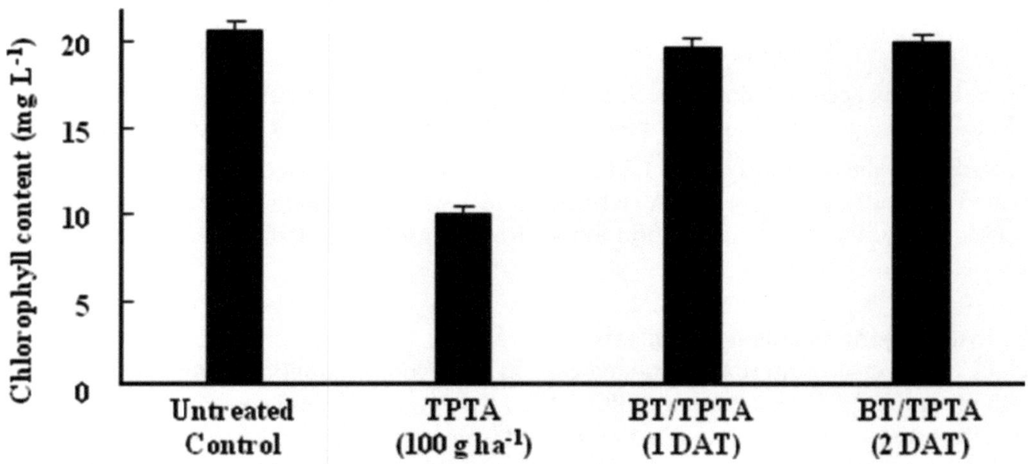

Fig. 8. **Reversal of *A. thaliana* growth inhibition with biotin supplement.** TPTA, Triphenyltin acetate; BT, Biotin; BT/TPTA, BT treatment followed by TPTA; DAT, day after treatment.

5.1 L-alanine accumulation in plants treated with TPTA

Alanine was determined from a detection system of copper complex with L-alanine described by Nakao et al. (1986) with some modification (Lin and Wu, 2005; Weinstein, 1984) from aqueous fraction of extractions. 40-day-old of the Arabidopsis plants grown as reported above was treated with TPTA (200 g·ha^{-1}) by foliar application with laboratory spray gun (spray volume of 1000 L·ha^{-1}). Plant leaves were harvested at 3 days after TPTA application. Ten grams of plant leaves were homogenized with 100 mL of distilled water and filtered with 2 layers of Mira cloth. The filtrates were separated with equal volume of ethyl acetate. The water fraction was concentrated by vacuum rotary evaporator. The solution (1 mL) was centrifuged at 1000g for 10 min, and chloroform (67 µl) was added to the supernatants, and then centrifuged at 1000g for 10 min, repeatedly. After reaction with 1 mg of NaN_3 and 67 mg of $Cu(OH)_2$ and standing for 20 min at room temperature. The copper complex of L-alanine was determined by the optical density at 620 nm of the supernatant (200 µl) using a microplate spectrophotometer (Benchmark Plus, Bio-Rad, USA). The concentration of L-alanine was determined by standard curve prepared from the same method with various concentrations of L-alanine. The standard curve was calculated as $Y = 0.4695X + 0.0146$, $r^2 = 0.9993$.

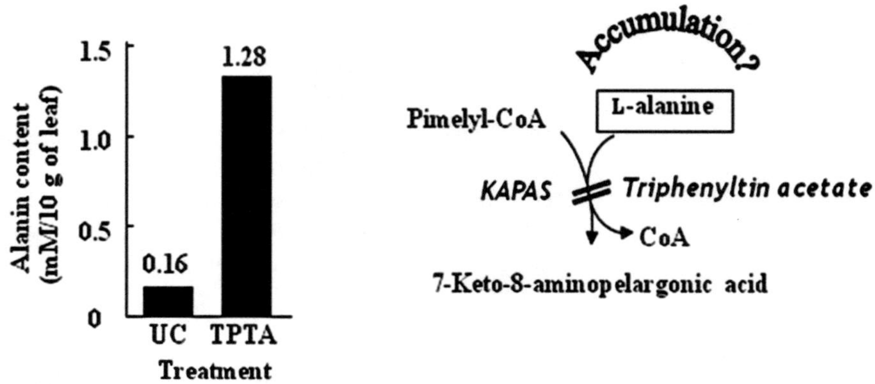

Fig. 9. L-alanin accumulation in *A. thaliana* plants treated with triphenyltin acetate. KAPAS, 7-Keto-8-aminopelargonic acid synthase; UC, untreated control; TPTA, triphenyltin acetate

According to the standard curve, 1.28 mM of L-alanine was detected from *A. thaliana* plants treated with 200 g $\cdot ha^{-1}$ of TPTA, whereas 0.16 mM of L-alanine from untreated plants. Consequently, the TPTA application induced 8-fold greater L-alanine accumulation in the plants (Fig 9).

5.2 KAPAS gene expression analysis

RT-PCR (Reverse transcription-polymerase chain reaction) amplifications were performed with an iCycler™ Thermal Cycler (BIO-RAD, http://www.bio-rad.com/), according to the manufacture's instructions. RNA was prepared from various tissues of Arabidopsis that had been immediately frozen in liquid nitrogen under RNase-free conditions. The RNA was isolated with the Qiagen RNeasy Plant Mini Kit (Qiagen, http://www.giagen.com/) for subsequent reverse transcription reactions. First-strand cDNA was synthesized with 1 µg of total RNA using the Oligo(dT)12–18 primer and the SuperScript™ III Reverse Transcriptase (Invitrogen, http://www.invitrogen.com/), following the manufacturer's instructions. One microliter of cDNA was used for PCR reactions. The PCR conditions were as follows: an initial denaturation at 94°C for 5 min, followed by 26 cycles of 94°C for 2 min, 55°C for 40 s and 72°C for 1 min. KAPAS-specific primers for RT-PCR were: KAPAS-F, 5′-GCTGAACGACAAGGAA ATGTTG-3′; KAPAS-R, 5′-GAGTGGCTGTGTTGTCAAAG-3′. Primers for amplification of reference gene, tubulin was: TUB-F, 5′-CTCAAGAGGTTCTCAGCAGTA-3′; TUB-R, 5′-TCACCTTCTTCATCCGCAGTT-3′.

To expand our understanding on the role of TPTA, the expression of KAPAS gene in the root, leaf, stem, and whole plant of *A. thaliana* was analyzed by RT-PCR at 1 day after treatment with or without 100 $g \cdot ha^{-1}$ TPTA (Fig. 8). KAPAS was expressed in most tissues, with the highest levels either in stems or roots of the untreated plants, and tubulin was also showed good reference gene expression in *A. thaliana*. However, RNA expression of KAPAS band was indeed fainter or disappeared in the lane representing leaf tissue of TPTA (+) plants. Also, slightly less RNA appeared to the tubulin band than in the other lane. This result implying that the TPTA treatment is severely subjected to protein KAPAS translation and/or post translational regulation in the leaf within 1 day of treatment like as *bio 1* mutants.

Fig. 8. Semi-quantitative RT-PCR analysis of KAPAS gene expression in *A. thaliana* plants. TPTA(+/-), treatment with/without 100 g ha^{-1} triphenyltin acetate; W, whole plant; R, root; L, leaf; S, stem; KAPAS, 7-Keto-8-aminopelargonic acid synthase; TUB, tubulin; Analysis was conducted 1 day after TPTA treatment.

6. Summary

As a number of enzymes related in the metabolic pathways of plants are essential for the growth and development, those can be utilized as potential herbicide targets. We have performed molecular genetic dissection using reverse genetics of antisense approach to identify *AtKAPAS* gene encoding KAPA synthase in the pathway of biotin biosynthesis and to characterize the phenotypic consequences of loss-of-function mutations (Hwang et al., 2003; 2010).

Many researchers have investigated the KAPAS in microorganisms and the most of these reports were focused on the biosynthesis in microorganisms (Eisenberg and Star, 1968), purification and characterization (Ploux 1992, Stoner and Eisenberg, 1975a; 1975b), crystal structure (Alexeev et al., 1998, Kack et al., 1999), binding and kinetics (Ploux et al., 1999), point mutation (Andrew et al., 2002), and stereospecificity (Vikrant et al., 2006). Among them, Ploux et al. (1999) reported that the KAPAS catalyzes the first committed step of biotin biosynthesis in micro-organisms and plants, and suggested that the inhibitors of this pathway might lead to antifungal or herbicide agents. Webster et al. (2000) also reported that biotin is an essential enzyme cofactor for carboxylase and transcarboxylase reactions. The biosynthesis of biotin appears to follow similar pathways in both plants and microorganisms, and thus, inhibition of the enzymes involved in the pathway is potential and attractive target for both herbicide and antibiotics development. These evidences strongly support the hypothesis that inhibitors of microbial enzymes of the biotin biosynthesis pathway might exhibit herbicidal properties as a biofunctional inhibitor.

For instance, two natural compounds isolated form culture filtrates of *Streptomyces* species, actithiazic acid and amiclenomycin, are biotin synthase, final step of biotin biosynthesis pathway, inhibitors of mycobacteria and plants. Ashkenazi et al (2005, 2007) reported the analogs of KAPA and DAPA, possessing chain lengths of eight carbon atoms, 7-aminooctanoic acid hydrochloride, 7-allyloxy-6-oxo-octanoic acid, and 6,7-diaminooctanoic acid dihydrochloride displayed a inhibitors of biotin biosynthesis as potential herbicides. In biotin biosynthesis pathway, four steps of enzymes, KAPAS, DAPA amino transferase, DTBS, and biotin synthase, were working continuously. Among them, the first step of KAPAS inhibitors was not introduced until now.

From these backgrounds, we studied KAPAS inhibition using various commercialized compounds as pesticides. Among them, the chemical TPTA was selected *in vitro* assay with *AtKAPAS* over-expressed from transgenic *E. coli*. Also, we investigated a genetic and chemical validation of the compound as a potential lead compound for KAPAS inhibitors under greenhouse condition *in vivo*. KAPAS activity was completely inhibited by 100 μM of TPTA *in vitro* enzyme assay with the IC_{50} value of 19.85 μM. The germination of *A. thaliana* seeds was also completely inhibited when TPTA concentration was greater than 63 μM. The foliar-treatment of more than 125 g·ha^{-1} TPTA to the 40-day old *A. thaliana* plants has caused almost complete death. Also, foliar application of TPTA to the 10 weed species was showed good herbicidal activity under a greenhouse condition.

Abell (1996) and Pillmoor (1995) suggested that if a protein is a potential target, a 60~80% inhibition of its activity leads to a severe growth phenotype. In accordance with this standpoint, our results suggest that the KAPAS might be a good target enzyme for new herbicide development. However, these results were not sufficient to explain the exact mechanism of action of TPTA as one of the KAPAS inhibitors. It is important to emphasize that the correlation between *in vitro* and *in vivo* inhibition patterns could be measured reproducibly and confirmed with reversal effect with the supplement of biotin and intermediates in the biotin biosynthesis pathway and/or substrate accumulation and/or RNA expression pattern in plants.

The supplement of biotin or biotin biosynthesis intermediates, except KAPA, was induced the germination and growth rescue previously inhibited by TPTA. The KAPA, even one of the biotin biosynthetic precursors, supplement could not rescue the germination inhibited by the compound TPTA, but additional supplement of 0.5 mM SAM increased up to 91% of the germination inhibited by 0.05 mM TPTA. In the same way, the antisense auxotrophs were rescued by supplementing of biotin (Hwang et al., 2003; 2010). From these results, we firstly reported the SAM is essential donor of amino group for synthesis of the biotin precursor DAPA in plants. DAPA aminotransferase is a pyridoxal 5'-phosphate (PLP) enzyme that catalyzes the transamination of KAPA to yield DAPA (Eisenberg and Stoner, 1971, Stoner and Eisenberg, 1975). In *E. coli*, the amino donor in this reaction is SAM (Breen et al., 2003). The enzyme from *E. coli* has been well characterized and its 3D structure determined.

Detailed mutational analysis of auxotrophic mutants has led to an inclusive understanding of biotin synthesis and regulation. The *bio1* auxotroph of *Arabidopsis*, first identified among the collection of recessive embryo-defective mutants, was shown to be defective in the early step of biotin synthesis, the conversion of KAPA to DAPA (Breen et al., 2003). Mutant *bio1*, first plant auxotroph for biotin, has shown to result in embryonic lethality, and its embryos remain pale throughout development, typically arrested between germination and cotyledon stage of embryogenesis (Alban et al., 2000, Meinke, 1985). Plant growth was rescued by biotin, dethiobiotin, or DAPA, but KAPA supply, or by genetic complementation by *E. coli bioA* gene coding DAPA aminotransferase, demonstrating that mutant plants are defective in this enzyme (Alban et al., 2000; Meinke, 1985; Shellhammer and Meinke, 1990; Patton et al., 1996; 1998). Based on feeding studies, Shellhammer and Meinke (1990) suggested that *bio 1* was defective in the conversion of KAPA to DAPA, the enzymatic function of the *BioA* protein of *E. coli*. This is the reason of the conversion of KAPA to DAPA in plant needs SAM supplement for rescue in mutant *bio 1* after treatment with TPTA, and appears to follow the same pattern as identified for *E. coli* (Shellhammer and Meinke, 1990; Patton et al., 1996; 1998). With these results, we firstly reported the SAM is an essential donor of amino group for the conversion of KAPA to DAPA in plants (Fig. 9).

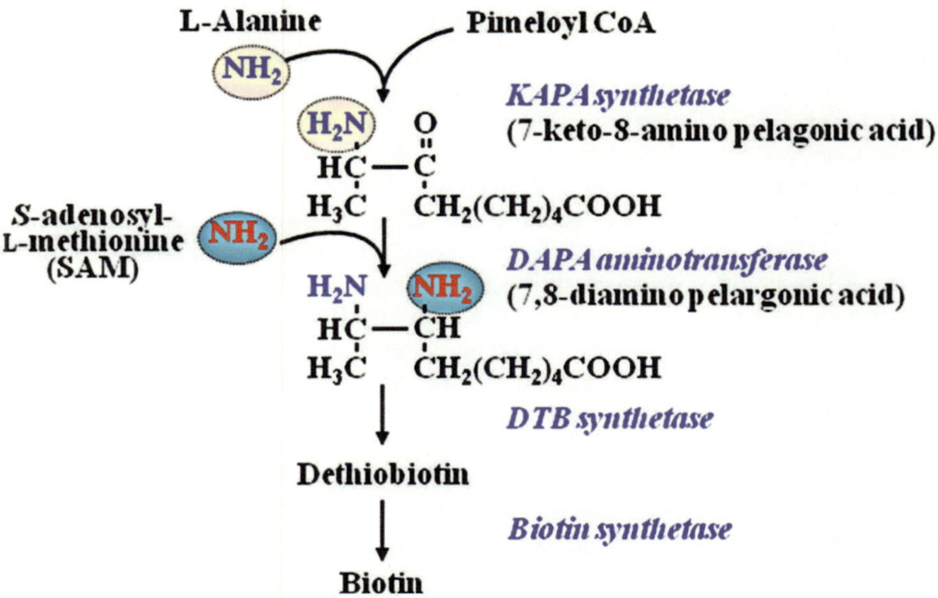

Fig. 9. Suggestion of S-adenoxyl-L-methione (SAM) is essential donor of amino group for the conversion of KAPA to DAPA in plants.

Furthermore, TPTA induced 8-fold greater accumulation of L-alanine, a substrate of KAPAS, in the foliar-treated plants. Also, RNA expression band for KAPAS was disappeared or indeed fainter in the lane representing leaf tissue treated with TPTA. This result suggested that the TPTA treatment is subjected to protein KAPAS translation and/or post translational regulation in the leaf like as *bio 1* mutants within 1 day of treatment. Also, TPTA showed slight inhibition to the tubulin translation and/or post translational regulation. Kourai et al. (1973) reported the mode of action of TPTA against *E. coli*. TPTA was only slightly inhibited respiration, permeability, protein synthesis and cell wall synthesis, but markedly inhibited RNA and DNA synthesis by *E. coli*. The antimicrobial action of TPTA was reversed by cysteine and 2-mercaptoethanol, and the active site of this compound is the metal atoms.

These results show that the action of TPTA was co-related with the enzyme activity of KAPAS in plants and RNA synthesis, coincidently. Because, TPTA inhibited the activity of KAPAS *in vitro*, germination of *A. thaliana* seeds *in vivo*, and the growth of weeds in a greenhouse condition. Also, TPTA inhibited the RNA expression in the leaf tissue of *A. thaliana*. This inhibition of seed germination was rescued by coincident treatment of KAPA and SAM, but could not rescue by supplement of KAPA only. It is not sure that the metal atoms of TPTA act on the active site or a cofactor was needed, but these results suggested that the KAPAS is a potential herbicidal target site in the biotin biosynthesis pathway, and TPTA is one of the KAPAS inhibiting chemicals even if the compound have been used as one of the fungicides.

Herbicidal symptoms after foliar treatment with TPTA were similar to herbicides targeting on the inhibition of fatty acid biosynthesis in grasses, leading to death of the susceptible plants. In this point of view, the mode of action of TPTA might be correlated with the fatty acid biosynthesis because the most important role of biotin is carboxylation of acetyl-CoA to give malonyl-CoA, which is the first step in fatty acid biosynthesis. Biotin is an essential vitamin and acts as cofactor for a number of enzymes involved in facilitation of CO_2 transfer during carboxylation, decarboxylation, and transcarboxylation reactions that are related to fatty acid and carbohydrate metabolism (Dakshinamurti and Bhagavan, 1985; Jelenska et al., 2002; Pinon et al., 2005; Nikolau et al., 2003).

These biotin-dependent carboxylases in plants include cytosolic acetyl-CoA carboxylase, chloroplastic geranyl-CoA and acetyl-CoA carboxylases, and mitochondrial methylcrotonoyl-CoA carboxylase (Alban et al., 2000; Nikolau et al., 2003). This complex contribution of biotin and biotin-mediated reactions in the plant cell implies an intracellular trafficking of biotin and precursors, thus requiring transport mechanisms. These transport steps include transfer of an intermediate, KAPA, DAPA, or dethiobiotin, between the cytosol and mitochondria was demonstrated by Pinon et al. (2005).

The reducing level of this enzyme activity required as a commercial herbicide target is hard to assume at present time. However, it appears that complete inhibition of enzyme activity at these targets is not necessary for plant death (Abell, 1996). In mutant plants with reduced amounts of glutamine synthetase activity, the target of glufosinate, reduction in glutamine synthetase activity of only 38% was sufficient to cause severe abnormalities (Blackwell et al., 1987). Antisense knock-out of acetolactate synthase (ALS), the target site of the sulfonylureas, imidazolinones, and triazolopyrimidines can produce plants displaying a range of ALS inhibitor-like symptoms such as growth retardation and necrosis (Blackwell et al., 1987; Höfgen, 1995). Such directed knock-outs allow the screening of enzymes whose inhibition might be expected to have catastrophic effects in the plant, based on knowledge of pathway dynamics. However, our knowledge of biochemical pathways in plants is incomplete and the next major herbicide target may lie in an unexpected area of plant

metabolism. Generally, it can be argued that we still do not know in detail how plants actually die as a result of inhibition of some known targets.
Even though whole plant screening will remain central to agrochemical discovery, high-throughput biochemical screening might be effective to accelerate the discovery of novel compounds. For example, these can allow the detection of hits that may be missed in glasshouse screens due to poor plant bioavailability, or the rapid and thorough evaluation of a target site by concerted screening against diverse sets of chemistry. Structure-activity relationships can provide inspiration for further chemical synthesis based on binding hypotheses or single parameter data not available from glasshouse screening. Further rationalization of activities and downstream of genomics such as high-throughput x-ray crystallography for three-dimensional analysis of protein-inhibitor interactions (structural genomics) will assist in developing 'virtual' or '*in silico*' screening of chemistry. Greater reliance on high-throughput biochemical screening will necessitate an improved ability to convert *in vitro* hits into biologically active molecules through a better understanding of whole plant-compound interactions and improved test systems would be confirmative for this speculation.

7. References

[1] Abell, L.M., J.V. Schloss, and A.R. Rendina (1993) Target-site directed herbicide design. Amer. Chem. Soc. Symp. Ser., 16~37.

[2] Abell, L.M. (1996) Biochemical approaches to herbicide discovery: advances in enzyme target identification and inhibitor design. Weed Sci., 44, 734-742.

[3] Alban, C., D. Job, and R. Dource (2000) Biotin metabolism in plants. Annu. Rev. Plant Physiol. Plant Mol. Biol., 51, 17~47.

[4] Arnon, D.I. (1949) Copper enzymes in isolated chloroplasts. Polyphenoloxidases in *Beta vulgaris*, Plant Physiol. 24, 1~5.

[5] Alexeev, D., M. Alexeeva, R.L. Baxter, D.J. Campopiano, S.P Webster, and L. Sawyer (1998) "The crystal structure of 8-amino-7-oxononanoate synthase: a bacterial PLP-dependent, acyl-CoA-condensing enzyme". J. Mol. Biol., 284, 401~19.

[6] Andrew, C.E., J. Sandmark, G. Schneider, and J.F. Kirsch (2002) The dual-specific active site of 7, 8-diaminopelargonic acid synthase and the effect of the R391A mutation. Biochemistry, 41(42), 12582~12589.

[7] Ashkenazi, T., A. Widberg, A. Nudelman, V. Wittenbach, and D. Flint (2005) Inhibitors of biotin biosynthesis as potential herbicides: Part 2. Pest Management Science, 61(10), 1024~1033.

[8] Ashkenazi, T., D. Pinkert, A. Nudelman, A. Widberg, B. Wexler, V. Wittenbach, D. Flint, and A. Nudelman (2007) Aryl chain analogs of the biotin vitamers as potential herbicides. Part 3. Pest Management Science, 63(10), 974~1001.

[9] Bachman, B.J. (1990) Linkage map of *Escherichia coli* K12. Microbiol. Rev., 54, 130~197.

[10] Baldet, P., H. Gerbling, S. Axiotis, and R. Douce (1993) Biotin biosynthesis in higher plant cells. Identification of intermediates. Eur. J. Biochem., 21, 479~485.

[11] Barker, D.F., and A.M. Campbell (1981) Genetic and biochemical characterization of the *birA* gene and its products: evidence for a direct role of biotin holoenzyme synthetase in repression of the biotin operon in *Escherichia coli*. J. Mol. Biol., 146, 469~492.

[12] Bartley, G.E., P.A. Scolnik, and P. Beyer (1999) Two *Arabidopsis thaliana* carotene desaturases, phytoene desaturase and ζ-carotene desaturase, expressed in *Escherichia coli*, catalyze a poly-*cis* pathway to yield pro-lycopene. Eur. J. Biochem., 259, 396~403.

[13] Berg, D, K. Tietjen, D. Wollweber, and R. Hain (1999) From genes to targets: impact of functional genomics on herbicide discovery. Brighton Conf-Weeds, 2, 491~500.

[14] Blackwell, R.D., A.J.S. Murray, and P. J. Lea (1987) Inhibition of photosynthesis in barley with decreased levels of glutamine synthetase activity. J. of Experimental Botany, 38, 1799~1809.

[15] Bower, S., J.B. Perkins, R.R. Yocum, C.L. Howitt, P. Rahaim, and J. Pero (1996) Cloning, sequencing, and characterization of the *Bacillus subtilis* biotin biosynthetic operon. J. Bacteriol., 178, 4122~4130.

[16] Breen, R.S., D.J. Campopiano, S. Webster, M. Brunton, R. Watt, and R.L. Baxter (2003) The mechanism of 7, 8-diaminopelargonate synthase; the role of *S*-adenosylmethionine as the amino donor. Org. Biomol. Chem., 1, 3498~99.

[17] Bult, C.J., O. White, G.J. Olsen, L. Zhou, R.D. Fleischmann, G.G. Sutton, J.A. Blake, L.M. FitzGerald, R.A. Clayton, J. Gocayne, and others (1996) Complete genome sequence of the methanogenic archaeon, *Methanococcus jannaschii*. Science, 273, 1058~1073.

[18] Coulter, C.V., J.A. Gerrard, J.AE Kraunsoe, and A.J. Pratt (1999) *Escherichia coli* dihydrodipicolinate synthase and dihydrodipicolinate reductase: kinetic and inhibition studies of two putative herbicide targets. Pestic. Sci., 55, 887~895.

[19] Cromartie, T.H., K.J. Fisher, J.N. Grossman (1999) The discovery of a novel site of action for herbicidal bisphosphonates. Pestic. Biochem. & Physiol., 63(2), 114~126.

[20] Dakshinamurti, K. and N.H. Bhagavan eds (1985) Biotin. Ann. NY Acad. Sci., 447, 1~441.

[21] Dayan, F.E., S.O. Duke, and K. Grossmann (2009) Herbicides as probes in plant biology. Weed Sci., In press.

[22] Eisenberg, M.A. and C. Star (1968) "Synthesis of 7-oxo-8-aminopelargonic acid, a biotin vitamer, in cell-free extracts of *Escherichia coli* biotin auxotrophs". J. Bacteriol., 96, 1291~7.

[23] Eisenberg, M.A. and G.L. Stoner (1971) Biosynthesis of 7, 8-diaminopelargonic acid, a biotin intermediate, from 7-keto-8-aminopelargonic acid and *S*-adenosyl-L-methionine. J. Bacteriol., 108(3), 1135~1140.

[24] Ficarelli, A., F. Tassi, and F.M. Restivo (1999) Isolation characterization of two cDNA clones encoding for glutamate dehydrogenase in *Nicotina plumbaginifolia*. Plant Cell Physiol., 40(3), 339~342.

[25] Fleischmann, R.D., M.D. Adams, O. White, R.A. Clayton, E.F. Kirkness, A.R. Kerlavage, C.J. Bult, J-F Tomb, B.A. Dougherty, J.M. Merrick, and others (1995) Whole-genome random sequencing and assembly of *Haemophilus influenzae* Rd. Science, 269, 496~512.

[26] Freund, A., U. Sonnewald, and L. Ding (2002) Dehydroquinate dehydrase/Shikimate dehydrogenase as an herbicide target. Patent No. WO/2002/004619.

[27] Gloeckler, R., I. Ohsawa, D. Speck, C. Ledoux, S. Bernard, M. Zinsius, D. Villeval, T. Kisou, K. Kamogawa, and Y. Lemoine (1990) Cloning and characterization of the

Bacillus sphaericus genes controlling the bioconversion of pimelate into dethiobiotin. Gene, 87, 63~70.

[28] Grossmann, J. and H. Schiffer (1999) Protoporphyrinogenoxidase- inhibiting activity of the new, wheat-selective isoindoldione herbicide, cinidon-ethyl. Pestic. Sci., 55, 687~695.

[29]Hiscox, J.D. and G.F. Israelstam (1979) A method for the extraction of chlorophyll from leaf tissues without maceration. Can. J. Bot., 57, 1332~1334.

[30] Höfgen, R., B. Laber, I. Schuttke, A.K. Klonus, W. Streber, and H.D. Pohlenz (1995) Repression of acetolactate synthase activity through antisense inhibition. Plant Physiology, 107, 469~477.

[31] Huang, W, J. Jia, K.J. Gibson, W.S. Taylor, A.R. Rendina, G. Schneider, and Y. Lindqvist (1995) Mechanism of an ATP-dependent carboxylase, dethiobiotin synthetase, based on crystallographic studies of complexes with substrates and a reaction intermediate. Biochemistry, 34, 10985~10995.

[32] Hwang, I.T., D.H. Lee, J.S. Choi, T.J. Kim, B.T. Kim, Y.S. Park, and K.Y. Cho (2001) A prognosis discovering lethal-related genes in plants for target identification and inhibitor design. The Korean J. of Pesticide Science, 5(3), 1~11.

[33] Hwang, I.T., D.H. Lee, J.S. Choi, Y.K. Min, T.J. Kim, J.H. Ko, T.H. Kim, Y.S. Park, K.Y. Cho, and S.W. Lee (2003) Novel polypeptide having function of 7-keto-8-aminopelargonic acid synthase of plant and method for inducing growth inhibition and lethality by suppressing expression of the polypeptide. Patent No. PCT/KR2003/001301.

[34] Hwang I.T., J.S. Choi, H.Y. Song, S.J. Cho, H.K. Lim, N.J. Park, and D.H. Lee (2010) Validation of 7-keto-8-aminopelargonic acid synthase as a potential herbicide target with lead compound triphenyltin acetate. Pestic. Biochem., and Physiol., 97, 24~31.

[35] Jelenska, J., A. Sirikhachornkit, R. Haselkorn, and P. Gornicki (2002) The Carboxyltransferase Activity of the Apicoplast Acetyl-CoA Carboxylase of *Toxoplasma gondii* Is the Target of Aryloxyphenoxypropionate Inhibitors. The J. of Biological Chemistry, 277(26), 23208~23215.

[36] Joseph, C. (2004) WORLD POPULATION TO 2300. United Nations Department of Economic and Social Affairs/Population Division, New York. P 240.

[37] Jun, J.H., C.S. Kim, D.S. Cho, J.M. Kwak, C.M. Ha, Y.S. Park, B.H. Cho, D.A. Patton, and H.G. Nam (2002) Random antisense cDNA mutagenesis as an efficient functional genomic approach in higher plants. Planta, 214, 668~674.

[38] Kack, H., J. Sandmark, K. Gibson, G. Schneider, and Y. Lindqvist (1999) Crystal structure of diaminopelargonic acid synthase: evolutionary relationships between pyridoxal-5'-phosphate-dependent enzymes. J. Mol. Biol., 291, 857~876.

[39] Kishore, G.M. and D.M. Shah (1988) Amino acid biosynthesis inhibitors as herbicides. Ann. Rev. Biochem., 57, 627~663.

[40] Kleier, D.A. and F.C. Hsu (1996) Phloem mobility of xenobiotics. VII. The design of phloem systemic pesticides. Weed Sci., 44, 749~756.

[41] Kourai, H., K. Takeichi, and I. Shibasaki (1973) Mode of action of triphenyltin acetate and tripropyltin phenyl ether. Hakko Kogaku Zasshi, 51(3), 197~204.

[42] Lin, T-Y. and C-H Wu (2005) Activation of hydrogen peroxide in copper (II)/amino acid/H_2O_2 systems: effects of pH and copper speciation. J. of Catalysis, 232, 117~126.
[43] Manjula, K., D. Wang, D.V. Quyen, T.K. Kim, K.S. Venkatesh, P. Srinidhi, and G.N. Rao (2010) Cyclin D1 Is a *Bona Fide* Target Gene of *NFATc1* and Is Sufficient in the Mediation of Injury-induced Vascular Wall Remodeling. J. of Biological Chemistry, 285(5), 3510~3523.
[44] Meinke, D.W. (1985) Embryo-lethal mutants of *Arabidopsis thaliana*: analysis of mutants with a wide range of lethal phases. Theor. Appl. Genet., 72, 543~552.
[45] Meinke, D.W. (1994) Seed development in *Arabidopsis. In* EM Meyerowitz, CR Somerville, eds, Arabidopsis. Cold Spring Harbor Laboratory Press, Cold Spring Harbor, NY, pp. 253~295.
[46] Nakao, Y., H. Nakamura, W. Mori, T. Sakurai, S. Suzuki, and A. Nakahara (1986) Azide-, thiocyanate- or cyanate-bridged di- and trinuclear copper (II) complexes with N,N,N',N'-tetramethylethylenediamine, N,N-diethylethylenediamine, 2,2'-bipyridyl or 1,10-phenanthroline. Bulletin of the Chemical Society of Japan, 59(9), 2755~2760.
[47] Nikolau, B.J., J.B. Ohlrogge, and E.S. Wurtele (2003) Plant biotin-containing carboxylases. Arch. Biochem. Biophys., 414, 211~222.
[48] Nudelman, A., D. Marcovici-Mizrahi, A. Nudelman, D. Flintd, and V. Wittenbache (2004) Inhibitors of biotin biosynthesis as potential herbicides. Tetrahedron, 60, 1731~1748.
[49] Ott, K-H, N. Aranibar, B. Singh, G.W. Stockton (2003) Metabolomics classifies pathways affected by bioactive compounds. Artificial neural network classification of NMR spectra of plant extracts. Phytochemistry, 62, 971~985.
[50] Pai, C.H. (1975) Genetics of Biotin Biosynthesis in *Bacillus subtilis*. J. Bacteriology, 121(1), 1~8.
[51] Palatnik, J.F., V.B. Tognetti, H.O. Poli, R.E. Rodriguez, N. Blanco, M. Gattuso, M. Hajirezaei, U. Sonnewald, E.M. Valle, and N. Carrello (2003) Transgenic tobacco plants expressing antisense ferredoxin-NADP(H) reductase transcripts display increased susceptibility to photooxidative damage. Plant J., 35, 332~341.
[52] Patton, D.A., S. Volrath, and E.R. Ward (1996) Complementation of an *Arabidopsis thaliana* biotin auxotroph with an *Escherichia coli* biotin biosynthetic gene. Mol. Gen. Genet. 251, 261~266.
[53] Patton, D.A., A.L. Schetter, L.H. Franzmann, K. Nelson, E.R. Ward, and D.W. Meinke (1998) An Embryo-Defective Mutant of Arabidopsis Disrupted in the Final Step of Biotin Synthesis. Plant Physiol., 116, 935~946.
[54] Pillmoor, J.B., S.D. Lindell, G.G. Briggs, and K. Wright (1995) The influences of molecular mechanisms of action on herbicide design. In: Proceedings of the Eighth International Congress of Pesticide Chemistry (Eds. N.N. Ragsdale, P.C. Kearney and J.R. Plimmer) America Chemical Society, Washington, D.C., pp. 292~303.
[55] Pinon, V., S. Ravanel, R. Douce, and C. Alban (2005) Biotin Synthesis in Plants. The First Committed Step of the Pathway Is Catalyzed by a Cytosolic 7-Keto-8-Aminopelargonic Acid Synthase Plant Physiol., 139(4), 1666~1676.

[56] Ploux, O. and A. Marquet (1992) The 8-amino-7-oxopelargonate synthase from *Bacillus ~sphaericus*. Purification and preliminary characterization of the cloned enzyme overproduced in *Escherichia coli*. Biochem. J., 283, 327~331.
[57] Ploux, O., O. Breyne, S. Carillon, and A. Marquet (1999) Slow-binding and competitive inhibition of 8-amino-7-oxopelargonate synthase, a pyridoxal-5-phosphate-dependent enzyme involved in biotin biosynthesis, by substrate and intermediate analogs kinetic and binding studies. Eur. J. Biochem., 259, 63~70.
[58] Rendina, A.R. and L.M. Abell (1994) Biochemical Approaches to Herbicide Discovery, enzyme target selection and inhibitor design. Amer. Chem. Soc. Symp. Ser., 407~424.
[59] Rüegg W.T., M. Quadranti, and A. Zoschke (2007) Herbicide research and development: challenges and opportunities. Weed Research, 47(4), 271~275.
[60] Saari, L.L. (1999) A prognosis for discovering new herbicide sites of action. In "Pesticide Chemistry and Bioscience: The Food-Environment Challenge" Ed. by G.T. Brooks and T.R. Robers Royal Society of Chemistry, Cambridge. pp. 207~220.
[61] Schloss, J.V. and A. Aulabaugh (1990) Acetolactate synthase and ketol- acid reductoisomerase: Targets for Herbicides Obtained by Screening and *de novo* Design. Z. Naturforsch, 45c, 544~551.
[62] Schulte, M. (2004) Transgene herbizidresistente Kulturen -Rückblicke und Ausblicke nach acht Jahren internationaler Anbaupraxis. Scientific colloquium. In: Weed Science on the Go (eds P. Zwerger & T. Eggers). 102~118. Universität Hohenheim, Stuttgart-Hohenheim, Germany.
[63] Shellhammer, J. and D. Meinke (1990) Arrested embryos from the *bio1* auxotroph of *Arabidopsis thaliana* contain reduced levels of biotin. Plant Physiol., 93, 1162~1167.
[64] Stitt, M. (1999) The first will be last and the last will be first: nonregulated enzymes call the tune? In Plant Carbohydrate Biochemistry. Edited by Burrell, J.A., Bryant, M.M., Kruger, N.J. Oxford: BIOS Scientific Publishing, 1~16.
[65] Stoner, G.L. and M.A. Eisenberg (1975a) Purification and properties of 7,8-diaminopelargonic acid aminotransferase. J. Biol. Chem., 250, 4029~4036.
[66] Stoner, G.L. and M.A. Eisenberg (1975b) Biosynthesis of 7,8-diaminopelargonic acid from 7-keto-8-aminopelargonic acid and *S*-adenosyl-L-methionine. The kinetics of the reaction. J. Biol. Chem., 250, 4037~4043.
[67] Subramanian, M.V., S.A. Brunn, P. Bernasconi, B.C. Patel, and J.D. Reagan (1997) Revisiting auxin transport inhibition as a mode of action for herbicides. Weed Sci., 45, 621~627.
[68] Toney, M.D., E. Hohenester, S.W. Cowan, J.N. Jansonius (1993) Dialkylglycine decarboxylase structure: Bifunctional active site and alkali metal sites. Science, 261, 756~759.
[69] Tissot, G., R. Pépin, D. Job, R. Douce, and C. Alban (1998) Purification and properties of the chloroplastic form of biotin holocarboxylase synthetase from *Arabidopsis thaliana* overexpressed in *Escherichia coli*. Eur. J. Biochem., 258, 586~596.
[70] Vikrant, M.B., D. Sagarika, G.R. Vasanthakumar, P. Kumar, S. Sinha, and A. Surolia (2006) Broad Substrate Stereospecificity of the *Mycobacterium tuberculosis* 7-Keto-8-aminopelargonic Acid Synthase Spectroscopic and Kinetic Studies. J. Biol. Chem., 281(35), 25076~25088.

[71] Webster, S.P., D. Alexeev, D.J. Campopiano, R.M. Watt, Alexeeva, M., L. Sawyer, and R.L. Baxter (2000) Mechanism of 8-Amino-7-oxononanoate Synthase: Spectroscopic, Kinetic, and Crystallographic Studies, Biochemistry, 39, 516~528.

[72] Weinstein, S. (1984) Resolution of optical isomers by thin layer chromatography. Tetrahedron Letters, 25(9), 985~986.

[73] Wolfgang, L., F. Börnke, A. Reindl, T. Ehrhardt, M. Stitt, and U. Sonnewald (2004) Target-based discovery of novel herbicides. Current Opinion in Plant Biology, 7(2), 219~225.

[74] Zhang, S., I. Sanyal, G.H. Bulboaca, A. Rich, and D.H. Flint (1994) The gene for biotin synthase from *Saccharomyces cerevisiae*: cloning, sequencing, and complementation of *Escherichia coli* strains lacking biotin synthase. Arch. Biochem. Biophys., 309, 29~35.

23

Possibilities of Applying Soil Herbicides in Fruit Nurseries – Phytotoxicity and Selectivity

Zarya Rankova
Fruit-Growing Institute, Plovdiv 4004, 12, Ostromila Str.
Bulgaria

1. Introduction

Fruit-growing is one of the major sub-branches of agriculture in Bulgaria and its development is enhanced by a number of positive prerequisites - favourable soil and climatic conditions, a rich genetic fund of local and introduced cultivars, production experience and national traditions. Production of grafted fruit tree planting material is an important starting point in developing modern fruit-growing. Requirements set by integrated fruit production and the needs of producing certified fruit planting material impose the development of ecologically sound integrated approaches for plant protection, including an efficient control of weed infestation in the fruit nursery, based on selective herbicides without any residual effects, as an element of high agrotechnical practice. In 1991 the European Plant Protection Organization (EPPO) published a scheme for production of certified virus-free fruit trees and rootstocks. The recommendations complying with the scheme of EPPO were included in Directive 92/34 of the Council of the European Economic Community of 28 April 1992 on the marketing of fruit plant propagating materials and fruit plants intended for fruit production. The aim of the Directive was to guarantee the quality of propagating material in the European Union countries after the borders have come down. Certified virus-free fruit planting material could only be produced under the conditions of very high agrotechnical background, including an efficient and ecologically sound control of weed infestation in the fruit nursery.

Weed vegetation is a serious problem in fruit crop nurseries. Weeds strongly suppress the growth of rootstocks and grafted trees in the process of planting material production. A direct damage caused by weed infestation is markedly expressed (weed-crop competition for moisture, light and nutrients from soil and introduced with fertilizers). Under the effect of weeds, growth and development of young trees is delayed, wood does not mature and the planting material obtained is non-standard. The indirect damage caused by weeds (dissemination of pests and diseases, including viral ones) in that case is quite strongly expressed, keeping in mind the modern issues to the production of certified, free of viral diseases fruit planting material.

In scientific literature there are data that a number of weed species could be also attacked by diseases, including viral ones, and thus weeds become the reason for their spread to the cultivated plants. It was established that PPV (plum pox potty virus) casing the economically most important Sharka disease in stone fruit species, could be hosted by a number of weed species contained in the weed association in the fruit nursery (Oosten, 1971; Rankova & Milusheva, 2001; Milusheva & Rankova, 2002; Milusheva & Rankova, 2006)

Under the conditions of modern fruit production, an efficient and ecologically sound weed control in the fruit nursery could hardly be carried out without a developed scientific chemical control system based on soil-applied and leaf herbicides with proven selectivity for fruit species.

There are data in literature about the different effects of some soil and leaf herbicides on growth of fruit species used as rootstocks - from lack of phytotoxicity and ability to produce good quality rootstocks suitable for grafting, to very strong toxicity after applying some active substances contained in herbicides, causing plant death. The effect of applying some soil herbicides in yellow plum seedling rootstocks was studied (Porterfield et al., 1993; Wazbinska, 1997; Kaufman & Libek, 2000 a; Kaufman & Libek, 2000 b; Rankova, 2004), in peach seedling rootstocks (Arenstein, 1980; Kuhns L., 1981; Lange, 1987; Lourens et al., 1989; Jankovic, et al., 1995; Abdul et al., 1998; Rankova, 2002; Rankova, 2004), in wild cherry seedling rootstocks (Crisp et al., 1984;Clay, 1984; Porterfield et al., 1993),in mahaleb (Rankova, 2006); in apricot seedling rootstocks (Arenstein, 1980; Mitchell & Abernethy, 1989).

The present work provides summarized data from studies carried out in the period 2001-2009 at the Fruit-Growing Institute - Plovdiv and it has set the aim of presenting the incidence of phytotoxicity and selectivity in some major seedling rootstocks after application of soil herbicides. The effect of a number of soil herbicides (napropamide, pendimethalin, metolachlor, oxyfluorofen, terbacil, linuron, oxadiargyl, etc.) on weed infestation, growth habits and physiological status of different seedling rootstocks for fruit species (yellow plum (*Prunus cerasifera*, Ehrh., Myrobolan), peach seedling rootstock (*Prunus persica* L., Batsch), wild cherry (*Prunus avium* L.), Mahaleb (*Prunus mahaleb* L.), apricot seedling rootstock (*Prunus armeniaca* L.), walnut (*Juglans regia* L.) was studied under the conditions of model pot experiments and field studies. Those seedling rootstocks have been widely used as rootstocks for plum, peach, sweet cherry, apricot, nectarine and walnut fruit cultivars, thanks to their good adaptability to the soil and climatic conditions and their excellent affinity to the range of cultivars grown in Bulgaria.

Probably, the use of the active substance terbacil in the studies will make an impression. After the accession of Bulgaria to the EU, its application was prohibited, despite the results of its excellent herbicide efficiency against a large number of annual grassy and broad-leaved weeds and its use in fruit-bearing orchards and in some nurseries (in rootstocks for apple cultivars and forest species). Although treatment with that soil herbicide is prohibited at present, the results of the carried out investigations showed that it could be applied in fruit tree nurseries for some fruit species (yellow plum, peach), because a depressing effect on the growth of the rootstocks was not reported. The other herbicides included in the study also have a comparatively broad spectrum of herbicide efficiency and persistence in soil for 2-3 months after the date of treatment. Their selection was made with the aim of providing herbicide efficiency during the first months after emergence of the seedlings, when the competition and the suppressing effect of the weed vegetation are most strongly expressed. Only preliminary stratified seeds (stones) were used during the implementation of the experiments for the reliable reporting of the soil herbicide effect on the process of plant emergency.

2. Model (pot experiments) for studying the effect of the soil herbicides on the vegetative habits of seedling rootstocks

2.1 Material and methods

Stratified seeds (stones) of yellow plum, peach, wild cherry, walnut of Kuklenski cultivar and apricot were planted (by 5 seeds) in pots of volume 1 kg sand and alluvial-meadow soil

(Fluvisol), pH 7,2 and content of mobile phosphorus P_2O_5 - 21,6 mg/100g of soil. Treatment with soil herbicides was applied immediately after seeding. Ten variants in five replications were set.

Variants: 1. Control (untreated); 2. Napropamide - Devrinol 4 F - 4,0 l/ha; 3. Pendimethalin - Stomp 33 EC - 4,0 l/ha; 4. Terbacil - Sinbar 80 WP - 1,0 kg/ha; 5. Oxadiargyl - Raft 800 WDG - 250 g/ha; 6. Metolachlor - Dual Gold 960 EC - 1,0 l/ha; 7. Isoxaflutole - Merlin 750 WG - 50 g/ha; 8. Linuron - Afalon 45 SC - 3,0 l/ha; 9. Acetochlor- Trophy- 3,0 l/ha; 10. Oxyfluorofen - Galigan 240 EC - 1,5 l/ha;.

The herbicide treatment rates were calculated according to the area of the cultivation vessel. The experimental plants were grown for 60 days under controlled conditions (temperature of 20-25^{0}C and relative air humidity 65-70%) in a glass-and-steel green house. During that period observations were made on seedling emergence, development and external symptoms of phytotoxicity. At the end of the period the following biometric indices were reported: stem height (cm) and above-ground mass (stem + leaves) - in average per plant (g). The results obtained were statistically processed following the standard methods.

2.2 Results and discussion

2.2.1 Effect of soil herbicides on the vegetative habits of yellow plum seedlings

The applied soil herbicides had a different effect on the emergence and development of yellow plum seedlings. No differences in the rate of seedling emergence were observed after treatment with napropamide (Devrinol 4 F - 4,0 l/ha) and terbacil (Sinbar 80 WP - 1,0 kg/ha). When treated with oxadiargyl (Raft 800 WG - 250 g/ha), the plants emerged at the same time with those of the control, however shortly after that symptoms of necrosis appeared on the cotyledons and the plants died about 10 days after emergence. Only single plants survived but white chlorosis developed on their leaves - the tissues around the central vein, and slight necrosis appeared at the leaf margins. Plant growth was suppressed.
The last to emerge (up to the 20th day after the emergence of the control seeds) were the plants treated with pendimethalin (Stomp 33 EC - 4,0 l/ha), (Var. 3). Their development was delayed and they had much shorter stems (rosettes) and smaller leaves (Fig. 1).

Fig. 1. Inhibiting effect of pendimethalin in yellow plum seedlings under sand culture conditions.

External symptoms of toxicity and growth suppression were not reported for the seedlings of the variants treated with napropamide and terbacil.
The plants treated with isoxaflutole (Merlin 750 WG - 50 g/ha) emerged at the same time with those of the control. Later their development was delayed, strongly expressed chlorosis appeared, the tissues about the central vein whitened (white chlorosis), a large number of the plants withered (Fig. 2). Again only single plants continued to develop in that variant.

Fig. 2. White chlorosis on the leaves of yellow plum after treatment with isoxaflutole (Merlin 750 WG).

The seedlings treated with metolachlor, linuron, acetochlor and oxyfluorofen emerged at the same time with the control plants but later their development was delayed and until the 20th day after emergence withering and dying of the plants was observed.
The least depressing effect on growth was exerted by the active substances napropamide and terbacil - Variants 2 and 4 (Fig. 3 and 4). After treatment of the seedlings with oxadiargyl - Raft 800 WG - 250 g/ha (Var. 5) a suppression of stem development was observed. The lowest values of both biometric characteristics were established after treatment with pendimethalin - Stomp 33 EC - 4,0 l/ha (Var. 3). The differences to the controlled variant were statistically highly significant.

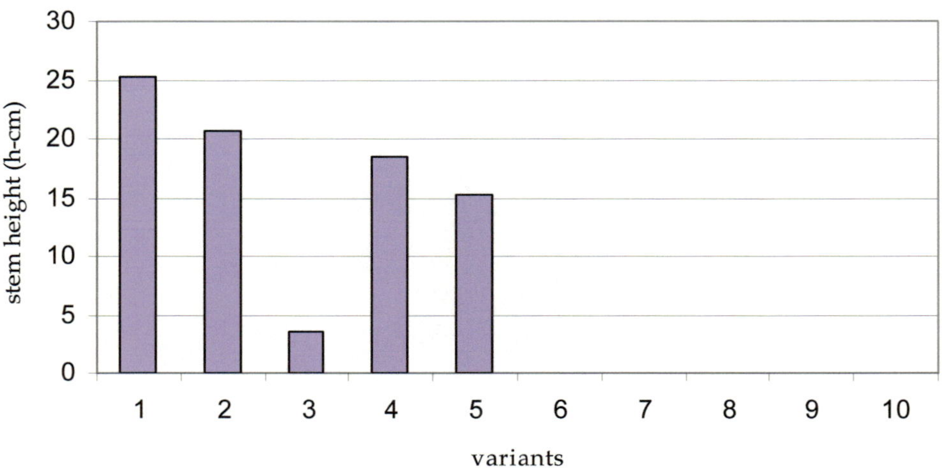

LSD 5%-0.87, 1%-1.27, 0.1%-1.90

Fig. 3. Stem height of yellow plum seedlings after treatment with soil herbicides (h-cm).

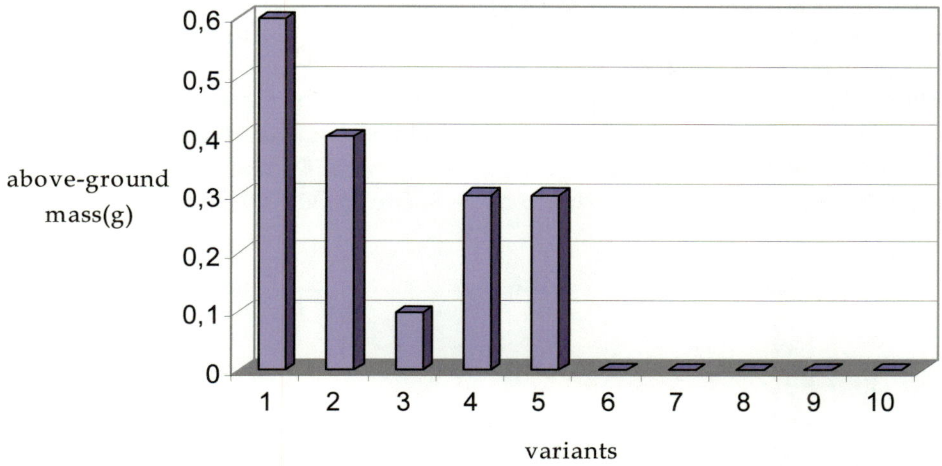

LSD 5%-0.13, 1%-1.20, 0.1%-1.29

Fig. 4. Effect of soil herbicides on the above-ground mass of yellow plum seedlings (g).

The strong inhibiting effect of pendimethalin under sand culture conditions could be explained by the physical basis of the herbicide selectivity and the possibility to induce phytotoxicity on light soils (sand) and in direct contact with the germinating seeds (stones).

The results obtained about the effect of the studied soil-applied herbicides on the vegetative habits of yellow plum seedlings under sand culture conditions gave the grounds to draw the following conclusions: 1. After treatment with napropamide and terbacil, toxicity was not observed and the growth habits of the plants were close to the untreated control; 2. Strongly suppressing effect was established after treatment with pendimethalin under sand culture conditions; 3. Strong phytotoxicity expressed in dying of the plants was reported after treatment with metolachlor, linuron, acetochlor and oxyfluorofen.

2.2.2 Effect of soil herbicides on the vegetative habits of peach seedlings

The obtained results showed that the applied herbicide active substances had a different effect on seedling development compared to sand culture. The plants of the variants treated with napropamide (Var. 2), pendimethalin (Var. 3), terbacil (Var. 4), oxadiargyl (Var. 8) and isoxaflutole (Var. 9) emerged at the same time as those of the control. They grew well without external symptoms of phytotoxicity. Later, in Variant 7 (Merlin 750 WG - 5 g/da) white chlorosis emerged along the leaf vein. The plants of the variants treated with metolachlor (Dual Gold 960 EC 1,0 l/ha), linuron (Afalon 50 WP - 3,0 l/ha), acetochlor (Trophy - 3,0 l/ha) and oxyfluorofen (Galigan 240 EC - 3,0 l/ha) emerged later than those of the control. When applying metolachlor (Var. 6) only single plants emerged and their growth was suppressed. Leaf withering and plant dying was observed until the 20th day after emergence. Analogous habits were established in the seedlings of Variant 8 (linuron - Afalon 50 WP - 3,0 l/ha), Variant 9 (acetochlor - Trophy - 3,0 l/ha) and Variant 10 (oxyfluorofen - Galigan 240 EC - 3,0 l/ha).

Single seeds emerged in the variants treated with linuron and acetochlor and their growth was strongly suppressed. Withering from the leaf tip was observed and the plants died until the 20th day after emergence. Phytotoxicity was also established in the plants treated with oxyfluorofen (Var. 10). The seedlings emerged later than those of the control. Later their growth was delayed, chlorosis appeared at the leaf margins, turning into necrosis and causing plant death in 20 - 25 days after emergence.

The slightest effect on the experimental plants was exerted by the active substances terbacil and pendimethalin (Fig. 5). Out of all the reported biometric characteristics, values closest to the control plants were established in the plants of Variant 4 - terbacil - Sinbar 80 WP - 1,0 kg/ha, followed by Variant 3 - pendimethalin - Stomp 33 EC - 3,0 l/ha.

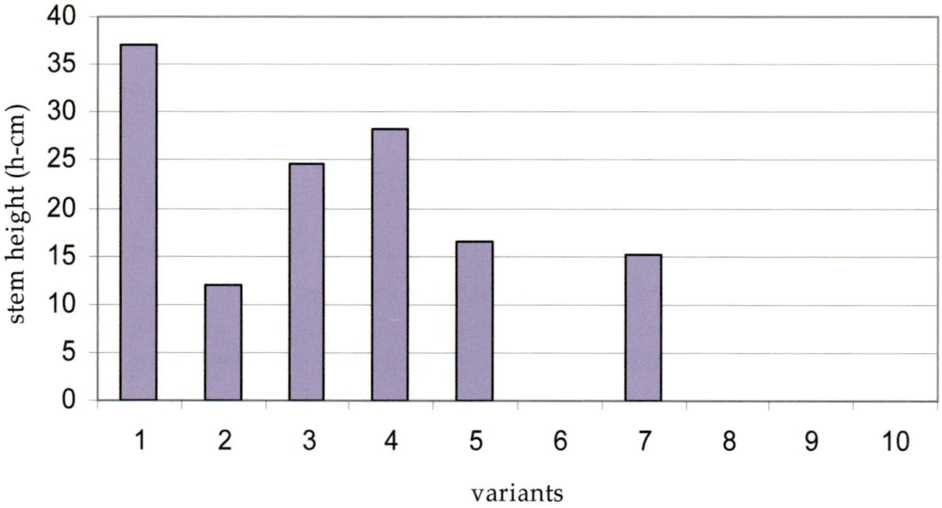

LSD $_{5\%}$-1,98, $_{1\%}$-2,65, $_{0.1\%}$-3,58

Fig. 5. Effect of soil herbicides on stem height of peach seedlings (h-cm)

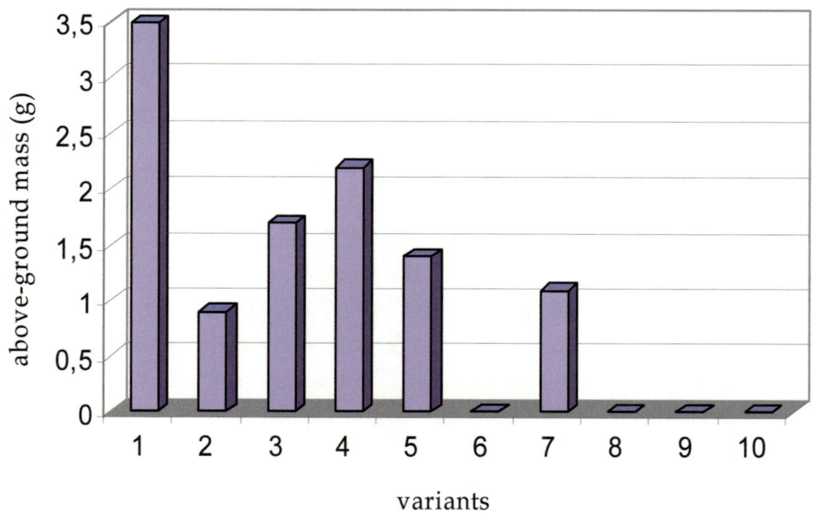

LSD $_{5\%}$-0,22, $_{1\%}$-0,30, $_{0.1\%}$-0,40

Fig. 6. Effect of soil herbicides on the above-ground mass of peach seedlings (g).

After application of napropamide (Var. 2), the strongest expression of the stem growth reduction in the seedlings was reported in comparison with the control. The differences were of high statistical significance.

There are data in literature about the response of some fruit species to soil herbicides when applying the sand culture method (Clay, 1984; Lourens, et al., 1989). Analogous results about the habits of peach seedlings after treatment with soil herbicides under sand culture conditions were obtained by Lourens, (1989). A slight phytotoxic effect was established after treatment of the experimental plants with pendimethalin, napropamide, oryzalin. The authors reported that even stronger phytotoxicity was observed after treatment with oxadiazon, alachlor, simazine.

The results obtained about the effect of the studied soil-applied herbicides on the vegetative habits of peach seedlings under sand culture conditions gave the grounds to draw the following conclusions: 1. Growth habits closest to those in the control, were established in the plants treated with terbacil (Var. 4) and pendimethalin (Var. 3);

2. Strong phytotoxic effect on peach seedlings expressed in dying of the plants, was exerted by linuron, metolachlor, acetochlor and oxyfluorofen.

2.2.3 Effect of soil-applied herbicides on the vegetative habits of wild cherry seedlings

Strong phytotoxicity under sand culture conditions was established after treatment with all the tested herbicides, expressed in blocking of seed germination or dying of the emerged plants. That was the reason to conclude that wild cherry as a species is strongly susceptible to the effect of soil-applied herbicides under sand culture conditions.

In the model pot experiment on alluvial-meadow soil (Fluvisol), the plants of the variants tested with napropamide (Var. 2), pendimethalin (Var. 3) and isoxaflutole (Var. 7) emerged at the same time with those of the control. External symptoms of phytotoxicity were not observed in the seedlings treated with napropamide and pendimethalin.

Fig. 7. White chlorosis in wild cherry seedlings after treatment with isoxaflutole (Var. 7)

After treatment with isoxaflutole (Var. 9), white chlorosis appeared in the leaves of the plants both along the leaf vein and between the leaf nerves. In those areas the chlorosis developed as white spots. Later (about a month) necrosis appeared in the white spots.

At a later stage (about 40 days after plant emergence), those symptoms were not observed in the newly formed leaves. The leaves were fresh, green, without obvious suppression of plant growth. That was the reason to accept that the phytotoxicity of isoxaflutole in wild cherry seedlings was overcome in 40-60 days after emergence.

The seeds of the other variants treated with herbicides (Var. 4, 5, 6, 7 and 8) did not emerge or only single plants developed. Withering of the plant tip was observed, followed by dying of the plants in about 20 days after emergence. Consequently, the active substances terbacil, metolachlor, linuron, oxyfluorofen and oxadiargyl have a strong phytotoxic effect on wild cherry seedlings, causing the plant death.

Fig. 8. Wild cherry seedlings with suppressed growth after treatment with terbacil (Var. 4) and metolachlor (Var. 6)

The results of the biometric analysis showed different effects of the soil herbicides on the vegetative habits of the plants. The least inhibiting effect on stem growth in height was observed in the variants treated with napropamide (Var. 2) and isoxaflutole (Var. 9), (Fig. 9). The lowest height was reported for the plants treated with pendimethalin (Var. 3). The differences to the control were statistically significant. After the application of isoxaflutole (Var. 7) the values of plant height were close to those in the control.

The results about the effect of the applied herbicides on the above-ground mass of the plants were analogous with those of plant height (Fig. 10). The results obtained after treatment of the seedlings with napropamide and isoxaflutole again showed values close to the control. Again a depressing effect on that characteristic was reported in the variant with application of pendimethalin (Var. 3). The differences were statistically significant.

The results obtained about the effect of the soil-applied herbicides on the vegetative habits of wild cherry seedlings gave the grounds to draw the following conclusions:

1. Strong phytotoxicity expressed in blocking the seed germination and dying of the emerged plants was exhibited after applying the soil herbicides napropamide, pendimethalin, terbacil, metolachlor, linuron, acetochlor, oxyfluorofen, oxadiargyl and isoxaflutole at the tested rates under sand culture conditions.
2. Under the conditions of alluvial-meadow soil (Fluvisol), external symptoms of phytotoxicity and growth depression was not observed in seedlings after treatment with napropamide (Devrinol 4 F - 400 ml/da).
3. Application of isoxaflutole led to incidence of white chlorosis in the plant leaves, however phytotoxicity was overcome in about 40 days after treatment with the herbicide and no suppression of the vegetative habits were observed.

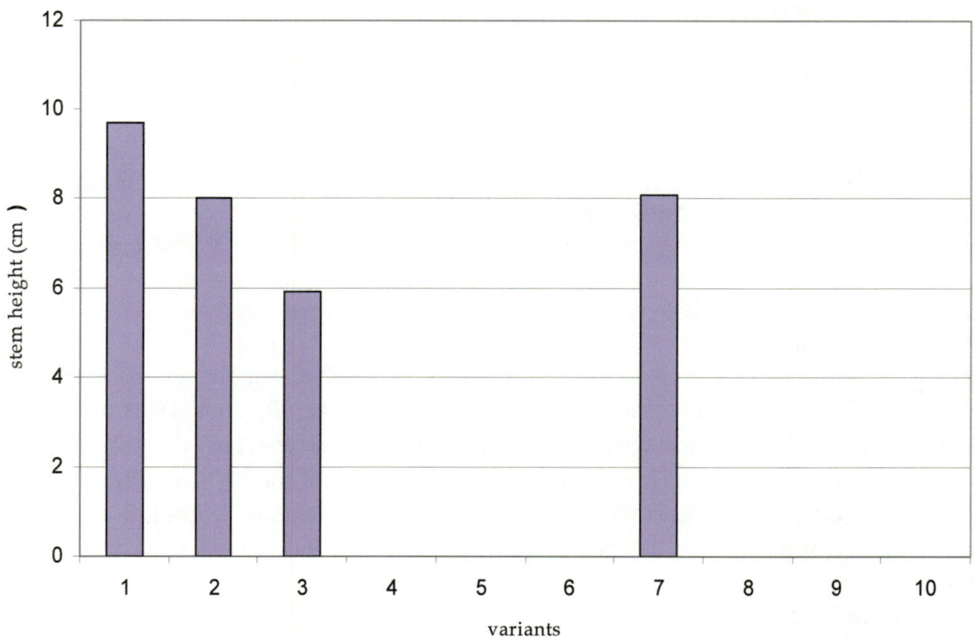

LSD 5%=1.87, 1%=2.72, 0,1 %=4.07

Fig. 9. Effect of soil herbicides on stem height (cm)

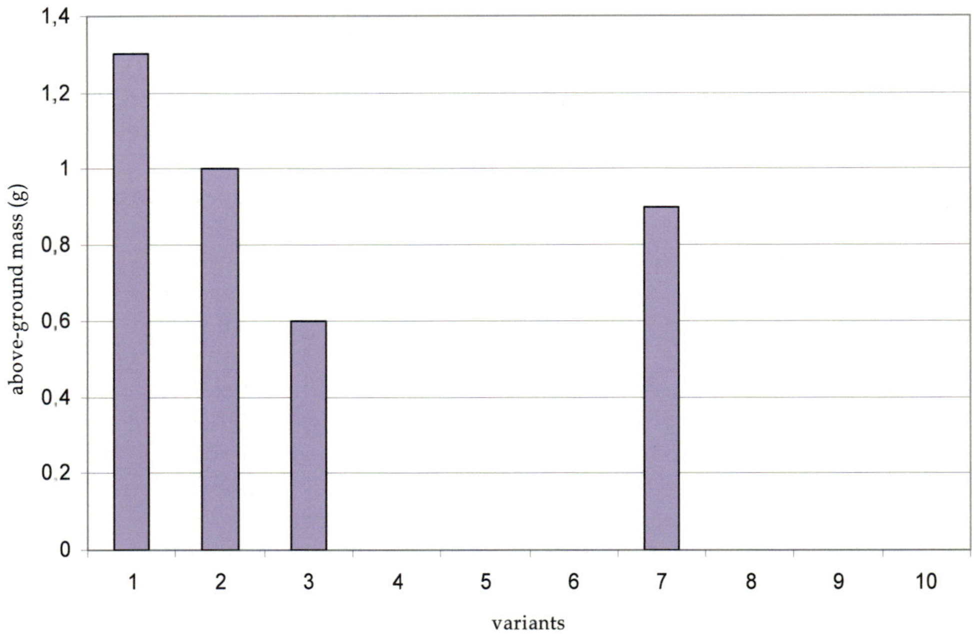

LSD 5%=0,36, 1%=0,53, 0,1 %=0,80

Fig. 10. Effect of soil herbicides on the above-ground plant mass (g)

4. Depressing effect on growth of wild cherry seedlings was established after treatment with pendimethalin - Stomp 33 EC - 4,0 l/ha.
5. Under the conditions of a model pot experiment on alluvial-meadow soil (Fluvisol) the active substances terbacil, metolachlor, linuron, acetochlor, oxyfluorofen and oxadiargyl had a strong phytotoxic effect on the seedlings, leading to plant death.

2.3 Effect of the soil-applied herbicides on the vegetative habits of walnut seedlings

2.3.1 Under sand culture conditions

The seeds treated with napropamide (Var. 2), pendimethalin (Var. 3) and terbacil (Var. 4) emerged at the same time with those of the control. External symptoms of phytotoxicity were not observed. Later, delayed development of the seedlings was reported for the plants of the Variants 2 and 3.

Delayed emergence compared to the control was established for the plants treated with metolachlor (Var. 6), linuron (Var. 8) and oxyfluorofen (Var. 10). In Variant 6 only single plants emerged. An incidence of necrosis in their leaves was reported. The seedlings of the variants treated with metolachlor and oxyfluorofen died until the 20th day after emergence. Similar symptoms of phytotoxicity were observed in the plants of Variant 8 - only single plants emerged and they had a seriously delayed development with necrosis at the margins of the apical leaves.

In the variants with application of oxadiargyl (Var. 10) and isoxaflutole (Var. 7) the seeds emerged at the same time as those of the control variant. External symptoms of phytotoxicity - chlorosis, necrosis, white chlorosis - typical of the species susceptible to isoxaflutole, were not observed.

The results of the biometric analysis showed that the soil-applied herbicides had an effect expressed in different ways on growth and development of walnut seedlings. The plants of the variants treated with oxadiargyl and isoxaflutole (Var. 5 and 7) had a bigger stem height compared to the control (Fig. 11).

That was the reason to conclude that those two active substances do not exert a depressing effect on seedling growth. Values of the stem height, close to those in the control, were established in the plants treated with terbacil, napropamide and linuron (Var. 4, 2 and 8). Strong inhibiting effect under sand culture conditions was established in the plants treated with pendimethalin (Var. 3). The differences to the control were of high statistical significance.

A similar effect of the soil-applied herbicides was observed on the other studied characteristic - the above-ground mass (Fig. 12). The plants treated with isoxaflutole (Var. 7) had a larger above-ground mass than those of the control variant. Consequently, the active substance isoxaflutole - Merlin 750 WG at the rate 50 g/ha did not suppress walnut seedling growth under sand culture conditions. A slighter depressing effect on that characteristic was established also in the plants treated with terbacil and linuron (Var. 4 and 8). The least above-ground mass was reported in the plants treated with pendimethalin (Var. 3) and napropamide (Var. 2). The differences to the control were again of high statistical significance.

Application of pendimethalin had a depressing effect on walnut seedling development under sand culture conditions. A similar result was observed in the other analogous studies about the effect of the soil-applied herbicides on the habits of peach and yellow plum seedling rootstocks.

The depressing effect of pendimethalin under sand culture conditions could be explained by the physical basis of the herbicide selectivity and the possibility to exhibit its phytotoxic effect on light soils (sand in the present case) and in a direct contact with the germinating seeds.

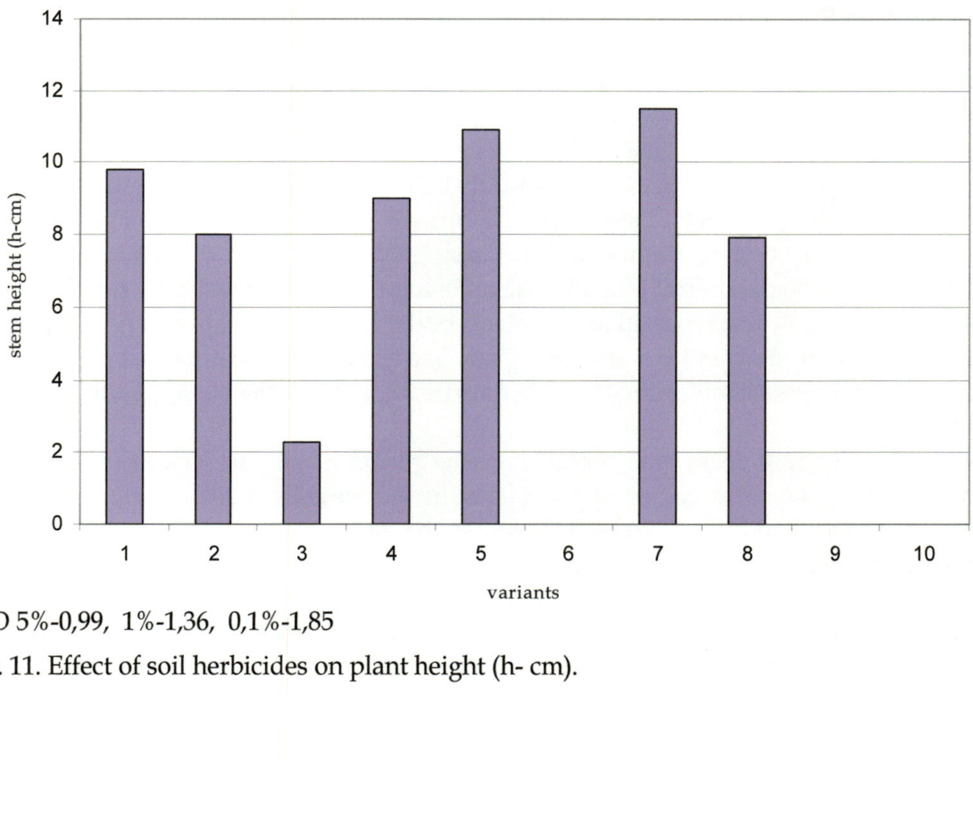

LSD 5%-0,99, 1%-1,36, 0,1%-1,85

Fig. 11. Effect of soil herbicides on plant height (h- cm).

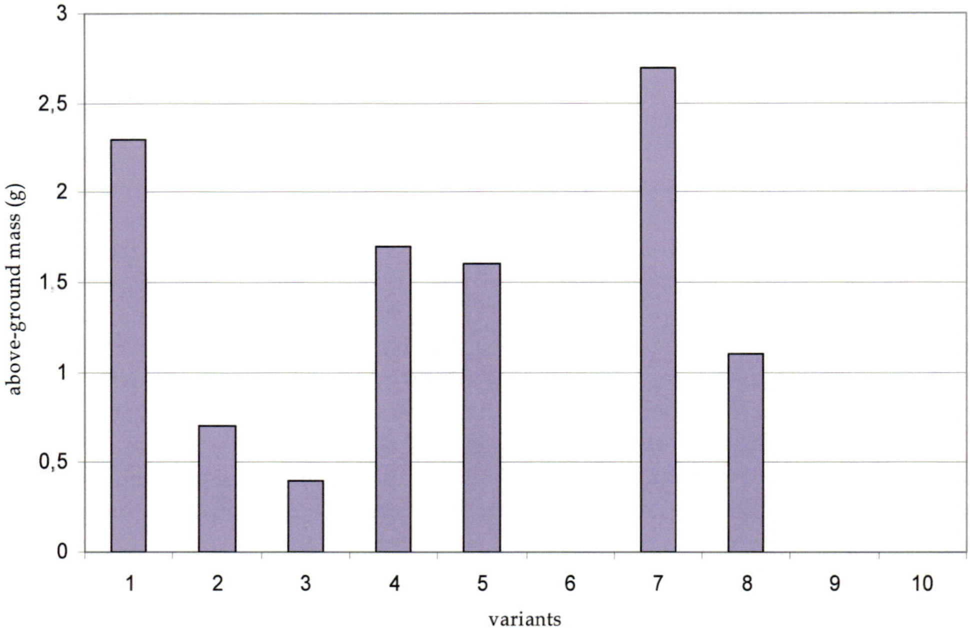

LSD 5%-0,32, 1%-0,43, 0,1%-0,59

Fig. 12. Effect of soil herbicides on the above-ground mass (g).

2.3.2 Under the conditions of a model pot experiment with alluvial-meadow soil (Fluvisol)

Differences in the rate of emergence of the seedlings in the variants treated with herbicides and those in the control were not observed. External symptoms of toxicity (chlorosis, necrosis, withering of the stem or leaves) did not appear. Later a delayed development of the plants in the variants treated with pendimethalin (Var. 3), metolachlor (Var. 6) and oxyfluorofen (Var. 10) was established.

The results of the biometric analyses showed that the applied soil herbicides had a different effect on growth and development of the walnut seedlings. Plants of the variants treated with napropamide, terbacil, linuron and isoxaflutole (Var. 2, 4, 8 and 7) had a closer to stem height compared to the control (Fig. 13). The differences were not statistically significant.

A lower stem height was reported for the plants of the variants with applied pendimethalin, metolachlor, acetochlor and oxyfluorofen (Var. 3, 6, 9 and 10). That gave the grounds to accept that the application of those soil herbicides suppressed the stem growth of walnut seedlings.

Lower above-ground mass was reported in the plants of all the variants treated with herbicides (Fig. 14). The values of the plants in the variants treated with terbacil and isoxaflutole (Var. 4 and Var. 7) were the closest to the control.

The results about the depressing effect of the soil herbicides pendimethalin, metolachlor, and oxyfluorofen on that characteristic were analogous.

Therefore, it could be admitted that those soil herbicides had a depressing effect on the growth of walnut seedlings.

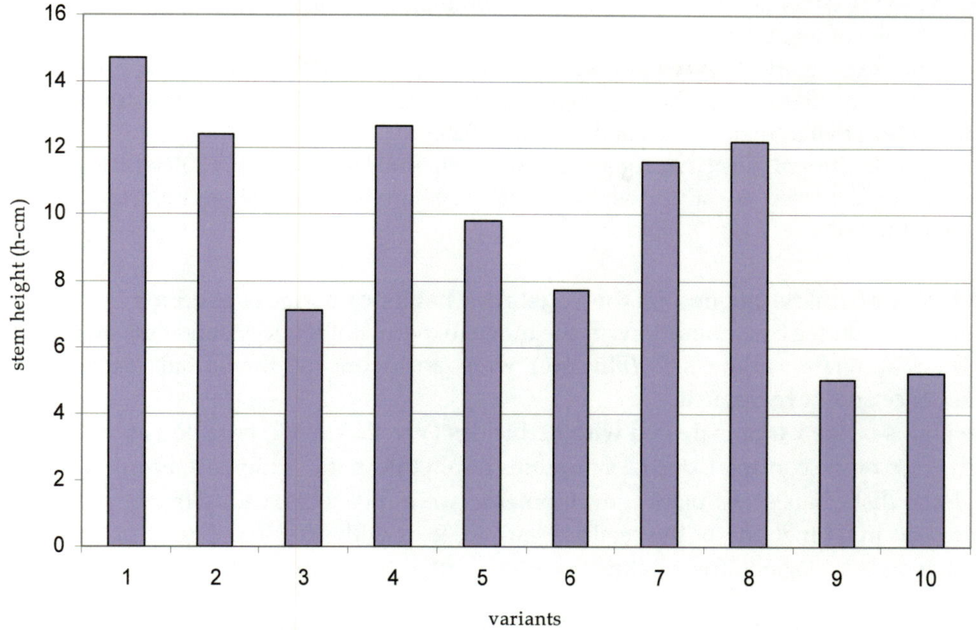

LSD 5%=4,44, 1%=6,08, 0,1 %=8,28

Fig. 13. Effect of soil herbicides on stem height of walnut seedlings (h - cm).

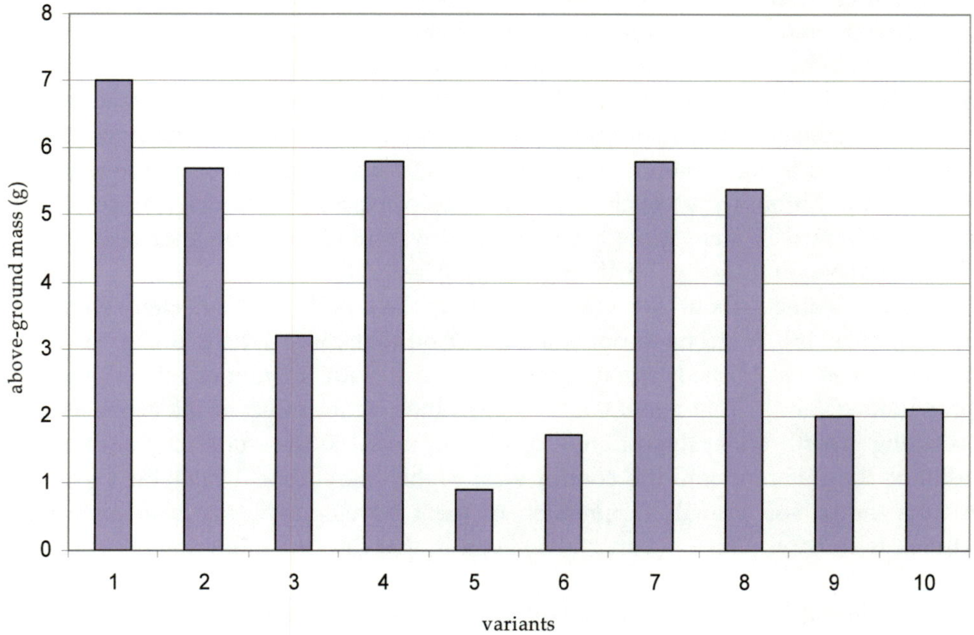

LSD 5%=1,51, 1%=2,07, 0,1 %=2,81

Fig. 14. Effect of soil herbicides on the above-ground plant mass (g)

The following conclusions could be drawn from the results about the effect of the herbicides on the habits of walnut seedlings:

1. A depressing effect on walnut seedling growth, expressed in growth suppression and a significant decrease of the above-ground plant mass, was exerted after treatment with pendimethalin, metolachlor and oxyfluorofen.
2. After treatment with napropamide, terbacil, linuron and isoxaflutole no phytotoxic effect expressed in suppression of the vegetative habits of walnut seedlings was established.

2.4 Effect of soil herbicides on the vegetative habits of apricot seedlings

Plant habits under both conditions of the model experiment - under sand culture conditions and on alluvial-meadow soil (Fluvisol) were analogous at the initial stages of plant emergence and development.

The plants of the variants treated with herbicides (Var. 2, 3 and 4) emerged at the same time with those of the control. External symptoms of phytotoxicity - chlorosis, necrosis, as well as obvious disturbance of plant development, were not observed. Strong phytotoxicity expressed in an inability of the seeds to germinate or withering and drying of the emerged plants was established after treatment with the other herbicides included in the study (Var. 5-10).

The results of the biometric analysis showed that after treatment with terbacil (Sinbar 80 WP - 1,0 kg/ha) the plants had a stem height close or bigger than those in the control variant. The differences to the control were statistically insignificant. Consequently, that active substance did not exert a depressing effect on stem growth (Fig. 15).

An inhibiting effect on growth was established after applying napropamide and pendimethalin under sand culture conditions. That could be explained by the mechanism of action of the active substances. It is known that napropamide stops the growth of the susceptible plants and pendimethalin inhibits cell division and elongation in the meristematic tissues of the stem (Tonev, 2000). Under sand culture conditions, when the effect of soil as a factor is eliminated, those characteristics of phytotoxicity in result of the herbicide application were much more obviously expressed. Under the conditions of alluvial-meadow soil, stem growth suppression was much weaker after applying those active substances. In that case the differences were insignificant.

The results obtained about the effect of the applied herbicides on stem weight were analogous (Fig. 16). A strong suppressing effect on that characteristic under sand culture conditions was established again after treatment with pendimethalin (Var. 3) and napropamide (Var. 2). That result could be explained by the effect of the active substances on seedling growth under the different conditions tested in the study. Under sand culture conditions the differences to the control were highly significant. Under the conditions of alluvial-meadow soil growth suppression of plant development was also reported after treatment with napropamide (Var. 2), however, in that case the inhibiting effect was much weaker.

The plants treated with terbacil under both experimental conditions had stem weight values close to that in the control. The differences to the control variant were statistically insignificant. That confirmed the results about the lack of phytotoxicity in the plants after applying Sinbar 80 WP - 1,0 kg/ha.

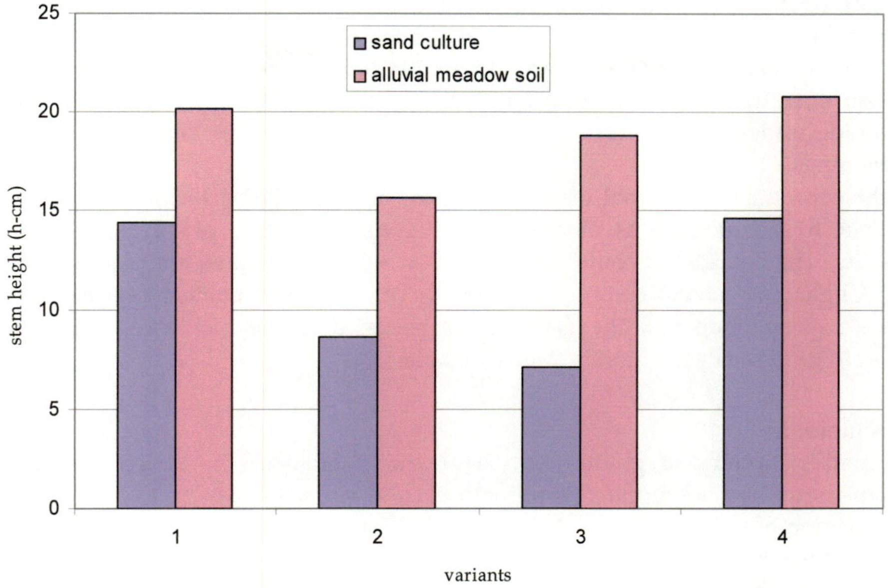

sand culture- LSD 5 %=4,72 alluvial meadow soil-n.s.
1%=6,87
0,1%=10,31

Fig. 15. Effect of soil herbicides on stem height (cm).

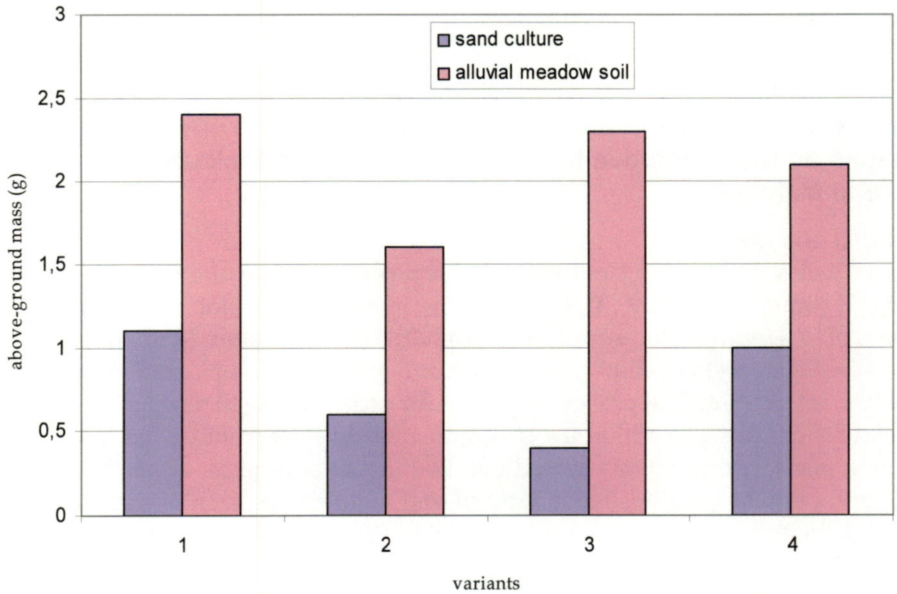

sand culture- LSD 5 %=0.53 alluvial meadow soil-n.s.
1%=0.78
0,1%=1.16

Fig. 16. Effect of soil herbicides on above-ground mass (g).

The results obtained gave the grounds to conclude that the application of terbacil, and pendimethalin under the conditions of alluvial-meadow soil did not cause phytotoxicity on apricot seedlings resulting in plant growth suppression. A negative effect on growth was established after treatment with napropamide. Similar results of phytotoxicity caused by napropamide under sand culture conditions were established in analogous studies with peach seedlings.
Under the conditions of alluvial-meadow soil, the effect of the herbicides was significantly weaker due to factors such as mechanical soil content, humus, pH, argillaceous (loamy) fraction, etc. The conclusion could be drawn that when the stones are planted shallow in soils of a light mechanical composition and there is a direct contact between the active substance napropamide and the germinating seeds, then the incidence of phytotoxicity expressed in growth suppression is quite probable.

2.5 Conclusions

1. External symptoms of phytotoxicity were not established in apricot seedlings after treatment with napropamide, pendimethalin and terbacil.
2. Application of terbacil (Sinbar 80 WP – 1,0 kg/ha) did not cause suppression of growth habits in the seedlings.
3. Suppression of plant growth and development was established after treatment with napropamide - Devrinol 4 F – 4,0 l/ha under sand culture conditions.

The results obtained in the model studies showed that the sand culture method could be applied as a rapid model system for testing the habits of fruit species in response to the soil-applied herbicides. Under sand culture conditions the soil type (mechanical composition, humus) is eliminated as a factor affecting the plant habits and the soil herbicides and it is possible to obtain information about the response of the cultural fruit species to the different active substances.

3. Field experiments for establishing the effect of some soil-applied herbicides on the vegetative habits of seedling rootstocks of yellow plum, peach and Mahaleb.

3.1 Material and methods

The field studies were carried out on the experimental site on the territory of the Fruit-Growing Institute - Plovdiv. The soil in the experimental plot was alluvial-meadow (Fluvisol), pH 7,4, with a good supply of phosphorus and potassium (Rankova, 2004,).
Stratified seeds (stones) of yellow plum, peach and Mahaleb were planted in the period 15 – 25 March on an experimental plot of a size 1 m^2 for each variant (10 seeds per 1 m^2) at 3-5 cm depth and 5-7 cm distance within the row. Immediately after planting the seeds, treatment with soil-applied herbicides was carried out. Four active substances of soil herbicides were used – napropamide, pendimethalin, terbacil and metolachlor, each of them used at three rates. The following variants were set: 1. Control (untreated); 2. Napropamide – Devrinol 4 F – 3,0 l/ha; 3. Napropamide – Devrinol 4 F – 4,0 l/ha; 4. Napropamide - Devrinol 4 F – 5,0 l/ha; 5. Pendimethalin – Stomp 33 EC – 3,0 l/ha; 6. Pendimethalin – Stomp 33 EC – 4,0 l/ha; 7. Pendimethalin – Stomp 33 EC – 5,0 l/ha; 8. Terbacil – Sinbar 80 WP – 750 g/ha; 9. Terbacil – Sinbar 80 WP – 1,0 kg/ha; 10. Terbacil – Sinbar 80 WP – 1,25 kg/ha; 11. Metolachlor – Dual Gold 960 EC – 1,125 l/ha; 12. Metolachlor – Dual Gold 960 EC – 1,5 l/ha; 13. Metolachlor – Dual Gold 960 EC – 1,875 l/ha.

The experiment was set by the standard chess-board method in 4 replications, the reporting area being 4 m^2. The control was maintained free of weeds by three weedings out by hand by 30-day intervals. During vegetation the rootstocks were grown following the standard technology.

Observations were made on plant growth and development during vegetation - emergence, external symptoms of phytotoxicity (chlorosis, necrosis, deformations of the plantlets).

In August (15-20 August) the rootstocks were qualified and the biometric characteristics stem height (h-cm) and thickness at the place of grafting (mm), were reported. Plant qualifying at that period coincided with the time of grafting, determined as the most suitable for grafting in Bulgarian fruit-growing practice.

3.2 Results and discussion

3.2.1 Effect of soil-applied herbicides on the vegetative habits of yellow plum seedling rootstocks

The plants of all the variants treated with herbicides emerged at the same time as those of the control. External symptoms of phytotoxicity (chlorosis, necrosis, deformations of the plantlets) were not observed. Later a slight delay of growth was reported in all the three variants treated with different rates of metolachlor /Dual Gold 960 EC/. Growth depression was more obviously expressed in the variants of the medium and the high rates of metolachlor - Dual Gold 960 EC - 1,5 l/ha and 1,875 l/ha (Var. 12 and Var. 13).

The measured biometric characteristics - thickness at the place of grafting and plant height during vegetation showed the different effect of the herbicides applied to soil at the respective rates on growth and development of the yellow plum seedling rootstocks. The results obtained in the different years showed the same tendency and they were discussed as averaged values.

The results obtained about the thickness at the place of grafting showed that the plants treated with the herbicides napropamide, pendimethalin and terbacil applied at the three studied rates, had a larger thickness compared to the rootstocks in the control variant (Fig. 17). That could be explained by the efficient control on weed vegetation exerted by the applied rates of herbicides during the first three months of seedling vegetation and the eliminated competition of weeds for moisture, nutrient substances and light. It created suitable conditions for the successful emergence, growth and maturing of the rootstocks. Thus they reached the thickness at the place of grafting permitting their inoculation in the same year of their planting.

After treatment with the mentioned herbicides, values closest to the control were reported in the variant with the low rate of napropamide - Devrinol 4 F - 3,0 l/ha (Var. 2) in all the three years of the study, which could be explained by the lower herbicide activity and efficiency against the weed vegetation. The differences were statistically insignificant - $LSD_{5\%}$ = 0,19 (2001); 0,39 (2002); 0,32 (2003).

During the three study years the largest thickness at the place of grafting was reported in the rootstocks of Variant 9 (terbacil - Sinbar 80 WP - 1,0 kg/ha) - 7,3 mm, of Variant 6 (pendimethalin - Stomp 33 EC - 4,0 l/ha) - 7,3 mm and of Variant 3 (napropamide - Devrinol 4 F - 4,0 l/ha) - 7,2 mm, versus 5,1 mm - the average thickness at the grafting zone in the control plants. Therefore, it could be admitted that the active substances napropamide, pendimethalin and terbacil applied at the tested rates, did not have a suppressing effect on growth and development of the seedlings.

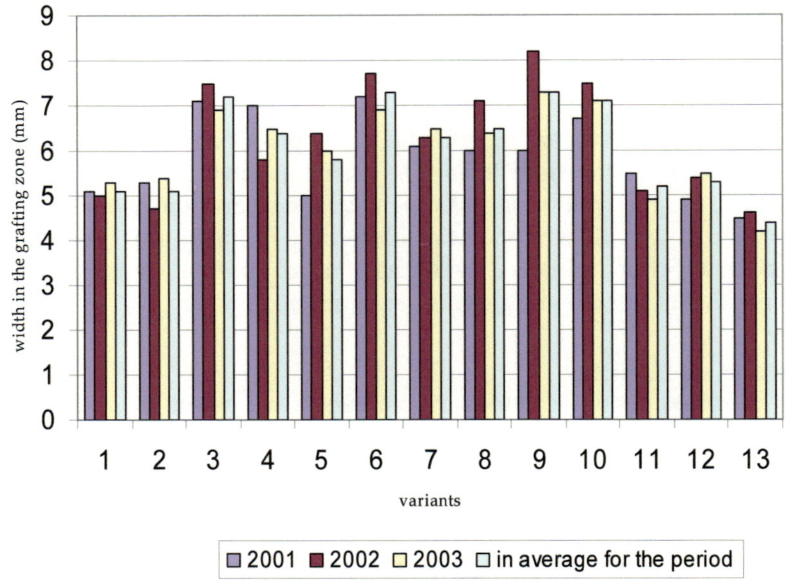

LDD5 % (2001) = 0,19 ; LSD 1 % = 0,25 ; LSD 0,1% = 0,32
LSD5 %(2002) = 0,39 ; LSD 1 % = 0,52 ; LSD 0,1% = 0,68
LSD5 % (2003) =0,32 ; LSD 1 % = 0,42 ; LSD 0,1% = 0,55

Fig. 17. Thickness at the place of grafting (mm) of the yellow plum seedling rootstocks in August

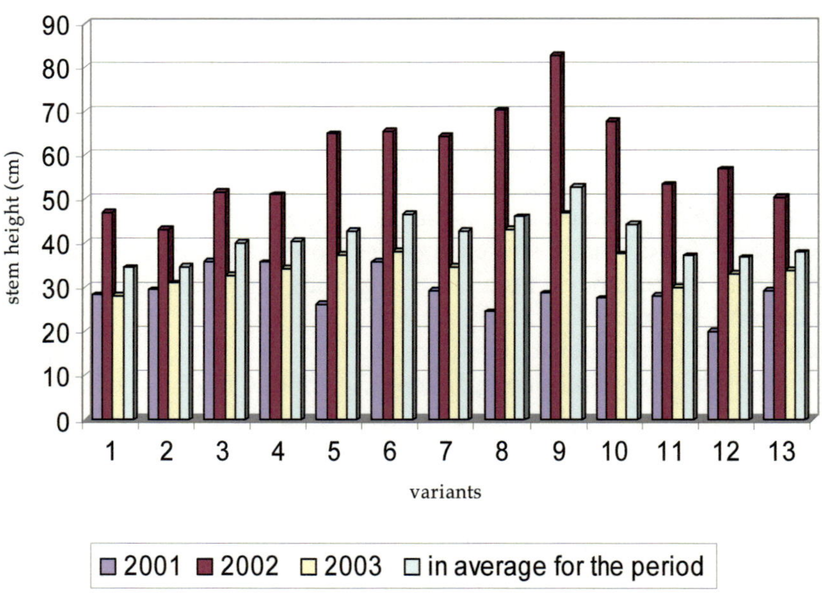

LSD5 % (2001г.) = 0,70 ; LSD 1 % = 0,93 ; LSD 0,1% = 1,21
LSD5 % (2002 г.) = 3,83 ; LSD 1 % = 5,10 ; LSD 0,1% = 6,63
LSD5 % (2003г.)= 3,60 ; LSD 1 % = 4,78 ; LSD 0,1% = 6,22

Fig. 18. Plant height (cm) of the yellow plum seedling rootstocks

With the increase of metolachlor rate, a tendency to a decrease of the thickness was observed (Var. 12-13).

That was also quite obviously expressed by the results about the effect of metolachlor on plant height, although the depressing effect of metolachlor on that characteristic was more weakly expressed. The obtained values were higher or close to those in the control (Fig. 18). In the three experimental years the highest values were obtained after applying the medium rate of: terbacil - Sinbar 80 WP - 1,0 kg/ha - 52,5 cm (Var. 9); pendimethalin - Stomp 33 EC - 4,0 l/ha - 46,5 cm (Var. 6) and the medium and the high rates of napropamide - Devrinol 4 F - 4,0 and 5,0 l/ha - 40,0 cm and 40,3 cm (Var. 3 and Var. 4), the plant height in the control being 34,5 cm.

In the experimental 2003, a variant with an unweeded and untreated control (K_0) was included in the study. Only single plants emerged in that variant, which were characterized by rather delayed development in result of the suppressing effect of the weed vegetation. When assessing their quality by reporting the stem height and thickness at the place of grafting, the advantages of the plants treated with herbicides were obvious - both concerning the quality of rootstocks and the number of the emerged seedlings (Fig. 19).

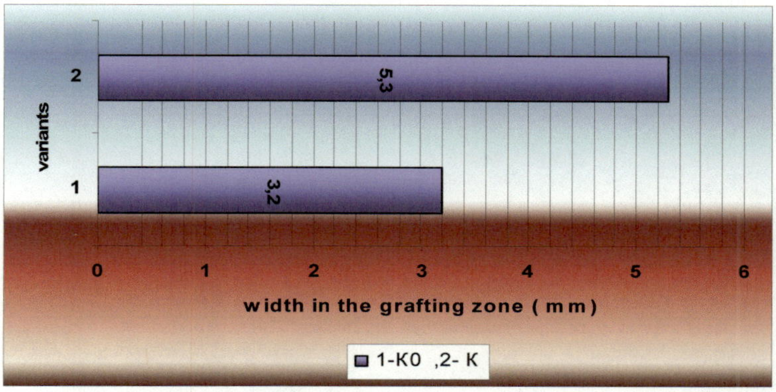

a)

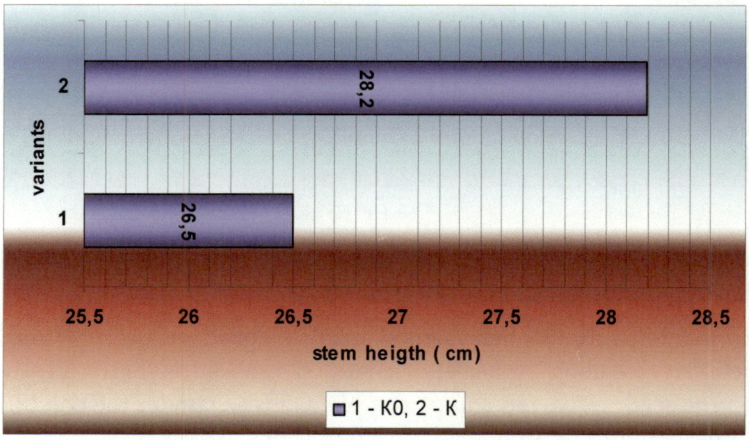

b)

Fig. 19. Yellow plum seedling rootstocks - comparison between unweeded (K_0) and weeded control (K) in August: a/ thickness at the place of grafting (mm); b/ plant height (cm)

a)

b)

Fig. 20. Yellow plum seedling rootstocks treated with: a) napropamid- Devrinol 4F-4.0 l/ha and b) pendimethalin - Stomp 33 EC - 4,0 l/ha

It is clear that a good quality planting material without agrotechnical practices for control of the weed vegetation (weeding by hand) or chemical substances (herbicides) would not be possible to be produced. However, comparing plants grown under the conditions of strong natural weed infestation to those treated with herbicides, makes it possible to evaluate the advantages of the chemical means of weed control. It is quite obviously expressed when comparing the qualitative characteristics of the plants grown under the conditions of different weed infestation background - unweeded control, weeded control and variants

with application of various rates of the respective soil herbicides . The measured thickness at the place of grafting in the rootstocks of the unweeded control (K_0) in August was small - 3,2 mm. It is obvious that planting material with such parameters of thickness at the place of grafting could not be inoculated in the year of seeding. Concerning the other characteristic - plant height, the depressing effect of weed infestation was less expressed. That was probably due to the attempts of the seedlings to overcome the weed competition for light.
Significant changes in the content of leaf pigments (chlorophyll a, b and a+b) and mineral elements in the leaves of the yellow plum seedling rootstocks were not established after treatment with the soil herbicides included in the study. There was a tendency to an increased content in the plants having higher values of the biometric characteristics (Rankova , 2004).
The following conclusions could be drawn from the results obtained about the effect of the soil-applied herbicides on the vegetative habits of yellow plum seedling rootstocks:

1. The following herbicides are recommended for realizing an efficient weed control in the production of planting material from yellow plum seeds: napropamide - Devrinol 4 F - 4,0 l/ha, pendimethalin - Stomp 33 EC - 4,0 ml/ha and terbacil - Sinbar 80 WP - 1,0 kg/ha.
2. After treatment of the seedling rootstocks with metolachlor - Dual Gold 960 EC, plant growth is suppressed. Consequently, the application of the active substance metolachlor should not be recommended in the integrated system of weed control in the production of yellow plum seedling rootstocks.

3.2.2 Effect of the soil-applied herbicides on the vegetative habits of peach seedlings

The plants of all the variants emerged simultaneously. External symptoms of phytotoxicity were not observed. Later a certain delay in plant growth and development was reported in the three variants treated with metolachlor - Dual Gold 960 EC (Var. 11 - 13). Growth suppression was very strong in the variant with the high rate of metolachlor - Dual Gold 960 EC - 1,875 l/ha (Var. 13), causing even the death of some plants.
The measured biometric characteristics in the middle of August showed the same tendency throughout the years of the study and they were discussed as averaged values. The data showed different effects of the soil-applied herbicides on rootstock development.
After treatment with the medium rates of napropamide - Devrinol 4 F - 4,0 l/ha (Var. 3), pendimethalin - Stomp 33 EC - 4,0 l/ha (Var. 6) and terbacil - Sinbar 80 WP - 1,0 kg/ha (Var. 9) the plants reached the highest values of thickness at the grafting zone - 7,0 mm, 7,2 mm and 7,2 mm, respectively (Fig. 21).
In all the three study years a lower value of thickness at the place of grafting was reported in the plants treated with metolachlor at the three applied rates (Var. 11 - 12) compared to that in the control. That was probably due to the depressing effect of metolachlor on plant development. It was most strongly expressed after applying the high rate of Dual Gold 960 EC - 1,875 l/ha (Var. 13). Taking into consideration that rootstocks having a stem thickness of 4,0 - 4,2 mm are not suitable for inoculation, it can be concluded that the herbicide metolachlor has a depressing effect on the growth of peach seedlings.
The results about the effect of the soil herbicides on plant height showed that the values were lower in the plants treated with the low rate of napropamide - Devrinol 4 F - 3,0 l/ha (Fig. 21).
That was observed in all the three years of the study and it was probably due to the weaker herbicide effect of the low rate of napropamide and the incidence of competition between

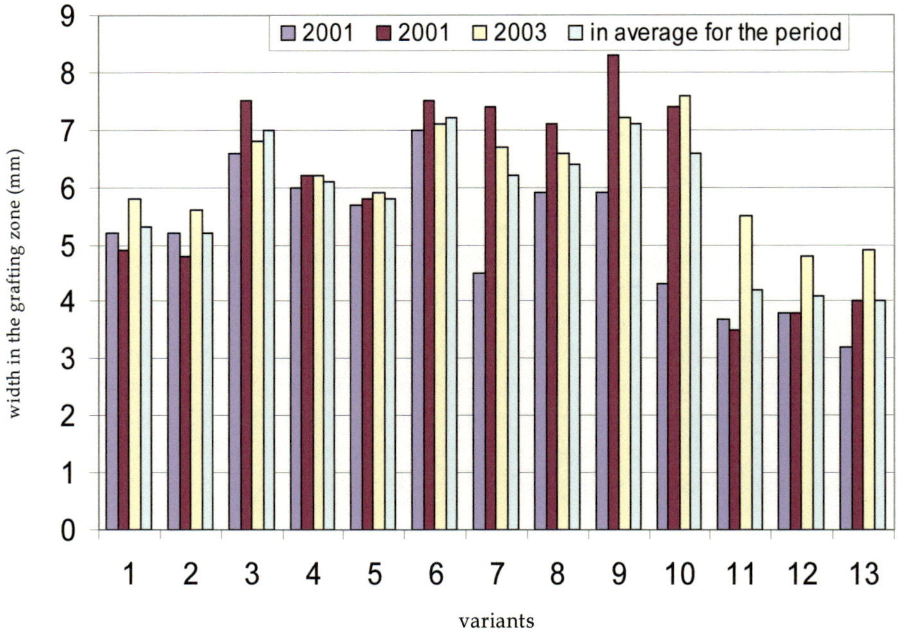

LSD5 % (2001г.) = 0,15; LSD 1 % = 0,20 ; LSD 0,1% = 0,26
LSD5 % (2002 г.) = 0,38;LSD 1 % = 0,50 ; LSD 0,1% = 0,75
LSD5 % (2003г.)= 0,33; LSD 1 % = 0,45 ; LSD 0,1% = 0,58

Fig. 21. Thickness at the place of grafting (mm) of peach seedling rootstocks

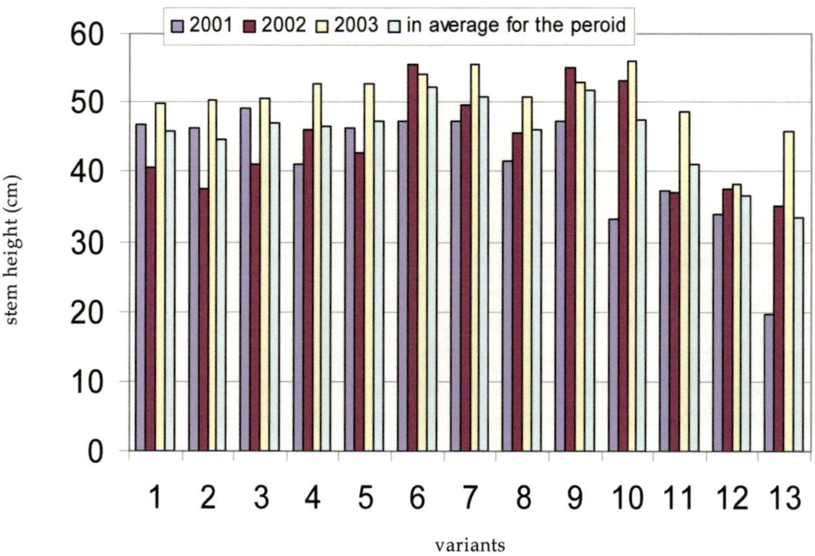

LSD5 % (2001г.) = 1,07 ; LSD 1 % = 1,43 ; LSD 0,1% = 1,87
LSD5 % (2002 г.) = 3,38 ; LSD 1 % = 5,10 ; LSD 0,1% = 6,68
LSD5 % (2003г.)= 3,34 ; LSD 1 % = 4,46 ; LSD 0,1% = 5,84

Fig. 22. Stem height (cm) of peach seedling rootstocks.

the seedlings and the developing weed vegetation for moisture and nutrient substances. However, the differences were statistically insignificant. Lower plant height was also reported for the plants of the variants treated with metolachlor at the three rates of Dual Gold 960 EC (Var. 11-13).

In average for the experimental period, it was most obviously expressed in the variant of the high rate of Dual Gold 960 EC - 1,875 l/ha (Var. 13) - 33,5 cm, the plant height in the control being 45,7 cm. Those differences were significant or highly significant by years.

However, when discussing the average values of plant height for the whole period of study, the medium rates of the soil-applied herbicides proved to be optimal. The highest values of plant height were reported in the variants treated with pendimethalin - Stomp 33 EC - 4,0 l/ha (Var. 6) - 52,2 cm, terbacil - Sinbar 80 WP - 1,0 kg/ha (Var. 9) - 51,7 cm and napropamide - Devrinol 4 F - 4,0 l/ha (Var. 3) - 47,0 cm, the plant height of the control being - 45,7 cm (Fig. 32).

On the basis of the obtained results about the effect of the soil herbicides on the biometric characteristics established at the moment of grafting, it could be concluded that the active substances napropamide, pendimethalin and terbacil at the three applied rates did not exert a negative influence on growth and development of peach seedlings. Eliminating weed - rootstock competition for moisture, nutrient substances and light in the first three months of seedling vegetation (the post-effect period of the applied herbicides) enables the production of plants suitable for grafting.

Significant changes in the content of chlorophyll and mineral elements in the leaves of the peach seedling rootstocks were not established after treatment with the soil herbicides included in the experiment. A tendency was established, similar to that in the yellow plum seedling rootstocks, that the content of chlorophyll and mineral elements increased in the plants having higher values of the biometric characteristics (Rankova, 2004).

In the experimental 2003, a variant with an unweeded and untreated control (K_0) was included in the study for establishing the effect of the soil-applied herbicides on growth habits of peach seedlings. In that variant only single plants emerged. They were characterized by a delayed growth and strongly suppressed development in result of the high competition of the weed vegetation for moisture, light and nutrients.

In August the rootstocks of that variant had 3,8 mm thickness at the place of grafting. Those rootstocks could not be grafted at that time (Fig. 23). The depressing effect of weed infestation had an obvious impact on the other characteristic - stem height (Fig. 23).

When comparing the qualitative characteristics of the rootstocks in that variant with the plants in the control weeded out by hand and the variants treated with herbicides, the results obtained in yellow plum rootstocks were confirmed. A good quality planting material suitable for grafting in the same year of seeding could not be produced without mechanical (weeding out by hand) or chemical methods of weed control. That shows the advantages of the chemical control of weeds in the production of peach seedling rootstocks.

1. The application of the following herbicides is recommended for the production of peach seedling rootstocks: pendimethalin - Stomp 33 EC - 4,0 l/ha, terbacil - Sinbar 80 WP - 1,0 g/da and napropamide - Devrinol 4 F - 4,0 l/ha.
2. Treatment with metolachlor showed an inhibiting effect on the growth habits of peach seedlings. The active substance metolachlor should not be applied for weed control in the production of peach seedling rootstocks.

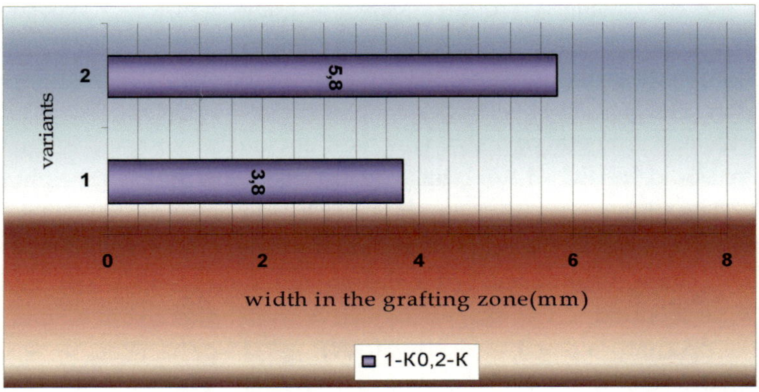

a)

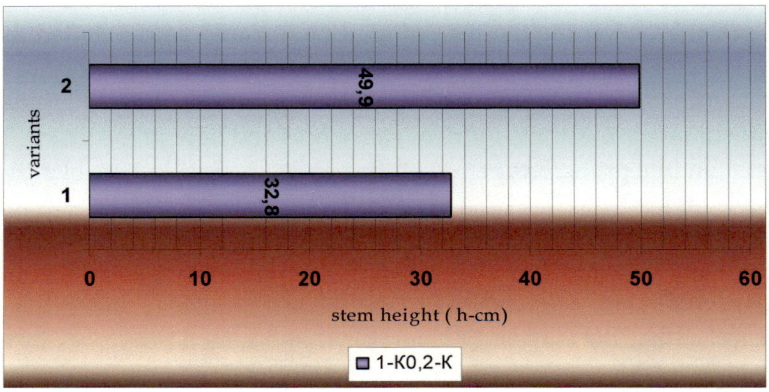

b)

Fig. 23. Peach seedling rootstocks - a comparison between unweeded control (K_0) and weeded out control (K) in August: a/ thickness at the grafting zone (mm); b/plant height (cm)

3.2.3 Effect of soil-applied herbicides on the vegetative habits of Mahaleb seedlings

Observations on the habits of Mahaleb seedlings treated with the soil-applied herbicides included in the study showed that they were highly susceptible to soil herbicides. Visual symptoms of phytotoxicity in the plants treated with napropamide, pendimethalin and metolachlor were not established. The seeds treated with those herbicides emerged at the same time as those in the control. Later, a certain delay in plant development was observed in the variant treated with the highest rate of metolachlor Dual Gold 960 EC - 1,875 l/ha (Var. 13).

In the three study years the plants treated with terbacil (Var. 8, 9, 10) responded in the same way - only single plants emerged and they had a very delayed development. Later chlorosis was detected, followed by withering and plant dying.

Consequently, it could be admitted that the soil herbicide terbacil was toxic for Mahaleb seedlings at all the three applied rates.

Data of the biometric analysis showed the same tendency throughout the years of the study and they were discussed as average values.

a)

b)

c)

Fig. 24. Peach seedling rootstocks treated with herbicides:
a) Var. 3 - napropamide – Devrinol 4 F – 4,0 l/ha; b) Var. 6 - pendimethalin – Stomp 33 EC – 4,0 l/ha; c) Var. 9- terbacil – Sinbar 80 WP – 1,0 kg/ha.

The biggest value of stem thickness at the place of grafting was reported in the plants treated with pendimethalin - Stomp 33 EC - 4,0 l/ha (Var. 4 and 5) and metolachlor - Dual Gold 960 EC - 1,5 l/ha (Var. 12), (Fig. 25). The differences to the control were of high statistical significance.

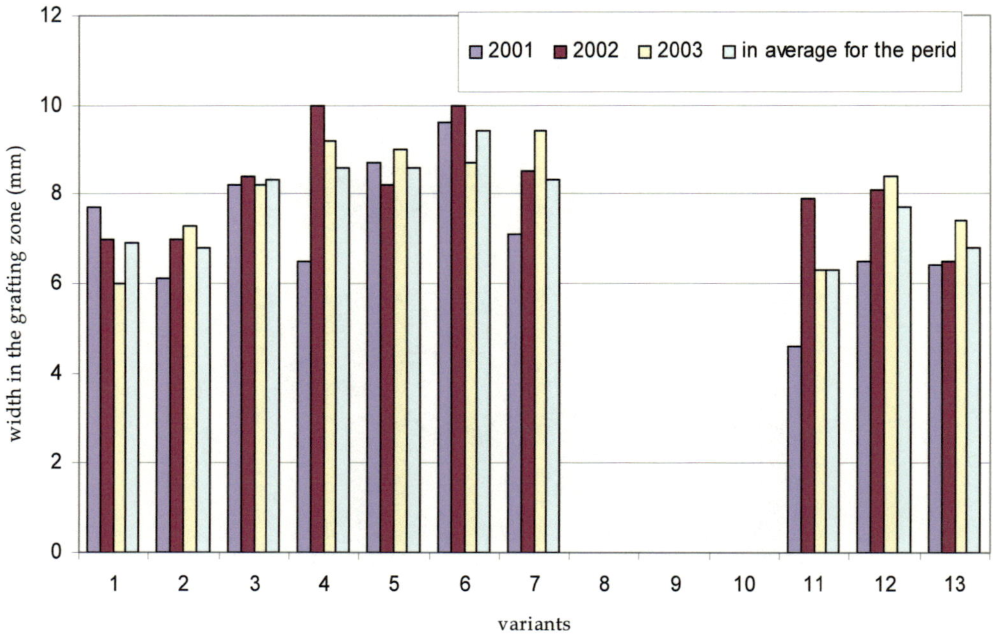

LSD5 % (2001г.) = 0,45 ; LSD 1 % = 0,60 ; LSD 0,1% = 0,80
LSD5 % (2002 г.) = 0,31 ; LSD 1 % = 0,42 ; LSD 0,1% = 0,55
LSD5 % (2003г.)= 0,55 ; LSD 1 % = 0,73 ; LSD 0,1% = 0,98

Fig. 25. Effect of soil-applied herbicides on thickness at the place of grafting (mm) in Mahaleb seedling rootstocks

Similar to the results in yellow plum and peach seedling rootstocks, values of thickness close to the control were also established in the plants of the variant treated with the low rate of napropamide (Var. 2). However, the differences were statistically insignificant. That could be explained by the poorer efficiency of the lower herbicide rate and the existing weed-rootstock competition for moisture, nutrients and light. Probably, due to the same reasons, a smaller thickness was established in the plants treated with the low rate of metolachlor (Var. 11).

Values of that biometric characteristic were also lower in the plants of the variant treated with the highest rate of metolachlor - Dual Gold 960 EC - 1,875 l/ha (Var. 13). That was probably due to the depressing effect of the high rate of metolachlor on rootstock development.

The results about the effect of the soil-applied herbicides on plant height were analogous (Fig. 26).

The plants of the variants with applied napropamide (Var. 2 - 4), pendimethalin (Var. 5 - 7) and metolachlor (Var. 11 and 12) had a bigger height compared to those in the control. The differences were statistically significant. Values close to or lower than the control variant were established again in the plants of variants 2 and 13. Probably that was due to the already mentioned poorer herbicide efficiency of the low rate of napropamide and the existing competition with the weeds (Var. 2) and the exerted depressing effect of the high rate of metolachlor (Var. 13).

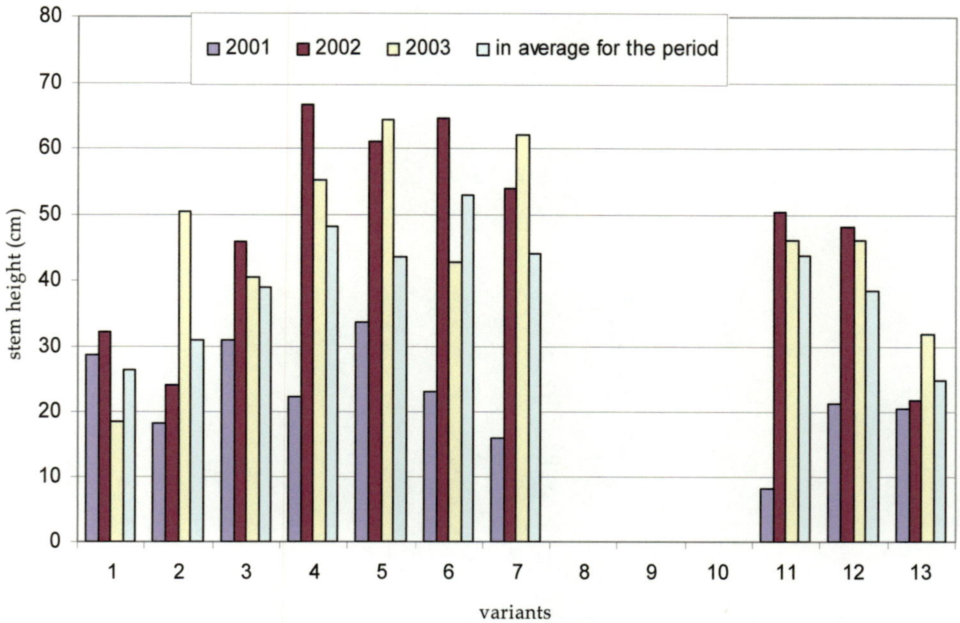

LSD5 % (2001г.) = 13,50 ; LSD 1 % = 18,11 ; LSD 0,1% = 23,91
LSD5 % (2002 г.) = 2,86 ; LSD 1 % = 3,83 ; LSD 0,1% = 5,06
LSD5 % (2003г.)= 1,09 ; LSD 1 % = 1,46 ; LSD 0,1% = 1,93

Fig. 26. Effect of soil herbicides on the height of Mahaleb seedling rootstocks (cm)

When including the variant with an unweeded control (K_0), all Mahaleb seedlings in that variant died in result of the strong competition with the weed vegetation.

Similar to the results obtained in the experiments with yellow plum and peach, significant changes in the content of chlorophyll and mineral elements in the leaves of Mahaleb seedling rootstocks were not established after treatment with the soil-applied herbicides included in the study. There was a tendency to an increase of their content in the plants having higher values of the biometric characteristics (Rankova, 2007).

3.3 Conclusions

1. The application of napropamide - Devrinol 4 F - 4,0-5,0 l/ha, pendimethalin - Stomp 33 EC - 4,0 l/ha and metolachlor - Dual Gold 960 EC - 1,5 l/ha is recommended in the production of Mahaleb seedling rootstocks.
2. The soil-applied herbicide terbacil (Sinbar 80 WP) had a strong toxic effect on Mahaleb seedling development and caused plant death.
3. The inhibiting effect of the soil herbicide metolachlor was exerted after treatment with the highest rate of Dual Gold 960 EC - 1,875 l/ha.

a)

b)

c)

Fig. 27. Mahaleb seedling rootstocks treated with: a) napropamide (Var. 5); b) pendimethalin (Var. 6) and c) metolachlor (Var. 13)

4. Discussion

The results obtained from the pot and field experiments confirmed the initial assumption that the stone fruit species are susceptible to the application of soil herbicides. In the production of seedling rootstocks risky for causing phytotoxicity proved to be the probable direct contact between the germ of the emerging plant and the soil herbicide. In parallel with those studies, model experiments under in vitro conditions were carried out for phytotoxicity caused by soil herbicides. Using the ebryoculture method, the different effect of the soil-applied herbicide pendimethalin on the development of the embryo root of yellow plum embryos was established, depending on its initial length at the time of treatment (Gercheva, et al., 2001). Phytotoxicity (inhibition of root meristem growth and browning of cotyledons) was established in the treatment of embryos with embryonic roots< 5mm in length. The embryos whose embryonic roots at the moment of herbicide application were longer than 5 mm did not show any symptoms of phytotoxicity (Fig.28). That allowed admitting that the selectivity of the active substance pendimethalin was of a physical character (a direct contact with the germinating seeds when they had been sown at a shallower depth) and of a physiological character (the type and the physiological stage of the plant development). Analogous results or results close to the model experiments with sand culture were obtained about the inhibiting effect or the lack of visual phytotoxicity of the soil herbicides napropamide, pendimethalin and terbacil under in vitro conditions in some vegetative rootstocks - GF-677, MM 106 and Wangenheims (*Prunus domestica*), (Rankova, et al., 2004; Rankova, et al., 2006a; Rankova, et al. , 2006b; Rankova, et al., 2009).

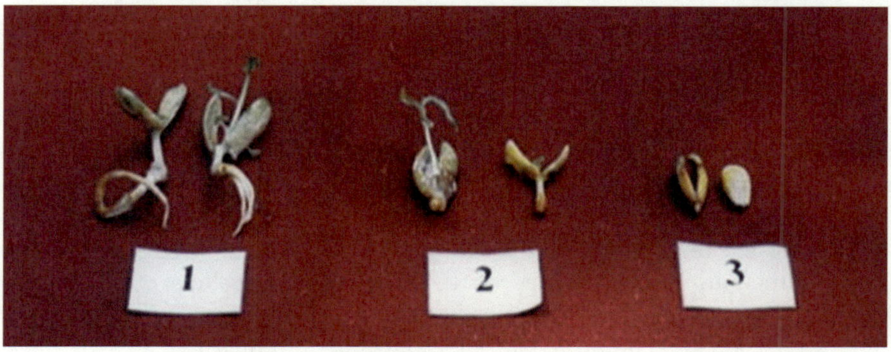

Fig. 28. Yellow plum embryos treated with pendimethalin

The results obtained from the field experiments allowed to accept the medium rates of napropamide - Devrinol 4 F - 4,0 l/ha, pendimethalin - Stomp 33 EC - 4,0 l/ha and terbacil - Sinbar 80 WP - 1,0 kg/ha as suitable rates for applying in fruit nurseries in yellow plum and peach seedling rootstocks, on the one hand, and, napropamide - Devrinol 4 F - 4,0-5,0 l/ha, pendimethalin - Stomp 33 EC - 4,0 l/ha and metolachlor - Dual Gold 960 EC - 1,5 l/ha - in Mahaleb seedlings. The highest values of biometric characteristics (thickness at the place of grafting and stem height) were reported in the plants of those variants. The major characteristic determining the planting material quality and its suitability for grafting is the thickness at the place of grafting. After treatment with herbicides applied at the rates mentioned, the highest values of the thickness at the place of grafting were obtained. That contributed to the production of high quality rootstocks suitable for inoculation in the year of seeding. The medium herbicide rates proved to be efficient against weed vegetation and they created good conditions for the development of the seedlings. It is known that eliminating the weed competition during the first three months of seedling vegetation (the period of seed emergence and the beginning of plant development) is a very important precondition for the normal growth and development of the plants. The results about the effect on the characteristics stem height and above-ground vegetative mass confirmed the incidence or the lack of a depressing effect on plant growth.

It is obvious that the low rates of the mentioned herbicides were less efficient against weed infestation. That was well expressed in the years of study under the conditions of higher weed density. The presence of weed plants in the plots of the variants treated with low rates created competition with the seedlings for moisture, light and nutrients. As a result of that lower values of the biometric characteristics were reported. Due to that only single cultural plants emerged with seriously delayed development in the variant of unweeded and untreated control.

In the variants with the high rates of napropamide (Var. 4), pendimethalin (Var. 7) and terbacil (Var. 10) the herbicide efficiency was also very well expressed. The differences in the values of the rootstock biometric characteristics in those variants compared to the values in the variants with the medium rates were small and statistically insignificant in most cases. External symptoms of phytotoxicity were not observed.

Metolachlor applied at the three rates had a suppressing effect on the growth habits of both the yellow plum and peach seedling rootstocks. That was quite obviously expressed when applying the high rate.

The results reported about the effect of the studied herbicides on growth habits of seedling rootstocks confirmed the results about the effect of some active substances obtained by other authors- (Hogue, 1983) established that napropamide (4 kg/ha) and the herbicide mixture napropamide + terbacil (4 + 2 kg/ha) were not toxic and did not cause disturbances in the development of peach seedlings. After treatment with trifluralin (1 kg/ha) referring to the group of nitroanilines, to which pendimethalin also belongs, no toxic effect on peach seedlings was established, as well. Strong toxicity of peach seedlings was observed after treatment with metolachlor (1,7 - 6,8 kg/ha) - suppression in seed emergence and growth disturbances were reported.

When comparing the responses of the seedling rootstocks of the three species, it could be concluded that the herbicide effect was weaker when treating yellow plum seeds (stones). Probably the good germination capacity of the yellow plum seeds and their easier adaptability to the soil and climatic conditions contributed to their much easier overcoming of the herbicide stress effect.

Comparing the bahaviour of Mahaleb seedlings to that of the seedling rootstocks of other fruit species allowed concluding that *Prunus mahaleb* L. species is more susceptible to the applied soil herbicides compared to yellow plum and peach. Similar habits were obtained after applying napropamide and pendimethalin at the studied rates, showing a lack of phytotoxicity in the rootstocks. A depressing effect of the active substance metolachlor in Mahaleb plants was established only after treatment with the highest rate of Dual Gold 960 EC (Var. 13).

The economic analysis of the chemical control of weeds in fruit nurseries showed that the application of herbicides in the production of yellow plum and peach seedling rootstocks led to 16 - 36 times higher return on investment and the efficiency coefficient was from 14 to 43 times higher compared to hand weeding out. Data about the economic effect of applying herbicides in the production of Mahaleb seedling rootstocks were similar - from 12 to 27 times higher return on investment and from 14 to 30 times higher efficiency coefficient in comparison with hand weeding out (Manolova & Rankova, 2005; Manolova & Rankova, 2007).

The recommended soil herbicides (with an exception of terbacil, which is prohibited for use nowadays) have about a three-month period of persistency, weak water solubility and they do not carry the risk of soil and ground water pollution with residues (Tonev, 2000). Due to the fact that usually soils of a light mechanical composition are used for establishing fruit nurseries - alluvial, alluvial-meadow, characterized by their weak absorption capacity - the risk of polluting such soils with residual amounts of herbicides was minimal (Bakalivanov, 1980). There are data that the herbicides recommended for application in fruit nurseries do not have an inhibiting effect on soil microflora (Bakalivanov, 1980). Consequently, their application does not have a negative effect on the biological activity of soil.

In conclusion, it should be mentioned that weed control in fruit nurseries should be carried out on the basis of a sound knowledge about the response of the separate rootstock species to the applied soil herbicides, looking for the point of intersection of the herbicide rate, so that it is efficient enough against the weeds and selective to the cultural plant. Thus, a good quality planting material for establishing new fruit orchards will be produced.

5. References

Abdul, B., Mohammad, K. & Tasleem, J. (1998): The effect of chemical weed control in peach nursery on germination and seedling growth. *Sarhad Journal of Agriculture*, 14, 1, pp. 43-47.

Arenstein ,Z. (1980). The control of annual weeds in young and matute orchards by means by terbutryne and terbutryne/simazine mixture. *British crop protection conference of weeds*, Nottingam, Proceedings, 1, pp. 159 - 163.

Bakalivanov, D.(1980). Soil- microbiological aspects on herbicide contamination. *Institute of soil science"Nikola Poushkarov'-Sofia*, pp. 254

Clay, D. V.(1984). Evaluation of the tolerance of cheery and plum trees to root application of herbicide using a sand culture method. *Aspects of applied biology*. Assoc of applied biologist, 8 , pp. 75 - 86.

Hogue E. (1983). Weed control in peach seedlings. *Exp. Committee on weeds. Western Canada Section*, Res. Rep., 3, pp. 145 - 146.

Jankovic, R., Stanojevic, V.& Rakicevic M. (1995): Herbicides in nursery tree production. *Jugoslovensko Vocarstvo*, 29, 1-2, pp. 93-101.

Gercheva, P., Rankova, Z. & Ivanova, K. (2002). In vitro Test System for Herbicide Phytotoxicity on Mature Embryos of Fruit Species, *Acta Horticulturae*, 577, ISSN 0567-7572, pp. 333- 336.

Kaufman, E.& Libek, A. (2000a). The application of herbicides in a fruit - tree nursery. *Proceedings of the International Conference Fruit Production and Fruit Breeding*, Tartu, Estonia,: 61-64.

Kaufman, E.& Libek, A. (2000b). Damages to cherry plum seedlings (Prunus cerasifera var. Daviricata Bailey) caused by herbicides. *Proceedings of the International Conference Fruit Production and Fruit Breeding*, Tartu, Estonia, pp. 132-137.

Lange, A.H.(1987). Comparative phytotoxicity studies in trees and vines. *Proceedings 39 th annual California weed conference*,pp. 210 - 212.

Lourens ,A.F.; Lange, A. & Calitz, F. (1989). Phytotoxicity of pre- emergence herbicides to peach seedlings (Prunus persica). *South African Journal of plant and soil*, 6 , 2 , pp. 97 - 102.

Manolova, V. & Rankova,Z. (2005). Economic results of using herbicides in the production of some seedling rootstoks, *Bulgarian Journal of Agricultural Science*, 2, pp. 159-163

Manolova, V. & Rankova, Z (2007). Comparative economic evaluation of herbicide use in the production of Prunus mahaleb seedling rootstocks. *Agricultural Economics and Managements*, 52, 2, pp. 56-58

Milusheva, Sn.& Rankova.Z. (2002). Plum pox poty virus Detection in Weed Species under Field Conditions. *Acta Horticulturae*, 577, ISSN 0567-7572, pp. 283 - 287.

Milusheva, Sn. & Rankova.Z. (2006). Serological Identification of Plum Pox Virus in some economic important weeds. *Agricultural Science* , ISSN 1313-3534, 4, 38-41

Mitchell, R.B.& Abernethy, R.J. (1989). The effect of weed removal on the growth of young apricot trees. *Proc.40 th New Zealand Weed and Pest control conf. Nelson* , Aug.11-13, Palmerston North, pp. 209 -212.

Oosten, Van H.J. (1971). Futher information about herbaceous host range of sharka /Plum pox/. *Ann. Phytopathologie*, pp. 195 -201.

Porterfield J.D., Odell J.D. & G.R. Huffman (1993). Effects of DCPA/ napropamide herbicide tank mix on germinants of seven hardwood species in tree nursery beds. *Tree Planters Notes*, 44, 4, pp. 149-153.

Rankova, Z. & Milusheva, Sn.(2001). Problems with the weed plants in the epidemiology of Plum Pox Virus. Fifth Scientific Practical conference "Ecological problems of Agriculture" AGROEKO 2001,Agricultural University-Plovdiv, Scientific Works, vol. XLVI, book1,pp. 381-384

Rankova, Z. (2002). Effect of some soil herbicides on growth habits of peach seedlings (*Persica vulgaris* L.) under the conditions of sand culture. *Proceedings of Papers from the Fourth Scientific and Technical Conference with International Participation "ECOLOGY AND HEALTH 2002"*, pp. 75-80.

Rankova, Z. (2004). Study on the effect of some soil herbicides on the vegetative habits of yellow plum and peach seedling rootstocks, *PhD Thesis.*,fruit growing Institute- Plovdiv, pp.156

Rankova, Z. , Gercheva, P. & Ivanova K. (2004). Screening of soil herbicides under in vitro conditions, *Acta Agriculturae Serbica*, vol. IX, 17,pp. 11-17

Rankova, Z(.2006). Effect of some soil herbicides on the vegetative habits of mahaleb cherry (*Prunus machaleb* L.) seedling rootstocks, *Bulgarian Journal of Agricultural Science*, 12,pp. 429-433

Rankova, Z., Nacheva L., Zapryanova K., Gercheva P. & Bozkova, V.(2006a). Effect of soil herbicides napropamid and pendimethalin on rooting and growth of the vegetative plum rootstock Pr. domestica Wangenheims under in vitro conditions. *Journal of mountain agriculture on the Balkans*, 9(3), pp. 349-359.

Rankova, Z., Nacheva L., Gercheva P, & Bozkova V.,(2006b). Vegetative habits of plum rootstock Wangenheims after treatment with terbacil under in vitro conditions. *VI National Conference "Ecology and helth" Plovdiv*, May 2006 pp. 339-344.

Rankova, Z (2007). Mineral composition and chlorophyll content in the leaf mass of mahaleb seedling rootstocks (Prunus mahaleb L.) after the application of some soil herbicides. *International scientific Conference-Plant Genetic Stoks-The basis of Agriculture of today. 13-14 June 2007*, Sadovo, ISBN 978-954-517-083-6, 2 ,pp. 599-601

Rankova, Z., Nacheva, L. & Gercheva P. (2009). Growth Habits of the Vegetative Apple Rootstock MM 106 After Treatment With Some Soil Herbicides Under *In Vitro* Conditions, *Acta Horticulturae 825*, ISSN 0567-7572, pp. 49-54

Tonev, T. (2000). Handbook of integrated weed control and culture of farming, Book 2, Higher Institute of Agriculture, Plovdiv, pp.275, pp 126-127

Wazbinska, J. (1997). Technological improvement of generative cherry plum rootstocks, one-year Wegierka Lowicka plum trees and apple seedlings. *Acta. Academiae Agriculturae ac Technicae, Olstenensis Agricultura*, 64 c, pp. 107.

24

Herbicide Sulcotrione

Nanxiang Wu
Department of Environmental Health, Institute of hygiene
Zhejiang Academy of medical Sciences
P.R.China

1. Introduction

Sulcotrione is a new kind of triketone-type herbicide, which inhibits the activity of plant 4-hydroxyphenylpyruvate dioxygenase (HPPD) [1]. It has been widely applied to control and prevent wild grass and weeds for crops such as barley, wheat and maize in Europe, the USA and many other countries, including China, since 2000 [2]. HPPD also exists in mammals for the catabolism of tyrosine. Although sulcotrione shows low toxicity in subchronic and chronic toxicity testing, it can also, through the food chain, be continuously transferred to and accumulate in the human body when it is widely used and applied and when part of its residue in the soil is absorbed by and aggregated in plants. Another part of its residue pollutes the water table through surface runoff and the underground permeable layer. Therefore, sulcotrione, as an HPPD inhibitor, has potential risks to human health, and its possible role as an environmental pollutant must raise attention and vigilance.

The residual level of sulcotrione in the soil directly affects its accumulation in agricultural products and is closely related to the level of human exposure. Currently, there is a standard for sulcotrione residue in the soil. Therefore, the investigation of the sulcotrione soil residual levels can provide scientific evidence for the rational application of sulcotrione and the establishment of pesticide residue standards. In addition, this investigation also can supply more data for soil pesticide monitoring databases and provide an informative source for scientific research.

HPPD, a target molecule of sulcotrione, exists universally in prokaryotes and eukaryotes. The catabolism of tyrosine is repressed when HPPD is inhibited by sulcotrione. Epinephrine is derived from tyrosine. The detection of tyrosine and epinephrine levels in the blood of rats exposed to sulcotrione can reflect the inhibiting effect of HPPD on epinephrine levels. The regulatory mechanisms for blood glucose are complicated. There are many possible pathways for sulcotrione to interfere with blood glucose levels. The rats were exposed to different doses of sulcotrione for different times to get blood glucose levels, enzyme activity, hormone levels, etc. The correlations between blood glucose levels and different parameters were analyzed in the rats exposed to sulcotrione to establish or exclude the possible interference of sulcotrione with blood glucose regulation.

In vitro experiments have already shown that sulcotrione can specifically, effectively and reversibly inhibit hepatic HPPD activity. However, there are few *in vivo* reports on this question. In this study, we investigate, the effects of sulcotrione on hepatic enzyme activity, tyrosinemia and cornea damage through subacute and chronic toxicity tests in rats.

2. Materials and methods

2.1 The investigation of sulcotrione soil residual levels

2.1.1 Soil sample collection

The soil samples were collected following the sampling rules of national farmland and the environmental standards for safety, high quality and pollution-free agricultural products GB/T18047.1-2001 from September 2007 to September 2008 within eight regions in the Zhejiang Province, P.R. China. The soil sampling depth was 0 to 20 cm. A single soil sample was a mixture of multiple (5) spots. A total of 200 soil samples were collected. At the same time, a total of 10 samples (from a mixture of 10 different regions) of cultured plants (corn) were collected in Quzhou and Dongyang in the Zhejiang Province.

2.1.2 Sample preparation

One gram samples (air-dried and passed through a 100 mesh sieve) were dissolved in 1 mL of methanol by mixing for 30 seconds. After incubation for 30 minutes, the sample was treated with ultrasound for 10 minutes. Then the sample was centrifuged for 5 minutes at 29000 g and the supernatant was collected and filtered through a membrane for detection.

2.2 Sulcotrione analysis

2.2.1 HPLC chromatographic conditions [3-4]

The mobile phase consisted of methanol:0.1M sodium dihydrogen phosphate triethylamine = 40:60 (v/v). The flow rate was 0.8 mL/min (A pump: 0.1M sodium dihydrogen phosphate triethylamine buffer 0.48 mL/min, B pumb: methanol 0.32 mL/min). The detection wavelength was 254 nm. The column temperature was room temperature. The sensitivity: was 0.001 AUFS. The injection volume was 10 μl.

2.2.2 Sulcotrione standard curve

2.2.2.1 Preparation of the standard buffer

A 0.0100 g (accuracy to 0.0001 g) sulcotrione standard sample was accurately weighed. It was dissolved in a small volume of methanol which was poured into a 100 mL volumetric flask. It was diluted in methanol to the final scale. It was the standard stock sulcotrione solution. Before use, the stock solution was diluted in methanol to 10 mg/L as the working solution. The sulcotrione standard HPLC profile was made according to the above mentioned chromatographic conditions. The sulcotrione standard analysis sample was provided by Sigma and had a purity of more than 98% (the number is 46318, which is valid until March 2013).

2.2.2.2 Calculating the results

The sample analytic results were extrapolated based on the area external standard method. The concentration calculation was: X = solution volume x C/sample weight, where X is the sulcotrione concentration (mg/kg) and C is the concentration calculated from the standard curve.

2.2.3 Statistical analysis

SPSS 15.0 version was used for the statistical analysis. The results are presented as mean ± SD. SD is the standard deviation.

2.3 Investigation of tyrosine levels in a population who might have had exposure to sulcotrione.

2.3.1 Determining the population

The population consisted of forty males and forty females (80 totals) who work in the sulcotrione industry and another forty males and forty females who do not work in the sulcotrione industry; ten laboratory staff who have had contact with sulcotrione and 10 staff who do not have had contact were also included.

2.3.2 Blood collection.

The blood was centrifuged at 3000 g for 15 minutes. Then the supernatant was collected and cryopreserved at -18°C. The blood sample was prepared for sulcotrione detection as follows: The frozen serum was thawed at room temperature. A certain volume of the serum sample was transferred and mixed with the same volume of methanol. After this, it was treated with ultrasound for 10 minutes. And then the sample was placed on table for 10 minutes. After centrifuging at 29000 g for 5 minutes, the supernatant was passed through the membrane for detection. The blood sample was prepared for tyrosine detection as follows: a certain volume of sample was dissolved in the same volume of 0.59M $HClO_4$. It was mixed for 30 seconds and centrifuged at 29000 g for 5 minutes. The supernatant was collected and passed through the membrane for detection.

2.3.3 Detection methods [5-7]

2.3.3.1 Standard materials

The analytical standard sulcotrione was purchased from Sigma as described as 2.2.2.1. The analytical standard tyrosine was provided by Dikma and had purity≥ 99.5% (the case No. is 0-110, which is valid until August of 2013).
The chromatographic conditions for tyrosine detection were: a mobile phase of acetonitrile:water = 5:95 (v/v); a 250 mm x 4.6 mm C_{18} column with particle size 5 μm; a flow rate of 0.8 mL/min; and a detection wavelength of 210 nm. The chromatographic conditions for sulcotrione detection were: a mobile phase of methanol:0.1 M sodium dihydrogen phosphate and triethylamine buffer = 40 : 60 (v/v); a 250 mm × 4.6 mm C_{18} column with particle size 5 μm; a flow rate of 0.8 mL/min; and a detection wavelength of 254 nm. Tyrosine and sulcotrione were quantified using the extrapolation method. Under the chromatographic conditions suitable for instrument characteristics, tyrosine and sulcotrione were obtained in a linear correlation curve. The linear regression equation for sulcotrione was $y = 20742.17x +655.12$, with a correlation coefficient of 0.9993. The linear regression equation for tyrosine was $y = 207412.2x-0.1092$, with a correlation coefficient of 0.9985.

2.3.3.2 The precision, accuracy and detection limit

The detection precision, accuracy and detection limit for the sulcotrione and tyrosine detection methods were:

parameters	Precision (RSD, %)	Accuracy (recovery, %)	Detection limit
Sulcotrione	4.83-6.79	94.65-109.23	0.044 mg.L^{-1}
Tyrosine	1.70-8.70	89.63-108.22	0.13 umol.L^{-1}

RSD: relative standard deviation.

According to the requirements of the testing methodology, the relative standard deviation (RSD) is less than 10% and the recovery rate is between 90-110%. The results show that HPLC method for sulcotrione and tyrosine detection meets the above requirement to ensure that the experimental data is accurate and reliable.

2.3.4 Statistical analysis

The statistical software SPSS15.0 was used for the one-way ANOVA analysis, least significant difference method (LSD) analysis and Dunnett's T test. The Pearson method is used for linear correlation analysis. The data are presented as mean ± SD.

2.4.The test of rats exposed to sulcotrione for 28 days to determine the time and response relationship between blood sulcotrione and tyrosine and main toxic response.

2.4.1 dose group [8-9]

The original sulcotrione (purity > 95%) was provided by a domestic corporation. The acceptable daily intake (ADI) of 0.005 mg/kg, extrapolated from the results of a sulcotrione rat chronic toxicity test for no observed adverse effect level (NOAEL), was designated as the low-dose group; the medium-dose group was 0.05 mg/kg and the high-dose group was 0.5 mg/kg. A group without sulcotrione exposure was the control group. There were 28 rats in each group (equal numbers of males and females).

2.4.2 Route of exposure, time and test indicators

The rats were fed with sulcotrione dissolved in cooking oil once a day for 28 continuous days. The control group was fed cooking oil only. Blood was collected from the tail vein at days 0, 7, 14, 21, and 28 after sulcotrione exposure. The rats were not fed on the days of blood collection. At day 28, half the male and female rats were sacrificed and the sulcotrione feeding was discontinued for the remaining rats. At days 35 and 42, blood was collected from the tail vein. During the feeding, the body weight, diet, hair and activity of the rats were routinely checked and recorded. After they were sacrificed, the liver, kidneys, adrenal glands and other organs were collected for pathologic study. The organ and body weight ratio was calculated.

2.4.3 The time of administration, sampling and sample preparation.

Sulcotrione was given to rats at the same time each day (around 16:00). The rats were fasted from 21:00 the day before blood collections. The blood was collected at 8:00 the next day. The blood was centrifuged at 3000 g for 15 minutes. The serum was collected and stored at -18°C for cryopreservation. The blood sample was prepared for sulcotrione detection as follows. The serum was thawed at room temperature. An aliquot of serum was added to the same volume of methanol and mixed for 30 seconds. The mixture was treated with ultrasound for 10 minutes and placed on table for another 10 minutes. Then the sample was centrifuged at 29000 g for 5 minutes. The supernatant was collected and filtered through the membrane for detection. The blood sample was prepared for tyrosine detection as follows. The serum was thawed at room temperature. An aliquot of serum was added to the same volume of 0.59 M $HClO_4$ and mixed for 30 seconds. The sample was centrifuged at 29000g for 5 minutes. The supernatant was collected and filtered through the membrane for detection. Blood glucose was directly measured from tail vein blood sampling.

2.4.4 The analysis method and quality control of detection indicators

The rat blood tyrosine and sulcotrione HPLC detection indicators were established. The accuracy, precision and detection limit for the detection method was under quality control (2.3.1). The blood glucose was detected by the Johnson & Johnson Rapid Blood Glucose Detector (USA). The instrument was calibrated with standard reference liquid before blood glucose detection. The original sulcotrione has more than 98.9% active component. The peak area extrapolation method was used for quantitative analysis of tyrosine and sulcotrione.

2.4.5 Statistical analysis

The statistical software SPSS15.0 was used for one way ANOVA analysis, least significant difference method (LSD) analysis and Dunnett's T test. The Pearson method was used for linear correlation analysis. The data are presented as mean ± SD.

2.5. Effects of sulcotrione on hepatic enzymes involved in tyrosine catabolism, tyrosinemia, and blood glucose in rat.

2.5.1 Animals and treatments

Sulcotrione with a purity of > 95.5% w/w, was supplied by Jia Hua Import & Export Co., Ltd (Zhejiang, China). male Alpk:APfSD (Wistar-derived) rats, aged from 5 to 6 weeks and obtained from Zhejiang Experimental Animal Center [SCXK (zhe) 2008-0033], were housed in stainless steel, wire bottom cages under standard housing conditions (controlled atmosphere with 12:12 h light/dark cycles, 55 ± 5% humidity and an ambient temperature of 22 ± 3°C). The rats were fed on a commercial powdered diet (GB 14924-2001) and given filtered water *ad libitum*. Rats were acclimatized for 3 days prior to the experiment. Groups of 8 sulcotrione-treated rats and 8 control rats were dosed with either corn oil alone or sulcotrione in corn oil at 5 ml/kg body weight at 0.1, 0.5, and 5 mg/kg/day for 90 days. Throughout the study, clinical signs were observed, body weight was recorded weekly, and food consumption was monitored twice weekly. Animal care and monitoring were carried out in accordance with strict guidelines issued by the P.R. China legislation. All animal procedures and treatments were performed according to our Institute Animal Care and Use Committee (Certificate No. IACUC-03-001) and animals were terminated when deemed to be under moderate stress or discomfort.

2.5.2 Hepatic enzymes involved in tyrosine catabolism assay

The rats were sacrificed after the 90 days test by inhalation of an overdose of halothane as previously described. Livers were removed and then homogenized with 10 up-down strokes in 30 ml of ice-cold 0.32 M sucrose, and then diluted to give a 20% (w/v) homogenate. The homogenate was then centrifuged at 105,000 g for 60 min at 4°C to remove particulate material. The supernatant (cytosol fraction) was stored in aliquots at -70°C prior to the assay of activities of TAT, HPPD and HGO. The protein content in the supernatant was measured using bovine plasma albumin (BSA) as the internal standard. TAT was assayed in liver cytosol by the method of Schepartz, 1969. HPPD and HGO were measured in liver cytosol by monitoring oxygen consumption after addition of the relevant substrate by the methods described by Ellis et al., 1995, respectively.

2.5.3 Serum epinephrine and tyrosine analysis

After oral glucose tolerance test, femoral artery blood was adopted and then separated the serum. Serum epinephrine and tyrosine were measured by rat epinephrine ELISA kit

(Uscnlife Science & Technology Company, USA) and High Performance Liquid Chromatography (HPLC, SHIMADZU, LC-20AD), respectively.

2.5.4 Fasting blood glucose test and Oral glucose tolerance test

During treatment fasting blood glucose of tail vein in rat were measured in 0, 30, 60, 90 day (Fasting time: from 08:00 to 14:00) using fast blood glucose meter (Johnson & Johnson Services Inc. USA). After final dosing, all rats were fasting for overnight (about 12 hours), and then oral glucose tolerance test was adopted. We first measured blood glucose of tail vein in rat in 0 min, after that immediately giving glucose water solution (2 g/kg b.w) , then blood glucose were measured in 30, 60, 120 min respectively.

2.5.5 Statistical analysis

The differences between sulcotrione-treated animals and controls were analyzed using SPSS Version 13.0 for Windows (SPSS Inc., Chicago, IL, USA). Changes in the concentration of tissue tyrosine or the enzymes involved in tyrosine catabolism in the time following sulcotrione treatment were analyzed using an analysis of variance followed by the Dunnett *t* test. *A* p-value below 0.05 was considered to be statistically significant between experimental groups

3.Results

3.1 Sulcotrione levels in the environment

3.1.1 Sulcotrione soil residual levels

Sulcotrione soil residual levels in the area where the sulcotrione was used in Zhejiang Province are summarized in table1-1 and figure1-1. The mean sulcotrione soil residue was between 0.19-0.47 mg/kg.

area	Quzhou	Wenzhou	Jiande	Dongyang	Haining	Ninghai	Longyou	Jinhua	Total
N	37	20	24	34	10	10	28	37	200
Mean ± sd	0.30±0.18	0.28±0.15	0.35± 0.19	0.33±0.19	0.19± 0.08	0.33±0.18	0.41±0.19	0.47±0.17	0.35±0.19

Table 1.1. Sulcotrione soil residual levels from the sampling area in Zhejiang province (mg/kg)

Figure 1-2 shows a plot of the distribution of the sulcotrione residual levels in the 200 soil samples against the expected normal probability distribution. The scatter graph follows an approximately straight line. So the test data follows a normal distribution. The arithmetic average of the sulcotrione soil residual levels in Zhejiang Province was 0.35 ± 0.19 mg/kg.

3.1.2 Sulcotrione corn residual levels

The sulcotrione residual levels in the 10 corn samples from the Quzhou and Dongyang areas are shown in Table 1-2. The detected values are between 0.02 to 0.10 mg/kg.

Fig. 1-1. The distribution of sulcotrione soil residual levels in Zhejiang Province (mg/kg, n: sample number)

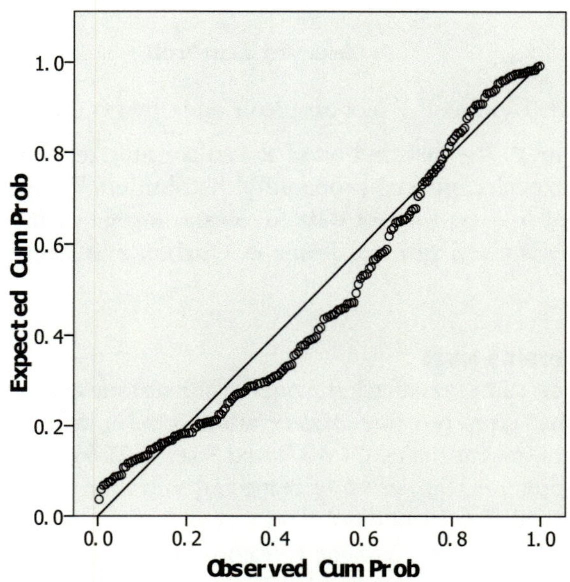

Fig. 1-2. The normal distribution P-P plot for sulcotrione soil residue.

Sample number	1	2	3	4	5	6	7	8	9	10
Average	0.03	0.06	0.05	0.04	0.02	0.08	0.07	0.10	0.05	0.07

Table 1-2. Sulcotrione corn residual levels (mg/kg)

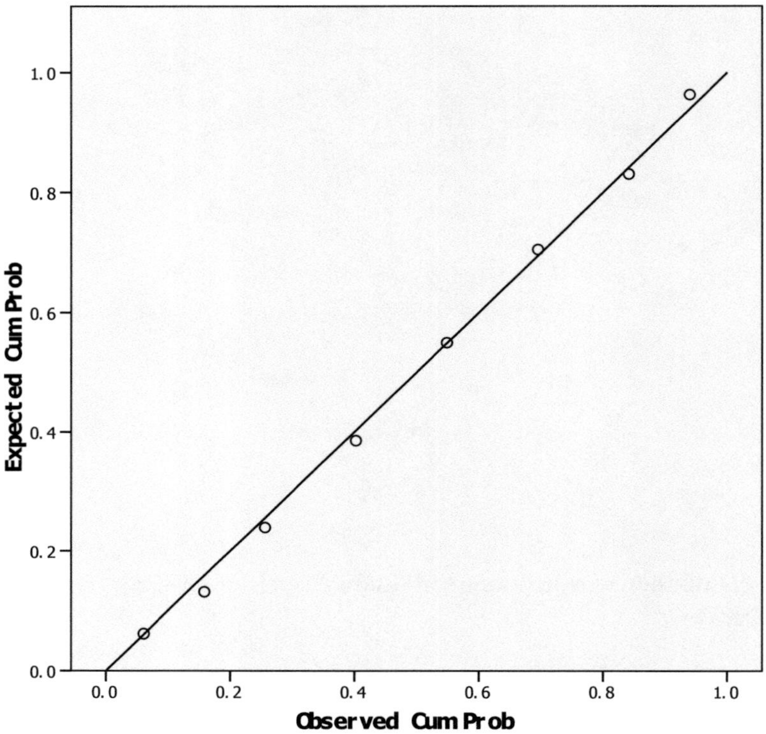

Fig. 1-3. The normal distribution P-P plot of sulcotrione coin residue.

Figure 1-3 shows a plot of the distribution of the sulcotrione residual levels in the 10 corn samples against the expected normal probability distribution. The scatter graph shows an approximately straight line. So the test data follows a normal distribution. The arithmetic average of the sulcotrione soil residual levels in Quzhou and Dongyang was 0.06 ± 0.02 mg/kg.

3.2 Human serum tyrosine level

We had 145 valid blood samples collected from the sulcotrione exposed population and the control population. The serum tyrosine concentration in the sulcotrione exposed population was 94.14 ± 19.67 nmol/ml for males (N = 37) and 98.85 ± 21.66 nmol/ml for females (N = 31). In the control population, the serum tyrosine concentration was 97.60 ± 16.27 nmol/ml for males (N = 35) and 100.2 ± 18.40 nmol/ml for females (N = 35). In laboratory staff who had contact sulcotrione, the serum tyrosine concentration was 82.02 ± 23.26 nmol/ml (male and female, N = 7), compared to 90.36 ± 19.27 nmol/ml (male and female, N = 10) for laboratory staff who did not have contact with sulcotrione.

3.3 The test of rats

exposed to sulcotrione for 28 days to determine the time and response relationship between blood sulcotrione and tyrosine and main toxic response.

3.3.1 The general situation of experimental rats

During the experiment, the body weight, diet, hair, activity in each dose group showed no significant difference from the control group.

3.3.2 The exposed rat liver, kidneys, adrenal glands and body weight ratio and pathological observations.

3.3.2.1 The exposed rat liver, kidney and adrenal gland and body weight ratios.

sex	group	Liver/body	Kidneys/body	Adrenal glands/body
female	control	2.98	0.69	0.03
	Low-dose	2.97	0.68	0.03
	Medium-dose	2.90	0.66	0.03
	High-dose	2.98	0.67	0.03
male	control	3.14	0.70	0.01
	Low dose	2.98	0.68	0.01
	Medium dose	3.00	0.71	0.02
	High dose	3.08	0.71	0.02

All have $p>0.05$ in the comparison with the control group.

Table 2-1. The exposed rat liver, kidney and adrenal gland and body weight ratios.

Table 2-1 indicates that in both male and female rats, the liver, kidneys and adrenal glands and the body ratio are not significantly different than those of the controls.

3.3.2.2 Pathological observations on the liver, kidney and adrenal gland in each dose group.

Although there were a few rats in each dose group that showed inflammatory lesions in the liver and kidneys, there was no significant difference in comparison with the control group.

3.3.3 The serum tyrosine level changes in the male and female rats in each dose group after different exposure times.

At different sulcotrione exposure times, the rat serum tyrosine levels in the medium- and high-dose groups were significantly higher than those in the low-dose and control groups. The serum tyrosine levels in the high-dose group were significantly higher than those in the medium-dose group. The serum tyrosine concentrations[figure3-1] in the medium- and high-dose groups increased with prolonged exposure before day 21 (the absorption phase). Then they remained relatively stable (the stable phase). After day 28, when exposure ceased, they began to decrease until day 42 (14 days of no exposure). However, they were still higher than the levels of the control at day 42. There were no significant differences in serum tyrosine levels between male and female rats.

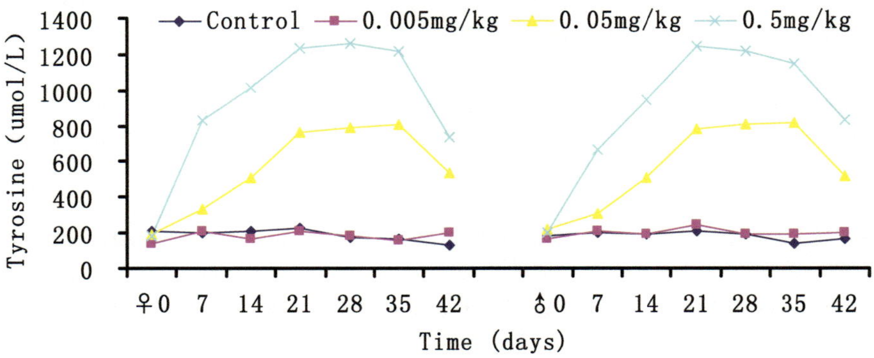

Fig. 3-1. The serum tyrosine level changes in the male and female rats in each dose group after different exposure times.

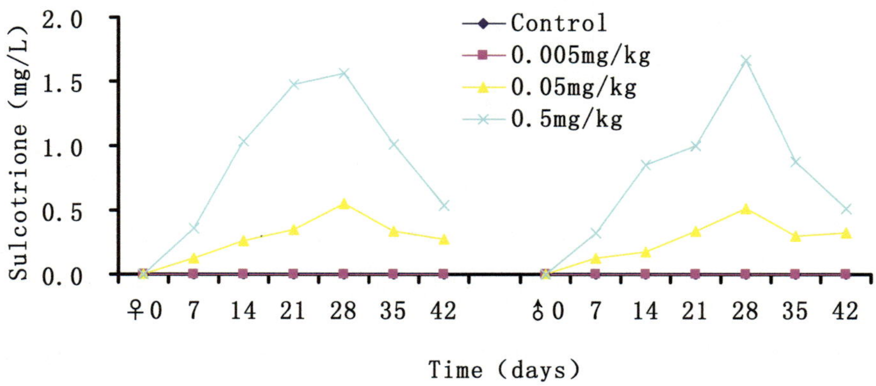

Fig. 3-2. The serum sulcotrione level changes in the male and female rats in each dose group after different exposure times.

The serum sulcotrione in the low-dose group was similar to that in the control group. The serum sulcotrione levels in the males and females were both below the detection limit. The serum sulcotrione was detected in the medium- and high-dose groups, and its level increased with increasing exposure over time and had a statistically significant difference from control. The serum sulcotrione concentration in the medium- and high-dose groups increased with prolonged exposure[figure3-2], but decreased significantly after no exposure. There were no significant differences between male and female rats.

Although there existed individual statistical differences in blood glucose levels between the dose groups and the controls, the values of the changes were all within the normal range [figure3-3]. There were no significant differences in blood glucose changes between male and female rats.

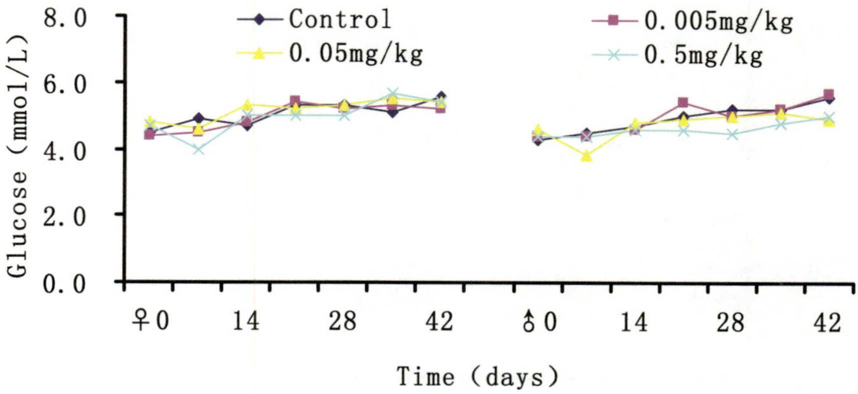

Fig. 3-3. The blood glucose level changes in the male and female rats in each dose group after different exposure times.

3.3.4 The relationship between serum sulcotrione and tyrosine

The scatter plots show that between day 7 and 28 of exposure, the serum tyrosine levels increased with the increasing serum sulcotrione in the medium- and high-dose groups. They show a dose-response relationship. There was a positive correlation [figure3-4,figure3-5]between serum sulcotrione and tyrosine that was statistically significant ($P<0.01$). This correlation is stronger in the female rats than in the males.

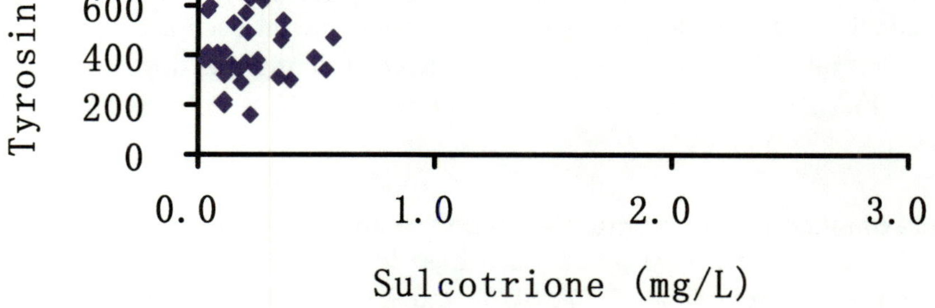

Fig. 3-4. The correlation between serum sulcotrione and tyrosine in female rats from the medium- and high-dose groups.

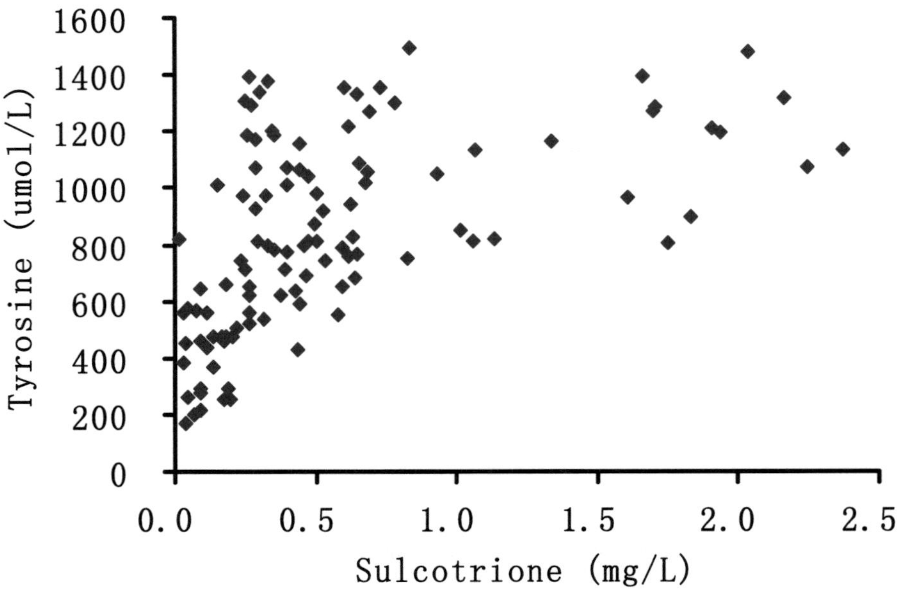

Fig. 3-5. The correlation between serum sulcotrione and tyrosine in male rats from the medium- and high-dose groups.

3.4. Effects of sulcotrione on hepatic enzymes involved in tyrosine catabolism, tyrosinemia, and blood glucose in rat.

3.4.1 Physical observation and mass gain

No sign of toxicity was observed in sulcotrione-treated animals until end of experiment and there were no changes in mass gain between controls and sulcotrione-treated animals (date not shown).

3.4.2 Effects of sulcotrione on hepatic enzymes involved in tyrosine catabolism

The activities of the hepatic enzymes TAT, HPPD and HGO examined after 90 days were shown at table 4-1. The activity of HPPD was dramatically reduced in liver cytosol by 90%, 92% and 95% at doses of 0.1, 0.5 and 5 mg/kg/day when compared with controls. Significantly higher activity of hepatic TAT (43%, 50%, 94%), the rate limiting enzyme in tyrosine catabolism, was evident in sulcotrione-treated rats at each dose as compared to controls. In contrast, the activity of hepatic HGO at 5 mg/kg/day was significantly decreased by 40% when compared to that of controls, and was not altered at 0.1 and 0.5 mg/kg/day.

3.4.3 Effects of sulcotrione on Serum epinephrine and tyrosine

Administration of sulcotrione to rats at each dose level for 90 days induced tyrosinemia in rats. There were approximately 717%, 817%, and 930% increases in serum tyrosine levels at 0.1, 0.5 and 5 mg/kg/day when compared to those of controls (Figure 4-2) and had no effect on their serum epinephrine concentration when compared to the control group (Figure 4-3).

Dose (mg/kg)	Enzyme activity[a]		
	Tyrosine aminotransferase (TAT) (nmol 4-hydroxyphenylpyruvate formed/min/mg protein)	4-Hydroxyphenylpyruvate dioxygenase (HPPD) (μl O_2 consumed/min/mg protein)	Homogentisic acid oxidase (HGO) (μl O_2 consumed/min/mg protein)
0	15.15 ± 1.22	1.55 ± 0 .07	1.14 ± 0.06
0.1	19.69 ± 0.44[b]	0.16 ± 0.02[b]	1.20 ± 0.05
0.5	24.53 ± 0.73[b]	0.12 ± 0.05[b]	0.96 ± 0.03
5	30.49 ± 0.61[b]	0.08 ± 0.03[b]	0.68 ± 0.05[b]

[a] Values are mean ± SE with at least six animals at each dose level.
[b] $P < 0.001$ compared to controls.

Table 4-1. Effects of sulcotrione on hepatic enzymes involved in tyrosine catabolism

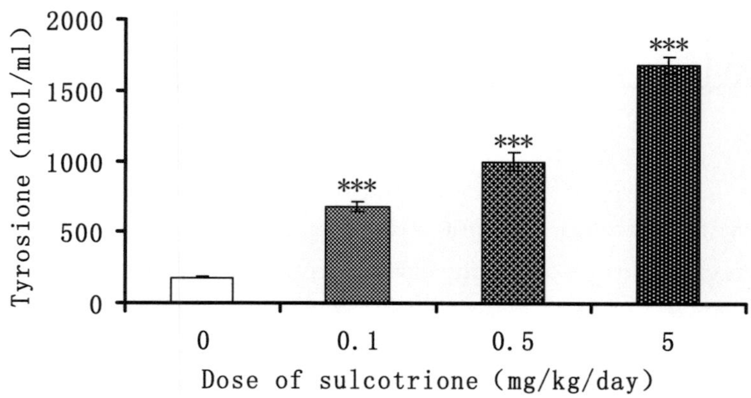

Values are mean ± SE with at least six animals at each dose level, *** $P < 0.001$ compared to controls.

Fig. 4-2. Effects of sulcotrione on Serum tyrosine

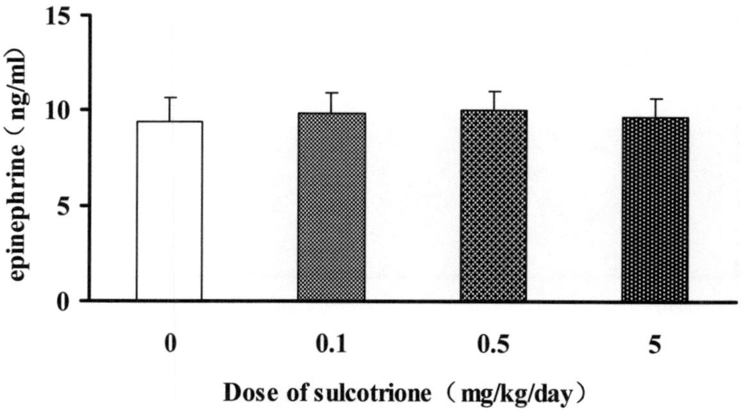

Values are mean ± SE with at least six animals at each dose level.

Fig. 4-3. Effects of sulcotrione on Serum epinephrine

3.4.4 Effects of sulcotrione on Fasting blood glucose and Oral glucose tolerance

Fasting blood glucose concentration and oral glucose tolerance concentration at each dose in the time as shown at Figure 4-5, and there were not significantly different between controls and sulcotrione-treated animals.

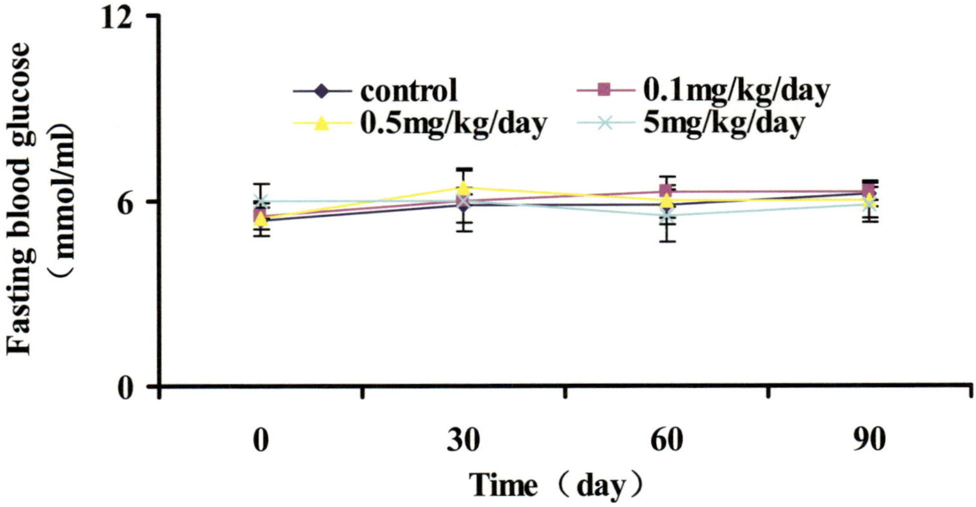

Values are mean ± SE with at least six animals at each dose level.

Fig. 4-4. Effects of sulcotrione on Fasting blood glucose

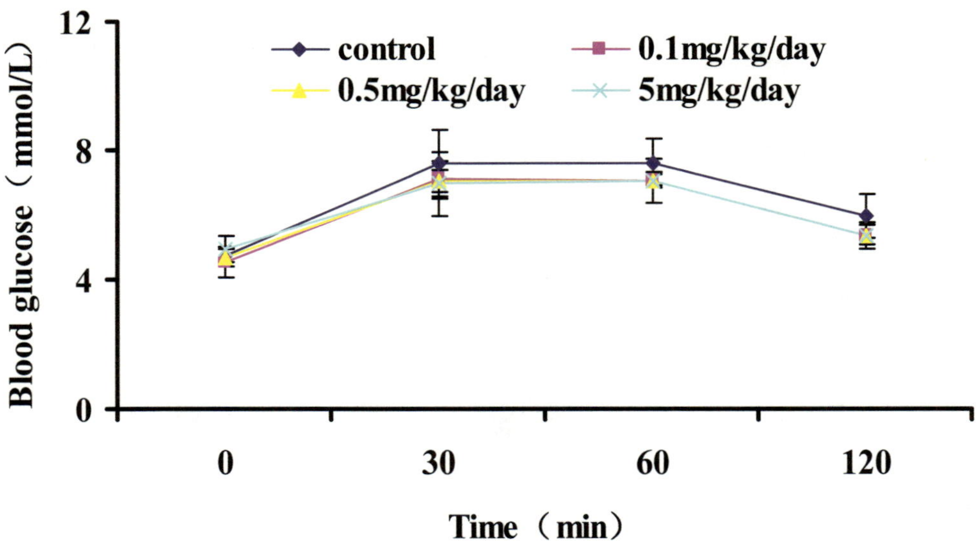

Values are mean ± SE with at least six animals at each dose level.

Fig. 4-5. Effects of sulcotrione on Oral glucose tolerance

3.4.5 Effects of sulcotrione on Serum tyrosine

Corneal lesions were observed in a few rats given sulcotrione at 5 mg/kg/day administration of sulcotrione for 90 days. The sulcotrione-treated rats showed corneal lesions that varied in severity from partial or hazy opacity to complete opacity, with edema and neovascularization evident in the more extensive lesions [Fig.4-6].

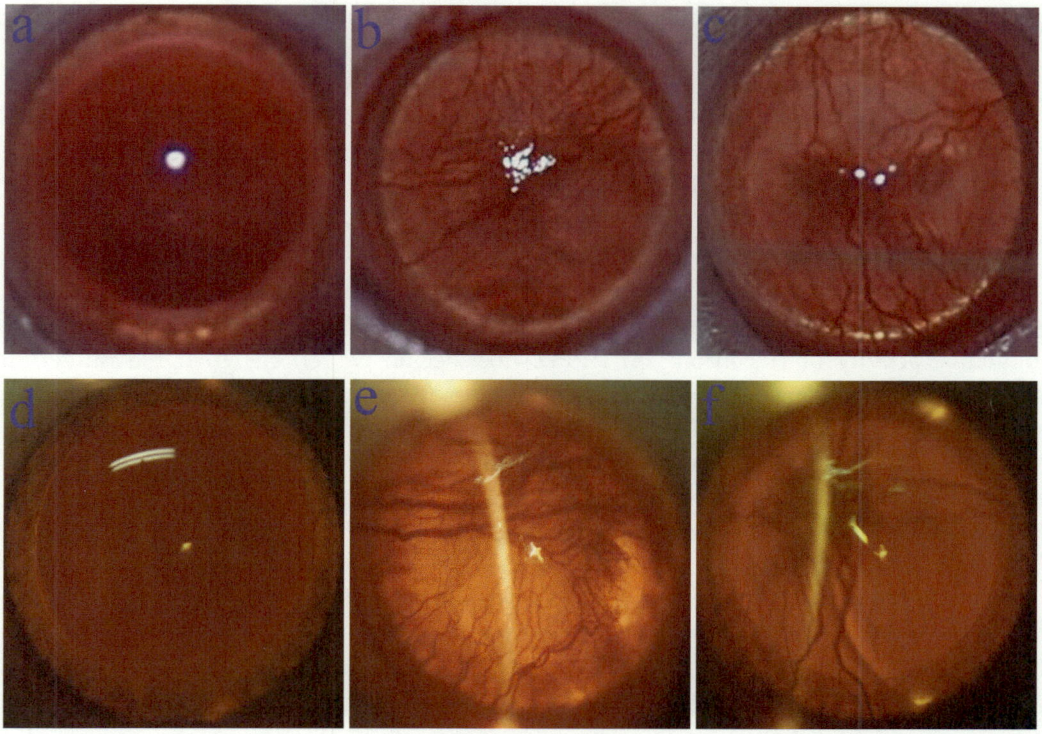

Fig. 4-6. Digital and slit lamp microscope photographs of normal and diseased rat eyes (5 mg/kg/day) after oral administration of sulcotrione for 90 days. a and d normal control rat cornea shows a uniformly bright surface and the pupil is clearly visible. b and e opacities over nearly the entire corneal surface and with neovascularization. c and f opacity of the corneal surface giving it a roughened appearance and edema of the stroma are maximal at the center and less so.

4. Discussion

The half-life of soil sulcotrione is up to 122 days [11-14]. The factors that affect the degradation of soil sulcotrione include soil pH, temperature, humidity and soil type. Sulcotrione has a relative shorter residence time in alkaline soil due to the very weak adsorption capacity of alkaline soil for sulcotrione [11]. The degradation rate is very slow in soil with low pH. Red to yellow soil is the main soil type and soil source in Zhejiang Province, located in the western Zhejiang Jinhua and Quzhou area. Cinnamonic soil is mainly located in the northern Zhejiang Jiaxing area. Red to yellow soil belongs to acidic soil, while cinnamonic soil belongs to neutral to slightly alkaline soil. Sulcotrione will be degraded quickly in soil in which microorganisms grow well. In low temperature and high humidity conditions, the soil adsorption of sulcotrione will increase and the degradation will decrease [15]. The soil adsorption capacity is correlated with soil particle size and organic matter content. Soil containing more organic content has greater herbicide adsorption capacity. Wilson and Foy indicated that the adsorption of sulcotrione was correlated with the organic content in the soil [16]. Sandy loam and clay has the strongest adsorption capacity for sulcotrione, followed by sandy clay and sandy soil. After adsorption by soil, sulcotrione is gradually released and does not disappear easily [17].

The sulcotrione soil residual level is affected by many factors. This investigation of the current status of sulcotrione soil residual levels aims to provide basic data for use in effective methods of mitigating the environmental damage caused by residual herbicide in the soil during the wide used of the new herbicide.

HPPD is the target molecule of sulcotrione. Its activity can directly reflect the inhibition strength of sulcotrione. At present, the enzymatic activity of HPPD is indirectly and quantitatively measured by the tissue oxygen consumption method, which is somewhat unstable. In this study, the detection of the serum tyrosine level can indirectly reflect the inhibitive activity of sulcotrione on HPPD. Lock, et al. [9] reported that the rat serum tyrosine concentration could be as much as 10 times the normal value after 24 hours when rats were given the sulcotrione analogue 2-nitro-4-trifluoromethylbenzoyl)-1,3-cyclohexanedione (NTBC) in a single dose of 0.5 mg/kg. It also caused cornea damage. In this study, corneal opacity in individual rats was also found, which might be related to the accumulation of excessive tyrosine in the anterior chamber of eye. There are three metabolic pathways for tyrosine in mammals: 1) Tyrosine is transferred into 4-hydroxyphenylpyruvic acid (HPPA), catabolized by tyrosine transaminase. It becomes homogentisate after decarboxylation, which is then broken down into acetyl acetate coenzyme A and fumaric acid for the TCA cycle. 2) Tyrosine, catabolized by tyrosine hydroxylase, is transferred into 3,4-dihydroxy phenylalanine (L-DOPA), which is then catabolized by DOPA decarboxylase and converted into dopamine. Dopamine forms norepinephrine after hydroxylation of a carbon atom by dopamine β hydroxylase. Finally, epinephrine is formed after methylation of norepinephrine. 3) Tyrosine, catabolized by tyrosine hydroxylase, is transferred into L-DOPA, which is then converted into dopaquinone by tyrosinase. The dopaquinone then spontaneously forms melanin. HPPD can transform HPPA into homogentisate by adding oxygen. The transformation from HPPA into homogentisate is inhibited by sulcotrione because it represses the activity of HPPD. Thus, sulcotrione inhibits the first metabolic pathway of tyrosine and induces the accumulation of serum tyrosine in the body, which constitutes the toxicity pathway of sulcotrione. Within the dosages used in this study, the tyrosine level is not high enough to affect blood glucose metabolism through the secondary pathway; or perhaps, even if all tyrosine is metabolized through the secondary pathway, it may still not be enough to affect glucose metabolism.

The results suggest that the metabolism of sulcotrione, although very fast, still has a sustainable impact on tyrosine. With the decreasing sulcotrione burden in the body, the repressed HPPD may be reactivated and the tyrosine metabolic pathway gradually becomes normal. This hypothesis is consistent with the results reported by Ellis et al. [18], who found that NTBC was a potentially reversible inhibitor of HPPD.

5. References

[1] Shaoquan Su, HPPD - New Target for Herbicide Development[J]. PESTICIDES, 2000, 39(5): 4-7

[2] Sehng Guo, Fuming Yang, Lin Zhang, The synthesis of herbicide sulcotrione [J]. PESTICIDES, 2001, 40(7): 20-22

[3] Zhizhong Sun, Wenyi Chu, Yanjun Hou, HPLC determination of sulcotrione[J]. JOURNAL OF NATURAL SCIENCE OF HEILONGJIANG UNIVERSITY, 2004, 21: 108-109

[4] JIN F, wu NX. Determination of sulcotrione in serum by HPLC[J]. Occup Health, 2008, 24(9): 843—844.

[5] Tang AG. Rapid high performance liquid ehromatography for determination of phenylalanine and tyrosine in serum[J]. Bull Hunan Med Univ, 2000, 25(2): 209—212.

[6] Mo XM, Tang AG. Determination of phenylalanine and tyrosine in finger blood by high-performance liquid chromatography with ultraviolet detection[J]. Chin J Birth Health Heredity, 2005, 13(3): 29—30.

[7] Zhang D, Chen H. RP—HPLC determination of glycyl-L-tyrosine and its related substance[J]. Chin J Pharm Anal, 2006, 26(5): 664—666.

[8] Hall MG, Wilks MF, Provan WM, Eksborg S, Lumholtz B. Pharmaeokineties and pharmacodynamics of NTBC(2-(2-nitro-4-fluoromethylbenzoy1)-1,3-cyclohexanedione) and mesotrione, inhibitors of 4-hydroxyphenyl pyruvate dioxygenase(HPPD)following a single dose to health male volunteers[J]. J Clin Pharmacol, 2001, 52(2): 169—177.

[9] Lock EA, Gaskin P, Ellis MK, Provan WM, Robinson M, Smith LL, et a1. Tissue distribution of 2-(2-nitro-4-trifluoromethylbenzoy1)cyclohexane-1-3-dione(NTBC): effect on enzymes involved in tyrosine catabolism andrelevance to ocular toxicity in the rat[J]. Toxieol Appl Pharmacol, 1996, 141(2): 439—447.

[10] Walker A, Brown PA. The relative persistence in soil of five acetanilide herbicides [J]. Bull Environ Contam Toxicol, 1985, 34: 143-149

[11] Allen R, Walker A. The influence of soil properties on the rates of degradation of metamitron, metazachlor and metribuzin[J]. Pestic Sci, 1987, 18: 95-111

[12] Rouchaud J, Neus O, Callens D et al. soil persistence and mobility in summer maize and winter wheat crops [J]. Weed Res, 1998, 38:361-371

[13] Baer U, Calvet R. Fate of soil applied herbicides: experimental data and prediction of dissipation kinetics [J]. J Environ Qual, 1999, 28: 1765-1777

[14] Beulke S, Malkomes HP. Effects of the herbicides metazachlor and dinoterb on the soil microflora and the degradation and sorption of metazachl or under different environmental conditions[J]. Biol Fertil Soils, 2001, 33:467-471

[15] Laure Mamy, Enrique Barriuso, Benoıt Gabrielle. Environmental fate of herbicides trifluralin, metazachlor, metamitron and sulcotrione compared with that of glyphosate, a substitutebroad spectrum herbicide for different glyphosate-resistant crops [J]. Pest Manag Sci, 2005, 61: 905-916

[16] Wilson S, Foy C. Influence of various soil properties on the absorption anddesorption of ICIA-0051 in five soils [J]. Weed Technol, 1992, 6: 583-586.

[17] Moshier LJ, Penner D. Factors influencing microbial degradation of 14C-glyphosate to 14CO2 in soil [J]. Weed Sci, 1978, 26: 686-691

[18] Ellis MK, Whitfield AC, Gowans LA, Auton TR, Pmvan WM, et al. Inhibition of 4-hydroxyphenylpyruvate dioxygenase by 2-(2-nitro-4-trifluommethylbenzoy1)cyclohexanel, 3-dione and 2-(2-nitro-4—methanesulphonyl benzoy1)-cy-elohexanel, 3-dione[J]. Toxicol appl Pharmacol, 1995, 133(1): 12—19.

25

The Hemodynamic Effects of the Formulation of Glyphosate-Surfactant Herbicides

Hsin-Ling Lee and How-Ran Guo
National Cheng Kung University
Taiwan

1. Introduction

1.1 Epidemiology of Glyphosate poisoning in Taiwan and other countries

Glyphosate ([N-(phosphonomethyl) glycine], CAS Number 1017-83-6) is the active ingredient of Roundup®, a common nonselective weed control agent. A variety of glyphosate-based formulations are registered in many countries under different brand names. The glyphosate–surfactant herbicide (GlySH) is usually a formulated commercial product containing glyphosate salts, such as isopropylamine, diammonium, potassium, trimesium, or sesquisodium salt. A GlySH commonly used in Taiwan contains 41% glyphosate as the isopropylamine salt (CAS Number 38641-94-0), water, and a variable amount of surfactant. The main surfactant used in GlySH products worldwide is polyoxyethyleneamine (CAS Number 61791-26-2). GlySH, an alternative to paraquat, has been used in suicide attempts in Taiwan and many countries in the Asia-Pacific region (Sawada et al., 1988; Menkes et al., 1991; Tominack et al., 1991; Talbot et al., 1991; Hung et al., 1997; Lee et al., 2000; Stella and Ryan, 2004; van der and Konradsen, 2006; Lee et al., 2008; Roberts et al., 2010). The case fatality rates were around 1.9 to 16 % (Sawada et al., 1988; Tominack et al., 1991; Talbot et al., 1991; Hung et al., 1997; Lee, et al., 2000; Suh et al., 2007), and a large study by the Poison Control Center (PCC) of Taiwan, which included 2186 cases of GlySH poisoning from 1986-2007, reported a case fatality rate of 7.2% (Chen et al., 2009). However, a much higher fatality rate up to 29.3% has been found in a recent study (Lee et al., 2008). Obviously, it continues to be a public health problem that calls for concerns.

1.2 Metabolism of glyphosate

Glyphosate is a nonselective herbicide that inhibits plant growth through interference with the production of essential aromatic amino acids by inhibition of the enzyme enolpyruvylshikimate phosphate synthase, which is responsible for the biosynthesis of chorismate, an intermediate in phenylalanine, tyrosine, and tryptophan biosynthesis (Williams et al., 2000). The absence of this biosynthetic pathway in mammals may explain the relatively low systemic toxicity of glyphosate (oral median lethal dose [LD50] for rats 4,320 mg/kg, rabbits 3,800 mg/kg) (Smith and Oehme, 1992). In the terrestrial environment, glyphosate is mainly biodegraded to aminomethylphosphonic acid (AMPA) when metabolized by bacterial in soils (Rueppel et al., 1977). According to the animal study in Sprague-Dawley rats, approximately 35-40% of the administered dose was absorbed from the gastrointestinal tract, and urine and feces were equally important routes of elimination after one oral dose (10 mg/kg) (Brewster et al., 1991). The animal study indicated that virtually no toxic metabolites of glyphosate were produced when it was administrated orally and that there was little evidence of metabolism (Müller et al., 1981). Essentially 100% of the body burden was the parent compound (Müller et al., 1981).

1.3 Systemic toxic syndrome of GlySH poisoning

Although GlySH is considered to be only slightly toxic to rats, ingestion of a substantial volume of GlySH has been reported to be associated with toxic effects, including gastrointestinal injury, laryngeal injury, pulmonary toxicity, impaired renal and liver functions, leukocytosis, impaired neurological function, dermatitis, metabolic acidosis, arrhythmias, myocardial depression, shock, and even death in humans (Sawada et al., 1988; Talbot et al., 1991; Tominack et al., 1991; Hung et al., 1997; Lin et al., 1999; Lee et al., 2000; Lee et al., 2008; Roberts et al., 2010). Although symptoms and signs of various organ systems could be seen clinically, the definite mechanism of systemic toxic syndrome in acute GlySH poisoning is still unclear. Aspiration pneumonitis and upper respiratory tract irritation are commonly reported findings (Tominack et al., 1991; Talbot et al., 1991; Hung et al., 1997). Hung et al. (1997) strongly suspected that severe laryngeal injury is the primary mechanism of respiratory aspiration and the leading cause of morbidity and mortality following GlySH intoxication. Previous animal studies in rats showed that intratracheal administration of GlySH produced more severe lung damages than oral administration (Martinez and Brown, 1991; Adam et al., 1997). They implied that at least some of the clinical manifestations are related to an aspiration complication. However, pulmonary hemorrhage and other systemic insults could also be seen in animals with oral administration of various components of GlySH (Martinez et al., 1990). Other mechanisms should be considered in explaining the impacts of GlySH on pulmonary and other systems.

1.4 The toxic mechanism of glyphosate and GlySH on mitochondria

Uncoupling of mitochondrial oxidative phosphorylation on rat liver mitochondria has been proposed as a lesion in glyphosate poisoning (Bababunmi et al., 1979; Olorunsogo et al., 1979a). These animal studies showed that the respiratory control ratios of liver mitochondria and state 3 respiration were significantly reduced. Enzyme inhibition of the Kreb's cycle and the uncoupling effect were also shown in the study of plant's mitochondria (Olorunsogo et al., 1979b; Olorunsogo et al., 1980). A study also showed that glyphosate enhanced mitochondrial ATPase with dose-dependent response (Olorunsogo et al., 1979b). The evidences suggested that glyphosate is an uncoupler of electron transport chain. In the study by Olorunsogo (1990), glyphosate significantly increased the permeability of the mitochondrial membrane to protons and to Ca^{2+} in liver mitochondria, and the author suggested that glyphosate may be able to act both as a chelator and a mild protonophore. The author also found that glyphosate had an inhibitive effect on energy-dependent transhydrogenase reaction in isolated rat liver mitochondria (Olorunsogo, 1982). In rats given glyphosate intragastrically for 2 weeks, glyphosate decreased the hepatic level of cytochrome P450 and monooxygenase activities, as well as the intestinal activity of aryl hydrocarbon hydroxylase (Hietanen et al., 1983). Even though most of the above studies claimed that glyphosate was tested, but actually used the isopropylamine salt of glyphosate (IPAG) (Bababunmi et al., 1979; Olorunsogo et al., 1979b; Olorunsogo, 1982; Hietanen et al.,

1983), those studies still implied that mitochondria may be a critical target in the toxic mechanisms of GlySH. However, the clinical significance of the relationship between these biochemical abnormalities and the systemic toxic syndrome is unclear. Further investigation should be conducted to clarify the possible toxic mechanisms in animal and human GlySH intoxication.

2. Studies for GlySH poisoning

It is within the context of the above background information that the two studies were undertaken. We first conducted a retrospective case-control study in a medical center to identify predictors of GlySH poisoning related fatality. On the basis of our data, among the clinical symptoms that GlySH intoxicated patients may present, the toxic symptoms on the cardiovascular system interested us. We then established an animal model to study the cardiovascular effects induced by each component of GlySH formulation, clarifying which one is responsible for the toxic symptoms.

3. Clinical outcomes and predictors of GlySH poisoning related fatality

In this section, we describe a retrospective case-control study accessing clinical outcomes and identifying the predictors of GlySH poisoning related fatality.

3.1 Study design

This was a retrospective study of patients with GlySH poisoning presenting to the emergency department (ED) of a referral center in a large agricultural area with approximately 2 million residents in southern Taiwan over a seven-year period. The ED's annual patient visits census is about 51,000. All the medical records of patients with GlySH poisoning following oral ingestion who presented to the ED of the referral center from June 1988 to December 1995 were reviewed.

3.2 Study protocol

We collected data on the date of admission, age, sex, estimated amount of GlySH ingested, co-ingestants of other agrochemicals, ethanol, or pharmaceuticals, suicide attempts, out-of hospital interval, initial clinical presentation, initial laboratory data in the ED, and clinical course. Laboratory variables that were reviewed included arterial blood gas (ABG), blood urea nitrogen (BUN), creatinine, alanine aminotransferase (ALT), aspartate aminotransferase (AST), bilirubin, sodium, potassium, calcium, phosphate, white blood cell (WBC) count, hematocrit, platelet, urine analysis, chest x-ray (CXR), and electrocardiogram (ECG). Only the laboratory studies done immediately upon the patients' arrival were taken into consideration. There were some patients who had received first aid and were then transferred from other EDs. For these patients, we used the primary data from those EDs. For clinical and statistical consideration, patients whose serum pH values < 7.35 on the ABG were considered to be "acidotic." Of note, the clinical practice at this hospital was to routinely obtain toxicological screens of other pesticides, such as paraquat and organophosphates, and screens of benzodiazepines.

We also performed specific tests according to the history offered by patients themselves, friends, or family members. The amount ingested was usually given in descriptive terms such as "a mouthful," "a small cup," or "half a bottle." For statistical purposes, we assigned a volumetric value to each description: 5 mL for "a little" or "a spoon," 25 mL for "a mouthful," and 100 mL for "a small cup." If the patient said "a bottle," the size was identified as being 150 mL, 300 mL, 500 mL, or 1 liter, according to the brand, empty bottles carried by family members or friends, or the description by family members or friends.

3.3 Data analysis

All analysis was performed using SPSS statistical software Version 6.03 (SPSS Inc., Chicago, IL). For univariate analysis, we used the Student t and Wilcoxon tests for continuous variables and the chi-square and Fisher's exact tests for categorical variables. We also calculated the odds ratio (OR) and associated 95% confidence interval (C.I.) for each variable. A p-value of less than 0.05 was considered statistically significant. Variables with ORs more than 5 were considered to be major prognostic predictors. All major prognostic variables were further evaluated by multiple logistic regression analyses with the stepwise approach. A patient's probability of survival (Ps) could then predicted using the logistic regression model $Ps = 1/(1 + e^{-b})$ where $b = b_0 + b_1 \times$ risk factor I + $b_2 \times$ risk factor II + $b_3 \times$ risk factor III ... + $b_N \times$ risk factor N.

3.4 Results

From June 1988 to December 1995, 131 patients presented to the hospital with GlySH ingestion, including 69 men and 62 women. There were 11 fatalities, yielding a fatality rate of 8.4%. The most common presentations included sore throat, nausea (with or without vomiting), and fever (Table 1).

Table 2 shows the initial laboratory data of patients. The most common laboratory abnormalities included leukocytosis (WBC count > 10^4/uL; 85/125, 68%), lowered bicarbonate (HCO_3^- < 22 mEq/L; 39/81, 48.1%), acidosis (serum pH < 7.35, 29/81, 35.8%), elevated AST (> 40 U/L; 32/108, 33.6%), hypoxemia (PO2 < 60 torr while breathing room air; 23/81, 28.4%), and elevated BUN (> 21 mg/dL; 21/123, 17.1%).

Of the 81 patients who had 12-lead electrocardiograms, 15 showed abnormal findings. The most frequent abnormalities were sinus tachycardia and nonspecific ST-T changes. Of 29 the patients who had serum pH < 7.35, 13 had metabolic acidosis, 1 had respiratory acidosis, and 15 had mixed-type acidosis. Of the 105 patients who had CXR, 22 revealed abnormal infiltrates or patches. Three patients had renal failure that necessitated hemodialysis, and all resulted in fatalities. Seven patients had co-ingestants, including sedative drugs (2), hypnotics (3), wine (3), and paraquat (1). The average survival time of the fatality cases was 2.8 ± 0.8 days.

Comparisons of clinical variables and laboratory data on arrival between survivors and fatalities are presented in Tables 1 and 2. The mean ± standard errors of the means (SEM) age of the survivors was 47 ± 2 years, while that of the fatalities was 60 ± 4 years (p = 0.02). No difference was found in the distributions of genders. The estimated amount of GlySH ingested averaged 122 ± 12 mL among the survivors and 330 ± 42 mL among the fatalities (p < 0.001). The mean out-of-hospital time among the survivors was longer than that in fatalities (Table 1), but the difference was not statistically significant.

Of the 17 variables identified as major prognostic predictors (Table 3), respiratory distress necessitating intubation, respiratory distress, renal dysfunction necessitating hemodialysis, abnormal CXR, shock, larger amount of ingestion (> 200 mL), altered consciousness, hyperkalemia, and pulmonary edema were associated with the largest ORs. Only the cases

Variable	Survivors (n=120) (%)	Fatalities (n=11) (%)	Total (n=131) (%)	p^b
Age (year)[a]	47 ± 2	60 ± 4	48 ± 2	0.02*
Gender (male/female)	62 / 58	7 / 4	69 / 62	0.47
Out-of-hospital interval (hr)[a]	4.0 ± 0.5	2.2 ± 0.4	3.8 ± 0.4	0.57
Estimated Ingested Amount (mL)[a]	122 ± 12	330 ± 42	138 ± 12	< 0.001*
Fever	48/120 (40.0)	6/11 (54.5)	54/131 (41.2)	0.36
Nausea and/or vomiting	88/118 (74.6)	5/8 (62.5)	93/126 (73.8)	0.43
Sore throat	96/118 (81.4)	5/9 (55.6)	101/127 (79.5)	0.08
Diarrhea	25/120 (21.0)	1/10 (9.1)	26/131 (19.1)	0.69
Respiratory distress	19/120 (15.8)	11/11 (100.0)	30/131 (22.9)	< 0.001*
Altered consciousness	19/120 (15.8)	10/11 (90.9)	29/131 (21.3)	< 0.001*
Respiratory distress necessitating intubation	7/120 (5.8)	11/11 (100.0)	18/131 (13.7)	< 0.001*
Pulmonary edema	2/119 (4.2)	4/11 (36.4)	6/130 (4.6)	< 0.001*
Abnormal CXR	15/98 (15.3)	7/7 (100)	22/105 (21.0)	< 0.001*
Shock	5/119 (4.2)	8/11 (72.7)	13/130 (10.0)	< 0.001*
Dysrrhythmia	9/71 (12.7)	6/10 (75.0)	15/81 (18.5)	< 0.001*
Renal dysfunction necessitating hemodialysis	0/120 (0.0)	3/11 (27.0)	3/131 (27.0)	< 0.001*
Suicide attempt	105/120 (17.5)	11/11 (100.0)	116/131 (88.5)	0.36

[a]Data are expressed as mean ± standard errors of the means.
[b]P values are for comparisons between survivors and fatalities.
*$p < 0.05$ is significant.
Data from Lee et al, 2000.

Table 1. Clinical variables on arrival at the emergency department among patients.

with complete data were used for the multiple logistic regression analysis, and we identified three significant independent predictors of survival, which could be applied to construct a logistic regression model as follows:

$$Ps = 1/(1+ e^{-b}) \quad (1)$$

$$b= -216.93 - 5.10 \times [\text{acute pulmonary edema}] - 1.80 \times [K] + 31.26 \times [pH] \quad (2)$$

Using Ps = 0.25 as the cutoff for predicting fatalities, we obtained a sensitivity of 100% and a specificity of 95.7%. Because pulmonary edema is a binary response, the above formula can be simplified as the following:

1. When pulmonary edema is absent, $31.26 \times [pH] - 1.80 \times [K] < 215.83$ predicts fatality.
2. When pulmonary edema is present, $31.26 \times [pH] - 1.80 \times [K] < 220.93$ predicts fatality.

Variables	Survivors (n=120)	Fatalities (n=11)	*p*
Complete blood count			
WBC (10^4/uL)	13.4 ± 0.5	18.5 ± 2.5	< 0.01*
Hematocrit (%)	42.0 ± 0.5	45.3 ± 1.5	0.07
Platelet count (10^3/cmm^3)	265 ± 9	239 ± 30	0.39
Biochemical data			
Urea nitrogen (mg/dL)	16 ± 1	19 ± 3	0.26
Creatinine (mg/dL)	1.0 ± 0.1	1.4 ± 0.2	< 0.01*
Sodium (mmol/L)	141 ± 1	141 ± 2	0.87
Potassium (mmol/L)	3.8 ± 0.1	4.7 ± 0.4	0.06
Chloride (mmol/L)	105 ± 1	103 ± 4	0.74
Total calcium (mg/dL)	9.1 ± 0.1	9.0 ± 0.2	0.79
Phosphate (mg/dL)	3.4 ± 0.1	3.9 ± 0.9	0.56
Total bilirubin (mg/dL)	1.0 ± 0.1	1.2 ± 0.4	0.99
ALT (U/L)	35 ± 3	64 ± 21	0.20
AST (U/L)	37 ± 3	110 ± 44	0.13
Arterial blood gases			
pH	7.39 ± 0.01	7.17 ± 0.05	< 0.001*
PO_2 (mmHg)	75.3 ± 2.6	48.2 ± 7.2	< 0.001*
PCO_2 (mmHg)	36.8 ± 0.8	41.8 ± 4.5	0.65
HCO_3^- (mEq/L)	22 ± 1	15 ± 2	< 0.001*

Data are expressed as means ± SEM, and $*p < 0.05$ is significant. WBC = white blood cell; ALT = alanine aminotransferase, AST = aspartate aminotransferase.
Data from Lee et al, 2000.

Table 2. Initial laboratory data of the patients.

3.5 Conclusion and discussion

3.5.1 Clinical presentations of GlySH poisoning

Clinical presentations of GlySH poisoning varied across studies (Sawada and Nagai, 1987; Kawamura et al., 1987; Sawada et al.,, 1988; Talbot et al., 1991; Tominack et al., 1991; Menkes et al. 1991). An analysis of three retrospective reviews of 246 cases (Sawada et al., 1988; Talbot et al., 1991; Tominack et al., 1991) revealed that patients most frequently presented with nausea and/or vomiting (40%), abdominal pain, and diarrhea (12%) initially, followed by sore throat (41–43%), fever (7%), gastrointestinal mucosal damage (7–43%), transient renal (10–14%) and hepatic (19–40%) dysfunction, metabolic acidosis, pulmonary edema (5–13%), shock (9%), and death (10.5–16.7%). In our study, nausea with or without vomiting (73.8%), sore throat (79.5%), and fever (41.2%) were the most common initial manifestations. We found leukocytosis (68.0%), low bicarbonate (48.1%), acidosis (35.8%), hepatic

Predictors	Fatalities (n = 11)	Survivors (n = 120)	Total (n = 131)	Odds Ratio (95% C.I.)
Respiratory distress necessitating intubation	11/11	7/120	18/131 (13.7%)	348.1 (98.8- ∞)*
Respiratory distress	11/11	19/120	30/131 (22.9%)	119.7 (29.6-484.6)*
Renal failure necessitating hemodialysis	3/11	0/120	3/131 (2.3%)	99.2 (26.4-372.4)*
Abnormal CXR	7/7	15/98	22/105 (21.0%)	80.8 (18.2-359.0)*
Shock (SBP < 90 mmHg)	8/11	5/119	13/130 (10.0%)	60.8 (10.1-435.8)†
Larger amount of ingestion (> 200 ml)	9/10	17/101	26/128 (20.3%)	53.5 (13.6-210.9)†
Altered consciousness	10/11	19/120	29/131 (22.1%)	53.2 (13.6-207.5)*
Hyperkalemia ([K] > 5.5 mmol/L)	4/10	2/118	6/128 (4.7%)	38.7 (4.6-398.6)†
Pulmonary edema	4/11	2/119	6/130 (4.6%)	33.4 (4.1-330.7)†
Elevated creatinine (> 1.5 mg/dL)	4/11	4/116	8/127 (6.3%)	16.0 (2.6-103.3)†
Lowered bicarbonate (HCO_3^- < 22 meq/L)	10/11	29/70	39/81 (48.1%)	14.1 (1.7-311.2)†
Acidosis (pH < 7.35)	9/11	20/70	29/81 (35.8%)	11.3 (1.98-83.3)†
Dysrrhythmia	6/10	9/71	15/81 (18.5%)	10.3 (2.0-56.5)†
Hyperphosphatemia ([P] > 5.0 mg/dL)	2/10	3/95	5/105 (4.8%)	7.7 (6.8-71.4)†
Elevated AST (> 40 U/L)	8/11	32/108	40/119 (33.6%)	6.3 (1.4-32.5)†
Hypoxemia (PO_2 < 60 mmHg)	7/11	16/70	23/81 (28.4%)	5.9 (1.3-28.2)†
Leukocytosis (WBC > 10^4/uL)	10/11	75/114	85/125 (68%)	5.2 (0.6-112.5)†

*Test-based 95% confidence interval for odds ratios.
†Cornfield's 95% confidence interval for odds ratios.
Data from Lee et al, 2000.

Table 3. Major predictors associated with poor patient outcome (odds ratio > 5).

dysfunction (33.6%), hypercapnea (30.9%), hypoxemia (28.4%), and renal insufficiency (17.1%) were the most common laboratory abnormalities. These findings were similar to previous reports of severe intoxications, except that our patients showed a higher prevalence of sore throat, nausea and/or vomiting, fever, acidosis, and diarrhea.

In this study, shock (8/11, 72.7%), respiratory distress necessitating intubation (11/11, 100%), pulmonary edema (4/11, 36.4 %), dysrrhythmia (6/10, 75%), altered consciousness (10/11, 90.9%), and renal dysfunction necessitating hemodialysis (3/11, 27.0%) were major predictors of fatality. Recent studies involving larger numbers of cases also showed that shock, respiratory failure, altered consciousness, and oligouria were more common in the fatal GlySH exposures (Roberts et al., 2010; Chen et al., 2009).

3.5.2 Predictors of GlySH poisoning

In this study, we identified acute pulmonary edema, hyperkalemia, and acidosis as major predictors of poor outcome, which are compatible with most of glyphosate studies in Taiwan. The risk factors of fatality or severity of GlySH exposure have been studied and discussed over the years, including the amount of exposure, hypovolemic shock, intractable shock, Acute Physiology and Chronic Health Evaluation II score, age, male gender, laryngeal injury with aspiration, abnormal chest X-ray, calendar time, reason for exposure, atropine therapy, elapsed time, delayed presentation, number of involved organs, metabolic acidosis, tachycardia, elevated serum creatinine, and high plasma glyphosate concentrations on admission (> 734 ug/mL) (Sawada et al., 1988; Tominack et al., 1991; Talbot et al., 1991; Hung et al., 1997; Lee et al., 2000; Lee et al., 2008; Chen et al., 2009; Roberts et al., 2010). Prognostic predictors can help emergency staff in identifying patients who are expected to deteriorate or die. We recommend that all the patients who are reported to have ingested large amounts of GlySH be carefully observed, especially for those who present with severe respiratory distress, unstable hemodynamics, requiring hemodialysis, pulmonary edema, and old age. The risk of immediate death is much less likely if the patient has no such risk factors on presentation.

4. Cardiovascular toxicity of GlySH poisoning

4.1 Presentation of cardiovascular toxicity in GlySH poisoning

Cardiovascular involvement in GlySH intoxicated patients may include ECG abnormalities such as sinus tachycardia, sinus bradycardia, first degree AV block, as well as shock (Sawada et al., 1988; Talbot et al., 1991; Tominack et al., 1991). Shock is one of poor prognostic signs in severely intoxicated patients (Tominack et al., 1991; Sawada et al., 1988). Sawada and Nagai (1987) reported that shock might be due to intravascular hypovolemia, which responds to fluid resuscitation and vasopressor agents. However, the study by Talbot et al. (1991) did not support the hypovolemic shock because they found shock developed after rehydration. Lin et al. (1999) reported one patient who presented with cardiogenic shock with left-ventricular hypokinesis after drinking about 150 mL of GlySH. Ventricular tachycardia was observed during resuscitation, and the blood pressure responded to neither vasopressor agents nor fluid resuscitation. The patient gradually recovered in the following 16 h, with the restoration of his left-ventricular function. In a beagle dog study, cardiac depression was observed by Roundup and surfactant injection (Tai et al., 1990). These data suggest that the suppression of the cardiac conduction system and contractility, rather than intravascular hypovolemia, plays an important role in the shock induced by acute GlySH poisoning in humans. However, the detailed mechanism of this cardiac involvement has not been demonstrated, not to mention the components responsible for these symptoms.

4.2 The hemodynamic effects of the formulation of GlySH

Because the acid form of glyphosate has low solubility in water (~12 g/L), commercial compositions of glyphosate generally contain glyphosate salts such as isopropylamine (IPA) (CAS Number 75-31-0), diammonium, potassium, trimesium, or sesquisodium salt, in which the acidic glyphosate is neutralized with a base to form the salt and becomes more water-soluble than the glyphosate acid. IPA is a colorless, flammable liquid with a tangy, ammonia-like odor (NFPA, 1997) and is usually used in the synthesis of dyes, pharmaceuticals, insecticides, rubber chemicals, textile-processing agents and other surface active agents (Harbison, 1998). Its oral LD_{50} for rats is 820 mg/kg (Bingham E et al., 2001). In a study of mongrel dogs, an IPA injection showed positive dose-dependent inotropic and chronotropic responses, with increasing myocardial contraction, arterial pressure, and pulse pressure, as well as significantly reduced vascular resistance in the hind leg (Ishizaki et al., 1974). Another study showed that infusion of IPA (2.5 mg/kg per min) produced an initial increase in arterial pressure and heart rate (HR), followed by prolonged hypotension and bradycardia, but lower doses produced only a hypotensive response (Privitera et al., 1982).
The surfactants commonly used in herbicide products serve several purposes, including acting as wetting agents, promoting uniform spread of the herbicide on the leaf surface, and assisting the penetration of glyphosate into the leaf (Bradberry et al., 2004). Polyoxyethyleneamine (POEA) is the surfactant commonly used in GlySH and has an oral LD_{50} of about 1200 mg/kg in rats (Williams et al., 2000), which is considerably more toxic than that of glyphosate itself (EPA, 1993). In human and animal studies, this nonionic polyoxyethylene alkyl group of surfactants is usually considered to be mainly or partly responsible for the toxic effects of various pesticides, inducing gastrointestinal tract, pulmonary, and depressive cardiac effects (Tai et al., 1990; Martinez and Brown, 1991; Koyama et al., 1994; Sawada et al., 1988; Adam et al., 1997). The clinical effects of other components used in GlySH, such as IPA or IPAG have rarely been studied and reported. Therefore, a study was conducted to characterize the major components leading to the cardiovascular failure in cases with GlySH poisoning.

5. The comparative effects of the formulation of GlySH on hemodynamics

In this section, we describe an animal experiment used for exploring the hemodynamic effects induced by the infusion of different components of GlySH formulation.

5.1 Animal model

We used male Landrace piglets (aged 6–8 weeks, body weight 8–15 kg) as the model for the study. The piglets were fasted for one day before surgery. Each piglet was initially sedated with an intramuscular injection of ketamine (20–30 mg/kg; Ketalar® 50 mg/mL, UBI Asia, Hsinchu, Taiwan) and atropine (0.05 mg/kg) and then placed in a supine position on a thermally controlled blanket on an operating table. A percutaneous venous cannula (24G) was placed into the piglet's marginal vein of the pinna, followed by an induction dose of propofol (0.5 mL/kg of 10 mg/mL; Propoful 1%, Fresenius Kabi, Austria) and pancuronium bromide (0.1 mg/kg; Pavulon® 4 mg/2 mL, Organon International, Oss, Netherlands). The piglet was then intubated with an appropriately sized endotracheal tube (4.5–5.0; Mallinckrodt® endotracheal tubes, Nellcor, Boulder, CO). Mechanical ventilation was initiated with an infant ventilator (North American Drager Narcomed 2A; DRE Inc., Louisville, KY) with oxygen gas (50% FiO_2) at a peak inspiratory pressure of 15 cmH_2O, inspiratory time of 0.75 s, a positive-end-expiratory-pressure of 5 cmH_2O, and a respiratory rate of 12 breaths per min. We measured the ABG intermittently and adjusted the peak pressure to maintain normocapnia ($PaCO_2$ 35–45 mmHg) during the baseline period. End-tidal CO_2 from the endotracheal humidity cuff was continuously monitored. Following intubation, the piglet was regularly paralyzed with intravenous pancuronium (100 μg/kg), and anesthesia was maintained with 2%–3% isoflurane (250 mL; Forane, Abbott Laboratories Ltd., Queenborough, Kent, UK). (Figure 1)

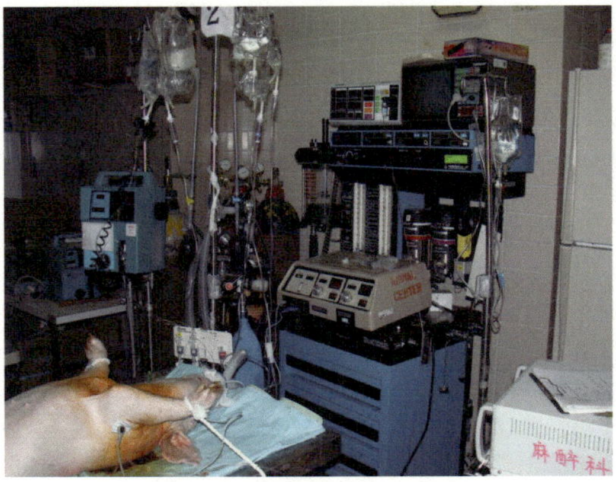

Fig. 1. Anesthesia and ventilator setting for experimental animals.

5.2 Monitoring physiological variables

We indwelled a rectal temperature probe for body temperature measurements and maintained the rectal temperature at 39.5–40.0 °C till the piglet was extubated. The left external jugular vein was aseptically exposed and cannulated with a 7F single-lumen central venous catheter (Arrow International Inc.) for chemical infusions. Normal saline with 5% glucose was given intravenously via the line in the piglet's marginal vein of the pinna by dripping at an hourly rate of 5 mL/kg. The right common femoral artery was exposed and cannulated with a 7F two-lumen central venous catheter (Arrow International, Inc.), and the catheter tip was advanced to lie in the proximal abdominal aorta for blood pressure measurements and blood sampling. We used a multiparameter physiological monitor (Hewlett Packard, 78399A) to monitor blood pressure, heart beats, and electrocardiography continuously. In addition, we inserted a 7.5F Swan-Ganz continuous cardiac output, mixed venous oxygen saturation monitoring (CCO/SvO_2) catheter (Edwards Lifesciences, 744H) via the right common femoral vein into the pulmonary artery and used a Vigilance monitor (Edwards Lifesciences) to monitor the pulmonary artery pressure (PAP), pulmonary capillary wedge pressure (PCWP), and central venous pressure (CVP) (Figure 2). The cardiac output (CO) was continuously measured using the thermodilution principle. The body surface area, cardiac index (CI), systemic vascular resistance index (SVRI), pulmonary vascular resistance index (PVRI), left-ventricular stroke work index (LVSWI), and right-ventricular stroke work index (RVSWI) were calculated for comparison.

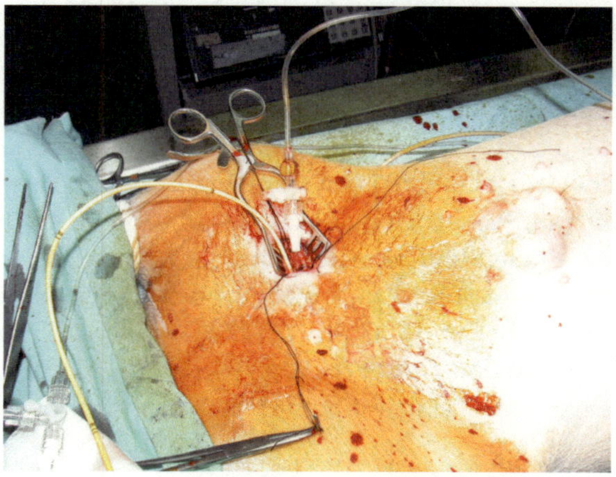

Fig. 2. Implantation of Swan–Ganz catheters during experiment.

5.3 Protocol for chemical infusion and data collection

After a stabilization period of approximately 20 min, we sampled blood for ABG, complete blood cell counts (CBC), and biochemistry, and recorded the mean arterial blood pressure (MABP), HR, CVP, MPAP, PCWP, and CO as baseline values. We separated piglets into five experimental groups: (1) control, receiving normal saline (NS), (2) G, receiving glyphosate ([*N*-(phosphonomethyl) glycine], Sigma-Aldrich, St. Louis, USA) 360 mg/mL in sodium hydroxide (NaOH) (~2.13 M, ~pH 5.7), (3) IPA, receiving IPA (CAS Number 75-31-0, Merck Schuchardt OHG, Hohenbrunn, Germany) 126 mg/mL in water (~2.13 M, ~pH 12.9), (4) IPAG group, receiving *N*-(phosphonomethyl) glycine, monoisopropylamine salt solution (Sigma-Aldrich) , 40 wt% (~2.13 M, ~pH 5.0), and (5) POEA group, receiving alkoxylated fatty amine (Kudos SL-101C, CAS Number 61791-26-2, Zhang Jia Gang Kudos Chemical Co. Ltd.) 15% in water, final ~pH 11.6. The concentration chosen for G, IPA, IPAG, and POEA were based on 40 wt % IPAG solution and 15% POEA.

In our preliminary study, we performed cardiographic examinations on piglets receiving different rates of IPAG infusions. We found that an infusion rate of 10 mL/h IPAG (~2.13 M) could result in slow reduction in blood pressure, and sudden death with ventricular arrhythmia or reversible depression of left-ventricular function may occur after discontinuing infusion right after the MABP decreased to 50% of the initial value. At an infusion rate higher than 10 mL/h, most piglets died soon after the IPAG infusion. For other chemicals, no obvious reduction in MABP values was noted within one hour of infusion at the rate of 10 ml/h. Therefore, we infused IPAG at 10 ml/hr and selected a 50% reduction in the MABP of the initial value (50% MABP) as the endpoint. The surviving piglets were then observed for up to 2 h from the beginning of the IPAG infusion. The NS, G, IPA, and POEA were infused at a rate of 10 mL/h for 1 h and then for another hour of observation. Temperature, HR, MABP, MPAP, CVP, PCWP, and CO values were recorded every 5 min. After the two hours of the experiments, the daily activities and urine amounts in the surviving piglets were observed and recorded for two days. Blood was sampled for ABG, CBC, biochemistry and serum glyphosate during the experiment and at 24 and 48 h after the chemical infusion began.

5.4 Serum levels of glyphosate analyzed by high-performance liquid chromatography (HPLC)

To explore the concentration change of glyphosate during infusion, serum concentrations of glyphosate were analyzed in the G and IPAG groups. We adopted HPLC method to measure serum levels of glyphosate, using a PerkinElmer LC 295 with a variable wavelength ultraviolet detector operated at a wavelength of 195 nm, and an anion-exchange column (4.6 mm × 250 mm, Partisil 10 μM SAX). Blood samples were centrifuged and the supernatants were then diluted and filtered through 0.2 μm nylon membranes before the analysis. The samples were dissolved in a mobile phase consisting of 0.05 M potassium dihydrogen phosphate (KH_2PO_4) in 60:40 KH_2PO_4: water, adjusted to pH 1.9 with phosphoric acid (H_3PO_4). The flow rate of the mobile phase was 1.0 ml/min. A sample of 20 μL was used for each injection. The detection limit is 1 ppm and the coefficient of variation was < 10%.

5.5 Statistical analysis

All numerical values are presented as means ± SEM. We used the general linear model (GLM) for repeated measures in comparing hemodynamic data, paired *t* test in comparing ABG data, and analysis of variance in comparing other data. One-compartment model intravenous infusion equations (Brewster et al., 1991; Bauer LA, 2006) were used for calculating the elimination rate constant (K_e), the half-life ($t_{1/2}$), and the volume distribution (V), which are:

$$t_{\frac{1}{2}} = \frac{0.693}{K_e} \tag{1}$$

$$K_e = -\frac{\ln C_1 - \ln C_2}{t_1 - t_2} \tag{2}$$

$$V = \frac{K_0(1 - e^{-K_e t'})}{K_e[C_{\max} - (C_{\text{predose}} e^{-K_e t'})]} \tag{3}$$

Where t_1/C_1 is the first time/concentration pair, t_2/C_2 is the second time/concentration pair, K_0 is the infusion rate, t' = infusion time, C_{max} is the maximum concentration at the end of infusion, and $C_{predose}$ is the predose concentration. All statistical tests were performed at the two-tailed significance level of 0.05.

5.6 Results

Table 4 shows the average infused dose of IPAG, G, IPA, and POEA in each group was 159.80 ± 15.79 mg/kg (piglet weight), 238.47 ± 17.49 mg/kg, 75.24 ± 4.51 mg/kg, and 0.0944 ± 0.00546 ml/kg. Both POEA and IPAG finally caused a fatality rate of 66.7% (4/6).

At the beginning of the experiment, we compared the MABP among all the groups. IPAG infusion reduced MABP from 89.17 ± 4.10 to 47.50 ± 6.02 mmHg, which reached 50% MABP at around 30.50 ± 1.67 min after the infusion began, and 50% (3/6) piglets died soon after that time point with the presentation of ventricular arrhythmia. After discontinuation, the MABP increased to the initial level in the piglets surviving after infusion. The IPA

Parameters	Control (N = 3)	Glyphosate[a] (N = 6)	Isopropylamine[a] (N = 6)	Isopropylamine[a] salt of glyphosate (N = 6)	Polyoxyethylene--amine[a] (POEA) (N = 6)
Body weight (kg)					
Mean ± SEM	15.57 ± 1.96	15.47 ± 1.02	17.08 ± 1.14	16.43 ± 1.43	16.17 ± 0.96
Body height (cm)					
Mean ± SEM	82.03 ± 4.08	81.07 ± 1.59	82.17 ± 0.40	79.40 ± 1.64	80.92 ± 1.18
Body surface area (m^2)					
Mean ± SEM	0.563 ± 0.052	0.558 ± 0.023	0.585 ± 0.018	0.565 ± 0.028	0.567 ± 0.021
Administered doses (mg/kg or mL/kg piglet weight)		238.47 ± 17.49 mg/kg	75.24 ± 4.51 mg/kg	159.80 ± 15.79 mg/kg	0.09 ± 0.01 ml/kg
Survival rate (%)					
No. surviving/total [no. (%)]	6/6 (100.00%)	6/6 (100.00%)	6/6 (100.00%)	2/6 (33.33%)*	2/6 (33.33%)*
Urine amount on postoperative day 1 (mL)					
Mean ± SEM	550.00 ± 180.28	345.00 ± 91.60	363.33 ± 40.79	140.00 ± 89.14[b]	191.67 ± 121.39[b]
Urine amount on postoperative day 2 (mL)					
Mean ± SEM	533.33 ± 169.15	545.00 ± 64.43	451.67 ± 32.09	160.00 ± 101.32[b]	208.33 ± 135.66[b]

SEM, standard error of the mean.

[a]The administered concentration for glyphosate, isopropylamine, IPAG, and polyoxyethyleneamine were calculated based on 40 wt % IPAG solution and 15% polyoxyethyleneamine, equal to 0.296 g/g (isopropylamine salt of solution), 0.104 g/g, 0.40 g/g, and 0.15 mL/mL ethoxylated tallowamine in water.

[b]Only two surviving piglets were counted.

*$p < 0.01$ by Pearson's χ^2 test.

Data from Lee et al, 2009.

Table 4. Values of body weight, body height, body surface area, survival rate, average survival time, and urine amount at postoperative days 1 and 2 in the five groups.

infusion led a marked increase in MABP. In all the other experimental groups, no significant changes in the MABP during the chemical infusion were observed. The average infused dose of IPAG, G, IPA, and POEA was 159.80 ± 15.79 mg/kg (piglet weight), 238.47 ± 17.49 mg/kg, 75.24 ± 4.51 mg/kg, and 0.0944 ± 0.00546 ml/kg. Although HR decreased gradually in the IPAG and POEA groups (10–30 min in the IPAG group and 35–100 min in the POEA group, $p < 0.05$), there was no significant difference in HR between these groups.

Compared to NS and G, IPAG and POEA had markedly decreased the CI after the initiation of infusion. Contrarily, the PCWP increased markedly in the IPAG and POEA groups. No significant changes in the CI or PCWP were noted in the G or IPA group. IPAG also increased the CVP and MPAP, but only a temporary increase in MPAP was noted.

The LVSWI, RVSWI, SVRI, PVRI calculated from MAP, PCWP, the stroke volume index (SVI), PAP, and CVP, were compared among the groups. IPAG infusion significantly reduced the LVSWI values, which subsequently stabilized after the discontinuation of the treatment. POEA also gradually reduced LVSWI during and after its infusion. These two chemicals also increased the values of PVRI, which were significantly different from those in the G group ($p < 0.05$). Whereas IPAG had no effect on the RVSWI, it increased the SVRI values after the discontinuation of infusion. POEA had no effect on the RVSWI or SVRI. Although IPA only transiently increased the RVSWI values during the infusion period (15–60 min), it significantly increased the PVRI values, which were higher than those of the G group. In contrast, G had no effect on the LVSWI, RVSWI, SVRI, or PVRI.

Table 5 shows the analysis of blood gas during the experiment. The initial mean pH was 7.45–7.51 in all experimental groups. The inhalation of oxygen during anesthesia caused elevated arterial blood P_{O2} initially, ranging from 186.50 to 210.17 mmHg, and the P_{CO2} were maintained around 35.83–41.33 mmHg. The initial lactate and base excess (BE) concentrations were similar across the groups. No significant changes in the arterial blood pH, P_{O2}, P_{CO2}, lactate, or BE occurred in the control group. POEA caused a reduction in the pH at the end of experiment ($p < 0.01$), accompanied by a gradual increase in lactate ($p < 0.01$) and a reduction in BE, which is compatible with the process of metabolic acidosis. Similar results were observed in the IPAG group, which also had an increase in lactate and a reduction in BE during and after infusion ($p < 0.01$), with a slight reduction in the P_{CO2} value during infusion. The G group had a reduction in pH and BE, with no changes in the other parameters during or after infusion. Unlike POEA, G, and IPAG, IPA caused a gradual increase in the BE.

A glyphosate standard and serum glyphosate concentrations were analyzed by HPLC as described in the Methods. Under the conditions employed in our study, glyphosate had a retention time of 10-11 min. The blood samples at different time points had retention time similar to parent glyphosate. The dose used in the G group produced an average glyphosate concentration of 166.54 ± 63.96, 236.47 ± 83.15, and 180.27 ± 33.19 ppm at 30, 45, and 60 min after its administration, and the chemical was barely detectable after nearly 48 h; while in the IPAG group, an average glyphosate concentration of 731.28 ± 151.38 ppm was detected at 50% MABP (around 30.5 min, averagely), and it could be detected with an average of 148.74 ± 73.36 ppm after nearly 48 h. The glyphosate concentration detected in IPAG infusion was four times higher than that in G infusion at ~30 min. We observe no plateau

Chemicals	pH	P_{O2} (mmHg)	P_{CO2} (mmHg)	Lactate	BE (mEq/L)
Control (normal saline)					
Initial (mean ± SEM)	7.47 ± 0.01	205.33 ± 1.21	41.33 ± 3.71	1.47 ± 0.37	6.27 ± 1.69
60 min (mean ± SEM)	7.47 ± 0.02	198.67 ± 7.86	41.33 ± 4.71	1.43 ± 0.09	6.47 ± 2.32
Final (mean ± SEM)	7.46 ± 0.02	227.00 ± 35.64	40.67 ± 3.93	1.23 ± 0.07	5.27 ± 1.68
Glyphosate (NaOH base)					
Initial (mean ± SEM)	7.51 ± 0.02	210.17 ± 6.96	35.83 ± 2.86	1.37 ± 0.12	5.55 ± 1.43
60 min (mean ± SEM)	7.45 ± 0.02**	193.83 ± 6.46	37.83 ± 3.43	1.55 ± 0.17	2.03± 1.00**
Final (mean ± SEM)	7.47 ± 0.02*	193.83 ± 9.89	37.10 ± 2.58	1.63 ± 0.25	3.37 ± 1.59*
Isopropylamine					
Initial (mean ± SEM)	7.45 ± 0.02	193.00 ± 11.72	39.67 ± 2.35	1.83 ± 0.39	4.17 ± 2.24
60 min (mean ± SEM)	7.46 ± 0.03	179.83 ± 8.75	43.33 ± 2.32	1.81 ± 0.55	7.43 ± 2.46*
Final (mean ± SEM)	7.48 ± 0.03	201.17 ± 25.9	43.17 ± 2.27	1.43 ± 0.16	8.82 ± 2.24**
Polyoxyethyleneamine					
Initial (mean ± SEM)	7.48 ± 0.03	196.67 ± 10.55	38.83 ± 3.99	1.67 ± 0.26	4.72 ± 1.00
60 min (mean ± SEM)	7.46 ± 0.04	196.17 ± 12.86	31.17 ± 3.64*	3.97 ± 0.62**	-1.42 ± 1.45*
Final (mean ± SEM)	7.23 ± 0.06**	167.83 ± 25.09	38.83 ± 3.82	7.58 ± 1.04**	-9.41 ± 2.62**
Isopropylamine salt of glyphosate					
Initial (mean ± SEM)	7.49 ± 0.01	186.50 ± 12.03	40.67 ± 1.96	1.33 ± 0.12	7.65 ± 1.76
50% of MABP (mean ± SEM)	7.50 ± 0.02	189.67 ± 10.50	30.83 ± 2.06**	2.12 ± 0.20**	0.90 ± 1.08**
Final (mean ± SEM)	7.42 ± 0.05	124.50 ± 30.68	34.01 ± 5.51	3.78 ± 0.67**	-3.01 ± 2.59**

SEM, standard error of the mean.
*$p < 0.05$ vs. Initial; **$p < 0.01$ vs. Initial.
Data from Lee et al., 2009.

Table 5. Arterial blood gas analysis at 60 min after control (normal saline), glyphosate, isopropylamine, or polyoxyethyleneamine injection and at 50% of the mean arterial blood pressure (MABP) after treatment with isopropylamine salt of glyphosate.

concentration for each piglet and therefore used the average concentrations for calculating pharmacokinetic parameters. For G infusion, the $t_{1/2}$ of glyphosate was 1.52 h, the K_e was 0.46 h^{-1}, and the V was 16.05 liter (L); for IPAG infusion, they were 1.46 h, 0.47 h^{-1}, and 3.92 L, respectively.

5.7 Conclusion and discussion

5.7.1 Infusion of IPA

In our study, the persistent elevated MABP and PVRI and the reversible RVSWI during IPA infusion suggest an inotropic effect of IPA. The lower dose used in our study (1.2–1.4 mg/kg per min *vs.* 2.5 mg/kg per min) may account for the differences observed between our and the other study (Privitera et al., 1982).

5.7.2 Infusion of IPAG

In contrast to G and IPA, POEA and IPAG infusions introduced high death rates. IPAG infusion lowered cardiac contractility and the MABP, accompanied by increases in the MPAP and vascular resistance, which caused heart failure. A 66.7% fatality rate and blood lactate formation with lowered BE values were noted following its infusion with ~50% of the dose in the concentration similar to other chemicals. No pulmonary rales were detected by auscultation during the experiments, and no hypoxemia, severe acidosis or alkalosis, or obvious pH changes that could result in changes in pulmonary vascular resistance or cardiac dysfunction were noted during the experiments. Uncoupling mitochondrial oxidative phosphorylation and reduced the respiratory control ratios of mitochondria have been reported as the possible toxic mechanism of glyphosate, IPAG or GlySH (Bababunmi et al., 1979; Olorunsogo et al., 1979a; Peixoto, 2005), which may be one of the reasons used for the explanation of lactate formation and acidosis; nevertheless, back to the level of more complex organisms with effective buffering capacities, we could not see severe acidosis with huge pH changes that could sufficiently lead to hemodynamic dysfunction. Therefore, the changes in the cardiovascular parameters in our study imply direct depressive cardiovascular and vasoactive effects exerted by IPAG.

5.7.3 Infusion of POEA

In our study, although POEA did not significantly affect MABP during the infusion period, it progressively depressed left-side ventricular function (decreased the CI and LVSWI and increased the PCWP and CVP), and increased pulmonary vasoconstriction effects (increased the MPAP and PVRI) during and after its infusion, leading to metabolic acidosis with the accumulation of lactate noted at 60 min and at the end of the experiment. In the POEA group, 66.7% (4/6) of the piglets died between 1 and 3 h after the discontinuation of this chemical. In a dog study, Tai et al. (1990) found that surfactant infusion decreased the MABP, CO, and LVSWI, and Koyama et al. (1994) reported similar effects in rats, when the surfactant polyoxyethylene alkylether produced negative chronotropic and inotropic responses. Reviewing the experimental records, we found that the increases in anal temperatures in the five groups, under the control of warm blanket, was no more than 1.4 °C, and the blood glucose levels, under the support of intravenous glucose/saline fluids, were kept around 100-200 mg/dL. The biochemistry data checked during one hour of chemical infusions showed no evidence of acute change in renal or liver function. The mild increase of lactate in the IPAG group might be induced by circulatory collapse or uncoupled oxidative phosphorylation. Because we found no report of uncoupled oxidative phosphorylation effects, the increase in lactate in the POEA group was most likely due to circulatory collapse which could worsen acidosis and lead to death. It is commonly assumed that acute acidosis could have adverse effects on hemodynamics. Therefore, it can be speculated that the deaths of our experimental animals from uncorrected metabolic acidosis was attributable to the infusion of POEA.

5.7.4 Infusion of glyphosate in NaOH base

The infusion of glyphosate in NaOH base had a reduction in pH and BE, with no significant hemodynamic changes during or after infusion.

5.7.5 Serum concentration of glyphosate during the infusion of glyphosate in NaOH and IPA base

According to the metabolic and pharmacokinetic studies, the vast majority of the body burden after the administration of glyphosate is unchanged parent glyphosate and no toxic metabolites are produced (Williams et al., 2000; Brewster et al., 1991). Human data on the kinetics of glyphosate are rare. The analysis of plasma concentration-time profiles in a prospective study of acute GlySH self-poisoning in adults suggested that the elimination of glyphosate is the first-order elimination and the best-fit apparent elimination $t_{1/2}$ of glyphosate is 3.1 h with a fairly narrow 95% C.I. of 2.7–3.6 h (Roberts et al., 2010). However, another study in rat showed after single 100 mg kg^{-1} intravenous (i.v.) and 400 mg kg^{-1} oral doses administration, plasma concentration–time curves were best described by a two-compartment open model; the elimination $t_{1/2}$ of α and β phase (distribution and elimination terminal phase) for glyphosate from plasma were 0.345 h and 9.99 h after i.v. and 4.17 h and 14.38 h after oral administration (Anadón et al., 2009). In our study, at the same infused concentration and infusion rate, the calculated $t_{1/2}$ and K_e values for glyphoaste in the G and IPAG infusion groups were relatively close (for G infusion, $t_{1/2}$ 1.52 h, K_e 0.46 h^{-1}; for IPAG infusion, 1.46 h, 0.47 h^{-1}, respectively). Distribution, elimination, and metabolism data are very important for being extrapolated from experimental animals to humans; however, they may vary across different study design in different experimental animals. In our piglet study, the elimination of glyphosate in intravenous infusion is described by a one-compartment model with the first-order elimination, which is similar to the report of Robert et al. in GlySH poisoning in humans.

In addition, a higher concentration of glyphosate was detected in the IPAG group than in the G group at the approximate time point (731 ppm vs. 167 ppm). This phenomenon could be explained by the different dissociation ability of IPA and NaOH salts. Since IPA is a weak base and NaOH is a strong base, in the environment of ~ pH 7.4 (blood), IPA salt would more easily dissociate than NaOH salt; thus, higher concentration of glyphosate in serum could be detected in the IPAG group. This might be one of the reasons that glyphosate in NaOH base with a pH of 5.7 had no obvious impact on hemodynamics during infusion, except for mild reductions in pH and BE values which were still within normal ranges. In contrast, glyphosate in the form of IPA salt produced more severe hemodynamic insults in our study.

6. Summary

GlySH has been commonly used in suicide attempt in Taiwan and other Asia countries. Case fatality rate ranged from 1.9 to 29.3% in Taiwan (Chen et al., 2009). The risk factors of fatality or severity of GlySH exposure identified are amount of exposure, hypovolemic shock, intractable shock, acute pulmonary edema, Acute Physiology and Chronic Health Evaluation II score, age, male gender, laryngeal injury with aspiration, abnormal chest X-ray, calendar time, reason for exposure, atropine therapy, elapsed time, delayed presentation, number of involved organs, hyperkalemia, metabolic acidosis, tachycardia, elevated serum creatinine, and high plasma glyphosate concentrations on admission (> 734 ug/mL) (Sawada et al., 1988; Tominack et al., 1991; Talbot et al., 1991; Hung et al., 1997; Lee et al., 2000; Lee et al., 2008; Chen et al., 2009; Roberts et al., 2010). All the patients who are reported to have ingested large amounts of GlySH should be carefully observed, especially for those who present with respiratory distress, unstable hemodynamics, and old age. In managing patients who have larger amount of GlySH ingestion, airway protection, early detection of pulmonary edema, and prevention of further pulmonary damage and renal damage appear to be of critical importance.

GlySH poisoning may induce severe cardiovascular symptoms in humans (Talbot et al., 1991; Lin et al., 1999). Animal and cell studies have also shown that GlySH are more toxic than POEA or glyphosate itself (Tai et al., 1990; Martinez and Brown, 1991; Richard et al., 2005; Peixoto, 2005; Marc et al., 2002), and therefore synergistic effects between the components of GlySH have been proposed (Peixoto, 2005; Marc et al., 2002). In the second study, we demonstrated that the negative cardiovascular effects seen in GlySH poisoning could be attributable to the surfactant POEA, IPAG, or both. Glyphosate in NaOH base or IPA alone had no similar cardiovascular effects. Here, we first demonstrated that IPAG has effects similar to POEA and provide further insight into the cardiovascular effects of different salts of glyphosate and the adjuvants used in GlySH on experimental animals under the circumstance of chemical infusion. Further studies that clarify more precisely the mechanisms of the synergistic effect of glyphosate and IPA are required.

In the evaluation of the toxicity of pesticides, the current practice is to evaluate the active ingredients. The current study shows that the adjuvant can be toxic. Therefore, the toxicity pattern related to the combination of active ingredients with adjuvants should be taken into consideration when evaluating the toxicity threshold of mixtures of pesticides. Furthermore, efforts should be taken to search for the safest formula in the development of commercially available pesticide products.

7. References

Adam A.; Marzuki A.; Abdul R.H.; Abdul A.M. (1997). The oral and intratracheal toxicities of ROUNDUP and its components to rats. *Veterinary & Human Toxicology*, Vol. 39, No. 3, pp. 147-151. ISSN: 0145-6296.

Anadón A.; Martínez-Larrañaga M.R.; Martínez M.A.; Castellano V.J.; Martínez M.; Martin M.T.; Nozal M.J. & Bernal J.L. (2009). Toxicokinetics of glyphosate and its metabolite aminomethyl phosphonic acid in rats. *Toxicology Letters*, Vol. 190, No. 1, pp. 91–95, ISSN: 0378-4274.

Bababunmi E.A.; Olorunsogo O.O. & Bassir O. (1979). The uncoupling effect of N-(phosphonomethyl) glycine on isolated rat liver mitochondria. *Biochemical Pharmacology*, Vol. 28, No. 6, pp. 925-927. ISSN: 0006-2952.

Bauer L.A. (2006) Clinical pharmacokinetic equations and calculations, In: *Clinical pharmacokinetics Handbook*, Brown M. & Edmonson K. (Ed.), pp. 19-32, McGraw-Hill Medical, ISBN: 007142542X.

Bingham E.; Cohrssen B. & Powell C.H. (2001). *Patty's Toxicology*, Vo. 4, 5th ed., John Wiley & Sons, Inc., ISBN: 047131935X, New York, NY 2001.

Bradberry S.M.; Proudfoot A.T. & Vale J.A. (2004). Glyphosate poisoning. *Toxicological Reviews*, Vol. 23, No. 3, pp. 159-167. ISSN: 1176-2551.

Brewster D.W.; Warren J. & Hopkins W.E. (1991). Metabolism of glyphosate in Sprague-Dawley rats: tissue distribution, identification, and quantitation of glyphosate-derived materials following a single oral dose. *Fundamental and Applied Toxicology*, Vol. 17, No. 1, pp. 43-51, ISSN: 0272-0590.

Chen Y.-J.; Wu M.-L.; Deng J.-F. & Yang C.-C. (2009). The epidemiology of glyphosate-surfactant herbicide poisoning in Taiwan, 1986-2007: a poison center study. *Clinical Toxicology (Phila)*, Vol. 47, No. 7, pp. 70-677, ISSN: 1556-3650.

EPA US. (1993). Glyphosate, In: *Reregistration Eligibility Decision Document*, Special Review and Reregistration Division, Office of Pesticide Programs, Washington, D.C.

Harbison R.D. (1998). *Hamilton and Hardy's Industrial Toxicology*, Mosby-Year Books, ISBN: 0815141815, St. Louis, MO.

Hietanen E.; Linnainmaa K. & Vainio H. (1983). Effects of phenoxyherbicides and glyphosate on the hepatic and intestinal biotransformation activities in the rat. *Acta Pharmacologica et Toxicologica (Copenh)*, Vol. 53, No. 2, pp. 103-112. ISSN: 0001-6683.

Hung D.-Z.; Deng J.-F. & Wu T.-C. (1997). Laryngeal survey in glyphosate intoxication: a pathophysiological investigation. *Human & Experimental Toxicology*, Vol. 16, No. 10, pp. 596-599. ISSN: 0960-3271.

Ishizaki T.; Privitera P.J.; Walle T. & Gaffney T.E. (1974). Cardiovascular actions of a new metabolite of propranolol: isopropylamine. *The Journal of Pharmacology and Experimental Therapeutics*, Vol. 189, No. 3, pp. 626-632. ISSN: 0022-3565.

Kawamura K.; Nobuhara H.; Tsuda K.; Tanaka A.; Matsubara Y. & Yamauchi N. (1987). Two cases of glyphosate (Roundup) poisoning. *Pharmaceuticals Monthly (Japanese)*, Vol. 29, pp.163–166, ISSN: 0016-5980.

Koyama K.; Goto K. & Yamashita M. (1994). Circulatory failure caused by a fungicide containing iminoctadine and a surfactant: a pharmacological analysis in rats. *Toxicology and Applied Pharmacology*, Vol. 126, No. 2, pp.197-201, ISSN: 0041-008X.

Lee C.-H.; Shih C.-P.; Hsu K.-H.; Hung D.-Z. & Lin C.-C. (2008). The early prognostic factors of glyphosate-surfactant intoxication. *The American Journal of Emergency Medicine,* Vol. 26, No. 3, pp. 75-281. ISSN: 0735-6757.

Lee H.-L.; Chen K.-W.; Chi C.-H.; Huang J.-J. & Tsai L.-M. (2000). Clinical presentations and prognostic factors of a glyphosate-surfactant herbicide intoxication: a review of 131 cases. *Academic Emergency Medicine,* Vol. 7, No. 8, pp. 906-910. ISSN: 1069-6563.

Lee H.-L.; Kan C.-D.; Tsai C.-L.; Liou M.-J. & Guo H.-R. (2009). Comparative effects of the formulation of glyphosate-surfactant herbicides on hemodynamics in swine. *Clinical Toxicology* (*Phila*), Vol. 47, No. 7, pp. 651-658. ISSN: 1556-3650.

Lin C.-M.; Lai C.-P.; Fang T.-C. & Lin C.-L. (1999). Cardiogenic shock in a patient with glyphosate-surfactant poisoning. *Journal of the Formosan Medical Association,* Vol. 98, No. 10, pp. 698-700. ISSN: 0929-6646.

Marc J.; Mulner-Lorillon O.; Boulben S.; Hureau D.; Durand G. & Bellé R. (2002). Pesticide Roundup provokes cell division dysfunction at the level of CDK1/cyclin B activation. *Chemical Research in Toxicology,* Vol. 15, No. 3, pp. 326-331. ISSN: 0893-228X.

Martinez T.T. & Brown K. (1991). Oral and pulmonary toxicology of the surfactant used in roundup herbicide. *Proceedings of the Western Pharmacology Society,* Vol. 34, pp.43-46. ISSN: 0083-8969.

Martinez T.T.; Long W.C. & Hiller R. (1990). Comparison of the toxicology of the herbicide roundup by oral and pulmonary routes of exposure. *Proceedings of the Western Pharmacology Society,* Vol. 33, pp. 193-197. ISSN: 0083-8969.

Menkes D.B.; Temple W.A. & Edwards I.R. (1991). Intentional self-poisoning with glyphosate-containing herbicides. *Human & Experimental Toxicology,* Vol. 10, No. 2, pp. 103-107. ISSN: 0960-3271.

Müller M.M.; Rosenberg C.; Siltanen H. & Wartiovaara T. (1981). Fate of glyphosate and its influence on nitrogen-cycling in two Finnish agriculture soils. *Bulletin of Environmental Contamination and Toxicology,* Vol. 27, No. 5, pp. 724-730. ISSN: 0007-4861.

NFPA. (1997). *Fire Protection Guide to Hazardous Materials.* National Fire Protection Association, Quincy, MA.

Olorunsogo O.O.; Bababunmi E.A. & Bassir O. (1979a). Effect of glyphosate on rat liver mitochondria in vivo. *Bulletin of Environmental Contamination and Toxicology,* Vol. 22, N0. 3, pp. 357-364. ISSN: 0007-4861.

Olorunsogo O.O.; Bababunmi E.A. & Bassir O. (1979b). Uncoupling of corn shoot mitochondria by N-(phosphonomethyl) glycine. *FEBS Letters,* Vol. 97, No. 2, pp. 279-282. ISSN: 0014-5793.

Olorunsogo O.O.; Bababunmi E.A. & Bassir O. (1980). Interaction of N-(phosphonomethyl) glycine with some respiratory chain enzymes of isolated corn-shoot mitochondria. *Archives of Environmental Contamination and Toxicology,* Vol. 9, No. 1, pp. 109-114. ISSN: 0090-4341.

Olorunsogo O.O. (1982). Inhibition of energy-dependent transhydrogenase reaction by N-(phosphonomethyl) glycine in isolated rat liver mitochondria. *Toxicology Letters,* Vol. 10, No. 1, pp. 91-95. ISSN: 0378-4274.

Olorunsogo O.O. (1990). Modification of the transport of protons and Ca2+ ions across mitochondrial coupling membrane by N-(phosphonomethyl) glycine. *Toxicology,* Vol. 61, No. 2, pp. 205-209. ISSN: 0300-483X.

Peixoto F. (2005). Comparative effects of the Roundup and glyphosate on mitochondrial oxidative phosphorylation. *Chemosphere,* Vol. 61, No. 8, pp. 1115-1122. ISSN: 0045-6535.

Privitera P.J.; Walle T. & Gaffney T.E. (1982). Nicotinic-like effects and tissue disposition of isopropylamine. *The Journal of Pharmacology and Experimental Therapeutics,* Vol. 222, No. 1, pp. 116-121. ISSN: 0022-3565.

Richard S.; Moslemi S.; Sipahutar H.; Benachour N. & Seralini G.E. (2005). Differential effects of glyphosate and roundup on human placental cells and aromatase. *Environmental Health Perspectives,* Vol. 113, No. 6, pp. 716-720. ISSN: 0091-6765.

Roberts D.M.; Buckley N.A.; Mohamed F.; Eddleston M.; Goldstein D.A.; Mehrsheikh A.; Bleeke M.S. & Dawson A.H. (2010). A prospective observational study of the clinical toxicology of glyphosate-containing herbicides in adults with acute self-poisoning. *Clinical Toxicology,* Vol. 48, No. 2, pp. 129-136. ISSN: 1556-3650.

Sawada Y. & Nagai Y. (1987). Roundup intoxication-its clinical status and participation with surface active agents. *Journal of Clinical and Experimental Medicine (Igaku No Ayumi),* Vol. 143, pp. 25-27. ISSN: 0039-2359.

Sawada Y.; Nagai Y.; Ueyama M. & Yamamoto I. (1988). Probable toxicity of surface-active agent in commercial herbicide containing glyphosate. *Lancet,* Vol. 1, No. 8580, pp. 299. ISSN: 0140-6736.

Smith E.A. & Oehme F.W. (1992). The biological activity of glyphosate to plants and animals: a literature review. *Veterinary and Human Toxicology,* Vol. 34, No. 6, pp. 531-543. ISSN: 0145-6296.

Stella J. & Ryan M. (2004). Glyphosate herbicide formulation: a potentially lethal ingestion. *Emergency Medicine Australasia,* Vol. 16, No. 3, pp. 235-239. ISSN: 1742-6731.

Suh J.H.; Oh B.J. & Roh H.K. (2007). Clinical outcomes after suicidal ingestion of glyphosate surfactant herbicide: severity of intoxication according to amount ingested. *Clinical Toxicology (Phila)*, Vol. 45, No. 6, pp. 641, ISSN: 1556-3650.

Tai T.; Yamashita M. & Wakimori H. (1990). Hemodynamic effects of Roundup, Glyphosate and Surfactant in dogs. *Japanese Journal of Toxicology (Chudoku Kenkyu)*, Vol. 3, pp. 63-68. ISSN: 0914-3777.

Talbot A.R.; Shiaw M.-H.; Huang J.-S.; Yang S.-F.; Goo T.-S.; Wang S.-H.; Chen C.-L. & Sanford T.R. (1991). Acute poisoning with a glyphosate-surfactant herbicide ('Roundup'): a review of 93 cases. *Human & Experimental Toxicology,* Vol. 10, No. 1, pp. 1-8. ISSN: 0960-3271.

Tominack R.L.; Yang G.-Y.; Tsai W.-J.; Chung H.-M. & Deng J.-F. (1991). Taiwan National Poison Center survey of glyphosate--surfactant herbicide ingestions. *Journal of Toxicology - Clinical Toxicology,* Vol. 29, No. 1, pp. 91-109. ISSN: 0731-3810.

van der Hoek W. & Konradsen F. (2006). Analysis of 8000 hospital admissions for acute poisoning in a rural area of Sri Lanka. *Clinical Toxicology (Phila),* Vol. 44, No. 3, pp. 225-231. ISSN: 1556-3650.

Williams G.M.; Kroes R. & Munro I.C. (2000). Safety evaluation and risk assessment of the herbicide Roundup and its active ingredient, glyphosate, for humans. *Regulatory Toxicology and Pharmacology,* Vol. 31, No. 2, pp. 117-165. ISSN: 0273-2300.

26

Herbicides and Protozoan Parasite Growth Control: Implications for New Drug Development

Ricardo B. Leite[1], Ricardo Afonso[1,2] and M. Leonor Cancela[1,2]
[1]*CCMAR – Centre of Marine Sciences and*
[2]*Department of Biomedical Sciences and Medicine, University of Algarve, 8005-139 Faro, Portugal*

1. Introduction

Modern chromalveolates were originated through a series of events of endosymbiosis of microalgae by an eukaryote, leading to the retention of its plastid by this new host. These complex events, which took place multiple times during the course of evolution, originated new organisms that either lost or retained parts of the metabolisms present in their ancestral microalgae symbionts, achieved by transferring some of their genes into the nucleus of the host cell or even maintaining a relic organelle, a plastid circular DNA. The presence of this relic DNA was discovered in some apicomplexas, and thus named apicomplast. It was later found that a large number of parasites belonging to Alveolata also contained a relic non-photosynthetic plastid, homologous to the chloroplast of plants and algae, thus expanding the apicomplast concept outside the apicomplexas. During the 90's, some important metabolic pathways just known to occur in plants, algae, fungus and some bacteria were also identified in a series of important human parasites like *Plasmodium falciparum* and *Toxoplasma gondii,* the agents of malaria and toxoplasmosis respectively, opening a new era in drug prevention/control. Some of those protozoan parasites are responsible for millions of disease cases worldwide and hence a major concern for human health. The growing understanding of the biology and biochemistry of protozoan parasites, which has considerably increased over the past two decades, has paved the way to the discovery of many potential targets for new parasite drugs. The decrypted genomes of several species and the new post-genomic tools considerably improved our ability to identify and study the different metabolic pathways at the molecular level, and consequently contribute to validate potential drug targets.

In 1998 the journal Nature published the discovery of a group of protozoan parasites that shared a metabolic pathway (shikimate pathway) essential for their survival with many plants, fungi and bacteria, but not found in mammals. Furthermore, researchers demonstrated that the herbicide glyphosate, known to interfere with this pathway in plants, could be used successfully to inhibit the *in vitro* growth of these parasites, opening the possibility of using available herbicides as a start point to develop new drugs to control parasite growth.

Various studies have since then highlighted the importance of plastid or algae-like pathways for other parasite metabolisms such as fatty-acid FAS II, heme and isoprenoid biosynthesis pathway, and since they are both essential for parasite survival and not present in their hosts, they have attracted considerable attention as targets for therapeutics. Furthermore, since they are already known as targets of existing herbicides, this should significantly reduce the time and cost of specific drug development.

Herbicides are a class of compounds known to produce a wide range of toxic side effects, thus posing a threat to several organisms including humans. However, and despite this toxicity, which represents the negative side of their use, we have to acknowledge that the development and use of pesticides and herbicides over the past decades has played an important role in increasing agricultural productivity and in controlling potential carriers of human diseases (Table 1). In the near future, the same herbicides may become the precursors of new drugs against protozoan parasite diseases.

2. Metabolic pathways as drug targets

Parasite cells as well as their corresponding hosts have hundreds of metabolic pathways that are vital for their normal function. Therefore, there is always a need to study host pathways and compare them with those of the parasite. Each pathway has a number of enzyme reactions that catalyze its different steps. Enzyme activity at every step is regulated to ensure that the final product of the pathway provides for the needs of the cell. With the recent completion of many genome projects, it became possible to provide provisional maps of the proteome of several parasites. A large portion of predicted or confirmed open reading frames result in unique proteins not found in the corresponding hosts, and these are good news when it comes to drug development since compounds that inhibit whatever function of these proteins are potentially less likely to cause severe side-effects to the host. The challenge resides in showing that a particular protein or pathway is essential for the survival of the parasite and thus if that protein can be a potential drug target. To find an answer to this question we either have high-throughput methods or can use a more traditional approach, based on some prior knowledge of candidate metabolic pathways or cellular processes.

2.1 Shikimate pathway

Biosynthesis of aromatic amino acids in plants, in many bacteria, and in microbes relies on the enzyme 5-enolpyruvylshikimate 3-phosphate (EPSP) synthase, a good target for several drugs including herbicides. Because the shikimate is absent in more complex organisms, EPSP synthase is an attractive target for the development of new antimicrobial agents effective against bacterial, parasitical, and fungal pathogens. A valuable lead compound and an exemple in the search for new drugs and herbicides is glyphosate. Glyphosate is a successfully used herbicide, being the active ingredient of the widely used weed control agent Roundup, and was shown to inhibit the growth of the parasites *P. falciparum*, *T. gondii*, *C. parvum* (Roberts et al. 1998) and *P. olseni* (Elandalloussi et al. 2008), among others.

2.2 Isoprenoid biosynthesis pathway

Isopentenyl diphosphate (IPP) is the central intermediate in the biosynthesis of isoprenoids, the most ancient and diverse class of natural products. Two distinct routes of IPP biosynthesis occur in nature, the mevalonate pathway and the recently discovered deoxyxylulose 5-phosphate (DOXP) pathway. The evolutionary history of the enzymes involved in both routes and the phylogenetic distribution of their genes across genomes suggests that the mevalonate pathway is unique to archaebacteria, as the DOXP is for

Herbicides	Organism	Target	References
Norflurazon, fluridone	*Plasmodium falciparum*	Carotenoid synthesis	US Patent 5859028
Diclofop, sethoxydim, tralkoxydim, alloxydim, clethodim and cycloxydim	*Plasmodium falciparum*	Fatty acid synthesis	US Patent 5877186
Fosmidomycin	*Plasmodium falciparum*	Isoprenoids pathway	(Lichtenthaler 2000)
Trifluralin, oryzalin and amiprophos-methyl	*Plasmodium falciparum*	Microtubule inhibitor	(Fennell et al. 2006)
Ethalfluralin, oryzalin and trifluralin	*Toxoplasma Gondii*	Microtubule inhibitor	(Stokkermans et al. 1996)
Dinitroaniline herbicides	*Trypanosoma cruzi*	Microtubule inhibitor	(Traub-Cseko et al. 2001)
Dinitroaniline herbicides	*Cryptosporidium parvum*	Microtubule inhibitor	(Arrowood et al. 1996)
Dinitroaniline and phosphorothioamidate herbicides	*Plasmodium falciparum*	Microtubule inhibitor	(Fennell et al. 2006)
Trifluralin, pendimethalin, oryzalin, and benfluralin (dinitroaniline herbicides)	*Plasmodium berghei*	Microtubule inhibitor	(Dow et al. 2002)
Dinitroaniline herbicides	*Entamoeba histolytica*	Microtubule inhibitor	(Makioka et al. 2000)
2,4,5 trichlorophenoxy acetic acid	*Tetrahymena pyriformis*	Mitochondria	(Silberstein and Hooper 1977)
Aryloxyphenoxypropionate herbicides	*Toxoplasma Gondii*	Plastid acetyl-CoA carboxylase	(Zuther et al. 1999)
Clodinafop-propargyl	*Babesia equi* and *B. caballi*	Plastid acetyl-CoA carboxylase	(Bork et al. 2003)
Flufenacet	*Perkinsus marinus*	PUFA pathway	(Venegas-Caleron et al. 2007)
Glyphosate	*Plasmodium falciparum*	Shikimate pathway	(Roberts et al. 1998)
Glyphosate	*Perkinsus olseni*	Shikimate pathway	(Elandalloussi et al. 2008)

Table 1. Herbicide derivatives used to control parasite proliferation, their metabolic targets and corresponding references

eubacteria, and that eukaryotes have inherited their genes for IPP biosynthesis from prokaryotes. The occurrence of genes specific to the DXP pathway is restricted to plastid-bearing eukaryotes, indicating that these genes were acquired from the cyanobacteria ancestor of plastids (Lim and McFadden 2010). The non-mevalonate isoprenoid biosynthesis pathway is essential for many protozoan parasites survival, since DOXP inhibitors like herbicide fosmidomycin strongly inhibit their *in vitro* proliferation (Lichtenthaler 2000; Wiesner et al. 2002), and thus is effective in managing the clinical symptoms of malaria that are associated with the intra-erythrocytic phase of the parasite cell cycle (Jomaa et al. 1999). Hence, all herbicides which are inhibitors of this pathway in plants are also potential drugs against all parasites bearing the same pathway.

2.3 Fatty acid biosynthesis

It was previously thought that apicomplexan parasites were incompetent for *de novo* fatty acid synthesis (Holz 1977; Matesanz et al. 1999), but recent work showed the presence of nuclear-encoded apicomplast-targeted genes for all enzymes of the fatty acid biosynthesis pathway in several apicomplexa parasites and this finding provided strong arguments in favor of the presence of a *de novo* fatty acid biosynthesis in this organelle (Surolia et al. 2004). The presence of highly conserved proteins known as Type II fatty acid synthase explains the susceptibility of *T. gondii* and *P. falciparum* to herbicides targeting plastidic Acetyl-CoA carboxylase, like the aryloxyphenoxypropionates. This pathway is seen as a promising drug target, mostly because it is structurally and functionally distinct from its equivalent pathway present in the vertebrate hosts (Goodman and McFadden 2008).

3. Perkinsus, a protozoa parasite of interest for pharmaceutical testing

Protozoa represents one of the earliest branches of eukaryotic organisms and the key to understand early global evolution. Inside protozoa, alveolata represents one of the classes better studied due to the presence, within this class, of the apicomplexa, which include parasites like the agents of malaria and toxoplasmosis (leading opportunistic infections often associated, among others, with AIDS and with congenital neurological birth defects), responsible for the infection of man and cattle, thus making this phylum particularly important for medical and veterinary reasons. Most of these organisms are a major cause of disease worldwide, but many of them have received little attention from pharmaceutical industry. This scenery is now changing due to the completeness of genome sequence of some of the most important protozoan parasites within this phylum. Comparative genomics using growing information provided by multiple genome sequencing efforts can now be used to help identify parasite-specific targets for drug development. However, and despite the increasing variety of genomes already sequenced, many possible applications resulting from their analysis are most likely not yet unveiled since, very often, knowledge of the genome alone is not sufficient to provide answers to many of the existing questions, and the unveiling of both transcriptome and/or proteome are also required.

3.1 The genus Perkinsus

The microorganisms of the genus Perkinsus are protist parasites responsible for important mortalities in different mollusc species. It was first described in 1946 as a spherical unknown organism found in moribund *Crassostrea virginica* oysters but not in healthy ones in Louisiana (USA) (Mackin et al. 1950). For the past two decades a severe mortality is affecting bivalve molluscs particularly in Portugal and Spain (Leite et al. 2004). This mortality was first associated with the parasite *Perkinsus* (P.) *atlanticus* in 1989 (Azevedo

1989). Recently, phylogenetic studies and the use of molecular data have shown that *P. atlanticus* and *P. olseni* are, in fact, the same organism (Murrell et al. 2002). On the other side of the Atlantic, oyster's mortalities are related with a parasite from the same genus, *P. marinus*. This parasite was first classified as a fungus, then as an Apicomplexa and now, a new taxonomic class (Perkinsea) was created inside Alveolata to place all Perkinsus.

Some authors suggest that Perkinsus represents an early branch between dinoflagellates and apicomplexa. These two groups of organisms are quite different, each containing unique characteristics, like the absence of histones and presence of photosynthesis in dinoflagellates and the presence of a circular DNA within a plastid in some apicomplexa like *P. falciparum* (Gardner et al. 1991) and *T. gondii*. Furthermore, it was recently suggested that Perkinsus also possesses both a non-photosynthetic plastid reminiscent of the apicomplexan relic plastid organelle, the apicomplast, (Teles-Grilo et al. 2007) and additional specific cell compartments (Fernández-Robledo et al. 2008) raising questions about the nature and origin of putative relic plastid/compartments in Perkinsus species.

Although the analysis of Perkinsus genes has only started very recently, some notorious findings are being made, in terms of characterization of metabolic pathways susceptible to play a critical role for drug targeting. Approaches to begin unveil Perkinsus transcriptome were already made by our group like the usage of Suppression Substractive Hybridization (SSH) to identify genes differentially expressed by *P. olseni* when exposed to hemolymph from Ruditapes decussatus (Ascenso et al. 2007) or the use of differential transcriptomic analysis to unveil Perkinsus transcripts resulting from gene transcription under differential conditions (Leite el al., unpublish data). But at present, as more and more genome sequence information for Perkinsus becomes available, the use of more high throughput techniques/tools such as microarrays is possible. There is currently ongoing a project conducting P. marinus genome (almost complete) sequencing by The Institute for Genomic Research (TIGR)/Center for Marine Biotechnology (COMB) (http://www.tigr.org/tdb/e2k1/pmg/). The strategy chosen for sequencing *P. marinus* was whole genome shotgun sequencing (8x coverage).

3.2 *In vitro* parasite cultures as tools for drug screening

In vitro cultures have been extensively used for screening chemotherapeutic agents. This technique is less costly than animal screening but depends on the ease of establishing laboratory cultures. The relationship between the parasite and the host is very intricate and in most cases disruptions of the host-parasite relationship *in vitro* leads to gradual death of the parasite. A frequent difficulty to establish continuous parasite cultures is their usually complex life cycle, in addition to some of them having intermediate hosts. An attempt to provide the parasite with a specific culture medium which is the most similar to its *in vivo* environment has been conducted for each parasite that can be cultured in vitro (Allen *et al.* 2005).

Various *in vitro* cultures of the protozoan parasite of genus *Perkinsus* (Figure 1) have already been developed (Gauthier et al. 1995; Robledo et al. 2002; Casas et al. 2008), and have the advantage of not requiring the presence of a host enhancing the capability of assessing the effects of drugs on parasite growth (Gauthier and Vasta 1994; Elandalloussi et al. 2005; Leite et al. 2008). The effects of a given drug can thus be determined by analyzing the survival of the parasites as a function of time in culture in the presence or absence of the drug to be tested (Figure 2) and analyzing the data using specific software (Figure 3).

Fig. 1. *Perkinsus olseni* cells under culture

3.3 Perkinsus proliferation can also be affected by herbicides derivatives

To demonstrate the advantages of the usage of Perkinsus as an alternative for the screening for new drug targets affecting viability/proliferation of the parasite and thus identification of possible drug precursors among known herbicides, a test using the most common herbicides was developed. Ten herbicides were chosen between the most frequently used worldwide (see Table 3). Other aspect taken into consideration was the choice of a non dinitroaniline, phosphorothioamidate or aryloxyphenoxypropionate herbicide because their mode of action is already known for several protists. The only exception to this criteria was the presence of Pendimethalin, which was used as positive control. A brief description of the main usages of the selected herbicides is presented below:

- 2,4-Dichlorophenoxyacetic acid (2,4-D), a broadleaf herbicide have been commercially available for over 50 years and are a widely used family of phenoxy herbicides worldwide, being a case study on agricultural chemicals. Now mainly used in a blend with other herbicides, it is the most widely used herbicide in the world, third most commonly used in the United States. It is an example of synthetic auxin (plant hormone) (Kennepohl et al. 2001).
- Atrazine, a triazine based herbicide is used in corn and. The low cost/good performance on a broad spectrum of weeds common in the U.S. is explains the widely usage of this pesticide. It is also commonly used with other herbicides to lower potential groundwater contamination. It is a photosystem II inhibitor (Lim et al. 2009).
- Dicamba another example of a synthetic auxin is a benzoic acid herbicide that acts by mimicking the effects of auxins (i.e., natural plant growth hormones), causing enhanced but uncontrolled growth rates, alterations in plant function homeostasis, and death. Dicamba is often combined with one or several other herbicidal agents including 2,4-D, 2,4-DP, atrazine, glyphosate, imazethapyr, ioxynil, and mecoprop. It is used to control a wide spectrum of annual and perennial broadleaf weeds and is effective in both pre- and post-emergence applications. A primary agricultural use is weed reduction in grain/cereal cropsand maintenance of pastures, forest lands, fence rows, and transportation and utility rights-of-way. (Harp et al. 2001).

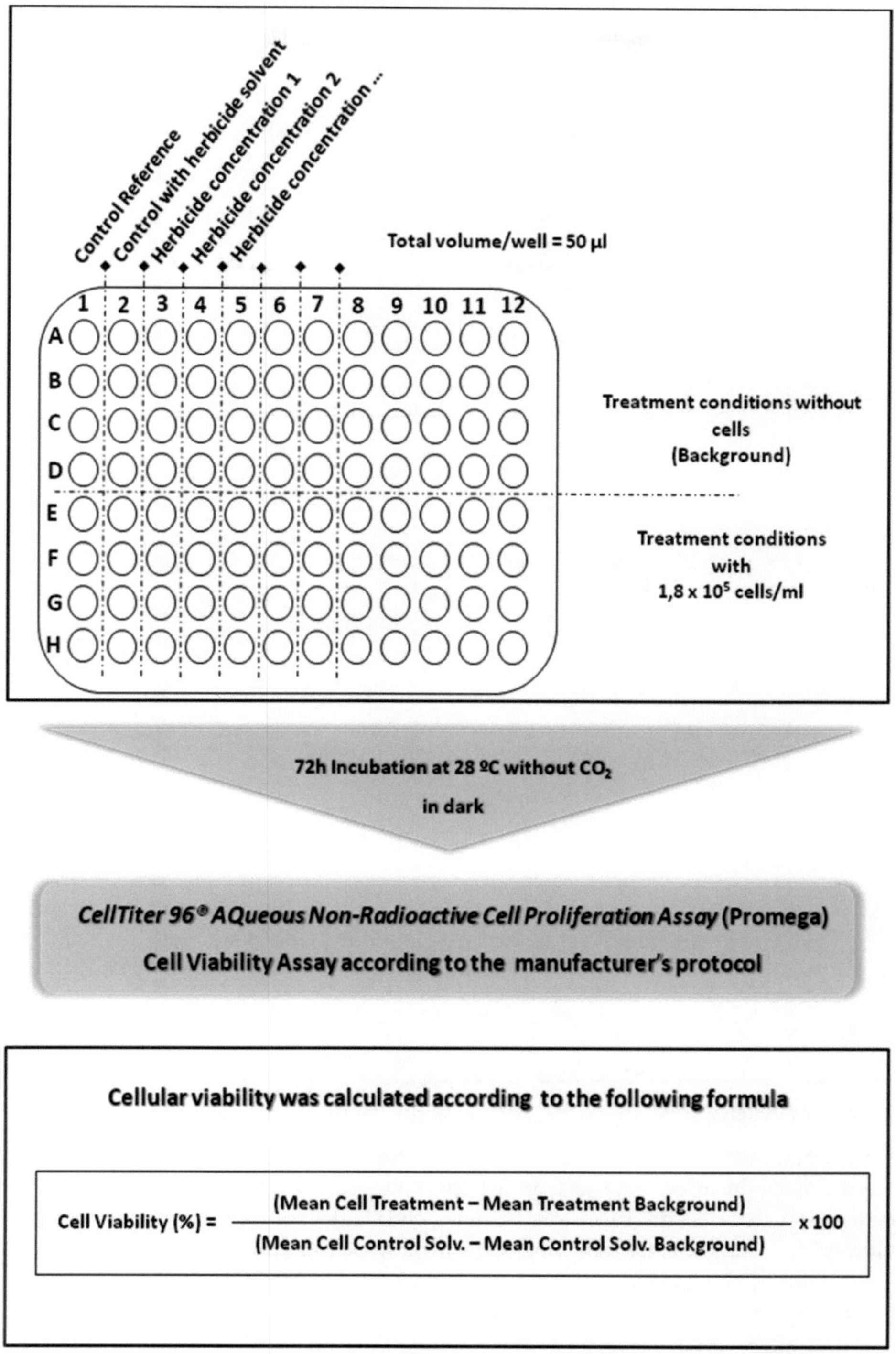

Fig. 2. Method to screen the effect of compounds in the proliferation of *Perkinsus olseni* using *in vitro* cultures (Elandalloussi et al. 2005; Leite et al. 2008).

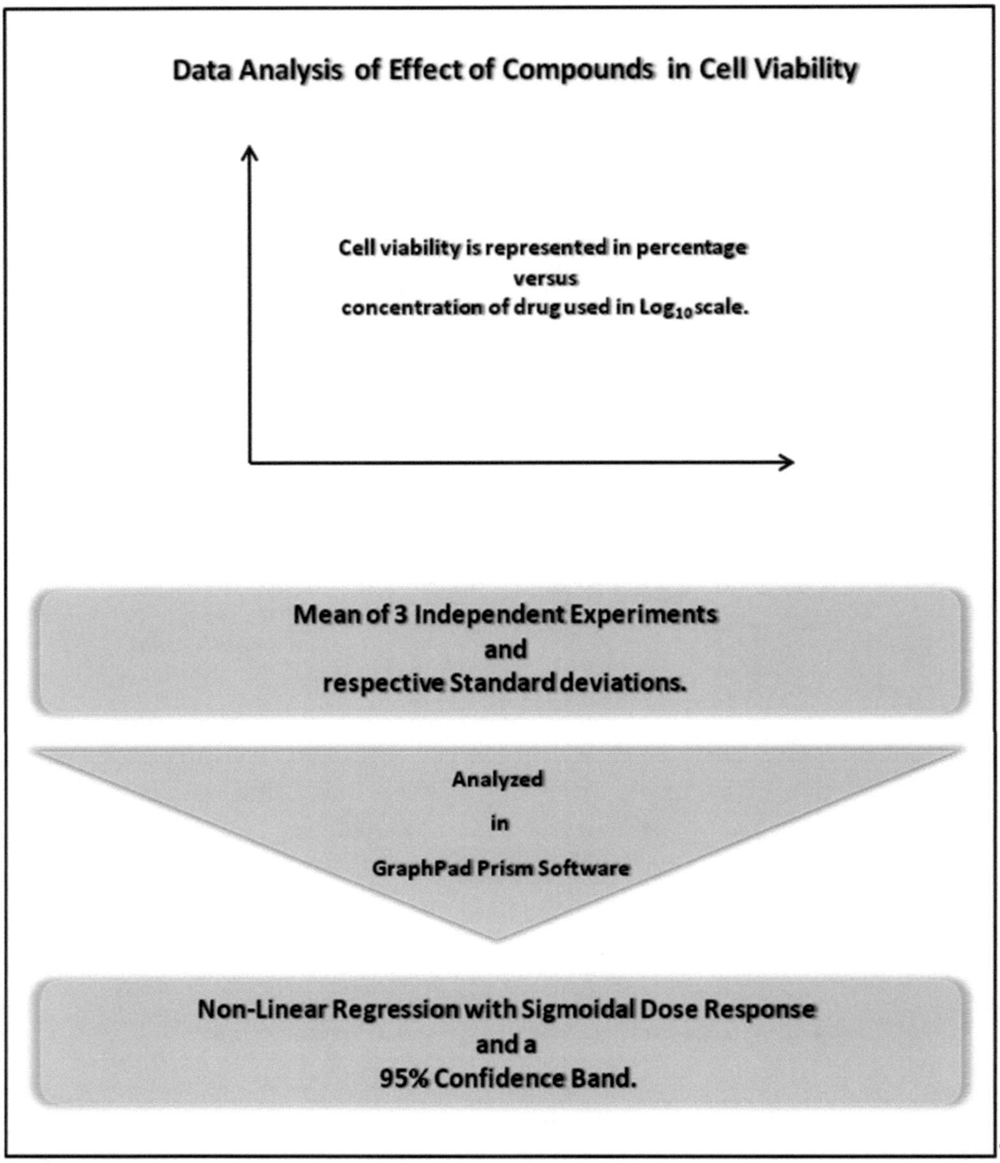

Fig. 3. Data analysis of *Perkinsus olseni* antiparasitic compounds screening

- Glufosinate ammonium, a broad-spectrum contact herbicide and is used to control weeds after the crop emerges or for total vegetation control on land not used for cultivation. It is a structural analogue of glutamate and acts in plants by inhibiting glutamine synthetase, thereby blocking synthesis of glutamine from glutamate and thus assimilation of NH_4 (Manderscheid and Wild 1986; Hack et al. 1994).
- Fluroxypyr, a systemic, selective herbicide is used for the control of broad-leaved weeds in small grain cereals, maize, pastures, range land and turf. It is a synthetic auxin. In cereal growing, fluroxypyr's key importance is in the control of cleavers, Galium aparine. Other key broad-leaved weeds are also controlled (Wu et al. 2009).

- Imazapyr is a non-selective herbicide used for the control of a broad range of weeds including terrestrial annual and perennial grasses and broadleaved herbs, woody species, and riparian and emergent aquatic species (Hess et al. 2010).
- Linuron is a non-selective herbicide used in the control of grasses and broadleaf weeds. It works by inhibiting photosynthesis (Snel et al. 1998).
- Metolachlor is a pre-emergent herbicide widely used to control annual grasses in corn and sorghum; it has partially replaced atrazine in these uses (Heydens et al. 2010).
- Pendimethalin, is a pre-emergent herbicide widely used to control annual grasses and some broadleaf weeds in a very wide range of crops, including corn, soybeans, wheat, cotton, many tree and vine crops, and many turfgrass species (Heydens et al. 2010).
- Picloram, is a pyridine herbicide mainly used to control unwanted trees in pastures and edges of fields. It is another synthetic auxin (Grossmann 2010).

All the herbicides were ordered from Sigma and the methodology followed was identical to that described in Figures 1 and 2. In order to test the range of concentrations to be used, a preliminary assay using all these herbicides in three different concentrations (1, 100 and 500µM) was conducted (table 2). Preliminary resulted suggested that only four herbicides demonstrated some effect on Perkinsus proliferation (Fig. 4) and thus a more extended test was performed using only these selected ones.

Coumpound	Chemical Formula	Group (herbicides)	Observed inhibition (Perkinsus)	IC50 (µM)
2,4-D	$C_8H_6Cl_2O_3$	phenoxy herbicides	Yes	ND
Atrazine	$C_8H_{14}ClN_5$	chlorotriazine	No	-
Dicamba	$C_8H_6Cl_2O_3$	benzoic acid	No	-
Fluroxypyr	$C_7H_5Cl_2FN_2O_3$	pyridine	No	-
Glufosinate-ammonium	$C_5H_{15}N_2O_4P$	organophosphorus	No	-
Imazapyr	$C_{13}H_{15}N_3O_3$	imidazolinone	No	-
Linuron	$C_9H_{10}Cl_2N_2O_2$	phenylurea	Yes	391,3
Metolachlor	$C_{15}H_{22}ClNO_2$	chloroacetanilide	Yes	193,2
Pendimethalin	$C_{13}H_{19}N_3O_4$	dinitroaniline	Yes	396,6
Picloram	$C_6H_3Cl_3N_2O_2$	pyridine	No	-
Glyphosate	$C_3H_8NO_5P$	organophosphorus	Yes	3400
Fosmidomycin	$C_4H_{10}NO_5P$	-	No	-

Table 2. Compound names and corresponding chemical formulas of herbicides derivatives shown to affect Perkinsus proliferation.

From the panel of herbicides tested, four of them had some effect on *Perkinsus olseni* proliferation, namely 2,4-D, Linuron, Metolachlor and Pendimethalin (Figure 4).

- The effect of Pendimethalin on Perkinsus proliferation was expected since an effect was already observed for trypanosomatids and Plasmodium falciparum (Chan and Fong 1994; Dow et al. 2002). It works by inhibiting microtubule disruption during parasite development.
- Metolachlor belongs to the class of chloroacetanilides herbicides responsible for inhibition of very-long-chain fatty acids (VLCFA) biosynthesis in plant and algal cells

(Böger et al. 2000; Trenkamp et al. 2004). It is already described in the literature that *Perkinsus marinus* possess some genes related with VLCFA like FAE-1. *P. marinus* FAE1-like elongating activity is also sensitive to the herbicide flufenacet, in accordance to some higher plant 3-ketoacyl-CoA synthases (Venegas-Caleron et al. 2007) and explain the results abtained with Metolachlor.

- The primary effect of linuron is the inhibition of photosystem II electron flow (Snel et al. 1998), resulting in damage and plant weakness. This result can be surprise but due to the lack of photosystem II in Perkinsus, but it can be related to the inhibition of electron flow in other systems.
- 2,4-D also revealed some effect but even at higher concentrations (500 μM) it inhibit less that 20% of the proliferation. Together with the pattern of the proliferation graphic, the results suggest that this herbicide has some cytotoxic effect on Perkinsus may be due to contaminations present in 2,4-D formulation.

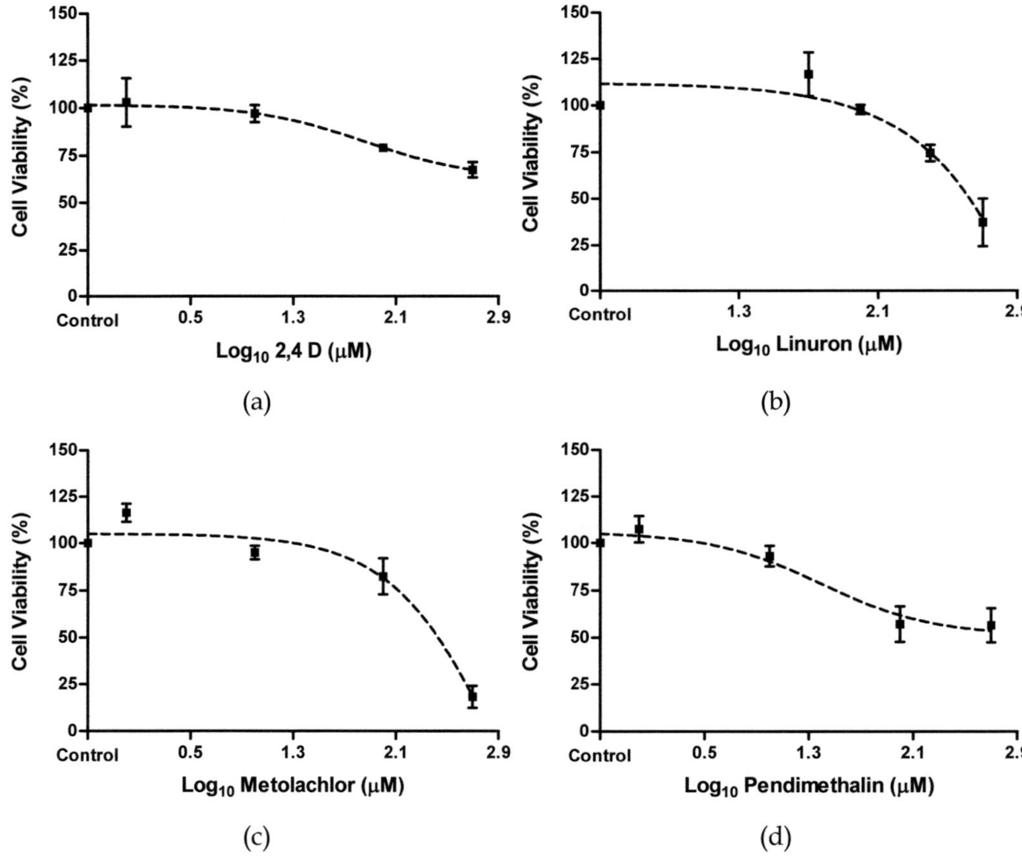

Fig. 4. Proliferation of *P. olseni* after 72 h of exposure upon different treatments.
Vertical axis stands for cell viability in percentage and horizontal axis for log_{10} concentration in μM (A) 2,4-Dichlorophenoxyacetic acid, (B) Linuron,(C) Metolachlor and (D) Pendimethalin. Percentage of proliferation is relative to normal (non-treated) conditions (100%).

4. Advantages of Perkinsus as model organism for new drug development against protozoa

Altogether, the available data suggest that Perkinsus can be a good alternative for herbicide screenings to detect potential drug precursors affecting parasite metabolic pathways not present in their hosts. Most of our currently used antiparasitic drugs have been identified as a result of random screening of a series of chemicals that are related to compounds with recognized therapeutic value. This is referred to as the empirical approach to drug discovery and remains a valid approach to the discovery of novel antiparasitic molecules. Despite not being comparable with true high-throughput screening, significant screening can be conducted against the parasite *in vitro* using direct approaches since all antiparasitic drugs must be tested against the parasite before advance to *in vivo* models (Woods and Knauer 2010).

Advantages of Perkinsus can be related to the fact that (i) it shares specific characteristics with algae, fungus and plants, in particular the presence of metabolic pathways that are not present in parasite hosts, (ii) these characteristic pathways are also present in several disease agents like plasmodium and toxoplasma and thus the response to specific drugs affecting these pathogenic parasites can first be tested in Perkinsus, (iii) it is harmless for mammals, (iv) it can be grown easily in vitro, (v) clonal cultures of the parasite have been developed and are available and (vi) it shares many similarities with highly pathogenic parasites, thus having the potential to became very useful for both the study of protozoan diseases and in pharmaceutical drug discovery and/or testing.

In addition, and because Perkinsus is an aquatic parasite and circulates and differentiates between the water column and the host, it can be an excellent biomarker of the degree of contamination by chemicals accumulated in the water or upon the host body and thus be very useful to detect effects of newly introduced chemicals and to be used to perform preliminary tests on aquatic bioaccumulation as quickly as possible, to avoid possible disasters.

5. References

Allen, D. D., Caviedes, R. l., Cardenas, A. M., Shimahara, T., Segura-Aguilar, J. and Caviedes, P. A. (2005). Cell Lines as In Vitro Models for Drug Screening and Toxicity Studies. *Drug Development and Industrial Pharmacy* 31(8): 757-768.

Arrowood, M. J., Mead, J. R., Xie, L. and You, X. (1996). In vitro anticryptosporidial activity of dinitroaniline herbicides. *FEMS Microbiology Letters* 136(3): 245-9.

Ascenso, R. M. T., Leite, R. B., Afonso, R. and Cancela, M. L. (2007). Suppression-subtractive hybridization: A rapid and inexpensive detection methodology for up-regulated *Perkinsus olseni* genes. *African Journal Microbiology Research* 1(3): 24-28.

Azevedo, C. (1989). Fine structure of *Perkinsus atlanticus* n. sp. (Apicomplexa, Perkinsea) parasite of the clam *Ruditapes decussatus* from Portugal. *Journal of Parasitology* 75(4): 627-635.

Böger, P., Matthes, B. and Schmalfuß, J. (2000). Towards the primary target of chloroacetamides -new findings pave the way. *Pest Management Science* 56(6): 497-508.

Bork, S., Yokoyama, N., Matsuo, T., Claveria, F. G., Fujisaki, K. and Igarashi, I. (2003). Clotrimazole, ketoconazole, and clodinafop-propargyl as potent growth inhibitors of equine Babesia parasites during in vitro culture. *Journal of Parasitology* 89(3): 604-6.

Casas, S. M., Reece, K. S., Li, Y., Moss, J. A., Villalba, A. and La Peyre, J. F. (2008). Continuous Culture of Perkinsus mediterraneus, a Parasite of the European Flat Oyster Ostrea edulis, and Characterization of Its Morphology, Propagation, and Extracellular Proteins in Vitro. *Journal of Eukaryotic Microbiology* 55(1): 34-43.

Chan, M. M. and Fong, D. (1994). Plant microtubule inhibitors against trypanosomatids. *Parasitology Today* 10(11): 448-451.

Dow, G. S., Armson, A., Boddy, M. R., Itenge, T., McCarthy, D., Parkin, J. E., Thompson, R. C. and Reynoldson, J. A. (2002). Plasmodium: assessment of the antimalarial potential of trifluralin and related compounds using a rat model of malaria, *Rattus norvegicus. Experimental Parasitology* 100(3): 155-60.

Elandalloussi, L., Leite, R., Rodrigues, P., Afonso, R. and Cancela, M. (2008). Effect of the Herbicide Roundup® on Perkinsus olseni in vitro Proliferation and in vivo Survival when Infecting a Permissive Host, the Clam Ruditapes decussatus. *Bulletin of Environmental Contamination and Toxicology* 80(6): 512-515.

Elandalloussi, L. M., Leite, R. B., Rodrigues, P. M., Afonso, R., Nunes, P. A. and Cancela, M. L. (2005). Effect of antiprotozoal drugs on the proliferation of the bivalve parasite *Perkinsus olseni. Aquaculture* 243: 9-17.

Fennell, B. J., Naughton, J. A., Dempsey, E. and Bell, A. (2006). Cellular and molecular actions of dinitroaniline and phosphorothioamidate herbicides on Plasmodium falciparum: Tubulin as a specific antimalarial target. *Molecular and Biochemical Parasitology* 145(2): 226-238.

Fernández-Robledo, J. A., Schott, E. J. and Vasta, G. R. (2008). Perkinsus marinus superoxide dismutase 2 (PmSOD2) localizes to single-membrane subcellular compartments. *Biochemical and Biophysical Research Communications* 375(2): 215-219.

Gardner, M. J., Feagin, J. E., Moore, D. J., Spencer, D. F., W.Gray, M., Williamson, D. H. and J.M.Wilson, R. (1991). Organisation and expression of small subunit ribosomal RNA genes encoded by a 35-kilobase circular DNA in Plasmodium falciparum. *Molecular and Biochemical Parasitology* 48(1): 77-88.

Gauthier, J. D., Feig, B. and Vasta, G. R. (1995). Effect of fetal bovine serum glycoproteins on the *in vitro* proliferation of the oyster parasite *Perkinsus marinus*: development of a fully defined medium. *Journal of Eukaryotic Microbiology* 42(3): 307-313.

Gauthier, J. D. and Vasta, G. R. (1994). Inhibition of in vitro replication of the oyster parasite perkinsus marinus by the natural iron chelators transferrin,lactoferrin, and desferrioxamine. *Development and Comparative Immunology*: 227-286.

Goodman, C. D. and McFadden , G. I. (2008). Fatty Acid Synthesis in Protozoan Parasites: Unusual Pathways and Novel Drug Targets. *Current Pharmaceutical Design* 14(9): 901-916

Grossmann, K. (2010). Auxin herbicides: current status of mechanism and mode of action. *Pest Management Science* 66(2): 113-120.

Hack, R., Ebert, E., Ehling, G. and Leist, K. H. (1994). Glufosinate ammonium--Some aspects of its mode of action in mammals. *Food and Chemical Toxicology* 32(5): 461-470.

Harp, P., Robert, I. K. and William, C. K. (2001). Dicamba. *Handbook of Pesticide Toxicology (Second Edition)*. San Diego, Academic Press: 1639-1640.

Hess, F. G., Harris, J. E., Pendino, K., Ponnock, K. and Robert, K. (2010). Imidazolinones. *Hayes' Handbook of Pesticide Toxicology (Third Edition)*. New York, Academic Press: 1853-1863.

Heydens, W. F., Lamb, I. C., Wilson, A. G. E. and Robert, K. (2010). Chloracetanilides. *Hayes' Handbook of Pesticide Toxicology (Third Edition)*. New York, Academic Press: 1753-1769.

Holz, G. G., Jr. (1977). Lipids and the malarial parasite. *Bull World Health Organ* 55(2-3): 237-248.

Jomaa, H., Wiesner, J., Sanderbrand, S., Altincicek, B., Weidemeyer, C., Hintz, M., uuml, rbachova, I., Eberl, M., Zeidler, J., Lichtenthaler, H. K., Soldati, D. and Beck, E. (1999). Inhibitors of the Nonmevalonate Pathway of Isoprenoid Biosynthesis as Antimalarial Drugs. *Science* 285(5433): 1573-1576.

Kennepohl, E., Munro, I. C., Robert, I. K. and William, C. K. (2001). Phenoxy Herbicides (2,4-D). *Handbook of Pesticide Toxicology (Second Edition)*. San Diego, Academic Press: 1623-1638.

Leite, R. B., Afonso, R. and Cancela, M. L. (2004). *Perkinsus* sp. infestation in carpet-shell clams, *Ruditapes decussatus* (L), along the Portuguese coast. Results from a 2-year survey. *Aquaculture* 240: 39-53.

Leite, R. B., Brito, A. B. and Cancela, M. L. (2008). An Oxygen Molecular Sensor, the HIF Prolyl 4-Hydroxylase, in the Marine Protist Perkinsus olseni. *Protist* 159(3): 355-368.

Lichtenthaler, H. K. (2000). Non-mevalonate isoprenoid biosynthesis: enzymes, genes and inhibitors. *Biochemical Society Transactions* 28(6): 785-789.

Lim, L. and McFadden, G. I. (2010). The evolution, metabolism and functions of the apicoplast. *Philosophical Transactions of the Royal Society B: Biological Sciences* 365(1541): 749-763.

Lim, S., Ahn, S. Y., Song, I. C., Chung, M. H., Jang, H. C., Park, K. S., Lee, K.-U., Pak, Y. K. and Lee, H. K. (2009). Chronic Exposure to the Herbicide, Atrazine, Causes Mitochondrial Dysfunction and Insulin Resistance. *PLoS ONE* 4(4): e5186.

Mackin, J. G., Owen, H. M. and Collier, A. (1950). Preliminary note on the occurrence of a new protistan parasite, *Dermocystidium marinum* n. sp. in *Crassostrea virginica* (Gmelin). *Science* 111: 328-329.

Makioka, A., Kumagai, M., Ohtomo, H., Kobayashi, S. and Takeuchi, T. (2000). Effect of dinitroaniline herbicides on the growth of Entamoeba histolytica. *Journal of Parasitology* 86(3): 607-10.

Manderscheid, R. and Wild, A. (1986). Studies on the mechanism of inhibition by phosphinothricin of glutamine synthestase isolated from *Triticium aestivum*. *Journal of Plant Physiology* 123: 135-142.

Matesanz, F., Durán-Chica, I. and Alcina, A. (1999). The cloning and expression of Pfacs1, a Plasmodium falciparum fatty acyl coenzyme A synthetase-1 targeted to the host erythrocyte cytoplasm. *Journal of Molecular Biology* 291(1): 59-70.

Murrell, A., Kleeman, S. N., Barker, S. C. and Lester, R. J. G. (2002). Synonymy of *Perkinsus olseni* Lester & Davis, 1981 and *Perkinsus atlanticus* Azevedo, 1989 and an update on the phylogenetic position of the genus *Prekinsus*. *Bulletin of the European Association of Fish Pathologists* 22: 258-265.

Roberts, F., Roberts, C. W., Johnson, J. J., Kyle, D. E., Krell, T., Coggins, J. R., Coombs, G. H., Milhous, W. K., Tzipori, S., Ferguson, D. J., Chakrabarti, D. and McLeod, R. (1998). Evidence for the shikimate pathway in apicomplexan parasites. *Nature* 393: 801-805.

Robledo, J. A. F., Nunes, P. A., Cancela, M. L. and Vasta, G. R. (2002). Development of an In Vitro Clonal Culture and Characterization of the rRNA Gene Cluster of *Perkinsus atlanticus,* a Protistan Parasite of the Clam *Tapes decussatus. Journal of Eukaryotic Microbiology* 49(5): 414-422.

Silberstein, G. B. and Hooper, A. B. (1977). The effect of the herbicide 2,4,5 trichlorophenoxy acetic acid (245T) on the growth and metabolism of Tetrahymena pyriformis. *J Cell Physiol.* 85: 331-8.

Snel, J., Vos, J., Gylstra, R. and Brock, T. (1998). Inhibition of photosystem II (PSII) electron transport as a convenient endpoint to assess stress of the herbicide linuron on freshwater plants. *Aquatic Ecology* 32(2): 113-123.

Stokkermans, T. J. W., Schwartzman, J. D., Keenan, K., Morrissette, N. S., Tilney, L. G. and Roos, D. S. (1996). Inhibition ofToxoplasma gondiiReplication by Dinitroaniline Herbicides. *Experimental Parasitology* 84(3): 355-370.

Surolia, A., Ramya, T. N. C., Ramya, V. and Surolia, N. (2004). 'FAS't inhibition of malaria. *Biochemical Journal* 383(3): 401-412.

Teles-Grilo, M. L., Tato-Costa, J., Duarte, S. M., Maia, A., Casal, G. and Azevedo, C. (2007). Is there a plastid in *Perkinsus atlanticus* (Phylum Perkinsozoa)? *European journal of protistology* 43(2): 163-167.

Traub-Cseko, Y. M., Ramalho-Ortigão, J. M., Dantas, A. P., de Castro, S. L., Barbosa, H. S. and Downing, K. H. (2001). Dinitroaniline herbicides against protozoan parasites: the case of Trypanosoma cruzi. *Trends in Parasitology* 17(3): 136-141.

Trenkamp, S., Martin, W. and Tietjen, K. (2004). Specific and differential inhibition of very-long-chain fatty acid elongases from Arabidopsis thaliana by different herbicides. *Proceedings of the National Academy of Sciences of the United States of America* 101(32): 11903-11908.

Venegas-Caleron, M., Beaudoin, F., Sayanova, O. and Napier, J. A. (2007). Co-transcribed Genes for Long Chain Polyunsaturated Fatty Acid Biosynthesis in the Protozoon Perkinsus marinus Include a Plant-like FAE1 3-Ketoacyl Coenzyme A Synthase. *Journal of Biological Chemistry* 282(5): 2996-3003.

Wiesner, J., Henschker, D., Hutchinson, D. B., Beck, E. and Jomaa, H. (2002). In Vitro and In Vivo Synergy of Fosmidomycin, a Novel Antimalarial Drug, with Clindamycin. *Antimicrobial Agents and Chemotherapy* 46(9): 2889-2894.

Woods, D. J. and Knauer, C. S. (2010). Discovery of veterinary antiparasitic agents in the 21st Century: A view from industry. *International Journal for Parasitology* 40(10): 1177-1181.

Wu, G., Cui, J., Tao, L. and Yang, H. (2009). Fluroxypyr triggers oxidative damage by producing superoxide and hydrogen peroxide in rice (*Oryza sativa*). *Ecotoxicology* 19(1): 124-132.

Zuther, E., Johnson, J. J., Haselkorn, R., McLeod, R. and Gornicki, P. (1999). Growth of Toxoplasma gondii is inhibited by aryloxyphenoxypropionate herbicides targeting acetyl-CoA carboxylase. *Proceedings of the National Academy of Sciences of the United States of America* 96(23): 13387-13392.

27

Synthesis and Evaluation of Pyrazine Derivatives with Herbicidal Activity

Martin Doležal[1] and Katarína Kráľová[2]
[1]*Faculty of Pharmacy in Hradec Králové, Charles University in Prague*
[2]*Faculty of Natural Sciences, Comenius University in Bratislava*
[1]*Czech Republic*
[2]*Slovak Republic*

1. Introduction

The pyrazine ring is a part of many polycyclic compounds of biological and/or industrial significance; examples are quinoxalines, phenazines, and bio-luminescent natural products pteridines, flavins and their derivatives. All these compounds are characterized by a low lying unoccupied π-molecular orbital and by the ability to act as bridging ligand. Due to these two properties 1,4-diazines, and especially their parent compound pyrazine, possess a characteristic reactivity. Pyrazine is a weak diacid base (pK_1 = 0.57; pK_2 = -5.51), weaker than pyridine, due to the induction effect of the second nitrogen (Bird, 1992). Its inherent bifunctionality and the low lying unoccupied molecular orbital permit pyrazine to form coordination polymers having unusual electrical and magnetic properties (Brown & Knaust, 2009). 1,4-Diazines may be employed to study inter- and intramolecular electron transfer in organic, inorganic and biochemical reactions. Autocondenzation of α-aminocarbonyle compounds to the dihydropyrazine derivative, which is followed by oxidation on the final substituted pyrazine, or the condenzation of α,β-dicarbonyle and α,β-diamino compounds forming during the fermentation of saccharides and peptides are the main routes of pyrazine ring building. Pyrazines are found mainly in processed food, where they are formed during dry heating processes *via* Maillard reactions (Maillard, 1912). They are also found naturally in many vegetables, insects, terrestrial vertebrates, and marine organisms, and they are produced by microorganisms during their primary or secondary metabolism (Adams et al., 2002; Beck et al., 2003; Wagner et al., 1999; Woolfson & Rothschild, 1990). The widespread occurrence of simple pyrazine molecules in nature, especially in the flavours of many food systems, their effectiveness at very low concentrations as well as the still increasing applications of synthetic pyrazines in the flavour and fragrance industry are responsible for the high interest in these compounds (Maga, 1992). Certain pyrazines, especially dihydropyrazines, are essential for all forms of life due their DNA strand-breakage activity and/or by their influencing of apoptosis (Yamaguchi, 2007). Synthetic pyrazine derivatives are also useful as drugs (antiviral, anticancer, antimycobacterial, etc.), fungicides, and herbicides (Doležal, 2006a). Furthermore, a simple pyrazine compound, 3-amino-6-chloro-pyrazine-6-carboxylic acid, has shown anti-auxin behaviour (Camper & McDonald, 1989). The importance of the pyrazine (1,4-diazine) ring for the biological

activity can be evaluated primarily according to the size of the studied molecules. In relatively small compounds, the pyrazine ring is necessary for biological action due to its resemblance (bioisosterism) to the naturally occurring compounds (e.g. nicotinamide, or pyrimidine nucleic bases). In bulky compounds the introduction of the pyrazine ring brings specific chemical and physicochemical properties for the molecule as a whole, such as basic and slightly aromatic character (Doležal, 2006a). A fully comprehensive study of the pyrazines including reactivity and synthesis is beyond the scope of this work but can be found in the literature (Brown, 2002; Joule & Mills, 2010).
Herbicides are generally considered as growth inhibitors, thus their different inhibitory responses have been studied in various culture systems. Plant tissue and cell cultures provide model systems for the study of various molecular, physiological, organism and genetic problems. These systems have been used in the study of herbicides and other xenobiotics (Linsmaier & Skoog, 1965).

2. Pyrazine herbicides

The most successful pyrazine derivative was diquat-dibromide (see Fig. 1, the structure I). This non-selective, contact herbicide has been used to control many submerged and floating aquatic macrophytes which interferes with the photosynthetic process, releasing strong oxidizers that rapidly disrupt and inactivate cells and cellular functions (at present banned in many EU countries). Severe oral diquat intoxication has been associated with cerebral haemorrhages and severe acute renal failure (Peiró et al., 2007). Also quinoxaline herbicides (containing the pyrazine fragment) are very useful herbicides. Among them propaquizafop (Fig. 1, II) and quizalofop-ethyl (Fig. 1, III) are the most important derivatives (Frater et al., 1987; Sakata et al., 1983).

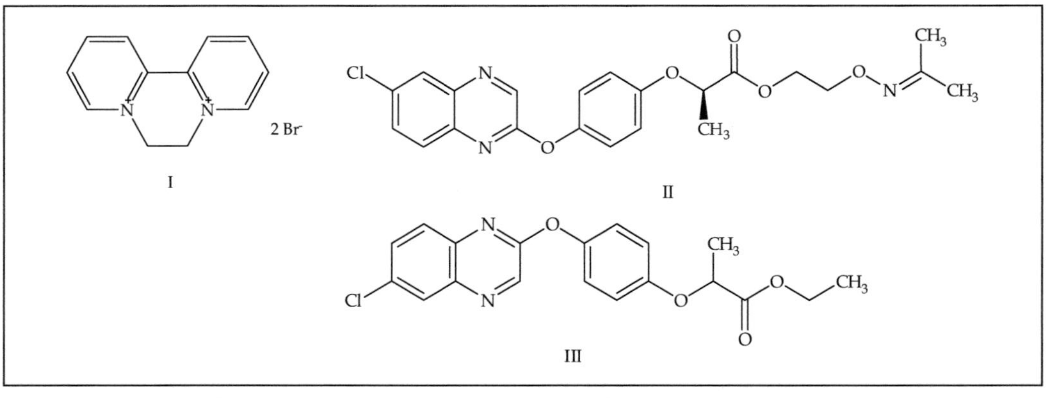

Fig. 1. Structures of diquat-dibromide (I), propaquizafop (II) and quizalofop-ethyl (III).

2.1 Diquat

Diquat-dibromide (6,7-dihydrodipyrido[1,2-a:2',1'-c]pyrazinediium-dibromide; for the structure see Fig. 1, I) is a quaternary ammonium salt used as a non-selective contact herbicide and desiccant, absorbed by the foliage with some translocation in the xylem. It is used for preharvest desiccation of many crops, as a defoliant on hops, for general weed control on non crop land etc. (Ritter et al., 2000; Ivany, 2005). It is applied as an aquatic herbicide in many countries since the late 1950s for control of emergent and submerged aquatic weeds (Ritter et al., 2000). According to Massachusetts Department of Agricultural Resources (2010) following weeds are controlled by diquat: *i)* submersed aquatics: *Ultricularia, Ceratophyllum demersum, Elodea* spp., *Najas* spp., *Myriophyllum* spp., *Hydrilla verticillata, Potamogeton* spp.; *ii)* floating aquatics: *Salvinia* spp., *Eichhornia crassipes, Pistia Stratiotes, Lemna* spp., *Hydrocotyle* spp.; *iii)* marginal weeds: *Typha* spp. ; *iv)* algae: *Pithophora* spp. , *Spyrogyra* spp. (filamentous algae). Diquat is stable in neutral and acidic solutions but unstable in alkaline medium. It breaks down by the UV radiation and the degradation increases with pH > 9 (Diaz et al., 2002). It is also biodegraded in water by microorganisms that uses this herbicide as a source of carbon or nitrogen (Petit et al., 1995).

Trade names for diquat-dibromide formulations included Desiquat®, Midstream®, Reglone®, and Reglex®. Mixtures of diquat with another quaternary herbicide paraquat (1,1'-dimethyl-4,4'-bipyridinium-dichloride) were sold under trade names including Actor®, Dukatalon®, Opal®, Pathclear® (also includes simazine and aminotriazole), Preeglox®, Preglone®, Seccatutto®, Spray Seed®, and Weedol® (Lock & Wilks, 2001).

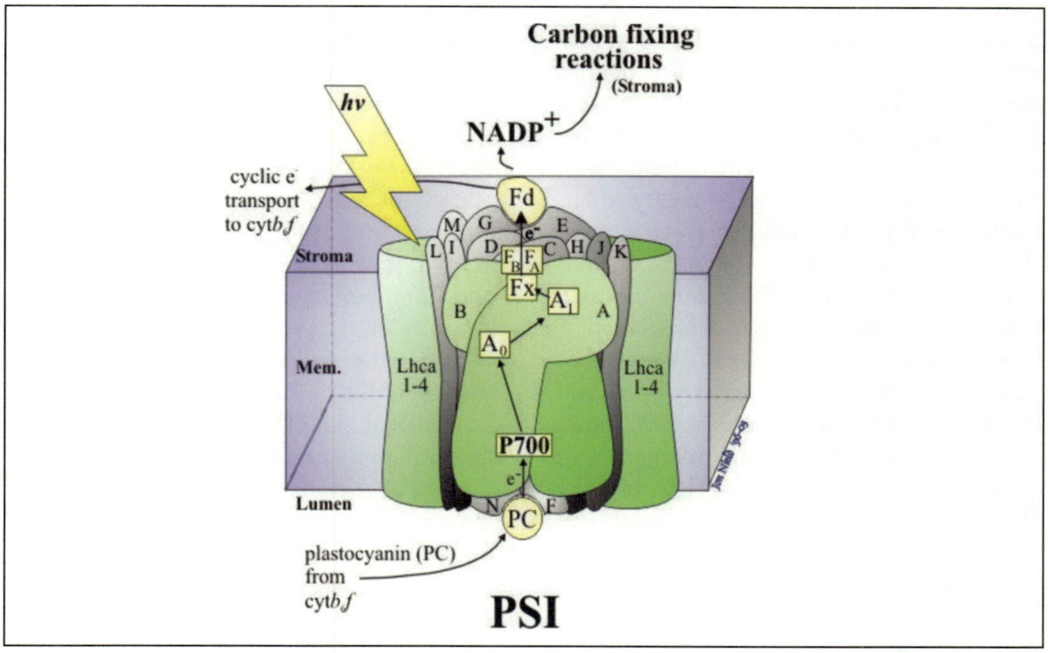

Fig. 2. Scheme of the photosynthetic electron transport in photosystem I (PS I). (Figure taken from http://www.bio.ic.ac.uk/research/barber/psIIimages/PSI.jpg with permission of Prof. Barber, Imperial College London).

The first paper dealing with the mode of action of diquat was published in 1960 by Mees who indicated that oxygen and light were essential for its herbicidal effect. Later Zweig et al. (1965) found that diquat caused a deviation of electron flow from photosystem (PS) I what resulted in an inhibition of $NADP^+$ reduction and the production of a reduced diquat radical. In Fig. 2 is shown scheme of the photosynthetic electron transport (PET) in PS I.

In plants, the PS I complex catalyzes the oxidation of plastocyanin and the reduction of ferredoxin (F_d). From the primary donor, P700, electrons are transferred to the primary acceptor, A_0 and then to phylloquinone (A_1) operating as a single electron acceptor. From A_1 electrons are transferred to a 4Fe-4S cluster (F_X) and subsequently to two 4Fe-4S clusters, F_A and F_B, located on the stromal side of the reaction center close to F_X. PS I produces a strong reductant that transfers electrons to F_d. Ferredoxin, one of the strongest soluble reductants found in cells, operates in the stromal aqueous phase of the chloroplast, transferring electrons from PS I to ferredoxin-$NADP^+$ oxidoreductase. The final electron acceptor in the photosynthetic electron transport chain is $NADP^+$, which is fully reduced by two electrons (and one proton) to form NADPH, a strong reductant which serves as a mobile electron carrier in the stromal aqueous phase of the chloroplast (Whitmarsch, 1998).

Due to deviation of electron flow from F_d, an inhibition of $NADP^+$ reduction occurs and a reduced diquat radical is formed. Davenport (1963) found that in the presence of oxygen the reduced diquat free radical was reoxidized with the production of hydrogen peroxide. Thus, an one-electron reduction of diquat results in a cation free radical that reacts rapidly with molecular oxygen and generates reactive oxygen species such as the superoxide anion radical (Mason, 1990). Reactive oxygen species cause oxidative stress in the cell with consecutive damage of biological membranes. In herbicide classification diquat, similarly to paraquat, is classified as HRAC Group D herbicide causing PS I electron diversion (HRAC 2005). Injury to diquat-treated crop plants occurs in the form of spots of dead leaf tissue wherever spray droplets contact the leaves indicating that this herbicide belongs to membrane disruptors. The use of diquat for the control of aquatic weeds is widespread in the US (US Environmental Protection Agency, 1995) whereas it is forbidden in the EU (European Commission, 2001, 2002).

As mentioned above, diquat toxicity to both aquatic plants and animals originates from the formation of reactive oxygen species in both chloroplasts and mitochondria (Cedergreen et al., 2006; Sanchez et al., 2006). The field effects of diquat to natural strands of aquatic vegetation were studied by Peterson et al. (1997) and Campbell at al. (2000). The filamentous cyanobacteria were slightly less tolerant than the unicellular cyanobacteria and the most sensitive was genus *Anabena* (Peterson et al., 1997). Gorzerino et al. (2009) showed that diquat, used as the commercial preparation Reglone 2®, inhibited the growth of *Lemna minor* in indoor microcosms. According to findings of Campbell et al. (2000) diquat has a minimal ecological impact to benthic invertebrates and fish; on the other hand, aquatic plants in the vicinity of application to surface waters appear to be at risk (nevertheless this is expected, as diquat-dibromide kills aquatic plants). However, Koschnick et al. (2006) observed that the accession of *Landoltia* from Lake County (Florida) had developed resistance to diquat and the resistance mechanism was independent of photosynthetic electron transport.

2.2 Patented pyrazine herbicides

The control of unwanted vegetation by means of chemical agents, *i.e.* herbicides, is an important aspect of modern agriculture and land management's. While many chemicals that are useful for the control of unwanted vegetation are known, new compounds that are more effective generally, are more effective for specific plant species, are less damaging to desirable vegetation, are safer to man or the environment, are less expensive to use or have other advantageous attributes, are desirable (Benko, 1997). Many structural variations of pyrazine compounds with herbicidal properties can be found in the patent literature.

Several thiazolopyrazines exhibited pre-emergent herbicidal activity when applied as aqueous drenches to soil planted with seeds of certain plants. For example, application of 4000 ppm of compound IV (Fig. 3) resulted in emergence inhibition of crabgrass (50% of the control) and barnyard grass (***Echinochloa crus-galli*** (L.) P. Beauv.) (45% of the control). Due to the treatment with a dose of 2 lb per acre of compound V (Fig. 3), the emergence of cotton reached only 30% of the control (Tong, 1978).

Böhner & Meyer (1989a, 1989b, 1990) prepared a set of aminopyrazinones (Fig. 3, VI) and aminotriazinones and tested these compounds for their herbicidal action before emergence of the plants. It was found that application of 70.8 ppm of some compounds on the substrate vermiculite resulted in very potent inhibition of seed germination of *Nasturtium officinalis, Agrostis tenuis, Stellaria media* and *Digitaria sanguinalis*. Due to the treatment with compound where $R^1 = CH_3$, $R^2 = OCH_3$, $R^3 = H$, $R^7 = H$, $R^8 = COOCH_3$, $X = O$ plants have not germinated and completely died. After spraying of 21 days old spring barley (*Hordeum vulgare*) and spring rye (Secale) plants shoots with an active substance VI (up to 100 g per hectare) new additional growth of plants reached only 60-90% of the control. For grasses *Lolium perenne, Poa pratensis, Festuca ovina, Dactylis glomerate* and *Cynodon dactylon* sprayed with the same dose of an active substance (Fig. 3, VII) reduction in new additional growth in comparison with the untreated control (10-30% of control) was observed, too (Böhner & Meyer, 1989a, 1989b, 1990).

Benko et al. (1997) patented a series of *N*-aryl[1,2,4]triazolo[1,5-a]pyrazine-2-sulfonamides as good pre- and post-emergence selective herbicides with good growth regulating properties. Excellent pre-emergence activity against pigweed and morning glory and very good post-emergence herbicidal activity against morning glory and velvet leaf (*Abutilon theophrasti*) have been exhibited by the title compounds.

Dietsche (1977) patented as herbicides a group of substituted 6,7-dichloro-3,4-dihydro-2*H*-pyrazino(2,3-*b*)(1,4)oxazines showing hundred-percent inhibitory effectiveness when applied as pre- as well as post-emergence herbicides (4000 ppm) for pigweeds.

Shuto et al. (2000) patented as useful active ingredients of herbicides a series of pyrazin-2-one derivatives (Fig. 3, VIII, IX) where R^1 is hydrogen or alkyl, R^2 is haloalkyl, R^3 is optionally substituted alkyl, alkenyl or alkynyl and Q is optionally substituted phenyl. Some compounds showed superb effectiveness against *Abtutilon theophrasti* and *Ipomoea hederacea* when applied as foliar or soil surface treatment on upland fields (2000 g/ha).

Griffin et al. (1990) patented alkylpyrazine compounds (Fig. 3, X) with plant growth regulating activity, where R^1 is C_1-C_4 alkyl optionally substituted with halogen or cyclopropyl, optionally substituted with C_1-C_4 alkyl; R^2 is C_1-C_8 alkyl, C_2-C_8 alkenyl, or C_2-C_8 alkynyl optionally substituted with halogen; C_3-C_6 cycloalkyl, C_3-C_6 cycloalkenyl. C_3-C_6 cycloalkylalkyl, C_3-C_6 cycloalkenylalkyl, phenylalkenyl or phenylalkynyl each optionally substituted on the ring group; R^3 is hydrogen or C_1-C_4 alkyl; R^4 is hydrogen, C_1-C_4 alkyl, halogen, alkylamino, cyano, or alkoxy; n is 0 or 1; and salts, ethers, acylates and metal complexes therof. The treatment of plants with these compounds can lead to the leaves developing a darker green colour. In dicotyledonous plants such as soybean and cotton, there may be promotion of side shooting. The compounds may be useful in rendering plants resistant to stress since they can delay the emergence of plants grown from seeds, shorten stem height and delay flowering. Engel et al. (1999) patented herbicidal pyrazine derivatives (Fig. 3, XI) which are suitable very effectively control weeds and grass weeds mainly in crops such as wheat, rice, corn, soybean and cotton, without significantly damaging the crops. It could be stressed that this effect occurs in particular at low application rates. In addition, these compounds can also be used in crops which have been made substantially resistant to the action of herbicides by breeding and/or by the use of genetic engineering methods.

N-pyrazinyl-haloacetamides (Fig. 3, XII) where R is hydrogen, hydrocarbonyl, halogen, epoxy, hydroxy, alkoxy, mercapto, alkylsulfanyl, nitro, cyano or amino, R´ is hydrogen or hydrocarbonyl, X is halogen, m is integer from 1 to 4 and n is 0, 1 or 2 showed herbicidal activity. For example, spraying of the 2,2,2-trichloro-*N*-pyrazinyl acetamide on the soil resulted in 100% growth inhibition of wild oats (dosage 1.12 g m^{-2}) and yellow foxtail or cultured rice (dosage 1.12 g m^{-2}) (Fischer, 1988).

Novel pyrazine-sulfonylcarbamates and thiocarbamates (Fig. 3, XIII) (where Z is oxygen or sulfur and R is C_1-C_4 alkyl, phenyl or benzyl; whereas the pyrazine ring may be variously further substituted) have been found to be good selective herbicides and therefore they are suitable for use in crops of cultivated plants. Moreover, these compounds can damage problem weeds which till then have only been controlled with total herbicides (Böhner et al., 1987). By means of surface treatment it is possible to damage perennial weeds to their roots. Moreover, the compounds are effective when used in very low rates of application and they are able to potentiate the phytotoxic action of other herbicides against certain noxious plants and to reduce the toxicity of such herbicides to some cultivated plants. These compounds can be used also as plant growth regulators causing inhibition of vegetative plant growth what results in substantial increase of the yield of plants. Böhner et al. (1987) synthesized and patented also a set of novel pyrazinyl sulfonamides of the formula Q-SO_2-NH_2 where Q is substituted pyrazine group which could be useful in controlling weeds and are suitable for selectively influencing plant growth. The compounds can be used as pre- and post-emergence herbicides and as plant growth regulators for growth inhibition of cereals (e.g. *Hordeum vulgare* or summer rye (Secale)) and grasses (e.g. *Lolium perenne, Poa partensis, Festuca ovina, Cynodon dactylon*). Selective inhibition of the vegetative growth of many cultivated plants permits more plants to be grown per unit of crop area, resulting in significant increase in yield with the same fruit setting and in the same crop area.

Zondler et al. (1989) prepared a set of 2-arylmethyliminopyrazines (Fig. 3, XIV) and tested them for their pre-emergent and post-emergent herbicidal action, as well as for their plant growth regulating activity. Compounds with R^5 = 4-Cl, R^6 = 2-Cl, R^7 = H and R^1 = SCH_3H_7(n) or $SCH_2CH{=}CH_2$ showed excellent pre-emergent effect (dose 4 kg/ha) against *Echinochloa crus-galli* and *Monocharia vag*. The last compound was active already at application rate of 500 g/ha. The 2-arylmethylimino-pyrazines were found to be also effective post-emergence herbicides and can be used for growth inhibition of tropical leguminous cover crops (e.g. *Centrosema plumieri* and *Centrosema pubescens*), growth regulation in soybeans and growth inhibition of cereals, too.

Cyanatothiomethylthiopyrazines have been found to be active as pesticides and find particular usage as fungicides, bactericides, nematocides and herbicides (Mixan et al., 1978).

Arylsulfanylpyrazine-2,3-dicarbonitriles have high herbicidal activity (Takematsu et al., 1984; Portnoy, 1978). Takematsu et al. (1981) patented 2,3-dicyanopyrazines (Fig. 3, XV) as compounds with high herbicidal activity as well as useful active ingredients of herbicides. The compounds have ability to inhibit the germination of weeds and/or wither their stems and leaves, and therefore exhibit an outstanding herbicidal effect as an active ingredient of pre-emergence and/or post-emergence herbicides in submerged soil treatment, foliar treatment of weeds, upland soil treatment, etc.

Compounds where A represents a phenyl group which may have 1 or 2 substituents selected from the class consisting of halogen atoms and lower alkyl groups containing 1 to 3 carbon atoms and B represents an ethylamino, *n*-propylamino, *n*- or *iso*-butylamino, 1-carboxyethylamino, 1-carboxy-*n*-propylamino, 1-carboxy-*iso*-butylamino, 1-carboxy-*n*-pentylamino or allylamino group have the property of selectively blanching (causing chlorosis, *i.e.* inhibiting the formation of chlorophyll and/or the acceleration of its decomposition) of weeds without chlorosis of useful crops. Hence, these compounds are most suitable as high selective herbicides of chlorosis type.

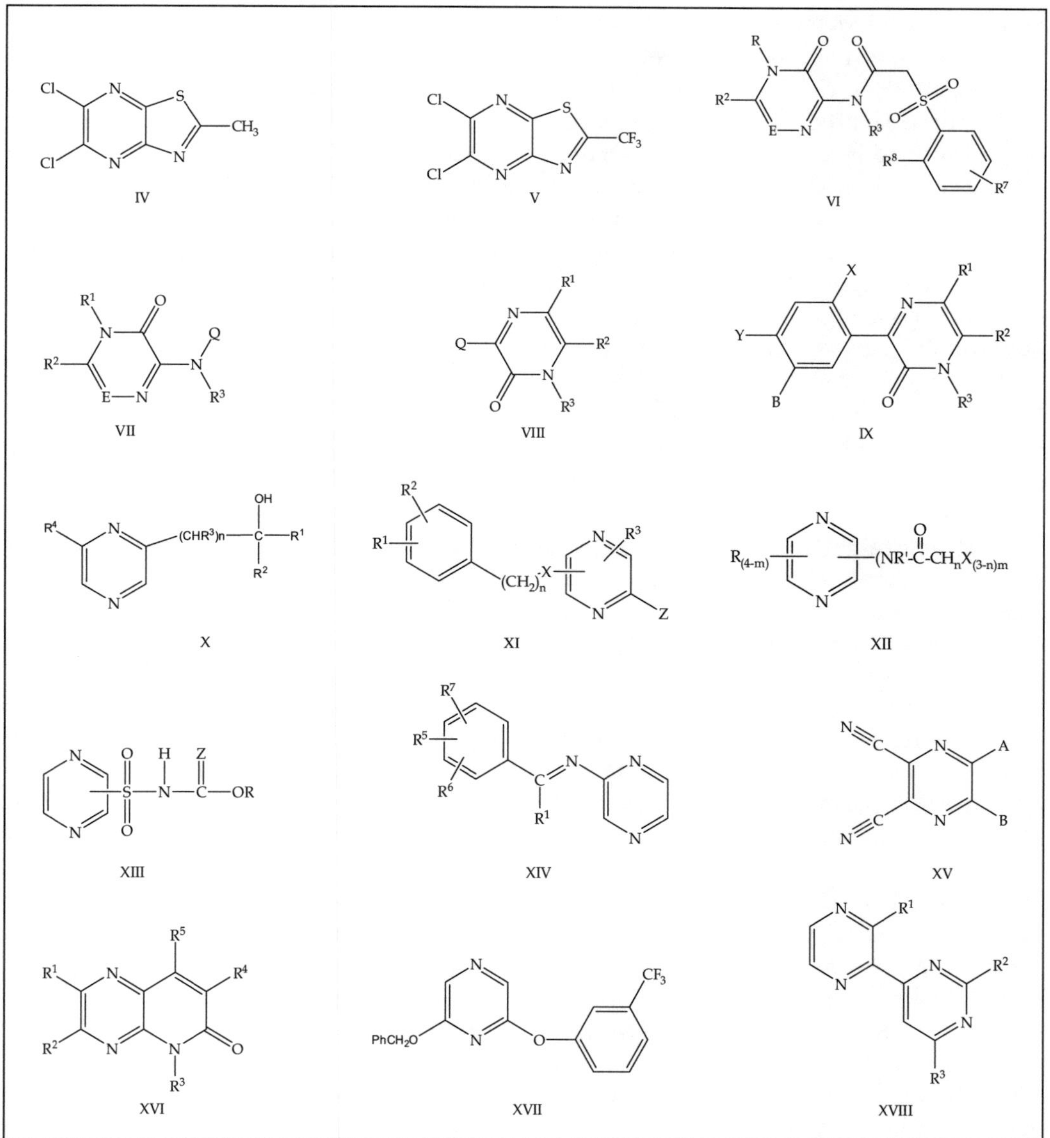

Fig. 3. Structures of patented thiazolopyrazines (IV,V), aminopyrazinones (VI,VII), substituted pyrazin-2-ones (VIII,IX), arylalkylpyrazines (X, XI), *N*-pyrazinyl-haloacetamides (XII), pyrazine-sulfonylcarbamates and thiocarbamates (XIII), 2-arylmethyliminopyrazines (XIV), substituted 2,3-dicyanopyrazines (XV), pyridopyrazines (XVI), aryloxopyrazines (XVII) and pyrimidinopyrazines (XVIII).

Takematsu et al. (1984) also patented a set of 2,3-dicyano-6-phenylpyrazine herbicides with outstanding herbicidal activities on paddy weeds in submerged soil treatment. Because they are not phytotoxic to rice, they can effectively control weeds in paddies. The compounds exhibited herbicidal activity against important upland weeds such are *Digitaria adscendens, Polygonum persicaria, Galinsoga ciliata, Amaranthus viridis, Chenopodium album, Chenopodium ficifolium, Echinochloa crus-galli* (without damaging upland crops) as well as against a very broad range of other upland weeds including *Galium aparin, Rumex japonicus, Erigeron philadelphicus, Erigeron annuus*, and *Capsella bursapastoria*.

Cordingley et al. (2008) prepared herbicidal effective pyridopyrazines (Fig. 3, XVI) with R^1,R^2 independently = H, alkyl, halo, CN, aryl, etc.; R^3 = H, (halo)alkyl, alkenyl, etc.; R^4 = (un)substituted heteroaryl; and R^5 = OH or group metabolizable to OH) or a salt or *N*-oxide thereof. XVI applied post-emergence at 1000 g/ha completely controlled *Solanum nigrum* and *Amaranthus retroflexus*. Also substituted aryloxopyrazines (Fig. 3, XVII) possess interesting herbicidal effect (Niederman & Munro, 1994). For example, in tests against 8 plants, title compound XVII at 5 kg/ha (foliar spray) gave complete kill of *Echinochloa crus-galli* with no damage to rice. Test data include foliar, pre-emergence, and soil drench applications against the 8 plants for most compounds. Sato et al. (1993) patented pyrimidinopyrazines (Fig. 3, XVIII) (R^1 = H, halo, alkoxy, alkylamino, alkyl, haloalkyl; R^2 = Ph, substituted Ph, benzyl, pyridyl, thienyl, furyl; R^3 = SR^4, OR^5, NR^6R^7; R^4,R^5,R^6,R^7 = H, alkyl, alkenyl, alkynyl; NR^6R^7 may form 3-7 membered ring), useful as herbicides, were prepared and showed herbicidal activity against *Stellaria neglecta* at 0.63 kg/ha.

2.2.1 Structure-activity relationships in series of herbicidal 2,3-dicyanopyrazines

Nakamura et al. (1983) synthesized sixty six 2,3-dicyano-5-substituted pyrazines and measured their herbicidal activities against barnyard grass in pot tests to clarify the relationship between chemical structure and activity. The activity of 59 derivatives showed parabolic dependence on the hydrophobic substituent parameter at the 5-position of the pyrazine ring, indicating that the compounds should pass through a number of lipoidal-aqueous interfaces to reach a critical site for biological activity. It was found that the moiety of 2,3-dicyanopyrazine is essential for herbicidal activity, and the 5-substituent on the pyrazine ring plays an important role in determining the potency of this activity and that *para*-substituted phenyl derivatives show undesirable effects on the potency of the activity at the ultimate site of herbicidal action.

Nakamura et al. (1983a) also synthesized sixty eight 6-substituted 5-ethylamino and 5-propylamino-2,3-dicyanopyrazines and tested their herbicidal activities against barnyard grass using pot tests. In general, these compounds induced chlorosis against young shoots of barnyard grass and inhibited their growth. The most active compound was 2,3-dicyano-5-propylamino-6-(*m*-chlorophenyl)-pyrazine. The results indicated that the structure of the 5-ethylamino and 5-propylamino-2,3-dicyanopyrazine moieties is an important function for the herbicidal activity and that the potency of activity of these two series of compounds is determined by the hydrophobic and steric parameters of substituents at the 6-position of the pyrazine ring.

3. Design, synthesis and evaluation of the pyrazinecarboxamides with herbicidal activity

The structural diversity of organic herbicides continues to increase; therefore classification of herbicides should be based on their chemical structure. The chlorinated aryloxy acids dominated for long period, later were replaced by chemicals of many distinct chemical classes, including triazines, amides (haloacetanilides), benzonitriles, carbamates, thiocarbamates, dinitroanilines, ureas, phenoxy acids, diphenyl ethers, pyridazinones, bipyridinium compounds, ureas and uracils, sulfonylureas, imidazolinones, halogenated carboxylic acids, and many other compounds. Carboxamide or anilide moieties are present in many used herbicides, *i.e.* alachlor, acetochlor, benoxacor, butachlor, diflufenican, dimethenamid, diphenamid, isoxaben, karsil, napropamide, pretilachlor, propyzamide, dicryl, diflufenican, flufenacet, mefenacet, mefluidide, metolachlor, naphtalan, picolinafen, propachlor, propanil, propham, solan (The Merck Index, 2006). Carboxamide or anilide herbicides are nonionic and moderately retained by soils. The sorption of several carboxamide herbicides has been investigated (Weber & Peter, 1982). The *N*-substituted phenyl heterocyclic carboxamides are an important class of herbicides as protoporphyrinogen-IX oxidase inhibitors with advantages such as high resistance to soil leaching, low toxicity to birds, fish, and mammals, and slow development of weed resistance (Hirai, 1999).

We have designed and prepared a series of 113 carboxamide herbicides derived from pyrazinecarboxylic acid and various substituted anilines. The final compounds XIX were prepared by the anilinolysis of substituted pyrazinoylchlorides (Doležal, 1999, 2000, 2002, 2006b, 2007, 2008a, 2008b). Their chemical structure, hydrophobic parameters (log *P* calculated by ACD/logP ver. 1.0, 1996), and photosynthesis-inhibiting activity, structure-activity relationship (SAR) were studied. We synthesized in preference: *i)* the compounds with the lipophilic and/or electron-withdrawing substituents on the benzene moiety (R^3), *ii)* the compounds with the hydrophilic and/or electron-donating groups on the benzene part of molecule (R^3), and finally *iii)* the compounds with the lipophilic alkyl (R^2), *i.e.* methyl (-CH_3) or *tert*-butyl (-$C(CH_3)_3$) and/or halogen (chlorine) substitution (R^1) on the pyrazine nucleus, for their synthesis and structure see Fig. 4 and Table 1.

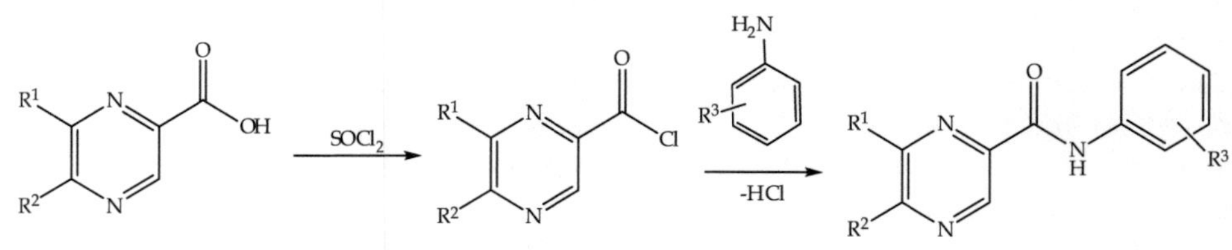

XIX

Fig. 4. Synthesis and structure of substituted *N*-phenylpyrazine-2-carboxamides (XIX).

3.1 Inhibition of photosynthetic electron transport by substituted *N*-phenylpyrazine-2-carboxamides

3.1.1 Photosynthetic electron transport in photosystem II

Photosystem II uses light energy to drive two chemical reactions: the oxidation of water and the reduction of plastoquinone. Five of redox components of PS II are known to be involved in transferring electrons from H_2O to the plastoquinone pool: the water oxidizing manganese cluster $(Mn)_4$, the amino acid tyrosine (Y_z), the reaction center chlorophyll (P680), pheophytin, and two plastoquinone molecules, Q_A and Q_B (Fig. 5). Tyrosine, P680, pheophytin (Pheo), Q_A, and Q_B are bound to two key polypeptides (D_1 and D_2) that form the reaction center core of PS II and also provide ligands for the $(Mn)_4$ cluster (Whitmarsh,

1998). After primary charge separation between P680 (chlorophyll *a*) and pheophytin (Pheo), $P680^+/Pheo^-$ is formed. Then electron is subsequently transferred from pheophytin to a plastoquinone molecule Q_A (permanently bound to PS II) acting as a one-electron acceptor.

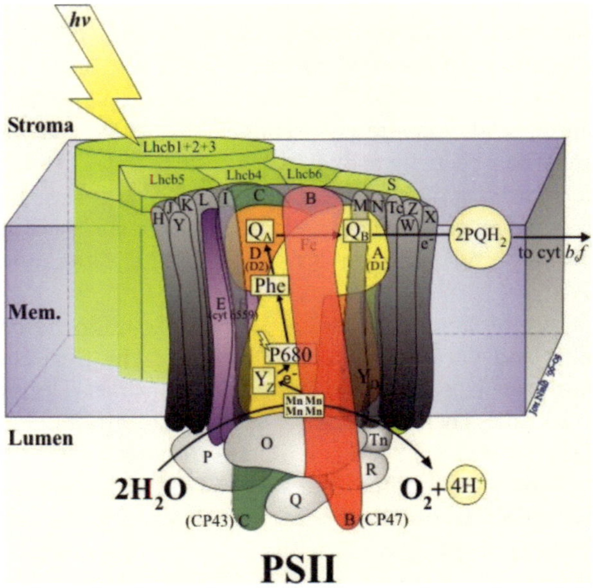

Fig. 5. Scheme of the photosynthetic electron transport in photosystem II (PS II). (Taken from Photosystem II in http://www.bio.ic.ac.uk/research/barber/psIIimages/PSII.jpg with permission of Prof. Barber, Imperial College London).

From Q_A^- the electron is transferred to another plastoquinone molecule Q_B (acting as a two-electron acceptor); two photochemical turnovers of the reaction centre are necessary for the full reduction and protonation of Q_B. Because Q_B is loosely bound at the Q_B-site, reduced plastoquinone then unbinds from the reaction centre and diffuses in the hydrophobic core of the membrane and Q_B-binding site will be occupied by an oxidized plastoquinone molecule (Whitmarsh, 1998). Several commercial herbicides inhibit Photosynthetic elektron transport (PET) by binding at or near the Q_B-site, preventing access to plastoquinone (*e.g.* Oettmeier, 1992). Photosystem II is the only known protein complex that can oxidize water, which results in the release of O_2 into the atmosphere. Oxidation of water is driven by the oxidized primary electron donor, $P680^+$ which oxidizes a tyrosine on the D_1 protein (Yz) and four Mn ions present in the water oxidizing complex undergo light-induced oxidation, too. Water oxidation requires two molecules of water and involves four sequential turnovers of the reaction centre whereby each photochemical reaction creates an oxidant that removes one electron. The net reaction results in the release of one O_2 molecule, the deposition of four protons into the inner water phase, and the transfer of four electrons to the Q_B-site (producing two reduced plastoquinone molecules) (Whitmarsh & Govindjee, 1999).

PET in chloroplasts can be estimated by electrochemical measurements of oxygen concentration using Clark electrode (PET through the whole photosynthetic apparatus is registered) or by spectrophotometric methods enabling the monitoring of PET through individual parts of photosynthetic apparatus. The site of action of PET inhibitors can be more closely specified by the use of chlorophyll fluorescence (*e.g.* Joshi & Mohanty, 2004) or by electron paramagnetic resonance (EPR) (*e.g.* Doležal et al., 2001a).

3.1.2 Hill reaction activity of *N*-phenylpyrazine-2-carboxamides

The Hill reaction is formerly defined as the photoreduction of an electron acceptor by the hydrogens of water, with the evolution of oxygen. *In vivo,* or in the organism, the final electron acceptor is $NADP^+$, in isolated chloroplasts an artificial electron acceptor that changes colour as it is reduced, is applied. We tested a large series of pyrazinecarboxamides (XIX) for their activity related to oxygen evolution rate (OER) using spinach chloroplasts and 2,6-dichlorophenol-indophenol (DCPIP) as an electron acceptor what intercepts the electrons before they transfer to cytochrome *bf* complex. Because the site of DCPIP action is plastoquinone pool (PQ) on the acceptor side of PS II (Izawa, 1980) this method is suitable for PET monitoring through PS II. The PET-inhibiting activities of the studied compounds XIX (expressed as IC_{50} values) are summarized in Table 1.

No.	R^1	R^2	R^3	IC_{50}	Ref.	No.	R^1	R^2	R^3	IC_{50}	Ref.
1	Cl	H	2-Br	334	a	**58**	Cl	tBu	2-Cl,5-OH	652	i
2	H	tBu	2-Br	171	a	**59**	H	CH_3	3-Br	648	b
3	Cl	tBu	2-Br	315	a	**60**	H	CH_3	3-C≡CH	668	b
4	Cl	H	3,5-Br-4-OH	995	a	**61**	Cl	tBu	3-C≡CH	385	b
5	H	tBu	3,5-Br-4-OH	404	a	**62**	Cl	tBu	3-C≡N	375	b
6	Cl	tBu	3,5-Br-4-OH	590	a	**63**	H	CH_3	3-Cl	174	b
7	Cl	H	3-OCH_3	500	a	**64**	H	CH_3	3-NO_2	402	b
8	H	tBu	3-OCH_3	800	a	**65**	H	CH_3	2-C≡N-4-NO_2	550	b
9	Cl	tBu	3-OCH_3	644	a	**66**	H	CH_3	3-I-4-CH_3	317	b
10	Cl	H	3,5-OCH_3	533	a	**67**	H	CH_3	2-COOH	75	b
11	H	tBu	3,5-OCH_3	317	a	**68**	Cl	tBu	3-F	262	c
12	Cl	tBu	3,5-OCH_3	435	a	**69**	H	tBu	3-OH-4-Cl	105	c
13	Cl	H	5-Br-2-OH	146	a	**70**	Cl	tBu	3-OH-4-Cl	44	c
14	H	tBu	5-Br-2-OH	80	a	**71**	Cl	tBu	2-Cl	43	c
15	Cl	tBu	5-Br-2-OH	42	a	**72**	H	tBu	2-Cl	371	c
16	Cl	H	3,4-Cl	105	a	**73**	H	H	2-Cl	47	c
17	H	tBu	3,4-Cl	1525	a	**74**	Cl	H	2-CH_3	1072	e
18	Cl	tBu	3,4-Cl	130	a	**75**	H	tBu	2-CH_3	440	e
19	Cl	H	3-F	565	d	**76**	Cl	tBu	2-CH_3	244	e
20	Cl	H	2,4-F	539	d	**77**	Cl	H	3-CH_3	486	e
21	Cl	H	4-Cl	486	d	**78**	H	tBu	3-CH_3	148	e
22	Cl	H	4-$CH(CH_3)_2$	118	d	**79**	Cl	tBu	3-CH_3	118	e
23	H	tBu	3-F	313	d	**80**	H	tBu	2-OCH_3	286	e
24	H	tBu	2,4-F	371	d	**81**	Cl	tBu	2-OCH_3	97	e
25	H	tBu	4-Cl	1502	d	**82**	Cl	H	3-Br	313	e
26	H	tBu	4-$CH(CH_3)_2$	110	d	**83**	H	tBu	3-Br	81	e
27	Cl	tBu	3-F	129	d	**84**	Cl	tBu	3-Br	107	e
28	Cl	tBu	2,4-F	106	d	**85**	Cl	H	3,5-CF_3	26	e

No.	R1	R2	R3	IC50	Ref.	No.	R1	R2	R3	IC50	Ref.
29	Cl	tBu	4-Cl	43	d	**86**	H	tBu	3,5-CF_3	114	e
30	Cl	tBu	4-$CH(CH_3)_2$	52	d	**87**	Cl	tBu	3,5-CF_3	241	e
31	Cl	H	2-OH	66	f	**88**	Cl	H	2,6-CH_3	649	e
32	Cl	H	3-OH	2288	f	**89**	H	tBu	2,6-CH_3	229	e
33	Cl	H	4-OH	3322	f	**90**	Cl	tBu	2,6-CH_3	242	e
34	Cl	H	2-OH-5-Cl	8	f	**91**	H	H	2-Cl-5-OH	722	g
35	H	tBu	2-OH	205	f	**92**	H	H	4-F	480	g
36	H	tBu	3-OH	431	f	**93**	H	H	2-CF_3	376	g
37	H	tBu	4-OH	314	f	**94**	H	H	3-CF_3	130	g
38	H	tBu	2-OH-5-Cl	465	f	**95**	H	H	4-CH_3	1475	g
39	Cl	tBu	2-OH	435	f	**96**	Cl	H	2-Cl-5-OH	624	g
40	Cl	tBu	3-OH	262	f	**97**	Cl	H	4-F	384	g
41	Cl	tBu	4-OH	43	f	**98**	Cl	H	2-CF_3	557	g
42	Cl	tBu	2-OH-5-Cl	105	f	**99**	Cl	H	3-CF_3	229	g
43	Cl	H	4-Cl-3-CH_3	595	h	**100**	Cl	H	4-CH_3	1524	g
44	Cl	H	3-I-4-CH_3	51	h	**101**	H	tBu	4-F	524	g
45	H	tBu	4-Cl-3-CH_3	190	h	**102**	H	tBu	2-CF_3	55	g
46	Cl	tBu	2-F	69	h	**103**	H	tBu	3-CF_3	283	g
47	Cl	tBu	4-CF_3	184	h	**104**	H	tBu	4-CH_3	164	g
48	H	H	4-F	480	i	**105**	Cl	tBu	2-Cl-5-OH	625	g
49	Cl	H	4-F	384	i	**106**	Cl	tBu	4-F	103	g
50	H	tBu	4-F	524	i	**107**	Cl	tBu	2-CF_3	205	g
51	Cl	tBu	4-F	103	i	**108**	Cl	tBu	3-CF_3	173	g
52	H	H	3-Cl	290	i	**109**	Cl	tBu	4-CH_3	73	g
53	Cl	H	3-Cl	262	i	**110**	Cl	H	2,4,6-CH_3	495	j
54	H	tBu	3-Cl	47	i	**111**	H	tBu	2,4,6-CH_3	434	j
55	H	tBu	3-Cl	103	i	**112**	Cl	tBu	2,4,6-CH_3	195	j
56	H	H	2-Cl-5-OH	722	i	**113**	H	tBu	4-$COCH_3$	664	j
57	Cl	H	2-Cl-5-OH	624	i	-	-	-	-	-	-

Table 1. IC_{50} values (in μmol dm^{-3}) related to PET inhibition in spinach chloroplasts by substituted pyrazinecarboxamides XIX (Ref. Doležal et al., 2006b[(a)], 2008a[(b)], 2001b[(c)], 2000[(d)], 2002[(e)], 1999[(f)], 2008b[(g)], 2007[(h)], 2004[(i)], 2001a[(j)]).

The compounds **1-18** inhibited PET in spinach chloroplasts; however the inhibitory activity of the majority of these compounds was relatively low. The IC_{50} values varied in the range from 42 to 1589 μmol dm^{-3}, the most efficient inhibitors was 5-*tert*-butyl-6-chloro-*N*-(5-bromo-2-hydroxyphenyl)-pyrazine-2-carboxamide (**15**, Table 1). The dependence of PET-inhibiting activity of compounds **1-18** on the lipophilicity of the compounds (log *P*) is shown in Fig. 6, A. Markedly lowered solubility of **4-6** as well as **17** due to insertion of two halogen atoms (Br or Cl) in R^3 substituent resulted in decreased inhibitory activity of these compounds. Based on the dependence of PET-inhibiting activity on log *P* of the rest compounds, these can be divided into two groups. In both groups increase of compound activity with increasing lipophilicity can be observed. Thus, with the exception of compounds **14** and **15** (R^2 = 5-Br-2-OH) it can be assumed, that the introduction of lipophilic

R^1 (Cl) and R^2 (*tert*-butyl, tBu) substituents, respectively, can result in partial decrease of the aqueous solubility and so in reduced inhibitory activity.

In other set of studied compounds **19-30**, compound **25** exhibited very low activity due to its low aqueous solubility (Table 1). As shown (Fig. 6, B), the PET-inhibiting activity of other compounds from the set expressed as log ($1/IC_{50}$) increased linearly with increasing compound lipophilicity (log *P*). The most active compounds from the set were 5-*tert*-butyl-6–chloro-*N*-(4-chlorophenyl)-pyrazine-2-carboxamide (**29**, IC_{50} = 43 μmol dm^{-3}) and 5-*tert*-butyl-6–chloro-*N*-(4-isopropylphenyl)-pyrazine-2-carboxamide (**30**, IC_{50} = 52 μmol dm^{-3}).

The inhibitory activity of the compounds **31-42** (Table 1) was affected not only by the lipophilicity of the compounds but also by the value of Hammett's constants of R^3 substituents. Very low activity of compounds **32** and **33** was connected with their low aqueous solubility. The most active compounds from this set were 6–chloro-*N*-(5-chloro-2-hydroxyphenyl)-pyrazine-2-carboxamide (**34**, IC_{50} = 8 μmol dm^{-3}) and 5-*tert*-butyl-6–chloro-*N*-(4-hydroxyphenyl)-pyrazine-2-carboxamide (**41**, IC_{50} = 43 μmol dm^{-3}), the activity of rest compounds from the set varied between 66 (**31**) and 465 μmol dm^{-3} (**38**).

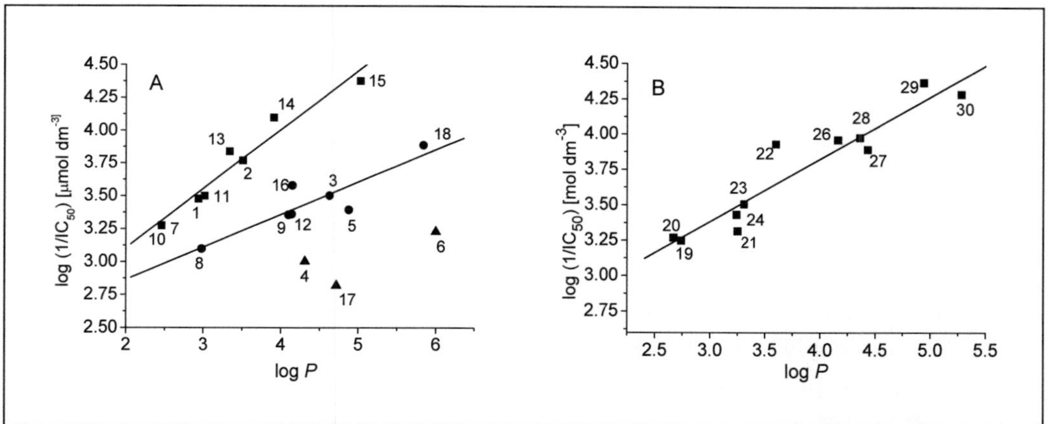

Fig. 6. The dependence of PET-inhibiting activity of compounds **1-18** (A) and compounds **19-30** (B) on the lipophilicity of the compounds (log *P*).

It was found that from the aspect of inhibitory activity it is much more favourable when on the phenyl ring (R^3 substituent) halogen atom occurs in *meta* and methyl moiety in *para* position (**44**, IC_{50} = 51 μmol dm^{-3}) in comparison with compound **43** where R^3 =4-Cl-3-CH_3 (IC_{50} = 595 μmol dm^{-3}). However, the inhibitory activity of the above mentioned compound **43** can be increased by introduction of *tert*-butyl substituent instead of H in R^2 (**45**, IC_{50} =190 μmol dm^{-3}). The IC_{50} values related to PET-inhibiting activity of compounds **48-58** varied in the range from 47.0 (**54**) to 722 μmol dm^{-3} (**56**). The inhibitory activity of majority of these compounds was relatively low, the most efficient inhibitors were 5-*tert*-butyl-6–chloro-*N*-(4-fluorophenyl)-pyrazine-2-carboxamide (**51**), *N*-(2-chloro-5-hydroxyphenyl)-pyrazine-2-carboxamide (**55**, both IC_{50} = 103.0 μmol dm^{-3}), and especially 5-*tert*-butyl-6–chloro-*N*-(3-chlorophenyl)-pyrazine-2-carboxamide (**54**, IC_{50} = 47.0 μmol dm^{-3}). Their log *P* values calculated ranged between 3.28 and 4.18.

In the set of compounds **59-67** the PET-inhibiting activity of compounds **61, 62, 63, 66** and **67** (Fig. 7, A) expressed as log ($1/IC_{50}$) showed a linear decrease with increasing values of lipophilicity parameter (log *P*). On the other hand, the biological activity of compounds **59,**

60, **64** and **65** was significantly lower and linear decrease of PET-inhibiting activity with increasing log *P* values was less sharp indicating that the biological activity of compounds **59-67** depended both on the compound lipophilicity as well as on Hammett's constants σ of the substituent R^2. The most active PET inhibitor from this set was found to be 2-(5-methylpyrazine-2-carboxamido)-benzoic acid (**67**, IC_{50} = 75.0 μmol dm^{-3}) (Doležal et al., 2008a).
From the set of compounds **68-73** the most active inhibitors with comparable inhibitory activity were compounds 5-*tert*-butyl-6–chloro-*N*-(3-chloro-4-hydroxyphenyl)-pyrazine-2-carboxamide (**70**, IC_{50} = 44 μmol dm^{-3}), 5-*tert*-butyl-6–chloro-*N*-(2-chlorophenyl)-pyrazine-2-carboxamide (**71**, IC_{50} = 43 μmol dm^{-3}) and *N*-(2-chlorophenyl)-pyrazine-2-carboxamide (**73**, IC_{50} = 47 μmol dm^{-3}).

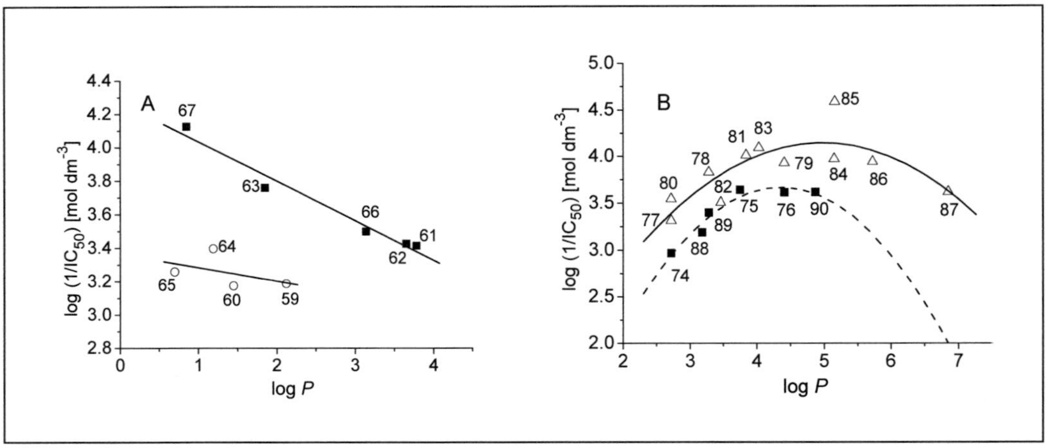

Fig. 7. The dependence of PET-inhibiting activity of compounds **59-67** (A) and compounds **74-90** (B) on the lipophilicity of the compounds (log *P*).

In the set of compounds **74-90** the IC_{50} values related to PET inhibition varied in the range from 26 (**85**) to 1072 μmol dm^{-3} (**74**), see Table 1. In general, the inhibitory activity of these compounds depended on their lipophilicity showing a quasi-parabolic trend (Fig. 7, B). However, the studied compounds could be divided into two groups. The compounds with 2-CH_3 substituents on the phenyl ring (**74**, **75**, **76**, **88**, **89** and **90**, squares in Fig. 7, B) had lower biological activity than the other investigated compounds with comparable log *P* values. Consequently, it can be assumed that the methyl substituent in *ortho* position of the benzene ring is disadvantageous from the viewpoint of interactions with the photosynthetic apparatus. On the other hand, compound **85** (6–chloro-*N*-(3,5-trifluoro-methylphenyl)-pyrazine-2-carboxamide) exhibited higher inhibitory activity than expected.
The majority of compounds **91-109** inhibited PET in spinach chloroplasts; however their inhibitory activity was rather low. From the obtained results it can be concluded that the activity depended on the lipophilicity and also on the electron accepting or withdrawing power of R^3 substituent(s). The most effective inhibitor was compound **102** (5-*tert*-butyl-*N*-(2-trifluoromethylphenyl)-pyrazine-2-carboxamide, IC_{50} = 55 μmol dm^{-3}). Among the three most active compounds **102**, **109** and **106** the optimal values of lipophilicity ranges from log *P* = 4.02-4.41. On the other hand, for the group of compounds **105**, **108** and **107** with the highest lipophilicity, the PET-inhibiting activity showed a decrease with increasing compound lipophilicity. The most effective inhibitor from the compounds with R^3= 2,4,6-

CH_3 was 5-*tert*-butyl-6-chloro-*N*-(2,4,6-methylphenyl)-pyrazine-2-carboxamide (**112**, IC_{50} = 195 μmol dm^{-3}) (Doležal et al., 2001a).

3.1.3 Determination of the site of inhibitory action of *N*-phenylpyrazine-2-carboxamides in the photosynthetic electron transport chain by electron paramagnetic resonance spectroscopy and chlorophyll a fluorescence measurements

The site of inhibitory action of some *N*-phenylpyrazine-2-carboxamides XIX in the photosynthetic electron transport chain was investigated using spinach (*Spinacia oleracea* L.) chloroplasts. For this purpose electron paramagnetic resonance spectroscopy (EPR) and measurement of chlorophyll *a* fluorescence were used.

Intact chloroplasts of algae and vascular plants exhibit EPR signals in the region of free radicals (g = 2.00), which are stable during several hours (Hoff, 1979) and could be registered at laboratory temperature by conventional continual wave EPR apparatus. These signals were denoted as signal I (g = 2.0026, ΔB_{pp} = 0.8 mT) and signal II (g = 2.0046, ΔB_{pp} = 2 mT) indicating their connection with photosystem (PS) I and PS II, respectively (Weaver, 1968). Signal II consists from two components, namely signal II_{slow} which is observable in the dark and signal $II_{very\ fast}$ which occurs at irradiation of chloroplasts by visible light and represents intensity increase of signal II at irradiation of chloroplasts by the visible light. It was found that signal II_{slow} belongs to the intermediate $D^{\bullet}$ and signal $II_{very\ fast}$ belongs to the intermediate $Z^{\bullet}$. Intermediates $Z^{\bullet}$ and $D^{\bullet}$ are tyrosine radicals which are situated at 161st position in D_1 and D_2 proteins which are located on the donor side of PS II (Svensson et al., 1991). The EPR signal I is associated with cation radical of chlorophyll *a* dimmer situated in the core of PS I (Hoff, 1979).

Using EPR spectroscopy it has been found that the studied compounds XIX affect predominantly the intensity of EPR signal II, mainly the intensity of its constituent signal II_{slow}. As mentioned above, the signal II_{slow} is well observable in the dark (see Fig. 8, full line) and it belongs to the $D^{\bullet}$ intermediate, *i.e.* tyrosine (Tyr_D or Y_D) radical which is located on the donor side of PS II in the 161st position in D_2 protein (Svensson et al., 1991; see Fig 5). From Fig. 8 it is evident that the intensity of signal II_{slow} has been decreased by the studied compounds (see Fig. 8, B and C, full lines). That means that in the suspension of spinach chloroplasts the 5-*tert*-butyl-6-chloro-*N*-(3-fluorophenyl)-pyrazine-2-carboxamide (**68**) and 5-*tert*-butyl-*N*-(3-hydroxy-4-chlorophenyl)-pyrazine-2-carboxamide (**69**) interact with the $D^{\bullet}$ intermediate. Due to this interaction of the studied anilides with this part of PS II, the photosynthetic electron transport from the oxygen evolving complex to the reaction centre of PS II is impaired. Consequently, the electron transport between PS II and PS I is inhibited as well and a pronounced increase of signal I intensity in the light can be observed (see Fig. 8, B and C, dashed lines). The signal I (g = 2.0026, ΔB_{pp} = 0.8 mT) belongs to the cation radical of chlorophyll *a* dimmer in the reaction centre of PS I (Hoff, 1979).

Similar site of action in the photosynthetic apparatus of spinach chloroplasts was confirmed for 2-alkylsulfanylpyridine-4-carbothioamides (Kráľová et al., 1997) and substituted benzanilides and thiobenzanilides (Kráľová et al., 1999). From Fig. 8 it is evident that the decrease of signal II_{slow} is greater in the presence of compound **69** (Fig. 8, B) than in presence of compound **68** (Fig. 8, C). These results are in agreement with those obtained for OER inhibition in spinach chloroplasts (Table 1, IC_{50} = 105 μmol dm^{-3} for **69** and 262 μmol dm^{-3} for **68**).

1,5-Diphenylcarbazide (DPC) is an artificial electron donor acting in Z•/D• intermediate (Jegerschöld & Styring, 1991). By addition of DPC to chloroplasts inhibited by PET inhibitors the supply of electrons to P680 is secured. However, the complete restoration of the electron transport to PS I occurs only in the case that photosynthetic electron transport chain between Z•/D• and plastoquinone is not damaged. After addition of DPC to chloroplasts inhibited by the studied anilides up to 70-80%, the OER in the suspension of spinach chloroplasts was not completely restored. It was restored only up to 55-75% of the untreated control sample what indicated that also some member of the photosynthetic electron transport chain between Z•/D• intermediate and plastoquninone is partially damaged by the studied compounds in the light (dashed lines).

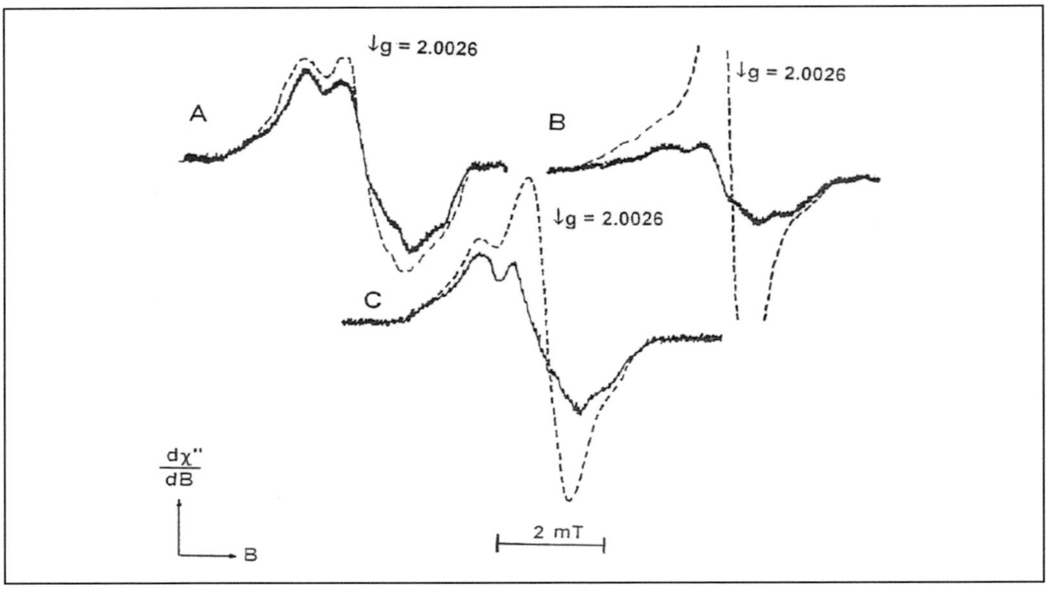

Fig. 8. EPR spectra of the untreated spinach chloroplasts (A) and in the presence of 0.05 mol dm^{-3} of 5-*tert*-butyl-*N*-(3-hydroxy-4-chlorophenyl)-pyrazine-2-carboxamide (**69**, B) and 5-*tert*-butyl-6-chloro-*N*-(3-fluorophenyl)-pyrazine-2-carboxamide (**68**, C) registered in the dark (full lines) and in the light (dashed lines). (Ref. Doležal et al., 2001a; reprinted with permission of editor).

The effects of *N*-phenylpyrazine-2-carboxamides XIX on the photosynthetic centres of spinach chloroplasts were investigated by studying chlorophyll *a* fluorescence. Fluorescence emission spectra of spinach chloroplasts were recorded on fluorescence spectrophotometer F-2000 (Hitachi, Japan) using excitation wavelength λ_{ex} = 436 nm for monitoring fluorescence of chlorophyll *a* and the samples were kept in the dark 10 min before measuring (Doležal et al., 2001a). When chloroplasts were irradiated with the light of λ_{ex} = 436 nm, an emission band with the maximum at λ = 686 nm was observed. This band belongs to the pigment-protein complexes present mainly in photosystem II (Govindjee, 1995). It was found that chloroplasts treated with the studied compounds exhibited quenching of the emission of Chl *a* molecules. Fig. 9 presents the dependence of F/F_{contr} in the suspension of spinach chloroplasts (F_{contr} - fluorescence intensity at λ = 686 nm in the control, F - fluorescence intensity at λ = 686 nm in the presence of the studied compound) on the concentration of 5-*tert*-butyl-*N*-(3-hydroxy-4-chlorophenyl)-pyrazine-2-carboxamide (**69**), 5-*tert*-butyl-6–chloro-*N*-(3-hydroxy-4-chlorophenyl)-pyrazine-2-carboxamide (**70**), 5-*tert*-butyl-6–chloro-*N*-(2-chlorophenyl)-pyrazine-2-carboxamide (**71**), and 5-*tert*-butyl-(2-chlorophenyl)-pyrazine-2-carboxamide (**72**). The greater is the fluorescence quenching, the more efficient is the interaction of the inhibitor with pigment-protein complexes in photosystem II. For the investigated compounds the intensity of this interaction showed a decrease in the following order: **70** > **69** > **72** > **71** (Fig. 9).

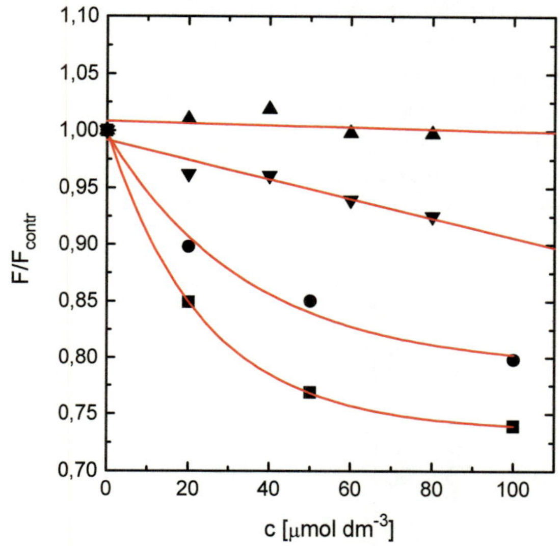

Fig. 9. Dependence of the fluorescence quenching on the concentration of 5-*tert*-butyl-6–chloro-*N*-(3-hydroxy-4-chlorophenyl)-pyrazine-2-carboxamide (**70**, squares), 5-*tert*-butyl-*N*-(3-hydroxy-4-chlorophenyl)-pyrazine-2-carboxamide (**69**, circles), 5-*tert*-butyl-(2-chlorophenyl)-pyrazine-2-carboxamide (**72**, down triangles) and 5-*tert*-butyl-6–chloro-*N*-(2-chlorophenyl)-pyrazine-2-carboxamide (**71**, up triangles) ($F_{contr.}$ = fluorescence of the untreated suspension of spinach chloroplasts; F = fluorescence of anilide treated suspension of spinach chloroplasts; λ = 686 nm). (Ref. Doležal et al., 2001a; reprinted with permission of editor).

The most effective compounds (**70** and **71**) contained two Cl substituents in their molecules. The results of fluorescence study obtained for compounds **70**, **69** and **72** are in agreement with those obtained for OER evolution in spinach chloroplasts (Table 1; IC_{50} = 44 (**70**), 105 (**69**) and 371 μmol dm^{-3} (**72**)). However, the fluorescence of the chloroplast suspension was not affected by 5-*tert*-butyl-6–chloro-*N*-(2-chlorophenyl)-pyrazine-2-carboxamide (**71**) which can be considered as relatively effective inhibitor of OER (IC_{50} = 43 μmol dm^{-3}). This can be explained with the decreased aqueous solubility of this compound. Whereas in the OER experiments the investigated compounds were dissolved in dimethyl sulfoxide, in fluorescence experiments ethanolic solutions were used and after evaporation of the solvent the compound was dissolved directly in the aqueous chloroplast suspension. Consequently, it can be assumed that the fluorescence was not affected due to insolubility of compound **71** in this suspension. The quenching of the fluorescence intensity at λ = 686 nm produced by the studied compounds suggested PS II as the site of action of the studied compounds.

3.1.4 Inhibition of oxygen evolution rate in suspensions of *Chlorella vulgaris* by *N*-phenylpyrazine-2-carboxamides

The inhibition of oxygen evolution rate (OER) in the suspension of *Chlorella vulgaris* was investigated with two model inhibitors (compounds **69** and **72**). The dependences of OER (expressed as the percentage of the untreated control sample) on the concentrations of compounds 5-*tert*-butyl-*N*-(3-hydroxy-4-chlorophenyl)-pyrazine-2-carboxamide (**69**) and 5-*tert*-butyl-*N*-(2-chlorophenyl)-pyrazine-2-carboxamide (**72**) are shown in Fig. 10. It is evident that both investigated compounds inhibited OER in the suspension of *Chlorella vulgaris* algae. Compound **69** was more effective inhibitor than compound **72** what is reflected in the corresponding IC_{50} values (99 μmol dm^{-3} for **69** and 329 μmol dm^{-3} for **72**). These results are in good agreement with those obtained for inhibition of OER in spinach chloroplasts (Table 1). The introduction of hydroxyl moiety in compound **69** enhanced its photosynthesis-inhibiting activity with respect to that of compound **72** approximately threefold.

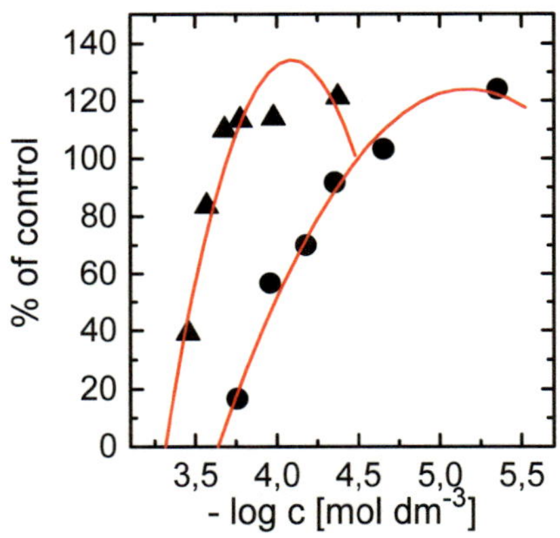

Fig. 10. Dependence of OER in the suspension of *Chlorella vulgaris* (expressed as the percentage of the control) on the concentration of 5-*tert*-butyl-*N*-(2-chlorophenyl)-pyrazine-2-carboxamide (**72**, triangles) and 5-*tert*-butyl-*N*-(3-hydroxy-4-chlorophenyl)-pyrazine-2-carboxamide (**69**, circles). (Ref. Doležal et al., 2001a; reprinted with permission of editor).

3.1.5 Reduction of chlorophyll content in *Chlorella vulgaris* by *N*-phenylpyrazine-2-carboxamides

Toxic effects of environmental pollutants on algae which are essential components of aquatic ecosystems can directly affect the structure and function of ecosystem (Campanella et al., 2000). Herbicides can alter species composition of an algal community what could result in modified structure and function of aquatic communities. Ma et al. (2000) examined the effects of 40 herbicides (belonging to 18 different chemical classes with nine different modes of action) on the green alga *Raphidocelis subcapitata* (formerly named *Selenastrum capricornutum*) and found that the highest acute toxicity exhibited herbicides acting as photosynthesis inhibitors. Photosynthetic pigments have often been used as biomarkers of exposure to different classes of herbicides in autotrophic plants including algae (Blaise, 1993;

Sandmann, 1993). The inhibitory effectiveness of some substituted pyrazinecarboxamides related to reduction of chlorophyll content in *Chlorella vulgaris* expressed by IC_{50} values is summarized in Table 2. The dependence of log (1/IC_{50}) on the compound lipophilicity (log *P*) showed a quasi-parabolic course (Fig. 11).

No.	R^1	R^2	R^3	IC_{50}	Ref.
43	Cl	H	4-Cl-3-CH_3	80	h
44	Cl	H	3-I-4-CH_3	44	h
45	H	tBu	4-Cl-3-CH_3	89	h
79	Cl	tBu	3-CH_3	63	e
84	Cl	tBu	3-Br	67	e
85	Cl	H	3,5-CF_3	125	e
86	H	tBu	3,5-CF_3	208	e
87	Cl	tBu	3,5-CF_3	356	e
88	Cl	H	2,6-CH_3	79	e
95	H	H	4-CH_3	71	b
97	Cl	H	4-F	32	b
100	Cl	H	4-CH_3	37	b
102	H	tBu	2-CF_3	33	b

Table 2. IC_{50} values (in μmol dm^{-3}) related to reduction of chlorophyll content in *Chlorella vulgaris* of some substituted pyrazinecarboxamides XIX. Cultivation conditions: 7 days; photoperiod 16 h light/8 h dark; photosynthetic active radiation 80 μmol m^{-2} s^{-1}; pH = 7.2; Chl content in the suspensions at the beginning of the cultivation was 0.01 mg dm^{-3}. (Ref. Doležal et al., 2007[(h)], 2002[(e)], 2008a[(b)]).

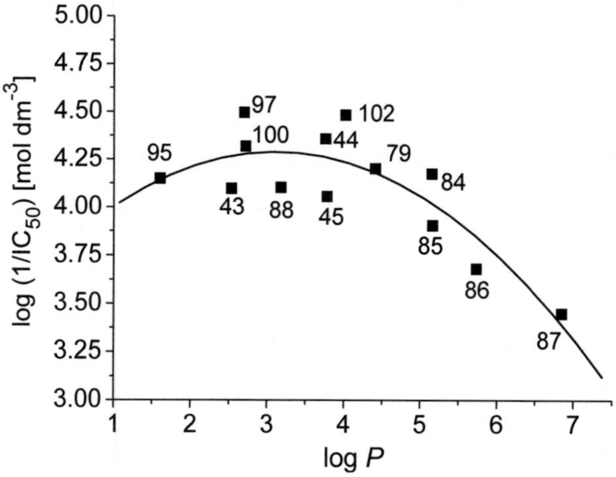

Fig. 11. The dependence of antialgal activity expressed as log (1/IC_{50}) on the lipophilicity (log *P*) of some substituted pyrazinecarboxamides XIX.

However, differences in IC_{50} values of compounds with comparable lipophilicity indicate that the biological activity is affected beside of lipophilicity also by the electronic properties of R^3 substituent(s). Because of too low aqueous solubility of many compounds from the tested set of pyrazinecarboxamides XIX the compounds fall out during experiment (7 days) and the corresponding IC_{50} values could be determined only for limited number of compounds.

4. Photosynthesis-inhibiting pyrazine analogues of chalcones

Chalcones and related compounds "chalconoids" are aromatic ketones containing two aromatic rings linked with three carbon chain. The presence of an unsaturated double bound is typical for chalcones. Hence, chalcones are 1,3-diarylprop-2-ones. They show antibacterial, antifungal, antitumor and anti-inflammatory properties (Dimmock et al., 1999). The aim of our project was the isosteric replacement of a phenyl moiety in chalcones with the pyrazine ring to form some pyrazine analogues of chalcones ("diazachalcones"). Several series (thirty two compounds) of ring substituted (*E*)-3-phenyl-1-(pyrazin-2-yl)-prop-2-en-1-ones XX (Fig. 12) were prepared in our laboratories by means of modified Claisen-Schmidt condensation of acetylpyrazines with aromatic aldehydes (Opletalová et al., 2002, Opletalová et al., 2006, Chlupáčová et al., 2005).

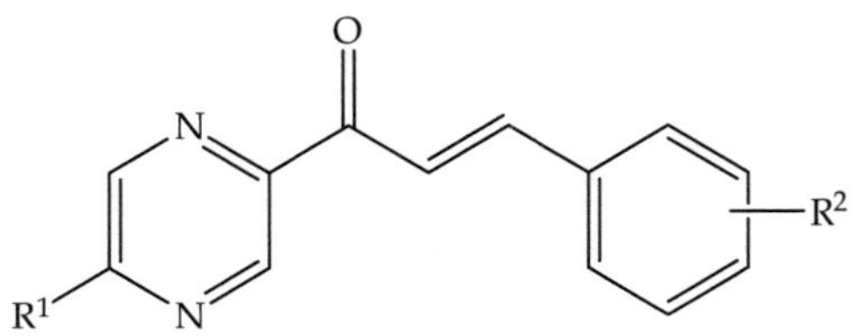

XX

Fig. 12. Pyrazine analogues of chalcones XX (R^1 = H, alkyl; R^2 = OH, NO_2, Cl).

Ring substituted (*E*)-3-phenyl-1-(pyrazin-2-yl)-prop-2-en-1-ones XX were tested for their activity related to OER inhibition in spinach chloroplasts and *Chlorella vulgaris* as well as reduction of chlorophyll content in statically cultured suspensions of freshwater alga *Chlorella vulgaris*. The corresponding IC_{50} values are summarized in Tables 3 and 4.

No.	R^1	R^2	OER inhibition/IC_{50}	
			S. oleracea	*C. vulgaris*
114	tBu	2-OH	167	78
115	isoBu	2-OH	144	63
116	nBu	2-OH	184	147
117	nPro	2-OH	187	100
118	tBu	4-OH	315	279
119	isoBu	4-OH	235	232
120	nBu	4-OH	306	265
121	nPro	4-OH	399	514

Table 3. IC_{50} values (in μmol dm^{-3}) related to OER inhibition in spinach chloroplasts and *Chlorella vulgaris* by diazachalcones XX. (Ref. Opletalová et al., 2002).

The inhibition of OER in spinach chloroplasts by substituted diazachalcones XX (**114-121**) (Fig. 12) has been investigated spectrophotometrically, using DCPIP as an electron acceptor (Kráľová et al., 1992). For the study of OER inhibition in the algal suspensions a Clark type electrode has been used. The IC_{50} values of compounds **114-121** related to OER inhibition varied in the range of 144-399 μmol dm^{-3} for spinach chloroplasts) and 63-514 μmol dm^{-3} for algal suspension of *Chlorella vulgaris* (Table 3). 2-Hydroxy substituted derivatives were found to be more effective inhibitors of photosynthesis than the 4-hydroxy substituted ones. The inhibitory activity of 2-hydroxy substituted derivatives was affected also by the branching of R^1 substituent: OER inhibition in photosynthesizing organisms by the isomers with branched alkyl chain (*tert*-butyl, *iso*butyl) was more pronounced than by the isomer with unbranched alkyl substituent (*n*-butyl) (Opletalová et al., 2002).

No.	R^1	R^2	IC_{50} PET inhibition *S. oleracea*	IC_{50} Chl. content reduction *C. vulgaris*	Ref.
122	H	2-NO_2	ND	70.6	k
123	tBu	2-NO_2	325.0	ND	k
124	isoBu	2-NO_2	ND	118.0	k
125	nBu	2-NO_2	393.0	585.0	k
126	nPro	2-NO_2	ND	123.0	k
127	H	3-NO_2	658.0	19.6	k
128	tBu	3-NO_2	461.0	ND	k
129	isoBu	3-NO_2	340.0	62.8	k
130	nBu	3-NO_2	236.0	ND	k
131	nPro	3-NO_2	ND	18.6	k
132	H	4-NO_2	ND	44.9	k
133	tBu	4-NO_2	ND	ND	k
134	isoBu	4-NO_2	ND	ND	k
135	nBu	4-NO_2	706.0	ND	k
136	nPro	4-NO_2	ND	238.3	k
137	H	3-OH	877.0	32.5	l
138	tBu	3-OH	105.0	238.3	l
139	isoBu	3-OH	256.0	65.5	l
140	nBu	3-OH	ND	95.9	l
141	nPro	3-OH	ND	69.9	l
142	H	4-Cl	ND	24.5	l
143	tBu	4-Cl	181.0	ND	l
144	isoBu	4-Cl	246.0	ND	l
145	nBu	4-Cl	374.0	ND	l

Table 4. IC_{50} values (in μmol dm^{-3}) related to PET inhibition in spinach chloroplasts and IC_{50} values (in μmol dm^{-3}) related to reduction of chlorophyll content in statically cultivated *Chlorella vulgaris* determined for diazachalcones XX. (Ref. Opletalová et al., 2006[k], Chlupáčová et al., 2005[l]), ND – not determined.

The effects of substituted diazachalcones XX (**114-121**) on the photosynthetic centres of chloroplasts were investigated by studying chlorophyll *a* fluorescence. The decreased intensity of the emission band at 686 nm, belonging to the pigment-protein complexes in photosystem (PS) II, suggested PS II as the site of action of the studied compounds (Kráľová et al., 1998).

Using EPR spectroscopy it has been found that in spinach chloroplasts the intensity of EPR signal II, mainly the intensity of its constituent signal II_{slow}, showed a decrease by the studied compounds **114-121**. Consequently it can be concluded that the studied compounds, similarly to *N*-phenylpyrazine-2-carboxamides (Doležal et al., 2001a), interact with $D^{\bullet}$ intermediate, *i.e.* with the tyrosine radical in 161st position (Tyr_D; Y_D) which is located in D_2 protein on the donor side of PS II (Fig. 5). Due to interaction of the studied compounds with $D^{\bullet}$ intermediate PET from the oxygen evolving complex to the core of PS II is impaired. A pronounced increase of EPR signal I intensity in the light belonging to the cation-radical of chlorophyll *a* dimmer in the core of PS I indicated that the electron transport between PS II and PS I is inhibited as well. However, addition of DPC to chloroplasts inhibited by the studied compounds completely restored the reduction of DCPIP indicating that the core of PS II (P680) and a part of the electron transport chain - at least up to plastoquinone - remained intact. These results are in accordance with those obtained with 2-alkylsulfanylpyridine-4-carbothioamides (Kráľová et al., 1997). Similar study with anilides of 2-alkylpyridine-4-carboxylic acids has shown that also the core of PS II was partially impaired by these inhibitors of photosynthetic electron transport (Kráľová et al., 1998a). On the other hand, after addition of DPC to chloroplasts inhibited by the studied *N*-phenylpyrazine-2-carboxamides **68** and **69** up to 70-80%, the OER in the suspension of spinach chloroplasts was restored only up to 55-75% of the untreated control sample indicating that also some member of the photosynthetic electron transport chain between $Z^{\bullet}/D^{\bullet}$ and plastoquininone was partially damaged by the these compounds (Doležal et al., 2001a).

In general, in the series of diazachalcones **122-145** the most effective reduction of chlorophyll content in the suspensions of *C. vulgaris* showed compounds with R^1 = H (Table 4): **127** (R^2 = 3-NO_2; IC_{50} = 19.6 μmol dm^{-3}), **142** (R^2 = 4-Cl; IC_{50} = 24.5 μmol dm^{-3}), **137** (R^2 = 3-OH; IC_{50} = 32.5 μmol dm^{-3}), **132** (R^2 = 4-NO_2; IC_{50} = 44.9 μmol dm^{-3}) and **122** (R^2 = 2-NO_2; IC_{50} = 70.6 μmol dm^{-3}). However, the highest anti-algal activity from this series showed compound **131** (R^1 = $CH_3CH_2CH_2$, R^2 = 3-NO_2; IC_{50} = 18.6 μmol dm^{-3}). On the other hand, the most effective inhibitors of PET in spinach chloroplasts were found to be two compounds with R^1 = $C(CH_3)_3$, namely **138** (R^2 = 3-OH; IC_{50} = 105 μmol dm^{-3}) and **143** (R^2 = 4-Cl; IC_{50} = 181 μmol dm^{-3}) whereby IC_{50} values for several compounds could not be determined due to too low solubility of these compounds.

5. Conclusion

Pyrazines are a class of compounds that occur almost ubiquitously in nature. The worldwide distribution of pyrazines in plants, insects, terrestrial vertebrates, marine organisms, fungi and bacteria, their specific properties, including their using as drugs, fungicides and herbicides invite reasonable attention. Our review brings the basic information about some commercially produced pyrazine herbicides including their mechanism of action as well as survey of patented herbicidal pyrazine derivatives. Special attention was paid to the original compounds from series of 113 substituted *N*-

phenylpyrazine-2-carboxamides XIX and 32 diazachalcones XX prepared and evaluated in our laboratories. In first series, pyrazinecarboxamides XIX connected via -CONH- bridge with substituted anilines can form centrosymmetric dimer pairs with the peptidic carboxamido group of some peptides, needed for binding to the receptor site, possibly by formation of hydrogen bonds. All compounds were tested as potential inhibitors of the photosynthetic electron transport in spinach chloroplasts. Based on the obtained results it could be assumed that the biological activity of the studied substituted pyrazinecarboxamides did not depend exclusively on the compound lipophilicity but it was also affected by electron accepting or withdrawing power of the substituents on the aromatic benzene ring. The site of action of some substituted *N*-phenylpyrazine-2-carboxamides XIX in the photosynthetic apparatus of spinach chloroplasts was studied using fluorescence and EPR spectroscopy. It was found that the studied compounds cause quenching of the chlorophyll *a* fluorescence at 685 nm belonging mainly to the pigment—protein complexes in photosystem (PS) II. The extent of the fluorescence quenching correlated with the effectiveness of the compounds concerning inhibition of oxygen evolution rate (OER) in spinach chloroplasts. Using EPR spectroscopy it was confirmed that the title compounds interact with the intermediate $D^•$ (Tyr_D), *i.e.* with the tyrosine radical, which is situated on the donor side of PS II at the 161th position of D_2 protein. It was found that the studied compounds inhibit OER not only in the suspension of spinach chloroplasts but also in the suspensions of *Chlorella vulgaris*. Introducing of Cl substituents into aromatic ring as well as pyrazine moiety of the studied molecules enhanced the effectiveness of OER—inhibiting activity. Some *N*-phenylpyrazine-2-carboxamides XIX reduced chlorophyll content in *Chlorella vulgaris* whereby their biological activity was affected beside of lipophilicity also by the electronic properties of R^3 substituent(s). The most effective inhibitor from the series XIX was 6-chloro-*N*-(5-chloro-2-hydroxyphenyl)-pyrazine-2-carboxamide (**34**, IC_{50} = 8 μmol dm^{-3}; Doležal, 1999).
The studied pyrazine analogues of chalcones, diazachalcones XX also reduced the rate of oxygen evolution in spinach chloroplasts and *C. vulgaris*, whereby the inhibitory activity of *ortho*-hydroxyl substituted derivatives XX was greater than that of *para*-hydroxyl substituted ones. The lowest IC_{50} values were found with compounds having a branched alkyl group on the pyrazine ring. The photosynthesis-inhibiting activity of nitro derivatives was lower than that of the corresponding hydroxylated analogs. In general, in the series of diazachalcones with R^2 = 2-NO_2; 3-NO_2; 4-NO_2; 3-OH and 4-Cl, the most effective reduction of chlorophyll content in the suspensions of *C. vulgaris* showed compounds with R^1 = H. It was confirmed that studied diazachalcones interact with $D^•$ intermediate, *i.e.* with the tyrosine radical in 161st position (Tyr_D) which is located in D_2 protein on the donor side of PS II and that they do not damage the core of PS II (P680) and a part of the electron transport chain - at least up to plastoquinone.

6. Acknowledgments

We dedicate the present article to Assoc. Prof. Jiří Hartl, former leader of our research team. Thanks to all our colleague and numerous graduate and undergraduate students who participated on the synthesis of biologically active pyrazines. The work was partly supported by the Ministry of Education of the Czech Republic (MSM002160822) and Sanofi Aventis Pharma Slovakia.

7. References

ACD/logP, ver. 1.0 (Advanced Chemistry Development Inc., Toronto, 1996)

Adams, T. B.; Doull, J.; Feron, V. J.; Goodman, J. I.; Marnett, L. J.; Munro, I. C.; Newberne, P. M.; Portoghese, P. S.; Smith, R. L.; Waddell, W. J. & Wagner, B. M. (2002). The FEMA GRAS assessment of pyrazine derivatives used as flavor ingredients. *Food Chem. Toxicol.*, Vol. 40, No 4. , 429-451, ISSN 0278-6915

Blaise, C. R. (1993). Practical laboratory applications with micro-algae for hazard assessment of aquatic contaminants. In: Richardson, M. (ed.), *Ecotoxicology Monitoring*, pp. 83–107, VCH, ISBN 3-527-28560-1, Weinheim

Beck, H. C.; Hansen, A. M. & Lauritsen, F. R. (2003). Novel pyrazine metabolites found in polymyxin biosynthesis by *Paenibacillus polymyxa*. *FEMS Microbiol. Lett.*, Vol. 220, 67-73, ISSN 0378-1097

Benko, Z.; Jachetta, J. J.; Costales, M. J. & Arndt K. E. (1997). N-*aryl[1,2,4]triazolo[1,5-a]pyrazine-2-sulfonamide herbicides.* US Patent 5602075, February 11, 1997

Bird, C. W. (1992). Heteroaromaticity, 5, a unified aromaticity index. *Tetrahedron*, Vol. 48, No. 2, 335-340 , ISSN 0040-4020

Böhner, B.; Fory, W.; Schurter, R. & Pissiotas, G. (1987). *Pyrazine-sulfonylcarbamates and thiocarbamates.* US Patent 4684730, August 4, 1987

Böhner, B.; Fory, W.; Schurter, R. & Pissiotas, G. (1987a). *Pyrazinyl sulfonamides useful as intermediates for herbicides.* US Patent 4707551, November 17, 1987

Böhner, B. & Meyer, W. (1989a). *Aminopyrazinones.* US Patent 4871853, October 3, 1989

Böhner, B. & Meyer, W. (1989b). *Aminopyrazinones and aminotriazinones.* US Patent 4795486, January 3, 1989

Böhner, B. & Meyer, W. (1990). *Aminopyrazinones and aminotriazinones.* US Patent 4920223, April 24, 1990

Brown, D. J. (2002). *The Pyrazines, Supplement I. The Chemistry of Heterocyclic Compounds.* John Wiley & Sons, Inc., ISBN 9780471403821, New York, 576 p.

Brown III, C. J. & Knaust, J. M. (2009). 2,5-Dimethylpyrazine 1,4-dioxide. *Acta Cryst. Sect. E – Struct. Reports Online*, Vol. 65 (December), No. 12, o3052-sup-9, ISSN 1600-5368

Campanella, L.; Cubadda, F. Sammartino, M. P. & Saoncella, A. (2000). An algal biosensor for the monitoring of water toxicity in estuarine environments. *Water Res.*, Vol. 25, No. 1, 69–76, ISBN 0043-1354

Campbell, K. R.; Bartell, S. M. & Shaw, J. L. (2000). Characterizing aquatic ecological risks from pesticides using a diquat dibromide case study. II. Approaches using quotients and distributions. *Environ. Toxicol. Chem.*, Vol. 19, No. 3, 760–774, ISSN 0730-7268

Camper, N. D. & McDonald, S. K. (1989). Tissue and cell cultures as model system in herbicide research. *Rev. Weed Sci.*, Vol. 4, No. 2, 169-190, ISSN 1042-4148

Cedergreen, N.; Kamper, A. & Streibig, J. C. (2006). Is prochloraz a potent synergist across aquatic species? A study on bacteria, daphnia, algae and higher plants. *Aquat. Toxicol.*, Vol. 78, No. 3, 243–252, ISSN 0166-445X

Chlupáčová, M.; Opletalová, V.; Kuneš J. & Kráľová, K. (2005). Synthesis of 3-alkyl- and 3-arylsulfanyl-1,3,-diphenylpropane-1-ones and their effects on two photosynthesizing organisms. *Folia Pharm. Univ. Carol.*, Vol. 31-32, 109-115, ISSN 1210-9495

Cordingley, M. R.; Turnbull, M. D.; Carter, N. B. & Crowley, P. J. (2009). *Herbicidal pyridopyrazines and their preparation and use in controlling plants.* WO 2009090402, January 16, 2009

Davenport, H. E. (1963). The mechanism of cyclic phosphorylation by illuminated chloroplasts. *Proc. R. Soc. B*, Vol. 157, March 26, pp. 332-345, ISSN 1471-2954

Diaz, T. G.; Merás, I. D. & Franco, M. F. (2002). Stopped flow kinetic-spectrophotometric determination of Diquat in waters. *Water Res.*, Vol. 36, No. 3, 783–787, ISSN 0043-1354

Dietsche, T. J. (1977). *6,7-Dichloro-3,4-dihydro-2H-pyrazino(2,3-b)(1,4) oxazines.* US Patent 49029657, June 14, 1977

Dimmock, J. R.; Elias, D. W.; Beazely, M. A. & Kandepu, N. M. (1999). Bioactivities of chalcones. *Curr. Med. Chem.*, Vol. 6, No. 12, 1125-1149, ISSN 0929-8673

Doležal, M.; Hartl, J; Miletín. M.; Macháček, M. & Kráľová, K. (1999). Synthesis and photosynthesis-inhibiting activity of some anilides of substituted pyrazine-2-carboxylic acids. *Chem. Pap.*, Vol. 53, No. 2, 126-130, ISSN 0366-6352

Doležal, M.; Vičík, R.; Miletín, M. & Kráľová, K. (2000). Synthesis and antimycobacterial, antifungal, and photosynthesis-inhibiting evaluation of some anilides of substituted pyrazine-2-carboxylic acids. *Chem. Pap.*, Vol. 54, No. 4, 245-249, ISSN 0366-6352

Doležal, M.; Kráľová, K.; Šeršeň, F. & Miletín, M. (2001a). The site of action of pyrazine-2-carboxylic acids in the photosynthetic apparatus. *Folia Pharm. Univ. Carol.*, Vol. 26, 13–20, ISSN 1210-9495

Doležal, M., Miletín, M.; Hejský, R.; Kráľová, K. & Kuneš, J. (2001b). Synthesis of some amides of substituted pyrazine-2-carboxylic acids and their photosynthesis-inhibiting activity. *5th International Electronic Conference on Synthetic Organic Chemistry* (ECSOC-5), C009, Seijas, J. A. (Ed.), ISBN 3-906980-06-5, CD-ROM edition, September 1– 30, 2001, MDPI, Basel

Doležal, M.; Miletín, M.; Kuneš, J. & Kráľová, K. (2002). Substituted amides of pyrazine-2-carboxylic acids: Synthesis and biological activity. *Molecules*, Vol. 7, No. 7, 363 – 373, ISSN 1420-3049

Doležal, M.; Palek, L.; Röslerová, Z.; Kuneš, J.; Buchta, V. & Kráľová, K. (2004). Substituted pyrazinecarboxamides: Preparation and biological activity. *8th International Electronic Conference on Synthetic Organic Chemistry* (ECSOC-8), C001, Seijas, J. A. & Tato, M. P. V. (Ed.), ISBN 3-906980-15-4, CD-ROM edition, November 1–30, 2004, MDPI, Basel

Doležal, M. (2006a). Biological active pyrazines of natural and synthetic origin. *Chem. Listy* Vol. 100, No. 11, 959-966, ISSN 0009-2770

Doležal, M.; Palek, L.; Vinšová, J.; Buchta, V.; Jampílek, J. & Kráľová, K. (2006b). Substituted pyrazinecarboxamides: Synthesis and biological evaluation. *Molecules*, Vol. 11, No. 4, 242-256, ISSN 1420-3049

Doležal, M.; Tůmová, L.; Kešetovičová, D.; Tůma, J. & Kráľová, K. (2007). Substituted *N*-phenylpyrazine-2-carboxamides, their synthesis and evaluation as herbicides and abiotic elicitors. *Molecules*, Vol. 12, No. 12, 2589-2598, ISSN 1420-3049

Doležal, M.; Čmedlová, P.; Palek, L.; Vinšová, J.; Kuneš, J.; Buchta, V.; Jampílek, J. & Kráľová, K. (2008a). Synthesis and antimycobacterial evaluation of substituted pyrazinecarboxamides. *Eur. J. Med. Chem.*, Vol. 43, No. 5, 1105-1113, ISSN 0223-5234

Doležal, M.; Lešňanská, K.; Peško, M.; Jampílek, J. & Kráľová K. (2008b). Pyrazinecarboxamides, their synthesis and evaluation as potential herbicides. *12th International Electronic Conference on Synthetic Organic Chemistry* (ECSOC-12), C0004, Seijas, J. A. & Tato, M. P. V. (Ed.), ISBN 3-906980-19-7, CD-ROM edition, November 1–30, 2008, MDPI, Basel

Engel, S.; Nubling, C.; Kardorff, U.; Kast, J.; von Deyn, W.; Plath, P.; Walter, H.; Westphalen, K. & Gerber, M. (1999). *Herbicidal pyrazine derivatives.* US Patent 5939359, August 17, 1999

European Commission, (2001). Commission Directive 2001/21/EC of 5 March 2001 amending Annex I to Council Directive 91/414/EEC concerning the placing of plant protection products on the market to include amatol, diquat, predate and thiabendazole as active substances. *Off. J. Eur. Communities L69*, 17–21, May 10, 2001.

European Commission, (2002). Opinion of the scientific committee on plants on an additional question from the commission concerning the evaluation of diquat in the context of Council Directive 91/414/EEC, *Scientific Committee on Plants SCP/DIQUAT-bis/002-Final 14 January 2002*. European Commission, Belgium, 12pp.

Fischer, G. C. (1987). *Biologically active* N-*pyrazinyl-haloacetamides*. US Patent 4742167, May 3, 1988

Frater, G.; Suchy, M.; Wenger, J. &Winternitz, P. (1987). *[(Isopropylideneamino)oxy]-ethyl-2-[[6-chloroquinoxalinyl)oxy]phenoxy]propionate postemergent herbicide.* US Patent 4687849, August 18, 1987

Gorzerino, C.; Quemeneur, A.; Hillenweck, A.; Baradat, M.; Delous, G.; Ollitrault, M.; Azam, D.; Caquet, T. & Lagadic, L. (2009). Effects of diquat and fomesafen applied alone and in combination with a nonylphenol polyethoxylate adjuvant on *Lemna minor* in aquatic indoor microcosms. *Ecotox. Environ. Saf.*, Vol. 72, No. 3, 802-810, ISSN 0147-6513

Govindjee (1995). Sixty-three years since Kautsky: Chlorophyll *a* fluorescence. *Aust. J. Plant Physiol.*, Vol. 22, No. 4, 131-160, ISSN 0310-7841

Griffin, D.; Rice, M. & Elliott R. (1990). *Pyrazine compound useful as plant growth regulators.* US Patent 4975112, December 4, 1990

Hirai, K. (1999). Structural evolution and synthesis of diphenylethers, cyclic imides and related compounds. In: *Peroxidizing Herbicides*. Böger, P.; Wakabayashi, K., Eds.; pp. 15-71, Springer, 1999; ISBN 3540645500, Berlin

HRAC: *Classification of Herbicides According to Mode of Action*, 2005 http://www.hracglobal.com/Publications/ClassificationofHerbicideModeofActio n/tabid/222/Default. aspx (23 June 2010)

Hoff, A. J. (1979). Application of ESR in photosynthesis. *Phys. Rep.*, Vol. 54, No. 2, 75-200, ISSN 0370-1573

Ivany, J. A. (2005). Response of three potato (*Solanum tuberosum*) cultivars to pyrafluen-ethyl used as desiccant in Canada. *Crop Protect.*, Vol. 24, No. 9, 836–841, ISSN 0261-2194

Izawa, S. (1980). Acceptors and donors for chloroplast electron transport. *Methods Enzymol.*, Vol. 69, Part C, pp. 413-434. Edited by Anthony San Pietro, Academic Press, ISBN 0121819698, New York - London

Jegerschöld, C. & Styring, S. (1991). Fast oxygen-independent degradation of the D1 reaction center protein in photosystem-II. *FEBS Lett.*, Vol. 280, No. 1, 87-90, ISSN 0014-5793

Joshi, M. K. & Mohanty, P. (2004). Chlorophyll *a* fluorescence as a probe of heavy metal ion toxicity in plants. In: *Chlorophyll a fluorescence: A signature of photosynthesis. Advances in Photosynthesis and Respiration.* ed. Papageorgiou, G. C. & Govindjee, Vol. 19, pp. 637-661, Springer, ISBN 1-4020-3217-X, Dordrecht

Joule, J. A. & Mills, K. (2010). *Heterocyclic chemistry.* John Wiley and Sons, ISBN 9781405133005, pp. 249-288, Chichester

Linsmaier, E. M. & Skoog, F. (1965). Organic growth factor requirements of tobacco tissue cultures. *Physiol. Plant*, Vol. 18, No. 1, 100-127, ISSN 0031-9317

Lock, E. A. & Wilks, M. F. (2001). *Diquat.* In: *Handbook of Pesticide Toxicology, Volume 2. Agents*, Krieger R. (ed.), 1605-1622, Academic Press, San Diego, USA, ISBN 0124262600

Koschnick, T. J.; Haller, W. T. & Glasgow, L. (2006). Documentation of landoltia (*Landoltia punctata*) resistance to diquat. *Weed Sci.*, Vol. 54, No. 4, 615-619, ISSN 0043-1745

Kráľová, K.; Šeršeň, F. & Sidóová, E. (1992). Photosynthesis inhibition produced by 2-alkylthio-6-R-benzothiazoles. *Chem. Pap.*, Vol. 46, No. 5, 348-350, ISSN 0366-6352

Kráľová, K.; Šeršeň, F.; Klimešová, V. & Waisser, K. (1997). Effect of 2-alkylthio-4-pyridinecarbothioamides on photosynthetic electron transport in spinach chloroplasts. *Collect. Czech. Chem. Commun.*, Vol. 62, No. 3, 516-520, ISSN 0010-0765

Kráľová, K.; Šeršeň, F. & Opletalová, V. (1998a). Inhibition of oxygen evolution in spinach chloroplasts and *Chlorella vulgaris* by 2',5'-diazachalcones. *Folia Pharm. Univ. Carol.*, Vol. 23 Suppl., pp. 71-72, ISSN 1210-9495

Kráľová, K.; Šeršeň, F.; Miletín, M. & Hartl, J. (1998b). Inhibition of photosynthetic electron transport by some anilides of 2-alkylpyridine-4-carboxylic acids in spinach chloroplasts. *Chem. Pap.*, Vol. 52, No. 1, 52-55, ISSN 0366-6352

Kráľová, K., Šeršeň, F., Kubicová, L., & Waisser, K. (1999). Inhibitory effects of substituted benzanilides on photosynthetic electron transport in spinach chloroplasts. *Chem. Pap.*, Vol. 53, No. 5, 328-331, ISSN 0366-6352

Ma, J. Y.; Wang, S. F.; Wang, P. W.; Ma, L. J.; Chen, X. L. & Xu, R. F. (2006). Toxicity assessment of 40 herbicides to the green alga *Raphidocelis subcapitata. Ecotox. Environ. Saf.*, Vol. 63, No. 3, 456–462, ISBN 0147-6513

Maga, J. A. (1992). Pyrazine update. *Food Rev. Inter.*, Vol. 8, No. 4, 479-558, ISSN 8755-9129

Maillard, L. C. (1912). Action des acides aminés sur les sucres: formation des mélanoïdines par voie méthodique. *Compt. Rend.*, Vol. 154, 66-68, ISSN 1631-0748

Mason, R. P. (1990). Redox cycling of radical anion metabolites of toxic chemicals and drugs and the Marcus theory of electron transfer. *Environ. Health Perspect.*, Vol. 87, No. 7, 237-243, ISSN 0091-6765

Massachusetts Department of Agricultural Resources, Aquatic Vegetation Management. *Appendix III – Diquat*, 19 pp., http://www.mass.gov/agr/pesticides/aquatic/docs/diquat.pdf (June 23, 2010)

Mees, G. C. (1960). Experiments on the herbicidal action of 1,1'-ethylene-2,2'-dipyridylium dibromide. *Ann. Appl. Biol.*, Vol. 48, No. 3, pp. 601-612, ISSN 0003-4746

The Merck Index, 14th Edition, 2006 by Merck & Co., Inc., Whitehouse Station, ISBN 0-911910-00-X, NJ, USA

Mixan, C. E. & Goralski, C. T. (1978). *Cyanatothiomethylthio pyrazines.* US Patent 4101546, July 18, 1978

Nakamura, A.; Ataka, T.; Segawa, H.; Takeuchi, Y. & Takematsu, T. (1983). Studies on herbicidal 2,3-dicyanopyrazines. 1. Structure-activity relationship of herbicidal 2,3-dicyano-5-substituted pyrazines. *Agric. Biol. Chem.*, Vol. 47, No. 7, pp. 1555-1560, ISSN 0002-1369

Nakamura, A.; Ataka, T.; Segawa, H.; Takeuchi, Y. & Takematsu, T. (1983a). Studies on herbicidal 2,3-dicyanopyrazines. 2. Structure-activity relationships of herbicidal 5-ethylamino- and 5-propylamino-2,3-dicyanopyrazines. *Agric. Biol. Chem.*, Vol. 47, No. 7, pp. 1561-1567, ISSN 0002-1369

Niedermann, H. P. & Munro, D. (1994). *Herbicidal pyrazines and pyrimidines.* GB 2285045, December 12, 1994

Oettmeier, W. (1992). Herbicides of Photosystem II. In: *Topics in photosynthesis*, Vol. 11: The Photosystems: Structure, Function and Molecular Biology (Barber, J. ed.) pp. 349-408, Elsevier Science Publishers B. V., ISBN 0444894403, Amsterdam, Netherlands.

Opletalová, V.; Hartl, J.; Patel, A.; Palát, Jr. K. & Buchta, V. (2002). Ring substituted 3-phenyl-1-(2-pyrazinyl)-2-propen-1-ones as potential photosynthesis-inhibiting, antifungal and antimycobacterial agents. *Farmaco*, Vol. 57, No. 2, pp. 135–144, ISSN 0014-827X

Opletalová, V.; Pour, M.; Kuneš, J.; Buchta, V.; Silva, L.; Kráľová, K.; Chlupáčová, M.; Meltrová, D.; Peterka, M. & Posledníková, M. (2006). Synthesis and biological evaluation of (*E*)-3-(nitrophenyl)-1-(pyrazin-2-yl)prop-2-en-1-ones. *Collect. Czech. Chem. Commun.*, Vol. 71, No. 1, pp. 44–58, ISSN 0010-0765

Peiró, A. M.; Zapater, P.; Alenda, C.; Ramírez, A.; Gutiérrez, A.; Pérez-Mateo, M. & Such, J. (2007). Hepatotoxicity related to paraquat and diquat absorption through intact skin. *Dig. Dis. Sci.*, Vol. 52, No. 11, pp. 3282–3284, ISSN 0163-2116

Peterson, H. G.; Boutin, C.; Freemark, K. E. & Martin, P. A. (1997). Toxicity of hexazinone and diquat to green algae, diatoms, cyanobacteria and duckweed. *Aquat. Toxicol.*, Vol. 39, No. 2, pp. 111–134, ISSN 0166-445X

Petit, V.; Cabridenc, R.; Swannell, R. P. J. & Sokhi, R. S. (1995). Review of strategies for modelling the environmental fate of pesticides discharged into riverine systems. *Environ. Int.*, Vol. 21, No. 2, pp.167–176, ISSN 0160-4120

Portnoy, R. C. (du Pont de Nemours, E. I., and Co.) (1978). *Arylthiodicyanopyrazines as plant disease control agents.* US Patent 4113724, September 12, 1978

Ritter, A. M.; Shaw, J. L.; Williams, W. M. & Travis, K. Z. (2000). Characterizing aquatic ecological risks from pesticides using a diquat bromide case study. I. Probabilistic exposure estimates. *Environ. Toxicol. Chem.*, Vol. 19, No. 3, 749–759, ISSN 0730-7268

Sakata, G.; Makino, K.; Kawamura, Y.; Ura, Y.; Ikai, T. & Kawamura, Y. (1983). NCI-96683, a new selective herbicide for annual and perennial grass weed control in broad leaf crops. *10th Int. Congr. Plant Prot., Proc. Conf.*, Vol. 1, pp. 315-23. ISBN 978-

0901436795, Brighton, England, 20-25 November, 1983, British Crop Protection Council

Sanchez, W.; Palluel, O.; Lagadic, L.; Ait-Aissa, S. & Porcher, J. M. (2006). Biochemical effects of nonylphenol polyethoxylate adjuvant, Diquat herbicide and their mixture on the three-spined stickleback (*Gasterosteus aculeatus* L.). *Mar. Environ. Res.*, Vol. 62, Suppl. S, pp. S29–S33, ISSN 0141-1136

Sandmann, G. (1992). Spectral determination of carotenoid precursors in *Scenedesmus* cells treated with bleaching herbicides. In: Böger, P. & Sandmann, G. (eds.), *Target Assays for Modern Herbicides and Related Compounds*, pp. 3–8. Lewis Publishers, ISBN 0-87371-539-X, Boca Raton, Florida

Sato, J.; Kondo, Y.; Kudo, Y.; Nawamaki, T.; Watanabe, S.; Ishikawa, K. & Ito, Y. (1993). *Preparation of pyrazine derivatives as herbicides.* JP 06192252, September 21, 1993

Shuto, A.; Hoshi, H. & Sanemitsu, Y. (2000). *Pyrazin-2-one derivatives, their use, and intermediates for their production*. US Patent 6015774, January 18, 2000

Svensson, B.; Vass, I. & Styring, S. (1991). Sequence analysis of the D_1 and D_2 reaction center proteins of photosystem II. *Z. Naturforsch.*, Vol. 46c, No. 9-10, 765-776, ISBN 0939-5075

Takematsu, T.; Segawa, H., Takamoro, M.; Ataka, T.; Chatani, M. & Nakamura, A. (1981). *2,3-Dicyanopyrazines.* US Patent 4259489, March 31, 1981

Takematsu, T.; Segawa, H.; Miura, T.; Ataka, T.; Chatani, M. & Nakamura, A. (1984). *2,3-Dicyano 6-phenyl pyrazine herbicides.* US Patent 4460403, July 17, 1984

Tong, Y. C. (1978). *Microbicidal/fungicidal thiazolopyrazines.* US Patent 4075207, February 21, 1978

US Environmental Protection Agency, 1995. *Reregistration Eligibility Decision (RED) Diquat Dibromide, EPA738-R-95-016.* Office of Prevention Pesticides and Toxic Substances, Washington, DC, 303 pp., October 17, 2007, http://www.epa.gov/oppsrrd1/REDs/0288.pdf (June 23, 2010)

Wagner, R.; Czerny, M.; Bielohradsky, J. & Grosch, W. (1999). Structure-odour-activity relationships of alkylpyrazines. *Z. Lebensm. Unters. Forsch.* A, Vol. 208, No. 5-6, 308-316, ISSN 1431-4630

Weaver, E. C. (1968). EPR studies of free radicals in photosynthetic systems. *Annu. Rev. Plant. Physiol.*, Vol. 19 (June), 283-294, ISSN 0066-4294

Weber, J. B. & Peter, C. J. (1982). Adsorption, bioactivity and evaluation of soil tests for alachlor, acetochlor and metolachlor. *Weed Sci.*, Vol. 30, No. 1, 14–20, ISSN 0043-1745

Whitmarsh, J. (1998). Electron Transport and Energy Transduction. In: *Photosynthesis: A Comprehensive Treatise*, Raghavendra A. S., (ed.), pp. 87-110, Cambridge University Press, ISBN 0-521-57000-X, Cambridge

Whitmarsh, J. & Govindjee (1999). The Photosynthetic Process. In: *Concepts in Photobiology: Photosynthesis and Photomorphogenesis*, G. S. Singhal, G. Renger, S. K. Sopory, K-D. Irrgang and Govindjee (eds.), pp. 11-51, Narosa Publishing House, ISBN 81-7319-151-4, New Delhi, India

Woolfson, A. & Rothschild, M. (1990). Speculating about pyrazines. *Proc. R. Soc. Lond. B*, Vol. 242, 113-119, ISSN 0962-8452

Yamaguchi, T. (2007). The relationship between the chemical structures of dihydropyrazine derivatives and DNA strand-breakage activity. *Chem. Pharm. Bull.*, Vol. 55, No. 4, 532-536, ISSN 0009-2363

Zondler, H.; Eckhardt, W. & Nyfeler R. (1989). *2-Arylmethyliminopyrazines, pyrazines compositions and method of combating undesirable plant growth and of regulating plant growth.* US Patent 4849009, July 18, 1989

Zweig, G.; Shavit, N. & Avron, M. (1965). Diquat (1,1'-ethylene-2,2'-dipyridylium dibromide) in photoreactions of isolated chloroplasts. *Biochim. Biophys. Acta*, Vol. 109, No. 2, 332-346, ISSN 0006-3002